Stefan Kubik

Supramolecular Chemistry

Also of Interest

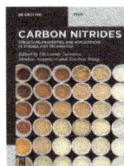

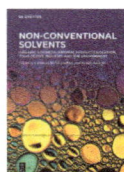

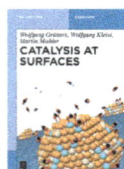

Stefan Kubik

Supramolecular Chemistry

From Concepts to Applications

2nd, Revised and Extended Edition

DE GRUYTER

Author
Prof. Dr. Stefan Kubik
Fachbereich Chemie – Organische Chemie
Rheinland-Pfälzische Technische
Universität Kaiserslautern-Landau
Erwin-Schrödinger-Str. 54
67663 Kaiserslautern
Germany
stefan.kubik@rptu.de

ISBN 978-3-11-131507-2
e-ISBN (PDF) 978-3-11-131517-1
e-ISBN (EPUB) 978-3-11-131576-8

Library of Congress Control Number: 2024934083

Bibliographic information published by the Deutsche Nationalbibliothek
The Deutsche Nationalbibliothek lists this publication in the Deutsche Nationalbibliografie;
detailed bibliographic data are available on the internet at http://dnb.dnb.de.

© 2024 Walter de Gruyter GmbH, Berlin/Boston
Cover image: Stefan Kubik
Typesetting: Integra Software Services Pvt.

www.degruyter.com

For my family
Daniela, Miriam, and Jakob
and my teachers
Peter Klein, Günter Wulff, and Julius Rebek Jr.

Preface to the first edition

The year 2017 marked the 50th anniversary of supramolecular chemistry, whose origin is generally considered to date back to 1967, when Charles Pedersen's first paper on crown ethers was published. One could argue that the field is, in reality, much older because supramolecular aspects were investigated even before 1967. Nonetheless, Pedersen's publication undoubtedly served as a starting point for supramolecular chemistry to develop into the prominent research field it is today. With the almost explosive developments over the last five decades, supramolecular chemistry has played an important role in shaping the face of modern chemistry, partly because of its multidisciplinary character that allows bridging many scientific disciplines. A modern education in chemistry would therefore be incomplete without including some relevant aspects. I wrote this book with the intention to provide this basic knowledge.

One of my objectives was to outline the differences between supramolecular chemistry and the chemistry taught in the early courses of chemistry studies. Supramolecular systems are, for example, typically held together by weak interactions, which causes entropy to have a profound influence on their thermodynamic stability. Moreover, the interactions are reversible, and most systems are therefore highly dynamic, which requires a special view to understand their behavior. In addition to these fundamental aspects, I have devoted a large fraction of the book to the many fascinating topics that are associated with supramolecular chemistry today. Readers will thus obtain an overview of the field, but can also use the book as a reference since key concepts are easily identified by the questions distributed among the chapters. The first one is the following:

> How is this book organized?

The book starts with an introduction in which the term "supramolecular chemistry" is defined, and the historical development of the field and its current relevance explained. The next two chapters are then devoted to the fundamentals of molecular recognition processes. These chapters introduce the formalism of describing and characterizing the interactions of molecules and the different natures of these interactions. Many key concepts are explained in this context, and a good grasp of these aspects helps understanding the behavior of the systems described in later chapters.

The fourth chapter focuses on the most important classes of receptors, their structures, syntheses, and binding properties. These receptors allow the complexation of a variety of different substrates, and the chapter thus concludes with a classification indicating which receptor is used for which substrate type.

The following chapters then deal with various specific topics of supramolecular chemistry, with the fifth chapter introducing strategies to assemble molecules and the sixth chapter concentrating on interlocked molecules. Chapters 7–10 address how to control molecular motion or develop supramolecular catalysts, carriers, or probes.

https://doi.org/10.1515/9783111315171-202

These chapters also give first insight into applications of supramolecular systems. This aspect is examined in more detail in the final chapter to also outline potential directions into which the field might head in the future.

Who could benefit from this book?

Although the book primarily addresses readers who are only starting to familiarize themselves with supramolecular chemistry and would like to obtain an accessible introduction, it may also be useful for those who already have a good understanding of the basic concepts and are interested in specific aspects. There is therefore no single way to read this book. Starting with the first three chapters is likely helpful to understand the concepts and terminology. All other chapters can be read independently, and additional information is hopefully located quickly with the available cross-references.

Are all aspects of supramolecular chemistry treated?

The short answer is no. The field is so large that I had to omit certain topics to retain the introductory character of the book. I decided to place the focus on supramolecular chemistry in solution, concentrating on discrete and structurally characterized complexes or assemblies, but to exclude or only briefly mention aspects relating to the solid state (clathrates, crystal engineering, and molecular tectonics), materials (polymers, foldamers, gels, and nanoparticles), or other larger assemblies (micelles and vesicles). I made this selection not because the latter aspects are less relevant, but because I felt that they are outside the scope of an introductory textbook. I am aware that this view may not be shared by everyone, and I therefore apologize to all those who miss certain topics. I also apologize to all whose work or names I did not mention, although they made important contributions to the field. I had to make choices, which meant that I had to leave out many fine examples, even from friends and colleagues whose work I admire.

I thank the people who supported me during the time of writing, most importantly my family. I thank Julia Bartl, Arne Lützen, and Konrad Tiefenbacher, who read parts of the book and made many helpful suggestions. I am also grateful to the team at De Gruyter for motivating me to embark on this venture and for their continued support. I hope that this book will turn out to be a valuable companion for many.

Kaiserslautern, June 2020

Preface to the second edition

Writing a book is like watching a child grow up. The beginnings are small, and the end is hard to predict. There are frustrating times when you struggle with every word, and there are harmonious times when everything falls into place. Progress is slow, until you reach a point where you can release the final product into the world, hoping it will behave. This one apparently did, at least I heard no complaints.

After publication, however, the more I looked at what I had written, the more I wished I had explained things differently, made fewer mistakes, or included things I had originally left out. Fortunately, you can do something with a book that you can never do with a child: you can improve it in a second edition. So, when the team at De Gruyter approached me and asked if I would be willing to prepare a new edition of this book, I immediately agreed. In preparing this edition, I carefully reviewed the entire text, improved the style, revised many of the explanations to make them clearer, and updated the references. In some places, I added details where concepts were too briefly explained, information was missing, or new developments needed to be mentioned. The most notable revision is the new chapter on supramolecular polymerization, which explains how to move from discrete self-assembled systems to infinite polymeric assemblies, focusing on the main mechanisms of supramolecular polymerization and their mathematical treatment. Because of this new chapter, I also had to restructure the final chapter on applications.

In addition, I felt that the book would benefit from information in many places that may not be essential for an introductory textbook, but that might still be of interest to some readers. Since I did not want to make the book significantly longer, I added short "Further reading" sections that briefly explain these aspects and provide one or two references. Finally, I added a selection of Python scripts to the appendix to illustrate how many of the graphs in the book were computed, and I hid an Easter egg in one of the figures.

With these changes, the book has hopefully become not only an updated version of the first edition, but also a much improved one that should be useful to all those who teach supramolecular chemistry or learn about this exciting field. I am very grateful to the entire team at De Gruyter who has supported me over the years. I also thank my family for their patience and for putting up with me when I was once again mentally absent, staring into the void, trying to figure out how best to explain enthalpy–entropy compensation or other strange concepts.

Kaiserslautern, March 2024

https://doi.org/10.1515/9783111315171-203

Contents

Questions

https://doi.org/10.1515/9783111315171-205

1 Introduction and overview

A significant part of the training in chemistry focuses on molecular chemistry. We learn about the relationship between the structure of a molecule and its reactivity, and the synthetic methods available to form covalent bonds. The corresponding theoretical framework allows us to predict how molecules will react or how they can be synthesized just by looking at their line drawings. As an example, let us look at the reaction in Figure 1.1.

We see that catechol and bis(2-chloroethyl) ether, when treated with sodium hydroxide, give a macrocyclic product with six ether groups along the ring. Considering the intrinsic reactivity of phenols and alkyl halides under basic conditions, one can attribute the formation of each of the four new C–O bonds to the initial deprotonation of a catechol hydroxy group. The nucleophilic phenolate formed then reacts with an electrophilic carbon atom in the reaction partner, which simultaneously releases a chloride ion. Thus, each bond formation involves a nucleophilic substitution and is associated with the formal formation of one equivalent of water and one equivalent of NaCl.

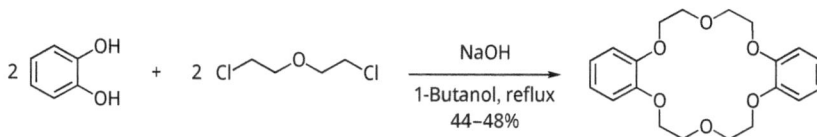

Figure 1.1: Reaction between catechol and bis(2-chloroethyl) ether under basic conditions that ultimately affords the depicted macrocyclic oligoether.

This mechanism adequately describes the formation of the C–O bonds, but does it also explain the formation of the macrocyclic product? Perhaps if only two molecules of catechol and two molecules of bis(2-chloroethyl) ether were present, but molecules are rarely as lonely as they appear in reaction schemes. Even reactions performed on a small scale involve the participation of an extremely large number of molecules when all the reaction partners, reagents, and solvent molecules are taken into account. Moreover, these molecules are constantly moving and bumping into each other, making reaction mixtures very crowded and dynamic environments. As a result, there is always a risk that a phenolate ion may not meet the correct partner in the above reaction, which could cause the product distribution to become very complex. Experimentally, however, the reaction proceeds quite selectively and in good yields, even at a catechol concentration of 1.3 M, far from the high dilution conditions often used in cyclization reactions to

https://doi.org/10.1515/9783111315171-001

favor ring formation over unwanted chain elongation. The mechanism of ether formation, that is, molecular chemistry, does not provide a straightforward explanation for
this observation, suggesting that effects beyond the actual bond formations operate in
this synthesis.

These effects are related to the presence of sodium ions, which, although unimportant for C–O bond formation, are not innocent bystanders but play an active role.
Initially, these ions interact preferentially with the solvent molecules, but with the
gradual appearance of linear oligoethers they prefer to bind to these more potent
binding partners. The interactions are mediated by the oxygen atoms, resulting in
complexes with an oligoether chain wrapped around a cation. The resulting proximity
of the phenolate group to the electrophilic carbon atom at the opposite end of the
chain, shown schematically in Figure 1.2 for the immediate precursor of the product,
explains why the cyclization is so efficient.

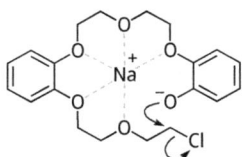

Figure 1.2: Structure illustrating how a sodium ion preorganizes the linear
precursor for the cyclization shown in Figure 1.1.

The principles of sodium ion complexation will be explained in later chapters, where
many other systems involving similar interactions will also be presented. In this introduction, the above example should serve only to illustrate that noncovalent interactions can exert characteristic effects on the outcome of a reaction. They are also
responsible for the stereoselectivity of transformations mediated by suitable catalysts,
but their relevance extends far beyond reaction control. In biological systems, for example, such interactions govern protein folding, the substrate selectivity of enzymes,
signal transduction, transport, the conservation and transmission of the genetic code,
and many other fundamental processes. Noncovalent interactions are also important
in several other areas of chemistry, such as medicinal and materials chemistry, but to
molecular chemistry, which is primarily concerned with forming covalent bonds,
they are not central. Instead, intermolecular interactions lie at the heart of supramolecular chemistry, a field of chemistry whose name was coined by one of its pioneers,
Jean-Marie Lehn, who chose the Latin prefix *supra* to indicate that supramolecular
chemistry transcends molecular chemistry. Lehn wrote in 1995, "Beyond molecular
chemistry based on the covalent bond there lies the field of supramolecular chemistry
whose goal is to gain control over the intermolecular bond. It is concerned with the
next step in increasing complexity beyond the molecule towards the supermolecule
and organized molecular systems [...]" [1]. This definition implies that supramolecular
systems consist of structurally defined assemblies whose formation is controlled by

organizational principles encoded in the structures of the individual interacting molecules. Figure 1.3 illustrates this idea.

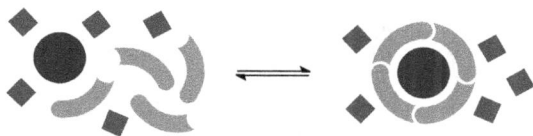

Figure 1.3: Schematic representation of the formation of a supramolecular complex from shape-complementary objects or molecules.

The curved building blocks in Figure 1.3 are ideally shaped to form a ring around the circle, forming a complex that must be stabilized by attractive interactions between the individual components. These components thus *recognize* each other, but ignore the square components that cannot be incorporated into the complex. An important requirement for this process to work is that any errors that occur on the way to the final structure, caused by incorrectly connected components, are continuously corrected. The interactions that stabilize the assembly must therefore be reversible, making the whole system dynamic and subject to thermodynamic control. Based on these considerations, we arrive at the following definition of supramolecular chemistry, which focuses on the two fundamental principles of recognition and reversibility.

Supramolecular chemistry is a field in chemistry that deals with molecular recognition phenomena mostly under thermodynamic control.

Although this definition covers most of the systems and processes discussed in this book, it could be argued that it does not sufficiently consider kinetically controlled processes that are sometimes used in supramolecular chemistry to drive interactions out of equilibrium. Nevertheless, it is preferable to broader definitions that associate supramolecular chemistry primarily with the design of functional molecules. While supramolecular systems are certainly functional, as we shall see, it is questionable whether the reverse is always true. For example, acid–base indicators are functional because their optical properties depend on the pH of the solution, but this property and its use have nothing to do with supramolecular chemistry. A narrower view is therefore preferred in this book.

How did supramolecular chemistry emerge and develop?

The advent of supramolecular chemistry is closely associated with the reaction shown in Figure 1.1. This reaction occurred as an unwanted side reaction in the synthesis of bis[2-(2-hydroxyphenoxy)ethyl] ether, performed in 1962 by Charles J. Pedersen at DuPont in Wilmington, Delaware (Figure 1.4).

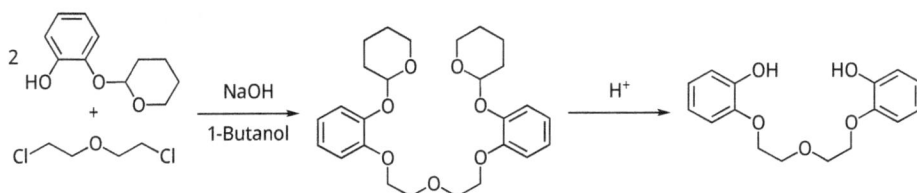

Figure 1.4: Synthesis of bis[2-(2-hydroxyphenoxy)ethyl] ether performed by Charles J. Pedersen.

Pedersen was interested in this reaction because he wanted to use the VO$^+$ complex of the product as a catalyst for olefin polymerization. He was aware that the tetrahydropyranyl-protected starting material he was using contained 10% unprotected catechol, but he did not consider this impurity to be problematic because he expected it to yield easily separable oligomeric and polymeric byproducts. Unexpectedly, the presence of catechol also afforded the macrocycle shown in Figure 1.1 in a yield of 1%, which might have gone unnoticed had the unusual properties of the cyclic product not sparked Pedersen's curiosity. For example, he observed that this compound was insoluble in methanol but readily dissolved in the presence of sodium hydroxide, which he correctly attributed to the binding of the sodium cation in the center of the ring. Pedersen went on to study the synthesis and properties of such cyclic oligoethers, for which he coined the term *crown ethers*, and in 1967 published a detailed account of nearly 50 derivatives in the *Journal of the American Chemical Society* [2]. This paper is generally considered to mark the birth of supramolecular chemistry.

Pedersen's work served as an inspiration for other groups to invent molecules that have cavities for accommodating suitable substrates. The next major step in this direction was made by Lehn with the development of cryptands, bi- or tricyclic analogs of crown ethers, that typically have a much higher affinity for alkali metal ions than their monocyclic counterparts. The work on cryptands was only the first of numerous contributions from Lehn to the field of supramolecular chemistry. His interests ranged from the development of various receptors and catalysts to polymetallic coordination compounds and organic materials. Accordingly, entire areas of supramolecular chemistry and the concepts and terms associated with them, including the term supramolecular chemistry itself, can be traced back to him.

The third pioneer in the field, Donald J. Cram, introduced the first chiral crown ethers and showed that these compounds were capable of enantioselective substrate recognition. Cram thus transferred a fundamental concept of nature, which is a consequence of the homochirality of biomolecules, to synthetic systems. He then turned his attention to the development of supramolecular catalysts that mimic not only the substrate affinity of enzymes but also their ability to transform the bound substrate. Cram also developed several new classes of receptors, introduced basic concepts of supramolecular chemistry such as that of preorganization, and promoted the use of the term host–guest chemistry when referring to receptor–substrate interactions. This term, which was actually invented before the advent of supramolecular chemistry [3], is still in use because it aptly illustrates the hosting of the substrate by the receptor (although it could be argued that the receptor is a *hostel* rather than a *host*). However, it refers to only part of the much broader field of supramolecular chemistry.

In 1987, the Royal Swedish Academy of Sciences awarded the Nobel Prize in Chemistry to Pedersen, Lehn, and Cram for their pioneering work, particularly for "their development and application of molecules with highly selective structure-specific interaction, i.e. molecules that can 'recognize' each other and choose with which other molecules they will form complexes" [4]. It was furthermore emphasized that the molecules developed by Cram, Lehn, and Pedersen are highly relevant to the life sciences, as they make it possible to mimic processes that were previously the exclusive domain of biomolecules. Yet, it would be wrong to say that supramolecular chemistry did not exist before the work of the three Nobel laureates. In fact, several relevant concepts were established long before Pedersen's seminal paper, some by groups working on what was later called host–guest chemistry. Table 1.1 gives an overview of selected concepts and discoveries that preceded supramolecular chemistry and helped lay the foundation for the field.

Table 1.1: Concepts and discoveries reported prior to 1967 relevant to the field of supramolecular chemistry.

1893	Alfred Werner describes the concepts of coordination chemistry.
1894	Emil Fischer introduces the *lock-and-key* concept to rationalize the substrate selectivity of enzymes.
1911	Hans Pringsheim describes the complexation of organic compounds by cyclodextrins, natural macrocyclic receptors that were originally discovered by Antoine Villiers in 1891.
1931	Linus C. Pauling publishes his seminal paper in the *Journal of the American Chemical Society* entitled "The Nature of the Chemical Bond" in which he also alludes to the hydrogen bond.
1953	James D. Watson and Francis H. C. Crick describe the structure of the DNA double helix.
1958	Daniel E. Koshland Jr. introduces the *induced fit* concept, which refers to the conformational changes that proteins undergo upon substrate binding.
1964	Daryle H. Busch proposes a classification for template effects.

Despite this important early work, the design and development of synthetic supramolecular systems really began only after 1967. A further rapid increase in worldwide research activity can be noted after the 1987 Nobel Prize, when many creative scientists joined the field. At the same time, the research became increasingly diverse, involving aspects of organic, inorganic, physical, theoretical, materials, and analytical chemistry, often in combination with biochemistry. This interdisciplinary work has led not only to a better understanding of the thermodynamic and kinetic aspects of molecular recognition phenomena but also to the development of a wide variety of novel supramolecular systems. In many cases, the inspiration came from nature, resulting in the transfer of biochemical concepts such as catalysis, allosteric control of substrate binding, induced fit, cooperativity, multivalency, information storage, replication, transport, motion, and so on, to synthetic systems. A better understanding of these principles then made it possible to move further and further away from natural models and to make new discoveries.

These extensive research activities caused supramolecular chemistry to develop into a prestigious and influential field of research. Its importance is also reflected in the fact that approximately 5,400 publications containing the term *supramolecular* were published in 2023, which is almost 15 articles per day. The number of articles related to supramolecular chemistry is probably even higher, as the term supramolecular is often not mentioned today. Clearly, supramolecular chemistry has contributed significantly to shaping the face of modern chemistry and will continue to do so. It is therefore worth taking a closer look at the various facets of this fascinating field, but this is not possible without a sound understanding of the basics. The next two chapters aim to provide this knowledge.

Bibliography

[1]   Lehn JM. Supramolecular Chemistry – Concepts and Perspectives. VCH: Weinheim, 1995, p. 2.
[2]   Pedersen CJ. Cyclic polyethers and their complexes with metal salts. J. Am. Chem. Soc. 1967, 89, 7017–36.
[3]   Crowley P. On the origins of the host–guest terminology. Cryst. Growth Des. 2023, 23, 8469–73.
[4]   The Nobel Prize in Chemistry 1987 (Accessed February 01, 2024, https://www.nobelprize.org/nobel_prizes/chemistry/laureates/1987/press.html).

2 Analyzing complex formation

CONSPECTUS: *Before we can discuss the actual binding forces that hold supramolecular systems together, we must first look at how complex formation is described mathematically and analyzed experimentally. Knowledge of these strategies will facilitate understanding of the concepts and methods used to characterize the supramolecular complexes presented later in the book.*

2.1 Thermodynamic and kinetic aspects

How are 1:1 complexation equilibria formally described?

Molecular recognition processes occur between structurally complementary molecules and result in complexes that are characterized by their stability and by the number and arrangement of interacting components. In the simplest case, a complex C consists of only two molecules, one of which, the so-called receptor R, has a cavity or cleft to accommodate the typically smaller substrate S (Figure 2.1).

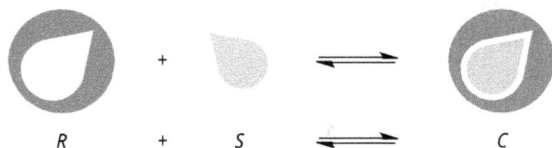

Figure 2.1: Schematic representation of the complexation of a substrate S by a structurally complementary receptor R and the corresponding reaction scheme.

Complex formation is reversible, as indicated by the equilibrium arrow in Figure 2.1, and R, S, and C therefore coexist in solution. The greater the extent to which R and S are bound and the smaller the amounts of uncomplexed species, the more efficient the interactions. This efficiency is expressed quantitatively by K_a, the stability or association constant, which is given by the law of mass action according to the following equation:

$$K_a = \frac{c_C}{c_R \, c_S} \tag{2.1}$$

Note that equation (2.1) specifies the amounts of R, S, and C in concentrations (c_R, c_S, c_C) instead of dimensionless activities. This approach is not only practical, since activity constants are generally not available for the species involved in binding equilibria, but also justified to some extent for the low concentrations typically used in binding

https://doi.org/10.1515/9783111315171-002

studies. However, the use of concentrations causes K_a to have units of M^{-1} ($L\ mol^{-1}$) because the denominator in equation (2.1) contains a product of two concentrations.

Supramolecular chemists prefer to use stability constants, perhaps because there is a direct correlation between magnitude and stability: the larger the K_a, the more stable the complex. In biochemistry, binding efficiency is usually expressed in terms of dissociation constants K_d, which are the reciprocal values of stability constants ($K_d = K_a^{-1}$). Accordingly, K_d has units of M ($mol\ L^{-1}$) and decreases with increasing complex stability. Whichever value one prefers, K_a and K_d have characteristic values for each receptor–substrate combination, but are strongly dependent on external influences such as temperature or solvent. Therefore, the binding properties of different receptors can only be compared if they have been characterized under identical conditions.

While equation (2.1) does not give direct access to the concentration of C because c_R and c_S are unknown, the equilibrium concentrations are related to the initial concentrations of receptor (c_R^0) and substrate (c_S^0) by the complex concentration (c_C) according to the following mass balances:

$$c_R = c_R^0 - c_C \tag{2.2a}$$

$$c_S = c_S^0 - c_C \tag{2.2b}$$

Combining equation (2.1) with (2.2a) and (2.2b) leads to (2.3), which is a quadratic equation in c_C that can be solved in a straightforward manner:

$$K_a = \frac{c_C}{\left(c_R^0 - c_C\right)\left(c_S^0 - c_C\right)} \tag{2.3}$$

$$c_C = \frac{c_R^0 + c_S^0 + K_a^{-1}}{2} - \sqrt{\frac{\left(c_R^0 + c_S^0 + K_a^{-1}\right)^2}{4} - c_R^0\, c_S^0} \tag{2.4}$$

Of the two possible solutions, only equation (2.4) with the minus sign in front of the square root correctly describes how c_C depends on c_R^0, c_S^0, and K_a. The following arguments explain why: when $K_a < 1\ M^{-1}$, K_a^{-1} becomes large, so that c_R^0 and c_S^0 have almost no influence on the expression $c_R^0 + c_S^0 + K_a^{-1}$. Consequently, the two terms before and after the minus sign on the right-hand side of equation (2.4) are approximately equal to $K_a^{-1}/2$, and since c_C is close to zero under these conditions, the equation is satisfied only when the terms are subtracted.

Equation (2.4) allows the equilibrium concentration of a 1:1 complex to be estimated from its stability constant and the initial concentrations of R and S. To illustrate the dependence, Figure 2.2a shows the changes in complex concentration as increasing amounts of S are added to a solution of R for complexes with stability constants of 10^2, 10^3, 10^4, and $10^6\ M^{-1}$.

In all cases, the complex concentration increases progressively with increasing substrate concentration. The black curve, representing the most stable complex, ex-

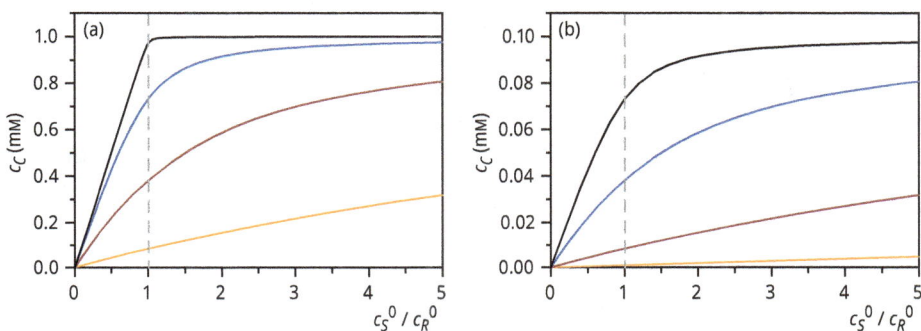

Figure 2.2: Graphs showing how the complex concentration changes when increasing amounts of a substrate are added to a receptor solution of given concentration. The initial receptor concentration amounts to $c_R^0 = 10^{-3}$ M (= 1 mM) in diagram (a) and to $c_R^0 = 10^{-4}$ M (= 0.1 mM) in diagram (b). The curves represent complexes of different stability with the K_a amounting to 10^2 M^{-1} (orange), 10^3 M^{-1} (red), 10^4 M^{-1} (blue), and 10^6 M^{-1} (black). The dotted lines mark the 1:1 receptor–substrate ratios in both graphs.

hibits an almost linear increase until a 1:1 substrate–receptor ratio is reached, indicating that each substrate molecule added to the solution is consumed in the complex. Once saturation is reached, further substrate molecules have no effect because there are no unused receptor molecules left. The shape of this curve is consistent with equation (2.4) as K_a becomes large and K_a^{-1} therefore negligible. Assuming that $c_S^0 = x c_R^0$, where x is the receptor–substrate ratio and $K_a^{-1} \sim 0$, the rearrangement of equation (2.4) gives $c_C = 0.5[1 + x - \text{abs}(1 - x)]c_R^0$. The complex concentration c_C thus increases linearly between $0 \le x \le 1$ and remains equal to c_R^0 when $x > 1$.

The curves describing the less stable complexes in Figure 2.2a become progressively shallower as K_a decreases. As a result, substantial amounts of uncomplexed receptor are still present even at high substrate–receptor ratios. For comparison, 97% of R is complexed at a 1:1 ratio for the complex with a K_a of 10^6 M^{-1}, whereas the corresponding fractions are 73%, 38%, and 8% for the complexes with stability constants of 10^4, 10^3, and 10^2 M^{-1}, respectively. Figure 2.2b shows that decreasing the receptor concentration also causes the curves to become shallower. As a consequence, higher amounts of S are required to reach saturation compared to the situation in more concentrated solutions. We conclude that receptor saturation can be achieved either by adding an excess of S, the exact amount depending on the stability of the complex, or by increasing the concentrations of R and S.

What happens when binding equilibria become more complex?

For complexes that do not have a 1:1 stoichiometry, equation (2.1) is no longer valid. However, the general strategy to mathematically describe such higher equilibria is not very different from that explained earlier. Let us assume that the receptor binds a second substrate molecule, resulting in the formation of a 1:2 receptor–substrate complex. In this case, complex formation is a stepwise process, involving the initial formation of the 1:1 complex, which is subsequently converted to the 1:2 complex. The overall reaction thus combines two equilibria, each of which is associated with an individual equilibrium constant. The following reaction schemes illustrate the two steps, and the corresponding laws of mass action are given by equations (2.5a, b):

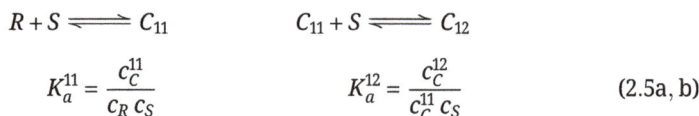

$$R + S \rightleftharpoons C_{11} \qquad\qquad C_{11} + S \rightleftharpoons C_{12}$$

$$K_a^{11} = \frac{c_C^{11}}{c_R\,c_S} \qquad\qquad K_a^{12} = \frac{c_C^{12}}{c_C^{11}\,c_S} \qquad\qquad (2.5\text{a, b})$$

In addition to these laws of mass action, we again need mass balances to relate the initial concentrations of the substrate c_S^0 and the receptor c_R^0 to the concentrations of the 1:1 complex c_C^{11}, the 1:2 complex c_C^{12}, the substrate c_S, and the receptor c_R in equilibrium. These mass balances are given by equations (2.6a) and (2.6b). Note the factor 2 in equation (2.6b), which reflects the fact that the 1:2 complex contains two substrate molecules.

$$c_R = c_R^0 - c_C^{11} - c_C^{12} \qquad\qquad (2.6\text{a})$$

$$c_S = c_S^0 - c_C^{11} - 2\,c_C^{12} \qquad\qquad (2.6\text{b})$$

The combination of equations (2.5) and (2.6a) and (2.6b) ultimately gives a cubic equation that can be solved in c_C^{11} and c_C^{12}. This solution is derived in Appendix 13.1. Here we will only qualitatively assess how the concentrations of the different complex species vary as a function of the receptor–substrate ratio and the stepwise binding constants using the two examples in Figure 2.3.

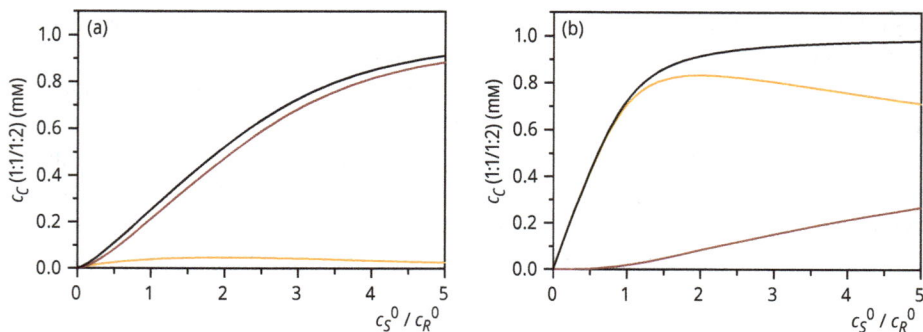

Figure 2.3: Graphs showing the concentrations of the 1:1 complex (orange) and the 1:2 receptor–substrate complex (red) when different amounts of a substrate are added to a 10^{-3} M (= 1 mM) receptor solution. The black curves denote the sum of the concentrations of the 1:1 and the 1:2 complexes. In diagram (a) $K_a^{11} = 100$ M^{-1} and $K_a^{12} = 10{,}000$ M^{-1} and in diagram (b) $K_a^{11} = 10{,}000$ M^{-1} and $K_a^{12} = 100$ M^{-1}.

The graphs in Figure 2.3a are associated with a stepwise equilibrium where K_a^{11} is smaller than K_a^{12}. In this case, the concentration of the 1:1 complex remains low even at high substrate concentrations because it is almost completely converted to the 1:2 complex. Accordingly, the equilibrium is dominated by the 1:2 complex. In addition, the curve describing the formation of this complex has a distinct sigmoidal shape. Figure 2.3b shows the graphs for a two-step equilibrium where $K_a^{11} > K_a^{12}$. In this case, the 1:1 complex is almost the only species in solution until the substrate is present in excess, at which point the 1:2 complex is formed at the expense of the 1:1 complex.

The shapes of the curves in Figures 2.2 and 2.3 are therefore influenced in characteristic ways by the strength and stoichiometry of the binding process. Accordingly, complexation equilibria can be characterized by following how the concentration of the complex changes as the receptor–substrate ratio is varied. Fitting the obtained binding isotherms to a suitable mathematical model then allows one to assess whether the assumed stoichiometry is correct, and to obtain the stability constants of the investigated system. An obvious mismatch between the experimental and theoretical isotherms immediately indicates that assumptions made in the selection of the binding model, such as the expected complex stoichiometry, were incorrect. For example, sigmoidal binding isotherms should not be evaluated on the basis of 1:1 equilibria.

Before we look at the analytical techniques available for following binding equilibria in Section 2.2, other pertinent thermodynamic parameters associated with the stability of a receptor–substrate complex will be introduced.

How does complex stability relate to binding enthalpy and entropy?

The thermodynamic driving force of complex formation is described quantitatively by the following equation:

$$\Delta G^0 = -RT \ln K_a \qquad (2.7)$$

This equation establishes a correlation between the association constant K_a of a complex and the Gibbs free energy of its formation ΔG^0. The other parameters are the gas constant R and the temperature T. Complexation processes are thus exergonic, that is, associated with a negative Gibbs free energy, when $K_a > 1$. These reactions occur spontaneously, with the extent of complex formation depending on the difference between the Gibbs free energy at the start of the reaction and at equilibrium. The more negative ΔG^0, the more stable the complex.

Note that ΔG^0 describes the thermodynamic driving force of complex formation when all binding partners are in their standard states, which for dissolved species means that their concentrations are 1 M. Relating the thermodynamics of complex formation to this standard state has the advantage that different systems can be compared, making ΔG^0 a useful alternative parameter to K_a for quantifying complex stability. However, unlike K_a, ΔG^0 refers to a single and in most cases relatively unrealistic situation. Experimentally, complex formation is rarely studied in 1 M solutions, and even if both binding partners are (or can be) mixed at this concentration, their interaction causes the system to immediately leave the standard state because the formation of the complex causes the concentrations of the free binding partners to gradually decrease. As a consequence, the absolute value of ΔG (note the absence of the index ^{0}, which indicates that this value refers to states of the system other than the standard state) decreases until the equilibrium is reached, where ΔG equals zero.

The effect of external conditions on ΔG can be qualitatively deduced by comparing the graphs in Figure 2.2. The blue curve in Figure 2.2a indicates that complex formation is 73% complete at equimolar concentrations of the binding partners, whereas the extent of complex formation decreases to 38% as the initial concentrations of receptor and substrate are reduced from 10^{-3} to 10^{-4} M (Figure 2.2b). The thermodynamic driving force for complex formation, that is, the exergonicity of the reaction, is thus lower in the dilute solutions although the K_a of the complex is unchanged. There are even ratios of R, S, and C that favor the reverse reaction, that is, complex dissociation. We must therefore carefully distinguish between ΔG^0, which refers to the standard state and allows the comparison of different systems, and ΔG, which refers to the thermodynamics of the system under specific experimental conditions. The latter not only has a different absolute value than ΔG^0 but can even have a different sign.

It must be emphasized that the term *system* refers to more than just the receptor, substrate, and complex. Complex formation is influenced by all interacting molecules, most importantly the solvent molecules, whose concentration greatly exceeds that of the solutes. The degree to which these solvent molecules stabilize the binding partners by solvation directly affects the strength of the interaction. For example, if the Gibbs free energy required to desolvate the binding partners cannot be compensated by the Gibbs free energy gained during complex formation, the corresponding complex will not form, even though it may be very stable in an environment where solvation is weaker. It is therefore useful to separate the total Gibbs free energy of binding ΔG^0 into the intrinsic free energy of complexation in the gas phase ΔG^0_{intr} and the free energies of solvation of the receptor $\Delta G^0_{solv}(R)$, the substrate $\Delta G^0_{solv}(S)$, and the complex $\Delta G^0_{solv}(C)$. The corresponding treatment gives equation (2.8), which shows that the binding process becomes exergonic only if the intrinsic binding strength overcompensates the difference between the free energies of solvation of the complex and its components $(\Delta G^0_{solv}(C) - \Delta G^0_{solv}(R) - \Delta G^0_{solv}(S))$. We will return to the influence of the solvent on the thermodynamics of complex formation in Section 3.3.

$$\Delta G^0 = \Delta G^0_{intr} + \Delta G^0_{solv}(C) - \Delta G^0_{solv}(R) - \Delta G^0_{solv}(S) \tag{2.8}$$

The Gibbs free energy of complex formation ΔG^0 can be further broken down into binding enthalpy and entropy. The corresponding underlying formalism is based on the Gibbs–Helmholtz equation (2.9):

$$\Delta G^0 = \Delta H^0 - T\Delta S^0 \qquad\qquad (2.9)$$

The enthalpy change ΔH^0 is defined as the heat change of the system during complex formation at a constant pressure. Again, this energy term refers to the standard state in which R and S are present at 1 M concentrations. Heat changes during a molecular recognition process are, to a first approximation, related to the direct interactions between the molecules involved in the complexation process, with an attractive receptor–substrate interaction producing a favorable exothermic ΔH^0. However, the receptor–substrate interactions are not the only factors influencing ΔH^0. Further contributions come from solvent effects, which can adversely affect ΔH^0 if the free binding partners are more strongly solvated than the complex. Moreover, unfavorable enthalpic contributions also result from strained receptor or substrate conformations in the complex. The overall binding enthalpy associated with complex formation therefore depends on the balance of a variety of factors and may ultimately be exothermic ($\Delta H^0 < 0$) or endothermic ($\Delta H^0 > 0$).

In the latter case, complex formation must be associated with a sufficiently large positive entropy value to become overall exergonic. Entropy refers to the order of a system, with a positive ΔS^0 denoting an increase in disorder, which promotes complex formation. A number of factors contribute to entropy. A fundamental one is that any binding process in which two or more molecules come together to form a complex is necessarily entropically unfavorable because the individual components lose degrees of freedom, the most important being translational and rotational mobility. The question then arises as to whether it is at all possible for a complexation process to be entropically favored. The answer, once again, lies in the solvent contributions, that is, the release of solvent molecules from the solvation shells of R and S when they interact. As a consequence, complexation processes, which in themselves lead to ordered assemblies, increase the disorder of the overall systems by allowing solvent molecules to gain freedom. Global disorder can therefore induce local order, which is a very important concept in supramolecular chemistry in general and self-assembly in particular, as we will see in Chapter 5. More detailed aspects of solvent effects are discussed in Section 3.3.

Note that entropy is temperature-dependent, which has implications when studying complex formation at different temperatures. Another important aspect is that the entropy of a solution increases with dilution. This effect explains why complexation equilibria shift toward the dissociated species as the concentration is reduced, as shown in Figure 2.2. It should also be noted that entropy has a greater effect on the formation of supramolecular complexes than on the formation of covalent bonds. The reason is that the interactions stabilizing supramolecular complexes are much weaker than covalent bonds and the corresponding ΔH^0 therefore smaller.

The substantial contribution of both enthalpy and entropy to the formation of a complex is the cause of a phenomenon that is almost universally observed when comparing the thermodynamic signatures of a series of related complexes: more favorable enthalpy terms are almost always associated with less favorable entropy terms.

This effect can have frustrating consequences when trying to optimize the performance of a receptor by increasing the binding strength, that is, by making the binding enthalpy more negative. Since the favorable effect on ΔH^0 is partially and sometimes even completely compensated by ΔS^0, the newly designed and perhaps laboriously synthesized receptor will be only marginally better than the original one. The correlation of ΔH^0 and $T\Delta S^0$ for a series of complexes is often linear and can thus be fitted to equation (2.10), where the parameter α denotes the extent to which an enthalpic gain is offset by an entropy loss in the respective complexes, while the intercept $T\Delta S_0$ describes the intrinsic entropy term for a complex whose formation is enthalpically neutral:

$$T\Delta S^0 = \alpha\Delta H^0 - T\Delta S_0 \qquad (2.10)$$

As an example, a plot of the ΔH^0 and $T\Delta S^0$ values associated with the formation of different crown ether complexes is shown in Figure 2.4. The straight line resulting from a regression analysis has a slope of 0.73 (an α value of 0.76 has been derived for a much larger data set [1]), showing that only 27% of the ΔH^0 increment contributes to the improvement in complex stability. A slope of 1 would indicate that a favorable effect of enthalpy is exactly compensated by an unfavorable entropic term.

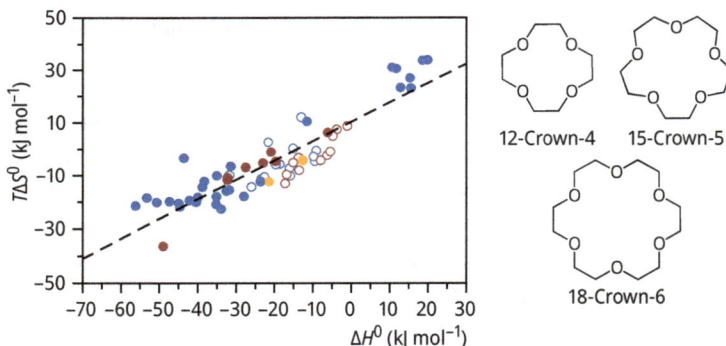

Figure 2.4: Plot of ΔH^0 versus $T\Delta S^0$ associated with the complexation of various cations by 12-crown-4 (orange), 15-crown-5 (red), and 18-crown-6 (blue) in water (open circles) and methanol (filled circles) at 298 K. The structures of the crown ethers are also depicted.

This so-called enthalpy–entropy compensation can be qualitatively explained as follows: the strengthening of the receptor–substrate interactions, manifested in the improvement of the binding enthalpy, simultaneously causes a tightening of the complex structure. As a consequence, conformational degrees of freedom are lost, which has an unfavorable effect on ΔS^0. Conversely, weak binding is entropically beneficial since the binding partners retain flexibility. In the case of complexes containing charged species, it must be considered that ion desolvation becomes enthalpically more costly but entropically more favorable as solvation becomes stronger. Accordingly, an adverse change in the binding enthalpy resulting from the exchange of a weakly solvated charged binding partner for a strongly solvated one is again offset by a more favorable entropy term. While enthalpy–entropy compensation is a rather general phenomenon, the reason why the gain in one parameter is often exactly cancelled out by a loss in the other is not yet fully understood.

Is there a relationship between complex stability and the rate of complex formation?

As we have seen, thermodynamics predicts that a negative ΔG^0 causes complex formation to proceed spontaneously. This does not necessarily imply that complexation is also fast, although many binding equilibria in supramolecular chemistry are indeed associated with low activation barriers and are therefore often (almost) diffusion-controlled. However, there are exceptions, one of which is illustrated in Figure 2.5.

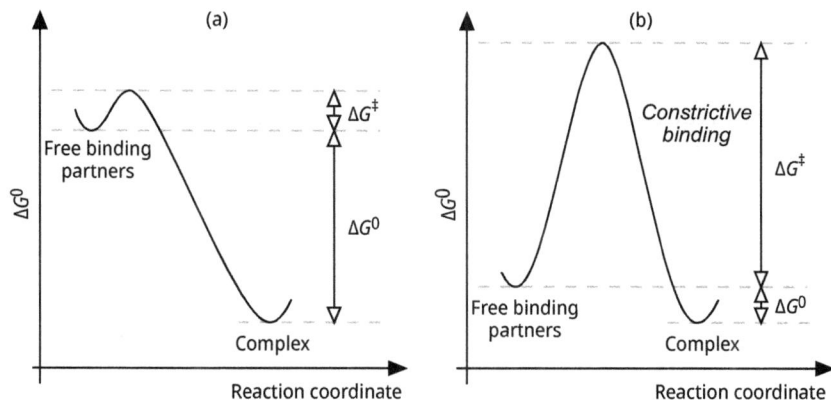

Figure 2.5: Energy profiles associated with complexation equilibria, of which one has a small activation barrier (small $\Delta G^\ddagger$) and leads to a stable complex (large ΔG^0) (a), and the other has a large activation barrier (large $\Delta G^\ddagger$) but leads to a thermodynamically not very stable complex (small ΔG^0) (b).

Figure 2.5a shows the energy profile of a complexation reaction in which binding is associated with a substantial exergonic stabilization of the complex and a small activation barrier $\Delta G^\ddagger$ that is easily overcome. In such a case, the equilibrium is strongly in favor of the complex and the system is dynamic, which means that complexes are constantly forming and dissociating. The situation shown in the energy profile in Figure 2.5b is different. In this case, the complex is thermodynamically not significantly favored over the free binding partners. However, complex formation and dissociation must overcome substantial activation barriers. Such a situation arises when the receptor and the substrate have to adopt strained conformations to allow substrate exchange, as in the hemicarcerands described in Section 4.1.9. Such complexes can be inert (no exchange at room temperature), although their thermodynamic stability is low. To describe the behavior of these systems in quantitative terms, Cram introduced the term *constrictive binding*, which relates the Gibbs free energy of the transition state to the Gibbs free energy of the binding partners (Figure 2.5b) [2]. In other words, constrictive binding is the free energy that must be invested, in addition to the ΔG^0 associated with complex stability, to reach the transition state from the complexed state. It is therefore a measure of the energy contribution that holds molecules together, not because they interact so strongly, but because they are so difficult to separate.

The reaction profile in Figure 2.5a exhibits a low activation barrier and we therefore expect complex formation to be fast. But what exactly does fast mean in this context? In order to obtain information in this respect, it is useful to correlate the stability constant K_a with the rate constants of complex formation k_{on} and dissociation k_{off}. This treatment is based on the rate equations (2.11a, b):

$$R + S \underset{k_{off}}{\overset{k_{on}}{\rightleftharpoons}} C$$

$$\frac{dc_C}{dt} = k_{on}\, c_R\, c_S \qquad\qquad -\frac{dc_C}{dt} = k_{off}\, c_C \qquad\qquad (2.11a, b)$$

Once the reaction reaches the thermodynamic equilibrium, k_{on} and k_{off} are equal (*steady state*), which allows us to write the following expression:

$$k_{on}\, c_R\, c_S = k_{off}\, c_C \qquad\qquad (2.12)$$

Rearrangement yields equation (2.13), which shows that the stability constant K_a is directly related to the ratio of the rate constants:

$$\frac{k_{on}}{k_{off}} = \frac{c_C}{c_R\, c_S} = K_a \qquad\qquad (2.13)$$

Stable complexes therefore form faster than they dissociate. Note that this statement also applies to the steady state, where the rates of complex formation ($k_{on}\, c_R\, c_S$) and dissociation ($k_{off}\, c_C$) are the same. Since the rate constants are multiplied with concentrations, a large k_{on} is multiplied with two small concentrations in the case of a stable complex, while a smaller k_{off} is multiplied with a large c_C, with both products ultimately yielding the same result.

The correlation in equation (2.13) allows us to estimate the lifetime of typical supramolecular complexes in solution. Assuming that complex formation is diffusion controlled ($k_{on} \sim 10^9$ M^{-1} s^{-1}) and that the complex has a K_a of 10^6 M^{-1}, which is a rather stable complex according to Figure 2.2a, we arrive at a k_{off} of 10^3 s^{-1}. The complex therefore has a lifetime on the order of milliseconds, which is short on the human timescale. The important lesson is that even highly stable complexes form and dissociate rapidly in solution and that the static picture suggested by reaction schemes, such as that in Figure 2.1, is misleading.

2.2 Analytical strategies and techniques

2.2.1 Strategies

The only value needed to derive the stability constant from the law of mass action in equation (2.1) is the complex concentration c_C (in the case of higher complexes, the concentrations of all complexes present), since the concentrations of the receptor c_R and substrate c_S are accessible from the mass balances (equations (2.2a, b)). Unfortunately, most of the analytical techniques used in binding studies do not yield absolute concentrations, but rather provide information on the extent to which the complex concentration increases or decreases when the concentrations of one or both binding partners are changed. For this reason, binding studies generally involve titrations in which the concentrations of the binding partners are varied and a physical property that correlates with c_C is measured. These titrations yield binding isotherms, such as those shown in Figures 2.2 and 2.3, from which K_a is determined by fitting them to the mathematical model underlying complex formation.

How can complex stoichiometry be determined?

A crucial aspect in this context is choosing the correct model because meaningful stability constants will only be obtained if all aspects of complex formation, in particular complex stoichiometry, have been correctly taken into account. It is therefore helpful to know the composition of the complex prior to recording binding isotherms, although information in this respect can sometimes be derived from the isotherms themselves. For example, the sharp bend at a receptor–substrate ratio of one, which characterizes the binding isotherm of the most stable complex in Figure 2.2a, clearly indicates that this complex has a 1:1 stoichiometry. Conversely, the black isotherm in Figure 2.3b reaches saturation at approximately a 1:2 receptor–substrate ratio, indicating that two substrate molecules are bound to the receptor. However, once the complexes become less stable, these correlations are no longer reliable. In addition, while sigmoidal shapes of binding isotherms suggest the presence of higher complexes, they give little information about the actual stoichiometry. Therefore, independent methods are usually required to determine the composition of a complex.

A classical approach is based on Job's method of continuous variations, introduced by Paul Job in 1928 to determine the stability and stoichiometry of coordination complexes [3]. The underlying strategy involves the preparation of a series of solutions containing R and S at the same total concentration but in different ratios, followed by measuring the amount of the complex (or a property that correlates with the complex concentration) in each solution. The idea behind this method is illustrated for 1:1 and 1:2 complexes in Figure 2.6.

Figure 2.6a shows that the concentration of a 1:1 complex should be highest in the solution containing the binding partners in equal amounts. The absolute amount of

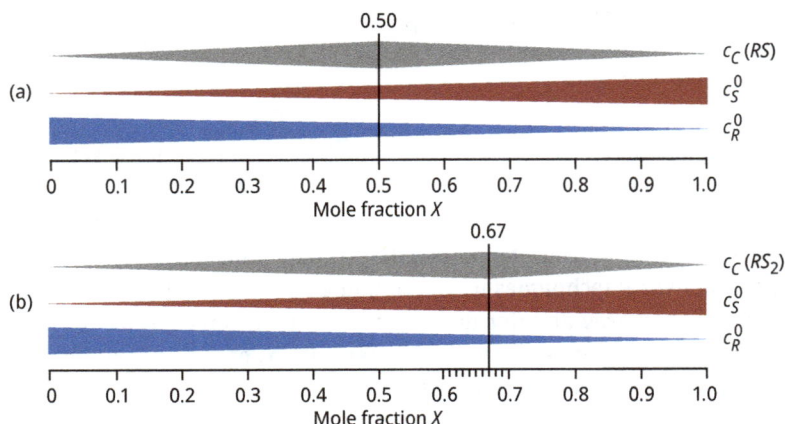

Figure 2.6: Schematic representation of the changes in complex concentration in solutions containing the receptor and the substrate at a constant total concentration but in different ratios for a 1:1 complex (a) and a 1:2 receptor–substrate complex (b). The mole fraction X is defined as $c_S^0 / (c_R^0 + c_S^0)$.

the complex at this mole fraction depends on its stability, but any reduction of either S or R will limit the maximum amount of C that can be formed, causing the complex concentration to decrease. The situation is similar for complexes of higher stoichiometry. For example, the maximum amount of the complex is formed at a mole fraction of 0.67 in the case of a 1:2 receptor–substrate complex (Figure 2.6b). In general, the mole fraction X at which the amount of complex is highest can be calculated by using equations (2.14a) or (2.14b), depending on how X is defined:

$$X = \frac{\text{number of substrate molecules in the complex}}{\text{total number of binding partners in the complex}} \quad \text{if} \quad X = \frac{c_S^0}{c_R^0 + c_S^0} \tag{2.14a}$$

$$X = \frac{\text{number of receptor molecules in the complex}}{\text{total number of binding partners in the complex}} \quad \text{if} \quad X = \frac{c_R^0}{c_R^0 + c_S^0} \tag{2.14b}$$

Thus, a plot of the mole fraction *versus* the amount of complex should yield a bell-shaped curve with the position of the extremum providing information about the complex composition. The actual equations describing these curves are based on the equations we derived earlier (e.g., equation (2.4) for a 1:1 complex). Examples of such Job plots are shown in Figure 2.7.

Figure 2.7 shows that, like the binding isotherms in Figure 2.2, the Job plots become shallower as the complex stability or the total initial concentrations of the binding partners decrease, reflecting the simultaneous decrease in the extent of complex formation. Since a reliable determination of the position of the extremum is not easy for shallow curves, carefully selected experimental conditions are required to obtain meaningful Job plots. In addition, other factors sometimes make Job plots problem-

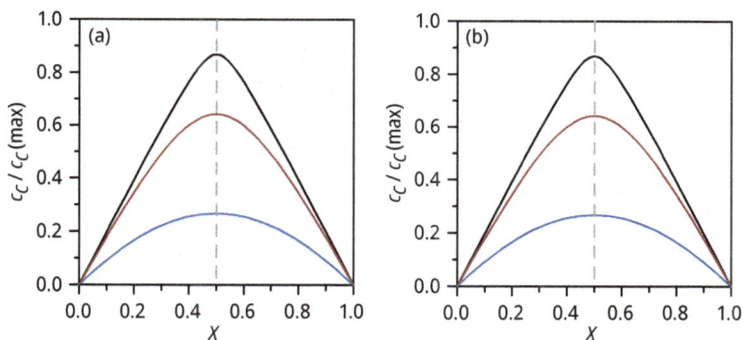

Figure 2.7: Job plots for 1:1 binding equilibria. The total concentration $c_S^0 + c_R^0$ amounts to 10^{-2} M in graph (a), with the curves representing complexes with stability constants K_a of 10^4 (black), 10^3 (red), and 10^2 M^{-1} (blue). The K_a is 10^4 M^{-1} in graph (b) and the total concentration $c_S^0 + c_R^0$ is varied between 10^{-2} (black), 10^{-3} (red), and 10^{-4} M (blue). The concentration $c_C(\mathrm{max})$ equals $(c_R^0 + c_S^0)/2$.

atic. One of them is illustrated in Figure 2.8, which shows examples of Job plots for 1:2 receptor–substrate binding equilibria, calculated by using the equations derived in Appendix 13.1.

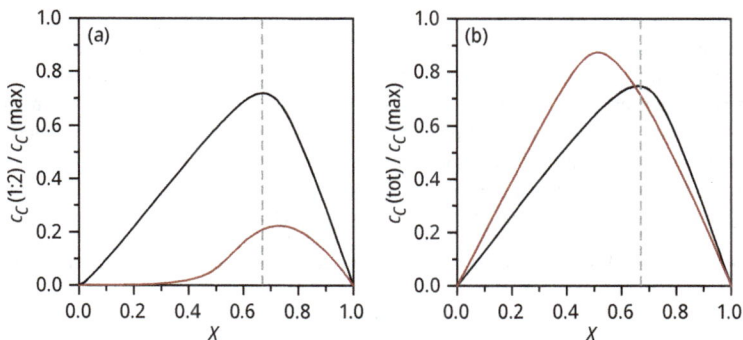

Figure 2.8: Job plots for 1:2 receptor–substrate binding equilibria. The total concentration $c_S^0 + c_R^0$ amounts to 10^{-2} M in both graphs. The black curves denote a complex with $K_a^{11} = 100$ M^{-1} and $K_a^{12} = 10{,}000$ M^{-1}, and the red curves a complex with $K_a^{11} = 10{,}000$ M^{-1} and $K_a^{12} = 100$ M^{-1}. Graph (a) shows the concentrations of only the 1:2 complexes $c_C(1{:}2)$ and graph (b) the total concentrations $c_C(\mathrm{tot}) = 2/3\, c_C(1{:}1) + c_C(1{:}2)$. The concentration $c_C(\mathrm{max})$ equals $(c_R^0 + c_S^0)/3$.

The black curves in Figure 2.8 show Job plots of a complex whose stepwise formation is associated with a K_a^{12} that is two orders of magnitude greater than K_a^{11}. Evidently, substantial amounts of the 1:2 complex are present at each mole fraction so that the maximum of the Job plot ends up at the expected value of 0.67, regardless of whether the formation of the 1:2 complex is followed individually or in combination with the 1:1

complex. However, if K_a^{12} is significantly smaller than K_a^{11}, as in the red curves, the 1:1 complex dominates. Consequently, the maximum of the Job plot is located either at a too high or too low mole fraction, depending on whether the plot shows only the concentration of the 1:2 complex, as in graph (a), or the combined concentrations of the 1:1 and 1:2 complexes as in graph (b). Neither Job plot reflects the true receptor stoichiometry, demonstrating that Job plots are unsuitable for assessing complex composition in certain situations. This problem becomes more pronounced when complex formation goes beyond 1:2 equilibria. Moreover, even if the maximum of the respective Job plot is located at the correct mole fraction, it is impossible to distinguish between complexes that have the same ratio of binding partners, for example 1:1 and 2:2 complexes, and distinguishing between maxima at 0.67 (1:2 complex) or 0.75 (1:3 complex) is experimentally challenging. These aspects have led some authors to proclaim "the death of the Job plot" [4]. While there are indeed situations where Job plots are useful, it is advisable to consider additional and ideally independent means of determining the binding model. Such methods may include careful statistical analysis of how well the fitted binding isotherm describes the experimental results to detect deviations from the chosen model, the results of crystallographic analyses, or mass spectrometric investigations. The latter in particular provide detailed structural information if the ionization method is mild enough to transfer the complexes to the gas phase without decomposition.

Further reading

Molecular recognition is mostly studied in solution. It is therefore not immediately obvious how gas phase studies can be useful for studying noncovalently stabilized complexes whose properties are normally sensitive to the environment. However, the use of mass spectrometry actually goes far beyond determining the ratio of the binding partners in a complex. Christoph Schalley and his group have shown that cleverly designed mass spectrometric experiments, including ion mobility studies and collision-induced fragmentation reactions, can be used to characterize supramolecular systems in terms of structure, reactivity, and stability. For a review of the use of mass spectrometry in supramolecular chemistry, see:

Baytekin B, Baytekin HT, Schalley CA. Mass spectrometric studies of non-covalent compounds: why supramolecular chemistry in the gas phase? Org. Biomol. Chem. 2006, 4, 2825–41.

How can complex stability be determined?

Once the composition of a complex has been determined, its stability is derived by fitting the experimental data to the corresponding theoretical binding isotherm by varying the value of the stability constant and any additional parameters associated with the technique used for the measurement until convergence is reached, that is, until the calculated and experimental binding isotherms are in best agreement [5].

At times when nonlinear regressions were difficult to perform because computers were not readily available, linearized forms of the binding isotherms were popular. All linearization methods are based on approximations, however, and therefore only yield meaningful results if the approximations are valid. In the case of the Benesi–Hildebrand method, for example, one binding partner must be present in excess so that its concentration can be assumed to remain constant over the course of the titration. These methods are now obsolete and should no longer be used in view of the many programs available to easily perform nonlinear regressions. There are even freely available online tools that conveniently allow the calculation of stability constants from binding isotherms without having to make approximations or assumptions [6].

A few guidelines must still be followed to ensure that the results of regression analyses are accurate. Reliability will suffer, for example, if binding isotherms are too steep, such as the black curve in Figure 2.2, because their nonlinear part, which mainly encodes for complex stability, is too small. Such isotherms sometimes allow deriving a lower limit of complex stability, but not an exact value. Conversely, binding isotherms that end before a substantial amount of the complex is present, such as the orange curves in Figure 2.2, contain too little information to allow calculating K_a. As a rule of thumb, nonlinear regressions give the most reliable results when the isotherms cover complex formation up to 80%.

An aspect that complicates the evaluation of higher equilibria is that the corresponding binding isotherms must be fitted to more than two independent parameters. Nonlinear regressions therefore sometimes yield different solutions depending on the parameters chosen as starting points. To avoid this problem, it is helpful to have estimates for one or more of these unknowns from independent measurements. Moreover, one can try to simplify the binding model by reducing the concentrations of the interacting species, thus suppressing the formation of higher complexes.

Because stability constants are so sensitive to external parameters, only constants determined under analogous conditions are comparable. That the measurements should have been performed in the same solvent or at the same temperature is trivial, but what about stability constants determined by different methods under comparable conditions? In principle, the value of a stability constant should be independent of how it was determined. However, different analytical techniques usually require different concentrations. For example, NMR titrations are performed at millimolar concentrations, whereas UV–vis titrations can be performed at concentrations one to three orders of magnitude lower. Accordingly, the corresponding binding isotherms cover different regions of the equilibrium and therefore vary. Different techniques also probe different properties, which can also affect the results. However, the deviations should be reasonably small and consistent with the different experimental conditions.

Finally, it is impossible to compare the binding constants of complexes that do not have the same composition because they have different units and are therefore incommensurable. Unfortunately, structural changes in a receptor sometimes lead to changes in the binding model, with one receptor forming a higher complex than a closely related other receptor, making comparison difficult. In some cases, it is useful to relate only the binding constants associated with the formation of the 1:1 complex. However, this is neither a general solution nor necessarily helpful, as it neglects the differences in the binding models. To solve this problem, the BC_{50} parameter has been introduced [7]. This so-called *median binding concentration* is defined as the total concentration of a reagent A (e.g., the receptor) required to bind 50% of a reagent B (e.g., the substrate). No further assumptions are made about the binding model underlying complex formation, making BC_{50} a general descriptor for quantifying affinity: the lower BC_{50}, the higher the affinity of A. While BC_{50} is a conditional parameter that varies with the total concentration of B and temperature, it approaches a constant minimum value when the total concentration of B becomes negligible. The corresponding concentration is the *intrinsic median binding concentration* BC_{50}^0, which depends on all relevant equilibrium constants, but not on specific conditions such as concentrations. BC_{50}^0 can therefore be regarded as the intrinsic maximum binding ability of A toward binding partner B. For simple 1:1 complexes, BC_{50}^0 is equal to the dissociation constant K_d.

The following example illustrates the usefulness of the BC_{50}^0 parameter. The two receptors shown in Figure 2.9 bind chloride ions in $CDCl_3/CD_3CN$, 80:20 (v/v). Under these conditions, other processes also take place, including receptor dimerization, pairing of the anions and the counterions, and binding of whole ion pairs to the receptors. Although the individual equilibrium constants of these processes are known, their comparison does not clearly indicate which of the receptors is more efficient. However, a clear distinction can be made on the basis of the BC_{50}^0 values. This parameter amounts to 4.6×10^{-3} M for **2.1** and 3.7×10^{-4} M for **2.2**, indicating that the pyrrole-derived receptor **2.2** has an intrinsic chloride affinity more than one order of magnitude higher than the urea-derived receptor **2.1**.

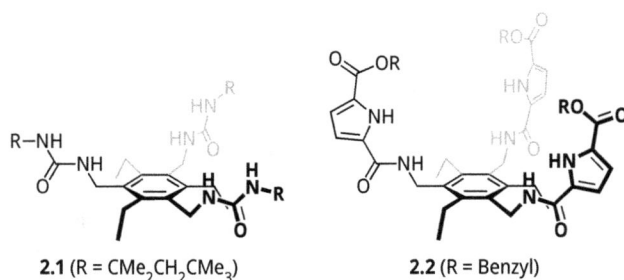

2.1 (R = CMe$_2$CH$_2$CMe$_3$) **2.2** (R = Benzyl)

Figure 2.9: Tripodal chloride receptors **2.1** and **2.2** whose chloride affinities are due to the cyclically arranged NH groups that can form hydrogen bonds with the anion.

2.2.2 NMR spectroscopy

How can binding equilibria be analyzed?

Complex formation usually produces characteristic changes in the NMR spectra of the receptor and/or the substrate, making NMR spectroscopy a very useful tool for studying molecular recognition processes [8]. ^{1}H NMR spectroscopy, in particular, is one of the most commonly used methods because it has several advantages over other techniques. One is that the extent to which a signal shifts or changes its integral upon complex formation correlates directly with the complex concentration without the need to consider parameters that relate the measured physical property to the concentration, such as the molar absorptivity coefficients in UV–vis spectroscopy (Section 2.2.3). ^{1}H NMR spectroscopy is also quite versatile as it only requires protons in the binding partners, while specific structural elements such as chromophores do not need to be present. Finally, since the spectroscopic changes produced by complex formation reflect the environment of the receptor and substrate protons in the complex, NMR spectroscopy also provides structural information, unlike many other techniques.

An important aspect to be considered in the context of NMR spectroscopy is the timescale of the method in relation to the lifetime of the complex. In this respect, two cases must be distinguished: equilibria that are slow on the NMR timescale and those that are fast. What does this mean and what are the consequences?

The NMR timescale determines whether protons whose environment changes in a dynamic process appear as two individual signals in the spectrum or collapse into one averaged signal. Examples of dynamic processes include conformational changes in a molecule, the exchange of acidic protons, or complexation equilibria. Whether signals from dynamic systems are resolved is determined by Heisenberg's uncertainty principle and depends on the excitation energy and the lifetime τ of the equilibrium state(s). This lifetime, or the corresponding rate constant of interconversion, is estimated by using equation (2.15), which strictly applies only to dynamic processes that follow first-order rate laws and involve the exchange of nuclei producing signals of equal intensity in the NMR spectrum:

$$k_c = \frac{\pi}{\sqrt{2}} \Delta \nu = 2.22 \, \Delta \nu = \frac{1}{\tau_c} \tag{2.15}$$

In this equation, $\Delta \nu$ refers to the distance in Hertz between the signals of the two exchanging protons if the exchange is slow. The parameter τ_c specifies the lifetime that just causes the signals to become indistinguishable, that is, to coalesce. If the actual lifetime of the system is greater than τ_c (or the rate constant of interchange is smaller than k_c), two signals will be observed, and if it is smaller, only an averaged signal will result.

Let us consider a specific example. Assuming that complex formation causes a receptor signal to shift by 1 ppm in the ^{1}H NMR spectrum when using a 300 MHz NMR spectrometer, $\Delta\nu$ is 300 Hz. Therefore, coalescence will occur if the exchange process has a lifetime of 1.5 ms (k_c = 666 s^{-1}) at the temperature at which the NMR spectrum is recorded. Processes with longer lifetimes (slower processes) at the same temperature will lead to two signals, and those with shorter lifetimes (faster processes) will give only one averaged signal. We have seen previously that the slow step in complexation equilibria is complex dissociation (which also follows a first-order rate law) and that a complex with a K_a of 10^6 M^{-1}, whose formation is diffusion controlled, has a k_{off} of 10^3 s^{-1} (Section 2.1). We would therefore expect averaged signals for the complexed and uncomplexed binding partners in this case. More stable complexes should produce separate signals, and the appearance of separate signals for free and complexed binding partners in the NMR spectrum indeed often indicates the formation of a very stable complex. Note, however, that the rate of complex formation also depends on activation barriers. Consequently, equilibria will be slow if complex dissociation is associated with a large activation energy, even if the thermodynamic stability is not very high (Section 2.1).

It should be emphasized that the above estimate is based on the assumption that $\Delta\nu$ amounts to 300 Hz. If the extent to which signals are shifted in the spectrum upon complex formation is smaller or larger, different lifetimes will result. It is even possible that complex formation causes some signals of the free and complexed binding partners to appear separately in the NMR spectrum, while others are averaged, if the extent to which complexation affects the chemical environment of different nuclei varies greatly. Another important aspect is that $\Delta\nu$ depends on the frequency of the NMR spectrometer. A difference of 1 ppm between two signals correlates to 300 Hz on a 300 MHz instrument, but to 800 Hz on an 800 MHz machine. Thus, NMR timescales depend on the spectrometer, and processes that are fast on one instrument may be slow on a more powerful instrument.

Deriving stability constants from NMR spectra for equilibria that are in slow exchange is relatively straightforward. A prototypical example of a series of NMR spectra is depicted in Figure 2.10. The signal in the lower spectrum should correspond to a proton in the free receptor. This signal gradually disappears as the substrate concentration increases, while a new signal appears, which is produced by the same proton now residing in the complex. At a certain substrate concentration, no signal from the free receptor is observed anymore.

Since the initial concentrations of receptor and substrate c_R^0 and c_S^0 are known and the complex concentration c_C is linked to these concentrations by mass balances, it is possible to quantify c_C, c_R, and c_S by signal integration. In the above example, c_C is calculated by using equation (2.16), which further allows the calculation of c_R and c_S based on equations (2.2a, b) if a 1:1 complex is formed:

$$c_C = \frac{\int \text{complex signal}}{\int \text{complex signal} + \int \text{receptor signal}} c_R^0 \qquad (2.16)$$

Once these concentrations are known, K_a is calculated by using the law of mass action in equation (2.1). Although a single spectrum is in principle sufficient to determine K_a,

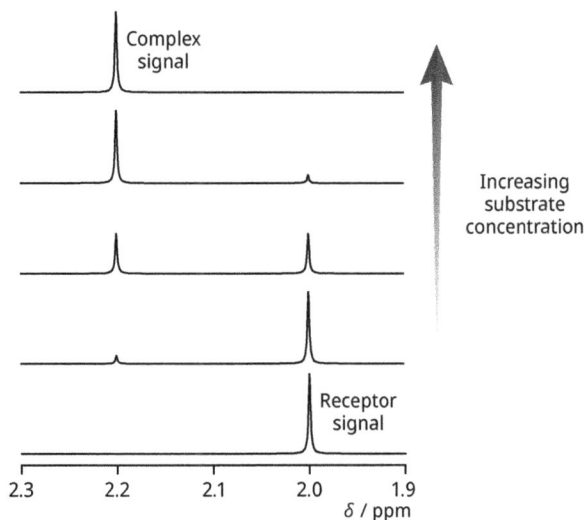

Figure 2.10: Exemplary ^{1}H NMR spectra illustrating the effect of complex formation on a receptor signal if the free and complexed species are in slow exchange with respect to the NMR timescale. The amount of substrate increases from the bottom to the top spectrum.

it is better to analyze a series of spectra with different receptor–substrate ratios in order to minimize the error. This method is only applicable if the complexes are not too stable. In a situation where the added substrate is completely consumed in the complex, c_S and/or c_R are almost zero, making it impossible to calculate a binding constant.

If complex formation is fast on the NMR timescale, the estimation of K_a involves recording spectra of a series of solutions containing the receptor and substrate in different ratios. Typically, the concentration of the component whose signal shift is being followed is held constant while the concentration of the other component is increased stepwise. Under these conditions, the equilibrium gradually shifts toward the complex, which is reflected in the continuous shift of a diagnostic signal in the NMR spectrum. Figure 2.11 shows an example of such a titration during which increasing amounts of the substrate are added to a receptor solution with the concentration c_R^0.

In this case, c_C is proportional, according to equation (2.17), to the extent to which the receptor signal at δ_0 is shifted toward the corresponding signal of the complex, which resides at $\delta_{\max}$:

$$c_C = \frac{\delta - \delta_0}{\delta_{\max} - \delta_0} c_R^0 \tag{2.17}$$

If a signal from the substrate is followed in the NMR titration, c_R^0 in equation (2.17) must be replaced by c_S^0. Combining equation (2.17) with equation (2.4) gives an expres-

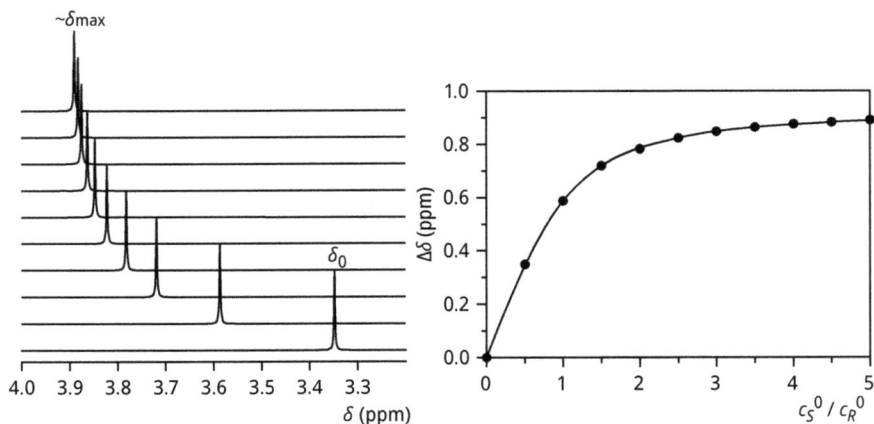

Figure 2.11: Exemplary ^{1}H NMR spectra illustrating the effect of complex formation on a receptor signal if the free and complexed species are in fast exchange with respect to the NMR timescale. The amount of substrate increases from the bottom to the top spectrum. The right graph shows a plot of the receptor–substrate ratio *versus* the relative shift changes $\Delta\delta = |\delta - \delta_0|$. The chemical shift δ_0 corresponds to the signal of the free receptor and δ to the chemical shifts of the signals in the mixtures. The line in the plot shows the binding isotherm obtained by fitting the individual data points to a 1:1 binding equilibrium.

sion that relates the experimentally determined chemical shift δ to the stability constant K_a. The K_a is estimated by fitting the observed binding isotherm to equation (2.18). The only other parameter that is typically unknown is δ_{max}, which therefore must also be fitted during the nonlinear regression:

$$\delta = \delta_0 + \frac{\delta_{max} - \delta_0}{c_R^0}\left(\frac{c_R^0 + c_S^0 + K_a^{-1}}{2} - \sqrt{\frac{\left(c_R^0 + c_S^0 + K_a^{-1}\right)^2}{4} - c_R^0\, c_S^0}\right) \tag{2.18}$$

We have seen in Section 2.2.1 that the reliability of such a nonlinear regression depends on the shape of the binding isotherm. Since NMR spectroscopy is not a very sensitive technique, binding studies are usually performed at millimolar concentrations of the binding partners. Consequently, the upper limit of binding constants that can be reliably determined is 10^4 M^{-1}, in some cases 10^5 M^{-1}. This estimate is consistent with the binding isotherms in Figure 2.2, which show that the binding isotherm of a complex with a K_a of 10^4 M^{-1} is almost too steep to allow an accurate estimation of complex stability if $c_R^0 = 1$ mM. One way of determining higher binding constants is to perform competitive titrations, for example by adding increasing amounts of a substrate whose affinity to the receptor is to be determined to a complex of known stability. These titrations afford a ratio of two stability constants from which the unknown K_a is calculated by using the known one.

Information on binding enthalpy and entropy can be obtained by performing NMR titrations at different temperatures and using van't Hoff plots to determine ΔH^0 and ΔS^0. The underlying equation (2.19) results from combining equations (2.7) and (2.9):

$$\ln K_a = -\frac{\Delta H^0}{R}\,T^{-1} + \frac{\Delta S^0}{R} \qquad (2.19)$$

Thus, plotting T^{-1} *versus* $\ln K_a$ should yield a straight line with a slope of $-\Delta H^0/R$ and an intercept of $\Delta S^0/R$. This method has several drawbacks. One is that ΔH^0 and ΔS^0 are not necessarily independent of temperature. Another is that, for practical reasons, temperature-dependent measurements can often only be made over a limited range of temperatures. As we will see later, isothermal titration calorimetry (ITC) allows the direct determination of the enthalpy and entropy in a single titration (Section 2.2.6), making this method much more suitable for precisely analyzing the thermodynamics of complex formation.

Further reading

Another NMR spectroscopic technique that is widely used to characterize supramolecular systems is *diffusion-ordered spectroscopy* or DOSY. DOSY gives rise to two-dimensional spectra in which the ^{1}H NMR spectrum of the sample is depicted along the *x*-axis and each signal in this spectrum is assigned a diffusion coefficient D on the *y*-axis. Spectra of pure compounds will therefore contain a single horizontal trace at a certain D value, whereas those of mixtures will contain a series of horizontal traces that allow the individual compounds, whose structures can be deduced from the corresponding signal patterns in the ^{1}H NMR spectrum, to be sorted according to their hydrodynamic radius R_H. Assuming a spherical shape, R_H is calculated by using the Stokes–Einstein equation $D = kT/6\pi\eta R_H$, where k is the Boltzmann constant, T is the temperature, and η is the viscosity of the medium. Accordingly, the smaller D, the larger R_H.

Since the formation of a supramolecular complex usually involves the binding of a smaller substrate molecule to a larger receptor, resulting in a complex similar in size to the receptor itself, DOSY spectra of receptor–substrate mixtures contain two horizontal traces if complex formation is slow on the NMR timescale and the substrate not fully bound. Of these traces, the one corresponding to the larger D value represents that of the free substrate and the other that of the complex. The latter trace contains signals in the ^{1}H NMR spectrum of the receptor and the substrate, both of which apparently have the same D value and therefore belong to the same species. If, on the other hand, complex formation is fast on the NMR timescale, the extent to which the diffusion coefficient of the substrate decreases with increasing receptor concentration correlates with the extent of complex formation, as in the case of NMR titrations. Binding isotherms can thus be obtained from which the stability of the complex can be determined. For an overview of the use of DOSY in supramolecular chemistry, see:

Cohen Y, Avram L, Frish L. *Diffusion NMR spectroscopy in supramolecular and combinatorial chemistry: an old parameter – new insights.* Angew. Chem. Int. Ed. 2005, 44, 520–54.

2.2.3 UV–vis spectroscopy

UV–vis spectroscopy can be used for binding studies if the receptor and/or the substrate contain suitable chromophores whose optical properties are affected by complex formation. This is usually the case when the environment of the chromophores differs in the complexed and uncomplexed states. A typical effect is an increase or decrease in the intensity of an absorption band, which is followed as the receptor–substrate ratio is changed. Since the electronic excitations that give rise to these absorption bands are fast, UV–vis spectroscopy can in principle distinguish between the free and the complexed binding partners. In reality, however, the corresponding bands overlap and differ mainly in their intensity. In addition, the molar absorptivity coefficient of the complex is usually unknown. Therefore, UV–vis spectroscopic binding studies must also rely on titrations and the fitting of binding isotherms.

A typical strategy is to prepare a series of samples containing a fixed concentration of one component, for example the receptor, and varying concentrations of the respective binding partner, followed by measuring the absorbance of the solutions A_{obs} at a given wavelength. For the mathematical treatment of the resulting binding isotherm, A_{obs} must be correlated with the complex concentration c_C. Provided that the Beer–Lambert law is valid under the conditions of the measurement, A_{obs} is expressed as the sum of the absorptions of all species in solution, which for 1:1 equilibrium are R, S, and C, according to the following equation:

$$A_{obs} = A_R + A_S + A_C \tag{2.20}$$

The mathematical treatment is facilitated by choosing the conditions such that only two of the interacting species absorb at the wavelength at which A_{obs} is recorded. Assuming that the substrate is UV–vis silent under the measurement conditions and that 1 cm cuvettes are used, equation (2.20) leads to the simpler equation (2.21):

$$A_{obs} = A_R + A_C = \varepsilon_R c_R + \varepsilon_C c_C \tag{2.21}$$

This equation can be rearranged to afford equation (2.22) by using the mass balance for c_R. The combination of equations (2.4) and (2.22) then gives (2.23), which correlates the measured absorption A_{obs} with the K_a of a 1:1 complex. The term $\varepsilon_R c_R^0$ is the absorption of the free receptor in the absence of the substrate, whereas $\Delta\varepsilon = \varepsilon_C - \varepsilon_R$ is usually unknown. Therefore, K_a and $\Delta\varepsilon$ must be fitted during the nonlinear regression.

$$A_{obs} = \varepsilon_R \left(c_R^0 - c_C \right) + \varepsilon_C c_C = \varepsilon_R c_R^0 + (\varepsilon_C - \varepsilon_R) c_C \tag{2.22}$$

$$A_{obs} = \varepsilon_R c_R^0 + (\varepsilon_C - \varepsilon_R) \left(\frac{c_R^0 + c_S^0 + K_a^{-1}}{2} - \sqrt{\frac{\left(c_R^0 + c_S^0 + K_a^{-1} \right)^2}{4} - c_R^0 \, c_S^0} \right) \tag{2.23}$$

The major difference between an NMR and a UV–vis titration is that the latter requires much lower concentrations. As a result, binding constants between 10^6 and 10^7 M^{-1} can be easily determined, depending on the molar absorptivity coefficient of the participating binding partners. Another advantage of UV–vis titrations is that lower amounts of receptor and substrate are required than for NMR titrations. A disadvantage is that the changes in the UV bands usually provide little structural information about the complex formed.

Another strategy for obtaining binding constants from UV–vis spectrometric measurements is the picrate extraction method developed by Cram to determine the stability of crown ether complexes. Rather than recording binding isotherms, this method measures the extent to which a metal salt partitions between an aqueous solution and an organic solution containing a receptor. The concentrations of the metal ions in both phases are estimated indirectly by UV–vis spectroscopy from the intensities of the absorption band of the picrate counterion. The mathematical treatment of the measurements is based on the following equations:

$$(M \cdot \text{Pic})_{\text{org}} + R_{\text{org}} \rightleftharpoons (M \cdot R \cdot \text{Pic})_{\text{org}} \qquad K_a = \frac{c_{(M \cdot R \cdot \text{Pic})}}{c_{(M \cdot \text{Pic})} \, c_R} \qquad (2.24a)$$

$$M_{\text{aq}} + \text{Pic}_{\text{aq}} \rightleftharpoons (M \cdot \text{Pic})_{\text{org}} \qquad K_{\text{distribution}} = \frac{c_{(M \cdot \text{Pic})}}{c_M \, c_{\text{Pic}}} \qquad (2.24b)$$

$$M_{\text{aq}} + \text{Pic}_{\text{aq}} + R_{\text{org}} \rightleftharpoons (M \cdot R \cdot \text{Pic})_{\text{org}} \qquad K_{\text{extraction}} = \frac{c_{(M \cdot R \cdot \text{Pic})}}{c_M \, c_{\text{Pic}} \, c_R} \qquad (2.24c)$$

$$K_a = \frac{K_{\text{extraction}}}{K_{\text{distribution}}} \qquad (2.25)$$

Equation (2.24a) specifies the binding constant of the complex between the receptor and a metal picrate when both reside in the organic phase. The distribution coefficient of the metal salt between the aqueous and the organic phases is given by equation (2.24b), and the corresponding extraction coefficient by equation (2.24c). Combining all three equations yields equation (2.25), which shows that the stability of the complex results from the ratio $K_{\text{extraction}}/K_{\text{distribution}}$.

These measurements involve two extractions, one in which the metal salt is extracted from the aqueous to the organic phase in the absence of the receptor, and an analogous one in which the receptor is present. The respective concentrations of the metal salt in the organic layers are quantified by measuring the absorption of the organic solutions and calculating the picrate concentrations using the known molar absorptivity coefficient. The concentration in the aqueous phase is calculated from the mass balance and the initial metal salt concentration. The receptor is considered to be insoluble in the aqueous phase, while the salt is considered to be ion-paired in the organic phase.

2.2.4 Fluorescence spectroscopy

Because of the sensitivity of fluorescence spectroscopy and the correspondingly low concentrations required, fluorescence titrations allow the determination of very high binding constants between 10^6 and 10^{10} M^{-1}. This technique requires the presence of a fluorescent chromophore in at least one of the binding partners. In the absence of quenching, the mathematical treatment for deriving stability constants from the binding isotherms of fluorescence titrations is very similar to that of UV–vis titrations. If there is a linear correlation between the observed fluorescence F_{obs} and the concentrations of the emitting compounds, and if one of the binding partners is nonfluorescent, we start from an expression closely related to equation (2.21):

$$F_{obs} = F_R + F_C = k_R c_R + k_C c_C \tag{2.26}$$

Equation (2.26) can then be transformed in a similar way as before to yield equation (2.27), which, together with the expression for c_C in equation (2.4), forms the basis of the nonlinear regression analysis of a 1:1 binding isotherm:

$$F_{obs} = k_R c_R^0 + (k_C - k_R) c_C \tag{2.27}$$

Note that if only the complex is fluorescent, that is, if fluorescence is turned on during complex formation, equation (2.27) simplifies further to give $F_{obs} = k_C c_C$. Modified versions of these equations must be used if F_{obs} is affected by quenching [5].

2.2.5 Potentiometry

Potentiometric titrations provide information about the strength of receptor–substrate interactions in systems that involve protonation equilibria. A typical example is the polyazamacrocycle **2.3** with seven amino groups along the ring, whose protonation is characterized by the equilibria specified in Figure 2.12. The corresponding cumulative protonation constants $\log \beta$ are determined electrochemically by measuring the change in pH when a solution of the fully protonated form of **2.3** is titrated with NaOH followed by fitting the resulting titration curve.

Substrates interacting with the receptor influence these protonation constants. If, for example, the substrate prefers to interact with the cationic form of the receptor, complex formation will stabilize the protonated state. Conversely, if the free electron pair of a nitrogen atom forms a coordinate bond to a metal ion, the protonation of this nitrogen atom becomes more difficult. Recording the protonation curves in the presence of the substrate, fitting them to the underlying equilibria, and comparing the $\log \beta$ values obtained with those of the free receptor thus provides information about the efficiency of substrate binding. If the substrate is also involved in protonation equilibria, the corresponding protonation constants must be determined independently.

As an example, Figure 2.12 summarizes the cumulative protonation constants obtained by titrating **2.3** alone and together with benzene-1,3,5-tricarboxylate [9]. The table shows that each $\log \beta_n^A$ value is larger than the corresponding $\log \beta_n^H$ value of the free receptor, demonstrating that the interaction of the tricarboxylate with the macrocycle stabilizes the respective protonated species. The stepwise binding constants $\log K_n$, that is, the differences $\log \beta_n^A - \log \beta_n^H$, specify the extent of this stabilization. These constants are equivalent to the association constants obtained by other methods. In the example above, the $\log K_n$ values show that the complex becomes more

2.3 (= L)

Benzene-1,2,3-tricarboxylate
(= A^{3-})

Equilibria:

$$\log \beta_n^H \quad L \quad + \quad n\,H^+ \quad \rightleftharpoons \quad H_nL^{n+}$$

$$\log \beta_n^A \quad L \ + \ n\,H^+ \ + \ A^{3-} \quad \rightleftharpoons \quad H_nLA^{(n-3)+}$$

$$\log K_n \quad H_nL^{n+} \quad + \quad A^{3-} \quad \rightleftharpoons \quad H_nLA^{(n-3)+}$$

n	$\log \beta_n^H$	$\log \beta_n^A$	$\log K_n$
3	27.7	30.9	3.2
4	34.1	39.2	5.1
5	37.8	46.1	8.3
6	40.0	51.0	11.0
7	41.9	54.6	12.7

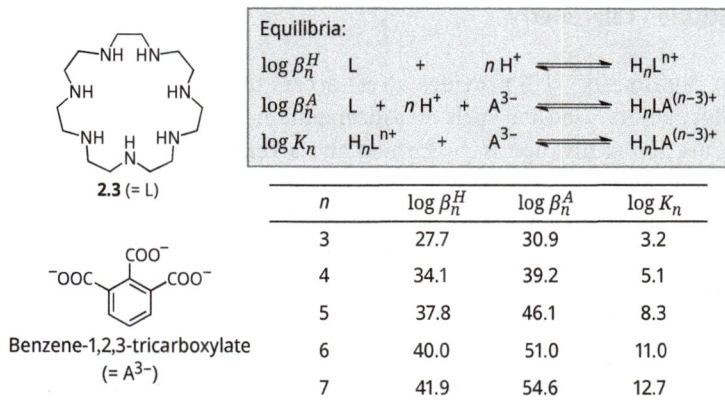

Figure 2.12: Structures of the polyazamacrocycle **2.3** (L) and benzene-1,3,5-tricarboxylate (A^{3-}), definition of the cumulative protonation constants $\log \beta_n^H$ and $\log \beta_n^A$, and the stepwise binding constant $\log K_n$ on the basis of the corresponding equilibria. The data in the table summarize selected cumulative and stepwise equilibrium constants of complexes between **2.3** and benzene-1,3,5-tricarboxylate with different degrees of protonation. The $\log K_n$ values are the differences $\log \beta_n^A - \log \beta_n^H$.

stable with increasing degree of protonation, which is consistent with the strengthening of electrostatic interactions as the charge of the receptor increases (Section 3.1.2). The pH-dependent distribution of the species present in such equilibria is often depicted graphically using speciation diagrams, such as the one in Figure 2.13.

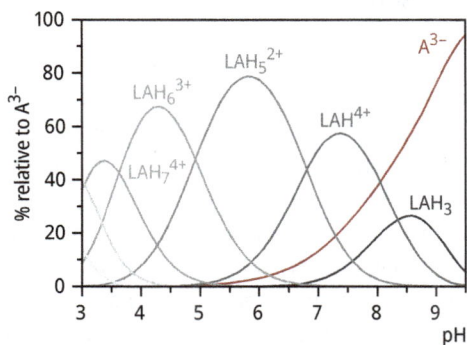

Figure 2.13: Speciation diagram of complexes of **2.3** (L) and benzene-1,3,5-tricarboxylate (A^{3-}), differing in the degree of protonation as a function of pH in 0.15 M aqueous $NaClO_4$ at 298 K. The concentration of the free anion is shown in red.

The stability constants determined by potentiometry depend on a number of factors, including the temperature, the concentrations of the binding partners, the concentration and type of the background electrolyte, and the ionic strength of the medium. Therefore, only results measured under identical conditions are comparable.

2.2.6 Isothermal titration calorimetry

Isothermal titration calorimetry (ITC) provides direct access to the enthalpy of complex formation. Calorimeters used for such measurements have two thermostatted cells, one in which the actual titration is performed and a reference cell containing only the solvent. During the measurement, a solution of one binding partner is added in several injections from a syringe to the solution of the other component of the complex, usually the receptor, and the heat generated or consumed during complex formation is recorded. This is done by measuring either the resulting temperature difference between the two cells or the energy input required to bring the cells back to the same temperature. Very small heat flows can thus be accurately determined.

Calorimetry is widely used in the life sciences to study biological systems operating in water. With the availability of calorimeters that allow the use of organic solvents, this technique has also become popular in supramolecular chemistry. Its major advantage over other analytical methods is that calorimetric measurements deliver the full thermodynamic profile of complex formation in a single titration and provide information about complex stoichiometry. ITC is also very versatile because it does not require the presence of specific structural elements in the binding partners. Only when complex formation is entirely entropy driven, lacking a measurable heat change, can ITC not be used.

Figure 2.14a shows the primary result of an ITC titration, consisting of a series of peaks reflecting the temperature change during each titration step. The direction of the peaks indicates that complex formation in this case is exothermic, whereas an endothermic reaction would produce peaks in the opposite direction. Integration of the peaks gives the heat of reaction Q associated with each addition, which correlates with the amount of complex formed in each step. The first additions usually produce large heat changes because the binding partner in the cell is present in excess, leading to the full complexation of the added component. The heat changes then become progressively smaller as the amount of free binding partner in the cell decreases. At the end of the titration, only the heat of dilution of the added compound is measured. These dilution effects are usually accounted for by titrating the solution in the syringe to the pure solvent and subtracting the measured heats from those of the actual titration. Because ITC titrations do not involve a series of individual measurements, but rather the stepwise addition of small aliquots of one binding partner to a solution of the other, the resulting binding isotherms ideally have a sigmoidal shape, as shown in Figure 2.14b, similar to the pH curves of acid–base titrations.

The mathematical formalism for deriving the binding constant from such an isotherm is based on the following considerations. The heat produced or generated during each addition is proportional to the total enthalpy ΔH^0 of the process, the amount of complex formed, and the cell volume V_0 according to the following equation:

$$Q = V_0 \, \Delta H^0 \, c_C \tag{2.28}$$

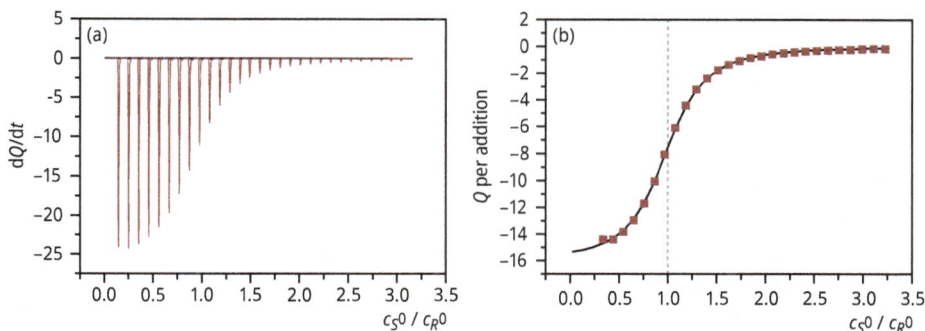

Figure 2.14: Thermogram resulting from an ITC titration (a) and binding isotherm reflecting the normalized heats per peak as a function of the molar ratio of the receptor and the substrate (b).

This equation establishes the correlation between the measured parameter Q and the complex concentration c_C. However, in contrast to the techniques discussed earlier, the total concentrations $c_R^{0,i}$ and $c_S^{0,i}$ of receptor and substrate are different for each injection. These concentrations are given by equations (2.29a, b), where v is the injection volume and the expression $(1 - v/V_0)$ specifies the reduction of the concentration in the cell after i injections. The volume V_0 remains constant throughout the titration. These equations are valid if the cell contains the receptor and the syringe contains the substrate.

$$c_R^{0,i} = c_R^0 \left(1 - \frac{v}{V_0}\right)^i \tag{2.29a}$$

$$c_S^{0,i} = c_S^0 \left[1 - \left(1 - \frac{v}{V_0}\right)^i\right] \tag{2.29b}$$

Since each injection is associated with a consumption or release of heat proportional to the change in complex concentration, the heat change between injection $i-1$ and i is given by the following equation:

$$Q = \frac{1}{v c_S^0} \left[Q_i - Q_{i-1}\left(1 - \frac{v}{V_0}\right)\right] = \frac{V_0 \Delta H^0}{v c_S^0} \left[c_C^i - c_C^{i-1}\left(1 - \frac{v}{V_0}\right)\right] \tag{2.30}$$

Combining equations (2.30) and (2.4) affords an expression that is used for the fitting of the binding isotherm to obtain the two unknowns ΔH^0 and K_a. Based on these parameters, ΔG^0 is calculated from K_a (equation (2.7)) and ΔS^0 from ΔG^0 and ΔH^0 (equation (2.9)).

Another useful parameter resulting from an ITC titration is the stoichiometry factor n, which is the receptor–substrate ratio at the inflection point of the binding isotherm and provides information about the composition of the complex. For a 1:1 complex, the inflection point occurs at an equimolar ratio of the binding partners

($n = 1$), whereas n values of 0.5 or 2 are consistent with 1:2 or 2:1 complexes, respectively. Values of n that deviate significantly from a reasonable binding stoichiometry often indicate that the actual concentration of one or both binding partners is different from that assumed, for example due to impurities in the samples.

A single ITC titration thus provides information on the binding model as well as the full thermodynamic profile of the binding process, making ITC a powerful method for binding studies. Stability constants between 10^1 M^{-1} and $10^8 M^{-1}$ can be determined, depending on the concentrations chosen. This range can be extended beyond 10^8 M^{-1} by performing competitive titrations. The reliability of K_a is mainly determined by the accuracy with which the binding isotherms can be fitted, as we have already seen for other methods. If the stability of the complex is too low, the curves will be flat and reliable analysis may not be possible. On the other hand, if the stability is too high, the curves do not have a pronounced curvature. To estimate the conditions that give rise to optimal S-shaped isotherms, the c parameter (Wiseman parameter) has been introduced. This parameter is given by the following equation:

$$c = n_R \; c_R^0 \; K_a \tag{2.31}$$

Accordingly, the Wiseman parameter represents the product of the number of binding sites per receptor n_R, the stability constant K_a, and the receptor concentration c_R^0 originally present in the cell. If this product is between 10 and 500, the curvature of the binding isotherm is sufficiently well defined to allow fitting. If $c < 10$, no reliable fitting is possible. If $c > 500$, only the binding enthalpy and complex stoichiometry can be accurately determined. Estimating c thus helps choose the optimal c_R^0 for a given (or assumed) K_a. Figure 2.15 shows a series of ITC binding isotherms to illustrate how their shapes change for different values of c.

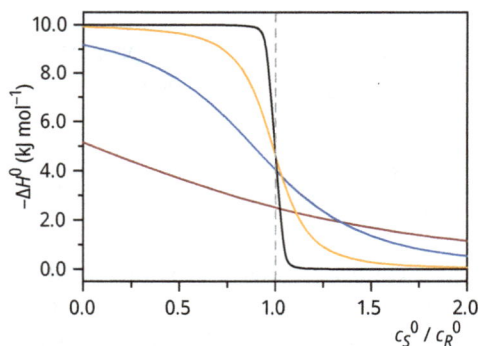

Figure 2.15: Examples of ITC binding isotherms corresponding to different c values. The concentration c_R^0 amounts to 10^{-2} M for all curves and K_a is varied between 10^2 M^{-1} ($c = 1$, red), 10^3 M^{-1} ($c = 10$, blue), 10^4 M^{-1} ($c = 100$, orange), and 10^6 M^{-1} ($c = 10,000$, black). The ΔH^0 amounts to -10 kJ mol^{-1} in all cases.

Bibliography

[1]   Inoue Y, Hakushi T. Enthalpy–entropy compensation in complexation of cations with crown ethers and related ligands. J. Chem. Soc., Perkin Trans. 1985, 2, 935–46.

[2]   Cram DJ, Blanda MT, Paek K, Knobler CB. Constrictive and intrinsic binding in a hemicarcerand containing four portals. J. Am. Chem. Soc. 1992, 114, 7765–73.

[3]   Job P. Formation and stability of inorganic complexes in solution. Ann. Chim. Appl. 1928, 9, 113–203.

[4]   Hibbert DB, Thordarson P. The death of the Job plot, transparency, open science and online tools, uncertainty estimation methods and other developments in supramolecular data analysis. Chem. Commun. 2016, 52, 12792–805.

[5]   Thordarson P. Determining association constants from titration experiments in supramolecular chemistry. Chem. Soc. Rev. 2011, 40, 1305–23.

[6]   Online tools for supramolecular chemistry research and analysis (Accessed February 01, 2024, http://supramolecular.org).

[7]   Vacca A, Francesconi O, Roelens S. BC_{50}: a generalized, unifying affinity descriptor. Chem. Rec. 2012, 12, 544–66.

[8]   Cohen Y, Slovak S, Avram L. Solution NMR of synthetic cavity containing supramolecular systems: what have we learned on and from? Chem. Commun. 2021, 57, 8856–84.

[9]   Bencini A, Bianchi A, Burguete MI, Dapporto P, Doménech A, García-España E, Luis SV, Paoli P, Ramírez JA. Selective recognition of carboxylate anions by polyammonium receptors in aqueous solution. Criteria for selectivity in molecular recognition. J. Chem. Soc., Perkin Trans. 1994, 2, 569–77.

3 Understanding molecular recognition

CONSPECTUS: When molecules meet in solution and experience attractive interactions, they stay together for some time. This chapter deals with the forces that mediate these interactions. First, however, a number of general factors that contribute to the efficiency of complex formation are outlined.

3.1 Modes of binding

3.1.1 General considerations

What general parameters influence complex formation?

A key parameter affecting the stability of a complex is the size of the interface shared by the binding partners, with larger areas of contact typically allowing for more efficient interactions (Section 3.5.3). This aspect is illustrated in Figure 3.1.

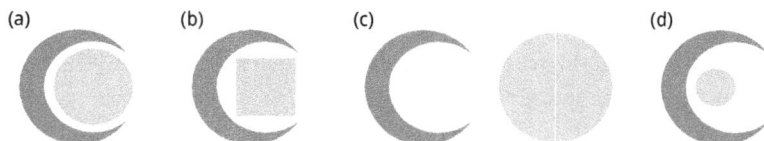

Figure 3.1: Dependence of the extent to which molecules come into contact during complex formation on their shape and size. The case of an optimal shape and size complementarity is shown in (a). In (b), the binding partners are not shape complementary, while in (c) and (d) they are shape complementary but not size complementary.

Assuming that the two binding partners are rigid, the spherical substrate in Figure 3.1a is clearly better suited to efficiently approach the inner surface of the receptor binding site than the square-shaped substrate in Figure 3.1b. However, shape selectivity only leads to strong binding if the binding partners are also complementary in size. Substrates that are too large cannot enter the receptor cavity (Figure 3.1c), while those that are too small fill only part of the available space, significantly reducing the size of the contact interface (Figure 3.1d).

To a first approximation, one would therefore expect the situation shown in Figure 3.1a to be the most favorable, but this is only true if the interfaces that come into contact in the complex are also complementary in their electronic properties, that is, if an electron-rich region in one binding partner comes close to an electron-poor region in the other binding partner. Characteristic distributions of (partial)

https://doi.org/10.1515/9783111315171-003

charges along the molecular backbones of the receptor and substrate thus create attractive forces that hold the complex together. These electrostatic interactions arise from polar bonds between carbon and heteroatoms and are therefore often classified according to the functional group or structural unit responsible for the bond polarization (hydrogen bonding, aromatic interactions, etc.). A more general approach to assess these interactions is based on the electrostatic potential surfaces of the binding partners. These surfaces are generated by mapping the potential energy of a virtual charged probe onto a surface of constant electron density (usually 0.002 electrons Å^{-2}), followed by color coding to indicate where along the surface a binding partner will find regions of positive or negative electrostatic potential. An example is shown in Figure 3.2.

(a)   (b)   (c)

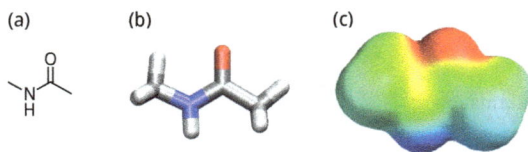

Figure 3.2: Molecular structure (a), stick model (b), and electrostatic potential surface (c) of *N*-methyl acetamide. The color coding covers a potential range from –120 to +120 kJ mol^{-1}, with red and blue indicating values greater than or equal to the absolute maximum in negative and positive potentials, respectively.

The red area surrounding the carbonyl oxygen atom of *N*-methyl acetamide in Figure 3.2 indicates that this atom has a negative electrostatic potential, as expected for an electronegative atom. The blue color surrounding the proton on the nitrogen atom indicates a positive electrostatic potential in this region. These areas are therefore available for electrostatic interactions with a positively or negatively charged or polarized binding partner, respectively.

Apparently, electrostatic potential surfaces are useful for estimating the ability of molecules to interact and for predicting the preferred binding mode (and we will see in Section 3.4 that they even allow prediction of binding strength). However, reliable predictions based on visual inspection can only be made if the scales of the color coding are chosen correctly. Very large ranges between the minimum and maximum electrostatic potentials are inappropriate because they cause molecular surfaces to appear nonpolar (green) when they could actually have a substantial positive or negative value, while ranges that are too small lead to an overestimation of the actual electrostatic potentials. Therefore, whenever electrostatic potential surfaces are used in this book to illustrate binding properties, the minimum and maximum values are specified. In addition, comparable scales are chosen in most cases to allow for comparison.

It must be emphasized that although the rationalization of noncovalent interactions based solely on electrostatic forces is straightforward and usually sufficient for supramolecular complexes, other binding mechanisms sometimes cannot be neglected. An important one involves orbital interactions, that is, the overlap of a filled orbital of a donor with an empty orbital of an acceptor. These interactions are based on the transfer of electrons from one binding partner to the other, hence the term charge-transfer interactions. The participating orbitals can be σ, π, or n orbitals. Of particular importance are charge-transfer interactions between the highest occupied molecular orbital (HOMO) of electron-rich aromatic systems and the lowest unoccupied molecular orbital (LUMO) of electron-poor systems, where orbital mixing leads to the appearance of a new band in the UV–vis spectrum, the so-called charge-transfer band. We will come back to this type of interactions in Section 3.1.10.

In addition to the actual binding forces, another aspect that influences complex stability is the flexibility of the binding partners. During complex formation, the receptor and/or the substrate often undergo a conformational reorganization to optimize the arrangement of functional groups in the contact regions, thereby enhancing their interactions. Conformational changes also allow the binding partners to correct a shape or size mismatch. For example, by contracting or expanding its cavity, a receptor can adapt to the size of the substrate. These effects typically have a favorable effect on binding enthalpy, but restricting the conformational mobility of the binding partners during complex formation also has entropic consequences. Entropy becomes unfavorable if the conformational changes associated with complex formation become too large or if the binding becomes too tight, potentially causing the situation in Figure 3.1a to be less beneficial than that in which the substrate retains some mobility. This relationship between receptor conformation and complex stability will be discussed in more detail in Section 3.5.1.

Finally, solvent effects have a profound influence on complex stability as already mentioned in Section 2.1 (see also equation (2.8)). The solvent promotes complex formation even in the case of weak direct interactions if the solvation of the complex is more favorable than that of the individual binding partners. Conversely, solvent molecules prevent intermolecular interactions if the binding partners are too strongly solvated. These effects will be further discussed in Section 3.3.

The stability of a complex therefore depends on the interplay of various factors. In the following, we will first look at the noncovalent interactions responsible for keeping the components of a supramolecular complex together. Subsequently, methods used to experimentally assess the strength of these interactions will be presented, and we will look at how conformational flexibility and the solvent influence complex stability. Coordinative interactions have a substantial covalent character and will therefore not be discussed here. However, several applications of coordinative interactions in supramolecular chemistry will be presented later.

3.1.2 Ion–ion interactions

What types of noncovalent interactions cause molecules to stay together?

The prototype of an electrostatic interaction is the attraction between a cation and an anion (Figure 3.3a). The potential energy E associated with this interaction is estimated by using equation (3.1), which is derived from Coulomb's law:

$$E = \frac{q_1 q_2}{4\pi\varepsilon\varepsilon_0 r} \tag{3.1}$$

In this equation, q_1 and q_2 denote the elementary charges ($\pm 1.602 \times 10^{-19}$ C per positive or negative charge), ε_0 the permittivity of the vacuum (8.854×10^{-12} C V^{-1} m^{-1}), ε the relative permittivity of the medium, and r the distance of the ions. The interaction becomes attractive ($E < 0$) when the ions are oppositely charged, that is, when q_1 and q_2 have different signs. Moreover, the inverse correlation of E with ε and r shows that the interaction strength becomes weaker as the distance of the ions increases or the medium becomes more polar. This dependence is illustrated in Figure 3.3b.

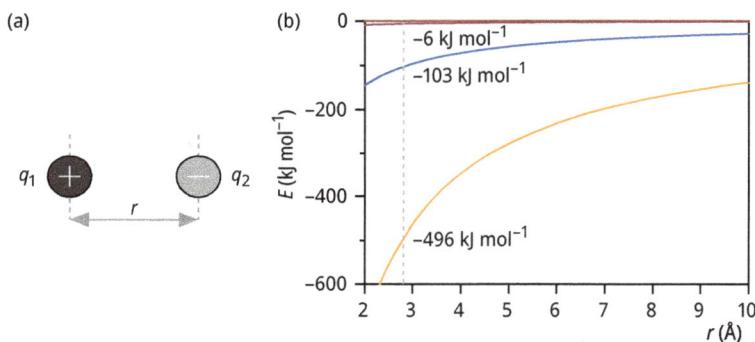

Figure 3.3: Schematic representation of two oppositely charged ions interacting at the distance r (a) and graphs showing the dependence of the potential energy E on the distance r in the gas phase ($\varepsilon = 1$, orange), in chloroform ($\varepsilon = 4.8$, blue), and in water ($\varepsilon = 78$, red) (b) (relative permittivities refer to 273 K). The dotted line denotes the distance between Na^+ and Cl^- ions in NaCl (2.8 Å).

According to Figure 3.3b, the strength of an ionic interaction in the gas phase easily reaches several hundred kJ mol^{-1}, comparable to covalent interactions. However, medium effects cause a substantial attenuation, reducing E by about two orders of magnitude when moving from the gas phase to water.

It must be emphasized that the estimates in Figure 3.3b do not reflect realistic binding strengths because equation (3.1) does not properly take into account atomic or molecular properties. First, this equation applies only to point charges, but ions have finite radii and therefore approach each other only within a certain distance. Once their electron clouds get too close, the interactions become repulsive, which is not considered in equation (3.1). Second, the effects of the medium on complex stability are only partly due to the relative permittivity of the solvent. Solvation, that is, the interaction between binding partners and solvent molecules, as well as the reorganization of the solvent molecules during complex formation (Section 3.3) also play a decisive role. Moreover, if the receptor cavity is shielded from the solvent, as in the case of deep cavitands (Section 4.1.9) or proteins, the permittivities experienced by the substrate inside and outside the binding site differ substantially. Finally, ions come in many different structures. Inorganic anions, for example, can be linear (N_3^-), trigonal planar (NO_3^-), tetrahedral (SO_4^{2-}), or octahedral (PF_6^-). Organic ions have an even greater structural variability, causing the strength of the interactions to depend on the direction from which the binding partner approaches. In addition, inductive and mesomeric effects lead to a distribution of the charge over relatively large regions of the molecules. This is shown in Figure 3.4a,b for the tetramethylammonium cation and the benzoate anion. The respective electrostatic potential surfaces illustrate that the positive charge of the tetramethylammonium cation is distributed across all four methyl groups, consistent with the higher electronegativity of nitrogen relative to carbon. Similarly, the negative charge of the benzoate anion is shared between the two oxygen atoms, and the phenyl ring also has a pronounced negative potential, which is the basis for a type of noncovalent interactions known as cation–π interactions (Section 3.1.8). Electrostatic interactions between charged molecules are therefore usually weaker and depend on many more factors than equation (3.1) suggests. However, this equation correctly predicts that the strength of the interactions is directly related to the number of charges on the binding partners.

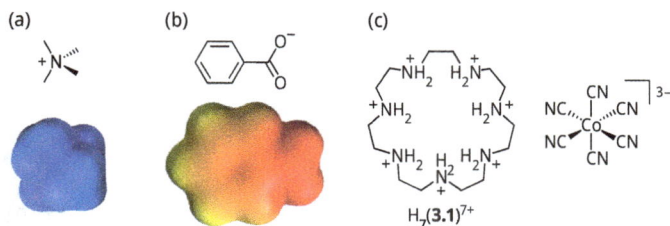

Figure 3.4: Molecular structures and electrostatic potential surfaces of the tetramethylammonium cation (a) and the benzoate anion (b). The color coding covers a potential range from –500 to +500 kJ mol^{-1}, with red and blue indicating values greater than or equal to the absolute maximum in negative and positive potentials, respectively. The heptaprotonated form of the polyazamacrocycle **3.1** and [Co(CN)$_6$]$^{3-}$ shown in (c) forms a complex with [Co(CN)$_6$]$^{3-}$ that is mainly stabilized by ion–ion interactions.

Macrocyclic polyammonium ions, such as the heptaprotonated form of the azacrown **3.1**, and complex anions, such as [Co(CN)$_6$]$^{3-}$ (Figure 3.4c), interact mainly by ionic interactions (Section 4.1.1).

3.1.3 Ion–dipole interactions

Ion–dipole interactions occur between permanently charged atomic or molecular species and polar molecules. In the electrostatic model, the strength of these interactions depends on the relative magnitude of the attractive and repulsive forces between the ion and the partial charges in the polarized molecule. This situation is illustrated in Figure 3.5, where a cation is oriented at a distance r from a diatomic polar molecule with the partial charges $\delta+$ and $\delta-$.

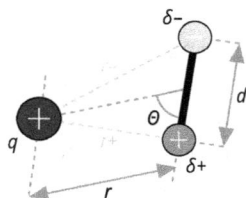

Figure 3.5: Schematic representation of a cation interacting with a polar diatomic molecule at a distance r and an angle θ. The distances between the ion and the individual partial charges $\delta+$ and $\delta-$ are denoted by $r+$ and $r-$, and d is the distance between the partial charges in the diatomic molecule.

The cation experiences an attractive interaction with $\delta-$, located at a distance of $r-$, while the interactions with $\delta+$, located at a distance of $r+$, are repulsive. The resulting balance of attraction and repulsion is given by equation (3.2a), which is derived from (3.1) and treats the electrostatic interactions of the ion with each partial charge separately. Taking into account the relevant geometric parameters and the dipole moment μ, which is defined by the magnitude and the distance of the charge separation in the polar molecule, equation (3.2a) transforms into equation (3.2b), which differs from (3.1) in several aspects:

$$E = \frac{q}{4\pi\varepsilon\varepsilon_0 r}\left(\frac{\delta+}{r+} + \frac{\delta-}{r-}\right) \tag{3.2a}$$

$$E = \frac{q\mu}{4\pi\varepsilon\varepsilon_0 r^2}\cos\theta \tag{3.2b}$$

First, the charge of one ion in equation (3.1) is replaced by the dipole moment of the polar molecule. Since μ is a vector pointing in the direction of the negative partial charge, ion–dipole interactions, unlike ion–ion interactions, are directional. This aspect is reflected in the $\cos\theta$ term. A cation ($q > 0$) oriented at an angle θ of 0° to the positive end of the dipole experiences maximum repulsion. At $\theta = 90°$, the attractive and repulsive interactions cancel so that E becomes zero, and the interactions become most attractive at $\theta = 180°$. Finally, E is proportional to the inverse squared distance as a consequence of the combination of attractive and repulsive interactions in the ion–dipole interactions. The strength of these interactions therefore decreases more rapidly than that of ion–ion interactions.

To estimate the extent to which the strengths of ion–ion and ion–dipole interactions differ, we must compare q and μ/r. Since dipole moments are in the order of 10^{-30}

C m and distances are in the Ångström range, μ/r is about 10^{-20} C. The elementary charge, on the other hand, is about 10^{-19} C, and a single ion–dipole interaction can therefore be expected if other parameters are comparable, to be at least one order of magnitude weaker than a single ion–ion interaction. Ion–dipole interactions can still become substantial when multiple dipoles are involved. A good example is the interaction of ions with the water molecules in their solvation shells. Hydration enthalpies therefore provide useful information on how the nature of the ion influences the binding strength. Selected values are given in Table 3.1 [1].

Table 3.1: Hydration enthalpies ΔH^0_{hydr} and ionic radii for a selection of inorganic cations and anions.

Cation	ΔH^0_{hydr}(kJ mol^{-1})	Ionic radius (Å)	Anion	ΔH^0_{hydr}(kJ mol^{-1})	Ionic radius (Å)
Li$^+$	−531	0.60			
Na$^+$	−416	0.95	F$^-$	−510	1.36
K$^+$	−334	1.33	Cl$^-$	−367	1.81
Rb$^+$	−308	1.48	Br$^-$	−336	1.95
Cs$^+$	−283	1.69	I$^-$	−291	2.16
Mg^{2+}	−1,949	0.65			
Ca^{2+}	−1,602	0.99			

According to Table 3.1, the strength of ion–dipole interactions correlates with the charge density of the ion, with smaller cations or anions interacting more strongly with water molecules than equally charged larger ions. This is because spreading the same charge over a larger volume weakens the electrostatic interactions, as illustrated by the electrostatic potential surfaces of the halides depicted in Figure 3.6a. Larger halides have less negative electrostatic potentials and therefore interact weaker with an oppositely charged or polarized binding partner.

(a)

(b)

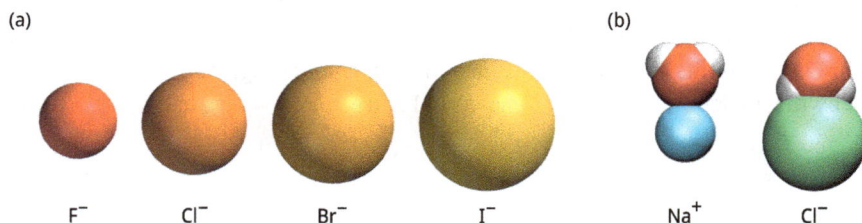

F$^-$ Cl$^-$ Br$^-$ I$^-$ Na$^+$ Cl$^-$

Figure 3.6: Electrostatic potential surfaces of the four halides fluoride, chloride, bromide, and iodide (a) and preferred orientation of a water molecule with respect to a cation and an anion (b). The potentials amount to −1,072.5 kJ mol^{-1} for F$^-$, −797.0 kJ mol^{-1} for Cl$^-$, −687.4 kJ mol^{-1} for Br$^-$, and −616.3 kJ mol^{-1} for I$^-$ according to the respective calculations.

When comparing ions of similar size but different charge, for example Li^+ and Mg^{2+}, the hydration of the doubly charged ion is stronger, in agreement with equation (3.2b). In addition, anions tend to be more strongly hydrated than their corresponding isoelectronic cations. This tendency may seem counterintuitive at first sight, because F^- is larger than Na^+ and Cl^- is larger than K^+, which should result in a lower charge density of the respective anions. However, it has to be considered that a water molecule approaches a cation with its oxygen atom, but an anion with its positively polarized protons, as shown schematically in Figure 3.6b. Since hydrogen atoms are smaller than oxygen atoms, water molecules approach anions at a smaller distance, making the respective ion–dipole interactions stronger than those between water molecules and cations. Typical complexes in supramolecular chemistry stabilized by ion–dipole interactions are the cation complexes of crown ethers (Section 4.1.1).

3.1.4 Dipole–dipole interactions

Two or more uncharged but polar molecules can engage in dipole–dipole interactions. These interactions occur, for example, in solvents with permanent dipole moments such as chloroform, acetone, or DMSO. They result in orientations of the individual dipoles in which their positive and negative ends preferentially approach each other. The strength of these interactions depends on the size of the dipole moments and the relative arrangement of the dipoles. Equation (3.3) mathematically describes the underlying relationship:

$$E = -\frac{\mu_1\mu_2}{4\pi\varepsilon\varepsilon_0 r^3}(2\cos\theta_1\cos\theta_2 - \sin\theta_1\sin\theta_2\cos\varphi) \tag{3.3}$$

This equation is closely related to equation (3.2b) but contains a product of two dipole moments $\mu_1\mu_2$. Since dipole moments are positive, the negative sign makes the interaction attractive if the dipoles are oriented in an antiparallel way. The last term in equation (3.3) accounts for the dependence of E on the arrangement of the dipoles, defined by the angles θ_1, θ_2, and φ, as shown in Figure 3.7.

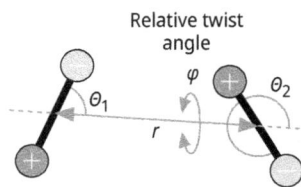

Figure 3.7: Schematic representation of the interaction of two dipoles at a distance r. The angles θ_1, θ_2, and φ appear in equation (3.3) and describe the relative arrangement of the dipoles.

As we have seen in the previous section, the formal replacement of one charge in equation (3.1) by μ/r causes ion–dipole interactions to be weaker than ion–ion interactions. With only dipole moments left in equation (3.3), dipole–dipole interactions are the weakest in this series. In addition, the interaction energy of dipole–dipole interactions falls off most rapidly with distance because E correlates with r^{-3} rather than r^{-2} or r^{-1}.

3.1.5 Hydrogen bonding

At this point, we turn to one of the most important types of noncovalent interactions, namely, hydrogen bonding. Hydrogen bonds contribute to the stabilization of protein structures and the DNA double helix, and are therefore crucial for the function of most biomolecules. Because of this importance, hydrogen bonds are generally treated as a distinct class of noncovalent interactions, but in terms of electrostatics, they are closely related to ion–dipole or dipole–dipole interactions, depending on whether one of the binding partners is an ion or a neutral polar molecule. The characteristic structural feature of hydrogen bonds is that the dipole is generated by a polar bond between a positively polarized proton and an atom or molecular fragment with a negative electrostatic potential. Since protons are small, two molecules interacting by hydrogen bonding can approach each other at a smaller distance than would be possible for binding partners involved in conventional dipole–dipole interactions. As a result, the electrostatic interactions are stronger, as already noted in Section 3.1.3, where we saw that the hydration of anions is more efficient than that of similarly sized cations.

It must be stressed that reducing hydrogen bonds to electrostatic interactions alone is a simplification, but one that is reasonable for the moderately strong hydrogen bonds found in the majority of supramolecular systems. Hydrogen bonds are actually more complex types of interactions involving a combination of electrostatics, polarization, and dispersion (Section 3.1.11), and very strong hydrogen bonds can even have a partially covalent character. This complexity may explain why it is not easy to give a clear-cut definition. In 2011, IUPAC recommended the following [2]:

> The hydrogen bond is an attractive interaction between a hydrogen atom from a molecule or a molecular fragment X–H in which X is more electronegative than H, and an atom or a group of atoms in the same or a different molecule, in which there is evidence of bond formation.

According to this definition, a hydrogen bond can be denoted in the following manner:

$$X - H \cdots Y(-Z)$$

X–H is the hydrogen bond donor because it carries the hydrogen atom. X is typically an electronegative heteroatom such as nitrogen, oxygen, or a halogen atom, but any molecular fragment that causes H to have a more or less pronounced partial positive charge is allowed. The hydrogen bond acceptor, that is, the binding partner of the proton, is a neutral or charged atomic species Y or a molecular species Y–Z. Since the above definition does not restrict the nature of Y to certain atoms or molecular fragments, any structural element that interacts with a positively polarized hydrogen atom is permitted as hydrogen bond acceptor – not only electronegative heteroatoms, but also, for example, π-systems.

The electrostatic attraction between the proton and Y causes the X–H bond to become longer and weaker in the hydrogen bond. These effects can be conveniently followed by different spectroscopic techniques. In the IR spectrum, for example, the vibrational band of the X–H group shifts to lower wavenumbers and becomes broader and more intense in the presence of a suitable hydrogen bond acceptor. At the same time, new vibrational modes belonging to the H···Y bond appear. In the ^{1}H NMR spectrum, the formation of a hydrogen bond is associated with a deshielding of the proton and the shift of the corresponding signal in the spectrum to higher ppm values.

The extent of proton transfer from the hydrogen bond donor to the hydrogen bond acceptor is influenced by the acidity of the donor and the basicity of the acceptor. Thus, there are donor–acceptor combinations where complete proton transfer is thermodynamically favored. However, proton abstraction results in an ion pair that forms a salt bridge, which is partly stabilized by hydrogen bonding, but also by a substantial contribution from ionic interactions. An example is the carboxylate–guanidinium ion pair shown in Figure 3.8a, which is formed from a carboxylic acid and a guanine derivative.

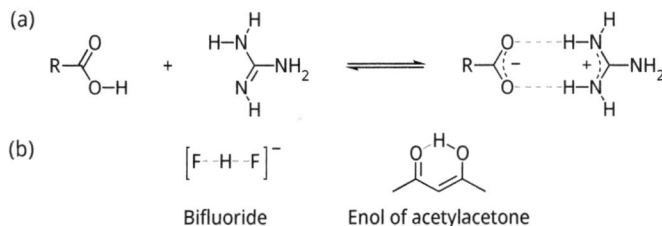

Figure 3.8: Structure and formation of the carboxylate-guanidinium ion pair (a) and examples of systems with very strong hydrogen bonds (b).

In special cases, the proton ends up being bound symmetrically by two short hydrogen bonds between X and Y, often with X = Y. These hydrogen bonds are very strong, potentially even stronger than weak covalent bonds (50–150 kJ mol^{-1} in the gas phase). Examples are the hydrogen bonds in the bifluoride anion or in the enol of acetylacetone (Figure 3.8b). Apart from these special cases, hydrogen bonds are often of intermediate strength (15–50 kJ mol^{-1}) and can also be very weak, not much stronger than dispersion interactions (<15 kJ mol^{-1}). Because of their importance not only in supramolecular chemistry but also in biochemistry, we will look in more detail at the key parameters that influence the strength of hydrogen bonds.

Nature of the hydrogen bond donor: Electrostatic arguments suggest that there should be a direct correlation between the strength of a hydrogen bond and the magnitude of the partial charge on the proton. Indeed, the donor strength of the hydrogen halides increases as the halide becomes more electronegative in the direction HI < HBr < HCl < HF. Note that this trend follows the size of the dipole moments and not the acidity of the hydrogen halides, which is highest for HI and lowest for HF (Figure 3.9a).

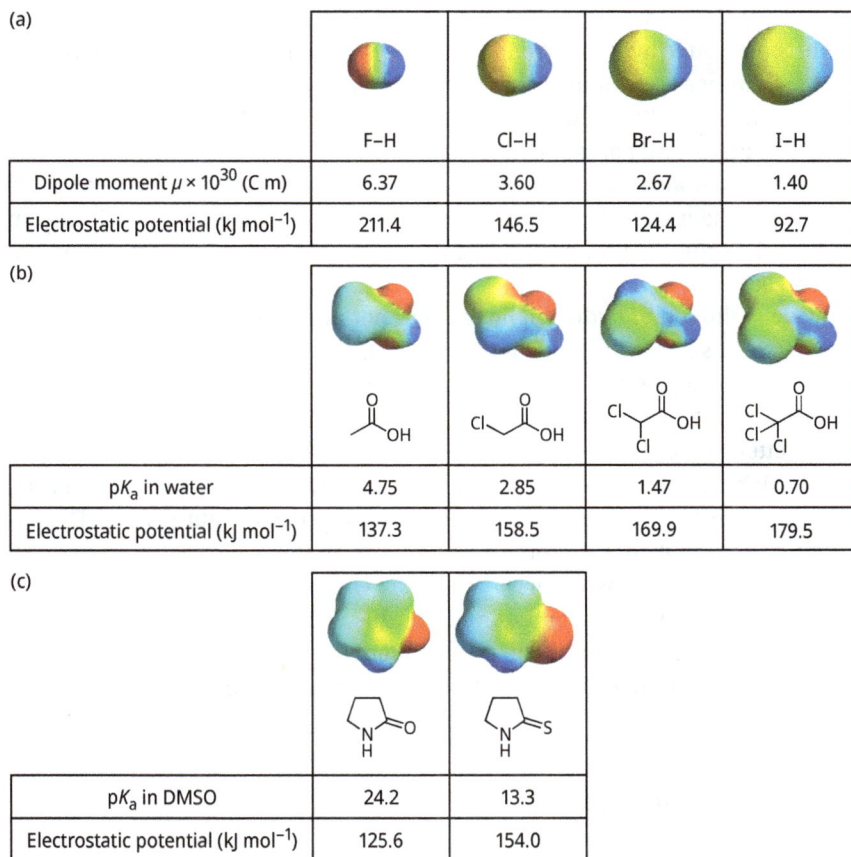

(a)

	F–H	Cl–H	Br–H	I–H
Dipole moment $\mu \times 10^{30}$ (C m)	6.37	3.60	2.67	1.40
Electrostatic potential (kJ mol^{-1})	211.4	146.5	124.4	92.7

(b)

pK_a in water	4.75	2.85	1.47	0.70
Electrostatic potential (kJ mol^{-1})	137.3	158.5	169.9	179.5

(c)

pK_a in DMSO	24.2	13.3
Electrostatic potential (kJ mol^{-1})	125.6	154.0

Figure 3.9: Electrostatic potential surfaces of the hydrogen halides HF, HCl, HBr, and HI together with their dipole moments (a), of acetic acid, chloroacetic acid, dichloroacetic acid, and trichloroacetic acid together with their pK_a values in water (b), and of N-methyl acetamide and N-methyl thioacetamide together their pK_a values in DMSO (c). In addition, the electrostatic potentials of the acidic protons involved in hydrogen bond formation are specified. The color coding covers a potential range from –120 to +120 kJ mol^{-1}, with red and blue indicating values greater than or equal to the absolute maximum in negative and positive potentials, respectively.

However, when the proton is bonded to the same atomic species in structurally related molecules, there is indeed a direct correlation between acidity and donor ability. This is because substituents that shift the electron density away from the proton, making it a better hydrogen bond donor, also stabilize the negative charge on the atom or group formed upon deprotonation. Good examples are acetic acid derivatives with electronegative substituents in the α-position, such as the series of chlorinated acetic acids shown in Figure 3.9b. In these compounds, the chlorine atoms shift the electron density away from the carboxyl group before and after deprotonation. These acids therefore become better hydrogen bond donors and stronger acids as the number of chlorine atoms increases. Another example is the hydrogen bonding ability of NH groups in amides and thioamides. Because sulfur atoms are larger than oxygen atoms, they can more easily accept a negative charge (recall in this context that thiols are more acidic than alcohols). As a result, not only are thioamides stronger acids than amides, but the greater charge-transfer from N to S in thioamides than to O in amides also causes the NH groups in thioamides to be the better hydrogen bond donors (Figure 3.9c).

The above definition of hydrogen bonds does not preclude weakly polarized X–H bonds from acting as hydrogen bond donors. In fact, even CH groups have frequently been observed to engage in hydrogen bonding. For example, electronegative substituents directly attached to the carbon atom increase the partial charge of the proton, making it a better hydrogen bond donor. Accordingly, α-CH groups in quaternary ammonium ions, which have a substantial positive potential, as shown in Figure 3.4a, form hydrogen bonds with acceptors such as carbonyl groups (Figure 3.10a). The hydrogen bond donor ability of CH groups also depends on the hybridization of the carbon atom and increases in the order $sp^3 < sp^2 < sp$, which explains why aromatic CH groups are versatile hydrogen bond donors. In this context, 1,4-disubstituted 1,2,3-triazole rings are particularly important binding motifs because they combine a relatively strongly polarized CH group with a large dipole moment, produced by the three nitrogen atoms opposite to the CH group (Figure 3.10b).

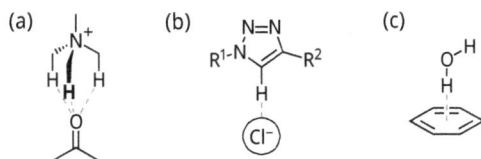

Figure 3.10: Examples of hydrogen bonds formed between the α-CH group in a quaternary ammonium ion and a carbonyl group (a), the CH group in a 1,4-disubstituted 1,2,3-triazole ring and an anion (b), and a water molecule with the surface of a π-system (c).

Nature of the hydrogen bond acceptor: Based on the same arguments used earlier, hydrogen bonding should be particularly strong to acceptors with a marked negative or partial negative charge. This tendency is indeed observed when the acceptor is an anion. In this case, anions with a high charge density are more potent hydrogen bond acceptors than larger, more charge-dispersed anions. The hydrogen bond acceptor ability increases in the order $I^- < Br^- < Cl^- < F^-$ for the halides, for example. We saw the same trend when discussing the hydration of anions by water molecules (Section 3.1.3), and the underlying interaction can indeed be regarded as hydrogen bond formation (but not the hydration of cations such as metal ions, which does not involve protons).

For neutral molecules containing hydrogen bond acceptors in the form of heteroatoms, the situation is more complex. Although oxygen atoms are usually better acceptors than nitrogen atoms, which is consistent with the higher electronegativity of oxygen, organic fluorides are poorer hydrogen bond acceptors because they are reluctant to share their electrons due to their high electronegativity. In haloalkanes, however, fluorine is a stronger hydrogen bond acceptor than the other halides [3]. Third row elements such as sulfur or phosphorous are usually weak acceptors, which is consistent with their size and associated low electron density.

Since the IUPAC definition of a hydrogen bond allows any "atom or group of atoms" to act as an acceptor as long as there is "evidence of bond formation," π-systems can also be considered as potential hydrogen bond acceptors. We will see in Section 3.1.8 that the quadrupole moment of π-systems and the associated negative potential along their faces do indeed give rise to polar interactions with cations (or anions in the case of electron-poor aromatic systems, as described in Section 3.1.9). Conversely, hydrogen bond donors also interact with π-systems. The interaction between a water molecule and benzene, for example, is worth about 8 kJ mol^{-1} in the gas phase, which qualifies as a hydrogen bond, albeit a weak one (Figure 3.10c).

In addition to the nature of the acceptor and the donor, there are a number of structural aspects that influence the strength of hydrogen bonds. We have seen that inductive effects of substituents mediate the partial charges in the donor and acceptor and therefore the strength with which they interact. Resonance has a similar effect. The conjugation between the carbonyl and the NH group affects the charge distribution in amides, for example, making them better hydrogen bond acceptors than ketones and better donors than amines (Figure 3.11). Such resonance-assisted hydrogen bonds also reinforce base pair interactions in double-stranded DNA.

The strength of hydrogen bonds also depends on whether the binding partners are already hydrogen bonded or not. This effect is best characterized for small clusters of water molecules where the binding of a water molecule to an already formed water dimer is stronger than the interaction between two isolated water molecules. The reason is that polarization effects cause the charge distribution in the water dimer to differ from that in an isolated water molecule. As shown schematically in Figure 3.11c, the existing hydrogen bond in the water dimer produces an increased electron density on the oxygen atom of the water molecule that donates the hydrogen bond, rendering it a better acceptor than the oxygen atom of an isolated water molecule. Conversely, the hydrogen atoms of the water molecule accepting the hydrogen bond in the dimer have reduced

(a)

(b)

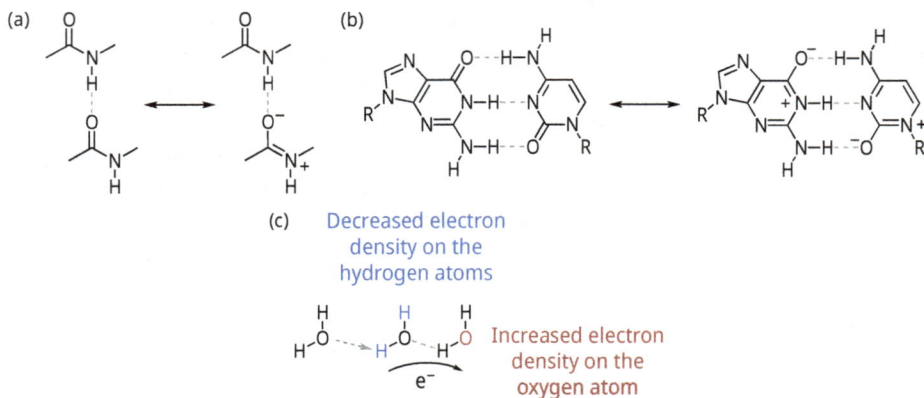

(c)   Decreased electron
       density on the
       hydrogen atoms

Increased electron
density on the
oxygen atom

Figure 3.11: Examples of resonance-assisted hydrogen bonds between amide groups (a) and a cytosine-guanine base pair (b), and schematic representation of the polarization enhancement of the binding of a water molecule to an already formed water dimer (c).

electron densities, improving their donor properties. Since the hydrogen bond in the dimer favors the formation of a new hydrogen bond, this effect is associated with the term cooperative hydrogen bonds, not to be confused with the strengthening of interactions by the formation of multiple hydrogen bonds (Section 3.5.3).

Geometry of the hydrogen bond: Interactions involving dipoles have a certain directionality, as reflected in the angular terms in equations (3.2b) and (3.3). The same is true for hydrogen bonds that are strongest when the proton approaches the point of highest electron density in the acceptor. The optimal arrangement of a hydrogen bond is therefore one where the donor is directly oriented in the direction of a free electron pair of the acceptor, the orientation of which can be derived from the valence shell electron pair repulsion (VSEPR) theory. Accordingly, amines and nitriles prefer linear hydrogen bonds, whereas X–H is arranged at certain angles to acceptors with more than one lone pair, as shown in Figure 3.12.

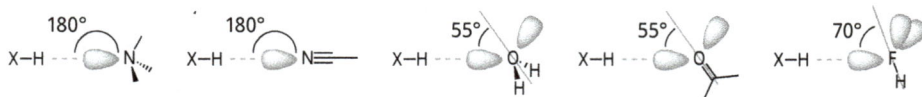

Figure 3.12: Optimal arrangement of hydrogen bond donors X–H with respect to acceptors having one, two, or three lone pairs.

Consistent with their primarily electrostatic nature, hydrogen bonds can tolerate significant deviations from the optimal arrangements shown in Figure 3.12 as a result of structural constraints or packing effects. Indeed, the orientation of hydrogen bond donors with respect to the axis of C=O groups varies within crystal structures between

+90° and −90°, with the optimal arrangements at ±55° dominating, but with many systems also exhibiting other angles.

Number of hydrogen bonds: So far, we have focused mostly on the structural aspects that control the formation of a single hydrogen bond. What happens when more than one hydrogen bond is formed between the binding partners? In the absence of structural constraints that prevent the efficient formation of multiple hydrogen bonds, it is generally expected that the stability of a complex will increase with the number of hydrogen bonds. The stabilization of complexes by multiple interactions is related to the concepts of multivalency and cooperativity, which will be discussed in more detail in Section 3.5.3. Note, however, that cooperativity caused by multiple interactions does not usually involve a direct influence of one interaction on the electronic factors controlling the next interaction, unlike the cooperativity in the polarization-enhanced hydrogen bonds discussed earlier.

Multiple hydrogen bonds also occur when two hydrogen bond donors bind to the same acceptor or *vice versa*. These so-called bifurcated hydrogen bonds do not necessarily lead to a stabilization beyond that of a single hydrogen bond if the geometries are not ideal. Examples of bifurcated hydrogen bonds are shown in Figure 3.13.

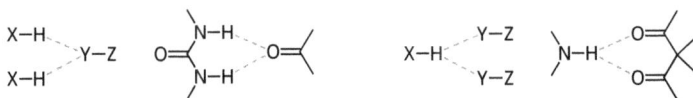

Figure 3.13: Examples of bifurcated hydrogen bonds.

A final case to consider is the formation of two or more *parallel* hydrogen bonds in close proximity. These arrangements are found, for example, between the complementary nucleobases in double-stranded DNA, and they also underlie many of the self-assembling systems presented in Section 5.3. Examples of hydrogen-bonded dimers are shown in Figure 3.14, each stabilized by three hydrogen bonds between the heterocyclic binding partners, but with a different pattern of the hydrogen bond donors and acceptors [4–6]. Despite the same number of primary interactions, the stabilities of these dimers vary considerably, indicating that the orientation of the hydrogen bonds also controls the binding strength.

The most stable complex in this series is the one in which one binding partner contains only donors and the other only acceptors. In this case, each proton finds acceptors in neighboring hydrogen bonds with which it can also engage in attractive interactions. These secondary interactions are shown as orange dotted lines in the schematic structures in the bottom row of Figure 3.14.

The exchange of the donor and acceptor in one of the lateral hydrogen bonds leads to the situation shown in the middle. While two attractive secondary interactions remain intact, the other secondary interactions become repulsive because two

--- Attractive primary interactions   --- Attractive secondary interactions   --- Repulsive secondary interactions

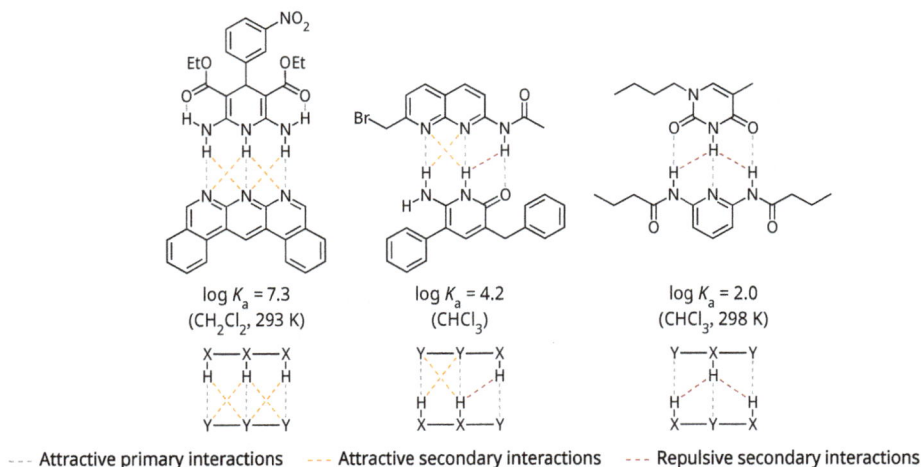

Figure 3.14: Examples of heterocyclic systems stabilized by three hydrogen bonds, stabilities of the three complexes, and schematic representation of the different hydrogen bond patterns with the assignment of the primary and secondary interactions.

positively polarized protons from different binding partners are arranged in close proximity. As a result, the stability of the dimer decreases by about three orders of magnitude compared to the dimer without repulsive secondary interactions. If the central hydrogen bond is oriented in the opposite direction to the two lateral ones, all secondary interactions are repulsive, which explains why the corresponding dimer is least stable. Secondary interactions thus cause the stability of the dimers in this example to differ by five orders of magnitude, although an exact comparison suffers from the fact that the stability constants in Figure 3.14 were determined under different conditions.

3.1.6 Halogen bonding

Halogen bonds are formally closely related to hydrogen bonds, except that the proton in the general expression of a hydrogen bond is replaced by a halogen atom:

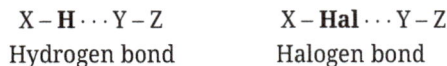

$$X-H \cdots Y-Z \qquad X-\mathbf{Hal} \cdots Y-Z$$

$$\text{Hydrogen bond} \qquad \text{Halogen bond}$$

Accordingly, the binding partner X–H containing the halogen atom bonded to an organic residue X is called halogen bond donor (as X–H is a hydrogen bond donor), while Y–Z is the halogen bond acceptor, with Y usually being an electronegative heteroatom. IUPAC defines this type of interaction as follows [7]:

A halogen bond occurs when there is evidence of a net attractive interaction between an electrophilic region associated with a halogen atom in a molecular entity and a nucleophilic region in another, or the same, molecular entity.

That these interactions should be attractive is not immediately obvious, as they appear to involve two electronegative and therefore partially negatively charged atoms. However, a closer look at the electrostatic potential surface of an organic halide reveals that the charge distribution around the halogen atom is anisotropic, with a belt of negative potential around the axis coinciding with the X–Hal bond and a small region of positive potential in the extension of the X–Hal bond, the so-called σ-hole (Figure 3.15).

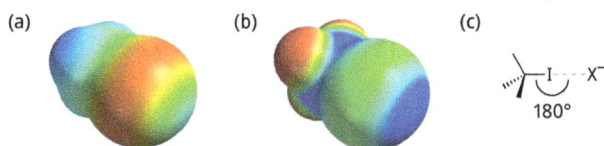

Figure 3.15: Electrostatic potential surfaces of methyl iodide (a) and trifluoromethyl iodide (b), showing the region of positive potential in the elongation of the C–I σ-bond. The color coding covers a potential range from –70 to +70 kJ mol^{-1}, with red and blue indicating values greater than or equal to the absolute maximum in negative and positive potentials, respectively. The preferred arrangement of a halogen bond is shown in (c).

The presence of this σ-hole is associated with the orbitals and hybridization of the halogen atom. Considering that the degree of hybridization is low for halogens, the free lone pairs occupy the s and the p_x and p_y orbitals when the p_z orbital is used for bond formation. The orthogonality of the p-orbitals and the depletion of the electron density of the p_z orbital caused by the covalent bond explain why there is a positive electrostatic potential at the end of the X–Hal bond. This view is consistent with the decrease in size and positive potential of the σ-hole when moving from iodide to fluoride in the halogen bond donor. Not only does fluoride have the highest electronegativity in this sequence, but the extent of sp^3-hybridization is also greatest, resulting in reduced anisotropy of the charge distribution around the halogen atom. The propensity of organic halogen compounds to act as halogen bond donors thus decreases in the order $I^- > Br^- > Cl^- > F^-$.

For donors containing the same halogen atom, the effectiveness improves as the electronegativity of the group X increases. This effect is due to a shift in the electron density along the X–Hal bond toward X, resulting in an increase in the size and positive potential of the σ-hole. This is illustrated in Figure 3.15 by the different electrostatic potentials of methyl iodide and trifluoromethyl iodide. The electrostatic model also explains the pronounced directionality of halogen bonds (Figure 3.15c). Since even small deviations from the optimal angle of 180° cause the halogen bond acceptor to experience a significantly smaller positive electrostatic potential at the halogen atom, halogen bonds are preferentially collinear with the X–Hal bond.

Again, this electrostatic model is insufficient to explain all aspects of halogen bonding. Weak interactions, for example, are mediated to a considerable extent by dispersion, while charge-transfer becomes important in the case of strong interactions. However, the electrostatic model helps to understand the situations that occur in most supramolecular systems. In this context, the following comparison with hydrogen bonding is useful:

- Focusing first on the hydrogen or halogen bond donor shows that any effect of X in X–H or X–Hal that increases the positive potential of the hydrogen atom or the σ-hole of the halogen atom is beneficial to the binding strength. Thus, the more electron-withdrawing X, the stronger the bond.
- While halogen bonding allows the binding strength to be tuned by varying the nature of the halogen atom, this is not possible for hydrogen bonds where only protons are allowed in the donor.
- The anisotropic charge distribution of halogen atoms causes halogen bonds to have a much greater degree of directionality than hydrogen bonds.
- Finally, protons are small and hard, whereas iodide, for example, is large and soft, causing halogen bond donors to favor softer acceptors. As a result, halogen bonding is less susceptible to solvent effects (halogen bond donors are less strongly solvated than hydrogen bond donors). Halogen bonding has therefore been called the "hydrophobic 'sister interaction' to hydrogen bonding" [8].

Halogen bonding has been extensively used in crystal engineering, that is, the programmed assembly of crystalline materials from building blocks that engage in defined and predictable interactions. An example is the cocrystal of 1,2-diiodotetrafluoroethane and N,N,N',N'-tetramethylethylenediamine, in which halogen bonds between the iodine and nitrogen atoms cause the two molecular building blocks to be arranged in linear chains in the crystal (Figure 3.16a). An example of a receptor whose binding in solution is based on halogen bonding is the tripodal benzene derivative **3.2** (Figure 3.16b) [9]. The three iodine atoms in the aromatic side chains of **3.2** interact with, for example, a chloride anion in acetone to form a 1:1 complex with a log K_a of 4.3. Harder oxoanions, such as hydrogensulfate or nitrate, interact much less efficiently with **3.2** (log $K_a < 1$). Other examples of receptors whose substrate recognition involves halogen bonding are presented in Sections 5.4 and 7.7.

(a)   (b)

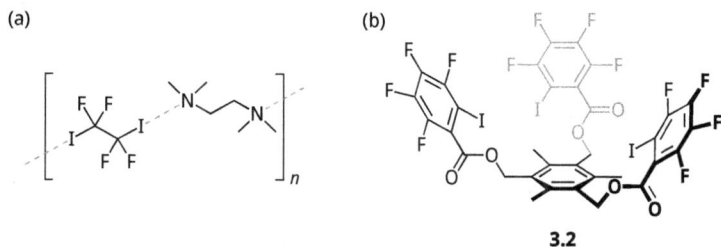

3.2

Figure 3.16: Schematic representation of the polymeric assembly resulting from the cocrystallization of 1,2-diiodotetrafluoroethane and *N,N,N',N'*-tetramethylethylenediamine (a), and structure of the tripodal receptor **3.2** with three aromatic residues containing iodine atoms as halogen bond donors (b).

Further reading

Work in the group of Orion B. Berryman has shown that the strength of the halogen bond is also influenced by hydrogen bond donors interacting with the belt of negative electrostatic potential of the halogen atom. Evidence for this effect was obtained by comparing the anion affinity of the two 1,3-bis(4-ethynylpyridinium) receptors **3.3a,b**, of which **3.3b**, with the additional amino group, binds halides by halogen bonding about an order of magnitude more strongly than **3.3b**. For example, chloride is bound by **3.3a** with a log K_a of 3.4 in $CDCl_3/CD_3NO_2$, 2:3 (v/v), while the chloride complex of **3.3b** has a log K_a of 4.4 under the same conditions.

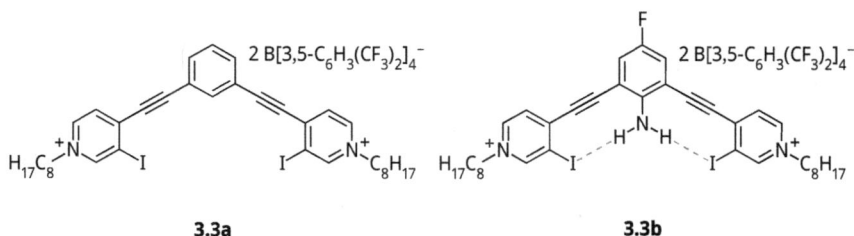

3.3a   **3.3b**

One reason for the higher anion affinity of **3.3b** is the better preorganization of this receptor, caused by hydrogen bonding between the protons of the amino groups and the negative electrostatic potential along the iodine atoms, which prevents rotation around the alkyne groups. In addition, there is computational evidence that these hydrogen bonds also increase the size and positive electrostatic potential of the σ-holes. This effect has been termed *hydrogen bond enhanced halogen bonding*. For more information and related studies by Pui Shing Ho using a biochemical system, see:

Riel AMS, Rowe RK, Ho EN, Carlsson ACC, Rappé AK, Berryman OB, Ho PS. Hydrogen bond enhanced halogen bonds: a synergistic interaction in chemistry and biochemistry Acc. Chem. Res. 2019, 52, 2,870–80.

3.1.7 Chalcogen bonding

Not only halogen bonding, but also other types of noncovalent interactions rely on σ-holes in one of the binding partners. Indeed, regions with an electron deficit can be found at the opposite end of the R–X bond on many atoms of group 14–17 elements when the R group is electron-withdrawing. Apart from the halogens, the most important elements in this context are the chalcogens sulfur, selenium, and tellurium, which we will focus on here. Early evidence that these elements can mediate noncovalent interactions came from crystal structures of chalcogen-rich compounds in which the chalcogen atoms were found to be arranged in close proximity. These short distances were attributed to attractive interactions, explaining why the term chalcogen–chalcogen interactions was originally used [10]. When it became clear that the binding partner of the chalcogen compound could also be an anion or a neutral molecule with an electronegative heteroatom, the more general term chalcogen bonding became more common.

To illustrate the arrangement of the σ-holes on a chalcogen atom, the electrostatic potential surface of bis(trifluoromethyl)tellurium is shown in Figure 3.17a. It can be seen that the tellurium atom, due to the two trifluoromethyl groups, has a larger and more extended region of positive electrostatic potential than halogen atoms. As a result, electrostatic interactions with an oppositely polarized binding partner are attractive at angles between 80° and 170° with respect to either of the R–X–R bonds (Figure 3.17b), causing chalcogen bonds to have a less pronounced directionality than halogen bonds. However, arrangements at the two extremes are favored, suggesting that the interaction is indeed controlled by the σ-holes.

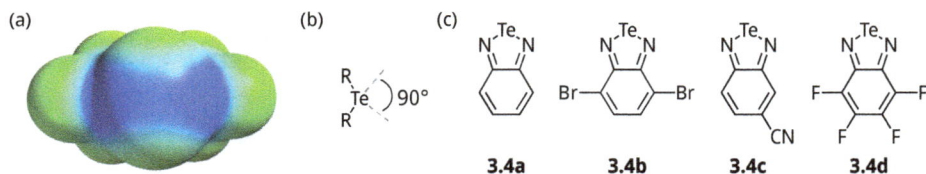

Figure 3.17: Electrostatic potential surface of bis(trifluoromethyl)tellurium (a), showing the regions of positive potential in the elongation of the two CF_3–Te σ-bonds. The color coding covers a potential range from −120 to +120 kJ mol^{-1}, with red and blue indicating values greater than or equal to the absolute maximum in negative and positive potentials, respectively. The preferred arrangement of a chalcogen bond is shown in (b) and the structures of the benzotelluradiazoles **3.4a–d** are shown in (c).

As in the case of halogen bonding, the strength of a chalcogen bond is expected to increase as the chalcogen atom becomes larger and the residues become more electron-withdrawing. While this trend is indeed observed, experimental and computational work has shown that chalcogen bonding, unlike other types of interactions, cannot simply be attributed to electrostatic attraction. Electrostatic and dispersion interactions (Section 3.1.11) contribute to the binding, but the most important contribution to the interaction often comes from orbital mixing, that is, the interaction of the empty σ* orbital of the chalcogen bond donor (the compound containing the chalcogen atom) and the filled n orbital of the chalcogen bond acceptor. As with electrostatic interactions, this interaction benefits from the increased electronegativity of the residues on the chalcogen atom, but the explanation in this case is the lowering of the energy level of the σ^* orbital, making it a better acceptor. Chalcogen bonds can be as strong as conventional hydrogen bonds, but are often less susceptible to solvent effects.

Despite these positive aspects, systems using chalcogen bonds are still somewhat underrepresented in supramolecular chemistry [11], perhaps because the introduction of the heavier chalcogens into a receptor is synthetically challenging. Examples of systems using chalcogen bonds are the self-assembling capsules discussed in Section 5.4.3, and the series of benzotelluradiazoles **3.4a–d** shown in Figure 3.17c. These compounds form 1:1 complexes with halides in tetrahydrofuran, with complex stability decreasing in the order of decreasing charge density of the anion from Cl^- to I^- [12]. The $\log K_a$ values of the chloride complexes increase from 3.0 for **3.4a** to 4.1 for **3.4b**, 4.6 for **3.4c**, and finally 5.1 for **3.4d**, as the electron-withdrawing nature of the aromatic subunit increases. The selenium analog of **3.4b** is a less efficient chalcogen bond donor and does not bind chloride under the same conditions.

3.1.8 Cation–π interactions

We have seen that molecules in which certain functional groups or structural elements induce an anisotropic charge distribution can interact with a complementary binding partner. In this and the following sections, we look at how π-systems participate in such interactions. Typical π-systems, such as those of alkenes or aromatic compounds, are usually considered to be nonpolar and it is therefore not immediately obvious how they promote binding. Again, looking at electrostatic potential surfaces helps to understand their behavior.

Figure 3.18 shows that benzene has a pronounced negative potential on either side of the ring and a positive potential along the edge due to the polarization of the six C–H bonds, which causes a shift in electron density from the peripheral protons to the center of the ring (recall that carbon is more electronegative than hydrogen, and that the electronegativity of carbon increases with the increasing *s*-character of the orbital used to form the C–H bond). The dipoles of the C–H bonds of benzene thus add up to produce a permanent quadrupole moment, characterized by two dipoles aligned end to end. Although these two dipoles cancel each other out, making benzene nonpolar overall, a cation approaching the center of one face of a benzene ring will experience an attraction. This electrostatic interaction is the basis of the cation–π interaction, which is not restricted to aromatic π-systems but is also observed for similar reasons in alkenes and other unsaturated compounds.

Cation–π interactions have been extensively studied in the gas phase, in the solid state, and in solution. The first evidence for their existence came from gas phase studies, which showed that the ΔH^0 of the interaction between benzene and a potassium ion in the gas phase is worth -80 kJ mol^{-1}, which is even slightly stronger than the interaction of K^+ with a water molecule (-75 kJ mol^{-1}) [13]. As expected, the binding strength increases with the charge density of the cation, reaching -117 kJ mol^{-1} for

(a) (b)

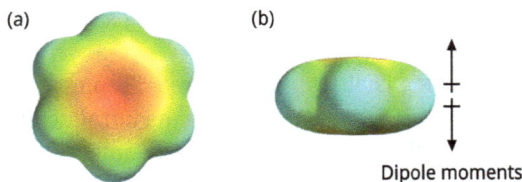

Dipole moments

Figure 3.18: Electrostatic potential surface of benzene viewed from the top (a) and from the side (b). The end-to-end aligned dipoles that make up the quadrupole moment are shown in the side view. The color coding covers a potential range from −100 to +100 kJ mol^{-1}, with red and blue indicating values greater than or equal to the absolute maximum in negative and positive potentials, respectively.

Na$^+$ and −160 kJ mol^{-1} for Li$^+$, but these energies only reflect the upper limit of the binding strength. As with ion–ion interactions, the presence of solvent molecules weakens cation-π interactions, and their contributions to a binding event in water are therefore typically only between 1 and 3 kJ mol^{-1} [14].

The structures of a number of complexes between alkali metal salts and so-called lariat ethers, which consist of a crown ether moiety with attached substituents (Section 4.1.1), provide strong evidence for the attractive nature of cation–π interactions [15]. An example is the lariat ether **3.5**, which adopts an extended conformation in the solid state, with the two phenol groups oriented away from the ring (Figure 3.19a,b). In the absence of cation–π interactions, a cation binding to the crown ether should not have a pronounced effect on this conformation and should remain ion-paired to its counterion. However, if cation–π interactions contribute to the stabilization of the complex, cation binding should induce a conformational reorganization. The crystal structure of the potassium iodide complex of **3.5** (Figure 3.19c), in which the potassium ion is sandwiched between the two phenol rings, illustrates that the latter is the case. Thus, the cation–π interactions in this complex are strong enough to outweigh the strength of the electrostatic interactions between the oppositely charged ions.

(a) (b) (c)

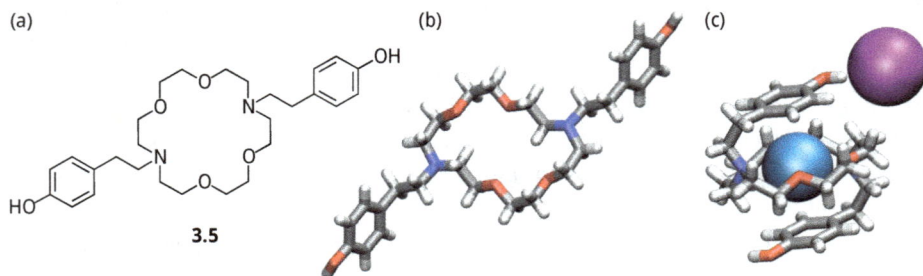

3.5

Figure 3.19: Molecular structure of lariat ether **3.5** (a) and crystal structures of the uncomplexed form of **3.5** (b), and of the corresponding potassium iodide complex (c). The potassium ion is shown in blue and the iodide ion in purple.

Similar close contacts between cationic and aromatic residues are found in the crystal structures of protein complexes, for example in the structure of the nicotinic acetylcholine receptor binding to acetylcholine. In this case, the cationic trimethylammonium head group of the substrate comes into close contact with a tryptophan indole moiety, providing evidence for the importance of cation–π interactions also in biological systems.

The dependence of the strength of cation–π interactions on the geometry of the complex and on the nature of the cation can be explained in terms of the electrostatic model. The interactions are strongest when the cation is positioned directly above or below the position of the aromatic system with the largest negative potential – in the case of benzene, the center of the ring. Deviations from this optimal arrangement lead to weakening of the interactions due to reduced electrostatic attraction. Furthermore, the binding strength in the gas phase correlates with the charge density of the ion, decreasing with increasing size of the ion in the order $Li^+ > Na^+ > K^+ > Rb^+$, as expected for electrostatic interactions. Solvent effects cause deviations from this trend. In water, for example, the two smallest and most strongly hydrated ions Li^+ and Na^+ rank at the end of the affinity scale and form complexes that are even less stable than those of Rb^+. Complex stability in solution thus depends on the balance between the intrinsic strength of the cation–π interaction and the degree to which the binding partners are solvated, with strong solvation causing the complex to become less stable. We have already seen this effect when discussing equation (2.8). Interestingly, cation–π interactions are not as strongly weakened by an increase in solvent polarity as other types of interactions. This is because the desolvation of one partner, namely the π-system, is usually not very solvent-dependent and therefore relatively easy, even in water. In contrast, when both binding partners are polar, they are both better solvated and therefore less likely to interact in more polar solvents.

Other factors affecting binding strength are the structure and substitution pattern of the aromatic system interacting with the cation. Again, the best way to assess these factors is to analyze the electrostatic potential surfaces of the respective aromatic compounds. Figure 3.20 shows these surfaces for a selection of substituted benzene derivatives and heterocyclic aromatic compounds.

At first glance, the influence of aromatic substituents on the strength of cation–π interactions appears to correlate roughly with the reactivity of the respective aromatic systems in electrophilic aromatic substitution reactions, with activating substituents strengthening and deactivating substituents weakening the interaction. On closer inspection, however, deviations become apparent. Phenol, for example, has an activated π-system but interacts less strongly with Na^+ than benzene. Furan is also much more reactive than benzene in S_EAr reactions, but is still a weaker cation binder. Indeed, the ability of aromatic systems to engage in cation–π interactions was found to correlate not with the polarization of the π-system, but with Hammett's σ_{meta} parameter, which describes the *inductive* effects of substituents on the aromatic ring [16]. Calculations confirmed that the effects of aromatic substituents arise primarily

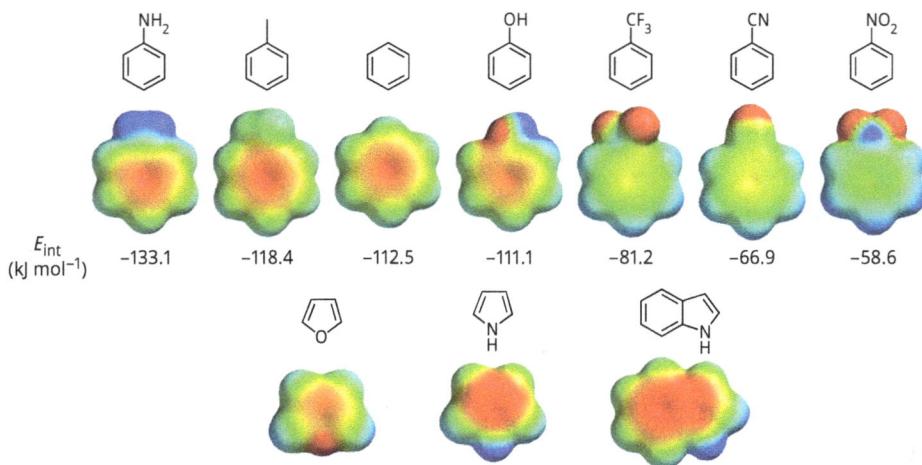

Figure 3.20: Electrostatic potential surfaces of aniline, toluene, benzene, phenol, trifluoromethylbenzene, benzonitrile, nitrobenzene, furan, pyrrole, and indole. The color coding covers a potential range from −100 to +100 kJ mol^{-1}, with red and blue indicating values greater than or equal to the absolute maximum in negative and positive potentials, respectively. The calculated energies E_{int} for the monosubstituted benzene derivatives, taken from ref. [17], describe the interaction of a sodium ion with the respective aromatic compound in the gas phase.

from the influence of their dipole moments on the electrostatic potential of the π-system [17].

Important information about the stabilizing effects of cation–π interactions came from the characterization of the binding properties of a family of cyclophanes, of which compound **3.6** (Figure 3.21) is a representative [18]. This compound was originally developed as a receptor for nonpolar substrates in water. However, with a log K_a of 8.4 in 10 mM borate buffer, the complex of **3.6** with the *N*-methylquinolinium cation is more than two orders of magnitude more stable than the complex with the neutral but otherwise similarly shaped 4-methylquinoline (log K_a = 5.9). The higher affinity for the positively charged substrate can be attributed to cation–π interactions, although additional contributions to complex stability from ionic interactions with the carboxylate groups cannot be completely excluded. Cation–π interactions also explain the cation affinity of many receptors with aromatic residues presented in Chapter 4.

3.1.9 Anion–π interactions

If cation–π interactions exist, what about anion–π interactions? Of course, aromatic systems are usually considered to be electron-rich because of the electrons delocalized in the π-orbitals, which should make anion–π interactions unlikely. However, we have seen that cation–π interactions are not governed by π-effects. In addition, nucleophiles

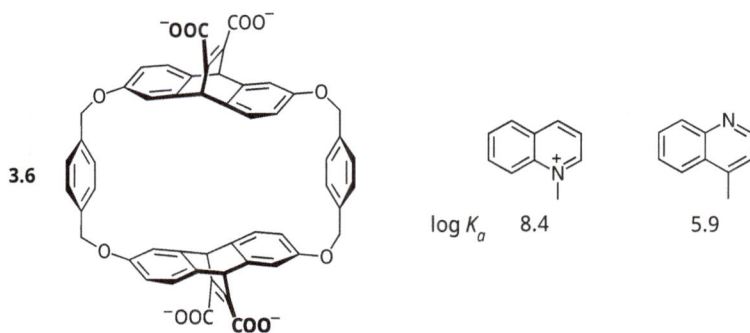

Figure 3.21: Structure of cyclophane **3.6** and binding constants log K_a of the complexes of **3.6** with the N-methylquinolinium cation and 4-methylquinoline in 10 mM aqueous borate buffer.

are known to react with electron-poor aromatic systems, suggesting that there must be a driving force that allows a nucleophile to approach a π-system. Indeed, electron-poor carbocyclic or heterocyclic aromatic compounds have a strong positive electrostatic potential, as shown in Figure 3.22, which allows them to attract anions.

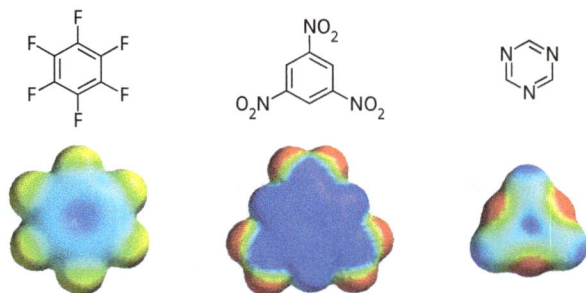

Figure 3.22: Electrostatic potential surfaces of hexafluorobenzene, 1,3,5-trinitrobenzene, and 1,3,5-triazine. The color coding covers a potential range from −100 to +100 kJ mol⁻¹, with red and blue indicating values greater than or equal to the absolute maximum in negative and positive potentials, respectively.

Computational work showed that anion–π interactions are mainly based on two effects. One is the electrostatic attraction between the anion and the positive potential of the electron-deficient aromatic system. This interaction is further enhanced by the ion-induced polarization of the π-system that arises as the anion approaches the aromatic ring. Once close enough, the anion causes a shift of the electron density to the opposite side of the ring, inducing a dipole moment whose positive end faces the anion.

Despite the computational evidence for the existence of anion–π interactions, the experimental proof has not been straightforward. In crystal structures, for example, anions are found in various arrangements with respect to nearby π-systems that are not always consistent with anion–π interactions. For example, anions approaching the substituted carbon atom of a benzene ring coincide with the trajectories of nucleophiles when forming Meisenheimer complexes during S_NAr reactions, providing no clear evidence that anion–π interactions are operative. Moreover, planar anions such as nitrate, carbonate, or azide can engage in π–π interactions with arenes. In certain systems, however, arrangements of anions with respect to electron-deficient aromatic rings have been observed that are consistent with even very strict views of anion–π interactions. Examples are the bromide salt of the quaternary ammonium ion **3.7** and the tetrabutylammonium bromide complex of benzamide **3.8a**, both of which contain the anion directly above the pentafluorophenyl moieties (Figure 3.23) [19, 20]. These structures thus provide similar evidence for the attractive nature of anion–π interactions as the lariat ether **3.5** does for cation–π interactions (Figure 3.19 in Section 3.1.8).

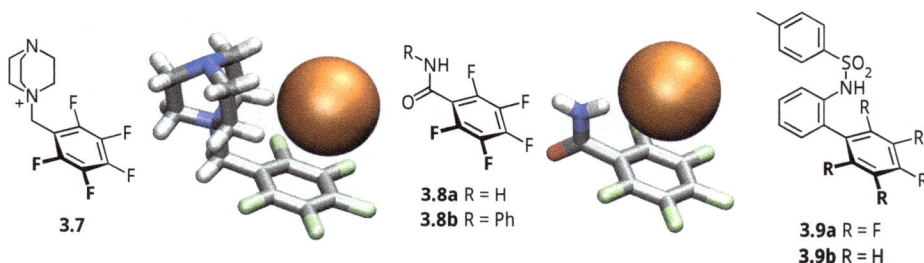

3.7

3.8a R = H
3.8b R = Ph

3.9a R = F
3.9b R = H

Figure 3.23: Molecular structures of the ammonium cation **3.7**, the benzamides **3.8a,b**, and the sulfonamides **3.9a,b**, and modes of interaction of **3.7** and **3.8a** with a bromide ion in the solid state. The tetrabutylammonium counterion in the complex of **3.8a** is omitted for reasons of clarity.

There is also evidence for the stabilizing effects of anion–π interactions in solution. An early example is the sulfonamide **3.9a** with an appended pentafluorophenyl ring (Figure 3.23) [21]. While the corresponding benzene derivative **3.9b** shows no measurable affinity for halides in chloroform, **3.9a** binds weakly to halides under the same conditions (K_a between 20 and 34 M^{-1}). According to the experimental results, the halide affinity is due to anion–π interactions and not to differences in the acidity of the sulfonamide NH groups of the two receptors. Following these early results, other systems were identified in which anion–π interactions play a role in binding, ion transport (Section 10.3.1), and catalysis.

3.1.10 Aromatic interactions

The fact that aromatic systems have regions of positive and negative electrostatic potential along their surfaces allows them to engage in electrostatic interactions not only with charged binding partners, as in cation–π and anion–π interactions, but also with themselves. For these interactions to be attractive, regions of opposite electrostatic potential should approach each other, as in the T-shaped or edge-to-face arrangement of two benzene molecules, in which positively polarized hydrogen atoms along the edge of one ring approach the face of the second ring, where the electrostatic potential is negative. This arrangement is the basis for the herringbone structure of crystalline benzene (Figure 3.24a,b).

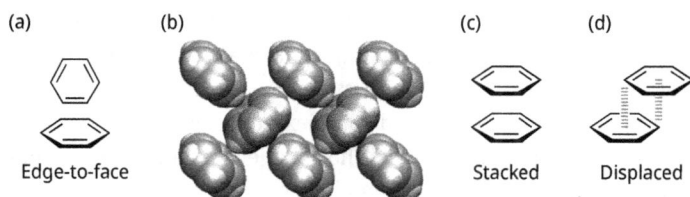

Figure 3.24: Possible arrangements of two aromatic rings. The T-shaped arrangement (a) and the displaced arrangement (c) are attractive. The stacked arrangement of two benzene rings is repulsive (d), but the alternating stacked arrangement of benzene and hexafluorobenzene is attractive (e). The herringbone structure of crystalline benzene (b) is characterized by multiple edge-to-face interactions.

Two benzene rings can also be arranged in parallel if structural effects within a molecule or in a receptor–substrate complex prevent the perpendicular orientation. A parallel arrangement can also be induced by solvent effects, as it is associated with more efficient shielding of hydrophobic surfaces than the T-shaped structure. When the rings are perfectly stacked, the dispersion interactions (Section 3.1.11) are most favorable (Figure 3.24c), but the electrostatic interactions between their quadrupole moments are repulsive. Consequently, a displaced arrangement is preferred in which the attractive dispersion interactions are largely preserved while the electrostatic repulsion is minimized (Figure 3.24d). This electrostatic argument is the basis of the Hunter–Sanders model of aromatic interactions [22] and the reason why the term π–π stacking, which implies that attractive interactions between π-systems could cause them to stack, has been considered misleading [23]. Computational work has subsequently challenged this view, suggesting that the displaced arrangement of aromatic rings can be rationalized without invoking electrostatics [24]. Rather, the interaction is due to a competition between Pauli repulsion (electrons must move to higher orbitals to ensure that each occupies an individual quantum state) and attractive dispersion interactions. This model explains not only the displaced arrangement of the two rings in the benzene dimer, but also that in the benzene–hexafluorobenzene dimer, where electrostatic arguments would predict a stacked arrangement. Thus, the term π–π stacking may be useful after all, describing a special type of dispersion interaction characteristic of π-systems.

The interaction between electron-rich and electron-poor π-systems goes beyond aromatic interactions if the HOMO of the donor and the LUMO of the acceptor are properly aligned and the energy gap between them is small enough to allow the orbitals to mix. In this case, charge-transfer interactions are possible, that is, the transfer of a fraction of the electronic charge from the donor to the acceptor. We have seen that orbital mixing contributes to many types of noncovalent interactions, so its role in aromatic interactions is not unique. Charge-transfer interactions between π-systems, however, often have the special feature of causing the solution to change color because the energy required to excite an electron from the HOMO to the LUMO of the complex is in the visible region of the electromagnetic spectrum. Complex formation can therefore be monitored by UV–vis spectroscopy. An example of a system in which charge-transfer interactions are important is the orange-red colored complex between the electron-rich 1,4-dimethoxybenzene and the receptor **3.10**, which has two electron-poor paraquat moieties [25] (Figure 3.25). We will return to this system in Sections 4.1.5 and 7.5.

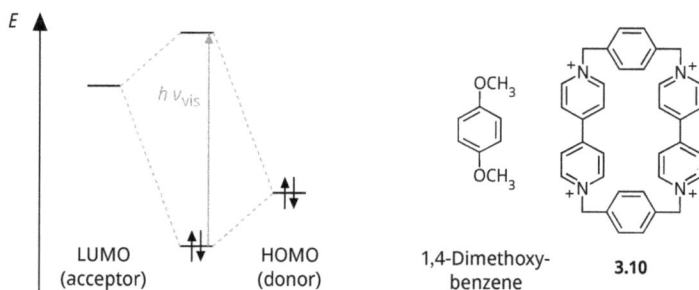

Figure 3.25: Schematic representation of the orbital mixing underlying charge-transfer interactions between an electron-rich and an electron-poor π-system, and structures of 1,4-dimethoxybenzene and receptor **3.10**, which form a complex stabilized by charge-transfer interactions.

3.1.11 Dispersion interactions

We now come to the weakest type of noncovalent interactions, dispersion interactions or London forces, which are the attractive component of van der Waals interactions. These interactions explain why atoms or molecules without anisotropic charge distributions, such as noble gases or alkanes, can be liquefied. To mathematically describe the dependence of the potential energy E on the distance r of the binding partners, the empirically derived Lennard–Jones potential in equation (3.4) is generally used:

$$E = \varepsilon \left[\left(\frac{\sigma}{r} \right)^{12} - \left(\frac{\sigma}{r} \right)^{6} \right]$$

(3.4)

According to equation (3.4), E is divided into a repulsive (positive sign) and an attractive (negative sign) term. The parameter σ in both terms corresponds to the distance at which the attraction and repulsion cancel each other out, so that E becomes zero ($\sigma = r$). The parameter ε describes the hardness of the interaction. This parameter is related to the polarizability of the binding partners, with soft–soft interactions having larger ε values than hard–hard interactions. The dependence of E on r is shown graphically in Figure 3.26a.

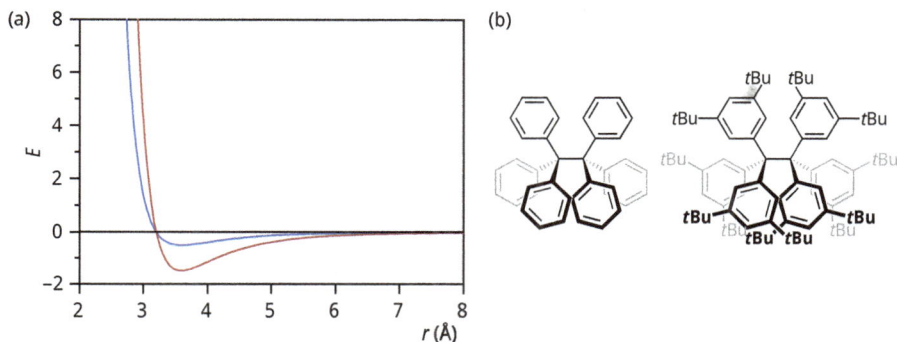

Figure 3.26: Lennard-Jones potential functions for $\sigma = 3.2$ Å and two different values of ε, with the red curve describing a soft–soft interaction (larger ε) and the blue one a hard–hard interaction (smaller ε) (a). In (b), the structures of 1,1,1,2,2,2-hexaphenylethane and 1,1,1,2,2,2-hexakis(3,5-di-*tert*-butylphenyl)ethane are shown, of which the former compound is unstable, while the latter is stabilized by the dispersion interactions mediated by the *tert*-butyl groups.

The curves in Figure 3.26a illustrate that the Lennard–Jones potential is repulsive at small distances, where the orbitals of the interacting molecules begin to overlap and Pauli repulsion occurs (electrons must move to higher orbitals to ensure that each occupies an individual quantum state). At distances greater than σ, there is a small region where the interactions are attractive. This attraction then quickly vanishes as r increases, with a distance dependence of r^{-6}. The attractive nature of the interactions within this small distance range is often explained by the model of oscillating dipoles: the motion of the electrons in a molecule causes a fluctuating dipole moment, which in turn induces a complementary dipole moment in the respective binding partner, resulting in a short-lived but attractive interaction. This view is consistent with the increasing strength of induced dipole-induced dipole interactions as the polarizability of the binding partners increases.

Dispersion interactions are weak when small atoms or molecules are involved (explaining the low boiling points of noble gases). However, they become stronger as the contact interfaces between the molecules increase in size. For example, the boiling points of alkanes increase with chain length. Intramolecular dispersion interactions are also responsible for the stabilization of sterically crowded molecules. 1,1,1,2,2,2-

Hexaphenylethane is, for example, unstable due to the steric crowding of the six phenyl groups. The much more crowded 1,1,1,2,2,2-hexakis(3,5-di-*tert*-butylphenyl)ethane (Figure 3.26b) should be even less stable, but the opposite is the case. The reason for this is that the interdigitation of the *tert*-butyl groups and the aromatic residues gives rise to *intramolecular* dispersion interactions that stabilize the molecule to the extent that it can be isolated and recrystallized. Since supramolecular complexes are often structurally quite complex, containing several polarizable groups in close proximity, *intermolecular* dispersion interactions lead to similar stabilization. Although these interactions are intrinsically weak and therefore difficult to quantify [26], they cannot be neglected.

3.2 Binding energies

3.2.1 General considerations

How strong are intermolecular interactions?

All the types of interactions discussed so far are associated with characteristic binding strengths, and we have seen equations in Section 3.1 that allow us to quantify the energetic gain of binding. These equations can also be used to rank the different interactions according to their strength. For example, an ion–dipole interaction should be about one order of magnitude weaker than an ion–ion interaction, due to the replacement of an elementary charge q in equation (3.1) by the term μ/r, as discussed in Section 3.1.3. Consequently, a dipole–dipole interaction should be even weaker. The same trend can be seen by looking at how the strength of the interactions varies with distance. While ion–ion interactions have a distance dependence of r^{-1}, ion–dipole and dipole–dipole interactions correlate with r^{-2} and r^{-3}, respectively, and their decrease in strength is therefore more pronounced as the distance between the binding partners increases. Accordingly, dispersive interactions are inherently the weakest. The different types of interactions were therefore roughly discussed in order of decreasing strength in the previous chapter.

Gas phase studies or calculations allow the quantification of intrinsic interaction energies, as shown for selected examples in Section 3.1. However, these energies typically overestimate how much the respective interaction is worth in solution. This is because, among other factors, the polarity of the medium and specific solvation effects strongly influence the interaction strength (Section 3.3). The high relative permittivity of water weakens ion–ion interactions, for example, and hydration can further reduce the electrostatic attraction to the point where it becomes negligible (Section 3.1.2). On the other hand, interactions between large hydrophobic surfaces are often enhanced in water, so that van der Waals interactions end up being apparently stronger than ion–ion interactions under certain conditions.

A statistical analysis of reported binding constants shows that typical receptor–substrate complexes in organic solvents have binding constants K_a between 1 and 10^7 M^{-1}, with an average of $10^{3.4 \pm 1.6}$ M^{-1}, corresponding to a ΔG^0 value of -19.8 ± 9.1 kJ mol^{-1} [27]. In water, the range of binding constants is similar, but the average is about 7 kJ mol^{-1} lower. Another literature survey found ΔG^0 values between -1 and -40 kJ mol^{-1} (reflecting the previously mentioned K_a values between 1 and 10^7 M^{-1}), with no clear correlation between the type of interaction and the strength [14]. Thus, not only are interaction energies in solution much lower than those calculated for the gas phase, but binding constants alone usually do not provide information about the actual driving force of complex formation. Therefore, special strategies are required to obtain information about the individual contribution to binding. These strategies must take into account that complex formation in solution is practically never due to a single type of interaction. For example, dispersion interactions almost always contribute to the stability of a complex, even if the main driving force of its formation is due to other effects. Therefore, strategies that aim to provide information about the effect of a particular type of interaction must be able to separate one contribution from another. Three methods that are useful in this context are presented below.

3.2.2 Trend analyses

The underlying notion of trend analyses is that each contribution to the stabilization of a complex is associated with a characteristic free energy value and that the overall stability of the complex is equal to the sum of all individual values according to the following equation:

$$\Delta G^0_{\text{total}} = \sum_{i=1}^{n} \Delta G^0_i \qquad (3.5)$$

This concept is illustrated in Figure 3.27a by using the example of the interaction of a macrocyclic receptor, containing a varying number of positively charged groups along the ring, with an anionic substrate. Although the complex of the singly charged receptor is mainly stabilized by a salt bridge (Coulomb attraction reinforced by a hydrogen bond), the Gibbs free energy of complex formation does not reflect the formation of this salt bridge alone, but also includes other factors, such as additional attractive or repulsive interactions, the entropic penalty associated with bringing the receptor and substrate together, and the loss of conformational flexibility. Therefore, a single ΔG^0 value is not sufficient to estimate the extent to which the salt bridge contributes to the stability of the complex. However, each additional salt bridge that is possible in the more highly charged receptors is expected to add a constant and characteristic energy term to ΔG^0 in the absence of cooperative effects that complicate the situation (Section 3.5.3). Thus, by comparing the stabilities of a series of complexes, it is possible to quantify how much a salt bridge is worth, in terms of energy.

(a)

(b)

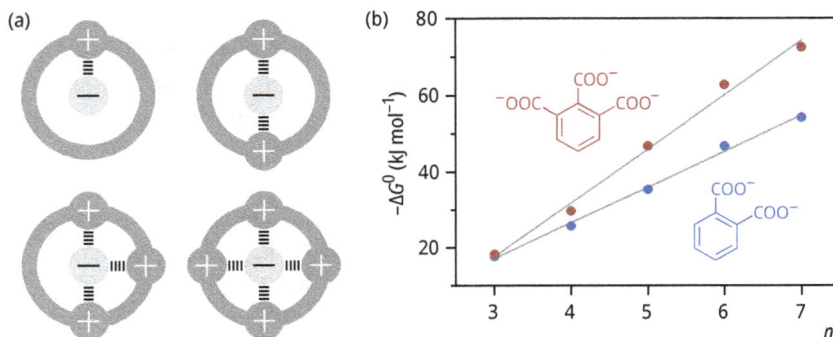

Figure 3.27: Schematic representation of the complexes of an anion with macrocyclic receptors containing one to four positive charges, which contribute to stability by the equivalent number of salt bridges (denoted by dotted lines) (a), and graph showing the dependence of the Gibbs free energy of formation of the complexes of terephthalate (blue) and benzene-1,2,3-tricarboxylate (red) on the degree of protonation of receptor **3.1** in Figure 3.4 (b).

Indeed, the logarithmic binding constants $\log K_a$ of the anion complexes of macrocyclic polyammonium receptors, for example receptor **3.1** shown in Figure 3.4, often increase linearly with the degree of protonation, obeying expression (3.6), where z_R and z_A are the charges of the receptor and the anion, respectively:

$$\log K_a = a \; z_A \; z_R + b \qquad (3.6)$$

Since $\log K_a$ values are directly correlated with ΔG^0 according to equation (2.7), plotting ΔG^0 versus the degree of protonation should yield a straight line with the slope $a \; z_A$, where the term a reflects the energetic contribution of each salt bridge to the overall stability.

That this treatment is indeed feasible is illustrated by the graphs in Figure 3.27b, which show the dependence of the ΔG^0 values associated with the terephthalate and benzene-1,3,5-tricarboxylate complexation by **3.1** on the degree of protonation (Figure 2.12). For both complexes, the free energies of complex formation increase linearly when going from the triprotonated to the fully protonated form of the receptor. The different charges of the anions cause the slopes of the lines to differ, but this effect can be eliminated mathematically according to equation (3.6). The resulting estimate of a indicates that, independent of the anion, each salt bridge contributes an energy term of -4.7 kJ mol^{-1} to the stability of the complex, a result consistent with the relatively constant value of $-(5 \pm 1)$ kJ mol^{-1} observed for other systems [14]. Estimates for other types of interactions, such as hydrogen bonding or van der Waals interactions, can be derived in a similar manner.

3.2.3 Double-mutant cycles

The idea of obtaining information about the contribution of a particular type of non-covalent interaction to binding strength by comparing the stability of a complex in which that interaction occurs with the stability of an analogous system in which the same interaction is absent is attractive, but often fails to yield meaningful results. The reason for this is shown schematically in Figure 3.28a. The complex **ABXY** is stabilized by pairwise interactions between the shape complementary subunits a and b, and x and y. However, secondary attractive or repulsive interactions between a and y and between b and x, as discussed in Section 3.1.5 and indicated by the red dotted lines in Figure 3.28a, cannot be completely ruled out. Replacing the groups x and y in the binding partners by the noninteracting *mutated* groups x' and y', thus yielding the double-mutant complex **AB**, causes not only the disappearance of the primary interaction, about which information should be obtained, but also of the secondary interactions. As a consequence, the difference $\Delta G^0_{ABXY} - \Delta G^0_{AB}$ cannot be used to determine how strongly x interacts with y. A workaround is to additionally include the single mutants **ABX** and **ABY** in the analysis.

The estimation of ΔG^0_{xy} is then based on the idea that all contributions to the binding strength occurring in the different complexes are additive and not affected by the structural changes. Consequently, the stabilities of the individual complexes can be calculated from the individual contributions ΔG^0_{ab}, ΔG^0_{xy}, ΔG^0_{ay} and ΔG^0_{bx}, allowing the estimation of ΔG^0_{xy} from the difference $\left(\Delta G^0_{ABXY} - \Delta G^0_{ABY}\right) - \left(\Delta G^0_{ABX} - \Delta G^0_{AB}\right)$ [or equivalently $\left(\Delta G^0_{ABXY} - \Delta G^0_{ABX}\right) - \left(\Delta G^0_{ABY} - \Delta G^0_{AB}\right)$] as derived in Figure 3.28b.

(a)

(b)

$$\Delta G^0_{ABXY} = \Delta G^0_{ab} + \Delta G^0_{xy} + \Delta G^0_{xb} + \Delta G^0_{ay}$$

$$\Delta G^0_{ABX} = \Delta G^0_{ab} + \Delta G^0_{xb}$$

$$\Delta G^0_{ABY} = \Delta G^0_{ab} + \Delta G^0_{ay}$$

$$\Delta G^0_{AB} = \Delta G^0_{ab}$$

$$\overline{\Delta G^0_{ABXY} - \Delta G^0_{ABY} = \Delta G^0_{xy} + \Delta G^0_{xb}}$$

$$\Delta G^0_{ABX} - \Delta G^0_{AB} = \Delta G^0_{xb}$$

$$\overline{\left(\Delta G^0_{ABXY} - \Delta G^0_{ABY}\right) - \left(\Delta G^0_{ABX} - \Delta G^0_{AB}\right) = \Delta G^0_{xy}}$$

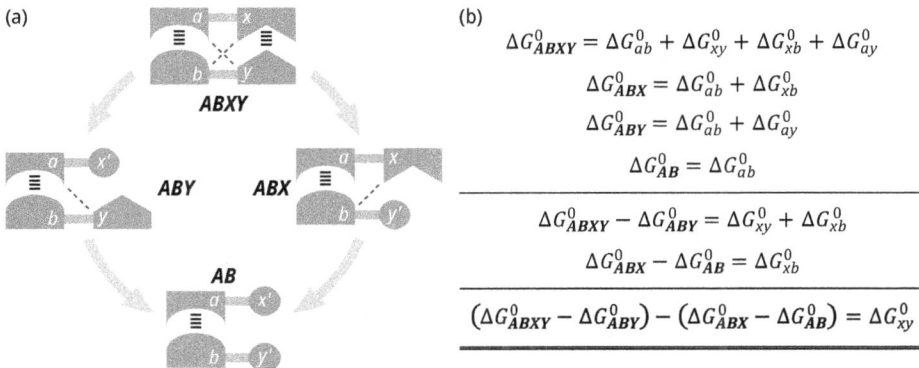

Figure 3.28: General strategy of a double-mutant cycle that provides quantitative information about the contribution of the interaction between x and y to the overall stability of the complex **ABXY** (a). The equations in (b) show that the Gibbs free energies ΔG^0_{ABXY}, ΔG^0_{ABX}, ΔG^0_{ABY}, and ΔG^0_{AB} of the individual complexes, which include the individual energy contributions ΔG^0_{ab}, ΔG^0_{xy}, ΔG^0_{ay}, and ΔG^0_{bx}, can be used to calculate ΔG^0_{xy}.

For this treatment to work, the following conditions must be met: (i) the structures of the different complexes must be very similar, (ii) the energetic contributions of the individual interactions must be additive, (iii) the interaction between a and b must be strong enough to ensure that the complex AB holds together even in the absence of the interacting groups x and y, and (iv) the mutated groups x' and y' must not interact.

Double-mutant cycles can provide information on the extent to which interactions between pairs of amino acid subunits along a protein backbone contribute to the stability of the folded state. In this case, the underlying strategy is to replace one or both of these amino acids with subunits that do not interact. A double-mutant cycle involving a synthetic system is shown in Figure 3.29 [28].

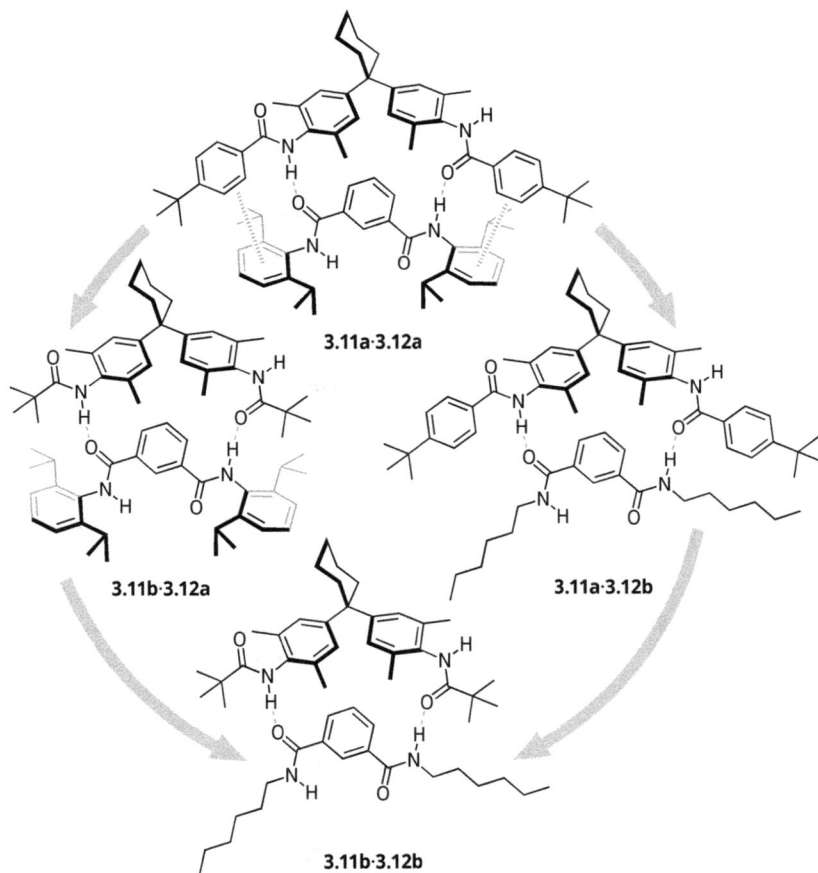

Figure 3.29: A double-mutant cycle that allows the quantification of the contribution of edge-to-face aromatic interactions in the complex between **3.11a** and **3.12a**.

In this system, the primary interaction between the binding partners **3.11a** and **3.12a** is based on hydrogen bonding. In addition, the complex is stabilized by edge-to-face aromatic interactions between the peripheral aromatic substituents. Replacement of these substituents by aliphatic residues yields the analogous compounds **3.11b** and **3.12b**, which allow the construction of a double-mutant cycle by individually estimating the stability of the four complexes **3.11a·3.12a**, **3.11b·3.12a**, **3.11a·3.12b**, and **3.11b·3.12b**. This treatment gives a ΔG^0 of –1.3 kJ mol^{-1} for the strength of the terminal edge-to-face aromatic interaction. The stabilization becomes stronger when one aromatic ring is electron-deficient and the other electron-rich, and repulsive when the interaction occurs between the edge of a phenyl ring and the face of a pentafluorophenyl ring, consistent with the electrostatic model.

The results depend sensitively on the arrangement of the interacting subunits in the complex, as shown by studies with binding partners derived from different core structures. However, structurally related complexes provide consistent information about the strength of certain types of interactions. In addition to aromatic interactions, double-mutant cycles have also been used to characterize cation–π interactions, CH–π interactions, and others.

3.2.4 Molecular balances

A third approach to characterizing the strength of different types of noncovalent interactions is based on compounds that adopt two well-defined conformational states, one in which the interaction is possible and another, in which it is not [29]. Such compounds typically contain the two groups x and y, whose interaction is being studied, attached to a core structure in such a way that rotation about a central bond orients these groups either close together (folded state) or too far apart to interact (unfolded state) (Figure 3.30). The rate of conformational interconversion must be slow enough to allow the two conformational states to be distinguished by NMR spectroscopy. In this way, the equilibrium constant K, which describes the extent to which one conformation is favored over the other, can be estimated by integrating selected signals from each conformer in the NMR spectrum. Based on this equilibrium constant, ΔG^0_{xy} can then be estimated by using equation (3.7). More reliable values for ΔG^0_{xy} are obtained by combining this concept with a double-mutant cycle, that is, by also considering the conformational equilibria of analogs in which either one or both substituents are replaced by noninteracting groups:

$$\Delta G^0_{xy} = -RT \ln K = -RT \ln \frac{c_{\text{folded}}}{c_{\text{unfolded}}} \tag{3.7}$$

A compound that allows such analyses is the Tröger's base derivative **3.13**, known as molecular torsion balance or molecular balance for short. This compound exists in two conformations, one of which is stabilized by an edge-to-face interaction between the 4-trifluoromethylbenzene unit and the benzene ring. The effect of this intramolecular interaction stabilizes the folded conformation in chloroform by -2.4 kJ mol^{-1} over the unfolded conformation, providing an estimate for the strength of the corresponding interaction in this system [30].

(a)

(b)

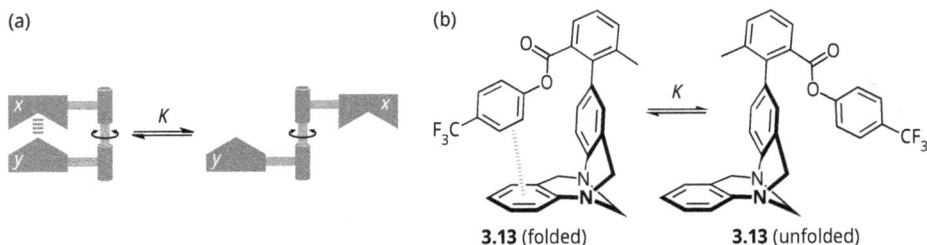

3.13 (folded)   3.13 (unfolded)

Figure 3.30: Working principle of a molecular (torsion) balance for estimating the strength of interactions between the groups x and y (a), and structure of the molecular balance **3.13**, derived from Tröger's base, which allows the strength of the edge-to-face interaction between the 4-trifluoromethylbenzene group and the benzene ring to be determined (b).

Molecular balances are attractive study objects, but there are several aspects to consider when using them. One is the influence of the solvent on the ratio of the folded and unfolded state. This influence is particularly pronounced when the conformational rearrangement is accompanied by a substantial reorganization of solvent molecules surrounding the subunits whose interaction is being probed. Another aspect is that the interaction energies depend on the exact arrangement of the substituents, which is defined by the core structure of the balance. Since this arrangement may be different from the one preferred by the same substituents in the absence of the scaffold, the extrapolation of the results to noncovalently assembled systems could be problematic. An advantage of molecular balances is that the interactions occur intramolecularly within a covalently assembled structure, so that bringing the interacting subunits together is not associated with a large entropic penalty. This allows even weak or repulsive interactions to be characterized. Several types of interactions have been studied using molecular balances, including CH–π, OH–π, NH–π, π–π, dispersion, and solvophobic effects [29].

3.3 Solvent effects

How does the solvent influence complex stability?

Imagine a complex that has been thoroughly characterized in the gas phase. How does the transfer of this complex into solution affect its structure and stability? To answer this question, one must evaluate the effects of the large number of surrounding solvent molecules. These molecules can influence the mutual arrangement of the binding partners, but these effects are difficult to predict or generalize. More obvious is that the solvent controls the relative permittivity ε of the medium. We have seen in the previous section that ε influences the strength of the electrostatic interactions between two particles, with higher relative permittivities causing the interactions to become weaker. Table 3.2 shows that the relative permittivities of typical solvents used in supramolecular chemistry vary over two orders of magnitude, so that correspondingly large variations in binding energies in different solvents can be expected as a consequence of this effect alone.

Table 3.2: Selection of parameters describing the properties of solvents typically used in binding studies [31].

Solvent	ε	$\mu \times 10^{30}$ a	π^*	α	β
Water	78.4	6.2	1.09	1.17	0.47
DMSO	46.5	13.5	1.00	0.00	0.76
DMF	36.7	12.7	0.88	0.00	0.69
Acetonitrile	35.9	13.1	0.66	0.19	0.40
Methanol	32.7	9.6	0.60	0.98	0.66
Acetone	20.6	9.0	0.62	0.08	0.48
Dichloromethane	8.9	3.8	0.82	0.13	0.10
Tetrahydrofuran	7.6	5.8	0.55	0.00	0.55
Chloroform	4.9	3.8	0.58	0.20	0.10
Benzene	2.3	0.0	0.55	0.00	0.10
1,4-Dioxane	2.2	1.5	0.49	0.00	0.37
Tetrachloromethane	2.2	0.0	0.21	0.00	0.10

a Dipole moment in C m.

Solvent effects include not only the relative unspecific influence on the medium, but also direct interactions between solvent molecules and solutes. For example, the dipole moments μ collected in Table 3.2 show that polar solvent molecules can bind to solutes through different types of electrostatic interactions (ion–dipole, dipole–dipole, or hydrogen bonding). The Kamlet–Taft parameters π^*, α, and β provide an even better insight into the propensity of solvents to engage in such interactions [32]. These parameters characterize solvents according to their polarizability (π^*), and their ability to donate (α) and accept (β) hydrogen bonds. The greater the values of α and β, for example, the better the corresponding solvent molecule is able to act as a binding partner in a hydrogen bond. Accordingly, water is a better hydrogen bond donor than acceptor (consistent with the stronger hydration of anions in water relative to cations of similar size and charge). DMSO is an even stronger hydrogen bond acceptor than water but cannot act as a donor. Tetrahydrofuran and 1,4-dioxane are also strong acceptors, while methanol is a good donor because of the acidic proton in the OH group.

Solvent molecules therefore have properties similar to those of the actual binding partners in a receptor–substrate complex, allowing them to participate in, and thus affect, the binding equilibrium. The equilibrium shown in Figure 3.31a, where a single solvent molecule Solv competes with the actual substrate S for the binding site in the receptor R, should serve as an example. At first sight, S should have little difficulty in displacing Solv from R if the interactions between R and S are stronger than those between R and the solvent molecule. However, one has to take into account the large excess of solvent molecules, which causes them to have a pronounced influence on the equilibrium, even in the case of weak solvation.

(a)

(b)

$R \cdots \text{Solv} + S \rightleftharpoons R \cdots S + \text{Solv}$

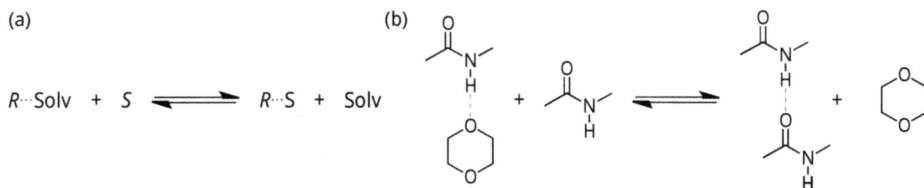

Figure 3.31: Reaction schemes showing the competition of a single solvent molecule Solv with the substrate S for the binding site of a receptor R (a), and the competition of 1,4-dioxane in the formation of the acetamide dimer (b).

This solvent effect is illustrated by the formation of an NH $\cdots$ O=C group hydrogen bond between two molecules N-methylacetamide (Figure 3.31b). In tetrachloromethane, the corresponding dimer has a K_a of 4.7 M^{-1} [33] – other authors have reported a K_a of 24 M^{-1} [34]. In contrast, practically no dimerization takes place in 1,4-dioxane. The absence of interactions in the latter solvent cannot be a medium effect, since both solvents have the same relative permittivity (Table 3.2). Instead, the stability of the N-methylacetamide dimer depends on the structural parameters of the solvent: 1,4-dioxane contains oxygen atoms that can interact with the NH group of N-methylacetamide by forming a hydrogen bond ($\beta = 0.37$), thus breaking up the N-methylacetamide dimer. Tetrachloromethane, on the other hand, does not have a strong tendency to act as a hydrogen bond acceptor ($\beta = 0.10$). Although the hydrogen bond between 1,4-dioxane and N-methylacetamide is not expected to be as strong as that between two amides (Section 3.1.5), the sheer number of solvent molecules causes the dimerization equilibrium to shift almost completely to the side of the dissociated species in 1,4-dioxane.

These considerations lead to the updated complexation equilibrium in Figure 3.32, which, unlike Figure 2.1, additionally shows the solvent molecules occupying the receptor cavity and surrounding (parts of) the substrate prior to complex formation. Enthalpically, complex formation has a good chance of outperforming solvation because the complex is stabilized by multiple interactions spread over large contact interfaces. However, when solvation becomes too strong, it impairs or even suppresses complex formation. A favorable entropic contribution results from the release of solvent molecules during complex formation, since fewer solvent molecules are needed to solvate the complex than the individual binding partners.

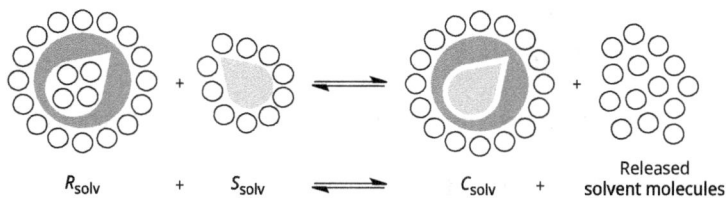

Figure 3.32: Schematic representation of a complexation process by explicitly considering the solvation of the substrate S, the receptor R, and the complex C. The white circles represent the solvent molecules.

The competition between complex formation and solvation is reflected in equation (2.8) (Section 2.1). In this equation, the total Gibbs free energy associated with complex formation ΔG^0 is equal to the sum of the intrinsic affinity of the binding partners in the gas phase ΔG^0_{intr} and the term $\Delta G^0_{solv}(C) - \Delta G^0_{solv}(R) - \Delta G^0_{solv}(S)$. This term attributes the energetic effect of solvent reorganization to the difference between the Gibbs free energy gained by solvating the complex and the energies required to desolvate its components. Several studies have confirmed the linearity of this relationship and have shown that the effect of the solvent on the Gibbs free energy of complex formation can often be estimated quantitatively using solvent parameters that correlate with $\Delta G^0_{solv}(C)$, $\Delta G^0_{solv}(R)$, and $\Delta G^0_{solv}(S)$. For example, according to the work of Hans-Jörg Schneider [35], the ΔG^0 of the K^+ complex of 18-crown-6, which varies between -11.3 kJ mol^{-1} (log $K_a = 2.0$) and -35.6 kJ mol^{-1} (log $K_a = 6.2$) in fourteen different solvents is linearly correlated with the β values of the solvent molecules, with binding becoming weaker as β increases. Thus, solvents that are strong hydrogen bond acceptors (or electron pair donors) solvate the cation efficiently and therefore compete in complex formation. There is an even better correlation between ΔG^0 and the Gibbs free energy ΔG^0_{tr} associated with the transfer of potassium ions from water to the organic solvent. This transfer energy is negative when the cation is more strongly solvated in the organic solvent than in water, or positive when solvation in the organic solvent is less efficient. The observed linear correlation between ΔG^0 and ΔG^0_{tr} indicates that, in this case, complex stability is mainly influenced by the solvation of the cation $\Delta G^0_{solv}(S)$, while the effect of the solvent on $\Delta G^0_{solv}(C)$ and $\Delta G^0_{solv}(R)$ is less pronounced. Such linear free energy relationships, which have often been observed [36], thus provide important insight into the effects of the solvent on binding.

For further discussion it is helpful to consider the enthalpic and entropic contributions to $\Delta G^0_{solv}(C)$, $\Delta G^0_{solv}(R)$, and $\Delta G^0_{solv}(S)$ separately, which leads to:

$$\Delta G^0 = \Delta G^0_{intr} + \left[\Delta H^0_{solv}(C) - \Delta H^0_{solv}(R) - \Delta H^0_{solv}(S)\right]$$

$$- T\left[\Delta S^0_{solv}(C) - \Delta S^0_{solv}(R) - \Delta S^0_{solv}(S)\right] \tag{3.8}$$

The enthalpy terms in this equation describe how the strength of the interactions between the solvent and the solute molecules affects the stability of the complex. In the case of weak solvation, the overall enthalpic effect of solvent reorganization is small. However, in solvents in which polar solutes are strongly solvated, $\Delta H^0_{solv}(C) - \Delta H^0_{solv}(R) - \Delta H^0_{solv}(S)$ can have a large positive value. This is because fewer molecules are generally needed to solvate the complex than its components, so that the enthalpy gained from solvating the complex $\Delta H^0_{solv}(C)$ cannot compensate the enthalpies $\Delta H^0_{solv}(R)$ and $\Delta H^0_{solv}(S)$ associated with the desolvation of the binding partners. The resulting positive enthalpy term thus reduces the absolute value of ΔG^0_{intr}. This effect is the reason why, as a rule of thumb, complex stability decreases as the polarity of the solvent increases or, more precisely, as the interactions of the solutes with the solvent molecules become stronger.

While solvent reorganization can reduce the enthalpic gain of complex formation, it is entropically advantageous. Figure 3.32 shows that complex formation is associated with the transfer of solvent molecules from the solvation shells into the bulk. Consequently, the unfavorable entropy associated with the solvation of the complex $\Delta S^0_{solv}(C)$ benefits from the favorable desolvation of the receptor and substrate. The term $\Delta S^0_{solv}(C) - \Delta S^0_{solv}(R) - \Delta S^0_{solv}(S)$ therefore usually returns a positive result, which is beneficial for the total ΔG^0. This effect may not be large in unstructured solvents, but becomes decisive when solvation has a strong effect on solvent organization. This is again the case in polar solvents, where a favorable entropic effect of solvent reorganization can compensate and sometimes even overcompensate an adverse enthalpic term.

The situation becomes even more complex when considering that the solvent molecules released during complex formation subsequently interact with other solvent molecules in the bulk, which adds further enthalpic and entropic contributions to the binding event (Section 3.4). This brings us to the special case of water, where the interactions between solvent molecules are sometimes more important than the direct interactions between the binding partners.

How does water mediate molecular recognition?

Based on the above considerations, it is not immediately obvious how molecular recognition should work in an environment with the high relative permittivity of water, where polar solutes are also strongly solvated due to the ability of water molecules to donate and accept hydrogen bonds. However, many compounds interact very efficiently in water, otherwise life would not have evolved and biological systems, for which molecular recognition is essential, would not operate. It could be argued that the comparison with nature is not entirely appropriate, because many binding events in biological systems take place inside the cavities of folded proteins or in the lipophilic environment of bilayer membranes, where the relative permittivity is lower than in the surrounding aqueous environment. In addition, the concentration of water in the cytosol is lower than normal. However, water molecules are still present under these conditions and can therefore potentially compete in binding events. To understand why even structurally relatively simple receptors are effective in water [37], not necessarily by using very strong types of interactions, we must first look at the properties and structure of water in the liquid phase.

Unlike many other solvent molecules, water molecules are both hydrogen bond acceptors and donors, allowing them to interact efficiently with each other. The underlying interactions are responsible for many properties of water, such as its unusually high boiling point compared to other hydrogen compounds such as ammonia or hydrogen sulfide. In ice, the formation of hydrogen bonds between individual water molecules leads to a highly ordered structure in which each oxygen atom is surrounded in a tetrahedral fashion by four hydrogen atoms, two of which are bound covalently and two by hydrogen bonds. This order is not completely lost in liquid water, where each water molecule forms an average of 3.6 hydrogen bonds with its neighbors at room temperature.

Water molecules also form hydrogen bonds with polar and especially ionic solutes. These interactions typically lead to an enthalpic stabilization ($\Delta H^0_{\text{hydr}} < 0$), but the effect on entropy is unfavorable because the water molecules in the hydration shell of these solutes are more ordered than in the bulk. Since enthalpy usually outweighs entropy, the Gibbs free energy of hydration ΔG^0_{hydr} is usually negative, explaining why the strong solvation of polar or ionic substances in water impairs molecular recognition.

However, this situation changes for solutes whose hydration has a different thermodynamic signature. For example, if the hydration of solutes has the usual entropic disadvantage but is enthalpically neutral, their interactions benefit from the entropically favorable release of water molecules. Conversely, if the hydration of solutes requires water molecules to give up hydrogen bonds, the release of these water molecules upon complex formation results in an enthalpy gain. The strong and defined interactions between water molecules in liquid water thus cause the effects of solutes on the solvent structure to have much larger enthalpic and/or entropic consequences than in other solvents. Not all the details are fully understood yet, but several models have emerged in recent years, also driven by intensive research activities in supramolecular chemistry, which help understand the underlying principles.

These models are based on concepts that correlate the influence of the dissolution of a nonpolar organic compound in water, with which water molecules do not form direct hydrogen bonds, with changes in the water structure. These changes depend sensitively on the size and shape of the dissolved species. Consequently, the recovery of the original water structure when these molecules form a complex is associated with a characteristic thermodynamic signature [38].

In the case of small (nearly) spherical compounds – examples are linear, branched, or cyclic alkanes or benzene derivatives – the creation of a cavity that allows these compounds to be incorporated into the water structure is enthalpically almost neutral or even slightly exothermic. This potentially counterintuitive thermodynamic signature is explained by the ability of water molecules near sufficiently small solutes to form the same number of hydrogen bonds, and even slightly stronger ones, as in the bulk. However, the increased strength and order of the water molecules lining the cavities causes the entropy to decrease. Thus, the merging of two cavities into a larger one, which accommodates two binding partners that were previously individually hydrated, does not cause a large change in enthalpy, but a pronounced gain in entropy because it is associated with the release of ordered water molecules. This entropy gain is the main reason why nonpolar solvents and water do not mix. The underlying process is associated with the term *hydrophobic effect t* (or solvophobic effect in general). Sometimes the term *hydrophobic interactions* can also be found in the literature, but this is misleading because it suggests that phase separation is caused by direct attractive interactions between nonpolar molecules, which is not the case. Once nonpolar solutes come together in water, they engage in van der Waals interactions, but the driving force for aggregation is predominantly the entropic gain associated with water reorganization.

This thermodynamic signature changes as the size of the solute increases because creating a cavity in water that can accommodate larger spherical or planar molecules does not allow all of the surrounding water molecules to maintain their hydrogen bonds. In this case, the hydration becomes enthalpically unfavorable but has a positive entropic component because the water molecules that give up hydrogen bonds gain translational freedom. The incorporation of these water molecules back into the water matrix causes the interaction of these solutes to be exothermic and entropically unfavorable, opposite to the hydrophobic effect discussed earlier. This combination of enthalpy and entropy is often referred to as the *nonclassical hydrophobic effect*.

The relationship between the size of the solute and the thermodynamic signature of hydration is summarized in Figure 3.33. The graph indicates that there should be a crossover point where complexation changes from entropically favored to enthalpically favored as the size of the binding partners increases and/or their shape changes.

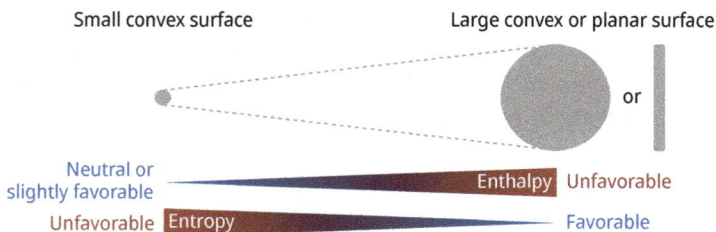

Figure 3.33: Dependence of the thermodynamic signature of the solvation of nonpolar solutes in water on size and shape.

A third case arises for solutes with concave surfaces, such as the inner surfaces of receptor cavities, which must be hydrated in water. Again, the ability of the solvating water molecules to retain all or most of their hydrogen bonds depends critically on the size and shape of the cavity. Large and shallow cavities can be hydrated without significant loss of hydrogen bonds, and the release of the trapped water molecules during complex formation is therefore mostly entropy driven. However, there are receptors with rigid cavities that hold only a few water molecules with a lower than optimal number of hydrogen bonds. When these water molecules are released, they regain hydrogen bonds, which is strongly exothermic but usually entropically unfavorable [39]. To describe the unfavorable situation of water in certain receptor cavities, the term *high-energy water* has been suggested. However, this term has been criticized, in part because it is not entirely clear that it refers only to enthalpy. Therefore, the more general and descriptive term *cavity water* is used throughout this book. In Section 4.1.11, we will take a closer look at receptors whose substrate recognition in water benefits from the release of water molecules.

Further reading

The group of Werner M. Nau showed that anionic borate clusters, such as $B_{12}Br_{12}^{2-}$, strongly interact with cyclodextrins (Section 4.1.4), with the complex between $B_{12}Br_{12}^{2-}$ and γ-CD having a log K_a of 6.0 in water, for example. The thermodynamic signature of complex formation resembles that of the nonclassical hydrophobic effect, with a pronounced negative ΔH^0 and unfavorable ΔS^0. However, borate clusters are well soluble in water and have significantly different hydration enthalpies and entropies than hydrophobic solutes. Nevertheless, they also cause water molecules in their solvation shell to have a less than optimal number of hydrogen bonds. The release of these comparatively unstructured water molecules into the structured bulk water allows them to regain hydrogen bonds, which is enthalpically favorable but entropically unfavorable. Thus, although the outcome of complex formation is similar to that of the nonclassical hydrophobic effect, the initial hydration state of these charge-dispersed anions differs from that of large convex hydrophobes. The term *chaotropic effect* has been associated with the complexation of these anions because, like typical chaotropes, they increase the solubility (*salting-in*) of proteins. For further information, see:

Assaf KI, Nau WM. The chaotropic effect as an assembly motif in chemistry. Angew. Chem. Int. Ed. 2018, 57, 13968–81.

Concluding, nonpolar solutes induce distinct changes in the bulk structure of water, which translate into characteristic effects on enthalpy and entropy. The thermodynamics of binding processes in water thus reflect, to a much greater extent than in other solvents, the contributions of solvent reorganization. These contributions can be so strong that they cause molecules to aggregate in water that otherwise have no great propensity to interact. Such binding processes range from the formation of structurally defined complexes between complementary binding partners to more complex self-assembled structures, such as micelles or vesicles, which are not observed outside the aqueous environment.

3.4 Predicting binding strength in solution

Can binding strength be predicted?

The above discussion shows that the stability of a complex in solution is determined by a complex interplay of various effects associated with direct interactions, solvation, and solvent reorganization. It could therefore be argued that any model that attempts to reliably predict binding strength in solution using a few readily available parameters is likely to fail. However, Christopher A. Hunter has shown that this is not the case [40]. His approach, assumptions, and main results are described below.

Molecular recognition events are treated in Hunter's model, as shown schematically in Figure 3.34a. Accordingly, the formation of a complex $R\cdots S$ between the receptor R and the substrate S requires the solvent molecules to leave the binding sites, in turn allowing them to interact with themselves (Solv$\cdots$Solv). The corresponding solvent–solvent interaction extends the reaction shown in Figure 3.31a by the influence of solvophobic effects, which cause the position of the binding equilibrium to depend not only on the relative strengths of the interactions in $R\cdots S$, $R\cdots$Solv, and $S\cdots$Solv, but also on how strongly the solvent molecules bind to each other. If the solvent molecules prefer to stay together rather than bind to R and S, for example, the solvent will promote the association of R and S even if the interactions in the complex are weak. In the case of water, this behavior gives rise to the hydrophobic effect.

Hunter argues that complex formation can be estimated solely on the basis of electrostatic arguments. Dispersion interactions, for example, tend to cancel each other out in solution, rendering their contribution to the overall binding affinity small. Binding strength can therefore be estimated by considering the electrostatic potentials of the binding partners in the regions of the molecules that come into contact in the complex. In the absence of steric constraints, these molecules are assumed to adopt the optimal mutual arrangement, so that the maxima and minima of the electrostatic potentials determine the interaction strength. All noncovalent interactions are therefore treated as electrostatic interactions between positively and negatively polarized regions of the interacting molecules. This approximation immediately reveals a limitation of the method: if similar values of the electrostatic potential are distributed over larger areas of a molecule, as in aromatic residues, a structurally less defined complex structure results, rendering the approach unreliable. However, in cases where the complexes are structurally well defined, it should be possible to estimate the binding strength from a set of universal parameters that describe the hydrogen bond donor and acceptor properties of the binding partners. These parameters must be chosen such that they allow the stability of each species involved in the equilibrium in Figure 3.34a to be estimated in terms of Gibbs free energies. Assuming further that these energies are additive, the overall $\Delta G^0_{\mathrm{H-bond}}$ of the equilibrium is given as follows:

$$\Delta G^0_{\mathrm{H-bond}} = -(\alpha\beta + \alpha_S\beta_S) + (\alpha\beta_S + \alpha_S\beta) = -(\alpha - \alpha_S)(\beta - \beta_S) \tag{3.9}$$

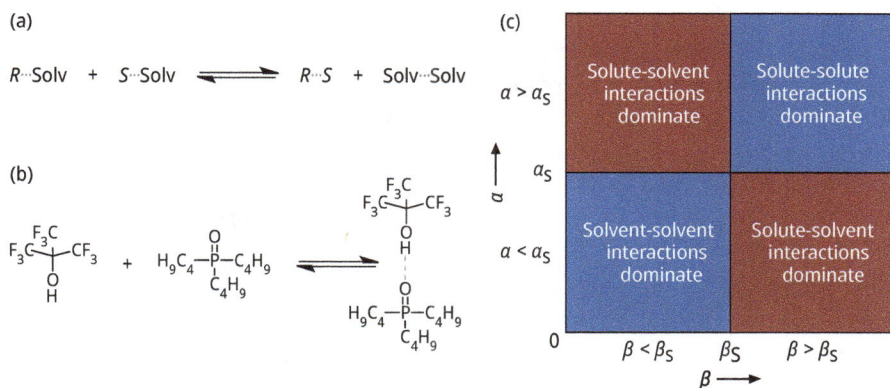

Figure 3.34: Competition between the solvation of the receptor R and the substrate S in a binding equilibrium with the formation of the complex $R{\cdots}S$, taking into account that the released solvent molecules can interact with themselves (a). The equilibrium describing the hydrogen bond formation between perfluoro-*tert*-butyl alcohol and tri-*n*-butylphosphine oxide is shown in (b), and (c) illustrates how the preferred interaction in solution depends on the hydrogen bond donating and accepting ability of the solutes (α and β, respectively) in relation to the corresponding parameters α_S and β_S of the solvent. The blue quadrants in the diagram show the regions where complex formation is favored, either because the interactions between R and S are stronger than those with the solvent, or because the solvent molecules interact so strongly that they promote the interaction of R and S. Conversely, the red quadrants indicate the regions where the solvation of R and S is so strong that complex formation does not occur.

The parameters α and β in this equation denote the hydrogen bond donor and acceptor strengths of the molecules involved in complex formation, respectively, and α_S and β_S stand for the corresponding parameters of the solvent molecules. Equation (3.9) thus relates the contribution of electrostatic interactions to complex formation to the strength of the interactions between R and S ($\alpha\beta$) and between two solvent molecules ($\alpha_S\beta_S$). Since all parameters are positive, the negative sign in front of the first sum in equation (3.9) is required to give a negative binding energy. The competition of the desolvation of R and S with the interactions in $R{\cdots}S$ and Solv${\cdots}$Solv is accounted for by the second term in equation (3.9), with the respective products $\alpha\beta_S$ and $\alpha_S\beta$ becoming larger as the solvation becomes stronger. It must be stressed that although the meaning of α and β is the same as that of the Kamlet–Taft parameters, the purpose of the two parameters is here is to provide information on binding energies rather than solvent properties. The values of α and β are therefore different from those given in Table 3.2.

Independent studies have shown that quantitative thermodynamic information can indeed be derived from parameters describing the hydrogen bond donor and acceptor properties of molecules. According to these studies, the association constants of many hydrogen-bonded complexes follow the general expression:

$$\log K_a = c_1 \alpha_2^H \beta_2^H + c_2 \tag{3.10}$$

In this equation, α_2^H and β_2^H again describe the hydrogen bond donor and acceptor properties of the interacting species. The factor c_1 reflects the solvent effect on the strength of the interactions. It increases with decreasing polarity of the medium. The constant c_2 typically amounts to -1.0 ± 0.1, independent of the solvent and the binding partners. The decrease in $\log K_a$ by about one order of magnitude caused by c_2, which translates into a ΔG^0 value of 6 kJ mol^{-1} at 298 K, thus appears to be a fundamental property of binding events and has been linked to the energetic cost of bringing two molecules together in solution to form a complex.

Although equations (3.9) and (3.10) describe similar aspects, they are not directly comparable. Equation (3.9) focuses only on the energetic contribution of the hydrogen bonding interactions ΔG^0_{H-bond}, but does not allow the calculation of the total binding strength because the effect of c_2 is missing. Conversely, equation (3.10) considers the solvent effect only in terms of the global constant c_1 and does not separate the hydrogen bond donor and acceptor properties of the solvent. Hunter's approach to combining the two equations is based on the experimental values of α_2^H and β_2^H. These values are converted to the α and β scale so that equation (3.11) allows the calculation of ΔG^0. The solvent parameters α_S and β_S are obtained in a similar way, assuming that these values do not change significantly when moving from the dilute solutions in which they are usually determined to the bulk. For compounds for which α_2^H and β_2^H are not available, α and β can be estimated computationally from the electrostatic potentials, allowing also new compounds to be included in the model.

$$\Delta G^0 = -(\alpha - \alpha_S)(\beta - \beta_S) + 6 \, \text{kJ} \, \text{mol}^{-1} \tag{3.11}$$

To confirm that equation (3.11) accurately predicts binding energies, the experimental and calculated stabilities of the complex between perfluoro-*tert*-butyl alcohol and tri-*n*-butylphosphine oxide in 13 different solvents were compared (Figure 3.34b) [41]. The agreement was mostly excellent with only one exception (*n*-decanol), which was attributed to the fact that the hydrogen bond parameters in this case did not properly reflect the solvent properties.

An illustrative example of the usefulness of the method is also the comparison of the experimental stability of the *N*-methylacetamide dimer in tetrachloromethane and 1,4-dioxane with the corresponding calculated values (Section 3.3). Using the hydrogen bonding parameters for an amide ($\alpha = 2.9$, $\beta = 8.3$), tetrachloromethane ($\alpha_S = 1.4$, $\beta_S = 0.6$), and an alkyl ether ($\alpha_S = 0.9$, $\beta_S = 5.3$), equation (3.11) affords a ΔG^0 of -5.6 kJ mol^{-1} for the stability of the *N*-methylacetamide dimer in tetrachloromethane, which corresponds to a K_a of 9 M^{-1}. In ether, a ΔG^0 of zero is obtained, in very good agreement with the experiments.

Equation (3.11) can also be used to derive general information about solvent effects on complex stability. The diagram in Figure 3.34c shows how the formation of $R \cdots S$ depends on the hydrogen bond donor and acceptor strengths of the solvent. The diagram is divided into four quadrants whose boundaries meet at the values corresponding to α_S and β_S, where no binding occurs ($\Delta G^0 = 6$ kJ mol^{-1}). In the region where solutes are better hydrogen bond donors ($\alpha > \alpha_S$) and better hydrogen bond acceptors ($\beta > \beta_S$) than the solvent, the interactions between R and S are stronger than solvation and complex formation dominates the equilibrium. Conversely, the complex is also formed in the region where the solvent–solvent interactions dominate ($\alpha < \alpha_S$, $\beta < \beta_S$). However, if the solutes are better hydrogen bond donors than the solvent ($\alpha > \alpha_S$) but worse hydrogen bond acceptors ($\beta < \beta_S$) or *vice versa* ($\alpha < \alpha_S$, $\beta > \beta_S$), they are too strongly solvated and complex formation is not possible. The exact shapes and sizes of the regions favoring (blue) and disfavoring (red) complex formation depend on the exact values of α_S and β_S as illustrated in Figure 3.35.

Figure 3.35a shows the hypothetical situation in the absence of a solvent ($\alpha_S = 0$, $\beta_S = 0$). As expected, all interactions are attractive under these conditions, but the actual binding strengths depend strongly on the hydrogen bonding properties of the interacting compounds. Selected hydrogen bond donors and acceptors are depicted along the axes of the diagram to illustrate the correlation between hydrogen bond strength and structure. Accordingly, weakly polarized compounds such as alkanes, alkenes, or aromatic compounds are located at the lower end of the scale, whereas compounds with strong dipole moments have large values of α and β and therefore interact more strongly.

For DMSO ($\alpha_S = 0.8$, $\beta_S = 8.9$), the diagram is dominated by the upper left quadrant (Figure 3.35b) because most of the compounds along the top axis are weaker hydrogen bond acceptors than DMSO itself ($\beta < \beta_S$). As a result, donors prefer to interact with the solvent. Accordingly, DMSO is a good solvent for compounds that form strongly hydrogen-bonded aggregates in the solid state but a poor solvent for promoting complex formation by hydrogen bonding.

Figure 3.35c shows that chloroform ($\alpha_S = 2.2$, $\beta_S = 0.8$) is much more suitable in this respect. Due to the low β_S value, chloroform does not compete with most hydrogen bond acceptors. Thus, hydrogen bond donors with an α value greater than 2.2 interact with almost all acceptors in chloroform, even the weakest. Weak donors, however, are well solvated in chloroform, giving rise to the red quadrant in the bottom right-hand corner of the diagram.

Finally, Figure 3.35d shows the situation in water ($\alpha_S = 2.8$, $\beta_S = 4.5$). The treatment of water is based on an adapted version of equation (3.11), which takes into account the fact that water does not interact directly with nonpolar solutes but prefers to entrap them in cavities lined by strongly hydrogen-bonded water molecules. The corresponding diagram is still closely related to the other three in that there are red regions in the upper left and the lower right corner where solvation of the binding partners opposes complex formation. Due to the large values of α_S and β_S, the blue region in the upper right corner is small, indicating that only the strongest hydrogen bond donors and acceptors are able to form complexes in water. In contrast to the

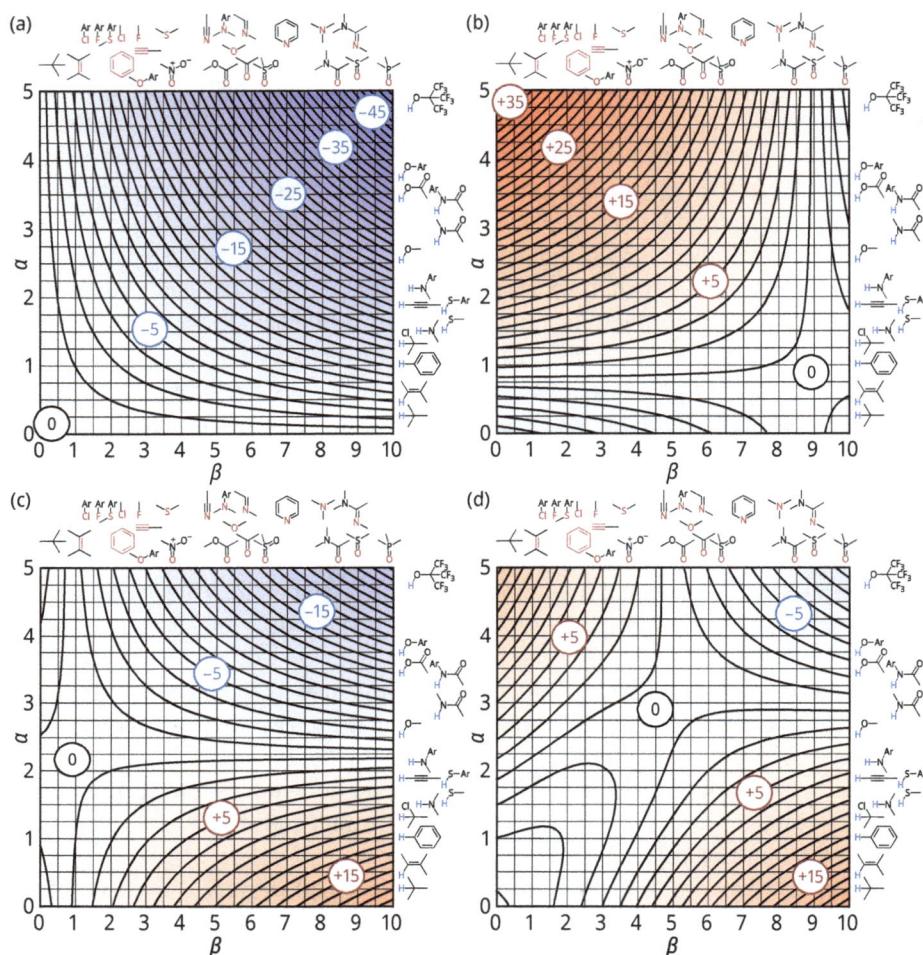

Figure 3.35: Functional group interaction profiles between solutes with different values of α and β in a hypothetical medium with $\alpha_S = \beta_S = 0$ (a), in DMSO ($\alpha_S = 0.8$, $\beta_S = 8.9$) (b), in chloroform ($\alpha_S = 2.2$, $\beta_S = 0.8$) (c), and in water ($\alpha_S = 2.8$, $\beta_S = 4.5$) (d). The contour lines indicate the values of ΔG^0 in kJ mol^{-1} at intervals of 1.25 kJ mol^{-1}, with blue regions denoting favorable and red regions unfavorable solute–solute interactions. Representative structures of hydrogen bond acceptors and donors are shown along the x- and the y-axis, respectively.

other plots, there is a substantial blue region in the lower left corner, consistent with the ability of water to induce the association of nonpolar molecules through the hydrophobic effect.

This model thus provides insight into some of the most important factors controlling complex formation in solution. The fact that even quantitative predictions of binding strength can be made by attributing complex formation to electrostatic interactions alone is the reason why the different types of interactions in Section 3.1 were discussed mainly in terms of these polar effects. With its recent extensions and refinements, such as its adaptation to charged binding partners, Hunter's model has become a valuable tool for understanding molecular recognition processes and how they are influenced by the solvent.

3.5 Guidelines for receptor design

How can strong binding be achieved?

So far, we have seen that different types of intermolecular interactions cause molecules to stay together, and we have learned about the influence of the solvent on complex formation. The examples discussed in this context already showed that receptors can have very different shapes and structures. However, despite these differences, receptor design is usually based on a number of general guidelines, the most important of which are outlined below.

3.5.1 Complementarity, preorganization, and induced fit

A good starting point to derive an important concept that often underlies receptor design is Figure 3.1, which shows schematic structures of receptor–substrate combinations. Of the four scenarios shown in this figure, the first is usually the most efficient because of the perfect size and shape complementarity of the receptor cavity and the substrate. If both binding partners are also complementary in their electronic properties, stronger binding is expected than in situations where the shape of the substrate does not match the shape of the cavity, or where the substrate is too large or too small. Size, shape, and electronic *complementarity* of the binding partners are therefore crucial factors for a stable complex.

The concept of complementarity has its roots in the *lock-and-key* principle formulated by Emil Fischer in 1894. Fischer used this analogy to rationalize the substrate selectivity of enzymes, suggesting that the substrate fits the enzyme's binding pocket like a key fits its lock. Fischer wrote "Um ein Bild zu gebrauchen, will ich sagen, dass Enzym und Glucosid wie Schloss und Schlüssel zu einander passen müssen, um eine chemische Wirkung auf einander ausüben zu können" ["To use an image, I would like to say that enzyme and glucoside must fit together like lock and key in order to exert a chemical effect on each other"] [42]. The idea is not restricted to biological molecular recognition processes but also applies to receptors in supramolecular chemistry. It is illustrated in Figure 3.36, which shows that the shape complementarity of the binding partners in (a) should allow complex formation, whereas the lack of complementarity in (b) prevents it.

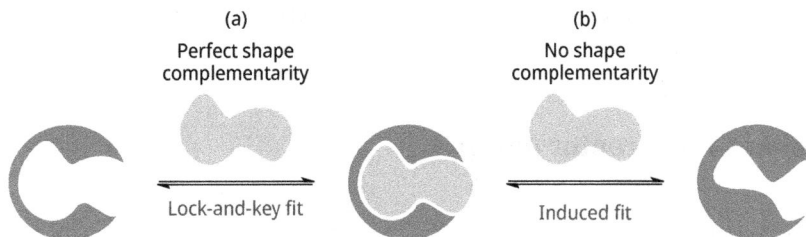

Figure 3.36: Schematic representation of the lock-and-key and induced fit concepts. The binding partners in (a) are perfectly complementary in shape and therefore form a complex according to the lock-and-key principle, whereas no binding is expected for the binding partners in (b). If the receptor in (b) can adapt its structure to accommodate the substrate, the complex can be formed by induced fit. However, the receptor is better preorganized for the substrate in (a) than in (b).

Although intriguing, the comparison of molecules to locks and keys is not entirely accurate, because molecules are never rigid. Indeed, Daniel E. Koshland Jr. showed that the active sites of enzymes do not have to be shape complementary to the substrates (or, more precisely, to the transition state of the reaction they catalyze) before binding. Instead, enzymes adapt to the structure of their substrate by conformational reorganization. Koshland compared this process to fitting a hand into a glove. An empty glove adopts all sorts of shapes but can still distinguish the right from the left hand when used [43]. The *induced fit* model introduced by Koshland therefore provides a more realistic view of a binding process, regardless of whether proteins or small receptors are involved. Figure 3.36b shows that it allows complex formation even when the empty cavity does not match the shape of the substrate.

The ability of receptors to adapt to the structural requirements of the substrate is actually an advantage. If the receptor is completely rigid, it must be perfectly shape complementary for optimal binding. Even small deviations from the optimal structure lead to a decrease in affinity. More flexible receptors, on the other hand, can respond to these deviations by adapting the receptor structure and thus strengthening the complex. If this conformational reorganization is not too extensive, the enthalpy gained by the structural adjustments easily overcompensates the adverse entropy term associated with the loss of conformational freedom. However, if the conformations of the receptor in the absence and presence of the substrate differ substantially, the stability of the complex suffers. The reason is mainly entropic, as complex formation is associated with a loss of conformational flexibility. An additional adverse enthalpic term arises if the conformation of the receptor in the complex is strained or otherwise unfavorable. Donald J. Cram summarized these effects in the principle of *preorganization*, which states that "the more highly hosts and guests are organized for binding [...] prior to their complexation, the more stable will be their complexes" [44].

Combining the three concepts of complementarity, induced fit, and preorganization provides the first guideline for receptor design:

> A receptor should be complementary to the substrate in size, shape, and electronic properties, and should also have some flexibility to optimize the interactions in the complex. However, a receptor should not be too flexible and sufficiently well preorganized to minimize the structural adjustments required for substrate binding.

3.5.2 Chelate effect and macrocyclic effect

Receptor–substrate interactions generally do not rely on a single point of contact, but on a combination of multiple interactions distributed over a larger contact area. The overall efficiency of binding benefits greatly from these multiple contacts, since a hypothetical complex in which the substrate interacts with an equivalent number of isolated binding partners is significantly less stable. An example from coordination chemistry illustrates this point. The formation of the tetraammine copper(II) complex $[Cu(NH_3)_4(H_2O)_2]^{2+}$ in water is a stepwise process in which water molecules are progressively replaced by ammonia molecules from the copper center. The overall equilibrium is characterized by a $\log K_a$ of 12.6. In comparison, the same reaction with the linear tetraamine **3.14** (Figure 3.37a) has an impressive eight orders of magnitude higher $\log K_a$ of 20.9 [45]. This difference is not due to the underlying metal–nitrogen interactions, the strength of which is more or less comparable in both complexes, but to entropy. In the case of $[Cu(NH_3)_4(H_2O)_2]^{2+}$, each step of complex formation involves the replacement of a water molecule from the hydrated copper center by an ammonia molecule. Therefore, the number of molecules does not change during the reaction. In contrast, the binding of the linear tetraamine results in the release of four water molecules, which is entropically favorable. This effect, termed *chelate effect*, generally causes complexes between metal centers and multidentate ligands to be more stable than complexes between the same metal ion and the corresponding number of monodentate ligands.

Figure 3.37 shows that the coordination of ligand **3.14** still suffers from the fact that this ligand must fold around the metal center. In the terminology of supramolecular chemistry, one could say that **3.14** is not well preorganized for complex formation. A way to address this issue involves using tetraamine **3.15** as ligand, the cyclic analog of **3.14**. In **3.15**, the cost of restricting conformational mobility and stabilizing the folded conformation required for metal coordination has already been paid for during the cyclization reaction, making the complex of this ligand with Cu^{2+} approximately four orders of magnitude more stable than that of **3.14** ($\log K_a = 24.8$) [45]. The stabilization of the complex associated with the use of a cyclic ligand is termed the

(a)

(b)

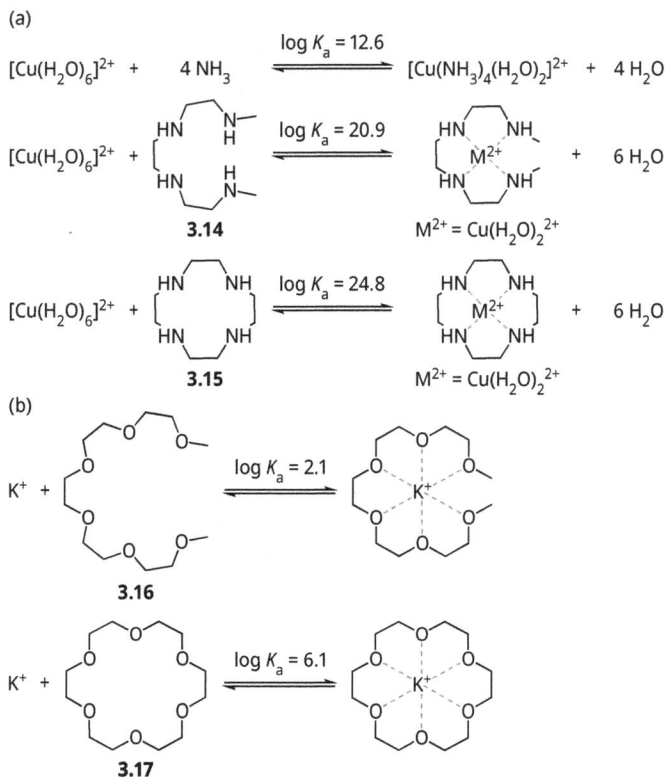

Figure 3.37: Comparison of the coordination of ammonia, the linear tetraamine **3.14**, and the cyclic tetraamine **3.15** to $[Cu(H_2O)_6]^{2+}$ (a), and of the podand **3.16** and crown ether **3.17** to K^+ (b). The binding constants of the podand and crown ether complexes refer to methanol at 25 °C [46].

macrocyclic effect. This effect is usually even more pronounced for macrobicyclic ligands (Section 4.1.2), where the term *macrobicyclic effect* is used.

These principles are not limited to coordination chemistry but also apply to supramolecular chemistry. For example, the complex of the linear podand **3.16** with K^+ is less stable than that of the crown ether **3.17** (Figure 3.37b) as a consequence of the macrocyclic effect. Although this macrocyclic effect is a rather general phenomenon, it is not always related to entropy as in the case of **3.15**. In Section 4.1.2, we will see that potassium binding in methanol is actually enthalpically more favorable for **3.17** than for **3.16**. Possible reasons for these enthalpy differences are a strained conformation of the acyclic receptor in the complex or a more difficult desolvation of the acyclic receptor during complex formation.

It should be emphasized that chelate effects operate in almost all supramolecular complexes because the formation of multiple contacts between the binding partners always also benefits from the entropically advantageous release of solvent molecules. We conclude:

Multidentate interactions within supramolecular complexes have a positive effect on stability, and the combination of this chelate effect with cyclic or bicyclic receptor structures is particularly beneficial.

3.5.3 Multivalency and cooperativity

Multivalency and cooperativity are concepts that refer to complexes stabilized by multiple receptor–substrate interactions. In contrast to complexes stabilized by multiple interactions at a common interface, multivalent systems contain at least two, but usually more, individual and typically distinct binding sites within each binding partner. Complex formation thus occurs between a receptor with multiple recognition units, which may or may not be identical, and a complementary polyfunctional substrate. This situation is illustrated in Figure 3.38.

(a)   (b)   (c)

Figure 3.38: Schematic illustration of a monovalent interaction (a) in comparison to divalent (b) and trivalent (c) interactions between receptors containing two or three binding sites and substrates with the equivalent number of complementary subunits.

Complexes stabilized by multivalent interactions are more stable than monovalent ones. This trend is closely related to the chelate effect, which also involves multiple interactions, but there is an important difference. The chelate effect describes the increase in binding strength caused by replacing *monodentate* binding partners in a complex with a *multidentate* analog that participates in an equivalent number of interactions. To assess the magnitude of this effect, we compare (somewhat inappropriately because of the different stoichiometries) the stability of a higher order complex such as $[Cu(NH_3)_4(H_2O)_2]^{2+}$ with that of the corresponding analog $[Cu(H_2O)_2 \cdot \mathbf{3.14}]^{2+}$. In contrast, in the case of multivalency, we compare the stability of a monovalent complex, shown schematically in Figure 3.38a, with that of a multivalent counterpart (Figure 3.38b,c). Since both complexes have the same stoichiometry, their binding constants are directly comparable. However, the binding constant of the multivalent complex must account for the fact that it is stabilized by multiple points of attachment, each of which contributes to the overall stability with an individual Gibbs free energy contribution. The overall stability therefore results from the sum of the individual binding energies $\sum_{i=1}^{N} \Delta G_i^0$ or the product of the respective association constants $\prod_{i=1}^{N} K_a^i$, so that the multivalent complex is always more stable than the monovalent one, even if each interaction does not contribute equally to the overall binding strength. The terms *affinity* and *avidity* are sometimes used to distinguish the stability of monovalent and multivalent complexes, respectively.

This brings us to the concept of cooperativity. To understand what cooperativity means, it is helpful to divide complex formation by multivalent interactions into a sequence of steps. The formation of the complex shown in Figure 3.38b, for instance, involves the initial binding of a subunit of the receptor RR to a subunit of the substrate SS, yielding the partially bound acyclic intermediate $(RR \cdot SS)_o$ (Figure 3.39). This intermediate then cyclizes to form the cyclic complex $(RR \cdot SS)_c$ and, if this cyclization is not efficient, higher complexes such as $RR \cdot (SS)_2$ are also formed. Polymeric complexes with multiple receptors binding to multiple substrates are possible under certain conditions, but the formation of these complexes can be neglected if SS is present in excess, simplifying the equilibrium to the situation shown in Figure 3.39.

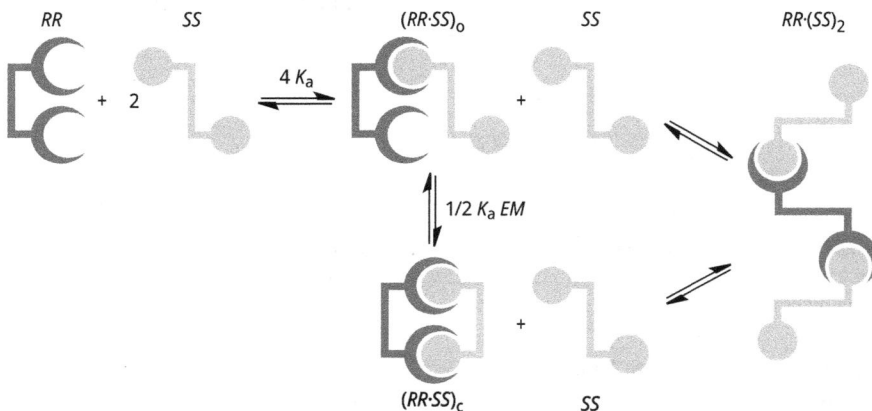

Figure 3.39: Stepwise formation of the cyclic complex $(RR \cdot SS)_c$ between a divalent receptor RR and a divalent substrate SS *via* the acyclic intermediate $(RR \cdot SS)_o$. In the case of high chelate cooperativity, the cyclic product dominates in the equilibrium. If cooperativity is low, however, higher complexes such as $RR \cdot (SS)_2$ are also formed to a considerable extent.

The ratio of $(RR \cdot SS)_o$ and $(RR \cdot SS)_c$ depends on how much the formation of the cyclic complex is favored over the intermolecular interactions. If the cyclic complex is formed very efficiently from $(RR \cdot SS)_o$, the cooperativity of the process is positive. It is negative if the cyclic complex does not form, for example because the tether connecting the two binding sites in the substrate does not allow the formation of $(RR \cdot SS)_c$. In this case, $RR \cdot (SS)_2$ becomes the favored product. To assess cooperativity quantitatively, the relative strengths of the intramolecular and intermolecular binding steps must be estimated. This approach is similar to the treatment of the chelate effect, which explains why the term chelate cooperativity (or intramolecular cooperativity) is used to describe the behavior of such systems [47, 48].

The effects of chelate cooperativity on complex formation can be understood quantitatively by considering the laws of mass action associated with the formation of the open complex $(RR \cdot SS)_o$ from its components RR and SS, and the formation of the cyclic complex $(RR \cdot SS)_c$ from $(RR \cdot SS)_0$. Equation (3.12) describes the first equilibrium, where K_a quantifies the stability of the complex between the monovalent receptor R and the substrate S:

$$4\ K_a = \frac{c_{(RR \cdot SS)0}}{c_{RR}\ c_{SS}} \tag{3.12}$$

According to this equation, the experimental equilibrium constant is four times larger than K_a, which is a statistical effect. Whereas only one product can be formed in the reaction between R and S, the first binding step between RR and SS leads to four different microspecies according to Figure 3.40a, giving it a statistical advantage over the corresponding monovalent interaction.

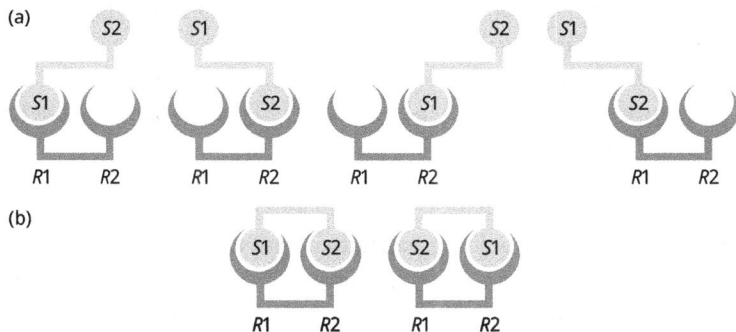

Figure 3.40: Possible interactions of a ditopic receptor RR with a ditopic substrate SS (b). The receptor subunits $R1$ and $R2$ are identical, as are the substrate subunits $S1$ and $S2$. The distinction is made only to illustrate that there are four ways in which these binding partners can interact, leading ultimately to indistinguishable complexes $(RR \cdot SS)_o$. The four complexes in (a) converge to give two possible arrangements of the closed complex $(RR \cdot SS)_c$ as shown in (b).

Such statistical factors always come into play when an equilibrium involving multiple interactions is compared with its monovalent counterpart. They depend on the structure of the binding partners and the binding mode, and can be estimated by calculating the ratio of the number of microspecies that can be formed on the product side n_{Prod} to the corresponding number of microspecies on the reactant side n_{React}. Since free components count as one microspecies, n_{Prod}/n_{React} in the above example is 4. In the following cyclization step, the four microspecies converge to give two microspecies of the product (Figure 3.40b), resulting in an n_{Prod}/n_{React} ratio of 1/2. Accordingly, the equilibrium between the open and the cyclized complex is described as follows:

$$\frac{1}{2} K_a \ EM = \frac{c_{(RR \cdot SS)c}}{c_{(RR \cdot SS)o}}$$

(3.13)

This equation contains an additional parameter, the effective molarity EM, which accounts for the fact that the formation of the cyclic from the open complex is an intramolecular process and the associated equilibrium constant is therefore dimensionless. The equilibrium constant of the monovalent interaction K_a must therefore be multiplied with a concentration to render the equation correct. The corresponding concentration can be viewed as the substrate concentration that must be reached for the intermolecular complex to outcompete the intramolecular complex. According to equation (3.13), the EM correlates with the extent to which the intramolecular process is favored over the intermolecular process for a given K_a. The higher the EM, the higher the concentration of the cyclic complex at equilibrium.

The product $K_a \ EM$ in equation (3.13) can be regarded as a measure of cooperativity. If $K_a \ EM \ll 1$, the partially bound complex is more stable than the cyclic one. In this case, the overall equilibrium is unaffected by the cyclic complex and the system behaves as if the substrate were monovalent. Positive cooperativity is observed when $K_a \ EM \gg 1$, resulting in $(RR \cdot SS)_c$ being the major species over a wide range of concentrations. Note that other ways of assessing chelate cooperativity that are independent of K_a have been proposed [48].

The strong coupling of the first and subsequent binding steps in the case of high chelate cooperativity thus leads to an all-or-nothing behavior, with the equilibrium either featuring the fully dissociated binding partners or the fully formed complex but no partially bound complexes. A typical example of such a system is the DNA double helix, which is stabilized by multiple interactions between the individual polynucleotide strands. Since each interaction strongly favors the next, only the fully formed double helix is present under normal conditions. Conversely, when DNA is heated in solution, strand dissociation occurs within a relatively narrow temperature range where all interactions break simultaneously. The structural integrity of multivalent complexes featuring chelate cooperativity still depends on the relative concentrations of the binding partners. If the substrate concentration exceeds EM, the 1:1 complexes dissociate and higher complexes are formed. With regard to receptor design, it can be stated that:

Multivalent interactions are always stronger than monovalent ones. Furthermore, when complex formation is characterized by positive chelate cooperativity, partially bound species are usually underrepresented in the binding equilibrium, while the fully bound complex dominates.

3.5.4 Allosterism and cooperativity

Allosteric behavior is also feature that is closely linked to receptors with more than one binding site. In this case, however, the receptor does not interact with a multivalent substrate but with several identical or different monovalent ones. The underlying concept comes from biochemistry where it describes the control of enzymatic activity by compounds called effectors that bind to enzyme pockets other than the actual active sites. Effectors typically trigger a conformational change in the enzyme, resulting in either activation or deactivation of enzymatic activity. In the first case, the effector molecule acts as an activator; in the second, it acts as an inhibitor.

In supramolecular chemistry, multivalent receptors exhibit allosteric behavior when the interaction with the first substrate affects the affinity of the other binding site(s). Such processes therefore also involve an element of cooperativity, which in this case describes the extent to which the different intermolecular binding steps influence each other. If the binding of the first substrates favors the subsequent binding steps, the system exhibits positive cooperativity, whereas negative cooperativity occurs when the first binding step weakens the subsequent interactions. The term allosteric cooperativity is used to distinguish this behavior from the earlier described chelate cooperativity [47, 48]. A prototypical example of a natural system exhibiting positive allosteric cooperativity is hemoglobin, where the binding of each oxygen molecule promotes the binding of the next until all four binding sites are occupied. Again, strong positive cooperativity leads to an all-or-nothing behavior characterized by a sharp transition from the fully dissociated to the fully bound complex.

In order to assess allosteric cooperativity, it is useful to distinguish between the monovalent association constants, which describe the strength of the individual interactions at each receptor subunit, and the experimental association constants, which additionally include the statistical degeneracy of the system in the case of identical binding sites. In this context, consider the example of a divalent receptor RR interacting with two substrate molecules S (Figure 3.41).

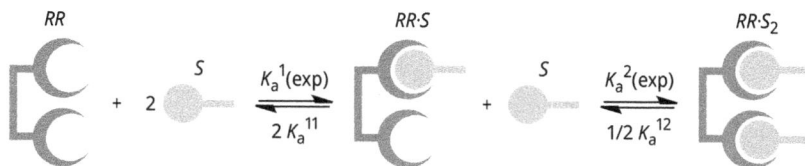

Figure 3.41: Stepwise complexation of two substrate molecules S by a divalent receptor RR. The microscopic association constants K_a^{11} and K_a^{12} describe the intrinsic strength of each binding step and the statistical factors 2 and 1/2 reflect the degeneracy of the partially bound intermediate.

The corresponding equilibrium comprises two steps, the initial formation of the partially bound intermediate $RR·S$, whose formation is associated with the microscopic association constant K_a^{11}, and the subsequent formation of the fully formed complex $RR{\cdot}S_2$, characterized by K_a^{12}. Since the interaction of a monovalent substrate S with the divalent receptor RR gives rise to two microspecies in which the substrate binds to one of the two binding sites of the receptor, a statistical factor of 2 must be introduced when correlating the experimental stability constant $K_a^1(\mathrm{exp})$ with K_a^{11} (Figure 3.41). Both microspecies will then give the same product, so that the $K_a^2(\mathrm{exp})$ of the second step is $1/2\ K_a^{12}$.

The cooperativity of the overall process depends on the direction and the extent to which the presence of S in the partially bound intermediate influences the strength of the interaction in the second step. The parameter used to describe this effect is the interaction parameter α, which is the ratio of the two microscopic association constants according to the following equation:

$$\alpha = \frac{K_a^{12}}{K_a^{11}} \tag{3.14}$$

In the absence of cooperativity, the intrinsic interactions at both binding sites have the same strength, which means that $K_a^{11} = K_a^{12}$ and $\alpha = 1$. Because of the statistical factors that correlate the microscopic and experimental binding constants, this relationship translates into $0.5\ K_a^1(\mathrm{exp}) = 2\ K_a^2(\mathrm{exp})$ or $K_a^1(\mathrm{exp}) = 4\ K_a^2(\mathrm{exp})$. The experimental stepwise binding constants of the equilibria shown in Figure 3.41 thus differ by a factor of 4 for a noncooperative system and any deviation from this ratio indicates cooperative behavior.

If $\alpha < 1$, the interactions in the intermediate $RR{\cdot}S$ are stronger than in the fully bound state $RR{\cdot}S_2$. $RR{\cdot}S_2$ is therefore only populated to a significant extent when S is present in excess. This case is illustrated by Figure 2.3b in Section 2.1, where binding isotherms are shown for a 1:2 receptor–substrate complex whose stepwise formation is associated with a K_a^{12} that is much smaller than K_a^{11}. In contrast, when complex formation is positively cooperative ($\alpha > 1$), there is a substantial thermodynamic driving force for the binding of the second substrate molecule once the intermediate complex has formed, as shown in Figure 2.3a. If α is very large, the intermediate state is never populated, and the system exhibits all-or-nothing behavior.

Whether a system exhibits allosteric cooperativity can be estimated using so-called Hill plots. This treatment is based on the idea that complex formation involves a single equilibrium, characterized by the binding of a multivalent receptor to n guests. The corresponding law of mass action is given by equation (3.15a) whose rearrangement leads to equation (3.15b):

$$K_a = \frac{c_C}{c_R\ c_S^n} \tag{3.15a}$$

$$K_a \, c_S^n = \frac{c_C}{c_R^0 - c_C} \tag{3.15b}$$

Expressing the ratio c_C/c_R^0 as the degree of saturation Θ_a then leads to equation (3.15c) after further rearrangement:

$$\log \left[\frac{\Theta_a}{1-\Theta_a} \right] = \log K_a + n \log c_S \tag{3.15c}$$

Plotting $\log[\Theta_a/(1-\Theta_a)]$ *versus* $\log c_S$ yields a graph, the Hill plot, whose slope n at 50% saturation, that is, at $\log[\Theta_a/(1-\Theta_a)] = 0$, provides information about the cooperativity of the reaction. If the Hill coefficient n is 1, complex formation is noncooperative. Negative cooperativity ($a < 1$) is reflected by a Hill coefficient $n < 1$ and positive cooperativity ($a > 1$) by $n > 1$. To illustrate the shapes of such graphs, Hill plots describing the complex formation of a ditopic receptor are shown in Figure 3.42 [47].

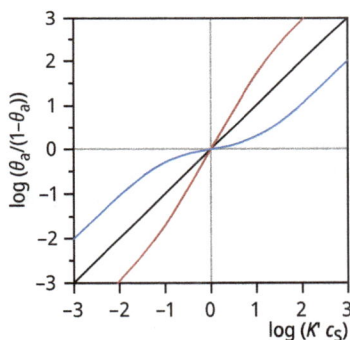

Figure 3.42: Hill plots describing the complex formation of a ditopic receptor when substrate binding proceeds with no cooperativity ($a = 1$) (black), with positive cooperativity ($a = 100$) (red), or with negative cooperativity ($a = 0.01$) (blue). The graphs show that positive cooperativity leads to a slope greater than unity ($n > 1$) at $\log[\Theta_a/(1-\Theta_a)] = 0$, while negative cooperativity leads to a slope smaller than unity ($n < 1$). Using $\log(K' c_S)$ as the x-axis, where $K' = \left(K_a^{11} K_a^{12} \right)^{0.5}$, causes the curves to pass through the origin for different values of a.

Allosteric behavior of supramolecular receptors can be realized in several ways. The one that most closely resembles the allosteric behavior of natural systems is based on the conformational coupling of two or more binding sites within a receptor [49]. Possible strategies are shown in Figure 3.43.

These examples illustrate the positive and negative allosteric behavior of receptors with different binding sites (*heterotopic* systems). If the binding of the first substrate causes a conformational reorganization of the receptor that improves the preorganization of the other binding site, the receptor exhibits positive cooperativity (Figure 3.43a). Conversely, the stabilization of a receptor conformation by the first substrate that is unsuitable for the uptake of the second substrate results in negative cooperativity (Figure 3.43b). Cooperativity can also be mediated by interactions between the bound substrate molecules. For example, if the two substrates are oppositely charged, attractive electrostatic interactions between the bound ions, as in some ion pair receptors (Section 4.2.3), stabilize the final complex, resulting in positive cooperativity. Since this mechanism of cooperativity differs from allosteric cooperativity, which involves confor-

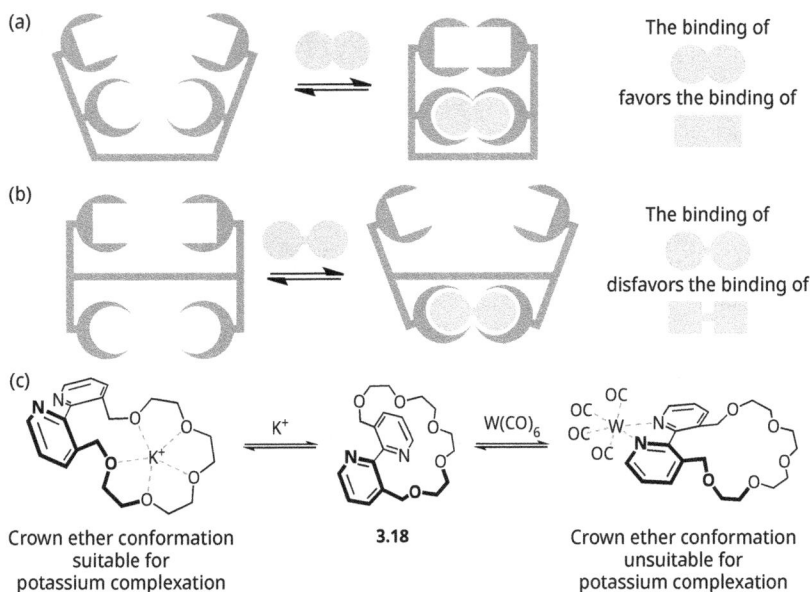

Figure 3.43: Schematic representation of the binding modes of allosteric receptors exhibiting positive heterotopic (a) and negative heterotopic (b) cooperativity. The allosteric receptor **3.18** shown in (c) has binding sites for a transition metal ion and for an alkali metal ion, with the conformational coupling of the two sites causing the coordination of the bipyridine moiety to a transition metal ion to weaken the binding of the potassium ion to the crown ether moiety.

mationally coupled binding sites, it is associated with the term intermolecular cooperativity [50]. *Homotopic* systems have two or more binding sites for the same substrate, but otherwise behave in a similar way.

Receptor **3.18** is a classic example of an allosteric system with negative heterotopic cooperativity [51]. This receptor has two distinct binding sites, the 2,2′-bipyridyl subunit for complexation of a transition metal ion and the crown ether moiety that can accommodate a potassium ion. In the absence of either substrate, the 2,2′-bipyridyl moiety in **3.18** prefers a conformation with divergent nitrogen atoms to minimize lone pair repulsion. As a result, the crown ether adopts a distorted conformation unsuitable for ion complexation. Addition of potassium ions induces a crown ether conformation in which the two nitrogen atoms converge, and a potassium ion can be bound. However, further rigidification of this arrangement by reaction with $W(CO)_6$ prevents the crown ether from forming a complex. The transition metal ion thus acts as an allosteric inhibitor. One can conclude:

The use of allosteric effects is an attractive means of controlling the substrate affinity of synthetic receptors with suitable effector molecules.

Bibliography

[1]   Marcus Y. Ion Properties. Marcel Dekker: New York, 1997.
[2]   Arunan E, Desiraju GR, Klein RA, Sadlej J, Scheiner S, Alkorta I, Clary DC, Crabtree RH, Dannenberg JJ, Hobza P, Kjaergaard HG, Legon AC, Mennucci B, Nesbitt DJ. Definition of the hydrogen bond (IUPAC Recommendations 2011). Pure Appl. Chem. 2011, 83, 1637–41.
[3]   Schneider HJ, Hydrogen bonds with fluorine. Studies in solution, in gas phase and by computations, conflicting conclusions from crystallographic analyses. Chem. Sci. 2012, 3, 1381–94.
[4]   Djurdjevic S, Leigh DA, McNab H, Parsons S, Teobaldi G, Zerbetto F. Extremely strong and readily accessible AAA-DDD triple hydrogen bond complexes. J. Am. Chem. Soc. 2007, 129, 476–77.
[5]   Kelly TR, Bridger GJ, Zhao C. Bisubstrate reaction templates. Examination of the consequences of identical versus different binding sites. J. Am. Chem. Soc. 1990, 112, 8024–34.
[6]   Hamilton AD, Van Engen D. Induced fit in synthetic receptors: nucleotide base recognition by a molecular hinge. J. Am. Chem. Soc. 1987, 109, 5035–36.
[7]   Desiraju GR, Ho S, Kloo L, Legon AC, Marquardt R, Metrangolo P, Politzer P, Resnati G, Rissanen K. Definition of the halogen bond (IUPAC Recommendations 2013). Pure Appl. Chem. 2013, 85, 1711–13.
[8]   Bulfield D, Huber SM. Halogen bonding in organic synthesis and organocatalysis. Chem. Eur. J. 2016, 22, 14434–50.
[9]   Sarwar MG, Dragisic B, Sagoo S, Taylor MS. A tridentate halogen-bonding receptor for tight binding of halide anions. Angew. Chem. Int. Ed. 2010, 49, 1674–77.
[10]   Gleiter R, Haberhauer G, Werz DB, Rominger F, Bleiholder C. From noncovalent chalcogen–chalcogen interactions to supramolecular aggregates: experiments and calculations. Chem. Rev. 2018, 118, 2010–41.
[11]   Biot N, Bonifazi D. Chalcogen-bond driven molecular recognition at work. Coord. Chem. Rev. 2020, 413, 213243.
[12]   Garrett GE, Gibson GL, Straus RN, Seferos DS, Taylor, MS. Chalcogen bonding in solution: Interactions of benzotelluradiazoles with anionic and uncharged Lewis bases. J. Am. Chem. Soc. 2015, 137, 4126–33.
[13]   Ma JC, Dougherty DA. The cation–π interaction. Chem. Rev. 1997, 97, 1303–24.
[14]   Biedermann F, Schneider HJ. Experimental binding energies in supramolecular complexes. Chem. Rev. 2016, 116, 5216–300.
[15]   Gokel GW, Barbour LJ, Ferdani R, Hu J. Lariat ether receptor systems show experimental evidence for alkali metal cation–π interactions. Acc. Chem. Res. 2002, 35, 878–86.
[16]   Mecozzi S, West Jr. AP, Dougherty DA. Cation–π interactions in simple aromatics: electrostatics provide a predictive tool. J. Am. Chem. Soc. 1996, 118, 2307–08.
[17]   Wheeler SE, Houk KN. Substituent effects in cation/π interactions and electrostatic potentials above the centers of substituted benzenes are due primarily to through-space effects of the substituents. J. Am. Chem. Soc. 2009, 131, 3126–27.
[18]   Kearney PC, Mizoue LS, Kumpf RA, Forman JE, McCurdy A, Dougherty DA. Molecular recognition in aqueous media. New binding studies provide further insights into the cation–π interaction and related phenomena. J. Am. Chem. Soc. 1993, 115, 9907–19.
[19]   Albrecht M, Müller M, Mergel O, Rissanen K, Valkonen A. CH-directed anion–π interactions in the crystals of pentafluorobenzyl-substituted ammonium and pyridinium salts. Chem. Eur. J. 2010, 16, 5062–69.
[20]   Giese M, Albrecht M, Krappitz T, Peters M, Gossen V, Raabe G, Valkonen A, Rissanen K. Cooperativity of H-bonding and anion–π interaction in the binding of anions with neutral π-acceptors. Chem. Commun. 2012, 48, 9983–85.

[21]   Berryman OB, Hof F, Hynes MJ, Johnson DW. Anion–π interactions augments halide binding in solution. Chem. Commun. 2006, 506–08.

[22]   Hunter CA, Sanders JKM. The nature of π-π interactions. J. Am. Chem. Soc. 1990, 112, 5525–34.

[23]   Martinez CR, Iverson BL. Rethinking the term "pi-stacking". Chem. Sci. 2012, 3, 2191–201.

[24]   Carter-Fenk K, Herbert JM. Reinterpreting π-stacking. Phys. Chem. Chem. Phys. 2020, 22, 24870–86.

[25]   Odell B, Reddington MV, Slawin AMZ, Spencer N, Stoddart JF, Williams DJ. Cyclobis(paraquat-*p*-phenylene). A tetracationic multipurpose receptor. Angew. Chem. Int. Ed. Engl. 1988, 27, 1547–50.

[26]   Gravillier LA, Cockroft SL. Context-dependent significance of London dispersion. Acc. Chem. Res. 2023, 56, 3535–44.

[27]   Houk KN, Leach AG, Kim SP, Zhang X. Binding affinities of host–guest, protein–ligand, and protein–transition-state complexes. Angew. Chem. Int. Ed. 2003, 42, 4872–97.

[28]   Cockroft SL, Hunter CA. Chemical double-mutant cycles: dissecting noncovalent interactions. Chem. Soc. Rev. 2007, 36, 172–88.

[29]   Mati IK, Cockroft SL. Molecular balances for quantifying noncovalent interactions. Chem. Soc. Rev. 2010, 39, 4195–205.

[30]   Fischer FR, Schweizer WB, Diederich F. Substituent effects on the aromatic edge-to-face interaction. Chem. Commun. 2008, 4031–33.

[31]   Marcus Y. The Properties of Solvents. John Wiley & Sons: Chichester, 1998.

[32]   Marcus Y. The properties of organic liquids that are relevant to their use as solvating solvents. Chem. Soc. Rev. 1993, 22, 409–16.

[33]   Klotz IM, Franzen JS. Hydrogen bonds between model peptide groups in solution. J. Am. Chem. Soc. 1962, 84, 3461–66.

[34]   Krikorian SE. Determination of dimerization constants of *cis*- and *trans*-configured secondary amides using near-infrared spectrometry. J. Phys. Chem. 1982, 86, 1875–81.

[35]   Solov'ev VP, Strakhova NN, Raevsky OA, Rüdiger V, Schneider HJ. Solvent effects on crown ether complexations, J. Org. Chem. 1996, 61, 5221–26.

[36]   Würthner F. Solvent effects in supramolecular chemistry: linear free energy relationships for common intermolecular interactions, J. Org. Chem. 2022, 87, 1602–15.

[37]   Esobar L, Ballester P. Molecular recognition in water using macrocyclic synthetic receptors. Chem. Rev. 2021, 121, 2445–514.

[38]   Hillyer MB, Gibb BC. Molecular shape and the hydrophobic effect. Annu. Rev. Phys. Chem. 2016, 67, 307–29.

[39]   Biedermann F, Nau WM, Schneider, HJ. The hydrophobic effect revisited – Studies with supramolecular complexes imply high-energy water as a noncovalent driving force. Angew. Chem. Int. Ed. 2014, 53, 11158–71.

[40]   Hunter CA. Quantifying intermolecular interactions: guidelines for the molecular recognition toolbox. Angew. Chem. Int. Ed. 2004, 43, 5310–24.

[41]   Cook JL, Hunter CA, Low CMR, Perez-Velasco A, Vinter JG. Solvent effects on hydrogen bonding. Angew. Chem. Int. Ed. 2007, 46, 3706–09.

[42]   Fischer E. Einfluss der Configuration auf die Wirkung der Enzyme. Ber. Dtsch. Chem. Ges. 1894, 27, 2985–93.

[43]   Koshland Jr. DE. The key-lock theory and the induced fit theory. Angew. Chem. Int. Ed. Engl. 1994, 33, 2375–78.

[44]   Cram DJ. Preorganization – From solvents to spherands. Angew. Chem. Int. Ed. Engl. 1986, 25, 1039–57.

[45]   Clay RM, Corr S, Micheloni M, Paoletti P. Non-cyclic reference ligands for tetraaza macrocycles. Synthesis and thermodynamic properties of a series of α,ω-di-*N*-methylated tetraaza ligands and their copper(II) complexes. Inorg. Chem. 1985, 24, 3330–36.

[46]   Haymore BL, Lamb JD, Izatt RM, Christensen JJ. Thermodynamic origin of the macrocyclic effect in crown ether complexes of sodium(1+), potassium(1+), and barium(2+). Inorg. Chem. 1982, 21, 1598–602.

[47]   Hunter CA, Anderson HL. What is cooperativity? Angew. Chem. Int. Ed. 2009, 48, 7488–99.

[48]   Ercolani G, Schiaffino L. Allosteric, chelate, and interannular cooperativity: a mise au point. Angew. Chem. Int. Ed. 2011, 50, 1762–68.

[49]   Kremer C, Lützen A. Artificial allosteric receptors. Chem. Eur. J. 2013, 19, 6162–296.

[50]   von Krbek LKS, Schalley CA, Thordarson P. Assessing cooperativity in supramolecular systems. Chem. Soc. Rev. 2017, 46, 2622–37.

[51]   Rebek Jr. J, Wattley RV. Allosteric effects. Remote control of ion transport selectivity. J. Am. Chem. Soc. 1980, 102, 4853–54.

4 Hosting ions and molecules

CONSPECTUS: Now that we understand the principles behind noncovalent interactions and the methods for characterizing noncovalently stabilized complexes, it is time to look at the actual types of receptors commonly used in supramolecular chemistry. Since the discovery of crown ethers, so many receptors have been described that it is impossible to give a comprehensive overview. Instead, we will focus on the most important receptor families and learn about their basic structures and properties. In the last part of this chapter, we will reverse the viewpoint and look at complex formation from the perspective of the substrate, which will allow us to derive guidelines on how to design a receptor for a given substrate.

4.1 Receptors

4.1.1 Crown ethers

Introduction: Crown ethers are macrocyclic polyethers typically, but not necessarily, derived from polyethylene glycol. They were discovered by Charles J. Pedersen, whose first paper on crown ethers covered the synthesis and properties of nearly 50 derivatives [1]. We saw in Chapter 1 that the first member of this family of receptors, dibenzo-18-crown-6 **4.1** (Figure 4.1), was a serendipitous discovery. After isolating a small amount of this byproduct from the synthesis of bis[2-(2-hydroxyphenoxy)ethyl] ether, Pedersen set out to elucidate its structure and properties. In one experiment, he checked whether the isolated compound contained free phenolic OH groups by measuring the UV–vis spectra of a solution in methanol before and after the addition of NaOH. To his surprise, both spectra were identical, indicating that the product contained no acidic protons. However, the addition of NaOH significantly improved the solubility of **4.1** in methanol, which Pedersen correctly attributed to complexation of the sodium ion. In his Nobel Prize lecture, he said: "It seemed clear to me now that the sodium ion had fallen into the hole in the center of the molecule and was held there by the electrostatic attraction between its positive charge and the negative dipolar charge on the six oxygen atoms symmetrically arranged around it in the polyether ring" [2].

At the end of his landmark publication, Pedersen mentioned that **4.1** also has "unusual physiological properties," including oral toxicity. While it was shown much later that the oral toxicity of crown ethers is actually not very high, with 15-crown-5 having an LD_{50} similar to that of aspirin [3], Pedersen's publication suggested a link between the ability to complex alkali metal ions and biological activity. The same had been observed around the same time for macrocyclic natural products such as the antibiotic valinomycin **4.2** (Figure 4.1), which stimulates mitochondrial potassium transport. In 1967, the same year as Pedersen's article on crown ethers was published, this biological

https://doi.org/10.1515/9783111315171-004

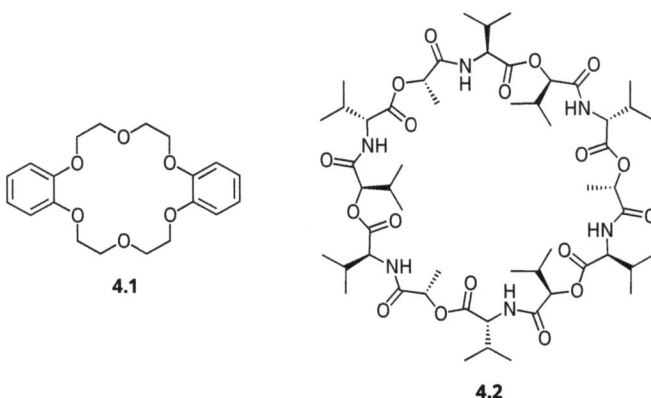

4.1

4.2

Figure 4.1: Structures of dibenzo-18-crown-6 **4.1** and valinomycin **4.2.**

activity was attributed to the ability of valinomycin to complex potassium ions and transport them across cell membranes. The relationship between valinomycin and crown ethers was quickly recognized, and several publications subsequently showed that crown ethers also mediate membrane transport of metal ions. It can be speculated whether the more or less simultaneous discovery of the cation-binding properties of crown ethers and natural ionophores such as valinomycin contributed to the immediate attention that Pedersen's paper received. In any case, this publication initiated intensive research in a field that would later become known as supramolecular chemistry.

Nomenclature: Pedersen suggested the name *crown* for macrocyclic polyethers because it adequately describes their shape when interacting with cations. Later, *crown* was replaced by *crown ether*, probably because the addition of the word ether helps clarify that the name refers to a chemical compound. The term *coronand* is used as a synonym for crown ethers, allowing their complexes to be referred to as *coronates*. Another way to denote a crown ether complex, also used for other types of complexes in supramolecular chemistry, is based on the use of the mathematical symbol of inclusion $\subset$. Accordingly, the abbreviation [K$^+$⊂dibenzo-18-crown-6] stands for the potassium complex of crown ether **4.1.**

Since the naming of crown ethers according to the systematic nomenclature would be too cumbersome, Pedersen invented a more convenient method. In this nomenclature, the word crown is used as the basis for the name. A preceding number indicates the total number of atoms in the ring, and the number of oxygen atoms is specified at the end of the name. For example, an 18-membered macrocycle with six oxygen atoms is called 18-crown-6. Prefixes are used to indicate the presence of other groups or heteroatoms (e.g., *benzo* for a benzene ring, *aza* for nitrogen, or *thia* for sulfur), hence the name dibenzo-18-crown-6 for crown ether **4.1.** To illustrate these rules, Figure 4.2 shows a selection of crown ethers along with their respective names.

This figure also shows that this nomenclature can readily be applied to macrocyclic polyethers with more than two carbon atoms between the heteroatoms.

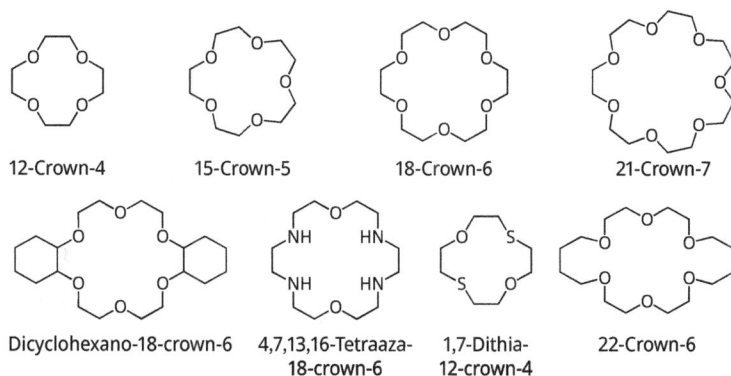

12-Crown-4   15-Crown-5   18-Crown-6   21-Crown-7

Dicyclohexano-18-crown-6   4,7,13,16-Tetraaza-18-crown-6   1,7-Dithia-12-crown-4   22-Crown-6

Figure 4.2: Examples of crown ethers together with their names according to the nomenclature introduced by Pedersen.

Synthesis: Crown ethers are usually prepared by treating a diol and a ditosylate in the presence of a base (Williamson ether synthesis). Dichlorides, dibromides, or diiodides can be used instead of ditosylates, but the latter usually afford the products in higher yields. An example is the synthesis of 18-crown-6 **4.3** shown in Figure 4.3a. This reaction gives the desired [1 + 1] macrocyclic product, but also 36-crown-12 as the [2 + 2] product, in addition to larger rings and polymeric byproducts. Product formation is irreversible, and the ratio of the different compounds is therefore kinetically controlled. In other words, the product whose formation is associated with the lowest Gibbs free energy of activation will dominate in the final mixture. If the macrocyclic product is unstrained, ring closure has no enthalpic disadvantage over the intermolecular reaction, which means that the preferred reaction pathway depends on entropy. To evaluate this influence, two opposing effects of the intramolecular and intermolecular reactions must be considered. Entropy generally favors intramolecular reactions because they do not involve a reduction in the number of molecules in solution. Therefore, cyclizations that produce small or medium-sized rings tend to be quite efficient. However, as the distance between two reacting groups in the linear precursor increases, the entropic term of the intramolecular reaction becomes increasingly unfavorable because the linear precursor loses more and more degrees of conformational freedom as it assumes the conformation required for the cyclization. Consequently, macrocyclizations such as the one shown in Figure 4.3a would normally be expected to produce a complex reaction mixture with significant amounts of polymeric materials, which is indeed the case when tetra-*n*-butylammonium hydroxide is used as the base.

However, in the presence of potassium *tert*-butoxide, the yield of **4.3** is high, exceeding 90% under optimal conditions [4]. This dramatic improvement over the reac-

(a)

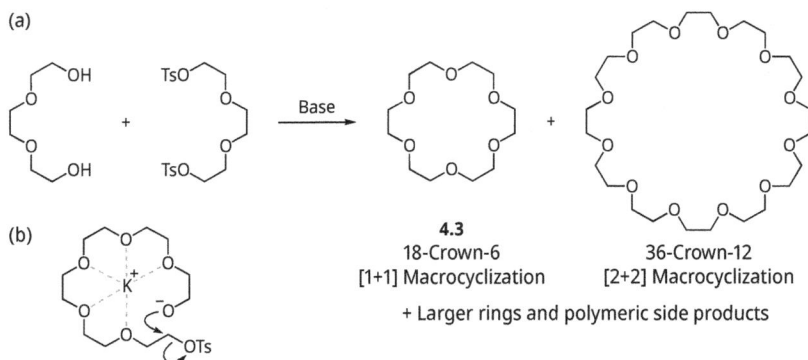

(b)

4.3
18-Crown-6
[1+1] Macrocyclization

36-Crown-12
[2+2] Macrocyclization

+ Larger rings and polymeric side products

Figure 4.3: Synthesis of 18-crown-6 from tetraethylene glycol and the ditosylate of tetraethylene glycol in the presence of a base (a) and structure illustrating how the binding of a potassium ion to the linear precursor of **4.3** mediates the ring closure (b).

tion mediated by tetra-*n*-butylammonium hydroxide is due to the ability of the potassium ions to interact with the linear intermediates formed during the reaction. These intermediates fold around the ions to maximize the ion–dipole interactions with the oxygen atoms, which in turn brings their end groups closer together and increases the probability that the ring can be closed (Figure 4.3b). The unfavorable activation entropy associated with the restriction of conformational mobility is thus paid for by complex formation, making the macrocyclization faster than the unwanted intermolecular reactions. In the terminology of supramolecular chemistry, the potassium ions act as *templates* that control the outcome of the reaction by favoring the [1 + 1] macrocycle over other products.

The term template was introduced in 1964 by Daryle H. Busch, who proposed the following definition [5]:

A chemical template organizes an assembly of atoms, with respect to one or more geometric loci, in order to achieve a particular linking of atoms.

Busch realized that there are two types of templates, those that influence the structure of the transition state of an irreversible reaction and those that thermodynamically stabilize a particular product of a reversible reaction. We will return to this classification in Section 5.1, where the differences between thermodynamic and kinetic templates will be explained in more detail. At this point, it is sufficient to note that potassium ions (or other metal ions) act as kinetic templates in crown ether syntheses. They promote the rate at which the desired macrocycle is formed and their role in the reaction is therefore similar to that of a catalyst. However, the cations must be present in stoichiometric amounts as they remain bound to the product. Since quaternary ammonium ions do not interact with the oxygen atoms of the reaction intermediates, they cannot act as templates, explaining why the synthesis of **4.3** does not proceed effectively in the presence of tetra-*n*-butylammonium hydroxide.

Binding properties: Crown ethers containing only oxygen atoms along the ring prefer to bind electropositive cations, such as alkali or alkaline earth metal ions. Larger crown ethers also interact with organic substrates containing hydrogen bond donors, such as protonated amino acids, but before discussing these systems, we will first look at the different binding modes available to crown ethers. These binding modes depend mainly on the ratio of the ring diameter to the diameter of the bound ion, as shown in the crystal structures in Figure 4.4.

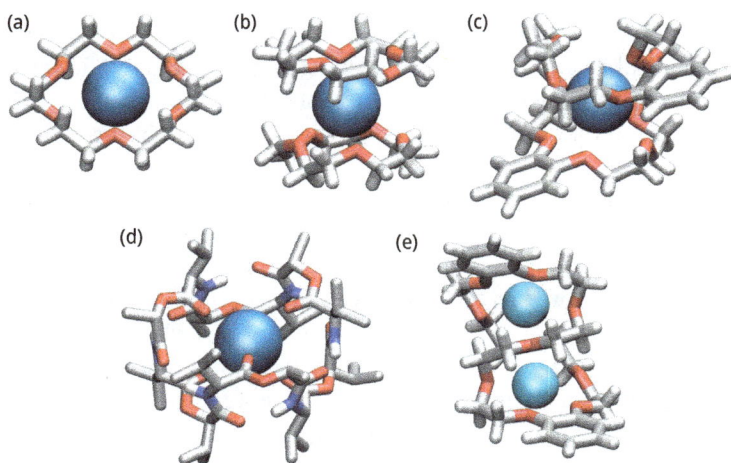

Figure 4.4: Crystal structures of [K$^+$⊂18-crown-6] (a), [K$^+$⊂(15-crown-5)$_2$] (b), [K$^+$⊂dibenzo-30-crown-10] (c), [K$^+$⊂valinomycin] (d), and [(Na$^+$)$_2$⊂ dibenzo-30-crown-10] (e). Only the acidic protons of valinomycin are shown in (d) for reasons of clarity.

If the diameter of the cation matches that of the available cavity, as in the case of [K$^+$⊂18-crown-6] (Figure 4.4a), complexes are formed with the metal ion located exactly in the center of the ring. If the ring is too small, the cation will remain perched above or below the ring and may recruit a second crown ether, leading to the formation of a sandwich complex, as in the K$^+$ complex of 15-crown-5 (Figure 4.4b). Conversely, very large rings either fold around the ion, as in [K$^+$⊂dibenzo-30-crown-10] (Figure 4.4c), or bind two metal ions simultaneously, as in the corresponding Na$^+$ complex (Figure 4.4e). In all cases, the oxygen atoms of the crown ethers approach the bound ion, demonstrating that complex formation is due to electrostatic interactions between the positively charged substrate and the negatively polarized oxygen atoms.

It is also instructive to examine how valinomycin manages to bind a potassium ion. The crystal structure of this complex is shown in Figure 4.4d. It illustrates that valinomycin adopts a folded bracelet-like conformation that is stabilized by six hydrogen bonds between each NH group and the carbonyl group of the following valine residue. Due to this intramolecular stabilization, the six remaining carbonyl groups of the α-hydroxycarboxylic acids face toward the center of the cavity, allowing them to coordinate to the metal ion in an almost perfect octahedral fashion. The cavity dimensions are optimal for K^+, but also allow the inclusion of larger metal ions such as Rb^+. However, the valinomycin backbone is too rigid to contract and bind Na^+ or even Li^+. The optimal size fit of the K^+ ion in the valinomycin cavity, the coordination of the cation by the six carbonyl oxygen atoms, and the simultaneous stabilization of the complex by six intramolecular hydrogen bonds results in a $\log K_a$ of 4.5 in methanol [6]. In comparison, the $\log K_a$ of the Na^+ complex of valinomycin under the same conditions is only 0.8 due to the smaller size of the Na^+ ion and its more difficult desolvation.

When comparing the binding properties of crown ethers of different sizes, the effects of conformational flexibility on binding affinity must be considered. While small crown ethers such as 12-crown-4 are relatively rigid and fixed in a conformation with convergent oxygen atoms, larger systems can equilibrate between conformations with convergent and divergent oxygen atoms, as shown for 18-crown-6 in Figure 4.5a. The major difference between these conformations is the arrangement of the oxygen atoms at the O–C–C–O bonds, which is synclinal in the first conformation (torsion angle ± 60°) and antiperiplanar in the second conformation (torsion angle 180°). Since the synclinal conformation is favored by the stereoelectronic *gauche* effect, the crown ether conformation with convergent oxygen atoms is the more stable one. Indeed, crown ethers do not undergo a substantial conformational reorganization upon cation binding [7], suggesting that even larger rings are well preorganized to form unstrained five-membered chelate rings at each ethylene glycol unit with a metal ion. In crown ethers with longer alkyl chains between the oxygen atoms, antiperiplanar conformations predominate at the C–C bonds, which are less suitable for complex formation, explaining why these systems typically exhibit reduced cation affinity.

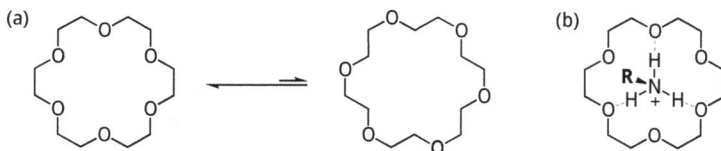

Figure 4.5: Schematic representation of the conformational reorganization of a crown ether upon metal ion complexation (a) and mode of ammonium ion binding to 18-crown-6 (b).

The structures of the complexes in Figure 4.4 suggest that their stability should correlate with the number of attractive contacts between the receptor and the substrate, a trend that is indeed observed in the gas phase, where the affinity for a given cation increases from 12-crown-4 to 15-crown-5 and on to 18-crown-6 [8]. With respect to the cation, the complexes become less stable in the order $Na^+ > K^+ > Rb^+ > Cs^+$, consistent with the decreasing charge densities in the same direction. The most stable complex is therefore the one between Na$^+$ and 18-crown-6. In solution, solvation effects cause deviations from this trend. To illustrate this effect, the stabilities of the alkali metal ion complexes of 12-crown-4, 15-crown-5, and 18-crown-6 in methanol are shown in Figure 4.6 [9].

Cation	$\log K_a$		
	12-Crown-4	15-Crown-5	18-Crown-6
Li$^+$	–	1.2	–
Na$^+$	1.5	3.3	4.4
K$^+$	1.6	3.5	6.1
Rb$^+$	1.7	2.8	5.4
Cs$^+$	1.6	2.7	4.6

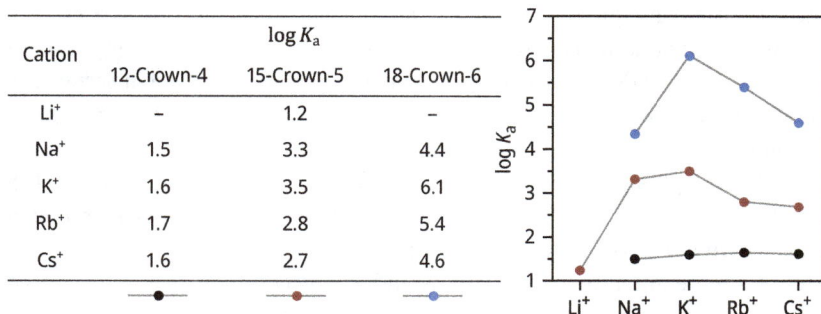

Figure 4.6: Dependence of the stability of crown ether complexes on the size of the ring and the size of the cation. All binding constants refer to methanol as the solvent.

Figure 4.6 shows that the correlation between complex stability and the number of oxygen atoms is maintained in solution. However, the stability of the complexes does not decrease progressively with increasing size of the cation, but shows a plateau for the Na$^+$ and K$^+$ complexes of 15-crown-5 and a pronounced peak for [K$^+$⊂18-crown-6]. Part of the reason for the high K$^+$ affinity of 18-crown-6 and the high Na$^+$ affinity of 15-crown-5 is the size match of the cation and the corresponding crown ether, leading to an efficient interaction. However, the results of the gas phase studies show that size matching is not a critical factor in determining complex stability. Moreover, if size matching were solely responsible for cation selectivity, the Li$^+$ or Na$^+$ affinity of 15-crown-5 should be much higher than experimentally observed according to the geometric parameters collected in Table 4.1.

Table 4.1: Geometric parameters of selected alkali metal ions and crown ethers.

Alkali metal ion	Ionic radius (Å)	Crown ether	Inner cavity radius (Å)
Li$^+$	0.60	12-Crown-4	0.60–0.75
Na$^+$	0.95	15-Crown-5	0.85–1.10
K$^+$	1.33	18-Crown-6	1.30–1.60
Rb$^+$	1.48	21-Crown-7	1.70–2.15
Cs$^+$	1.69		

Complexes of crown ethers with smaller ions are apparently less stable in methanol than expected. The reason for this is ion solvation, which can be best understood on the basis of equation (2.8) derived in Section 2.1 ($\Delta G^0 = \Delta G^0_{\text{intr}} + \Delta G^0_{\text{solv}}(C) - \Delta G^0_{\text{solv}}(R) - \Delta G^0_{\text{solv}}(S)$). Assuming that $\Delta G^0_{\text{solv}}(C)$ and $\Delta G^0_{\text{solv}}(R)$ do not change much for a range of crown ethers and their complexes (which is certainly an approximation due to the different structures), the overall stability of a crown ether complex ΔG^0 depends on the balance between ΔG^0_{intr} and $\Delta G^0_{\text{solv}}(S)$. Since cation–crown ether interactions and cation solvation are both mediated by ion–dipole interactions, cations that strongly bind to crown ethers are also strongly solvated. As a consequence, the desolvation of small ions with high charge density causes a more pronounced reduction in ΔG^0_{intr} than the desolvation of larger, more weakly solvated ions. Recall the solvent effect on complex stability discussed in Section 3.3, where we also saw the strong effect of $\Delta G^0_{\text{solv}}(S)$ on complex stability. This solvent effect causes the cation complexes of crown ethers to be typically at least two orders of magnitude less stable in water than in methanol, while the effect of the solvent on cation selectivity is small. For example, [K$^+$⊂18-crown-6] also has the highest stability in water among the complexes shown in Figure 4.6 (log K_a = 2.1) [9].

Consistent with the ion–dipole interactions underlying complex formation, crown ether complexes of doubly charged alkaline earth metal ions should be more stable than those of alkali metal ions, but this trend is again seen only for the weakly solvated Ba^{2+} ions. Ca^{2+} binding is even weaker than that of singly charged alkali metal ions due to the strong solvation of this small and doubly charged ion.

As mentioned earlier, ammonium ions are also suitable substrates for crown ethers. Because NH$_4^+$ and K$^+$ have similar ionic radii and a good symmetry match, 18-crown-6 is particularly suitable for such guests. Ammonium ions bind to crown ethers by forming hydrogen bonds between the NH groups and the crown ether oxygen atoms (Figure 4.5b), but these complexes are typically not very stable. For example, with a log K_a of 4.1, the association constant of the NH$_4^+$ complex of 18-crown-6 in methanol is two orders of magnitude lower than that of the corresponding K$^+$ complex [10]. Organic ammonium ions are bound even more weakly due to steric effects between their substituents and the surrounding 18-crown-6 backbone. Nevertheless, Cram could use the affinity of crown ethers for ammonium ions to show for the first time that a fundamental concept of nature, namely that of enantioselective substrate recognition, could be transferred to a synthetic receptor. The structure of this classic system, the chiral bis(1,1′-bi-2-naphthol)-derived crown ether (*R,R*)-**4.4**, is shown in Figure 4.7.

Receptor (*R,R*)-**4.4** allows the separation of the enantiomers of amino acid ammonium salts such as the perchlorate salt of D/L-phenylglycine methyl ester. These substrates can bind to either side of (*R,R*)-**4.4**, but since the receptor is C_2-symmetric, identical arrangements result. In the complexes, the orientation of the guest is determined by the steric effects of the flanking 1,1′-bi-2-naphthol (BINOL) groups, as shown in Figure 4.7. The methylated BINOL moiety exerts the greatest steric hindrance, allowing only the proton on the α-carbon atom of the bound amino acid to be in close proximity. The other substituents either occupy the cleft between the BINOL units or are arranged along the face of the other naphthyl moiety. Since the side chains of the substrates are typically larger than the ester groups, they prefer to occupy the larger cleft. As a result, (*R,R*)-**4.4** binds D-amino acids more strongly than the

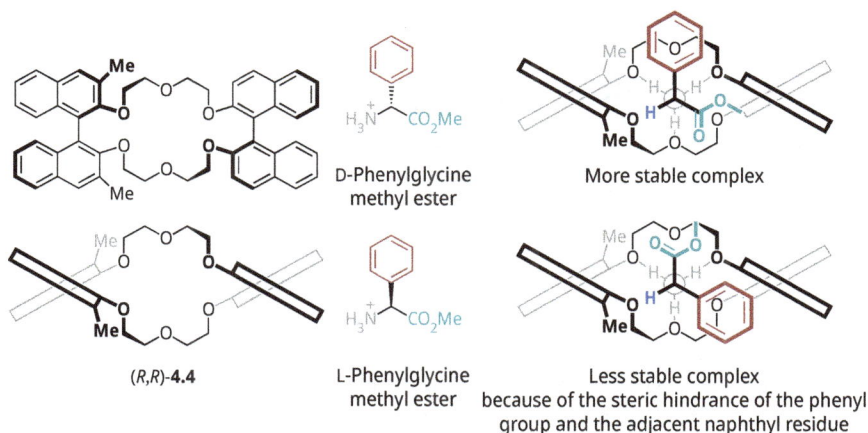

Figure 4.7: Structure of the chiral crown ether (*R,R*)-**4.4** (a) and preferred arrangements of D- and an L-phenylglycine methyl ester in the complexes with (*R,R*)-**4.4** (b). The complex of the D-enantiomer is more stable due to a greater steric hindrance in the complex with the L-amino acid.

corresponding L-enantiomers. Although the overall stability of these complexes is relatively low, (*R,R*)-**4.4** can be used to separate enantiomers on a preparative scale. To this end, Cram developed an ingenious method based on the enantioselective transport of an amino acid from an aqueous solution through an organic phase into a receiving aqueous solution. The principle is illustrated in Figure 4.8.

The experimental setup consists of two U-tubes, one containing a chloroform solution of (*R,R*)-**4.4** and the other containing a chloroform solution of the (*S,S*)-enantiomer of the receptor. Both tubes are in contact with an aqueous solution containing the racemic amino acid. Since the D-enantiomer is selectively transported by (*R,R*)-**4.4** into one of the receiving phases, whereas (*S,S*)-**4.4** mediates the transport of the other enantiomer, each receiving phase contains predominantly an enantiomerically pure amino acid at the end of the experiment (enantiomeric excess (*e.e.*) 86–90%). Other applications of crown ethers are described in Chapter 12.

Derivatives: The number of currently known crown ethers or cation receptors derived from crown ethers is enormous. Therefore, we will only focus on a few general strategies that have been used to modulate the binding properties of these compounds. Probably the most obvious one is to replace the oxygen atoms along the ring with other heteroatoms. The most commonly used elements in this context are nitrogen and sulfur, but other elements (e.g., Se and Te) have also been used. Heteroatoms less electronegative than oxygen generally interact more weakly with electropositive cations, so that the cation affinity decreases as the number of these heteroatoms increases. This effect is evident along the series of receptors shown in Table 4.2 [11]. Conversely, the affinity for softer transition metal ions increases in the same direction.

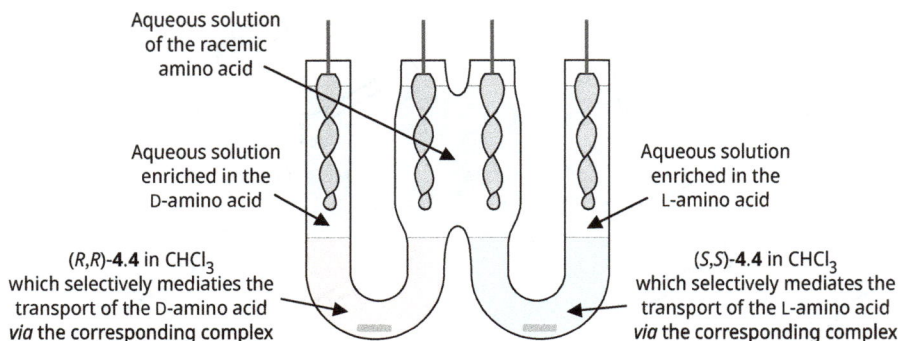

Figure 4.8: Method for separating the racemate of amino acid ammonium ions. Each receptor enantiomer in the chloroform solutions mediates the transport of one enantiomer of the substrate, so that each aqueous receiving phase contains predominantly an enantiomerically pure compound at the end of the experiment.

Table 4.2: Effect of the replacement of oxygen atoms in 18-crown-6 with nitrogen atoms on the log K_a values of the corresponding K^+ complexes in methanol and Ag^+ complexes in water at 298 K.

Cation			
K^+	6.1	3.9	2.0
Ag^+	1.6	3.3	7.8

The introduction of trivalent nitrogen atoms into the ring has the advantage of allowing the subsequent attachment of further substituents. For example, bridging two nitrogen atoms in a crown ether with a suitable chain leads to bicyclic cryptands, which will be discussed in the next section. Appending substituents to crown ethers leads to lariat ethers (one substituent) or bibrachial lariat ethers (two substituents). The term lariat, which usually refers to a rope in the form of a lasso, illustrates very well that these receptors consist of a linear chain attached to a loop. Lariat ethers combine the properties of cyclic crown ethers with those of their acyclic analogs, the so-called podands. In the next section, we will look in more detail at how the binding properties are affected when moving from a linear podand to a macrocyclic coronand and further to a lariat ether and a bicyclic cryptand. For now, just recall that lariat ethers have played a role in the characterization of cation–π interactions, as explained in Section 3.1.8.

Replacing most or all of the oxygen atoms of a crown ether with nitrogen atoms leads to polyazamacrocycles, another large family of receptors, whose members serve either as ligands for transition metal ions or, after protonation, as anion receptors. Two important macrocyclic tetraamines that have found widespread use in the coordination of transition metal ions are 1,4,7,10-tetraazacyclododecane **4.5** (cyclen) and 1,4,8,11-tetraazacyclotetradecane **4.6** (cyclam) (Figure 4.9a). Both compounds form highly stable complexes with, for example, Cu^{2+} and Zn^{2+} in water ([Cu⊂**4.5**]$^{2+}$, $\log K_a$ = 24.8; [Zn⊂**4.5**]$^{2+}$, $\log K_a$ = 16.2; [Cu⊂**4.6**]$^{2+}$, $\log K_a$ = 27.2; [Zn⊂**4.6**]$^{2+}$, $\log K_a$ = 15.5) [12]. The metal ions have square pyramidal coordination geometries with the nitrogen atoms at the base and the counterion at the apical position (Figure 4.9b). The lability of the bond to the counterion makes such metal complexes useful building blocks for the construction of anion receptors. Their chemistry otherwise falls within the realm of coordination chemistry and these systems will therefore not be discussed further here. Instead, we will concentrate on the properties of metal-free polyamines, an example of which is compound **4.7** in Figure 4.9c.

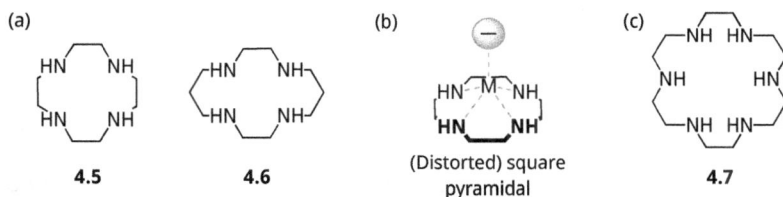

(a)

HN   NH
HN   NH
4.5

HN   NH
HN   NH
4.6

(b)

HN–M–NH
HN   NH
(Distorted) square
pyramidal

(c)

–NH HN–
NH   HN
–NH HN–
4.7

Figure 4.9: Structures of cyclen **4.5** and cyclam **4.6** (a), schematic structure of the metal complexes of these ligands (b), and structure of the polyaza analog **4.7** of 18-crown-6 (c).

These polyamines are the nitrogen-containing counterparts of crown ethers. Their degree of protonation is high in aqueous solution, making them well suited to interact with anions. Anion binding can be understood on the basis of a few guidelines. First, the degree of protonation of macrocyclic polyamines is pH dependent. Thus, the number of positively charged groups along the ring increases with decreasing pH, causing the anion affinity to increase in the same direction. With regard to the protonation sequence, ammonium groups in a partially protonated polyamine tend to be placed as far apart as possible to reduce charge repulsion. Thus, nonadjacent amino groups are protonated first and the least basic groups are protonated last as the pH of the solution is lowered. It is important to note that secondary amino groups have a greater propensity to accept a proton than tertiary amino groups. The introduction of tertiary amines into strategic positions of such polyamines therefore provides a means of controlling the protonation pattern. The distance of the amino groups along the receptor backbone also strongly influences the conditions required for protonation, with smaller distances making it more difficult to achieve a high degree of protonation. Protonation affects the conformation of the receptor because the ammonium groups tend to move away from each other, causing the molecular framework of the receptor to expand upon protonation. Finally, the ammonium groups preferentially adopt orientations in which the protons are outside the macrocyclic cavity. As a consequence, the receptor must undergo a conformational reorganization to allow the formation of hydrogen bonds with a bound anion.

The effects of these factors on anion affinity have been systematically investigated by using complex anions as substrates, such as the octahedral complexes $[Fe(CN)_6]^{4-}$ and $[Co(CN)_6]^{3-}$. Comparison of the stabilities of the corresponding complexes provides information, for example, on how the charge of the anion affects the stability of the complex, without having to consider many interferences from other factors. These studies showed that (i) the complexes of the more highly charged $[Fe(CN)_6]^{4-}$ anion are more stable than those of $[Co(CN)_6]^{3-}$; (ii) complex stability increases with increasing degree of protonation of the receptor; and (iii) complex stability increases with increasing charge density for a given degree of protonation, meaning that smaller macrocycles form more stable complexes than larger ones. These correlations may break down for other substrates, particularly oxoanions, whose binding not only involves ionic interactions but also hydrogen bonds.

4.1.2 Cryptands

Introduction: In 1969, 2 years after Pedersen's first publication on crown ethers, two successive short communications by Bernard Dietrich, Jean-Marie Lehn, and Jean-Pierre Sauvage appeared in *Tetrahedron Letters*, both written in French, dealing with the synthesis, structural characterization, and cation affinity of macrobicyclic analogs of crown ethers [13, 14]. It is worth noting here that not only Lehn but also Sauvage, who was Lehn's PhD student in 1969, received the Nobel Prize for his contributions to supramolecular chemistry, specifically for his work on molecular machines (Chapter 8). Representative structures of the family of macrobicyclic receptors introduced by Lehn are shown in Figure 4.10. They all have nitrogen atoms as bridgeheads, connected by three ethylene glycol chains of different lengths. They therefore feature three-dimensional cavities that allow a cation to be fully engulfed and surrounded from all sides. This mode of binding explains why the cation complexes of these re-

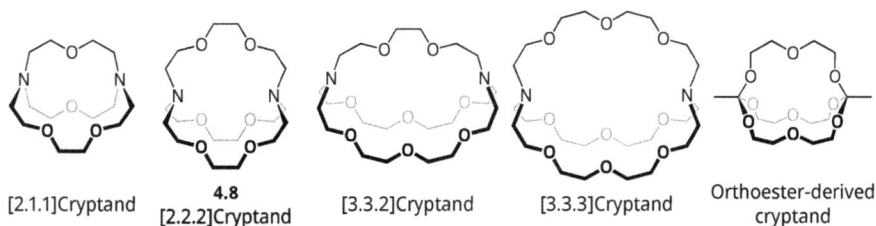

[2.1.1]Cryptand   **4.8** [2.2.2]Cryptand   [3.3.2]Cryptand   [3.3.3]Cryptand   Orthoester-derived cryptand

Figure 4.10: Representative structures of bicyclic crown ether derivatives, so-called cryptands, together with their respective names.

ceptors were called *cryptates*, a term that first appeared in the title of one of the 1969 publications ("Les Cryptates") and comes from the Latin *crypta* = vault, cave. The free receptors are, accordingly, called cryptands.

For a long time, the family of cryptands consisted mainly of compounds with nitrogen atoms as bridgehead atoms. It was not until 2015 that Max von Delius described cryptands in which carbon atoms connect the three straps [15]. An example is shown in Figure 4.10. These cryptands are synthesized by treating orthoesters with diols under suitable conditions (Section 5.7). Their binding properties range between those of crown ethers and classical cryptands of similar size.

Nomenclature: The archetypal cryptands with ethylene glycol chains between bridgehead nitrogen atoms are distinguished by the number of oxygen atoms in the straps. Accordingly, the smallest cryptand shown in Figure 4.10 is called [2.1.1]cryptand and the largest is called [3.3.3]cryptand. [2.2.2]Cryptand (**4.8**) is commercially available under the name Kryptofix® 222. Many other cryptands have been described, varying widely in the number and type of heteroatoms and in the structure of the bridging straps. Although a systematic way of naming them has been proposed [16], this nomenclature has not found widespread use.

Synthesis: The classical approach to the preparation of cryptands, introduced by Lehn, is illustrated in Figure 4.11a, using the original synthesis of **4.8** as an example. The first step is the reaction of a diamine with a diacid chloride to give a macrocyclic dilactam. This product is then reduced with lithium aluminum hydride to afford 4,13-diaza-18-crown-6 in a total yield of 75%. Treatment of this crown ether with another equivalent of the acid dichloride used in the first step produces the macrobicyclic product in 45% yield. The subsequent reduction with diborane proceeds almost quantitatively to afford **4.8** [13].

It should be noted that the transformations giving the macrocycle in the first step and the macrobicycle in the third step each involve a sequence of two separate reactions. The initial intermolecular step leads to an intermediate that still contains an unreacted acid chloride and an amino group. The subsequent step can proceed either intramolecularly, affording the desired product, or intermolecularly, yielding unwanted oligomeric byproducts. The question is how to favor the intramolecular reaction pathway.

What strategies exist to favor macrocyclization reactions?

One possibility is the use of suitable templates, as mentioned in Section 4.1.1 and discussed in more detail in Section 5.1. If template effects cannot be used or prove to be inefficient, the reaction is usually carried out at very low concentrations. Under these conditions, the probability that two molecules will meet is reduced and reactive groups will therefore preferentially react intramolecularly rather than intermolecularly. The disadvantage of such high dilution conditions is that they require large amounts of solvent and large reaction vessels if the syntheses are carried out on a

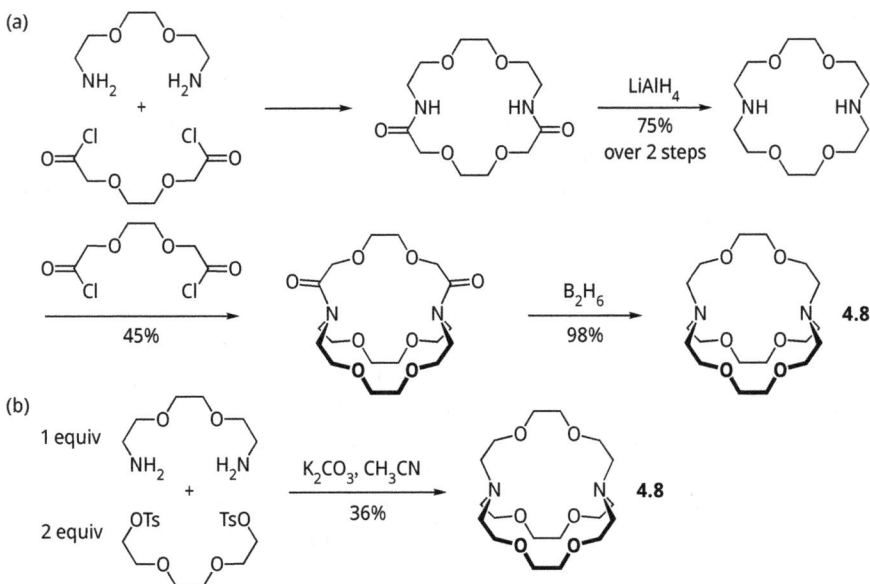

Figure 4.11: Syntheses of cryptand **4.8** *via* dilactams (a) or by using nucleophilic substitution reactions under the influence of a suitable template (b).

large scale. It is therefore often preferable to add one reaction partner very slowly to the solution of the second starting material by using a motorized syringe pump (pseudo-high dilution conditions). In this way, the reaction mixture will always contain one component at a very low concentration, while keeping the total volume easily manageable, and if the addition is slow enough, the product will be formed before the next drop of the reactant reaches the solution.

The synthetic approach outlined in Figure 4.11a gives access to a variety of cryptands, including cryptands that differ in the structures of all three bridges. However, such multistep syntheses are time-consuming, making simpler strategies more attractive. One possibility is shown in Figure 4.11b. In this reaction, the treatment of one equivalent of a diamine and two equivalents of a ditosylate in the presence of potassium carbonate in acetonitrile gives **4.8** in a single step with a good yield of 36% [17]. Characteristic effects of the cationic counterion of the base on the yield suggest that this synthetic strategy is mediated by template effects, as in the crown ether syntheses discussed earlier.

Binding properties: Figure 4.12 shows the crystal structures of the potassium and ammonium complexes of **4.8** to illustrate the binding modes. Both cations are fully incorporated into the three-dimensional cavity and interact with the heteroatoms by ion–dipole interactions and, in the case of the NH_4^+ ion, by hydrogen bonding. In addition to internal binding, cryptands have fewer binding modes available than crown ethers (Figure 4.4) due to their conformational rigidity. If the cation cannot enter the cavity because it is too large, it can also interact with two bridges from the outside, but internal binding leads to more stable complexes.

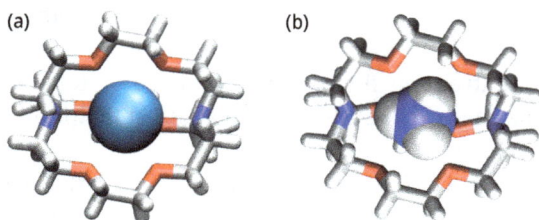

Figure 4.12: Crystal structures of the complexes of K^+ (a) and NH_4^+ (b) with **4.8.**

To illustrate the effect of receptor structure on cation affinity, the stabilities of the potassium complexes of six different polyethers are compared in Table 4.3 [11]. All binding constants were measured in methanol, but by different methods, so that only general trends should be derived.

Table 4.3: Comparison of the potassium affinity in methanol at 298 K of six receptor types differing in structure and the number of oxygen atoms along the cavity.

Receptor	log K_a	Receptor	log K_a	Receptor	log K_a
	2.1		6.1		3.9
	6.1		2.0		10.4

Table 4.3 shows that the potassium affinity of 18-crown-6 is four orders of magnitude higher than that of the corresponding acyclic podand with the same number of oxygen atoms. We have already seen this trend in Section 3.5.2, where the increase in binding strength caused by cyclization was linked to the macrocyclic effect: the podand must fold around the cation to maximize its interactions, and entropic and/or enthalpic penalties associated with this conformational reorganization are detrimental to the overall binding strength. In the case of the crown ether, these unfavorable

energetic contributions have already been paid for during macrocyclization, making cation binding stronger than that of the podand. According to the results of calorimetric studies, the lower binding affinity of the linear receptor in this particular case is almost exclusively due to the less favorable binding enthalpy with respect to the crown ether (podand: $\Delta H^0 = -36.4$ kJ mol^{-1}, $T\Delta S^0 = -24.3$ kJ mol^{-1}; crown ether: $\Delta H^0 = -56.1$ kJ mol^{-1}, $T\Delta S^0 = -21.4$ kJ mol^{-1}) [18]. Thus, the macrocyclic effect is not due to a more unfavorable binding entropy of the less preorganized system, as might be expected, but to the fact that the acyclic system has to adopt enthalpically unfavorable conformations during complex formation.

Table 4.3 shows that the transition from the all-oxygen crown ether to the derivative with one nitrogen atom leads to a pronounced decrease in affinity, as already explained in Section 4.1.1. Appending an ethylene glycol chain with two oxygen atoms to this nitrogen atom compensates for this effect, but only to the extent that the corresponding lariat ether has the same potassium affinity as the crown ether without the substituent. A further improvement in binding strength results when the substituent in the lariat ether is covalently connected to a second nitrogen atom in the ring. Importantly, the third bridge overcompensates for the detrimental effect of the second nitrogen atom on cation affinity, making the bicyclic cryptand by far the best potassium binder in this series of receptors. Similar trends are observed for other receptor–substrate combinations. The good preorganization of cryptands in combination with their ability to surround the cation from all sides, which are the main reasons for their high cation affinity, are associated with the term macrobicyclic effect.

Because cryptands are relatively rigid and cannot easily adapt to the size of the bound substrate, they exhibit pronounced size selectivity. If the cation is too small, the interactions cannot easily be optimized by reducing the size of the cavity. Conversely, large cations will not fit into a cavity that is too small. Cryptands therefore show peak selectivities, with the [2.1.1]cryptand having the highest affinity for Li$^+$ (log $K_a = 8.0$) despite the strong solvation of this cation. The alkali metal ion affinity of the [2.2.1]cryptand peaks at Na$^+$ (log $K_a = 9.7$) and that of **4.8** at K$^+$ (log $K_a = 10.4$) (Figure 4.13) [11]. Note also the much larger differences between the cation affinities of these cryptands (Figure 4.13) compared to the more flexible crown ethers (Figure 4.6). For example, cryptand **4.8** binds K$^+$ eight orders of magnitude more strongly than Li$^+$ and six orders of magnitude more strongly than Cs$^+$. For the corresponding 18-crown-6 complexes, the binding constants vary only by less than two orders of magnitude, illustrating the advantage of improved structural complementarity between the receptor and the substrate for binding selectivity.

The formation of these complexes is fast on the human timescale despite the conformational changes a cryptand must undergo to accommodate the cation. For example, the k_{on} rates when **4.8** binds an alkali metal ion are on the order of 10^8 M^{-1} s^{-1} [11]. The peak selectivities observed for cryptand complexes therefore result from the different rates of complex dissociation according to equation (2.12), with the dissocia-

Cation	log K_a		
	[2.1.1]Cryptand	[2.2.1]Cryptand	[2.2.2]Cryptand
Li$^+$	8.0	5.4	2.6
Na$^+$	6.1	9.7	8.0
K$^+$	2.3	8.5	10.4
Rb$^+$	1.9	6.7	9.0
Cs$^+$	<2.0	4.3	4.4

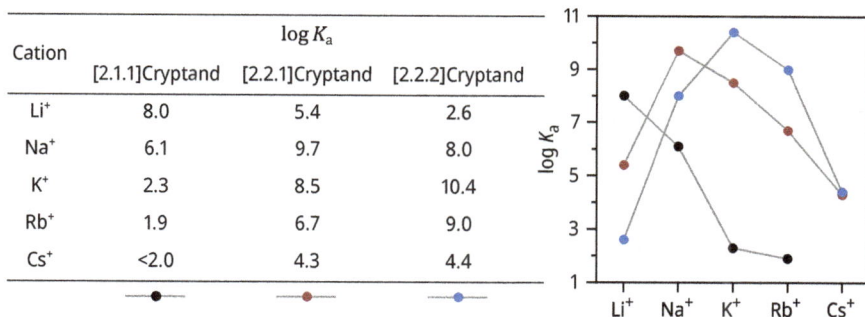

Figure 4.13: Dependence of the affinity of different cryptands for alkali metal ions of varying size. All binding constants refer to methanol as the solvent.

tion of the K$^+$ complex of **4.8** being much slower ($k_{off} = 1.8 \times 10^{-2}$ s^{-1}) than that of the other alkali metal ion complexes.

Derivatives: The concept of creating a three-dimensional cavity to accommodate a suitable guest has been extended to macrotricyclic systems, the most prominent of which is the so-called soccer ball **4.9** (Figure 4.14a) developed by Lehn [19]. The introduction of two bridges in a crown ether ring requires the presence of four nitrogen atoms, so that **4.9** ends up containing an equal number of oxygen and nitrogen atoms, with the nitrogen atoms occupying the vertices and the oxygen atoms lining the edges of a distorted tetrahedron.

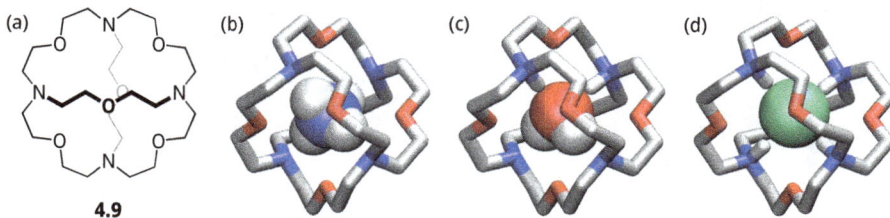

4.9

Figure 4.14: Structure of the macrotricyclic soccer ball **4.9** (a) and calculated structures of the NH$_4^+$ complex of deprotonated **4.9** (b), the water complex of diprotonated **4.9** (c), and the Cl$^-$ complex of tetraprotonated **4.9** (d). Only the acidic hydrogen atoms are shown in all three structures for reasons of clarity.

The nitrogen atoms lining the cavity of **4.9** are expected to disfavor interactions with hard alkali metal ions, and the K$^+$ affinity log K_a of **4.9** in water indeed only amounts to 3.4. For comparison, the K$^+$ complex of cryptand **4.8** has a log K_a of 5.4 under the same conditions. However, the soccer ball **4.9** has a high affinity for NH$_4^+$ in water (log $K_a = 6.4$) because its cavity perfectly matches the size and shape of the ammonium ion. The complex is stabilized by four hydrogen bonds between the bridgehead nitrogen atoms and the NH groups and by ion–dipole interactions between the cation and the surrounding oxygen atoms (Figure 4.14b).

Cation binding is only observed for the deprotonated form of **4.9**. Once the four basic nitrogen atoms accept protons, the binding properties of the soccer ball change profoundly, with the diprotonated form of **4.9** being perfectly complementary to a water molecule and the tetraprotonated form to a chloride anion. The structures in Figure 4.14c and d show that both complexes are stabilized by hydrogen bonds similar to those in the NH_4^+ complex. The difference is that in the water complex of **4.9**, two bridgehead NH groups act as hydrogen bond donors, interacting with the oxygen atom of the guest, whereas the chloride complex is stabilized by hydrogen bonds between the anion and four converging NH groups.

The tetramethylated analog of **4.9**, receptor **4.10** (Figure 4.15), has four permanent positive charges, which make anion binding independent of pH. This receptor, developed by Franz P. Schmidtchen, has an extended tetrahedral shape due to the charge repulsion of the four ammonium groups, and its methyl groups are oriented outwards. It cannot interact with halides by hydrogen bonding, which are held inside the cavity only by electrostatic interactions. Consequently, the halide affinity of **4.10** or the related analog **4.11** with all-carbon chains connecting the nitrogen atoms is relatively low in water, with typical log K_a values ranging between 1 and 2.5 [20].

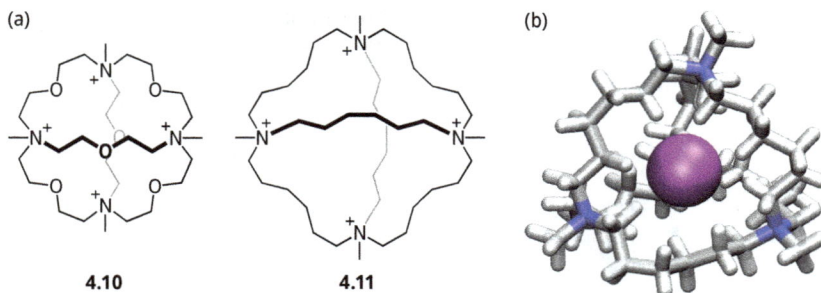

Figure 4.15: Structures of the macrotricyclic tetracationic anion receptors **4.10** and **4.11** (a) and crystal structure of the iodide complex of **4.11** (b).

Cryptands in which the number of nitrogen atoms exceeds the number of oxygen atoms or those with only nitrogen atoms also act as anion receptors. Like in the macrocyclic polyamines described earlier, anion affinity is strongly dependent on the degree of protonation and the structural complementarity between the receptor and the substrate. For example, the hexaprotonated form of the small polyazacryptand **4.12** (Figure 4.16a) interacts practically only with fluoride with a high log K_a of 10.6 [21]. Under the same conditions, chloride is bound with a log K_a of <2, resulting in an impressive F^-/Cl^- selectivity of 10^8. The hexaprotonated ellipsoidal receptor **4.13** prefers linear to spherical anions. Among singly charged anions, it has the highest affinity for azide (log K_a = 4.3) [22]. The arrangements of the anions in the cavities of these receptors are shown in Figure 4.16b and c.

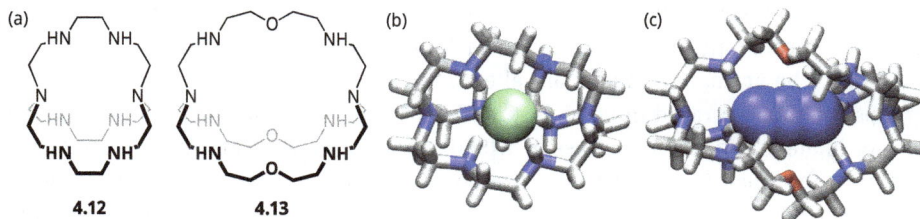

Figure 4.16: Structures of the polyazacryptands **4.12** and **4.13** (a) and crystal structures of the F⁻ complex of H₆**4.12**⁶⁺ (b) and the N₃⁻ complex of H₆**4.13**⁶⁺ (c).

Most polyazacryptands have tris(2-aminoethyl)amine (TREN) moieties at both ends. They are therefore also prone to coordinate to transition metal ions, leading to the formation of dimetallic complexes. In these complexes, the two metal ions are positioned at a certain distance from each other, allowing suitable substrates, typically anions, to be inserted between them [23]. As the formation of the final complex involves a cascade of complexation steps, first the coordination of the ligand to the metal ions and then the coordination of the actual substrate, such systems have been termed cascade complexes by Lehn [24]. An example is the dicopper(II) complex of the polyazacryptand **4.14** (Figure 4.17), which binds an azide anion with a log K_a of 4.8 in water.

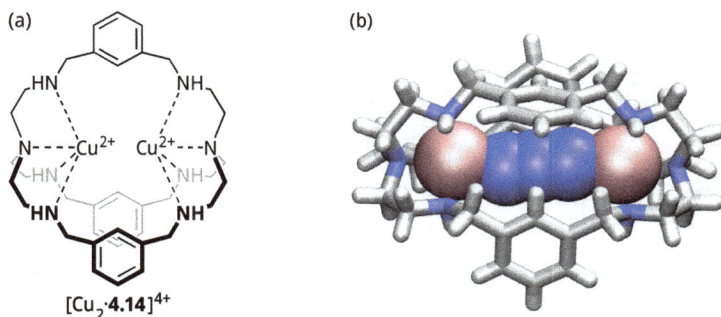

Figure 4.17: Structure of the dicopper(II) complex of polyazacryptands **4.14** (a) and crystal structure of the N₃⁻ complex of this receptor (b).

4.1.3 Spherands

Introduction: With the aim of demonstrating the concept of preorganization, Cram developed a series of receptors with exceptionally high cation affinity and selectivity, the spherands. The idea was to design a macrocyclic system that would undergo only small conformational changes during complex formation to minimize the entropic and/or enthalpic disadvantages of complex formation [25]. The structure of the archetypal spherand **4.15** developed in this context is shown in Figure 4.18a. It has six *p*-methylanisole units directly linked to each other in the *ortho* positions. Due to steric effects, the aromatic subunits are alternately oriented up and down. The methoxy groups are thus arranged in an almost perfect octahedral geometry that is well suited to coordinate to cations. In the case of **4.15**, the size of the cavity best matches the ionic radius of the small Li^+ ion. Na^+ can also be accommodated, but other cations are rejected. The perfect fit of the Li^+ ion is evident in the crystal structure of the corresponding complex (Figure 4.18b).

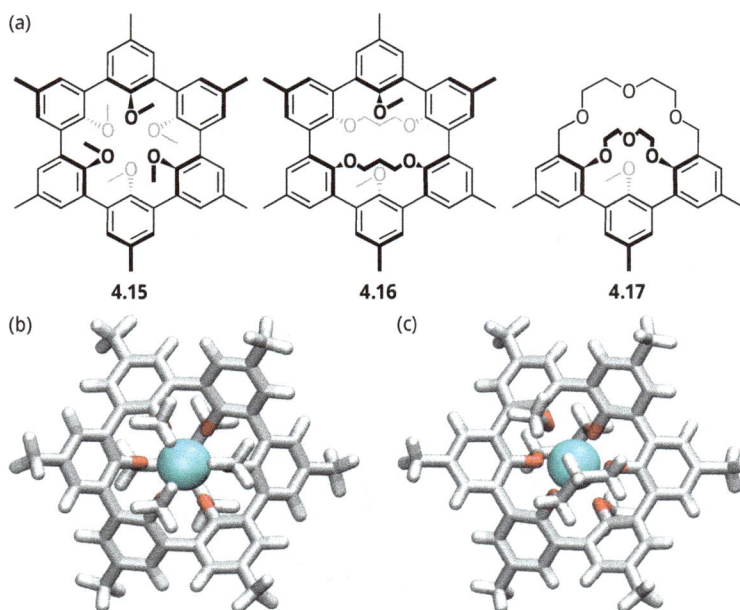

(a)

4.15   **4.16**   **4.17**

(b)   (c)

Figure 4.18: Structures of spherands **4.15**, **4.16**, and hemispherand **4.17** (a), and crystal structures of the Li^+ complexes of **4.15** (b) and **4.16** (c).

Synthesis: The synthesis of spherands involves several steps. It generally begins with the oxidative trimerization of three molecules of *p*-cresol to give one half of the macrocycle (Figure 4.19). The treatment of this compound with bromine, followed by methylation of the hydroxy groups, affords a dibromide that is oxidatively dimerized after lithiation in the presence of tris(acetylacetonato)iron(III). The lithium complex of **4.15** thus obtained is finally converted to the metal-free form by treatment with ethylenediaminetetracetic acid (EDTA) to remove the iron salt, followed by refluxing in methanol/water to remove the bound Li^+. The overall yield of this multistep synthesis is only about 6%, which explains why spherands have found only limited use in supramolecular chemistry despite their excellent binding properties.

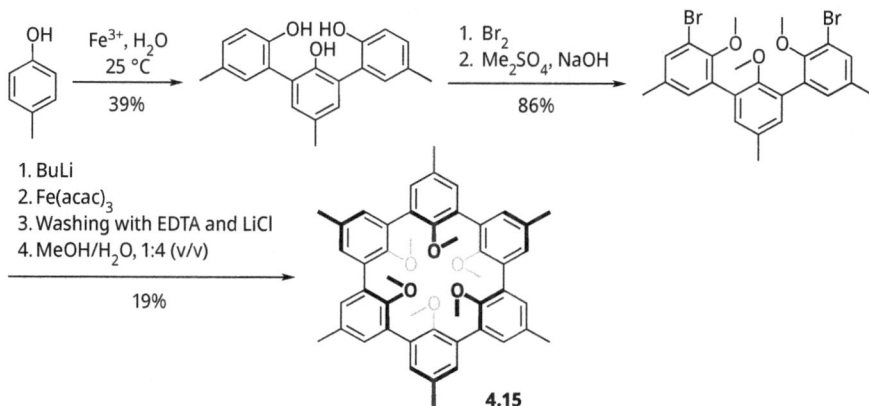

Figure 4.19: Synthesis of spherand **4.15**.

Binding properties: Spherands are hydrophobic molecules, soluble only in organic solvents and practically insoluble in water. Therefore, cation affinity has mostly been studied in (water-saturated) chloroform by assessing the extent to which picrate salts of different metal ions can be extracted from water into the organic phase (Section 2.2.3) [26]. Even with this method, the $\log K_a$ of the Li^+ complex could not be quantified for **4.15**, indicating only that it is greater than 16.8. The corresponding Na^+ complex has a $\log K_a$ of 14.1. Li^+ is thus bound at least two orders of magnitude more strongly than Na^+. Salts of larger metal ions cannot be extracted into the chloroform phase.

Spherand **4.16**, which contains propylene chains between two pairs of oxygen atoms, has an overall structure similar to **4.15**, but the crystal structure of the Li^+ complex in Figure 4.18c shows close contacts between the cation and the additional alkyl chains. As a result, the $\log K_a$ of the Li^+ complex is reduced to 12.3, which is still remarkable but almost identical to the stability of the Li^+ complex of [2.1.1]cryptand, which has a $\log K_a$ of 12.2 under the same conditions. Again, only Na^+ is also bound, in this case with a $\log K_a$ of 9.7. More flexible receptors, such as hemispherand **4.17** (Figure 4.18a), have a lower affinity for Li^+ ($\log K_a = 5.3$), demonstrating the beneficial effect of preorganization on the receptor properties and confirming Cram's design concept.

Derivatives: The structural motif of spherands has been combined with that of other receptors, such as crown ethers (Section 4.1.1) or calixarenes (Section 4.1.7). Spherands with more than six aromatic subunits have also been described that bind metal ions larger than Li^+ or Na^+, but these compounds are more flexible than **4.15** and therefore less well preorganized. Overall, the spherand family is smaller than other receptor families, probably because spherands are more difficult to synthesize. Therefore, only one other spherand derivative, compound **4.18** (Figure 4.20a), is presented here.

Due to the alternating arrangement of aromatic and urea subunits, this macrocycle has convergent carbonyl and methoxy oxygen atoms that induce an affinity for metal ions in chloroform. The cation selectivity is lower than that of conventional spherands because **4.18** is more flexible and can adapt to the size of the bound cation [27]. For example, the $\log K_a$ values of the alkali metal complexes of **4.18** differ by only two orders of magnitude, ranging from 8.9 for Li^+ to 11.3 for K^+. Importantly, this receptor also interacts with ammonium ions, which cannot be bound by the spherands discussed earlier. Complex formation involves hydrogen bonding between the NH groups of the substrates and the C=O groups of the urea moieties. In the complexes formed, the ammonium ions are perched on one side of the macrocycle, as shown in the crystal structure of the *tert*-butylammonium complex of **4.18** in Figure 4.20b. The stability of the corresponding complexes is high in chloroform ($\log K_a = 10.5$ for the NH_4^+ complex and 9.7 for the complex of $(H_3C)_3CNH_3^+$), which is why Cram used these receptors as scaffolds for the design of enzyme mimics as we will see in Section 9.1.

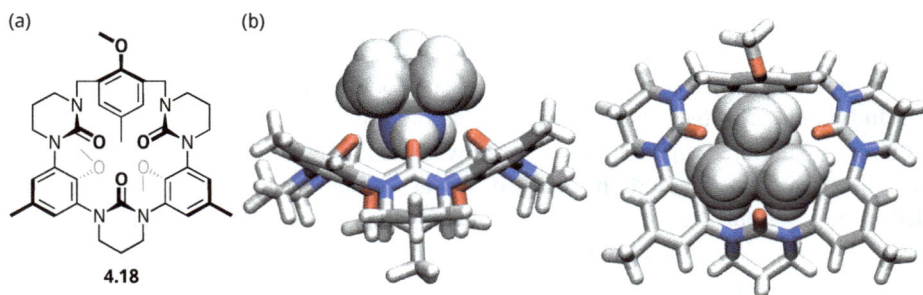

Figure 4.20: Structure of spherand **4.18** (a) and crystal structure of the *tert*-butylammonium complex of **4.18** in side and top view (b).

4.1.4 Cyclodextrins

Introduction: Cyclodextrins are among the most widely used receptors in supramolecular chemistry. Their origins date back long before Pedersen's discovery of crown ethers, with the first reports about cyclodextrins appearing in the literature as early as 1891. In that year, Antoine Villiers discovered new products in a mixture resulting from the bacterial degradation of potato starch. He noted their surprising resistance to acid hydrolysis, which is also a characteristic property of cellulose, and therefore suggested the name *cellulosine*. We now know that Villiers was the first to prepare cyclodextrins, but it would be many years before a clearer picture of the structure of these compounds emerged. In the years following Villiers' publication, several groups worked on cyclodextrins. Important contributions came from Franz Schardinger, who improved the original method of cyclodextrin preparation. Schardinger also isolated the microorganism that mediates the degradation of starch into cyclodextrins, *Bacillus macerans*, identified two fractions of the products (later shown to represent α-cyclodextrin and β-cyclodextrin), and described many of the fundamental properties of cyclodextrins, including their ability to interact with iodine. Although he was ultimately unable to elucidate the structure of cyclodextrins, his contributions were so significant that for many years cyclodextrins were known as *Schardinger dextrins*. In 1930, Hans Pringsheim found that cyclodextrins interact with organic molecules in water, a property that was later used to separate the different congeners. The exact structure of cyclodextrins remained unknown until the work of Dexter French, who deduced from the molecular weights of α- and β-cyclodextrin that they are macrocycles containing six and seven glucose units, respectively. French thus confirmed the hypothesis of Karl Johann Freudenberg, who first proposed cyclic structures for cyclodextrins. Freudenberg also discovered a third starch degradation fraction, later shown to be γ-cyclodextrin. Extensive work on the host–guest chemistry of cyclodextrins was then carried out by Friedrich Cramer, who did his PhD with Freudenberg and whose contributions laid the foundation for the use of cyclodextrins in supramolecular chemistry. Cramer also proposed the names α-, β-, and γ-cyclodextrin for the three most important congeners. The final structural proof came in the 1970s, when Wolfram Saenger described several crystal structures of cyclodextrins. Even before this work, the field of cyclodextrin chemistry was developing rapidly due to the many interesting properties of these compounds. Other eminent researchers were involved in these developments, whose contributions are acknowledged in a historical review [28].

It is worth noting that the field suffered a setback in 1957 when reports were published indicating that cyclodextrins are toxic. A few years later, it was shown that these claims were incorrect and probably due to impurities in the samples tested. Today, cyclodextrins are considered toxicologically safe and they and their derivatives are therefore permitted in drug formulations, as food additives, and for other applications (Chapter 12).

Cyclodextrins are structurally related to starch, from which they are produced. They have macrocyclic structure, with α-cyclodextrin, β-cyclodextrin, and γ-cyclodextrin containing, respectively, six, seven, and eight 1,4-linked α-D-glucopyranose units along the ring (Figure 4.21a). Larger cyclodextrins (δ-, ε-cyclodextrin, etc.) are known but do not (yet) play a major role in supramolecular chemistry.

Cyclodextrins are considered to have the shape of a truncated cone or torus due to the arrangement and orientation of their subunits. All of them adopt chair conformations in which the anomeric carbon atom is arranged below and the carbon atom in 4-position is above the plane formed by the other four ring atoms ($^4C_1$ conformation). The ring planes are arranged almost parallel to the axis of the torus, with the secondary and primary hydroxy groups oriented along the wider and narrower cavity openings, respectively, as illustrated by the β-cyclodextrin structure in Figure 4.21b. The thus resulting uneven distribution of the OH groups along the two cavity openings and the slightly tilted arrangement of the glucose units induce a dipole moment that influences complex formation.

The secondary OH groups form a seam of intramolecular hydrogen bonds, which is only uninterrupted in β-cyclodextrin. In α-cyclodextrin and the more flexible γ-cyclodextrin, not all OH groups form intramolecular hydrogen bonds along the wider rim. As a result, β-cyclodextrin has the least number of OH groups available for interactions with water molecules, which is one of the reasons why the seven-membered cyclodextrin has the lowest water solubility (Table 4.4). In contrast to the polar rims, the cyclodextrin cavities are less polar because they are surrounded by C–C, C–H, and C–O bonds. The CH groups additionally induce a slightly positive electrostatic potential, which explains why cyclodextrins prefer to bind neutral compounds and certain anions, but have no pronounced affinity for cations when the positively charged group ends up inside the cavity.

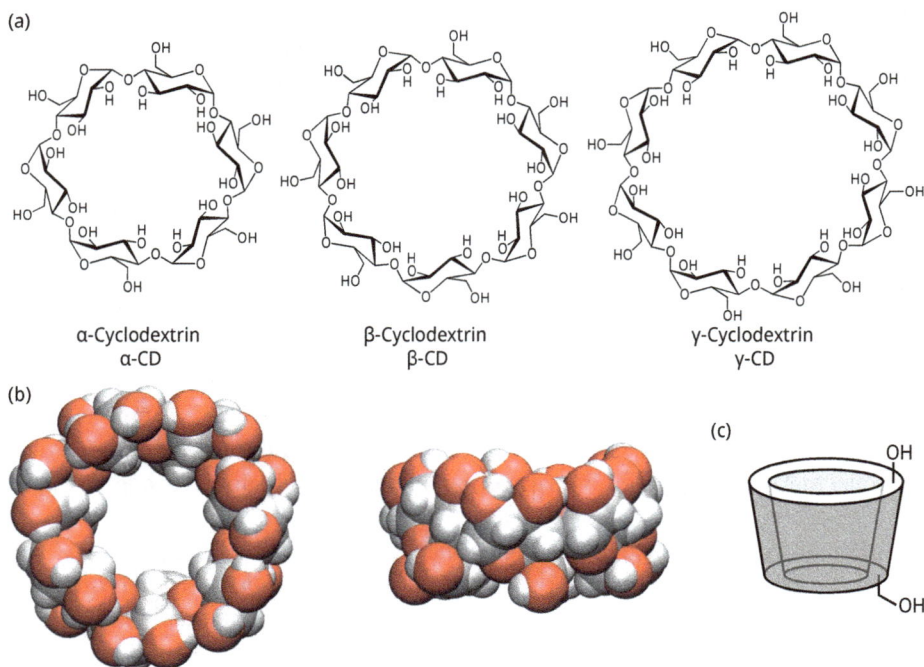

Figure 4.21: Structures of α-cyclodextrin, β-cyclodextrin, γ-cyclodextrin (a), top and side views of the space-filling model of β-cyclodextrin (b), and schematic representation of the cyclodextrin torus with the secondary and primary hydroxy groups lining the wider and the narrower rim, respectively (c).

The cavity dimensions of cyclodextrins correlate with the ring size (Table 4.4). Thus, cyclodextrins exhibit size selective complex formation. It should be noted that the geometric parameters summarized in Table 4.4 are derived from Corey–Pauling–Koltun (CPK) models under the assumption that the rings adopt rigid C_n-symmetric conforma-

Table 4.4: Properties of α-, β-, and γ-cyclodextrin.

	α-Cyclodextrin	β-Cyclodextrin	γ-Cyclodextrin
Number of glucopyranose units	6	7	8
Internal cavity diameter (Å)	4.7–5.2	6.0–6.4	7.0–8.3
External ring diameter (Å)	14.2–15.0	15.0–15.8	17.1–17.9
Height of rings (Å)	7.9–8.0	7.9–8.0	7.9–8.0
Cavity volume (Å^3)	174	262	427
Solubility in water at 25 °C (g L^{-1})	145	18.5	232
Typical substrates	Linear aliphatic compounds, benzene, phenol	4-*tert*-Butylphenol, naphthalene, ferrocene, adamantane, and other bulky alkanes	Perylene, cholesterol, crown ethers

tions [29]. However, cyclodextrins are much more flexible than often assumed. For example, in the crystal structure of the dihydrate of α-cyclodextrin, one glucose ring is oriented almost perpendicular to the main axis [30]. Certain derivatives of cyclodextrins even contain an inverted glucose unit whose primary hydroxy group is oriented toward the secondary OH groups of the other glucose units. Although these conformations are likely high energy conformations that are stabilized by interactions with solvent molecules or suitable substrates, they illustrate how flexible cyclodextrins are. The dimensions of cyclodextrins determined from rigid CPK models in Table 4.4 should therefore be regarded as averages that vary over a certain range.

Nomenclature: Cyclodextrins are classified according to ring size, with α-cyclodextrin containing six, β-cyclodextrin containing seven, and γ-cyclodextrin containing eight glucose subunits. Larger cyclodextrins are named analogously by using the corresponding Greek letters. In the older literature, one also finds the terms cyclohexaamylose, cycloheptaamylose, and cyclooctaamylose for α-, β-, and γ-cyclodextrin, respectively, and the term *Schardinger dextrins* is used for the whole family of receptors. The most common abbreviation for cyclodextrins is CD, which will be used from here on.

To distinguish the individual glucose units in a CD, the ring is viewed from the primary side and one glucose is labeled with the letter A. Starting from this subunit, the other glucose units are labeled clockwise with the following letters of the alphabet. Accordingly, two neighboring glucose units are labeled AB. This method is useful for distinguishing CD derivatives with substituents in more than one glucose unit. The ring atoms in the glucose units are numbered according to the conventions of carbohydrate chemistry, starting with 1 at the anomeric center.

Synthesis: The enzyme in *Bacillus macerans* responsible for the conversion of starch to CDs is cyclodextrin glycosyltransferase (CGTase or cyclodextrin glucano-transferase, EC 2.4.1.19). This enzyme cleaves an oligosaccharide from amylose, the linear component of starch, at a certain distance from the nonreducing end of the chain. The resulting product is then cyclized by joining the anomeric center and the 4-position of the glucose unit at the opposite chain end (Figure 4.22).

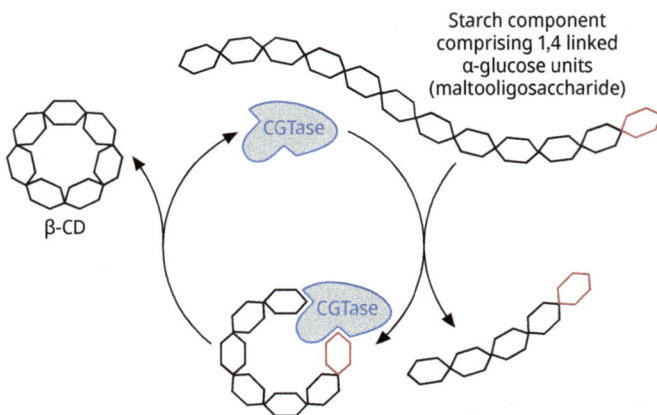

Figure 4.22: Schematic representation of the enzymatic CD synthesis from a linear maltooligosaccharide. The red rings denote reducing chain ends.

This transformation yields mixtures of CDs and acyclic products, with β-CD typically predominating. Pure CDs are isolated by repeated precipitation and crystallization steps. For example, β-CD forms insoluble complexes with toluene, *p*-xylene, trichloro-ethylene, or other organic compounds and can therefore be separated relatively selectively from a mixture *via* these complexes. Complex formation is also used to drive the enzymatic CD synthesis. For example, if the enzymatic reaction is carried out in the presence of toluene the [toluene⊂β-CD] complex precipitates from the mixture. α-CD is preferentially produced in the presence of 1-decanol and γ-CD in the presence of cyclohexadec-8-en-1-one.

The drawback of these procedures is that the organic and potentially toxicologically problematic precipitation agents must be carefully removed in the next step before the products can be used in certain applications. Indeed, residual organic impurities may have been the reason for the toxicity of CD samples in early investigations. To avoid extensive separation and purification steps and the use of precipitants, microbiologically modified variants of CGTases have been developed that selectively produce the three CDs. In combination with amylases that mediate the initial breakdown of starch into smaller linear maltooligosaccharides, these CGTases are now used to produce CDs on an industrial scale. With these methods available, CDs have become commodity chemicals, with an annual global production of several thousand tons. The bulk price for β-CD is no more than a few dollars per kilogram. α-CD and γ-CD are more expensive because they are produced in lower yields and are more difficult to purify.

Further reading

CGTases catalyze both the rapid, reversible intermolecular and intramolecular transglycosylation and the slow hydrolysis of the α-1,4-glucosidic linkages of amylose or shorter maltooligosaccharides, ultimately yielding glucose as the thermodynamic product. The cyclic intermediates formed on the way to the end point of the reaction persist in solution long enough to follow their evolution and transformation into each other. In their investigation of this reaction, the group of Sophie R. Beeren found that α-CD, β-CD, and γ-CD dominate the mixture after a few hours and that the ratio of the three CDs reflects their thermodynamic stability. The CDs interconvert rapidly, but their degradation to glucose is slow. This subsystem of macrocyclic products therefore has the typical characteristics of a dynamic combinatorial library (Section 5.7) as long as it remains in the kinetically trapped state. Consequently, the product distribution can be influenced by the addition of templates that interact preferentially with one of the library members. This strategy allowed the Beeren groups not only to selectively enrich the reaction mixture in α-CD, β-CD, or γ-CD, but also to produce large-ring cyclodextrins such as the nine-membered δ-CD in unprecedented yields. For more information, see:

Larsen D, Beeren SR. Enzyme-mediated dynamic combinatorial chemistry allows out-of-equilibrium template-directed synthesis of macrocyclic oligosaccharides. Chem. Sci. 2019, 10, 9981–7. Erichsen A, Peters GHJ, Beeren SR. Templated enzymatic synthesis of δ-cyclodextrin. J. Am. Chem. Soc. 2023, 145, 4882–91.

Binding properties: CDs bind their substrates within the cavity created by the cyclically arranged glucose subunits. Other modes of binding, such as the arrangement of the substrate along the cavity openings where it interacts with the cyclically arranged hydroxy groups, are known but are often unspecific and not very strong. Complex formation is most efficient in water, in which CDs are well soluble and in which complex formation is actively mediated and supported by the water molecules, as shown schematically in Figure 4.23.

Figure 4.23: Schematic representation of the inclusion of an organic guest molecule into the cavity of a CD. Water molecules are shown as circles.

Figure 4.23 shows that the substrate and the CD are surrounded by water molecules prior to complex formation. In addition, water molecules occupy the CD cavity. These water molecules cannot form the maximum number of hydrogen bonds and their release during complex formation therefore adds a characteristic signature to the overall thermodynamics of binding. While the desolvation of the substrate, which is usually a small organic molecule, should be enthalpically nearly neutral and entropically favorable in water, the release of the water molecules from the CD cavity should make the binding exothermic and entropically unfavorable, as discussed in Section 3.3. Both thermodynamic signatures are observed for CD complexes, as shown in Table 4.5 [31]. The negative ΔH^0 values of the α-CD and β-CD complexes indicate that the release of cavity water, whose contribution to complex formation was proposed as early as 1967 [32] and subsequently confirmed [33], is apparently more important for the smaller CDs than for γ-CD. However, there are also cases, especially for γ-CD, where complex formation is endothermic and driven entirely by entropy. The widely varying ΔH^0 and $T\Delta S^0$ values also show that complex formation is sensitively dependent on the structure of the substrate. For example, the complexation of benzene changes from exothermic to endothermic as the ring size increases, while the decrease in stability of the sodium naphthalene-2-sulfonate complex when moving from β-CD to the smaller or larger CD is due to a decrease in the enthalpic term in one case and an unfavorable entropic term in the other. Thus, in addition to stabilizing receptor–substrate interactions in the complex, both cavity and substrate desolvation contribute to the thermodynamics of binding. The important and probably decisive influence of water molecules on complex formation is clearly demonstrated by the negative effects of organic solvents on complex stability: in DMSO or aqueous solvent mixtures containing a significant amount of an organic cosolvent, CD complexes are usually not very stable or do not form at all.

Table 4.5: Thermodynamic parameters associated with the formation of selected α-CD, β-CD, and γ-CD complexes (ΔH^0 and $T\Delta S^0$ in kJ mol^{-1}).

Cyclodextrin	Substrate	log K_a	ΔH^0	$T\Delta S^0$
α-CD	Benzene	1.5	−13.1	−4.5
	Benzoic acid	3.0	−33.9	−16.7
	Sodium naphthalene-2-sulfonate	2.6	−3.3	11.3
	Adamantane-1-carboxylic acid	2.3	−13.5	−0.4
β-CD	Benzene	2.2	−1.9	10.8
	Benzoic acid	2.6	−18.0	−3.2
	Sodium naphthalene-2-sulfonate	5.4	−29.3	1.3
	Adamantane-1-carboxylic acid	4.3	−20.3	4.2
γ-CD	Benzene	1.0	14.2	19.7
	Benzoic acid	2.1	−4.5	7.6
	Sodium naphthalene-2-sulfonate	1.2	−17.5	−10.9
	Adamantane-1-carboxylic acid	3.5	5.0	25.2

Note that the binding of dodecaborate clusters, for example $B_{12}Br_{12}^{2-}$, to γ-CD is also strongly exothermic [34], but since water molecules do not have to give up many hydrogen bonds when hydrating the large γ-CD cavity, the negative ΔH^0 in this case results from chaotropic effect, that is, the thermodynamically favorable desolvation of the large anionic substrate, as discussed in Section 3.3.

Once the substrate is inside the CD cavity, further interactions enhance the binding strength. Table 4.5 shows, for example, that adamantane-1-carboxylic acid forms a more stable complex with β-CD than with the other CDs, which is due to the perfect fit of the adamantane group into the β-CD cavity. As a result, dispersion interactions with the surrounding cavity walls are more efficient than in the case of the smaller and larger CDs. Another factor influencing complex stability is the formation of hydrogen bonds between the OH groups lining the CD cavity and functional groups in the substrate. This effect could explain why benzoic acid forms a more stable complex with α-CD than benzene. Alternatively, dipole–dipole interactions may be responsible for this trend. The importance of the latter type of interactions is reflected in the effects of introducing nitro groups into benzoic acid and phenol on the stabilities of the corresponding β-CD complexes: while the $\log K_a$ of the benzoic acid complex decreases from 3.1 to 2.5 when 4-nitrobenzoic acid is used as substrate, the $\log K_a$ increases from 1.9 for the phenol complex to 2.5 for the complex of 4-nitrophenol. Accordingly, the reduction of the dipole moment of benzoic acid caused by the introduction of the nitro group has an unfavorable effect on the complex stability, whereas the complex becomes more stable when phenol is exchanged for a derivative with a higher dipole moment. Dipole–dipole interactions, which are most favorable when the dipoles are antiparallel, also determine the preferred orientation of the substrate in the CD cavity.

Unlike many other receptors, CDs are chiral. They are therefore potentially able to discriminate between enantiomers. Indeed, complexes of CDs with the two enantiomers of a chiral substrate often differ in their structures. However, these differences usually translate into only minor differences in complex stability [35]. For example, the β-CD complex of N-acetyl-D-tryptophan has a K_a of 12.7 M^{-1} and that of the corresponding L-enantiomer a K_a of 17.1 M^{-1}. The corresponding $\Delta \Delta G^0$ of 1.3 kJ mol^{-1} is still sufficient to allow CDs to be used as chiral stationary phases for chromatographic separations (Section 12.3.1).

Complex formation can cause CDs to precipitate from solution. This is usually the case when the cavity openings are poorly hydrated because they contain nonpolar residues from the substrate. In contrast, polar substituents in these positions lead to soluble complexes. The toluene complex of β-CD has a low solubility in water, for example, whereas the complex with benzyl alcohol is soluble. The effect of certain substrates on solubility is used to separate α-CD, β-CD, and γ-CD, as we have seen earlier. If large hydrophilic portions of the substrate remain outside the cavity after binding to a CD ring, the sandwiching of the substrate between two rings is also possible, leading to the formation of 2:1 complexes. An example is the complex between camphor and two molecules of α-CD.

Derivatives: CDs are polyfunctional molecules whose hydroxy groups are available for the formation of ethers or esters, but also for oxidations, deoxygenations, and other transformations. Special synthetic techniques are required to carry out these transformations selectively because even the introduction of only one substituent can lead to three products, depending on whether the 2-, the 3-, or the 6-OH group of a glucose unit reacts. The number of possible isomers increases rapidly with increasing degree of functionalization, reaching a maximum for CDs in which half of the hydroxy groups are modified [36]. Therefore, most synthetic techniques leading to structurally defined CD derivatives involve the modification of either all available hydroxy groups, or of only one or two. In addition, random modification is also possible.

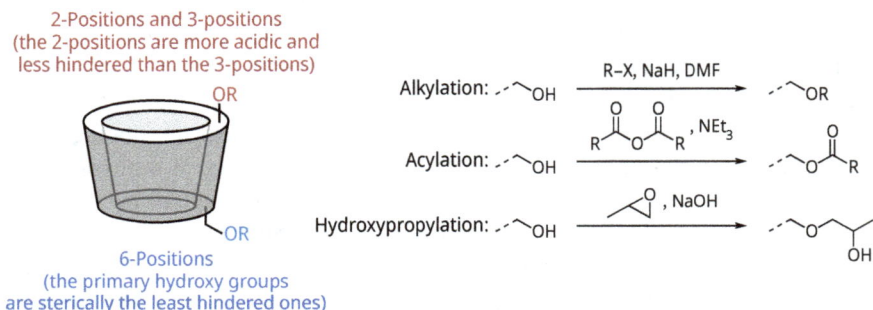

Figure 4.24: Strategies for CD modification.

Fully substituted CDs are obtained by alkylation or acylation. These reactions typically involve treating CDs under basic conditions with an excess of an alkyl halide, an acyl chloride, or an anhydride (Figure 4.24). The products are generally water-insoluble and have conformations and binding properties very different from those of native CDs. Notable exceptions are the permethylated CD derivatives, which are surprisingly soluble in water, even better soluble than native CDs. For example, up to ca. 200 g of permethylated β-CD can be dissolved in 1 L of water at 25 °C, whereas native β-CD has a water solubility of only 18.5 g L^{-1} under the same conditions. Both types of CDs also exhibit a different temperature dependence of water solubility. Native CDs behave as expected, becoming more soluble with increasing temperature, whereas permethylated CDs precipitate from water at higher temperatures. In the case of native CDs, the improved solubility at elevated temperatures is due to the increase in flexibility, which facilitates the interactions of the OH groups with water molecules. Methylated CDs, on the other hand, are flexible even at room temperature because they contain no intramolecular hydrogen bonds. Despite the presence of the methyl groups, they are efficiently hydrated in water, but the hydration entropy is negative. Hydration enthalpy and entropy thus act in different directions, causing the free enthalpy of hydration to change sign at a certain temperature according to the Gibbs–Helmholtz equation (equation (2.9)). Above this temperature, the entropic advantage of desolvation outweighs the negative hydration enthalpy and methylated CDs precipitate from solution in their dehydrated form.

Permethylated CDs are significantly weaker binders than native CDs due to the reduced preorganization resulting from the higher flexibility. This effect does not simply correlate with the number of methyl groups, but depends on the exact substitution pattern, as shown by comparing the 4-*tert*-butylbenzoic acid affinity of different methylated β-CD derivatives [37]. According to the data in Table 4.6, the introduction of methyl groups in the 2-position or in the 2- and 6-positions of the glucose units does not have a strong effect on the binding properties. However, once the 3-positions are methylated, the substrate affinity drops markedly. The reason for this effect is the influence of methylation on the intramolecular hydrogen bonding pattern along the cyclodextrin cavity. According to the structure of the unsubstituted β-CD in Figure 4.21b, the OH groups in the 3-positions act as hydrogen bond donors, binding to the oxygen atoms of the 2-OH groups of the neighboring glucose units. Methylation prevents the 3-OH groups from participating in hydrogen bonding, which has a negative effect on the preorganization of the β-CD ring. In contrast, the acceptor properties of the 2-OH groups are not strongly affected whether they are methylated or not. Interestingly, methylation of β-CD at the 6-positions leads to a small increase in complex stability compared to the native β-CD.

Table 4.6: Dependence of the affinity of β-CD for 4-*tert*-butylbenzoic acid in water on the degree of methylation and the substitution pattern.

Cyclodextrin	Substitution pattern	log K_a	Cyclodextrin	Substitution pattern	log K_a
Unsubstituted β-CD		4.2	6-O-Methyl-β-CD		4.5
2,6-Di-O-methyl-β-CD		4.2	2-O-Methyl-β-CD		4.1
2,3,6-Tri-O-methyl-β-CD		3.1	2,3-Di-O-methyl-β-CD		2.9

The higher water solubility of partially methylated β-CD derivatives and their complexes, combined with a binding affinity that is hardly reduced compared to native CDs, are attractive features for many practical applications. However, the synthetic procedures required to obtain the compounds shown in Table 4.6 are far too laborious for large-scale syntheses. Statistical modifications are therefore used instead. The resulting CD derivatives are characterized by their degree of substitution (d.s.), which specifies the average number of substituents per glucose unit. Thus, persubstituted CDs have a d.s. of 3, while a d.s. of 1 means that each glucose unit contains on average one substituent. Statistically substituted CDs are usually well soluble in water and still possess high substrate affinity. For example, a randomly methylated β-CD with a d.s. of 1.7–1.8 binds 4-*tert*-butylbenzoic acid with a $\log K_a$ of 4.1, almost as strongly as native β-CD itself (Table 4.6) [37]. Other popular derivatives for applications are statistically hydroxypropylated CDs, which are prepared by treating CDs with propylene oxide under basic conditions (Figure 4.24).

The substitution pattern resulting from the chemical modification of CDs is controlled by the reactivity of the different hydroxy groups. The primary OH groups are less hindered than the secondary OH groups and bulky groups such as trityl- or *tert*-butyldimethylsilyl groups are therefore preferentially introduced along the narrower rim of the cavity. The secondary OH groups are more acidic than the primary OH groups, making them more reactive in basic media. Of the secondary OH groups, those in the 2-positions are more accessible than those in the 3-positions, which are located between the glucose units. These intrinsic reactivities explain why statistical transformations have little effect on the binding properties: they occur preferentially in the 6- and 2-positions and not in the more problematic 3-positions.

Central starting points for the selective synthesis of monofunctionalized β-CD derivatives are the monotosylated derivatives **4.19** and **4.20** (Figure 4.25a), which carry the leaving group in the 6- or 2-position of a glucose unit, respectively. Both compounds are obtained from β-CD and 4-toluenesulfonyl chloride in about 15–30% yield, but under different conditions. The 6-substituted tosylate **4.19** is formed when β-CD and 4-toluenesulfonyl chloride are treated in water. Under these conditions, an inclusion complex is presumably formed in which the sulfonyl chloride group protrudes from the narrower cavity opening, explaining the regioselective modification of a primary OH group. The corresponding isomer **4.20** can be obtained from β-CD and 4-toluenesulfonyl chloride in several different ways (Bu$_2$SnO, NEt$_3$, DMF, 100 °C; NaH, DMF; Na$_2$CO$_3$, 1,4-dioxane/water, 1:1 (*v/v*), 50 °C), all of which have in common that no complexation of 4-toluenesulfonyl chloride occurs. In this case, the regioselectivity is due to the preferential deprotonation of the 2-OH group.

Both tosylates can be converted into a variety of products by reaction with nucleophiles. For example, direct treatment of **4.19** with thiols gives thioethers, and reaction with sodium azide gives an azide that can be coupled with terminal alkynes under the conditions of the copper(I)-catalyzed azide–alkyne cycloaddition to give

(a)

4.19   **4.20**

(b)

(c)

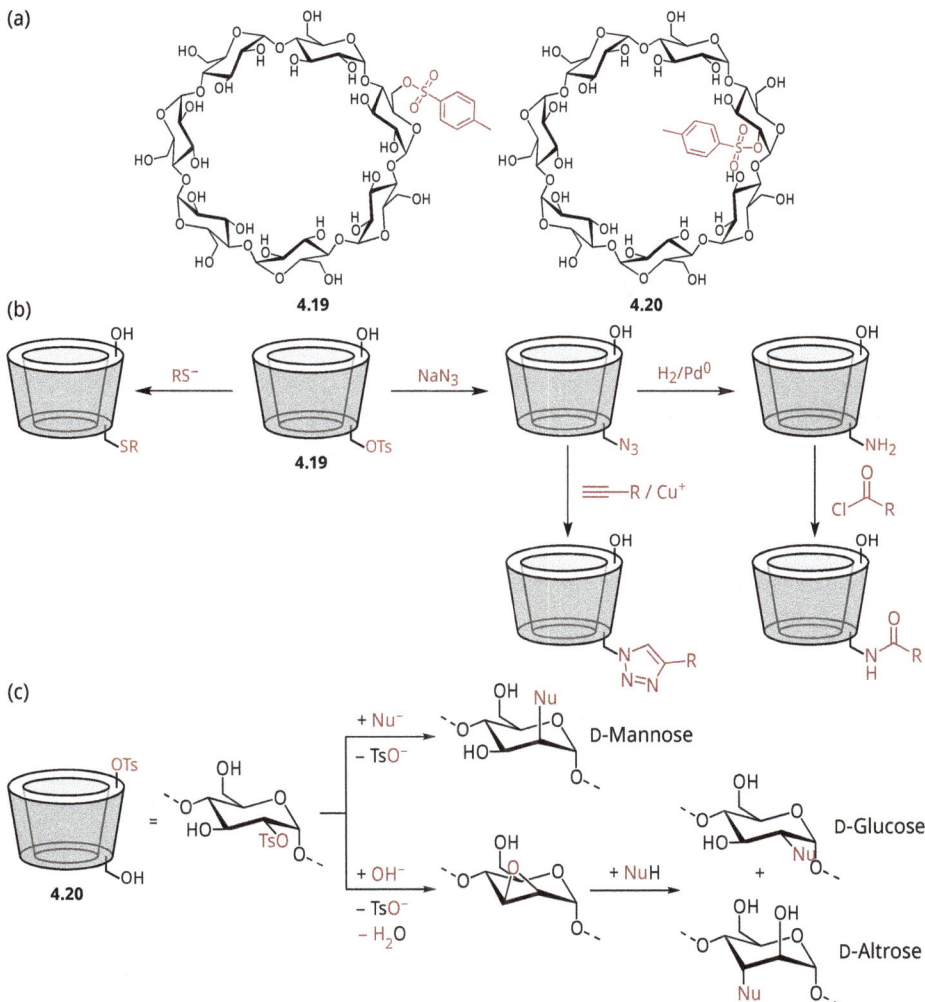

Figure 4.25: Structures of β-CD monotosylates **4.19** and **4.20** (a) and examples of reaction sequences available to convert **4.19** (b) and **4.20** (c) into corresponding β-CD derivatives.

1,2,3-triazoles or, after reduction to the corresponding amine, with acid chlorides to give amides (Figure 4.25b).

In the case of **4.20**, the nucleophilic substitution of the tosylate group proceeds under inversion of the reaction center (Figure 4.25c), leading to a β-CD derivative with the new substituent in a 2-deoxy-D-mannose unit. Alternatively, **4.20** can first be converted into the corresponding 2,3-epoxide by treatment with NaOH. A nucleophile which opens the epoxide group in the 2-position will give a β-CD with the substituent in the 2-position of a 2-deoxy-D-glucose unit. If the epoxide is opened at the 3-position, a substituted 2-deoxy-D-altrose subunit is formed, which distorts the CD cavity and therefore negatively affects the receptor properties.

Disubstituted CDs are prepared in a similar way, starting from suitable ditosylates. The relative positions of the two leaving groups in the products are controlled by performing the tosylation with appropriate bis(sulfonyl chlorides), which allow the regioselective modification of two glucose units in AB (directly adjacent), AC (one separating glucose unit), or AD (two separating glucose units) positions. In addition, methods are available to regioselectively introduce more than two substituents along the ring. The achievable degree of structural control is illustrated well by the elegant, albeit lengthy, synthesis of a β-CD derivative with a defined sequence of seven different substituents at the narrower opening [38].

Substituents in functionalized CDs can contribute to substrate binding, mediate the chemical transformation of a bound substrate (Chapter 9), report the binding event by an optical change (Chapter 11), or allow the immobilization of CDs on surfaces, nanoparticles, polymers, or other supports. In addition, bis(cyclodextrins) consisting of two covalently linked CD rings are also accessible. An example is the disulfide **4.21** (Figure 4.26), which is prepared from **4.19** by reaction with thiourea, followed by basic hydrolysis and oxidative dimerization.

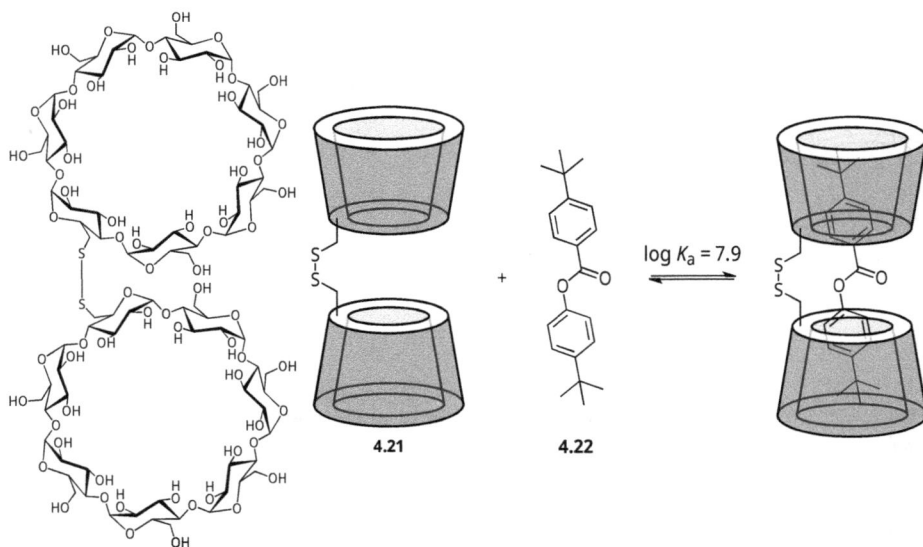

Figure 4.26: Structures of the bis(cyclodextrin) **4.21** and its substrate, the ester **4.22**.

Bis(cyclodextrin) **4.21** binds the ester **4.22** with a $\log K_a$ of 7.9 [39], which is approximately twice the $\log K_a$ of 4.2 associated with the *tert*-butylbenzoic acid complex of β-CD (Table 4.6). This and other bis(cyclodextrins) are therefore well suited for investigating concepts such as multivalency or cooperativity (Section 3.5.3), but also for designing high-affinity receptors.

4.1.5 Cyclophanes

Introduction: Cyclophanes are, by definition, cyclic compounds containing at least one, but usually several, disubstituted aromatic subunits. Of the many compounds in this category, only some play a role in supramolecular chemistry, namely those with a sufficiently large cavity into which a substrate can be included. The [2.2]paracyclophane shown in Figure 4.27 is a typical cyclophane, for example, but not a receptor. Cyclophanes useful for substrate binding include cyclotriveratrylenes, calixarenes, resorcinarenes, and pillararenes, major receptor families so important that they are discussed separately. Other cyclophane-based receptors comprise a structurally diverse group of macrocyclic compounds. They allow the binding of substrates that vary greatly in size, shape, and properties and since it is impossible to give a complete overview, we will only look at a few representative examples [40].

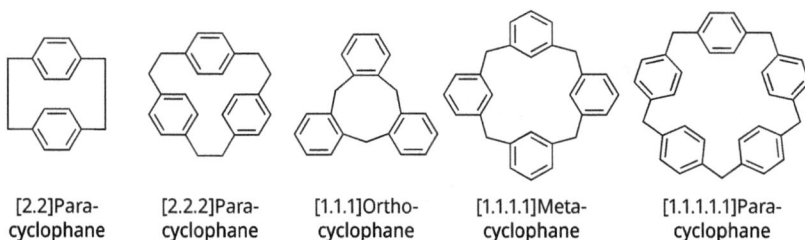

| [2.2]Para-cyclophane | [2.2.2]Para-cyclophane | [1.1.1]Ortho-cyclophane | [1.1.1.1]Meta-cyclophane | [1.1.1.1.1]Para-cyclophane |

Figure 4.27: Examples of cyclophanes and their names.

Nomenclature: Cyclophanes are systematically named in the same way as cryptands. The name begins with numbers in square brackets indicating the lengths of the bridges between the aromatic subunits, starting with the longest. The substitution pattern of the aromatic units is then denoted by using the descriptors *ortho*, *meta*, or *para*, immediately followed by the family name *cyclophane*. The presence of heteroatoms is indicated using the prefixes *aza* for nitrogen, *oxa* for oxygen, or *thia* for sulfur. These rules are illustrated in Figure 4.27 for a selection of cyclophanes. Naming derivatives with more complex structures requires additional rules that are not explained here.

Synthesis: The cyclophanes discussed in the following have carefully designed structures to allow binding of a complementary substrate. They vary widely in structure and contain different functional groups to mediate complex formation, with the underlying interactions typically involving a combination of solvophobic effects, aromatic interactions, cation–π interactions, charge-transfer interactions, and hydrogen bonding. There is no single synthetic approach to accessing these compounds. However, the procedures described often make use of important general synthetic concepts, such as the use of templates or high dilution conditions.

Binding properties: An early receptor that provided the first direct evidence for successful complexation of an aromatic guest in aqueous solution is the 1,6,20,25-tetraaza[6.1.6.1] paracyclophane **4.23** (Figure 4.28a) described by Kenji Koga [41]. This compound is somewhat reminiscent of the macrocyclic polyamines described in Section 4.1.1, but differs structurally by the presence of the diphenylmethane moieties. It is soluble in acidic aqueous media (pH < 2), where all amino groups are protonated. Under these conditions, **4.23** has an extended conformation with a defined cavity due to the charge repulsion of the ammonium groups. It hosts aromatic substrates such as 2,7-dihydroxynaphthalene or 1,2,4,5-tetramethylbenzene (durene). The crystal structure of the latter complex shows that the aromatic guest is fully included into the cavity, with the methyl groups oriented toward the cavity openings (Figure 4.28b). The aromatic CH groups of the substrate face benzene rings in the diphenylmethane subunits, indicating that the complex is stabilized by aromatic edge-to-face interactions (Section 3.1.10). Similar interactions are seen between the faces of the durene ring and CH groups of the alkyl linkers.

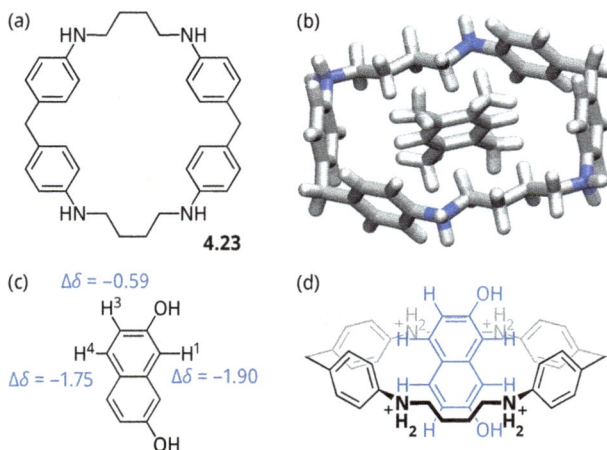

Figure 4.28: Structure of cyclophane **4.23** (a), crystal structure of the complex between **4.23** and durene (b), ^{1}H NMR chemical shift changes ($\Delta\delta$ in ppm) induced by the complexation of 2,7-dihydroxynaphthalene with **4.23** in DCl/D$_2$O (c), and schematic structure of the corresponding complex derived from the NMR spectroscopic results (d).

Similar complexes are formed in solution, as shown by the pronounced effects of **4.23** on the 2,7-dihydroxynaphthalene proton resonances in the NMR spectrum. Figure 4.28c shows that all the guest signals shift upfield, with the signals corresponding to the positions 1 and 4 undergoing larger shifts than the signal from the proton at position 3. The complexation-induced shielding of the guest protons suggests that the guest is included into the cyclophane cavity with the axis of the macrocycle and the longer axis of the guest aligned, causing the guest protons to be positioned within the shielding regions of the magnetically anisotropic π-systems of **4.23**. The relative magnitudes of the signal shifts show that the protons at C1 and C4 preferentially reside inside the cavity, while the protons at C3 are oriented toward the cavity openings, where the shielding effect is weaker (Figure 4.28d).

The group of François Diederich studied the binding of aromatic substrates to cyclophane **4.24** (Figure 4.29a), which also contains two diphenylmethane moieties [40]. The solubilizing ammonium groups are located at the periphery of this receptor, but its overall shape is similar to that of **4.23**. Therefore, **4.24** also binds aromatic compounds that can engage in aromatic edge-to-face interactions with the aromatic cavity walls. The most stable complexes are formed in water, and complex stability decreases progressively as the solvent becomes less polar [42]. The complex between **4.24** and 1,4-dimethoxybenzene has a $\log K_a$ of 4.0 in water, for example, and a $\log K_a$ of only 0.9 in methanol.

These results demonstrate the strong effect of solvation on binding strength. In nonpolar solvents, where the binding partners are well solvated, the stability of the complex is mainly determined by the direct receptor–substrate interactions, which are weak in this case. In more polar solvents, and especially in water, the release of solvent molecules from the receptor cavity and the solvation shell of the substrate is associated with a thermodynamic gain that adds to the Gibbs free energy of the direct receptor–substrate interactions. Importantly, the binding of neutral aromatic substrates to **4.24** is consistently exothermic with an unfavorable entropy term, suggesting that, as in the case of CDs, complex formation benefits from the ability of solvent molecules to interact with each other after leaving the receptor cavity [33]. This effect is strongest in water, but also contributes to some extent to complex formation in other solvents such as alcohols. Diederich therefore correlated complex stability with the cohesiveness of the solvent: the higher the solvent cohesiveness, the more stable the complex [42].

Not only aromatic interactions but also other types of interactions contribute to the substrate affinity of cyclophanes. We have already seen an example in Section 3.1.8. The corresponding receptor **4.25** (Figure 4.29a), containing two ethenoanthracene moieties and two xylylene-derived linkers, was designed by Dennis A. Dougherty to study the effect of cation–π interactions on binding strength [43]. This receptor binds to the neutral 4-methylquinoline with a $\log K_a$ of 5.9 in 10 mM borate buffer. The structurally related but cationic N-methylquinolinium cation is bound more strongly by more than two orders of magnitude ($\log K_a$ = 8.4) due to the reinforcement of complex stability by cation–π interactions and attractive interactions between the oppositely charged ions.

Receptors with electron-poor paraquat (N,N'-dimethyl-4,4'-bipyridinium) moieties were introduced almost simultaneously in 1988 by J. Fraser Stoddart [44] and Sieg-

(a)

4.24 **4.25**

(b)

4.26

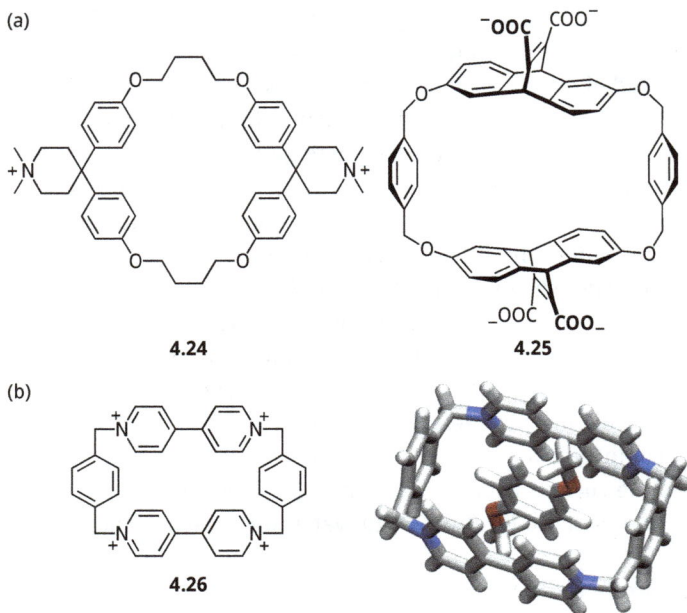

Figure 4.29: Structures of cyclophanes **4.24**, **4.25** (a) and of the blue box **4.26** together with the crystal structure of the 1,4-dimethoxybenzene complex of **4.26** (b).

fried Hünig [45]. An example is the tetracationic cyclophane **4.26** (Figure 4.29b), first mentioned in Section 3.1.10 and better known as the (little) *blue box* [46]. The reason for the term *box* is obvious – **4.26** has a rigid rectangular shape, with the planes of the aromatic subunits oriented parallel to the C_2 axis – but why *blue*? This attribute does not refer to the color produced by **4.26**, since solutions of this compound are nearly colorless, but to the color coding used by Stoddart in many of his publications, where the structural drawings of electron-poor and electron-rich aromatic rings contain filled blue and red circles, respectively. These color codes help identify complementary binding partners.

The blue box **4.26** binds aromatic guests through a combination of aromatic and charge-transfer interactions (Section 3.1.10), as illustrated by the crystal structure of the 1,4-dimethoxybenzene complex in Figure 4.29b. In this complex, the substrate is positioned within the cavity with the ring faces parallel to the paraquat moieties, allowing for orbital mixing and explaining the orange-red color of solutions of the complex. In addition, the proximity of the aromatic 1,4-dimethoxybenzene protons to the faces of the xylylene moieties leads to further stabilization. In solution, the complex geometry is similar, as indicated by the upfield shift of the signal of the 1,4-dimethoxybenzene protons in the ^{1}H NMR spectrum upon complex formation. The stability of this complex is not high (log K_a = 1.3) [44], but naphthalene derivatives are more strongly bound due to their larger π-surface areas, and changing the solvent from acetonitrile to water also increases the stability of the complex. The Stoddart group used such complexes exten-sively to construct rotaxanes and catenanes and molecular machines, as we will see in Section 7.5 and Chapter 8.

Receptors **4.27** and **4.28** (Figure 4.30) are further examples of cyclophane-derived receptors. In addition to the diphenylmethane motif, both contain NH groups in charac-teristic arrangements that enable them to form hydrogen bonds with suitable sub-strates. Cyclophane **4.27** was designed by Christopher A. Hunter as a receptor for *p*-benzoquinone [47]. The corresponding complex is stabilized not only by hydrogen bonding between the NH groups of **4.27** and the C=O groups of the substrate, but also by aromatic interactions, as shown schematically in Figure 4.30a. This binding mode is consistent with the complexation-induced signal shifts observed in the ^{1}H NMR spec-trum upon complex formation, which include an upfield shift of the *p*-benzoquinone signal by 2.36 ppm and a downfield shift of the signal of the NH protons of **4.27** by 1.03 ppm. The former is consistent with the positioning of the substrate protons close to the diphenylmethane sidewalls of the receptor, while the latter is a characteristic effect of hydrogen bonding. It results from the proximity of the proton in the hydrogen bond donor to the electron-rich hydrogen bond acceptor, causing a through-bond transfer of electron density from the proton to the attached residue. The resulting reduction in electron density explains the deshielding of the protons.

Receptor **4.27** is highly selective for *p*-benzoquinone. Tetramethyl-*p*-benzoquinone, for example, cannot be bound because there is not enough space in the cavity to accom-modate the larger guest. The $\log K_a$ of the *p*-benzoquinone complex is 3.1 in chloroform and decreases in mixtures containing methanol. In water, a water-soluble analog of **4.27** binds *p*-benzoquinone with a $\log K_a < 0.7$ in water [48]. Thus, the dependence of complex stability on solvent polarity is opposite to that observed for the cyclophanes discussed so far. The reason is that **4.27** contains polar groups whose desolvation is energetically costly in polar solvents. As a consequence, the stability of the complex decreases with increasing solvent polarity. In contrast, the substrate affinity of cyclophanes **4.24** and **4.26** benefits from the favorable energetic contributions of receptor desolvation in polar media.

Cyclophane **4.28** was described by Andrew D. Hamilton [49]. It is characterized by two 2,6-diaminopyridine residues attached to the central isophthalic acid. The conver-gent arrangement of hydrogen bond donors and acceptors mediates the binding of guests with a complementary hydrogen bonding pattern, especially barbiturates. Since this recognition motif has found widespread use after its introduction, it is known as the *Hamilton receptor*. The diphenylmethane moiety confines the cavity of **4.28**, further influencing the arrangement of the guest (Figure 4.30b). Consistent with this structure, complex formation causes the NH signals of the receptor and the sub-strate to shift downfield in the ^{1}H NMR spectrum, while the signals of the ethyl groups shift upfield due to their proximity to the diphenylmethane residue. The complex has a $\log K_a$ of 6.1 in chloroform. Barbiturates with phenyl groups in the 5-position are too large to fit into the cavity and therefore form much less stable complexes.

(a)

4.27

(b)

4.28

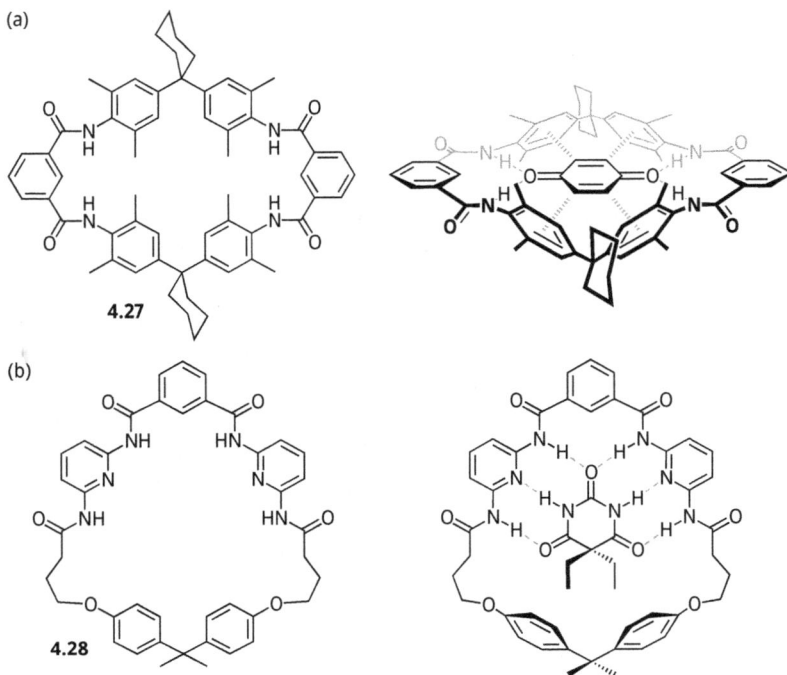

Figure 4.30: Structure of cyclophane **4.27** and schematic structure of its complex with *p*-benzoquinone (a) and structure of cyclophane **4.28** together with the schematic structure of its complex with barbital (b).

How does NMR spectroscopy help characterize receptor–substrate complexes?

We have now seen several examples where structural information about a complex has been derived from complexation-induced shifts of receptor and/or substrate signals in the ^{1}H NMR spectrum. Indeed, ^{1}H NMR spectroscopy is an important tool not only for the determination of complex stability, as outlined in Section 2.2.2, but also for the structural characterization of supramolecular complexes in solution. Several techniques are available, the simplest and most straightforward of which is to measure the NMR spectra of the complex and the free binding partners and analyze the differences.

Since NMR spectroscopy probes the electron density at the nuclei of a molecule, a complexation-induced increase or decrease in electron density at receptor and/or substrate protons (or other NMR-active nuclei), caused by the change in environment during complex formation, will result in shifts of the corresponding signals in the NMR spectrum. Downfield shifts (higher ppm values), indicating a decrease in electron density, usually occur when an electron-rich heteroatom is positioned close to the absorbing proton in the complex, for example due to the formation of a hydrogen bond. Thus, protons that are deshielded typically contribute to the stabilization of the complex as hydrogen bond donors. Note that hydrogen bonding can also cause upfield signal shifts if hydrogen bonds between the corresponding protons and solvent molecules prior to complex formation are replaced by weaker hydrogen bonds in the complex. More often, however, upfield signal shifts (lower ppm values) of NMR signals indicate that the corresponding proton in the complex is located above the plane of an aromatic ring, where it experiences the magnetic anisotropy of the π-system.

While simple ^{1}H NMR spectra often provide sufficient information about the orientation of the substrate in the receptor cavity or the mode of binding, a number of techniques are available for more detailed structural assignments. For example, when two or more protons in the complex are in close proximity, typically closer than 5 Å, they give rise to cross peaks in NOESY spectra. Alternatively, such nuclear Overhauser effects (NOEs) can be observed in one-dimensional spectra by irradiating at the frequency of one proton and detecting a signal for the nearby proton. NOEs are due to the transfer of spin polarization from one proton to another through space, with the distance of the magnetization transfer affecting the intensity of the resulting peak, which provides additional structural information when multiple cross peaks are observed in a NOESY spectrum. An application of this method is the structural assignment of CD complexes (Section 4.1.4), where the spatial proximity of substrate protons to the glucose protons lining the cavity wall is examined. Of these CH protons, those in the 5-position are close to the narrower cavity opening and those in the 3-position are close to the wider opening. A cross peak between a substrate signal and one of these signals therefore indicates, which way the substrate is incorporated into the CD cavity. In certain cases where NOESY spectroscopy does not provide meaningful information, rotating-frame nuclear Overhauser effect spectroscopy (ROESY) can be a useful alternative.

There are also a number of NMR spectroscopic techniques to determine the kinetic parameters of complex formation (Section 2.1). In cases where substrate exchange is slow on the NMR timescale, cross peaks are observed in NOESY spectra between the signals of the free binding partners and the corresponding signals of the complex. These exchange signals result from dynamic processes during the evolution period of the NMR experiment, most importantly the complexation and decomplexation of a substrate molecule, causing the spin polarization and relaxation to occur in different environments. The intensities of the corresponding cross peaks can be used to quantify the rate constants k_{on} and k_{off}. This technique is called exchange spectroscopy or EXSY. Applications of this method in supramolecular chemistry have been reviewed [50].

In cases where characteristic NMR signals of the complex cannot be detected, for example due to fast exchange, GEST NMR spectroscopy can be useful to obtain information about the complexation kinetics [51]. In this technique, NMR-active nuclei in the complexed substrate are magnetically labeled by saturation or an inverse pulse. Substrate exchange then results in the transfer of this magnetization to the large pool of free substrate molecules, whose signal intensity subsequently drops. Fitting the data to appropriate mathematical models yields the exchange rate of free and bound guest and the fractional occupancy of the host under the experimental conditions. This method is most convenient when the substrate contains NMR-active heteroatoms such as ^{19}F.

When exchange rates and the NMR timescale are comparable, broad signals are observed in the spectra that often do not provide much useful information. In these cases, it may be helpful to perform the measurement with a spectrometer operating at a different frequency. At lower frequencies, the guest exchange can become faster than the timescale of the NMR spectrometer, leading to a sharpening of the signals. Conversely, measuring the spectrum on a more powerful instrument can result in separate signals for free and complexed binding partners. The same effects can be achieved by increasing or decreasing the temperature at which the NMR spectrum is recorded. However, changes in temperature simultaneously affect the kinetics and thermodynamics of dynamic processes and therefore also cause a shift in the degree of complexation.

4.1.6 Cyclotriveratrylenes

Introduction: Cyclotriveratrylenes (CTVs) are formally [1.1.1]orthocyclophanes. The parent CTV **4.29** (Figure 4.31a) is derived from the acid-mediated macrocyclization of veratrole (1,2-dimethoxybenzene) in the presence of formaldehyde, hence the name cyclotriveratrylene. Today, the family name is used not only for CTVs containing veratrole subunits, but also for analogs with substituents other than methoxy groups. At least one substituent in each aromatic moiety is usually a free hydroxy or alkoxy group, while the other substituent may be the same or a different alkoxy group, a substituted nitrogen or sulfur atom, or a proton.

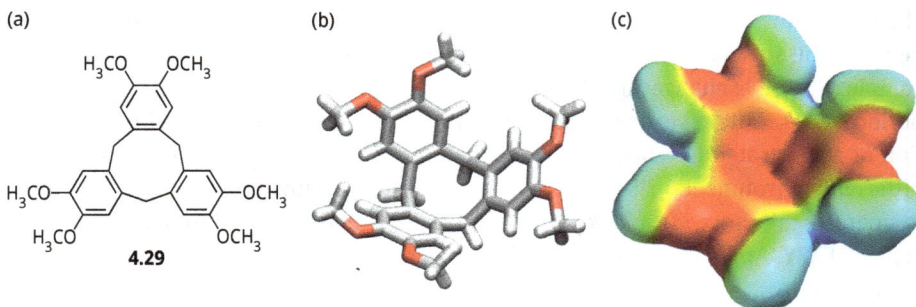

Figure 4.31: Molecular structure of the veratrole-derived CTV **4.29** (a), crystal structure of **4.29**, showing the thermodynamically most stable crown conformation (b), and electrostatic potential surface of **4.29** (c). The color coding covers a potential range from –100 to +100 kJ mol^{-1}, with red and blue indicating values greater than or equal to the absolute maximum in negative and positive potentials, respectively.

CTVs were first structurally characterized in the 1960s. Extensive work on the supramolecular chemistry of CTVs and their derivatives then followed in the 1980s, with many important contributions coming from the group of André Collet [52]. Much of the current research on CTVs builds on his important groundwork. The crystal structure in Figure 4.31b illustrates the preferred bowl-shaped structure of **4.29**. This conformation, known as crown, has a pronounced negative electrostatic potential along its inner surface due to the electron-rich subunits, which explains the binding properties of CTVs (Figure 4.31c).

CTV derivatives with six identical substituents are C_{3v}-symmetric, but different substituents along the ring cause a reduction in symmetry. For example, a CTV with an alternating sequence of substituents is C_3-symmetric and therefore chiral (Figure 4.32). Racemization involves the inversion of the three aromatic subunits, a process that has an activation barrier of 110.8 kJ mol^{-1} in the case of a chiral analog of **4.29** [53]. Thus, the racemization half-life is approximately 70 days at room temperature, which is long enough for isolation, structural characterization, and further derivatization of chiral CTVs.

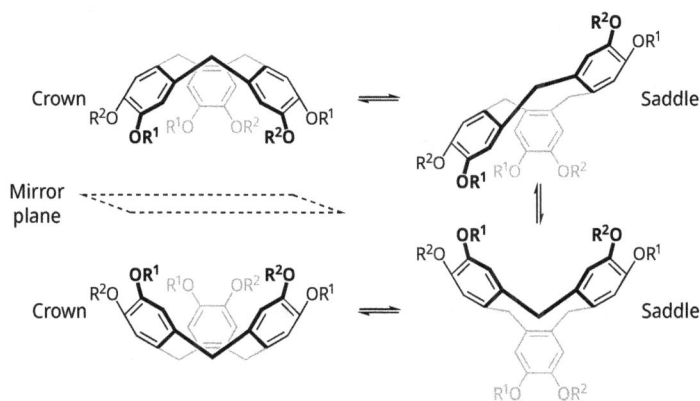

Figure 4.32: Proposed mechanism of the racemization of the two enantiomeric crown conformations of a chiral CTV *via* the respective saddle conformations.

The inversion of the crown conformations proceeds *via* saddle conformations in which one aromatic subunit is inverted with respect to the other two (Figure 4.32). The intermediate saddle conformer can be isolated by quenching a hot solution of **4.29** [54]. It represents a metastable state that slowly isomerizes in solution to the thermodynamically more stable crown. The activation barrier of this process is similar to that of the CTV racemization.

Synthesis: There are two general methods for the preparation of CTVs. The first involves the treatment of veratrole or a veratrole derivative with formaldehyde. Alternatively, benzyl alcohols can be cyclotrimerized. Both reactions are mediated by acids, such as perchloric acid, trifluoroacetic acid, sulfuric acid, or BF_3 etherate. If the starting material contains two identical alkoxy groups, both methods will give the same product, but the result will be different if the alkoxy groups are not the same. Figure 4.33a illustrates that treating a 1,2-alkoxybenzene derivative with formaldehyde will give a mixture of a C_1- and a C_3-symmetric product in a ratio of 3:1 because the initial reaction is not regioselective. In contrast, a benzyl alcohol with two substituents will afford only the C_3-symmetric product (Figure 4.33b).

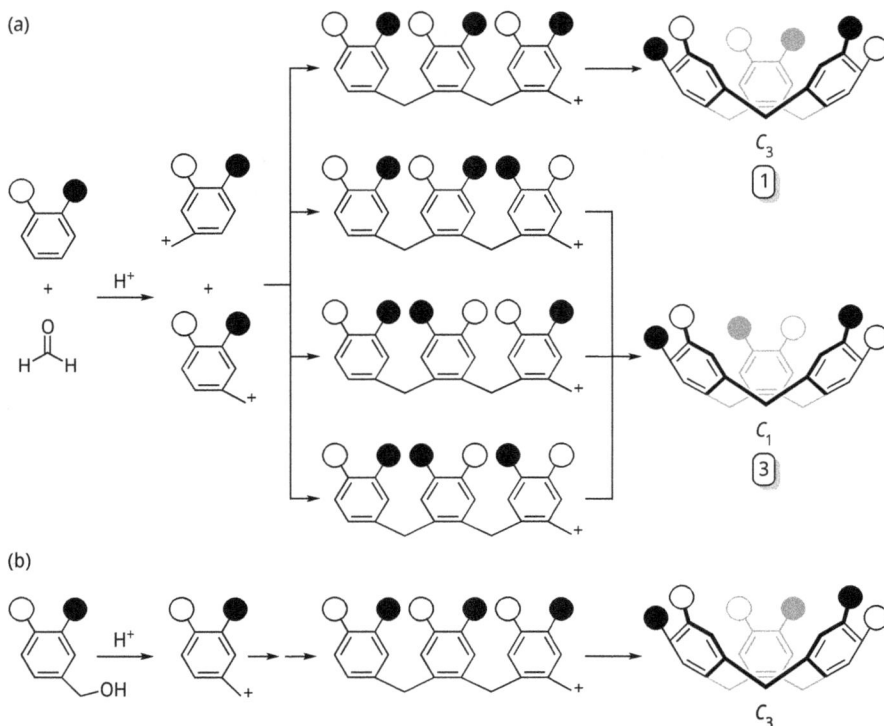

Figure 4.33: Synthesis of a CTV with different substituents in the aromatic subunits, shown as filled and unfilled circles, by reaction of the respective veratrole derivative with formaldehyde in the presence of an acid (a), or by acid-mediated cyclotrimerization of a 3,4-disubstituted benzyl alcohol (b). The first reaction gives a mixture of a C_1- and a C_3-symmetric product in a 3:1 ratio, whereas the second reaction gives only the C_3-symmetric CTV. The schemes show the formation of only one enantiomer of the products.

Once the CTV is obtained, it can be further modified. For this purpose, suitably protected CTV derivatives are useful. After deprotection, substituents can be introduced along the ring that contribute to substrate binding. Alternatively, the unprotected functional groups can be used to covalently link two CTV rings. In the latter case, so-called cryptophanes are obtained, whose synthesis and properties are discussed below. C_3-Symmetric CTVs can also serve to prepare enantiopure CTV derivatives.

Binding properties: CTVs themselves do not exhibit pronounced host–guest chemistry because the substrates included into the shallow cavity remain partially in contact with the surrounding solvent. The complexes are therefore not very stable unless strengthened by contributions from substituents along the ring. CTVs can interact with suitable guest molecules in the solid state and bind fullerenes in solution, but their receptor properties are otherwise unremarkable. More important as receptors are derivatives of CTVs, the so-called cryptophanes and hemicryptophanes, which are discussed below.

Derivatives: We will first consider cryptophanes, which were developed with the aim of constructing three-dimensional cavities in which substrates can be encapsulated and shielded from all sides. As shown in Figure 4.34a, these compounds consist of two CTV subunits covalently linked by three spacers. They therefore have a *roof* and a *floor* connected by several *bars*, making them members of a class of receptors commonly referred to as molecular cages.

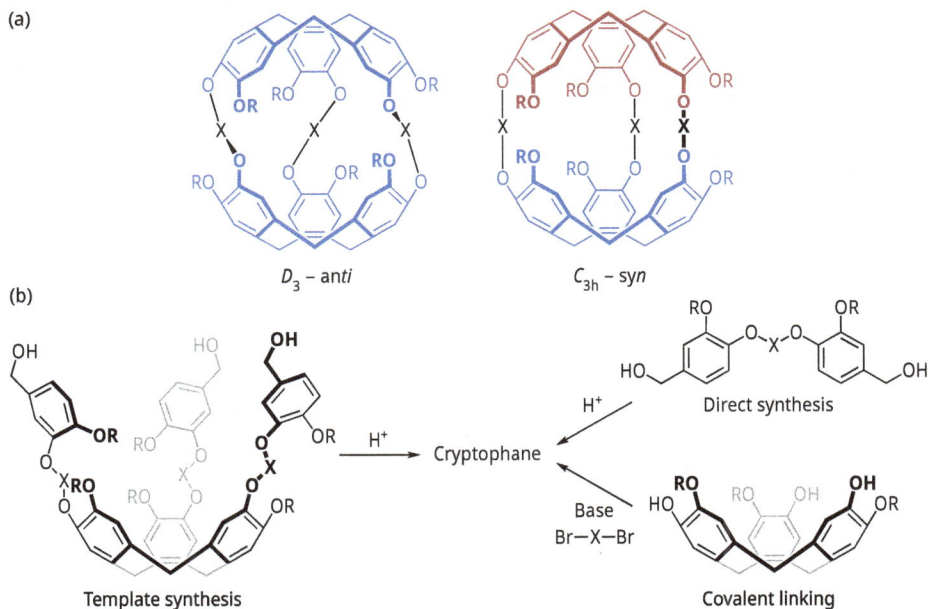

Figure 4.34: General structures of a D_3-symmetric chiral *anti*-cryptophane and the corresponding C_{3h}-symmetric achiral *syn*-isomer (a), and synthetic strategies for the preparation of cryptophanes (b).

Since cryptophanes contain the spacers in every other position of each CTV subunit, both subunits are chiral. Accordingly, the overall symmetry of a cryptophane depends on whether the two subunits have the same or a different absolute configuration (Figure 4.34a). In the first case, the two CTV rings in the cryptophane are staggered, giving rise to a tilted arrangement of the spacers with respect to the main axis. This so-called *anti*-isomer is D_3-symmetric and chiral. In a cryptophane derived from the two mirror images of a chiral CTV, the subunits are eclipsed, resulting in the C_{3h}-symmetric, achiral *syn*-isomer.

Cryptophanes can be synthesized in three different ways (Figure 4.34b). In the template synthesis, three veratrole-derived substituents are first attached to a preformed CTV, followed by intramolecular coupling to give the second CTV ring. The first ring thus directs the subsequent cyclization, with the directionality of the product formed in the first intramolecular coupling step determining whether an *anti*-cryptophane or a *syn*-cryptophane is formed. This approach is useful for the synthesis of cryptophanes that differ in the structure of the two CTV subunits. Due to the preorganization of the three substituents on a common scaffold, it is also often more efficient than the direct synthesis, which involves the simultaneous formation of both CTV rings by cyclotrimerization of a bis(veratrole) derivative. The advantage of the latter strategy is that it involves fewer steps than the template synthesis and requires no high dilution conditions. Both approaches usually differ in the ratio in which the *syn*- and *anti*-isomers are formed. The template synthesis gives cryptophane-E **4.30** (Figure 4.35a) with three propylene linkers in a *syn/anti* ratio of about 2:1, for example, whereas the direct synthesis leads almost exclusively to the *anti*-isomer. The final synthetic approach, the intermolecular linking of two CTV rings, is useful to prepare cryptophanes that are not accessible by the other methods.

Cryptophanes bind hydrocarbons, halogenated hydrocarbons, quaternary ammonium ions, and noble gases, especially xenon. These guests are surrounded in the cavity by the π-systems of two CTV subunits and by the alkyl groups of the linkers. As a result, the complexes are stabilized mainly by dispersion interactions, supplemented by cation–π interactions in the case of cation complexes. Cryptophanes are therefore useful tools for assessing the strength of these types of interactions. However, the outcome of such studies is strongly influenced by the solvent. In chloroform, for example, the binding of dichloromethane to **4.30** is qualitatively detectable but too weak to be quantified because the solvent and substrate molecules are similar in size and can both be bound. In 1,1,2,2-tetrachloroethane, where competition is less pronounced because the large solvent molecules are difficult to bind, the chloroform and dichloromethane complexes of **4.30** have binding constants of 100 M^{-1} and 470 M^{-1}, respectively. Both complexes are therefore quite stable, considering that they are mainly stabilized by dispersion interactions [55]. These stability constants still do not reflect the actual binding strength, since 1,1,2,2-tetrachloroethane is still not completely excluded from the cavity [56]. Therefore, even higher stability constants are obtained when further increasing the size of the solvent molecules. In 1,2-dichlorobenzene, for example, the chloroform complex of **4.30** has an impressive K_a of 26,000 M^{-1}. This extraordinary affinity could be due in part to the release of strain as the cavity is filled. However, calculations indicate that the contribution of dispersion interactions to the stability of the complex is substantial. We will see in other chapters (e.g., Section 4.1.9) that the use of solvents that are too large to enter a receptor cavity is generally useful to promote complex formation when electronic effects are insufficient to mediate the preferential binding of the substrate.

(a)

(b)

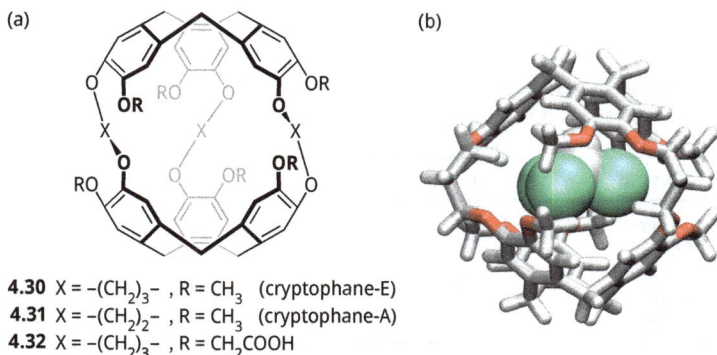

4.30 X = –(CH$_2$)$_3$– , R = CH$_3$ (cryptophane-E)
4.31 X = –(CH$_2$)$_2$– , R = CH$_3$ (cryptophane-A)
4.32 X = –(CH$_2$)$_3$– , R = CH$_2$COOH

Figure 4.35: Molecular structures of cryptophane-E **4.30**, cryptophane-A **4.31**, and the water-soluble analog of **4.30**, cryptophane **4.32** (a), and crystal structure of the chloroform complex of **4.30** (b).

Cryptophanes also exhibit a pronounced size dependence of complex stability. For example, cryptophane-A **4.31** (Figure 4.35a), in which ethylene residues bridge the CTV rings, has the highest affinity for dichloromethane, whereas the larger cryptophane-E **4.30** prefers chloroform, with the crystal structure in Figure 4.35b illustrating the structural complementarity between the receptor cavity and a chloroform molecule [57]. This seemingly simple correlation between cavity size and substrate affinity has both enthalpic and entropic reasons. Table 4.7 shows that the binding enthalpies are negative for all complexes and consistently more favorable for chloroform than for dichloromethane [58]. Thus, regardless of receptor size, direct receptor–substrate interactions favor the inclusion of the larger substrate, presumably because it interacts more efficiently with the cavity walls. The enthalpic stabilization is strongest for **4.31** and the enthalpy therefore favors binding of the larger guest to the smaller cryptophane. The fact that **4.31** still has higher overall affinity for dichloromethane must therefore be due to entropy. Indeed, the binding entropies are unfavorable for chloroform and favorable for dichloromethane, indicating that complexation of the larger substrate leads to a pronounced loss of conformational flexibility. The most unfavorable binding entropy is observed for the complex between chloroform and the **4.31** because this combination suffers most from the tight fit. The entropic disadvantage is so pronounced that it causes the chloroform complex of **4.31** to be less stable than the dichloromethane complex, despite the more favorable binding enthalpy. In the case of **4.30**, the entropic disadvantage of chloroform binding is smaller and does not reduce the corresponding binding enthalpy as much, causing the larger cryptophane to prefer chloroform to dichloromethane. The thermodynamic parameters observed for these complexes thus illustrate the interplay between binding strength (enthalpically favorable) and conformational constraints (entropically unfavorable) that underlies the phenomenon of enthalpy–entropy compensation (Section 2.1). They show that simple arguments to rationalize binding strength and selectivity based solely on structural complementarity can be misleading.

Table 4.7: Thermodynamic parameters associated with the dichloromethane and chloroform complexation of cryptophanes **4.30** and **4.31** in $C_2D_2Cl_4$ at 298 K (ΔH^0 and $T\Delta S^0$ in kJ mol^{-1}).

Cryptophane	Substrate	log K_a	ΔH^0	$T\Delta S^0$
4.30	CH_2Cl_2	2.0	−4.2	7.5
	$CHCl_3$	2.9	−25.1	−8.7
4.31	CH_2Cl_2	2.7	−13.8	1.2
	$CHCl_3$	2.5	−34.3	−19.9

The enantioselectivity of chiral cryptophanes has been studied using bromochlorofluoromethane CHFClBr as guest, one of the simplest chiral molecules with a single stereogenic center [59]. The ^{1}H NMR spectrum of a mixture of the two enantiomers of CHFClBr and an enantiomerically pure D_3-symmetric *anti*-cryptophane contains two sets of signals, demonstrating the formation of diastereomeric complexes. The CHFClBr signal appears as two doublets (the multiplicity is due to the H,F coupling) whose integrals provide information about the ratio of the two enantiomers.

Receptor **4.30** also binds quaternary ammonium ions, which can interact with the π-systems lining the inner surface through cation–π interactions. As a result, cations are typically bound more efficiently than neutral substrates. The cation affinity of cryptophanes is maintained in water, although complex stability is generally lower due to cation hydration. For example, while the tetramethylammonium complex of **4.30** has a log K_a of 4.4 of in 1,1,2,2-tetrachloroethane, the water-soluble analog of **4.30**, cryptophane **4.32** (Figure 4.35a), binds the same cation with a log K_a of 2.6 in water [60]. Smaller and larger ammonium ions are less efficiently bound. Increasing the linker length generally leads to an increase in complex stability, probably because the greater mobility of the guests within the larger cryptophane cavity reduces the entropic disadvantage of complex formation, as also observed for neutral substrates.

The log K_a of the complex between Xe and **4.31** is 3.6 in $C_2D_2Cl_4$ at 278 K [61], which is remarkably high for a neutral guest bound mainly by dispersion interactions. In water, the xenon affinity of water-soluble cryptophanes is even higher [62]. These complexes have the potential to be used as biosensors in magnetic resonance imaging, since the nucleus of the ^{129}Xe isotope is NMR-active with intermediate sensitivity and a spin of 0.5 (Section 12.2.4).

The other CTV-derived receptors that will be discussed are the hemicryptophanes. As the name suggests, these receptors contain only one CTV moiety rather than two, just as only half of the hemispherand **4.17** is derived from a spherand. The CTV ring is covalently linked *via* three spacers to another C_3-symmetric subunit that contributes to substrate binding. The presence of the two different recognition sites makes these receptors heteroditopic. Like cryptophanes, they are molecular cages.

The structural variability of the receptor subunits attached to the CTV ring is very large. Possible structural elements include crown ethers, calix[6]arenes, α-CDs, cyclopeptides, triamines, triamides, and many others [63]. These subunits either complement the binding properties of the CTV ring or allow hemicryptophanes to bind to two different structural elements in the substrate. The following examples illustrate this concept.

The hemicryptophane **4.33** (Figure 4.36) binds alkylammonium ions as picrate salts in CDCl$_3$/CD$_3$OD, 95:5 (*v/v*) [64]. Complex formation induces the shielding of the guest protons, indicating that their alkyl group is incorporated into the bowl of the CTV ring where it interacts with the aromatic rings through CH⋯π interactions. The ammonium group resides on the opposite side of the cavity where it forms hydrogen bonds to the amino groups. The stability constants log K_a range between 4 and 5.4, depending on the steric bulk of the alkyl group in the substrate. This receptor also binds the neurotransmitter dopamine under the same conditions.

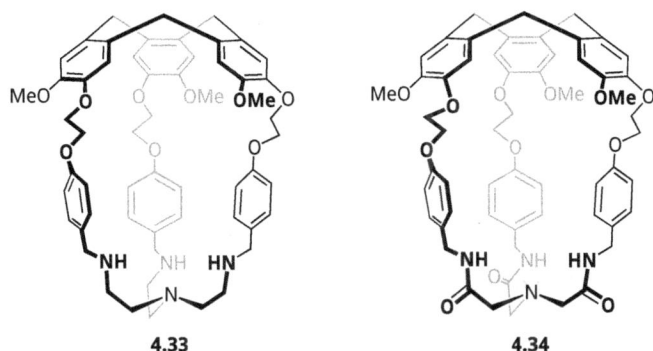

Figure 4.36: Molecular structures of hemicryptophanes **4.33** and **4.34**.

The triamide analog of **4.33**, hemicryptophane **4.34** (Figure 4.36), has a binding site for cations in the form of the CTV ring and for anions in the form of the three cyclically arranged amide NH groups. Accordingly, this receptor binds ion pairs or zwitterions (Section 4.2.3). The affinity of **4.34** for individual cations or anions can be estimated by using salts in the binding studies whose respective counterions are too large to fit into the receptor cavity. Specifically, tetrabutylammonium salts allow the assessment of anion affinity and picrate salts allow the assessment of cation affinity. According to these studies, the affinity of **4.34** for individual ions is moderate. For example, **4.34** binds chloride with a log K_a of 2.0 and the NMe$_4^+$ cation with a log K_a of 2.6 in chloroform [65]. However, the NMe$_4$Cl complex has a log K_a of 3.2, showing that simultaneous binding of both components of the ion pair leads to an increase in complex stability. The shielding of the cation protons, induced by the aromatic residues of the CTV, and the deshielding of the NH protons, induced by hydrogen bonding, confirm the expected arrangement of the ions within the receptor cavity.

The same hemicryptophane also interacts with zwitterions that differ in the distance and the nature of the two oppositely charged groups [66]. Taurine (2-aminoethane-1-sulfonic acid) forms the most stable complex ($\log K_a = 4.1$ in CD_3CN/D_2O, 9:1 (v/v)) because it contains the two oppositely charged head groups at a distance that is particularly well-matched to the distance of the binding sites in **4.34**.

4.1.7 Calixarenes

Introduction: We have seen in Section 4.1.6 that the acid-catalyzed reaction between veratrole and formaldehyde gives rise to cyclophanes with methylene groups as the bridging units. The reaction proceeds through a sequence of electrophilic aromatic substitutions, with the resulting substitution pattern of the aromatic subunits determined by the orienting effects of the veratrole methoxy groups. A number of other cyclophanes are accessible by similar approaches, including the calixarenes.

The chemistry on which the synthesis of calixarenes is based goes back to the work of Adolf von Baeyer, who described the reaction between phenol and formaldehyde in 1872. This reaction is mediated by acids or bases and produces a black tar whose structure von Baeyer was unable to determine. We now know that the product is a highly cross-linked polymer in which phenol rings are linked by methylene groups through their *ortho* and *para* positions (Figure 4.37a). In 1908, Leo Baekeland patented a process for converting the tar described by von Baeyer into a hard and brittle material. Baekeland then commercialized this phenol–formaldehyde resin under the name *Bakelite*, thus creating the first plastics industry.

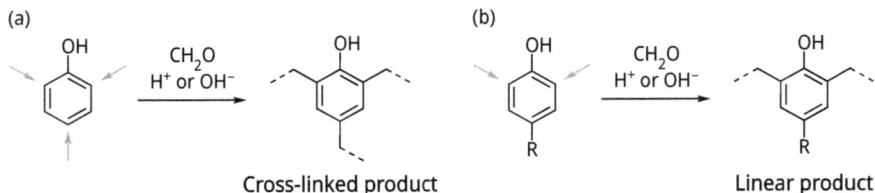

Figure 4.37: Reactions of phenol (a) and a 4-substituted phenol with formaldehyde (b).

The question arose as to whether a phenol derivative with only two reactive positions would yield linear rather than cross-linked polymers (Figure 4.37b). In this context, Alois Zinke studied the reaction of 4-*tert*-butylphenol and formaldehyde in the presence of sodium hydroxide at high temperatures. He obtained a crystalline product which he postulated in 1944 to be the cyclic tetramer **4.35** (Figure 4.38a) [67]. Although there is no doubt that Zinke was the first to hold a calixarene in his hands, it is possible that his sample might not have been pure. Indeed, when John Cornforth reproduced Zinke's procedure in 1952, he obtained two products, which he initially assumed to be two conformational isomers of **4.35** [68], but which later turned out to be the cyclic tetramer and octamer. Today, we know that the reaction between 4-*tert*-butylphenol and formaldehyde leads to a whole series of macrocyclic products, the most important of which are those containing four, six, and eight aromatic subunits. The influence of product distribution on the reaction conditions was systematically studied by C. David Gutsche, who published the first reliable procedures for the synthesis of calixarenes in 1978 [69]. Gutsche recognized the potential of calixarenes as receptors and continued to make important contributions. He also introduced the term *calixarene*, which combines the Greek word *calix* (κάλυξ) for chalice, symbolizing the shape of the calix[4]arene cone conformation (Figure 4.38b,c), with the word *arene*, denoting the presence of the aromatic subunits along the ring. The improved synthetic accessibility made calixarenes versatile study objects, and they soon became one of the most important receptor families in supramolecular chemistry. According to the cyclophane nomenclature, calixarenes are [1*n*]metacyclophanes, with **4.35**, for example, representing a [1.1.1.1]metacyclophane.

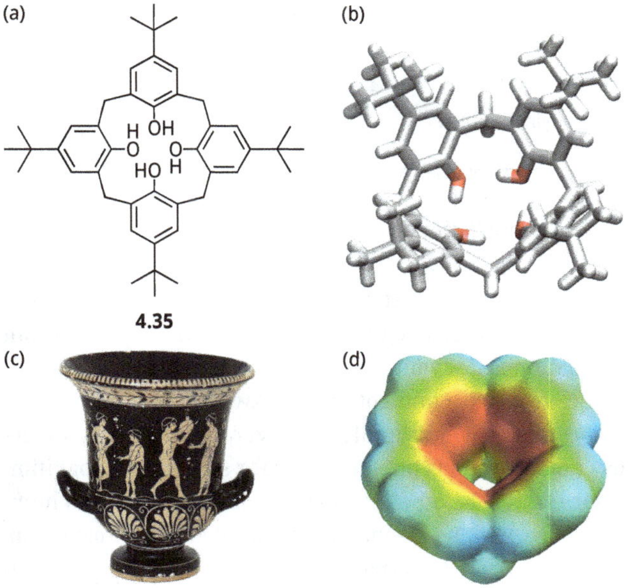

Figure 4.38: Molecular structure of the 4-*tert*-butylcalix[4]arene **4.35** (a), crystal structure of **4.35** in the cone conformation (b), image of a Greek vase to illustrate the origin of the name calixarene (c) [70], and electrostatic potential surface of a derivative of **4.35** lacking the *tert*-butyl groups (d). The color coding covers a potential range from −100 to +100 kJ mol^{-1}, with red and blue indicating values greater than or equal to the absolute maximum in negative and positive potentials, respectively.

The binding properties of calixarenes are determined in part by the size and shape of the cavity and the degree of preorganization. These parameters vary with the number of subunits and are best understood for the cyclic tetramers. In this case, four conformations can be distinguished, which differ in the relative orientation of the subunits. These conformations are shown in Figure 4.39.

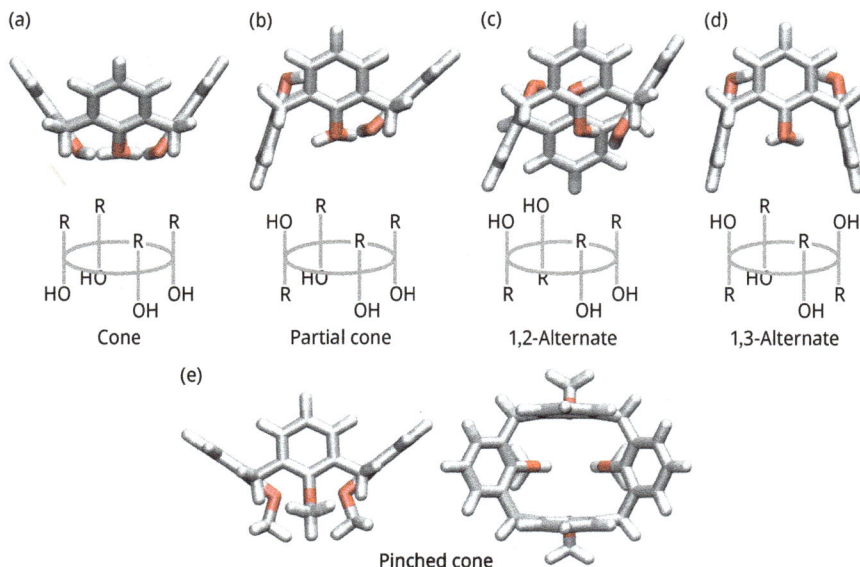

Figure 4.39: Calculated structures and schematic representations of the cone (a), partial cone (b), 1,2-alternate (c), and 1,3-alternate (d) conformations of a four-membered calix[4]arene, and side and top views of the pinched cone conformation of a calix[4]arene with methylated OH groups (e).

The cone conformation (Figure 4.39a) is the most stable conformation. It has all the hydroxy groups oriented along the narrower cavity opening where they form a seam of hydrogen bonds. This conformation is C_4-symmetric because of the directionality of the OH bonds. Assuming that the clockwise and counterclockwise orientations have equal probability and that they interconvert rapidly, the average symmetry of the cone conformation is C_{4v}. This conformation is often the best for substrate recognition because it has a deep and defined cavity. The other conformations have a reduced number of hydrogen bonds (Figure 4.39b–d). In the C_s-symmetric partial cone conformation, one aromatic ring is inverted, whereas the 1,2-alternate (C_{2v}) and 1,3-alternate (D_{2d}) conformations contain two adjacent or opposite inverted rings. Some calix[4]arene derivatives, such as those with alkylated OH groups, prefer the C_{2v}-symmetric pinched cone conformation, in which two opposite aromatic rings are nearly parallel to each other while the other two diverge. An example is shown in Figure 4.39e.

In solution, calix[4]arenes with free OH groups exist only in the cone conformation. They are still dynamic, undergoing a ring inversion similar to that of CTVs (Section 4.1.6, Figure 4.40a). This inversion causes the pseudoequatorial and pseudoaxial protons in the CH$_2$ groups to exchange position, and since the signal pattern produced by these protons in the ^{1}H NMR spectrum depends on the rate at which the exchange takes place, dynamic NMR spectroscopy can be used to characterize the conformational equilibrium. Below room temperature, the CH$_2$ protons give rise to an AB system of two doublets in the 100 MHz ^{1}H NMR spectrum of **4.35** in CDCl$_3$, showing that the cone-to-cone interconversion is slow on the NMR timescale [71]. Increasing the temperature causes the signals to broaden until they first coalesce at 52 °C and then sharpen again to give a broad singlet (Figure 4.40b). A Gibbs free energy of activation of 65.7 kJ mol^{-1} can thus be derived for the ring inversion in chloroform, which means that it occurs approximately 185 times per second at 52 °. The rate is lower at room temperature.

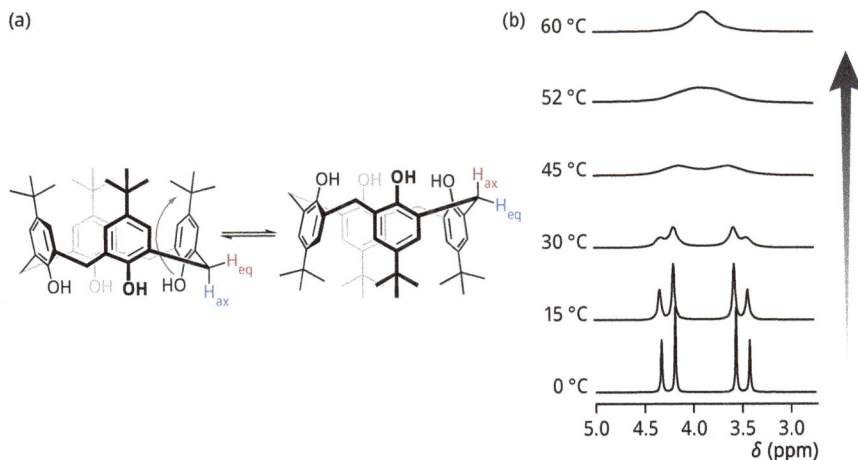

Figure 4.40: Schematic representation of the ring inversion of calix[4]arene **4.35**, which causes the interconversion of the pseudoequatorial and pseudoaxial protons of the methylene units (a), and sections of ^{1}H NMR spectra illustrating how the shape of the CH$_2$ signals changes with the temperature (b). Equivalent protons are shown in the same color and the arrow in the structure indicates the movement of one aromatic ring during the cone-to-cone interconversion.

It is not known whether the ring inversion proceeds in a concerted manner or stepwise *via* intermediate conformations. It is known, however, that the conformational interconversion of calix[4]arenes involves the passage of the hydroxy groups through the ring because the larger parts of the aromatic subunits cannot fit through the cavity opening. As a consequence, the rate of the cone-to-cone equilibrium is almost unaffected by substituents in the 4-position of the aromatic subunits, but strongly affected by substituents at the OH groups. These substituents cause the loss of the intramolecular hydrogen bonds and thus facilitate the conformational exchange if they are small enough to pass through the ring. This is the case for methyl groups, and the tetra-*O*-methylated analog of **4.35** indeed equilibrates much faster than **4.35** itself ($\Delta G^{\ddagger}$ = 20 kJ mol^{-1}). More bulky ethyl groups reduce the rate of the conformational interconversion because they cannot easily squeeze through the cavity opening, and once a calix[4]arene contains propyl or even larger alkyl groups attached to the phenolic oxygen atoms, conformational interconversion no longer takes place. Thus, conformations other than the cone conformation can be stabilized by introducing sufficiently large substituents along the narrower opening of the cavity. The solvent polarity also has a strong effect on the equilibrium, with polar solvents destabilizing the hydrogen bonds between the OH groups and thus facilitating the cone-to-cone interconversion.

From the pentamer on, calixarenes become large enough to allow both ends of the aromatic subunits to pass through the ring. In addition, the conformational equilibria become more complex because larger calixarenes not only adopt more than the four conformations shown in Figure 4.39, but these conformations also interconvert more readily. However, the relationship between interconversion rate and ring size is not straightforward. The eight-membered calixarene, for example, is almost as rigid as the tetramer because it is also stabilized by intramolecular hydrogen bonds. The corresponding conformation is highly symmetric, with all eight hydroxy groups involved in an almost planar cyclic array of hydrogen bonds [72]. This structure is known as *pleated loop* conformation. Other calixarenes with ring sizes between five and seven aromatic units can adopt (distorted) cone conformations that allow their use as receptors but also less symmetric conformations. Thus, receptors derived from these calixarenes often contain linkers or other structural elements to stabilize the conformation optimal for substrate recognition.

Nomenclature: The most important criterion for differentiating calixarenes is the ring size, which is specified by the number of aromatic subunits in square brackets between the two name components *calix* and *arene*. Thus, the tetramer **4.35** is a calix[4]arene, a pentamer would be a calix[5]arene, and so on. This name refers strictly to calixarenes without substituents. The names and positions of substituents are added as prefixes to the family name using the numbering schemes shown in Figure 4.41. Thus, the systematic name of **4.35** is 5,11,17,23-tetra-*tert*-butylcalix[4]arene-25,26,27,28-tetrol.

Calixarenes with four identical substituents in the *para* positions of the hydroxy groups are sometimes named in a simplified way by just stating the name of the substituents. Compound **4.35** could therefore be called *tert*-butylcalix[4]arene, which implies that the OH groups are unsubstituted.

Several descriptors are used to specify certain regions of the calixarene molecules or the relative positions of substituents. Specifically, the OH groups of calixarenes are arranged along the *narrower* rim of the calixarene cavity, while substituents in the *para* positions of the OH groups can be found along the *wider* rim, with the terms narrower and wider referring to the shape of the cone conformation. The terms *upper* and *lower*

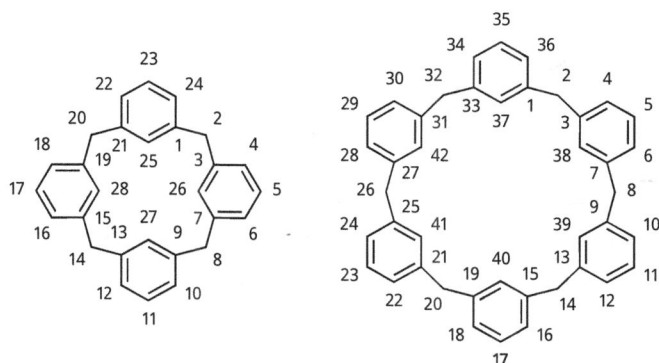

Figure 4.41: Numbering systems used to denote the positions of the substituents in calix[4]arene and calix[6]arene. Similar schemes are used for other calixarenes.

rim are used as synonyms for the wider and narrower rim, respectively. The individual aromatic subunits along the calixarene ring are numbered consecutively so that adjacent rings are designated 1,2, rings separated by one ring are designated 1,3, and so on. Adjacent aromatic rings are also referred to as *vicinal* or *proximal*, while the rings in the 1,3 positions of a calix[4]arene or in the 1,4 positions of a calix[6]arene are referred to as *distal* or *diametral*.

Synthesis: Calixarenes are synthesized by treating 4-*tert*-butylphenol with formaldehyde in alkaline media at high temperatures. Since the role of the substituent at the 4-position of the phenol component is to prevent cross-linking, the reaction works in principle with any *para*-substituted phenol. Indeed, a variety of substituted phenols can be used as starting materials, but 4-*tert*-butylphenol has the advantage that optimized methods are available for the selective preparation of calixarenes of different sizes and that the *tert*-butyl groups can be easily removed, allowing further functionalization.

The mechanism of calixarene formation is shown in Figure 4.42. It is best understood by assuming that a phenol represents the enol form of the corresponding ketone. The initial steps of the reaction can then be viewed as an aldol condensation, initially leading to an α,β-unsaturated ketone. The chain elongation represents a Michael addition between this ketone and the formal enolate of another aromatic building block. This reaction sequence is repeated until the oligomers formed are long enough to cyclize.

Optimized procedures are available for the selective preparation of the various calixarenes. The thermodynamically favored calix[4]arenes are prepared using 3 mol% of NaOH as the base in refluxing diphenyl ether (~260 °C). In this way, **4.35** is obtained in yields between 50% and 60%. The octamer is the kinetic product and is therefore formed at lower temperatures (refluxing xylene, ~ 140 °C), again with NaOH as base. This octamer can be converted to the tetramer by treatment in basic media at high temperatures,

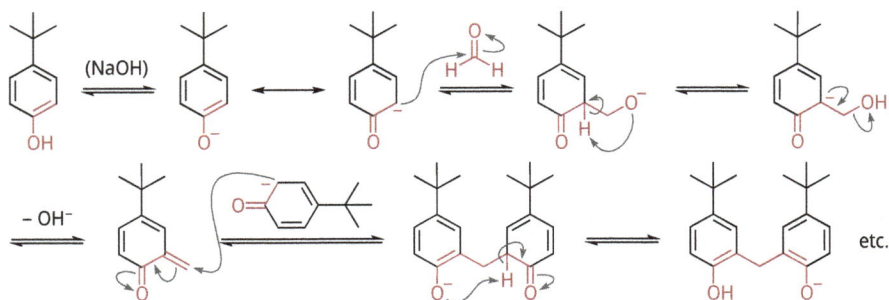

Figure 4.42: Mechanism of the base-catalyzed reaction between 4-*tert*-butylphenol and formaldehyde that underlies the reaction sequence leading to a calixarene. The substructures illustrating the relationship of this mechanism to an aldol condensation and a Michael addition are shown in red.

showing that all steps of calixarene formation are reversible and under thermodynamic control. When the reaction is carried out in xylene with KOH as base, the main product is the cyclic hexamer, probably due to a template effect of the potassium ions. Calixarenes with an odd number of aromatic subunits are prepared in a similar way, although the yields are usually lower. All these syntheses can be carried out on a large scale. In addition, most calixarenes are commercially available, making them attractive building blocks for receptor development.

Binding properties: Calixarenes combine two structural elements that act as binding sites, the oxygen atoms arranged along the narrower cavity opening and the aromatic moieties lining the cavity. Similar binding motifs are found in crown ethers (Section 4.1.1), where the oxygen atoms induce an affinity for electropositive metal ions, and in cyclophanes (Section 4.1.5), where the electron-rich aromatic systems serve as binding sites for softer cationic substrates. The combination of both structural motifs in calixarenes is expected to induce affinity for both types of substrates, which is indeed observed. Information on how the respective binding modes differ can be derived from the effect of complex formation on the conformational equilibrium of flexible calix[4]arenes, such as the tetramethylated derivative **4.36a** (Figure 4.43).

In chloroform, **4.36a** exists as a mixture of all four isomers with the partial cone conformation predominating (cone: 2%, partial cone: 92%, 1,2-alternate: 5%, 1,3-alternate: 1%) [73]. The high proportion of the partial cone conformation under these conditions is attributed to solvent effects: the large dipole moment of the cone conformation in which all aromatic subunits are equally aligned is unfavorable in nonpolar media, causing the less polar partial cone conformation to dominate. This assumption is consistent with the observed increase in the proportion of the cone conformation with increasing solvent polarity.

The presence of sodium perchlorate causes a shift in the conformational equilibrium in favor of the cone conformation to maximize the number of oxygen atoms coordinating to the cation. The extent of this stabilizing effect is greater in less polar

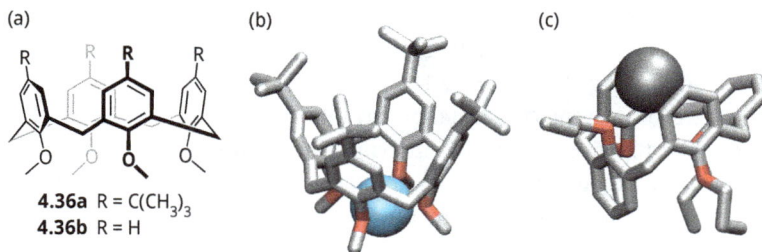

Figure 4.43: Molecular structure of the calix[4]arenes **4.36a** and **4.36b** (a), crystal structure of the Na$^+$ complex of **4.36a** (b), and crystal structure of the Ag$^+$ complex of the tetrapropylated analog of **4.36b** (c). Protons and counterions are not shown for reasons of clarity.

solvents where Na$^+$ binding is stronger. The crystal structure in Figure 4.43b illustrates the stabilization of the cone conformation by the Na$^+$ ion.

The same conformation is also stabilized when calix[4]arenes interact with quaternary ammonium ions. However, such bulky cations do not bind to **4.36a** due to the steric hindrance of the *tert*-butyl groups. Binding studies therefore require the use of **4.36b**, which lacks the alkyl groups. This calix[4]arene exists in CDCl$_3$/CD$_3$CN, 10:1 (*v/v*) at −50 °C as a mixture of cone (31%) and partial cone (69%). The addition of one equivalent of *N*-methyl-pyridinium iodide increases the proportion of the cone conformation to 67%, almost reversing the ratio observed in the absence of the substrate. That substrate binding in this case takes place within the cavity lined by the electron-rich aromatic subunits (Figure 4.38d) is reflected in the pronounced upfield shifts of the guest signals in the ^{1}H NMR spectrum upon complex formation. Thus, both small electropositive and soft charge-dispersed cations bind to calix[4]arenes, but at different positions.

When a metal cation is too large to bind simultaneously to all four oxygen atoms, the partial cone conformation becomes the preferred one. The binding of soft Ag$^+$ ions, for example, involves cation–π interactions combined with ion–dipole interactions, the latter involving the oxygen atom of the inverted aromatic residue. The crystal structure of the Ag$^+$ complex of the tetra-*O*-propylated analog of **4.36b** illustrates this arrangement (Figure 4.43c).

Derivatives: Additional substituents along the narrower or the wider rims of calixarenes can induce affinity for substrates not bound by unsubstituted calixarenes, such as anions. In addition, the substituents can mediate properties beyond molecular recognition, such as self-assembly or catalytic activity, with the calixarene ring serving only as a scaffold. Examples of such systems will be discussed in later chapters. Here, we will focus on the synthetic approaches available to structurally modify calixarenes and look at a selection of calixarenes to see how these modifications affect binding properties.

An important strategy to modify the hydroxy groups of calixarenes along their narrower rim involves treating them with an alkylating agent in the presence of a base (Williamson ether synthesis). Only ethoxy and methoxy groups are small enough to pass through the cavity opening (although ethyl groups already significantly slow down the conformational interconversion), while larger groups completely restrict the mobility of the ring. Sufficiently large substituents thus allow the stabilization of the cone conformation or other calix[4]arene conformations by taking advantage of the fact that the alkylation proceeds stepwise *via* flexible intermediates. Using substoichiometric amounts of the alkylating reagent (less than the minimum amount required to convert all OH groups) results in incompletely substituted products whose substitution patterns can sometimes be controlled.

The introduction of alkyl groups serves primarily to control the conformational mobility of the calixarene ring, while functionalized substituents may participate in substrate recognition. The calix[4]arene derivatives shown in Figure 4.44a and b illustrate this concept. In **4.37**, the four carbonyl oxygen atoms of the acetamide groups contribute to the binding of electropositive metal ions, as illustrated by the crystal structure of the corresponding potassium complex (Figure 4.44a) [74]. Cation affinity is similar to that of cryptands (Section 4.1.2), despite the flexibility of the attached residues. Furthermore, the selectivity of **4.37** is impressive. With a $\log K_a$ of 9.3, Na^+ is bound about 70 times more strongly than K^+. Increasing the ring size reduces the affinity for small alkali metal ions, while at the same time causing a pronounced increase in the affinity for alkaline earth metal ions.

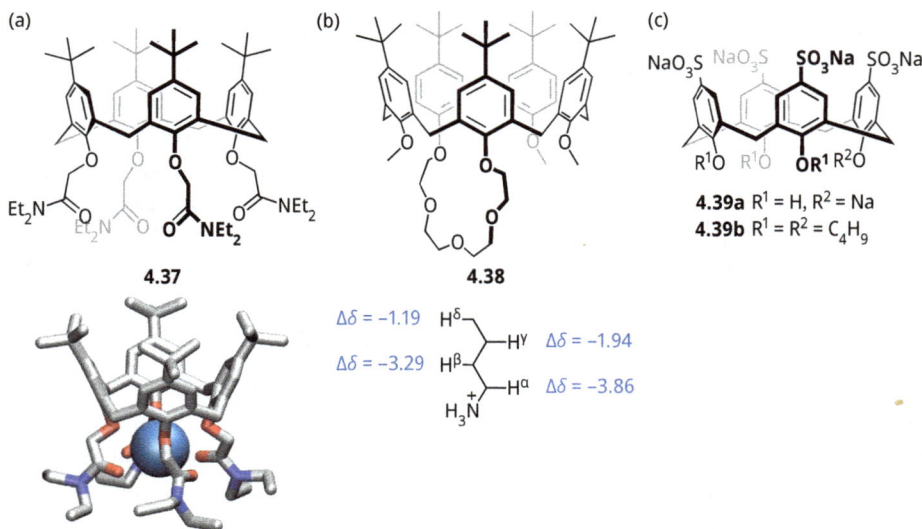

Figure 4.44: Molecular structure of the calix[4]arene **4.37** together with the crystal structure of its K^+ complex (a), of the crown ether-containing calix[5]arene **4.38** together with the ^1H NMR chemical shift changes ($\Delta\delta$ in ppm) induced when **4.38** binds to *n*-butylammonium picrate in $CDCl_3/CD_3OD$, 9:1 (*v/v*) (b), and of the sulfonatocalix[4]arenes **4.39a** and **4.39b** (c). Protons in the crystal structure and counterions are not shown for reasons of clarity.

The calix[5]arene derivative **4.38** contains a crown ether moiety linking the 1,3 rings, which stabilizes the cone conformation. This receptor recognizes protonated n-alkyl amines in CDCl$_3$/CD$_3$OD, 9:1 (v/v) by threading their linear alkyl chains through the ring to allow interactions between the cationic head groups of the guest and the oxygen atoms at the narrower cavity opening [75]. Pronounced upfield shifts of the guest signals in the ^{1}H NMR spectrum confirm this binding mode. The shifts are largest for the protons of the methylene unit immediately adjacent to the nitrogen atom and decrease with increasing distance from the cationic head group (Figure 4.44b), illustrating that the protons that penetrate deepest into the calix[5]arene cavity are also those that experience the greatest shielding. Substrate binding to the crown ether from the outside is prevented by steric effects of the methoxy groups at the narrow cavity rim. The recognition of linear ammonium ions is highly selective, as branched substrates can neither enter the calix[5]arene ring nor approach it from the outside. Accordingly, the corresponding receptor with free OH groups, in which these steric effects are eliminated, binds to both types of substrates with low selectivity.

Substituents can also be introduced along the wider rim of the calixarene cavity. It is quite fortunate in this context that the most readily available 4-tert-butylcalixarenes are also the most convenient precursors for such structural variations. This is because the tert-butyl groups are relatively easy to remove by retro-Friedel–Crafts reactions. Accordingly, treating tert-butylcalixarenes in toluene in the presence of aluminum trichloride gives the corresponding unsubstituted products, which then allow a number of further structural modifications (Figure 4.45). For example, calixarenes can be nitrated, the nitro groups reduced, and the resulting amino groups alkylated or acylated to add further substituents. Calixarenes can also be formylated or halogenated at the 4-position of the aromatic units. The halogenated calixarenes can be used as starting materials for various cross-coupling reactions, or can be converted into organometallic compounds, which can then be reacted with suitable electrophiles. Some of these reactions require the protection of the calixarene OH groups. In addition to fully substituted calixarenes, derivatives containing one or more substituents in a specific relative arrangement are also accessible by appropriate methods. Calixarenes can thus be structurally varied over a wide range, which explains their importance in supramolecular chemistry and the prominence of calixarenes in later chapters of this book.

Important water-soluble derivatives are sulfonatocalixarenes (Figure 4.44c), which are readily accessible by treating unsubstituted calixarenes with sulfuric acid. Sulfonatocalixarenes are nontoxic and capable of binding various biomolecules, including peptides and proteins, making them interesting candidates for biomedical applications (Section 12.2.1). They generally have a high affinity for ammonium ions, especially quaternary ammonium ions, whose binding involves ionic interactions and benefits from the release of cavity water. It is important to note that one phenolic OH group of **4.39a** is deprotonated even at neutral pH (the deprotonation of the first OH group is associated with a pK_a of 3.3, whereas the pK_a values of the remaining OH groups are approximately 11) and that the negative charge along the narrower cavity opening contributes

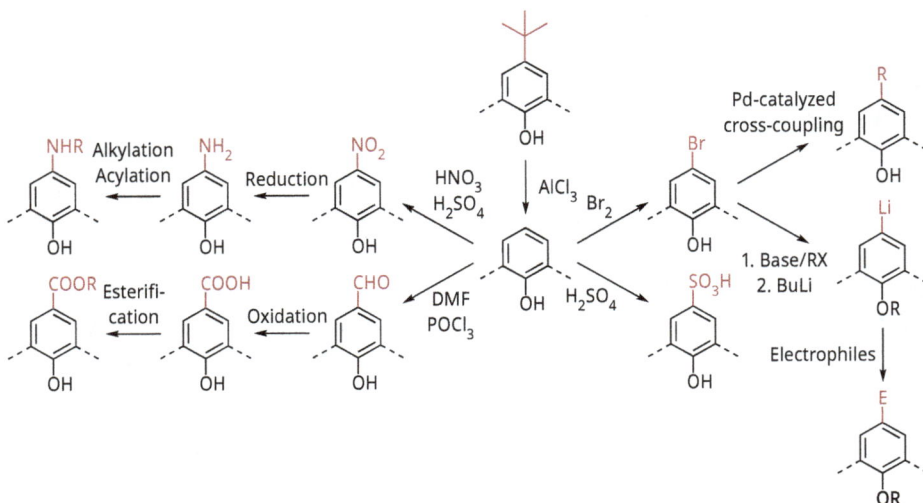

Figure 4.45: Examples of synthetic methods available to vary the substituents in calixarenes along the wider rim.

to cation binding. In addition, cation affinity is influenced by structural effects: the tetramethylammonium complex of **4.39a** has a $\log K_a$ of 5.6 in water, for example, whereas the tetra-*O*-butylated analog **4.39b** binds to the same cation with an approximately three orders of magnitude lower affinity [76]. This effect is due to the lower negative charge of **4.39b** and the preferred pinched cone conformation of alkylated calixarenes, which is not well suited to accommodate spherical substrates.

Further reading

In addition to calixarenes with methylene groups between the aromatic subunits, there are a number of related cyclophanes in which heteroatoms bridge the aromatic subunits. Because of their close structural relationship to conventional calixarenes, these compounds are called oxacalixarenes, azacalixarenes, and thiacalixarenes, depending on whether they contain oxygen, nitrogen, or sulfur as linking units, respectively. The properties of these cyclophanes are partly similar to those of calixarenes, but there are also differences. For example, three-membered oxacalixarenes and azacalixarenes are known, which do not exist for calixarenes. In the case of azacalixarenes, the conformational space is restricted to the 1,3-alternate conformation, and the heteroatoms in all these systems also mediate characteristic reactivities. Despite their promising properties, information on the host–guest chemistry of these compounds is relatively limited. For further information, see:

Morohashi N, Narumi F, Iki N, Hattori T, Miyano S. Thiacalixarenes. Chem. Rev. 2006, 106, 5291–316. Maes W, Dehaen W. Oxacalix[n](het)arenes. Chem. Soc. Rev. 2008, 37, 2393–402. Tsue H, Ishibashi K, Tamura R. Azacalixarene: a new class in the calixarene family. Top. Heterocycl. Chem. 2008, 17, 73–96.

4.1.8 Calixpyrroles

Introduction: Another electron-rich aromatic compound that readily reacts with carbonyl derivatives to give cyclophane-type receptors is pyrrole. The treatment of pyrrole with acetone affords the cyclic tetrapyrrole **4.40** (Figure 4.46a), which is called calix[4]pyrrole because of its structural relationship to calix[4]arenes [77]. This reaction was first reported by Adolf von Baeyer in 1886, fourteen years after he described the condensation of formaldehyde with phenol [78]. The correct structure of the product was proposed in 1916, after which calixpyrroles appeared regularly in the literature. However, it was not until the advent of supramolecular chemistry that they found wider applications. Important contributions in this context came from the group of Jonathan L. Sessler, who discovered that calixpyrroles are potent anion receptors.

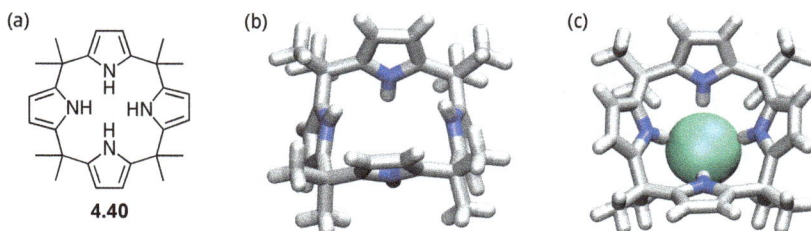

Figure 4.46: Molecular structure of calix[4]pyrrole **4.40** (a), crystal structure of this compound illustrating the preferred 1,3-alternate conformation (b), and crystal structure of the chloride complex of **4.40** in which the calix[4]pyrrole ring adopts the cone conformation (c).

Calix[4]pyrrole **4.40** is a flexible nonplanar molecule that can adopt conformations similar to those of calix[4]arenes, namely the cone, the partial cone, the 1,2-alternate, and the 1,3-alternate conformation. Of these four conformations, only the 1,3-alternate conformation, in which the NH groups are arranged alternately on opposite sides of the ring, is populated in the absence of suitable guests due to repulsive interactions between the positively polarized NH hydrogen atoms (Figure 4.46b). Recall that calix[4]arenes prefer to adopt the cone conformation because the interactions between the OH groups along the narrower cavity opening are attractive rather than repulsive.

Another difference between calix[4]pyrroles and calix[4]arenes is that calix[4]pyrroles prefer to bind anions rather than cations because the NH groups in the pyrrole moieties act as hydrogen bond donors. Typical substrates are halides, phenolates or the negatively polarized oxygen atoms of *N*-oxides. Complex formation proceeds by an induced fit in which the 1,3-alternate conformation is converted to the cone conformation to allow all four NH groups to participate in hydrogen bonding. The resulting arrangement is illustrated by the crystal structure of the chloride complex of **4.40** in Figure 4.46c. In this complex, the bowl-shaped cavity of the calixpyrrole ring provides a second binding site, in this case for cations, which can interact by cation–π interactions with the electron-rich aromatic subunits and by ionic interactions with the simultaneously bound anion. Typical cationic substrates are cesium or quaternary ammonium ions with at least one methyl group. The halide salts of these cations form complexes with calixpyrroles in which the entire ion pair is bound (Section 4.2.3) [79, 80].

Synthesis: Calixpyrroles are obtained by condensation of pyrrole with carbonyl compounds under the influence of a Brønsted or Lewis acid. The reaction involves repeated electrophilic aromatic substitutions, initially yielding oligomers with sp^3-hybridized carbon atoms between the aromatic moieties. Suitable linear intermediates then undergo cyclization, with both aldehydes and ketones preferentially leading to the thermodynamically most stable tetramers (Figure 4.47). In the case of aldehydes, the corresponding product is susceptible to oxidation and reacts further to give a conjugated planar porphyrin. Ketones, on the other hand, yield stable calixpyrroles that cannot be oxidized because their linking units are fully substituted. This synthesis is simple and high-yielding (up to 90%), involving only the heating of pyrrole and a ketone in the presence of an acid. Larger calixpyrroles are also known, but their synthesis requires special routes [81]. These compounds are less important than the tetramers and will not be discussed further.

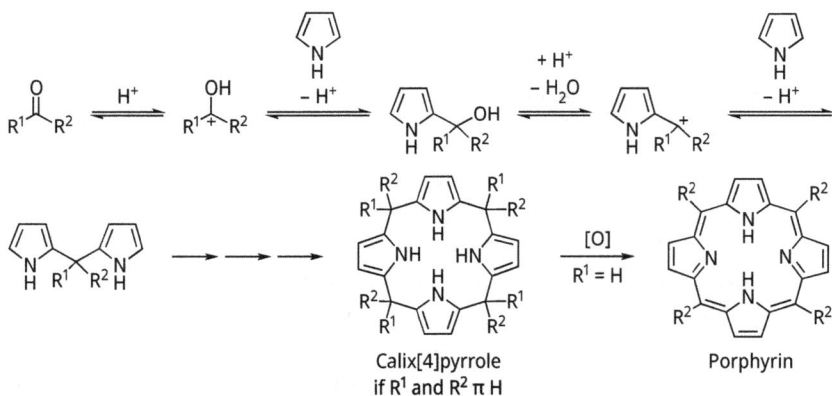

Figure 4.47: Mechanism of the acid-catalyzed reaction between pyrrole and a carbonyl compound, yielding a macrocyclic tetramer as the major product. When aldehydes are used as the carbonyl component, this product is unstable and oxidizes to give a porphyrin. The macrocyclic tetramer derived from ketones cannot oxidize and represents a stable calix[4]pyrrole.

Binding properties: According to early binding studies, small halides interact significantly more strongly with **4.40** in dichloromethane than oxoanions such as sulfate or hydrogen phosphate [82]. In the series of halides, complex stability increases progressively from the large iodide to the small fluoride, with a pronounced jump in affinity from chloride ($\log K_a = 2.5$) to fluoride ($\log K_a = 4.2$). This apparently straightforward trend was attributed to the good size fit and high charge density of the fluoride anion, making it the best guest. Subsequent work then showed that solvent and counterion effects can cause the stability of the chloride complex of **4.40** to approach or even exceed that of the fluoride complex under certain conditions. These results led to the conclusion that "it is recommended that anion binding studies involving new receptors be carried out in several different solvents and with several different countercations before a detailed understanding of the anion binding properties or receptor-based selectivities is claimed" [83]. The difficulty in correlating the anion affinity of calixpyrroles with the properties of the anion alone is partly due to the ability of calixpyrroles to simultaneously bind both components of an ion pair, with the individual equilibria involved in anion and cation binding not being independent. In addition to inorganic anions, calixpyrroles also bind organic hydrogen bond acceptors, the most important of which are phenolates and *N*-oxides.

Derivatives: Structural variations in the calixpyrrole core or the attachment of substituents allow fine tuning of the anion binding properties. For example, an analog of **4.40** with 3,4-difluoropyrrole units has a significantly higher anion affinity than the nonfluorinated parent compound because the electron-withdrawing nature of the fluorine atoms enhances the hydrogen bond donor properties of the pyrrole NH groups. Monosubstituted derivatives of calixpyrrole can also be prepared using appropriate protocols, allowing the synthesis of receptors with two or more covalently linked calixpyrrole rings or the attachment of further substituents [81].

Two special classes of calixpyrrole derivatives are worth mentioning. The first was introduced by Chang-Hee Lee and is characterized by a bridge linking two carbon atoms in the calixpyrrole ring [84]; compound **4.41** (Figure 4.48a) is an example. These so-called strapped calix[4]pyrroles are obtained by reacting acetone with an acyclic precursor containing two dipyrromethane units covalently linked by a spacer, which becomes the strap in the final product. This strap does not improve the preorganization – strapped calixpyrroles also require the presence of anions to adopt the cone conformation – but it can serve as an additional binding site. In **4.41**, for example, the CH group in the aromatic linker facing the binding site interacts with the substrate through an additional hydrogen bond. As a result, the chloride affinity in DMSO increases from a $\log K_a$ of 3.1 for **4.40** to a $\log K_a$ of 5.0 for **4.41** [85].

Other important calixpyrrole derivatives are the α,α,α,α-stereoisomers of calix[4]pyrroles, introduced by Pablo Ballester, which are synthesized using acetophenone derivatives as the carbonyl component. These calixpyrroles contain a methyl and an aryl group on each linking carbon atom, with the cone conformation of the α,α,α,α-stereoisomer having all the aryl groups in the axial positions and all the methyl groups in the equatorial positions, as shown for compound **4.42** in Figure 4.48b [86]. As a result, the binding site is surrounded by a wall of π-systems whose electronic effects influence complex stability. In the case of **4.42**, the interactions between the faces of the phenyl rings and an anion bound to the NH groups are repulsive, resulting in a reduction of the anion affinity compared to **4.40**. This destabilizing effect is less pronounced when the aryl groups con-

(a) (b)

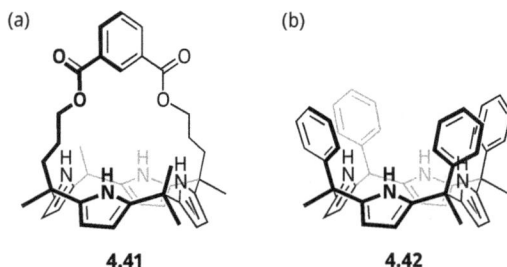

4.41   **4.42**

Figure 4.48: Molecular structures of the strapped calix[4]pyrrole **4.41** (a), and of the calix[4]pyrrole **4.42** with four aryl groups arranged around the binding site (b).

tain electron-withdrawing substituents in the 4-position, and the analog of **4.42** with four 4-nitrophenyl residues has an even higher chloride affinity than **4.40**. These receptors thus provide information on the contributions of anion–π interactions to anion binding. Based on the same scaffold, molecular capsules with inwardly projecting binding sites or cavitands that bind small neutral organic molecules in water are also accessible [87].

4.1.9 Resorcinarenes

Introduction: We have seen that veratrole, 4-substituted phenols, and pyrrole can be used as bifunctional monomers together with formaldehyde to prepare cyclophane-type receptors. Another suitable aromatic building block for this reaction is 1,3-dihydroxybenzene (resorcinol), where the 4- and 6-positions are available for electrophilic aromatic substitution due to the *ortho*- and *para*-directing effects of the two OH groups. The corresponding macrocyclic products have a 1,3-substitution pattern in the aromatic moieties similar to that of calixarenes (Figure 4.49a).

(a) (b)

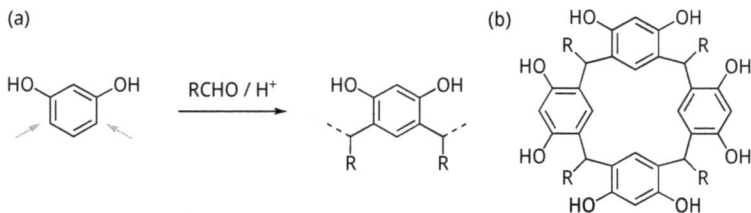

Figure 4.49: Reaction of resorcinol with an aldehyde (a) and structure of the resulting cyclic tetramer (b).

Under the normally used acidic conditions, formaldehyde itself is too reactive and gives only complex reaction mixtures, but the condensation between resorcinol and aldehydes works well. It was postulated in 1940 on the basis of molecular weight determinations and confirmed by X-ray crystallography in 1968 that cyclic tetramers are the major products [88]. The structures of these so-called resorcinarenes were studied in detail by A. G. Sverker Högberg [89], and their applications in supramolecular chemistry started to develop in the 1980s, initially mainly in the Cram group [90].

The general structure in Figure 4.49b shows that, like calix[4]arenes, resorcin[4]arene are [1.1.1.1]metacyclophanes. Analogies to calix[4]arenes are therefore to be expected, but there are also differences. One important difference is that resorcinarenes with more than four aromatic subunits are rare. Larger rings are formed as intermediates in the synthesis, but are ultimately converted to the thermodynamically favored cyclic tetramer. The group of Agnieszka Szumna showed that resorcin[5]arenes (and pyrogallol[5]arenes), with R = H, can be obtained in useful amounts from formaldehyde and resorcinol (or pyrogallol) under basic conditions [91], but the properties of these compounds are only beginning to be explored [92].

Another difference between calix[4]arenes and resorcin[4]arenes is that resorcin[4]arenes contain substituents in the linking units, resulting from the use of aldehydes as starting materials. These substituents give rise to four diastereomers (Figure 4.50), all of which are *meso*-forms and therefore achiral.

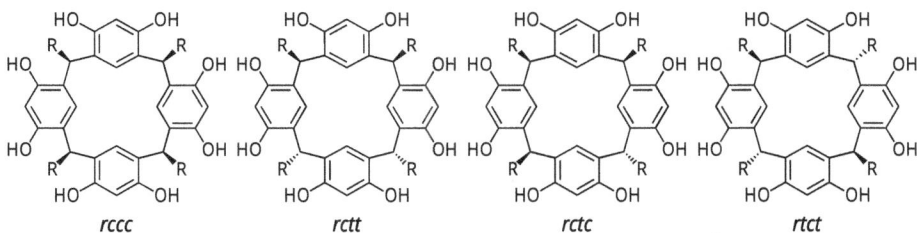

Figure 4.50: Possible stereoisomers of a resorcin[4]arene differing in the relative orientation of the substituents R and the corresponding stereochemical descriptors.

In general, only two of these diastereomers are formed in the synthesis, namely those with the *rccc* or the *rctt* configuration (*r* for *reference*, *c* for *cis*, and *t* for *trans*). Under certain reaction conditions, the *rctc*-isomer is also produced in small amounts [93]. The reason for the observed selectivity is the reversibility of product formation, which causes the thermodynamically favored stereoisomer(s) to dominate in the final mixture. In addition, certain products have a tendency to precipitate, preventing their further participation in the reaction. The exact product ratio therefore depends on the reaction conditions and the nature of R.

The relative configurations of the residues R along the ring determine the preferred orientation of the aromatic subunits (Figure 4.51). In the *rccc*-isomer, all aromatic rings are tilted in the same direction as in the cone conformation of calix[4]arenes. In the terminology of resorcinarene chemistry, this diastereomer is called the crown. The R groups can in principle occupy either the axial or the equatorial position, but in reality, there is not enough space between the aromatic subunits for the residues to be arranged equatorially. Therefore, only the conformation with axial residues exists and ring inversion is not possible.

Figure 4.51a shows that neighboring OH groups in the crown interact by hydrogen bonding, a feature that contributes to the thermodynamic stability of the *rccc*-isomer. This on average C_{4v}-symmetric conformation is often observed in the solid state. In solution, the *rccc*-isomer of resorcin[4]arenes also exists in boat conformations, especially when the OH groups are acetylated, which resemble the pinched cone conformation of calix[4]arenes (Figure 4.51b). Formally, two equivalent boat conformations are possible, the interconversion of which is associated with a Gibbs free energy of activation $\Delta G^{\ddagger}$ of 79.5 kJ mol^{-1} for the resorcin[4]arene with R = phenyl [89]. Thus, the boat-to-boat interconversion proceeds (at least in this resorcin[4]arene) at a slightly slower rate than the cone-to-cone interconversion of a calix[4]arene. The orientation of the substituents does not change during this process – all substituents remain in the axial orientations –but the aromatic residues change their relative tilt angles in a pseudorotation-like process, possibly *via* the symmetrical crown as the transition state.

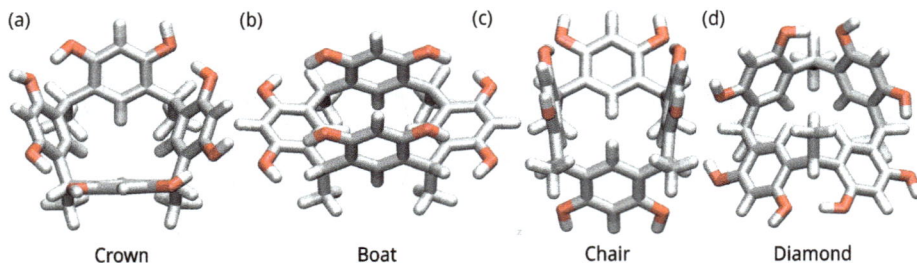

(a) Crown   (b) Boat   (c) Chair   (d) Diamond

Figure 4.51: Calculated structures of the crown (*rccc*-isomer, cone) (a), boat (*rccc*-isomer, pinched cone) (b), chair (*rctt*-isomer, partial cone) (c), and diamond (*rctc*-isomer, 1,2-alternate) (d) configurational isomers of a resorcin[4]arene with methyl groups as substituents R.

The resorcin[4]arene stereoisomer with the *rctt* configuration resembles the partial cone conformation of calix[4]arenes. It also contains all four substituents R in the axial positions, but with two pairs of neighboring residues on opposite sides of the ring (Figure 4.51c). The preferred conformation of the *rctc*-isomer is called the diamond. It corresponds to the 1,2-alternate conformation of calix[4]arenes (Figure 4.51d). The 1,3-alternate conformation is likely to have its counterpart in the *rtct*-isomer, but this isomer is unknown. As the crown, chair, and diamond structures are configurational isomers, they can only be converted into each other by breaking covalent bonds.

Nomenclature: The name resorcin[4]arene was introduced in 1994 by Hans-Jörg Schneider [94]. It is an abbreviation of the longer and more descriptive name calix[4]resorcinarene, which indicates that resorcin[4]arenes, like calix[4]arenes, have a chalice-like structure in the crown (cone) conformation with four aromatic subunits. Since there are (almost) no resorcinarenes other than the four-membered ones, the ring size is often not specified. However, the more precise name resorcin[4]arene is preferred, also because resorcin[5]arenes have recently become available [91]. To specify the nature of the substituents in the aliphatic bridging units, the prefix *C-(substituent name)* is added to the family name. Accordingly, the resorcin[4]arenes in Figure 4.51 are *C*-methylresorcin[4] arenes or *C*-methylcalix[4]resorcinarenes. Analogous cyclophanes derived from pyrogallol are called pyrogallol[4]arenes. They contain additional OH groups at the 2-positions of the aromatic subunits, which allow further functionalization, as we will see below.

Synthesis: Resorcin[4]arenes are prepared by treating an aliphatic or aromatic aldehyde with resorcinol under acidic conditions. The cyclic tetramer often crystallizes from the reaction mixture and can be easily isolated. Depending on the structure of the aldehyde, the ratio in which different product stereoisomers are formed varies. As a rule of thumb, aliphatic aldehydes typically lead to the preferential or even exclusive formation of the *rccc*-isomer, whereas aromatic aldehydes often lead to mixtures of the *rccc*- and *rctt*-isomers [95]. Mechanistically, the reaction proceeds through a series of Friedel–Crafts alkylations, initially leading to oligomers that cyclize once the appropriate chain length is reached (Figure 4.52).

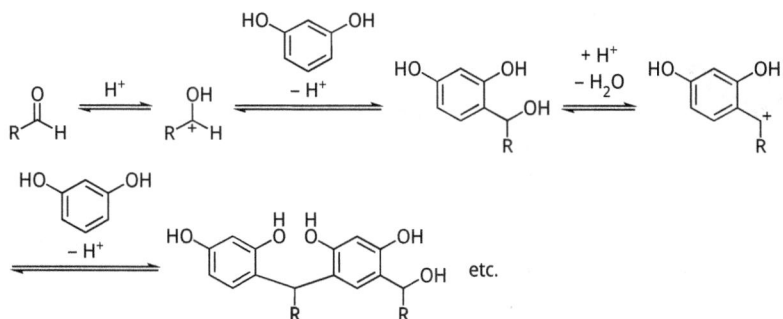

Figure 4.52: Mechanism of the acid-catalyzed reaction between resorcinol and an aldehyde that eventually leads to a resorcin[4]arene *via* corresponding linear oligomers.

The same reaction can also be carried out with 1,2,3-trihydroxybenzene (pyrogallol), 2,6-dihydroxypyridines, or related compounds as the aromatic component.

Binding properties: The *rccc*-isomer of a resorcin[4]arene has a bowl-shaped cavity lined by four electron-rich aromatic subunits. Resorcin[4]arenes are therefore able to bind cations, such as quaternary ammonium ions, by cation–π interactions. However, the interactions are often not very efficient because the resorcin[4]arene cavity is relatively shallow, leading to complexes with only moderate stability. Cation binding can be strengthened by deprotonation of the OH groups, allowing ion pairing to contribute to binding [94], but complex formation in this case requires relatively basic media.

The group of Yasohiro Aoyama investigated whether the OH groups along the resorcin[4]arene cavity can contribute to substrate binding. They found that resorcin[4]arenes efficiently solubilize polar organic molecules such as dicarboxylic acids, alcohols, carbohydrates, and also larger biomolecules (steroids, vitamins, or coenzymes). Complex formation takes place in nonpolar solvents, such as chloroform or tetrachloromethane, in which the substrates are normally insoluble. The fact that glutaric acid binds strongly to resorcin[4]arene **4.43** in CDCl$_3$, while dimethyl glutarate fails to form a complex, suggests that hydrogen bonds between the terminal carboxyl groups of glutaric acid and pairs of OH groups on opposite sides of the cavity are crucial for complex formation (Figure 4.53) [96]. Glutaric acid appears to be ideal for spanning the cavity, as dicarboxylic acids with shorter or longer distances between the head groups are less strongly bound. This mode of binding is further supported by the upfield shifts of the guest signals in the ^{1}H NMR spectrum induced by complex formation. Most complexes have a 1:1 resorcin[4]arene–substrate ratio, but a 2:1 is ratio is also possible. Finally, separate signals of free and bound guests can sometimes be observed in the NMR spectra, indicating that the corresponding complexes are in slow exchange.

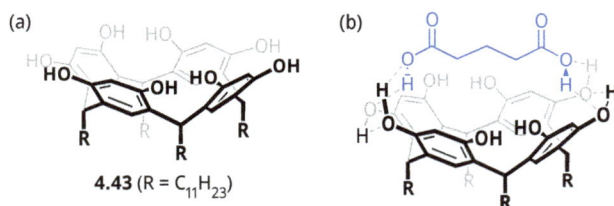

Figure 4.53: Molecular structure of resorcin[4]arene **4.43** (a) and binding mode proposed by Aoyama for the complex between **4.43** and glutaric acid (b).

Work by several groups later revealed that resorcin[4]arenes tend to form hollow aggregates in which six resorcin[4]arene moieties surround a large cavity. Julius Rebek Jr. subsequently showed that many of the complexes studied by Aoyama were also formed from such self-organizing capsules, and we will therefore return to these systems in Section 5.3.3 on self-assembled receptors. In the following, we will focus on special classes of resorcin[4]arene derivatives, such as cavitands, carcerands, and hemicarcerands, all of which were originally introduced by Cram.

Derivatives: One approach to improve the receptor properties of resorcin[4]arenes is to increase their rigidity and the size of the internal surface area that makes contact with the substrate. This can be achieved by introducing suitable spacers between the hydroxy groups of neighboring aromatic subunits. Cram introduced the term *cavitand* for such receptors, which have "enforced cavities to accommodate simple molecules and ions" [97]. While this definition applies to many types of receptors, some of which have already been introduced (e.g., CTVs in Section 4.1.6 or calixarenes in Section 4.1.7), we will restrict the use of the term cavitand to receptors derived from resorcinol.

The resorcin[4]arene **4.44** (Figure 4.54) is one of the first cavitands developed by Cram. Its methylene groups between the oxygen atoms, resulting from the treatment of C-methylresorcin[4]arene with bromochloromethane (CH_2BrCl) in the presence of a base [97], prevent **4.44** from adopting the boat conformation, thus stabilizing the C_{4v}-symmetric crown with a concave cavity well suited for substrate binding. Analogous cavitands with ethylene or propylene linkers have also been described, which differ from **4.44** in terms of flexibility and average tilt angle of the aromatic subunits. All of these compounds have a limited substrate scope and interact relatively weakly with small organic molecules such as CD_3CN. For example, the CD_3CN complex of an analog of **4.44** has a log K_a of only 2.5 in CCl_4 [98]. Aliphatic bridges between the OH groups are therefore not sufficient to significantly improve the receptor properties of resorcin[4]arenes.

(a)

(b)

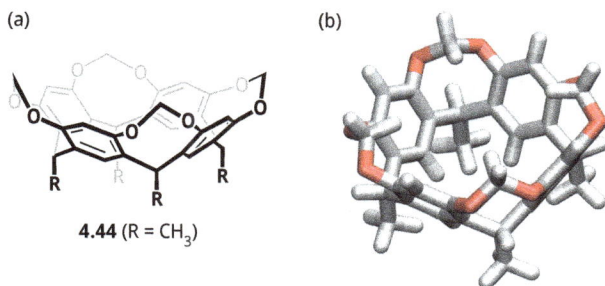

4.44 (R = CH_3)

Figure 4.54: Molecular structure and calculated preferred conformation of cavitand **4.44**.

The reason for the moderate substrate affinity of **4.44** is that its concave binding site is too shallow to make efficient contact with the substrate and shield it from the solvent. Resorcin[4]arenes with aromatic moieties as sidewalls, such as the deep cavitands **4.45** and **4.46** in Figure 4.55a, should not have this disadvantage. They are obtained from C-methylresorcin[4]arene by nucleophilic aromatic substitution using 2,3-dichloro-1,4-diazanaphthalene or 1,2-difluoro-4,5-dinitrobenzene as the reaction partners.

However, deep cavitands are generally not well preorganized for complex formation because they oscillate between a C_{4v}-symmetric and a C_{2v}-symmetric conformation (Figure 4.55b). In the first conformation, called vase, the aromatic sidewalls are parallel to the main axis, forming a cavity, which is approximately 7 Å wide and 8 Å deep in the case of **4.45**. This cavity has a large opening at one end, while the opening at the other end, where the resorcin[4]arene ring is located, is so narrow that only thin aromatic compounds such as alkynes can pass through. In the other so-called kite conformation, all the sidewalls are folded away, with the resorcin[4]arene ring adopting the boat conformation.

(a)

4.45 (R = CH₃)   **4.46** (R = CH₃)

(b)

Vase   Kite

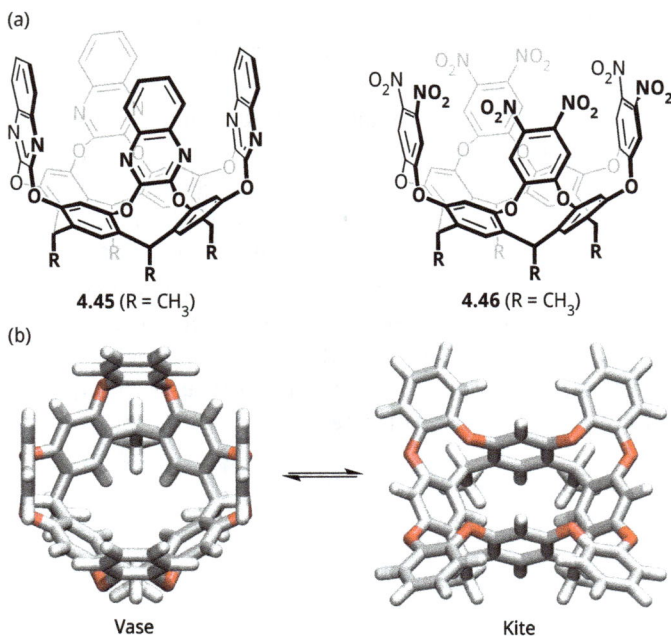

Figure 4.55: Molecular structures of the deep cavitands **4.45** and **4.46** (a), and calculated structures of the C_{4v}-symmetric vase and the C_{2v}-symmetric kite conformation of a derivative of **4.46** lacking the nitro groups (b).

According to dynamic NMR spectroscopy, the kite conformation is preferred at lower temperatures and almost completely dominates the conformational equilibrium at −62 °C in CDCl₃/CS₂, 1:1 (v/v) [99]. With increasing temperature, the fraction of the vase conformation increases and reaches 100% at temperatures above 45 °C. This behavior can be explained by the interplay between thermodynamic stability and solvation. The kite conformation is preferred at lower temperatures because it is thermodynamically more stable than the vase conformation. At higher temperatures the conformational equilibrium is more strongly controlled by entropic factors according to the Gibbs–Helmholtz equation (equation (2.9)), and since the transition from the kite conformation, with its large solvent-exposed surface, to the more compact vase is associated with an entropically favorable release of solvent molecules, the latter conformation dominates at higher temperatures.

Cavitands in the kite conformation are shape complementary and therefore tend to dimerize in solution, reducing the solvent-exposed surface area. This dimerization stabilizes the kite conformation to such an extent that the formation of the vase conformation can only be induced by the addition of suitable guests. If methyl groups or other small substituents are present at the 2-position of the resorcinol subunits, the vase conformation is completely prevented. Such cavitands form stable dimers in solution, especially in polar solvents such as methanol, where solvent effects induce a further stabilization [100]. Cram proposed the name velcrands for these compounds and velcraplexes for their complexes to describe their tendency to stick together like velcro.

As expected, deep cavitands in the vase conformation are able to bind larger guests than cavitand **4.44**, including substituted benzene derivatives. The binding equilibria are fast on the NMR timescale and the stability of the complexes is moderate due to the lack of preorganization. For example, the complex between **4.45** and N,N-dimethyl-4-nitroaniline in acetone-d_6 has a $\log K_a$ of 2.3 [101]. The larger upfield shift of the signal belonging to the protons *ortho* to the dimethylamino group with respect to the signal of the *meta* protons indicates that the substrate prefers to be arranged with the dimethylamino group at the bottom of the cavity, close to the resorcin[4]arene ring. This orientation is induced by the antiparallel arrangement of the dipole moment of the substrate and the dipole moments of the aromatic sidewalls surrounding the cavity. We have seen a similar influence of electronic factors on the structure of CD complexes (Section 4.1.4).

Work in the Rebek group has shown that the vase conformation of deep cavitands can be stabilized by a seam of intramolecular hydrogen bonds resulting from acetamide groups along the cavity opening [102]. Accordingly, the self-folding cavitand **4.47** (Figure 4.56a) is much better preorganized for substrate binding than **4.46**. The conformational stabilization is more pronounced in less competitive solvents such as CDCl$_3$, benzene, or xylene than in DMSO, consistent with the effect of solvent polarity on the strength of hydrogen bonding (Section 3.1.5).

Cavitand **4.47** binds cyclohexane and adamantane derivatives as well as macrocyclic lactams such as ε-caprolactam. Binding is strongest in xylene-d_{10}, where hydrogen bonding of the acetamide groups is most efficient. However, even under these conditions the stability of the complex is not very high, ranging between $\log K_a$ values of 1.6 and 2.9. The reason for this is that the exchange of a cavity-bound solvent molecule for the substrate is associated with only a small thermodynamic gain, since both guests are bound only by relatively weak dispersion and dipole–dipole interactions. The binding properties of **4.47** are therefore not remarkable from a thermodynamic point of view, and this receptor would not have been mentioned here were it not for the unique effect of the amide groups on the kinetics of complex formation.

In contrast to conformationally flexible deep cavitands, whose guest exchange is fast, separate signals of the free binding partners and the complex are visible in the ^{1}H NMR spectra of mixtures of **4.47** and suitable substrates in xylene-d_{10}. The complexation equilibrium is therefore slow on the NMR timescale, which normally accounts for a high stability due to fast complex formation (large k_{on}) and slow dissociation (small k_{off}), which is clearly not the case here (Section 2.2.2). Alternatively, a binding equilibrium is slow when complex formation *and* dissociation are associated with high activation barriers. In this case, separate signals from the free binding partners and the complex are also observed in the NMR spectra, but the ratio of k_{on}/k_{off} and, hence, K_a is small. An

(a)

4.47

(R¹ = $C_{11}H_{23}$, R² = C_7H_{15})

4.48

(b)

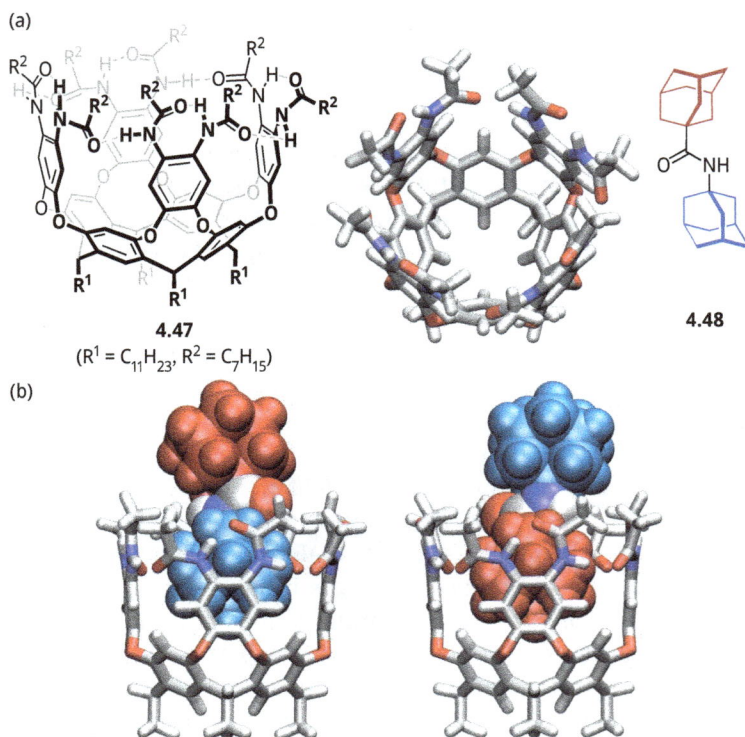

Figure 4.56: Molecular structure and calculated preferred conformation of the self-folding cavitand **4.47** (a) and structures of the two diastereomeric complexes of **4.47** with 1-[*N*-(1-adamantyl)] adamantanecarboxamide **4.48** (b). Equivalent adamantyl residues of the substrate are shown in the same color.

energy profile of such a case is shown in Figure 2.5b in Chapter 2. Receptors exhibiting this behavior typically surround the substrate on all sides and must undergo a substantial conformational reorganization to allow the substrate to enter or leave the cavity. Examples are the hemicarcerands described later. That receptor **4.47** with its wide opening behaves in a similar way is surprising, but a closer look at the guest exchange mechanism reveals the reasons.

An important aspect to consider in this context is that the cavity of **4.47** is open at only one end. Therefore, the substrate molecules entering and exiting the cavity must use the same opening. Since there is not much space between the substrate and the cavity walls, simultaneous exchange of two substrate molecules is impossible when the receptor is in its folded state. For the same reason, a suitably sized substrate cannot easily leave the cavity, as this would create a vacuum in the vacated space. Consequently, substrate exchange must involve folding away at least one sidewall of the receptor, which is only possible if hydrogen bonds along the cavity opening are broken. This process is energetically costly, which explains why it is slow. Considering that the vase-to-kite interconversion of **4.45** has an activation barrier of 48.5 kJ mol^{-1} [99], that breaking a hydrogen bond costs about 5 kJ mol^{-1} [103], and that a total of four hydrogen bonds must be broken to allow the vase-to-kite interconversion, the experimental Gibbs free energy of activation for the guest exchange of 70.7 kJ mol^{-1} is consistent with the idea that it involves opening of the cavity.

The fact that **4.47** is an unsymmetrical receptor with an open and a closed end also gives rise to interesting stereochemical aspects of complex formation. Although complexes that differ in the orientation of a guest in the receptor cavity are formally diastereomeric, it is usually not possible to distinguish the different stereoisomers because of the rapid tumbling of the guest in the cavities of most receptors. In the case of **4.47**, however, the movement of the guest is restricted because it can only rotate along the cavity axis. Since complex formation is also slow on the NMR timescale, complexes that differ in the guest orientation can be distinguished by NMR spectroscopy. Accordingly, the ^{1}H NMR spectrum of the complex between **4.47** and 1-[N-(1-adamantyl)]adamantanecarboxamide **4.48** contains two sets of signal, reflecting the two binding modes shown in Figure 4.56b [102].

In water, soluble derivatives of **4.47** usually exist in the kite conformation because the hydrogen bonding between the acetamide groups is not strong enough to stabilize the vase conformation under these conditions [104]. Two molecules in the kite conformation additionally dimerize to screen the hydrophobic surfaces of the folded aromatic sidewalls from the surrounding water molecules. However, in the presence of suitable substrates, the dimers dissociate and the conformational equilibrium shifts toward the vase to allow the binding of the guest. The hydrogen bonding along the edge of the receptor makes guest exchange slow on the NMR timescale even under these conditions.

Another example of a self-folding cavitand is **4.49**. In this case, the stabilization of the vase conformation does not rely on direct hydrogen bonding interactions between substituents along the cavity opening, but on water molecules bridging the four benzimidazole units in the sidewalls (Figure 4.57a). This receptor adopts the vase conformation even in water. It binds quaternary ammonium ions such as choline, typically with values of log K_a > 4 in D$_2$O [105]. Interestingly, dodecyltrimethylammonium salts bind to **4.49** with the alkyl chain and not with the cationic head group included into the cavity. This alkyl chain adopts an unfavorable helical conformation to efficiently fill the available space [104]. The energy required to adopt this conformation is paid for by the dehydration of the dodecyl chain during complex formation and the stabilizing dispersion interactions between the folded chain and the cavity walls.

The final example of a deep cavitand is the so-called octa acid receptor **4.50** developed by Bruce C. Gibb (Figure 4.57b). This receptor is structurally more closely related to **4.44** than to the deep cavitands **4.45** or **4.46** in that the aromatic subunits of the resorcin[4]arene core are bridged by benzal groups. These benzal groups form the sidewalls of an approximately 8 Å-wide and 8 Å-deep hydrophobic cavity, which is rigidified by the 3,5-dihydroxybenzoic acid moieties along the opening. The eight car-

(a)

4.49 (R = C$_2$H$_5$)

(b)

4.50

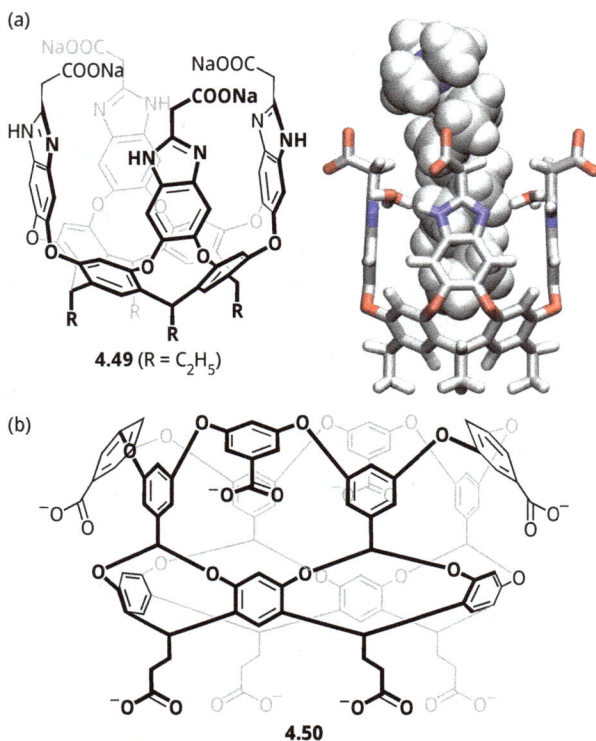

Figure 4.57: Molecular structure of **4.49** and calculated structure of its complex with a dodecyltrimethylammonium ion, showing the bridging water molecules and illustrating the helical arrangement of the dodecyl chain inside the receptor cavity (the ethyl groups in the resorcin[4]arene ring are replaced by methyl groups) (a), and structure of octa acid **4.50** (b).

boxylate groups along the *feet* of the resorcin[4]arene unit and the rim of the cavity render **4.50** water-soluble at pH > 7. Under these conditions, **4.50** hosts guest molecules such as alkanes or aliphatic carboxylic acids [106]. In the case of alkanes, the filling of the cavity creates a large hydrophobic surface at the cavity opening. As a consequence, the complexes tend to dimerize, driven by the hydrophobic effect, forming capsules containing the guest molecules inside the cavity. This mode of binding will be discussed in Section 5.2. Carboxylic acids are bound in a 1:1 fashion because their polar head groups mediate the hydration of the cavity entrance, thus weakening the hydrophobic effect and preventing self-assembly.

The octa acid **4.50** usually binds hydrophobic guests more strongly than the similarly sized β-CD. For example, adamantane-1-carboxylic acid is bound with a log K_a of 6.7 in 10 mM phosphate buffer at pH 11.3 [107], whereas the corresponding β-CD complex is more than two orders of magnitude less stable (Table 4.5). Complex formation is strongly exothermic with a small unfavorable entropy term, as observed for CDs. Interestingly, **4.50** also binds large weakly coordinating anions such as perchlorate, raising the question of why a negatively charged receptor should bind an anionic substrate. One aspect to consider is that the negatively charged groups in **4.50** are too far away to repel an anion incorporated into the cavity. In addition, the inner surface of the cavity has a positive electrostatic potential in the region of the inward-facing methine protons, which explains the affinity for anionic guests. Anion binding causes the interactions with hydrophobic guests in water to be influenced in a characteristic way by salts. Perchlorate, for example, reduces the stability of the adamantane-1-carboxylic acid complex because it also binds to the receptor cavity [107].

Instead of building a wall around the upper rim of resorcin[4]arenes to obtain receptors with deep cavities that are open at one end, it is also possible to prepare resorcin[4]arene-based receptors with completely closed cavities. The synthetic strategy to access these compounds involves covalently linking two resorcin[4]arene moieties *via* four spacers, resulting in molecular cages similar to the cryptophanes discussed in Section 4.1.6.

Such cages were first prepared in the Cram group on the basis of the conformationally rigid cavitand **4.44** (Figure 4.54) [108]. Since **4.44** lacks functional groups that would allow the bridging of two subunits, the pyrogallol derivative **4.51** (Figure 4.58a), whose free hydroxy groups are available for reactions with halocarbons under basic conditions, has to be used as starting material. Treatment of **4.51** with CH_2BrCl in dimethylacetamide (DMA), dimethylformamide (DMF), or dimethylsulfoxide (DMSO) yields **4.52**, which is one of the first resorcin[4]arene-derived molecular cages [109].

After isolation, **4.52** contains one solvent molecule in the cavity that cannot escape because the portals surrounding the cavity are too small for it to squeeze through. The presence of this guest can be inferred by various analytical techniques. For example, the ^{1}H NMR spectrum of the product not only contains the signals of the included solvent molecule, but these signals also appear at a significantly higher field than normal. This indicates that the guest experiences the shielding effect of the π-systems lining the cavity wall. In the mass spectrum, the product peaks have *m/z* ratios consistent with the mass of the cage plus the mass of the trapped guest molecule, demonstrating that the guest and the cavitand are part of the same entity. Finally, the crystal structure clearly shows that the cavity is occupied. To illustrate the binding mode, the structure of the DMF complex of an analog of **4.52** with shortened side chains is shown in Figure 4.58b. Because of the permanent confinement of the guests, Cram suggested the name *carcerand* for these molecular cages. The corresponding complexes are called *carceplexes*.

Carcerand syntheses are often surprisingly efficient considering that seven molecules are brought together and eight covalent bonds are formed. For example, the DMA, DMF, and DMSO complexes of **4.52** are isolated in yields of 49%, 54%, and 61%, respectively, using the synthesis shown in Figure 4.58 [109]. When the reaction is carried out in the bulky solvent *N*-formylpipyridine (NMP), no product is obtained, indicating that the solvent molecules act as templates. These templates influence the

(a)

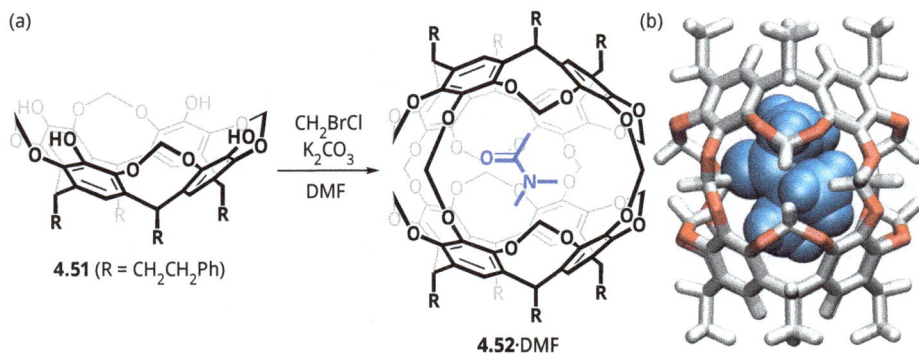

(b)

4.51 (R = CH$_2$CH$_2$Ph)

CH$_2$BrCl
K$_2$CO$_3$
DMF

4.52·DMF

Figure 4.58: Synthesis of **4.52** from the corresponding pyrogallolarene-derived cavitand **4.51** (a), and calculated structure of an analog of **4.52** with R = CH$_3$ containing a trapped DMF molecule (b). The DMF molecule is shown in blue.

course of the reaction once two cavitand molecules have been linked (Figure 4.59). Due to the chelate effect, the corresponding singly linked intermediate has a higher affinity for a solvent molecule than two separate receptors, an effect we also saw for the CD dimers discussed in Section 4.1.4. The interaction preorganizes the two receptor subunits so that the second linkage can only be formed between correct pairs of OH groups. Once this second connection has been made, all further linkages must lead to the desired product. The solvent molecule remains in the cavity, mediating its own incarceration.

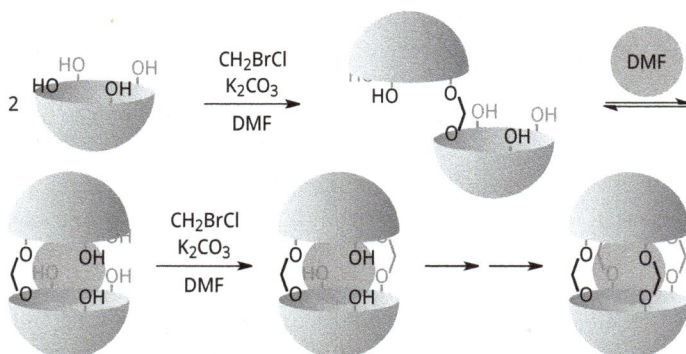

Figure 4.59: Schematic course of a carcerand synthesis, illustrating the template effect of a solvent molecule.

Carcerand syntheses in solvents such as NMP, where the solvent molecules are too large to act as templates, have been used to screen for templates that are particularly suitable for inducing product formation. A correlation between template structure and efficiency can also be established by using competing templates and analyzing the ratio at which the respective products are formed. According to these studies, pyrazine gives the highest yields because it induces an optimal arrangement of the two pyrogallol units during the synthesis. The authors concluded that templation is driven "by an optimum of van der Waals interactions and a minimum of steric interactions between the guest–template molecule and the interior of the forming shell" [110].

While carcerands and their complexes are conceptually interesting, representing "a new phase of matter" according to Cram [109, 111], they are of limited use because guest exchange is impossible. This disadvantage can be overcome by using hemicarcerands, which have larger portals along their surface, allowing guests to enter or leave the cavity. Note that the prefix *hemi* in the word hemicarcerand has a different meaning than in the terms hemispherand and hemicryptophane. The latter two receptor types contain only half a spherand or cryptophane. Hemicarcerands, on the other hand, contain two pyrogallol units but, unlike carcerands, do not permanently trap their guests. Hemicarcerands are, by definition, molecules with a closed surface and an enforced internal cavity, whose complexes are kinetically inert at ambient conditions but permit guest exchange at elevated temperatures.

Two strategies provide access to such receptors. The first is to reduce the number of linkers by using pyrogallolarenes as starting materials that lack one or two OH groups. An example is hemicarcerand **4.53** (Figure 4.60). Alternatively, the distance between the two macrocyclic subunits can be increased by using longer linkers, as in hemicarcerand **4.54**. The advantage of the latter strategy is that the corresponding hemicarcerands allow the complexation of larger guests.

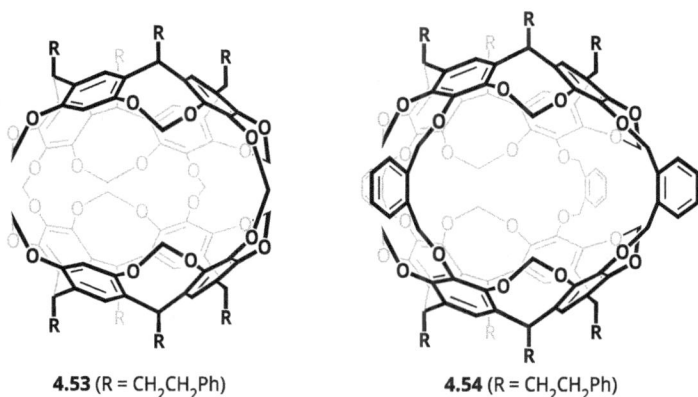

4.53 (R = CH₂CH₂Ph)   4.54 (R = CH₂CH₂Ph)

Figure 4.60: Molecular structures of hemicarcerands **4.53** and **4.54**.

These hemicarcerands also contain entrapped solvent molecules after isolation, which can be expelled from the cavity by prolonged heating in solvents that are too large to enter the cavity, such as xylene, 1,3,5-trinitrobenzene, or chlorobenzene. At 165 °C, the removal of DMF or DMA from the cavity of **4.53** takes 24 h and 12 h, respectively, illustrating how difficult it is for guests to leave a hemicarcerand, even if a suitable portal is available [112]. The filling of hemicarcerands is similarly achieved by heating in the presence of an excess of a guest.

The rates at which molecules enter and leave hemicarcerand cavities under defined conditions provide information about the activation barriers associated with these processes and the thermodynamic stability. For example, the free activation energies for a DMA molecule to enter and leave **4.54** in o-xylene-d_{10} at 100 °C are 98.4 kJ mol^{-1} and 113.9 kJ mol^{-1}, respectively [113]. The difference between these energy values reflects the thermodynamic stability of the complex in terms of ΔG^0. The corresponding –15.5 kJ mol^{-1} correlates with an association constant K_a of 150 M^{-1} (at 100 °C), showing that the stability of the complex is moderate. Thus, the fact that the complex is not dynamic at room temperature is entirely due to the high activation barriers associated with its formation and dissociation.

Breaking down the ΔG^0 of complex formation into the enthalpic and entropic contribution shows that both parameters contribute approximately equally to the overall complex stability ($\Delta H^0 = -6.3$ kJ mol^{-1}, $T\Delta S^0$, $= 9.2$ kJ mol^{-1}) [113]. The small ΔH^0 indicates that the direct interactions between the guest and the groups lining the inner surface of the hemicarcerand are weak. Small differences are observed for different guests, explaining the effect of guest structure on templation efficiency. The entropy gain during complex formation has several causes, including the desolvation of the guest, the emptying of the cavity, and the release of strain when filling it again.

To describe these kinetic effects quantitatively, Cram introduced the concept of constrictive binding [114], which was already mentioned in Section 2.1. This term describes the free energy of the transition state relative to the free energy of the separate binding partners (Figure 2.5). Constrictive binding therefore essentially describes how difficult it is, in terms of activation energy, for a substrate to enter the cavity of a hemicarcerand. This energetic barrier is mainly due to the strain induced in the hemicarcerand or in the guest as it squeezes through the portals. Calculations provided information on the possible pathways of these processes. For **4.53**, they showed that guest egress involves the folding away of the acetal CH$_2$ groups linking the two pyrogallol moieties [115]. This gating mechanism is analogous to the opening of a French door. Alternatively, the enlargement of a hemicarcerand side portal can also proceed *via* a sliding door mechanism. In this case, the entire molecular skeleton of the hemicarcerand is distorted during complex dissociation without a pronounced outward movement of the CH$_2$ groups (Figure 4.61).

A notable application of hemicarcerands is the stabilization of highly reactive substrates. The first example was reported by Cram and involved the *taming* of cyclobutadiene, which prior to Cram's work had only been characterized at 8 K in an argon matrix. In solution, cyclobutadiene is unstable and rapidly undergoes a series of transformations, starting with dimerization *via* a Diels–Alder reaction. In stark contrast, cyclobutadiene is stable even at room temperature when protected in the cavity of hemicarcerand **4.53** [116].

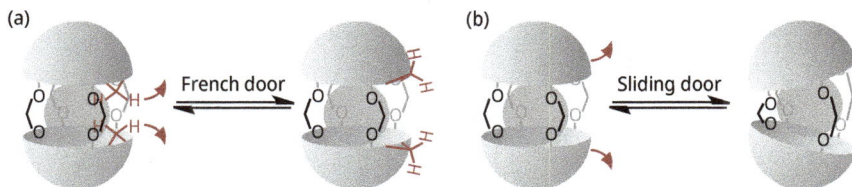

Figure 4.61: Schematic illustrations of the *French door* (a) and *sliding door* (b) gating mechanisms that allow the egress of a substrate from the cavity of a hemicarcerand.

The cyclobutadiene complex of **4.53** is obtained by first incorporating α-pyrone into the cavity. Irradiation in chloroform initiates a two-step reaction sequence, starting with the electrocyclic reaction of the 1,3-butadiene part of the guest, followed by the elimination of CO_2 to give the product (Figure 4.62). The CO_2 molecule leaves the cavity, but the cyclobutadiene remains inside. This isolated molecule cannot undergo any further reactions and can therefore be characterized spectroscopically. According to the results, cyclobutadiene has a singlet ground state and produces a signal in the ^1H NMR spectrum at 2.27 ppm, about three ppm upfield from the proton signal of an isolable cyclobutadiene derivative with three bulky alkyl groups.

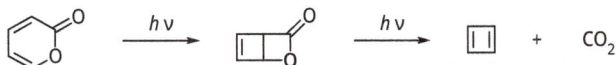

Figure 4.62: Reaction sequence performed inside the hemicarcerand **4.53** for the preparation of cyclobutadiene from α-pyrone.

This strategy can also be used to stabilize and characterize other reactive intermediates. In this context, the group of Ralf Warmuth described the generation of *o*-benzyne in the cavity of a hemicarcerand [117]. Irradiation of the hemicarceplex containing benzocyclobutanedione as the guest at 77 K induces CO extrusion, initially leading to a molecule of benzocyclopropenone within the cavity (Figure 4.63). This compound slowly converts to benzoic acid when the hemicarceplex is stored at room temperature in the presence of water. Alternatively, further irradiation at 77 K induces the extrusion of another CO molecule and the formation of *o*-benzyne. Again, ^1H NMR spectroscopy provides clear evidence for the presence of the product in the cavity. This guest is so reactive, however, that it has to be characterized at 77 K. At higher temperatures, it undergoes a Diels–Alder reaction with an aromatic ring lining the inner cavity.

Other reactive intermediates stabilized by hemicarcerands include carbenes [118]. The limitation of the concept is that the reactions must be initiated by triggers that reach the bound substrate. Light-induced reactions are useful, but small molecules such as O_2, CO_2, or H_2O can also pass through the portals, while larger reagents cannot be used to induce transformations.

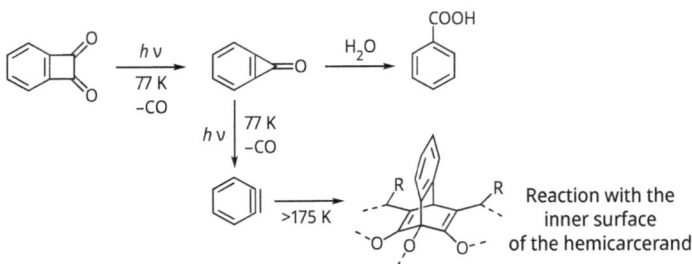

Figure 4.63: Reaction sequence performed inside a hemicarcerand for the generation of o-benzyne from benzocyclobutanedione. The benzocyclopropenone intermediate reacts with water to give benzoic acid (top right) or loses another molecule of CO. The o-benzyne thus formed adds to an aromatic subunit along the inner surface of the hemicarcerand, as shown in the lower right-hand corner of the reaction scheme.

4.1.10 Pillararenes

Introduction: Given the long history of CTVs, calixarenes, and resorcinarenes in supramolecular chemistry, it is surprising that a fourth family of receptors, also formed by treating a phenol derivative and an aldehyde, only entered the field in 2008 [119]. These cyclophanes were first described by Tomoki Ogoshi, who showed that the reaction between 1,4-dimethoxybenzene and paraformaldehyde in the presence of a Lewis acid yields a five-membered paracyclophane in which the aromatic subunits are linked by methylene groups in the 2- and 5-positions.

Reaction conditions are now known that also allow the preparation of larger rings (up to the 10-membered one), but yields are often low. The most widely used members of this receptor family are therefore those with five and six aromatic subunits, as they combine good synthetic accessibility with useful binding properties [120]. Representative crystal structures are shown in Figure 4.64. These structures differ from those of many of the cyclophanes discussed earlier in that the aromatic subunits are arranged almost parallel to the main axis of the ring, rather than in a tilted arrangement as in CTVs, calixarenes, and resorcinarenes. The resulting characteristic cylindrical shape is reminiscent of the columns that make up the Parthenon in Athens, which explains why this family of receptors has been named pillararenes.

The reason for the tubular structure of pillararenes is the substitution pattern of the aromatic subunits, which causes a linear arrangement of the CH_2–Ar–CH_2 bonds. In contrast, the CH_2–Ar–CH_2 subunits in ortho- or metacyclophanes are bent. The 1,4-disubstituted aryl groups connecting the CH_2 group can formally be regarded as extensions of the C–C bonds in cyclopentane, and since the angle at which they are arranged is almost the same as the internal angles in a pentagon (109° *vs.* 108°), the five-membered pillararene forms easily without producing strain. The six-membered analog is somewhat distorted, as shown by the slightly tilted arrangement of the aromatic sub-

(a)

(b)

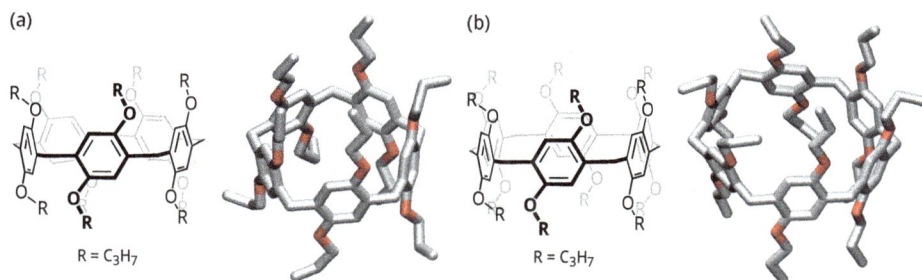

R = C₃H₇

R = C₃H₇

Figure 4.64: Molecular structures and crystal structures of the pillar[5]arene (a) and the pillar[6]arene (b) derived from 1,4-dipropoxybenzene. The protons in both structures are not shown for reasons of clarity.

units in the crystal structure shown in Figure 4.64b, but the corresponding derivative with free OH groups has the conformation of an almost ideal hexagon. Pillar[5]arene has a cavity diameter of about 5 Å, similar to α-CD, whereas pillar[6]arene has a cavity diameter of 7 Å, intermediate between β-CD and γ-CD.

The outer substituents in pillararenes are oriented at an angle to the main axis of the ring. As a result, the substituents can be arranged in two different ways, leading to a total of eight conformations for a pillar[5]arene in which all substituents are identical. All of these conformations are chiral, with the planar aromatic subunits representing the stereogenic elements. In terms of stereochemistry, four conformational diastereomers can be distinguished, each consisting of a pair of enantiomers. In the case of pillar[6]arene, three of the eight possible diastereomeric conformations are *meso*-forms, giving a total of 13 conformational isomers, but only a few of these conformations are relevant. In the case of pillararenes with OR groups, the most stable conformations are those with the highest symmetry, where all substituents are aligned (C_5 for pillar[5]arene and C_6 for pillar[5]arene). All other conformations have OR groups of adjacent subunits in close proximity, which is sterically unfavorable. The flipping of all aromatic rings leads to the interconversion of the two enantiomeric C_n-symmetric conformations (Figure 4.65a), with the rate of this conformational equilibrium depending on the size of the aromatic substituents. However, these substituents must be much larger than in calixarenes to completely prevent interconversion. Dodecyl chains, for example, are not yet large enough because they can wriggle through the cavity. Only when the pillar[5]arene carries bulky cyclohexylmethyl substituents is it conformationally stable, allowing chromatographic separation of the conformational isomers [121].

The situation is different for pillararenes with free OH groups. In the case of pillar[6]arene, the even number of subunits allows a conformation in which all OH groups form intramolecular hydrogen bonds. This conformation is a *meso*-form in which the orientations of the aromatic rings alternate (Figure 4.65b). In contrast, the unsubstituted pillar[5]arene cannot form a continuous seam of hydrogen bonds due to the odd number of subunits. The most stable conformation is therefore the one

(a)

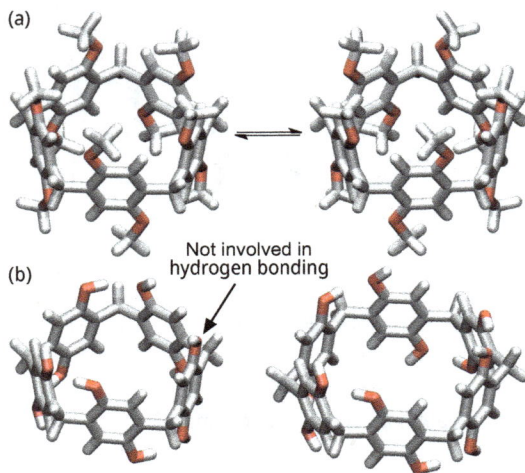

(b)

Not involved in
hydrogen bonding

Figure 4.65: Selected calculated conformations of pillar[5]arene and pillar[6]arene. In (a), the C_5-symmetric enantiomeric conformations of a pillar[5]arene with identical substituents are shown. Stereochemically, the left structure represents the (pR,pR,pR,pR,pR)-enantiomer and the right structure the corresponding (pS,pS,pS,pS,pS)-enantiomer. In (b), the preferred conformations of a pillar[5]arene (left) and a pillar[6]arene (right) with free OH groups are shown. The OH group that is not involved in a hydrogen bond in the pillar[5]arene is marked.

with the maximum number of hydrogen bonds, consisting of four alternating subunits. At room temperature, this compound is flexible, but the conformational equilibrium can be slowed down sufficiently by lowering the temperature to detect the unsymmetrical structure by ^{1}H NMR spectroscopy.

Nomenclature: The ring size of pillararenes is indicated in the same way as for calixarenes by specifying the number of aromatic subunits in square brackets between the two parts of the name. Thus, the five- and six-membered unsubstituted pillararenes are called pillar[5]arene and pillar[6]arene, respectively. The substitution pattern is indicated by using the names of the aromatic building blocks. For example, the pillar[5]arene derived from 1,4-dimethoxybenzene is called 1,4-dimethoxypillar[5]arene. Note that this name is technically incorrect because the molecule contains 10 methoxy groups and not two. Pillararenes with different substituents along the ring are sometimes called copillararenes. To indicate the position of the substituents in these compounds, each aromatic subunit is assigned a letter and the cavity openings are distinguished by using the numbers 1 and 2. Thus, an A1/B1 co-pillararene has two substituents on the same side of the ring in neighboring aromatic residues.

Synthesis: Most pillararene syntheses are based on the treatment of 1,4-dialkoxybenzenes with paraformaldehyde in the presence of a Brønsted or Lewis acid [120]. Mechanistically, the reaction resembles the formation of resorcinarenes. The first step is an electrophilic aromatic substitution leading to a benzyl alcohol derivative (Figure 4.66a). Since all positions in symmetrical 1,4-dialkoxybenzenes are equivalent, the product always contains the new substituent in the *ortho* position of an alkoxy group. The introduction of the next substituent, which can occur at any stage of the synthesis, is then controlled by a combination of steric and electronic effects: all free positions in the benzyl alcohol intermediate are activated by the adjacent alkoxy groups, but the position adjacent to the CH_2OH group is sterically hindered so that the second electrophilic substitution does not take place there. Of the two remaining positions, the one *para* to the CH_2OH group is more strongly activated and the final product therefore preferentially contains the linking CH_2 groups in positions 2 and 5. Note that the order of the electrophilic substitutions is not important. All steps are reversible and the formation of the preferred product is therefore subject to thermodynamic control.

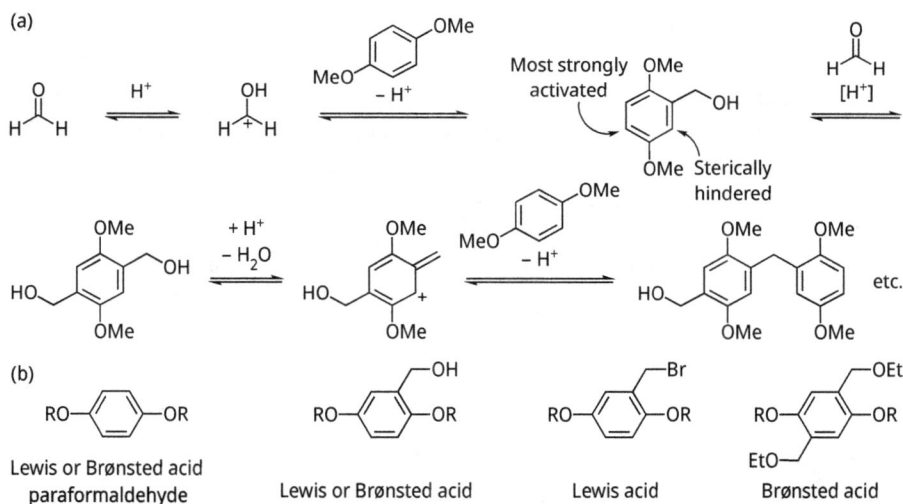

Figure 4.66: Mechanism of the acid-catalyzed reaction between 1,4-dimethoxybenzene and formaldehyde, explaining the preferred substitution pattern of the aromatic subunits in pillararenes (a), and overview of the reaction conditions for pillararene syntheses starting from different aromatic precursors (b).

In his first pillar[5]arene synthesis, Ogoshi used $BF_3 \cdot OEt_2$ as the Lewis acid and 1,2-dichloroethane as the solvent and isolated the product in a yield of 22% [119]. The influence of acid, solvent, and other parameters on product formation was then systematically studied, leading to optimized conditions that allow the synthesis of pillar[5]arene in yields above 80%. The correct choice of solvent is crucial. 1,2-Dichloromethane typically yields the five-membered ring in high yields, whereas the reaction is much less efficient in bulky solvents such as chloroform. This influence of the solvent is probably due to a template effect, where the linear 1,2-dichloroethane induces a conformation of the pillar[5]arene precursor that readily undergoes cyclization, whereas the conformation induced by the more bulky chloroform is unsuitable for ring closure.

Pillar[6]arenes are usually more difficult to access, but with the right choice of template they can also be synthesized on a preparative scale. For example, the reaction between 1,4-diethoxybenzene and paraformaldehyde in chloroform and in the presence of $FeCl_3$ as catalyst gives a mixture of the five- and six-membered rings, both in a yield of about 30%. However, when using 1,4-bis(cyclohexylmethoxy)benzene, paraformaldehyde, $BF_3 \cdot OEt_2$, and working in chlorocyclohexane as a large bulky template, the pillar[6]arene is obtained in a high yield of 87% with only 3% of pillar[5]arene as a byproduct.

In addition to 1,4-dialkoxybenzenes, other aromatic precursors can also be converted into pillararenes. When starting from 2,5-dialkoxybenzyl alcohols or 2,5-dialkoxybenzyl bromides, the presence of formaldehyde or paraformaldehyde is not required because the methylene groups are already present in the starting materials (Figure 4.66b). When 1,4-alkoxy-2,5-bis(ethoxymethyl)benzenes and a Brønsted acid are used, the reaction proceeds *via* an *ipso*-substitution accompanied by the release of diethoxymethane [122].

Binding properties: The intrinsic receptor properties of pillararenes are determined by the negative electrostatic potential of their inner cavity surfaces induced by the electron-rich aromatic subunits (Figure 4.67a). Accordingly, substrate binding relies mainly on charge-transfer interactions with electron-poor π-systems, CH–π interactions, and cation–π interactions, resulting in a substrate scope similar to that of many of the cyclophanes discussed earlier. However, unlike calixarenes and resorcinarenes, pillararenes have cavities that are open at both ends, allowing substrate molecules to thread through the ring, as in the case of CDs.

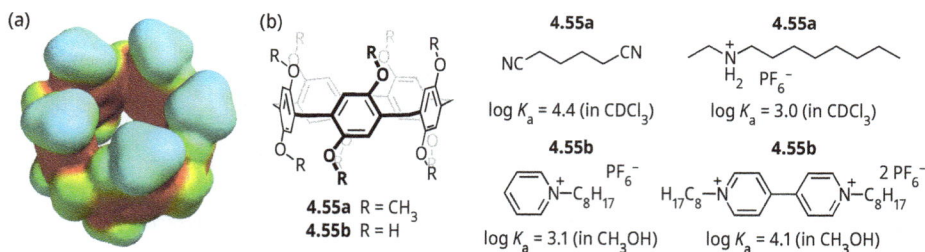

Figure 4.67: Electrostatic potential surface of the pillar[5]arene **4.55a** derived from 1,4-dimethoxybenzene (a), and representative substrates interacting with the pillararenes **4.55a** and **4.55b** together with the binding constants of the corresponding complexes (b). The color coding covers a potential range from −100 to +100 kJ mol^{-1}, with red and blue indicating values greater than or equal to the absolute maximum in negative and positive potentials, respectively.

Pillar[5]arene preferentially binds to linear aliphatic and monocyclic aromatic substrates, demonstrating that the comparable cavity diameter to that of α-CD results in analogous substrate selectivity. Examples of typical substrates for **4.55a** and the analog with free OH groups **4.55b** are collected in Figure 4.67b together with information on the stabilities of the corresponding complexes. The magnitudes of the $\log K_a$ values show that pillar[5]arenes are efficient receptors even for substrates whose binding involves relatively weak types of interactions. Due to the larger cavity diameter, pillar[6]arenes prefer to bind bulkier substrates, such as those containing bicyclic aliphatic residues or quaternary ammonium ions.

Derivatives: The structural variation of the side chains of pillararenes can be used to influence the binding properties, to link pillararenes to other receptors, or to immobilize them on surfaces. Many pillararenes are directly accessible by using suitable building blocks in the reaction, but the introduction of functional groups along the cavity openings by post-functionalization is also possible. These transformations often start with the deprotection of the OH groups in a permethylated or otherwise alkylated pillararene by treatment with boron tribromide. Modification of the free OH groups is then achieved in a number of ways, some of which are shown in Figure 4.68. Alkylation with propargyl bromide, for example, yields pillararenes with terminal alkyne groups in the side chains, which can be further converted to 1,4-disubstituted 1,2,3-triazoles under the conditions of the copper(I)-catalyzed azide–alkyne cycloaddition. Sulfonylation of the OH groups with triflic anhydride leads to valuable precursors for Pd(0)-catalyzed cross-coupling reactions such as the Sonogashira–Hagihara reaction. Carboxymethylation with chloroacetates gives esters, which can either be hydrolyzed to give water-soluble pillararene carboxylates, or converted to positively charged pillararenes containing quaternary ammonium ions by using the reaction sequence shown in Figure 4.68. In addition, several strategies are known to produce co-pillararenes with different substituents at defined positions along the ring [120, 123, 124]. Pillararenes can therefore be structurally varied over a wide range, which explains why they have found widespread use in supramolecular chemistry within just a few years. Applications in supramolecular chemistry are in the fields of self-assembly, sensing, or the development of interlocked molecules, including molecular machines.

In order to show the influence of the substituents on the binding properties, we focus here on two pillararene derivatives, namely the polyanionic and polycationic water-soluble ones containing carboxylate groups and quaternary ammonium ions in the periphery, respectively. The ionic groups in these compounds either enhance the intrinsic cation affinity of pillararenes or induce affinity for anions that would otherwise be rejected.

The cooperative effect of negatively charged groups on cation affinity is reflected in the stability of the paraquat (N,N'-dimethyl-4,4'-bipyridinium) complex of **4.56** (Figure 4.69), which has a $\log K_a$ of 4.9 in water. In contrast, the $\log K_a$ of the complex between the neutral pillararene **4.55b** and N,N'-dioctyl-4,4'-bipyridinium is only 4.1 in the less competitive methanol (Figure 4.67b) [125]. Increasing the number of carboxylate groups to twelve by using a substituted pillar[6]arene increases the $\log K_a$ of the paraquat complex to 8.0 in water [126], and the $\log K_a$ of the complex

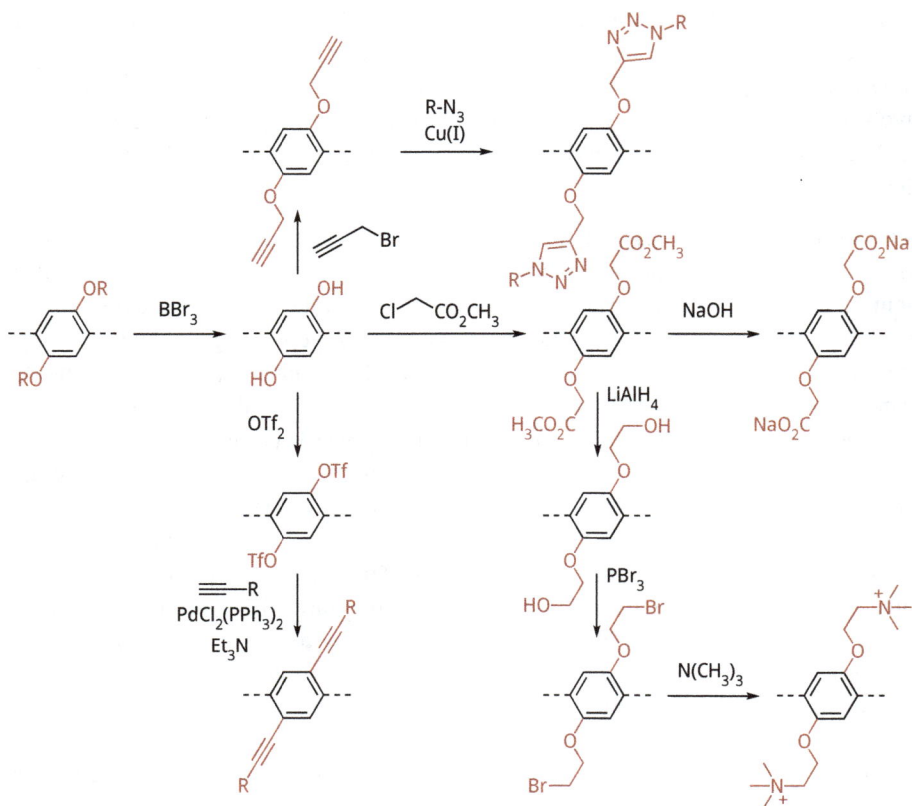

Figure 4.68: Examples of synthetic methods available to vary the substituents in pillararenes.

between paraquat and a pillar[7]arene with fourteen carboxylate groups even amounts to 9.5 [127]. This trend is due to a better fit of the substrate into the cavities of the larger pillararenes and a strengthening of the ionic interactions as the number of negatively charged substituents increases.

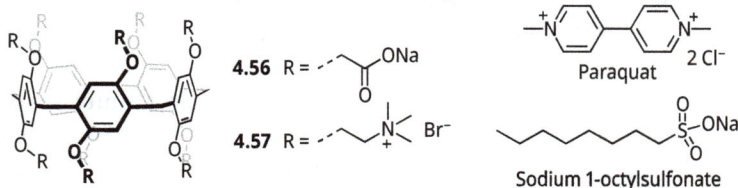

Figure 4.69: Molecular structures of the water-soluble pillar[5]arenes **4.56** and **4.57** and structures of the corresponding preferred guests.

An example of a pillar[5]arene that binds to anionic substrates is **4.57** (Figure 4.69). This receptor binds to 1-octanesulfonate with a $\log K_a$ of 4.1 [128]. Complex formation results in shielding of the protons in the octyl group, giving rise to eight separate signals in the ^{1}H NMR spectrum. The protons in the middle region of the octyl group are most strongly shielded, indicating that the linear guest is threaded through the cavity of **4.57** so that its negatively charged head is oriented close to the cationic groups at one opening.

Further reading

We have seen that several important families of cyclophane-derived receptors are obtained by treating an aromatic building block with formaldehyde (or another aldehyde) under acidic or basic conditions. At this point, the question may arise whether this approach is limited to the receptor families presented above or whether there are other cyclophanes that can be accessed in a similar manner. Indeed, the discovery of the pillararenes has triggered intense research activities to answer this question, which in turn have yielded numerous new receptor scaffolds. The prism[*n*]arenes derived from 2,6-dimethoxynaphthalene, the pagoda[*n*]arenes derived from 2,6-dimethoxyanthracene, and the helic[*n*]arenes derived from triptycene derivatives are just a few examples. Together with the many macrocyclic receptors prepared by using other synthetic strategies, there will be no shortage of receptors in the near future, as the following reviews demonstrate:

Shi Q, Wang X, Liu B, Qiao P, Li J, Wang L. Macrocyclic host molecules with aromatic building blocks: the state of the art and progress. Chem. Commun. 2021, 57, 12379–405. Han XN, Han Y, Chen CF. Recent advances in the synthesis and applications of macrocyclic arenes. Chem. Soc. Rev. 2023, 52, 3265–98. Zhang W, Yang W, Zhou J. Biphenarenes, versatile synthetic macrocycles for supramolecular chemistry. Molecules 2023, 28, 4422.

4.1.11 Cucurbiturils

Introduction: We have seen that the history of formaldehyde-derived cyclophanes goes back to Adolf von Baeyer, whose work laid the foundation for one of the first commercial plastic materials. Interestingly, the discovery of the cucurbiturils is also linked to the development of a synthetic polymer, namely the urea-formaldehyde or UF resins. In 1905, Rolf Behrend studied the reaction between a bicyclic bis(urea) derivative, glycoluril **4.58** (Figure 4.70a), and formaldehyde under the typical acidic polymerization conditions, and isolated a white crystalline material that proved to be very stable and insoluble in water, but formed complexes with a number of salts [129]. Since the structure of this product could not be determined at that time, it was initially referred to as Behrend's polymer.

In 1981, William L. Mock reproduced Behrend's original procedure and, using the modern analytical techniques available to him, showed that the product formed from glycoluril and formaldehyde has the structure of **4.59** (Figure 4.70a) [130]. Behrend's polymer thus turned out to be a macrocycle with six glycoluril subunits with methylene units connecting pairs of nitrogen atoms. According to the crystal structure, the glycoluril subunits are oriented with their protons pointing outward and their car-

(a)   (b)

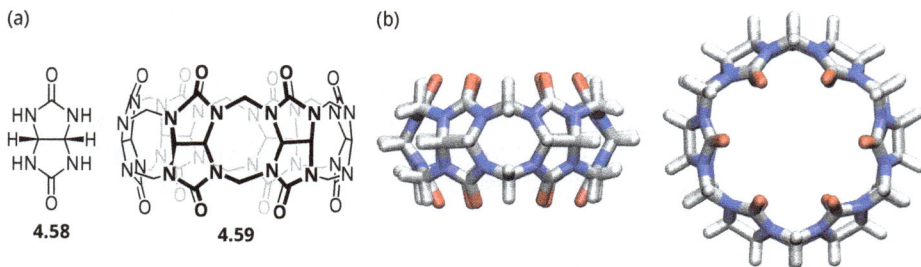

Figure 4.70: Molecular structures of glycoluril **4.58** and cucurbit[6]uril **4.59** (a), and side and top views of the crystal structure of **4.59** (b).

bonyl groups converging at the two cavity openings (Figure 4.70b). The ring thus has a convex shape around the equator and a cavity that is accessible from both sides. Mock suggested naming this compound *cucurbituril* because of its resemblance "to a gourd or pumpkin (family Cucurbitaceae), and by devolution from the similarly named (and shaped) component of the early chemists' alembic" (i.e., the round bottom flask in modern chemistry) [130]. The comparison of the cucurbituril structure with the shape of a flask or container suggests that Mock saw potential in cucurbiturils to act as receptors, and he (and several other groups) indeed continued to systematically study the binding properties of these compounds. Unfortunately, due to the low solubility of **4.59** in other media, many studies had to be carried out initially in rather acidic solvent mixtures, such as aqueous formic acid.

This situation changed when the groups of Kimoon Kim and Anthony Day succeeded in synthesizing and isolating other cucurbituril homologs, some of which are considerably more soluble than **4.59**. The cucurbituril family now extends from the five-membered to the ten-membered ring, and a 14-membered cucurbituril has also been described. With the availability of these compounds and the realization that some of them exhibit exceptional binding strength in aqueous media, cucurbiturils have gradually become a popular family of receptors with the potential to replace CDs in certain applications [131].

The structure of cucurbiturils does not change significantly with ring size. All homologs up to the octamer have a cylindrical shape, closely related to the structure of **4.59**. The height of the ring remains constant as it is determined by the distance between the carbonyl groups of the glycoluril units. The cavity diameter increases with the number of subunits and is larger at the equatorial region than at the cavity openings where the carbonyl groups are located. Figure 4.71 summarizes the structural parameters of the most important cucurbituril derivatives. This figure also shows the electrostatic potential surface of **4.59**. As can be seen, the cavity openings have a pronounced negative potential, indicating that they can act as binding sites for appropriately sized cations. The inner surface of the cavity is slightly negative and its concave shape makes it difficult for a guest molecule to make efficient contact. Note also the pronounced positive potential of the convex outer surface in the region of the methine protons, which explains the tendency of certain anions to associate with the outside of cucurbiturils and the anion affinity of macrocyclic receptors with inverted glycoluril units, such as the bambusurils discussed later.

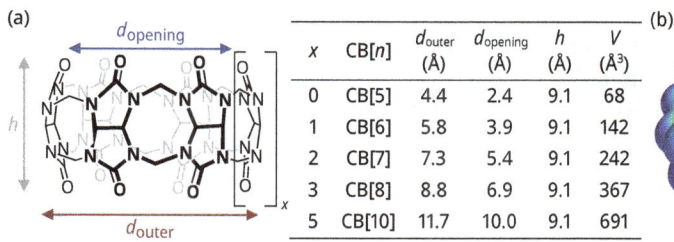

(a)

x	CB[n]	d_{outer} (Å)	$d_{opening}$ (Å)	h (Å)	V (Å³)
0	CB[5]	4.4	2.4	9.1	68
1	CB[6]	5.8	3.9	9.1	142
2	CB[7]	7.3	5.4	9.1	242
3	CB[8]	8.8	6.9	9.1	367
5	CB[10]	11.7	10.0	9.1	691

(b)

Figure 4.71: Molecular dimensions of glycolurils (a) and electrostatic potential surface of cucurbit[6]uril **4.59** (b). The color coding covers a potential range from −100 to +100 kJ mol^{-1}, with red and blue indicating values greater than or equal to the absolute maximum in negative and positive potentials, respectively. The values of d_{outer}, $d_{opening}$, and h take into account the van der Waals radii of the respective atoms.

Nomenclature: The family name cucurbituril refers to the compounds derived from unsubstituted glycoluril **4.58**. The ring size is indicated, as we have seen for cyclophanes, by specifying the number of glycoluril subunits in square brackets between the two parts of the name. Accordingly, compound **4.59** is a cucurbit[6]uril, or CB[6] for short. For cucurbiturils with substituents along the equator, the number, type, and positions of the substituents must be included in the name, and there are several ways of doing this. The naming of fully substituted cucurbiturils with only one type of substituent is straightforward. A CB[5] with methyl groups in all glycoluril subunits, for example, is a decamethylcucurbit[5]uril or permethylcucurbit[5]uril. Cucurbiturils with cyclohexane rings in the periphery are cyclohexanocucurbiturils.

Synthesis: The synthesis of cucurbiturils involves the treatment of glycoluril or glycoluril derivatives with formaldehyde under acidic conditions. The glycolurils required as starting materials are condensation products of 1,2-dicarbonyl compounds and urea. For example, unsubstituted **4.58** is obtained from glyoxal, dimethylglycoluril from butane-2,3-dione, and cyclohexanoglycoluril from cyclohexane-1,2-dione.

In the original CB[6] synthesis used by Behrend and Mock, the reaction between glycoluril and formaldehyde was carried out in hot concentrated H_2SO_4, yielding only the six-membered ring. Product formation involves the initial hydroxymethylation of one or more glycoluril nitrogen atoms (Figure 4.72). Under the acidic conditions, the hydroxymethyl groups eliminate water and form iminium ions. Their reaction with a nitrogen atom from another glycoluril molecule then leads to the bridging of two subunits. Cyclic ethers such as compound **A** can also be formed as reaction intermediates. All of these compounds are rapidly consumed as the reaction proceeds, initially yielding ribbon-like oligomers, which cyclize once a certain chain length is reached. As the underlying reactions are reversible, any errors in the synthesis can be corrected. For example, the bridging of two glycolurils can lead to a C-shaped product **B**$_{endo}$, with an *endo* arrangement of the two subunits, but also to the S-shaped isomer **B**$_{exo}$. Chain elongation of the latter does not give products with the curvature required for cyclization, but these products can isomerize during the reaction, ultimately leading to the macrocyclic product. Under the harsh conditions employed, CB[6] is likely to be thermodynamically preferred, although its formation may also be mediated by template effects of cationic species present in the reaction mixture.

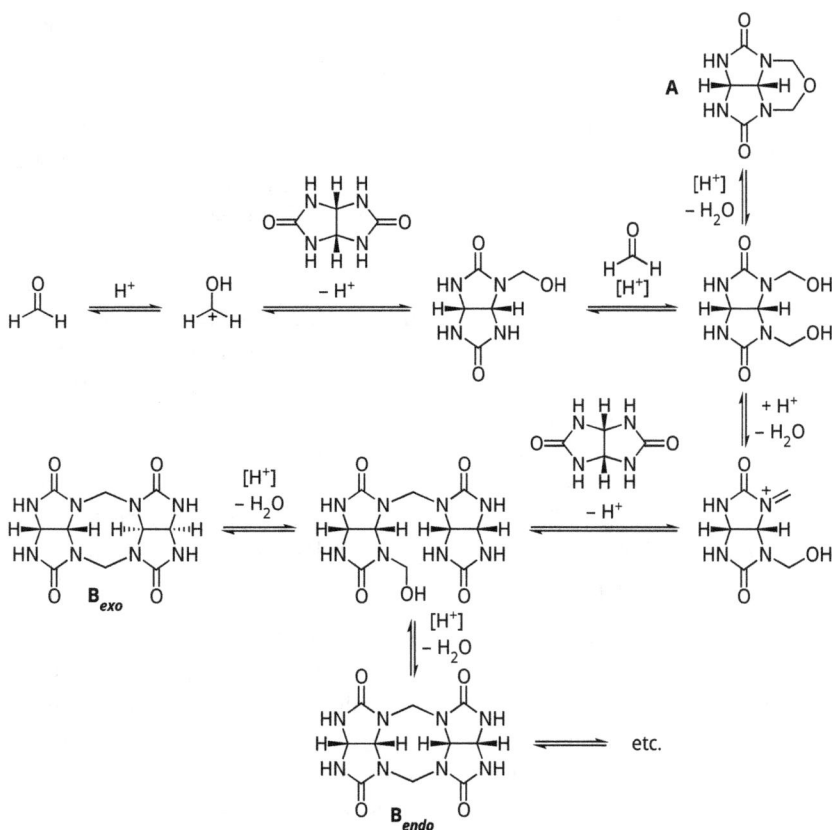

Figure 4.72: Mechanism of the acid-catalyzed condensation reaction between glycoluril and formaldehyde. Possible intermediates on the way to macrocyclic product, such as the diether **A** or the stereoisomers of the glycoluril dimer **B**$_{exo}$ and **B**$_{endo}$, are marked.

Kim and Day later showed that performing the reaction under milder conditions yields a mixture of different cucurbiturils, typically dominated by CB[6], but also containing CB[5], CB[7], and CB[8] in useful amounts. Interestingly, macrocycles containing an inverted glycoluril residue with its methine protons oriented toward the interior of the cavity are also isolated from these mixtures. These byproducts are thought to be kinetic intermediates, since they eliminate the mismatched subunit upon treatment with acid at higher temperatures, thus yielding the next smaller ring. The product ratios in these syntheses depend on the solvent, the temperature, the type of acid, and also on the presence or absence of templates such as metal ions [132]. In combination with appropriate separation techniques that take advantage of the different solubilities of the cucurbituril homologs, their receptor properties, and their elution behavior on stationary phases, the available procedures permit the synthesis of many members of this receptor family on a preparative scale.

Binding properties: Cucurbiturils have nonpolar cavities lined by cyclically arranged oxygen atoms at the two openings. These oxygen atoms converge, suggesting that they can interact with metal ions by ion–dipole interactions, like in crown ethers. Indeed, cucurbiturils bind efficiently to alkali and alkaline earth metal ions in water, as shown by the stability constants collected in Table 4.8. The binding selectivity correlates with the size of the ring and is most pronounced for the smallest CB[5], which binds most strongly to K^+, similar to 18-crown-6 [133]. The cation selectivity of the larger cucurbiturils is less pronounced due to the larger cavity openings, which prevent the oxygen atoms from making efficient contact with the metal ions [the largest ion has a radius of 1.69 Å (*cf.* Table 3.1), but the radius of the cavity opening of **4.59** is 1.95 Å ($d_{opening}$ = 3.9 Å)].

Table 4.8: Comparison of the affinity of cucurbiturils and 18-crown-6 to various metal ions in terms of log K_a values.

Receptor	Li$^+$	Na$^+$	K$^+$	Rb$^+$	Cs$^+$	Ca^{2+}
CB[5]	2.0	3.9	4.7	3.2	2.6	2.6
CB[6] (**4.59**)	2.4	3.9	3.8	4.3	5.3	4.2
CB[7]	2.3	3.4	3.5	3.4	3.5	4.3
CB[8]	1.7	2.5	2.7	2.6	2.6	3.3
18-Crown-6	–	0.8	2.0	1.6	1.0	1.3

The binding constants were measured in water at 298 K.

The cation affinity of cucurbiturils explains why they are generally much more soluble in aqueous Na_2SO_4, LiCl, KCl, CsCl, or $CaCl_2$ solutions than in water: the cations of these salts bind to the oxygen atoms at the cavity openings and thus mediate receptor hydration. As a result, the affinity of cucurbiturils for nonpolar guest molecules such as methane, ethane, or tetrahydrofuran is normally reduced. An exception is the improvement of the methane affinity of CB[5] by Li$^+$, Rb$^+$, and Cs$^+$, which is due to the enlargement of the cucurbituril cavity when these ions bind to the carbonyl groups and thereby remove solvating water molecules [134].

Early investigations by Mock also revealed a high affinity of **4.59** for primary alkylammonium salts or α,ω-diammonium alkanes (in H_2O/formic acid, 1:1 (v/v)), whose cationic head groups interact with the carbonyl groups at the cavity openings [135]. The stability of the complexes depends on the length of the alkyl chains in these substrates (Figure 4.73).

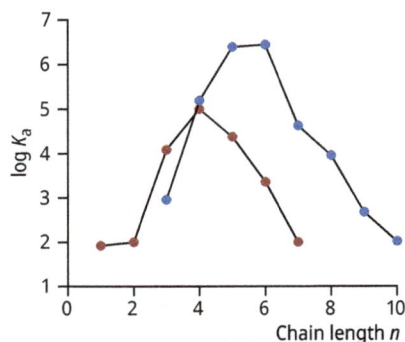

Figure 4.73: Dependence of the stability of the complexes of **4.59** in H_2O/formic acid, 1:1 (v/v) with alkylammonium ions $H-(CH_2)_n-NH_3^+$ (red circles) and α,ω-diammonium alkanes $^+H_3N-(CH_2)_n-NH_3^+$ (blue circles) on the chain length n.

The graphs in Figure 4.73 show that the stability of the alkylammonium complexes of **4.59** increases by about three orders of magnitude with increasing chain length up to the butyl derivative. A further increase in chain length then causes a marked decrease in affinity, with the heptyl derivative binding as weakly as the ethylammonium ion. The trend for α,ω-diammonium alkanes is similar, but in this case the maximum affinity is observed for the substrates with five or six carbon atoms between the cationic head groups. The maximum $\log K_a$ value is also almost two orders of magnitude larger than that of the most stable alkylammonium complex.

These trends reflect structural aspects of the complexes. Since guest protons oriented close to the cucurbituril carbonyl groups are deshielded and those included deep into their cavity are shielded, the signal shifts observed in the ^1H NMR spectra upon complex formation indicate that both types of substrates bind with their ammonium groups to the carbonyl groups of **4.59**, while the alkyl groups occupy the cavity. The higher stability of the complexes of dialkylammonium ions is due to the fact that they are stabilized by ion–dipole interactions at both cavity openings. Of the monoammonium ions, the butyl derivative binds most strongly because its alkyl chain has the optimal length to completely fill the cavity. Ammonium ions with shorter alkyl chains are too small to replace all the water molecules, while longer alkyl groups protrude from the cavity opening opposite to where the ammonium group is bound. Both situations are energetically unfavorable because they require hydration of nonpolar surfaces inside or outside the cavity. In the case of diammonium ions, chain lengths with five or six methylene units allow the two ammonium groups to bridge the carbonyl groups at both openings without creating strain.

According to calorimetric measurements, complex formation is exothermic with a favorable or unfavorable entropy term, depending on the actual substrate [136]. For example, the formation of the butylammonium complex is associated with a ΔH^0 of -26.8 kJ mol^{-1} and a $T\Delta S^0$ of -2.9 kJ mol^{-1}, whereas these parameters are -29.6 kJ mol^{-1} and 8.9 kJ mol^{-1}, respectively, for the 1,6-hexyldiammonium complex. We have seen similar thermodynamic signatures for the formation of CD, cyclophane, and cavitand complexes, where they have been attributed to the release of cavity water during complex formation. We will see below that cucurbiturils benefit even more from this effect.

Note also that the complexes of **4.59** with α,ω-diammonium alkanes approach stability constants $\log K_a$ of up to 7, which is unusually high compared to many other types of synthetic receptors, but characteristic of cucurbiturils. For example, CB[7] binds to ferrocene derivatives with $\log K_a$ values up to 15.5 in water [137]. With a record $\log K_a$ of 17.8, CB[7] has an attomolar affinity for a diamantane diammonium ion [138]. Figure 4.74a shows a selection of guests that form such high-affinity complexes, together with the corresponding thermodynamic parameters of complex formation.

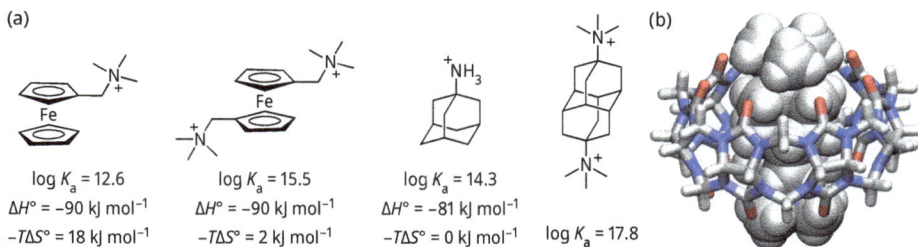

(a)

$\log K_a = 12.6$
$\Delta H° = -90$ kJ mol^{-1}
$-T\Delta S° = 18$ kJ mol^{-1}

$\log K_a = 15.5$
$\Delta H° = -90$ kJ mol^{-1}
$-T\Delta S° = 2$ kJ mol^{-1}

$\log K_a = 14.3$
$\Delta H° = -81$ kJ mol^{-1}
$-T\Delta S° = 0$ kJ mol^{-1}

(b)

$\log K_a = 17.8$

Figure 4.74: Structures of selected substrates that form highly stable complexes with CB[7], together with the thermodynamic parameters of complex formation (a), and crystal structure of the diamantane complex of CB[7] (b).

How do water molecules in a receptor cavity contribute to complex formation?

The extremely high affinities of these CB[7] complexes could result from the tight fit of the guest molecules into the rigid cucurbituril cavity, as illustrated by the crystal structure of the diamantane complex in Figure 4.74b. CB[7] is also very well preorganized for complex formation and the loss of flexibility associated with guest binding is therefore small. A further contribution to complex formation could come from the entropic gain associated with receptor dehydration. However, complex formation has a relatively small entropic component according to the thermodynamic parameters collected in Figure 4.74a, while the main thermodynamic driving force comes from the substantial negative enthalpy term. This strong exothermicity of complex formation is consistent with the tight interactions between receptor and substrate but it is striking that the binding enthalpy is relatively independent of the substrate structure. Therefore, Werner M. Nau and Frank Biedermann proposed another explanation for the exceptional binding properties of CB[7] in water [33, 139]. The underlying ideas were briefly presented in Section 3.3, where the favorable effect of cavity desolvation on complex formation was introduced. According to this concept, water molecules, which form an average of 3.62 hydrogen bonds with surrounding water molecules in bulk water, must give up hydrogen bonds when occupying the confined cavities of receptors lined with nonpolar subunits, with the size of the cavity determining how many water molecules can be accommodated and how many hydrogen bonds each water molecule must sacrifice.

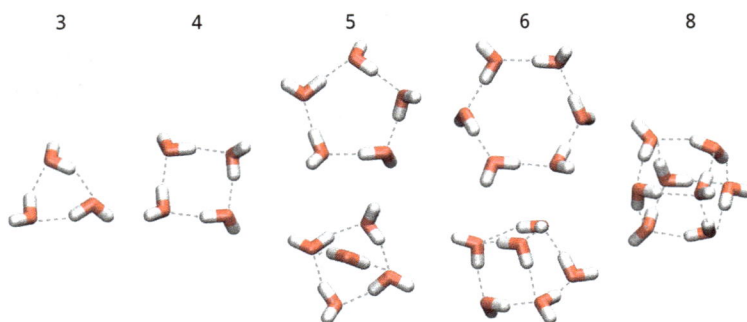

Figure 4.75: Examples of water cluster structures containing between three and eight water molecules.

This information was obtained for different receptors using molecular dynamics calculations. The results in Table 4.9 show that each receptor can accommodate a characteristic number of water molecules N, depending on its cavity size. In addition, the structural characterization of the corresponding water clusters provided information on how many hydrogen bonds m they contain per water molecule (Figure 4.75) [140].

Equation (4.1) can now be used to calculate the total number of hydrogen bonds regained when the receptor-bound water clusters are released into the bulk. In this equation, the number 3.62 represents the average number of hydrogen bonds per water molecule in bulk water:

$$Z = N(3.62 - m) \tag{4.1}$$

According to Table 4.9, the binding equilibria of all considered receptors should benefit to some extent from the release of cavity water. Since the associated recovery of hydrogen bonds represents an enthalpic gain, the dehydration of the receptor cavity contributes to complex formation with a negative enthalpy term. The total binding enthalpy may still end up being positive if other factors overcompensate this effect, but if not, complex formation should be exothermic, as indeed observed in many

Table 4.9: Parameters describing the cavity hydration of different receptors.

Receptor	N	m	Z
α-CD	3.6	2.86	3.1
β-CD	4.4	2.96	3.1
Calix[4]arene	0.8	2.15	1.2
Pillar[5]arene	0.5	1.24	1.2
CB[6]	3.3	1.71	6.3
CB[7]	7.9	2.52	8.7
CB[8]	13.1	3.06	7.3

N is the average number of water molecules in the respective cavity, m refers to the average number of hydrogen bonds per bound water molecule, and Z is the combined hydrogen bond deficit of these water molecules in comparison to bulk water. Z is derived from equation (4.1).

cases. Table 4.9 also shows that the enthalpy gain of complex formation should be largest for cucurbiturils because their cavities contain more water molecules than those of other receptors. CB[7], in particular, has the most unfavorable combination of N and m values. There are receptors where the cavity water molecules form fewer hydrogen bonds, but these receptors also contain less water so that Z, the number of hydrogen bonds regained when the cavity water is released, ends up being smaller than for CB[7]. Receptors with larger cavities than CB[7] bind more water molecules, but their overall hydrogen bond deficit is less pronounced. Thus, of the receptors collected in Table 4.9, CB[7] should benefit the most from cavity dehydration, which is consistent with the experimental results. Receptor dehydration thus provides a rationale for the extremely high stabilities of cucurbituril complexes in aqueous solution. Furthermore, this concept highlights the importance of solvent effects in binding equilibria.

The substrate scope of cucurbiturils is immense [141], ranging from noble gases and simple aliphatic hydrocarbons to biologically relevant molecules such as amino acids, peptides, neurotransmitters, hormones, drugs, and toxins. Complex formation generally benefits from the release of cavity water, but when ion–dipole interactions additionally contribute to stability, as in the case of cationic substrates, the complexes are typically 10–100 times more stable than those of neutral guests. Another interesting aspect is that the large CB[8] even allows the simultaneous complexation of two guests, either two individual substrates or two substituents of a larger molecule. These complexes are particularly stable when reinforced by interactions between the two guest molecules, as in complexes containing an electron-rich and an electron-poor aromatic π-system. Examples are the complexes between CB[8], a paraquat derivative, and an electron-rich aromatic guest such as 2,6-dihydroxynaphthalene, 1,4-dihydroxybenzene, tyrosine, dopamine, or tryptophan. To illustrate the binding mode, the crystal structure of a ternary CB[8] complex is shown in Figure 4.76 [142]. The formation of such complexes is accompanied by the development of a charge-transfer band in the UV–vis spectrum, which can be conveniently monitored by the associated color change of the solution. CB[8] acts as a *molecular glue* in these complexes, allowing large molecules such as polymers and peptides with suitable side chain substituents, to be linked together.

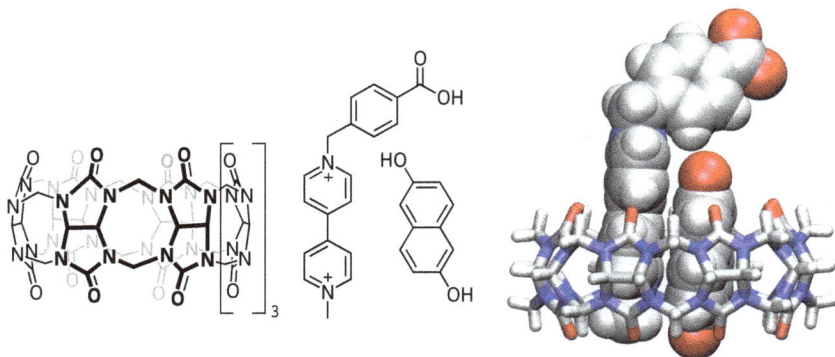

Figure 4.76: Molecular structures of CB[8], 2,6-dihydroxynaphthalene, and a paraquat derivative, and the crystal structure of the corresponding ternary complex.

Derivatives: A disadvantage of cucurbiturils is that they cannot be easily functionalized due to the lack of functional groups. Structural variations therefore usually involve changes at an early stage of the synthesis, for example by using glycoluril precursors with appropriate substituents. One method of postfunctionalization is the oxidation of the outward-facing methine protons with $K_2S_2O_8$, which converts **4.59** to the corresponding perhydoxylated derivative, whose OH groups are available for further modification. This oxidative method also allows the preparation of monohydroxylated cucurbiturils, which are valuable precursors for linking cucurbiturils to other molecules or supports. It is also possible to use a mixture of different glycolurils for macrocyclization, but this strategy results in mixtures of products that differ in the ratio and arrangement of the subunits along the ring. Overall, the methods for structural variation of cucurbiturils are more limited than those available for other macrocycles, and we therefore focus in the remainder of this section on selected cucurbituril analogs that have gained importance as receptors. Acyclic cucurbiturils have also been developed and will be discussed in Section 4.1.12.

Hemicucurbiturils are analogs of cucurbiturils with ethylene urea instead of glycoluril subunits. They lack the second row of methylene units, which makes them flexible. The preferred conformation has an alternating up-down arrangement of the carbonyl groups with a slight tilt of the individual subunits that causes the protons in the ethylene units to point into the cavity, as shown for the hexamer in Figure 4.77a. As a result, hemicucurbiturils differ from cucurbiturils in their binding properties, particularly in their ability to bind anions.

Another class of cucurbituril derivatives, this time with glycoluril subunits, has an even more pronounced anion affinity. In these compounds, each glycoluril residue contains substituted nitrogen atoms in one urea subunit, leaving only the unsubstituted nitrogen atoms to form the ring. The structure of hexamer **4.60** (Figure 4.77b) shows that, like hemicucurbiturils, these macrocycles have only one row of methylene groups and feature alternately oriented glycoluril subunits. In addition, the regions of positive electrostatic potential on the convex sides of the glycoluril subunits no longer face outward, as in cucurbiturils, but line the cavity (Figure 4.77c). Vladimír Šindelář, who first described this family of receptors, proposed the name bambus[6]urils because of their resemblance to the stem of a bamboo plant [143].

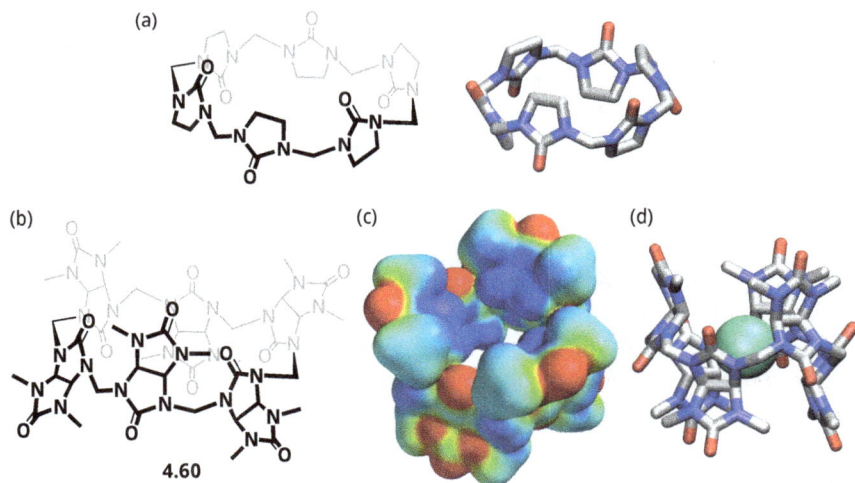

Figure 4.77: Molecular structure and calculated structure of a hexameric hemicucurbituril (a), molecular structure of bambus[6]uril **4.60** (b), electrostatic potential surface of **4.60** (c), and crystal structure of the chloride complex of **4.60** (d). The protons in both structures are not shown for reasons of clarity. The color coding covers a potential range from −100 to +100 kJ mol^{-1}, with red and blue indicating values greater than or equal to the absolute maximum in negative and positive potentials, respectively.

Complex formation involves the incorporation of the anion into the center of the cavity, where the positive electrostatic potential of the glycoluril moieties mediates the interaction with the oppositely charged ions, as illustrated by the crystal structure of the chloride complex of **4.60** (Figure 4.77d). The binding strength roughly correlates with the size of the anion, increasing as the anion becomes larger. Thus, a derivative of **4.60** with benzyl instead of methyl groups attached to the nitrogen atoms binds iodide ($\log K_a = 10.2$) and perchlorate ($\log K_a = 10.3$) with exceptional affinity in chloroform [144]. Appropriately substituted analogs of **4.60** have slightly lower but still remarkable anion affinities even in water, placing them among the most potent receptors for large, weakly coordinating anions in this solvent. The thermodynamic signatures of complex formation suggest that anion binding is mediated in part by the release of cavity water.

4.1.12 Clefts and tweezers

Introduction: So far we have mostly discussed cyclic receptors, and most receptors in supramolecular chemistry are indeed macrocyclic. This is because macrocyclic receptors are often better preorganized for substrate binding and therefore have a higher substrate affinity than their acyclic counterparts. Recall in this context the concept of the macrocyclic effect introduced in Section 3.5.2 and illustrated in Section 4.1.1 by comparing the binding properties of crown ethers and podands. Does the prominence of cyclic receptors mean that acyclic receptors are unimportant? The clear answer is no. Nature itself relies on acyclic systems for molecular recognition, the most important of which are proteins and polynucleotides. Moreover, synthetic acyclic receptors can be quite efficient, as we will see in this and the next section. Initially, we will focus on smaller systems in which suitably arranged functional groups produce well-preorganized binding sites. In Section 4.1.13, we will then discuss oligomeric receptors, whose folding, determined in a predictable way by the type and sequence of monomer units, leads to binding properties comparable to those of proteins.

The receptors discussed in this section have structures with concave regions where substrate binding takes place. Functional groups in the periphery of these binding sites participate in substrate binding or mediate substrate selectivity. Examples are the tripodal receptors derived from 1,3,5-trisubstituted triethylbenzene (Figure 4.78a) [145]. In these systems, the triethylbenzene core serves as a rigid platform to which binding sites are attached in the 1-, 3-, and 5-positions. Steric effects of the ethyl groups associated with the term *steric gearing* cause the six substituents along the ring to be alternately oriented up and down. As a result, the groups responsible for substrate binding are well preorganized to interact with a substrate. This receptor design is very versatile because it allows the recognition of substrates of very different nature by simply adapting the substituents along the ring. For example, the receptors **4.61** and **4.62** in Figure 4.78b have been designed to recognize cations (NH_4^+) and anions (citrate), respectively. We saw similar receptors in Section 2.1, where the BC_{50}^0 parameter was introduced (Figure 2.9). In Section 4.2.4 we will see that they are also useful for binding monosaccharides. A selection of other receptor families will be presented in more detail below. Various terms are used to refer to such receptors, examples are molecular clefts, tweezers, pincers, or clips, not all of which are well defined and are therefore sometimes used synonymously.

Rebek's clefts: Julius Rebek Jr. described a series of preorganized acyclic receptors whose binding properties benefit from the characteristic arrangement of the functional groups in *cis,cis*-1,3,5-trimethylcyclohexane-1,3,5-tricarboxylic acid or Kemp's triacid

(a)

(b)

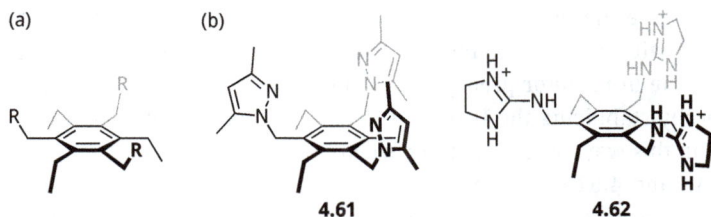

4.61

4.62

Figure 4.78: Preferred conformation of a 1,3,5-trisubstituted triethylbenzene derivative, illustrating the effect of steric gearing (a), and molecular structures of the tripodal receptors **4.61** and **4.62**, designed to recognize an ammonium ion and citrate, respectively (b).

[146, 147]. This cyclic tricarboxylic acid **4.63** (Figure 4.79) is locked into one ring conformation by the three methyl groups, which are bulkier than the planar carboxyl groups and therefore occupy the equatorial positions. As a result, the three carboxyl groups point into the same direction, rendering **4.63** an ideal basis for the design of receptors with convergent binding sites.

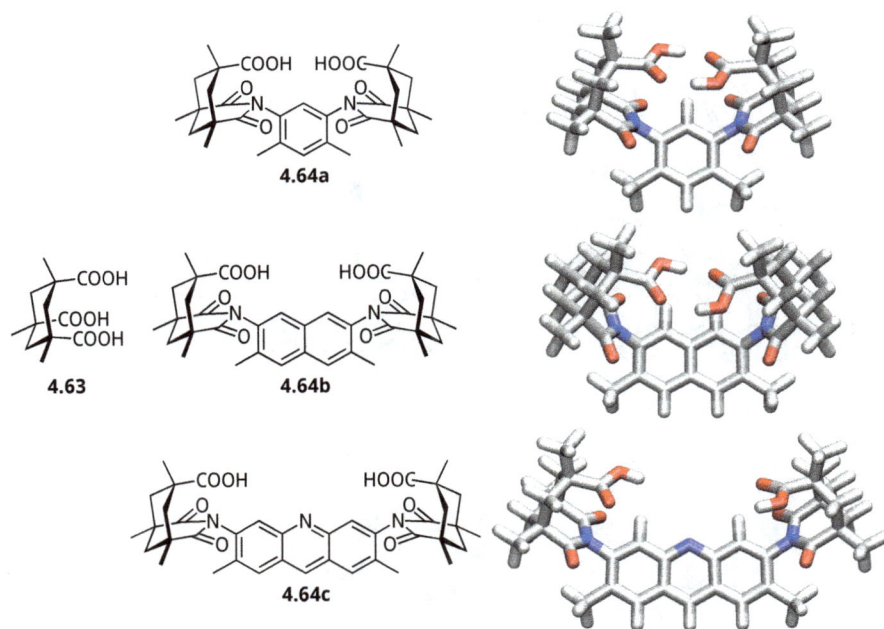

4.64a

4.63

4.64b

4.64c

Figure 4.79: Molecular structure of Kemp's triacid **4.63** and molecular and calculated structures of the clefts **4.64a**, **4.64b**, and **4.64c**.

Connecting two subunits of **4.63** to rigid aromatic scaffolds in such a way that their carboxyl groups face each other creates a binding site into which the substrate can be incorporated. Examples are the bis(imides) **4.64a**, **4.64b**, and **4.64c** (Figure 4.79), obtained by treating **4.63** with aromatic diamines. The methyl groups in the aromatic linkers play a crucial role in receptor preorganization by preventing rotation around the Ar–N bonds, thereby stabilizing the U-shaped structures. Because of the concave binding sites created in this way, these receptors are known as molecular clefts.

In receptors **4.64a** and **4.64b**, the carboxyl groups are close enough to interact intramolecularly by hydrogen bonding, thus requiring deprotonation for the coordination of divalent metal ions. Receptor **4.64c** has a richer host–guest chemistry due to the greater distance between the carboxyl groups. In its diprotonated form, **4.64c** binds to diamines either by hydrogen bonding as in the pyrazine complex or by salt bridge formation when the diamines are more basic. Both binding modes are shown in Figure 4.80a,b. If the aromatic diamine interacts with the acridine linker by additional aromatic interactions, the stability of the complex increases. The $\log K_a$ of the pyrazine complex of **4.64c** is 3.1 in chloroform, for example, whereas the quinoxaline (1,4-diazanaphthalene) complex has a $\log K_a$ of 4.4. Since salt bridges are stronger than hydrogen bonds, the 1,4-diazabicyclo[2.2.2]octane complex is even more stable ($\log K_a = 5.2$). These results demonstrate that the excellent preorganization of **4.64c** induces high substrate affinity even without the macrocyclic effect.

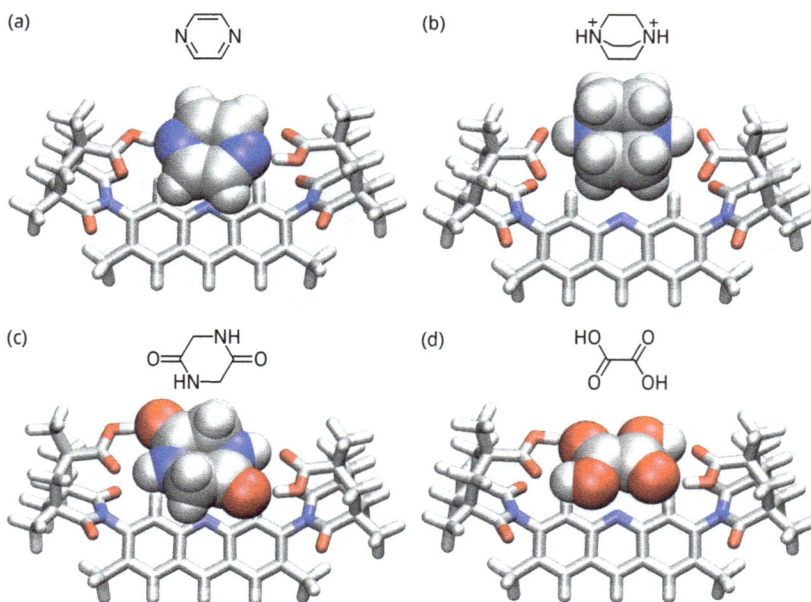

Figure 4.80: Binding modes in the complexes of **4.64c** with pyrazine (a), 1,4-diazabicyclo[2.2.2]octane (DABCO) (b), diketopiperazine (c), and oxalic acid (d). The binding of DABCO involves the transfer of two protons of the receptor to the substrate.

Cleft **4.64c** also forms complexes with diketopiperazines (Figure 4.80c) and with dicarboxylic acids such as oxalic or malonic acid (Figure 4.80d). Zwitterionic α-amino acids are bound by a combination of hydrogen bonding, ionic, and aromatic interactions (if the amino acid contains an aromatic side chain), but complex formation in this case involves the binding of two receptor molecules to a single substrate.

Another receptor type derived from Kemp's triacid is obtained by connecting two carboxyl groups in **4.63** *via* an NH group as in **4.65a** (Figure 4.81a). This receptor has an A-D-A hydrogen bonding pattern at the imide group, which induces affinity for complementary substrates such as 9-ethyladenine. The remaining carboxyl group can be used for further structural variation. In **4.65c**, **4.65d**, and **4.65e**, for example, the different aromatic residues enhance nucleobase binding by mediating additional aromatic interactions. Using the ΔG^0 associated with the formation of the 9-ethyladenine complex of methylamide **4.65b** as a reference, the affinities of the other receptors increase by a surprisingly constant energy contribution of about 1.8 kJ mol^{-1} per benzene ring, such that the anthryl moiety is responsible for about one-third of the total binding strength of **4.65e** [148]. This series of receptors thus provides quantitative insight into a type of interaction that contributes to the stabilization of the DNA double helix.

9-Ethyladenine binds to these receptors by either Watson–Crick or Hoogsteen-type base pairing (Figure 4.81b). Since both arrangements are possible, the U-shaped receptor **4.66**, in which two imide subunits of **4.65** are arranged in a convergent fashion, uses both binding modes for substrate recognition. The 9-ethyladenine complex of **4.66** has a log K_a of 4.0 in chloroform [149], again demonstrating how efficient well-designed acyclic receptors can be. We will return to Kemp's triacid-derived receptors in Chapter 9 on supramolecular catalysis.

Zimmerman's tweezers: The term *molecular tweezers* was introduced by Howard W. Whitlock for acyclic receptors with two aromatic substituents arranged in a parallel fashion at a distance that allows the intercalation of a suitable substrate [150]. The following examples of such receptors were developed in the group of Steven C. Zimmerman [151]. Receptor **4.67** consists of a 5,6,8,9-tetrahydrodibenz[*c,h*]acridine spacer to which two 9-anthryl groups are attached (Figure 4.82). The spacer arranges the substituents in a parallel fashion at a distance of about 7.2 Å as estimated from the distance of the respective protons in dihydrodibenz[*c,h*]acridine. While this distance is already well suited for the intercalation of planar aromatic systems (the inter-ring distance of π-systems in donor–acceptor complexes is typically between 3.2 and 3.5 Å), it can be further optimized, if necessary, by small adjustments of the angles between the anthryl units and the spacer in **4.67**. This receptor binds electron-poor aromatic guests, which are sandwiched between the anthryl groups. The log K_a of the 2,4,5,7-tetranitrofluorene (**4.68**) complex is, for example, 4.3 in chloroform. An analog of **4.67** in which one anthryl unit is replaced by an acridine moiety has an affinity approximately one order of magnitude lower, and the replacement of the second anthryl residue by an acridine moiety causes a further decrease in affinity. This trend shows that the stability of these complexes corre-

(a)

	4.65a	4.65b	4.65c	4.65d	4.65e
$\Delta G°$ (kJ mol^{-1})	−9.7	−11.4	−13.4	−15.1	
$\Delta G°_{Ar}$ (kJ mol^{-1})		0.0	−1.7	−3.7	−5.4

(b)

(c)

Watson–Crick   Hoogsteen

4.66

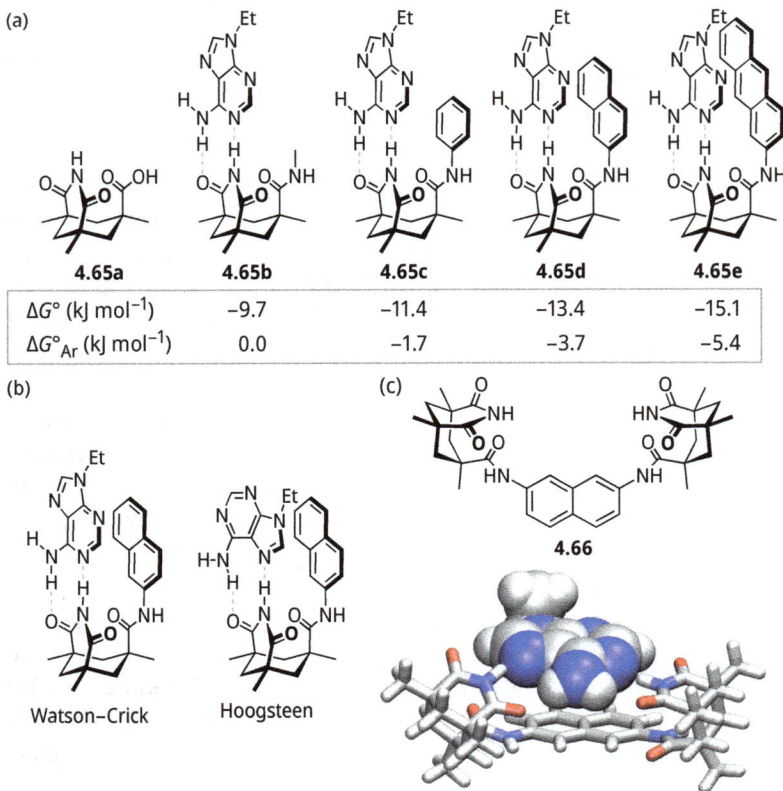

Figure 4.81: Molecular structures of imides **4.65a–e** together with the $\Delta G°$ values associated with their binding to 9-ethyladenine in chloroform (a). The $\Delta G°_{Ar}$ values are the differences between the $\Delta G°$ of **4.65b** and the $\Delta G°$ values of the other receptors. Structures of the Watson–Crick and Hoogsteen hydrogen bonding pattern between the A-D-A binding site of **4.64c** and 9-ethyladenine are shown in (b), and (c) shows the molecular structure of receptor **4.66** together with the calculated structure of its 9-ethyladenine complex.

lates with the electronic properties of their aromatic residues, becoming lower as electron-rich systems are replaced by electron-poor ones.

Receptor **4.69** (Figure 4.82) combines aromatic interactions and hydrogen bonding to the carboxyl group for substrate recognition. It binds 9-propyladenine with a $\log K_a$ of 4.4 in chloroform, whereas the complex of the dimethylated N,N-dimethyl-9-ethyladenine only has a $\log K_a = 2.6$. Accordingly, the 9-propyladenine complex benefits from the strengthening of the aromatic interactions by additional hydrogen bonding interactions.

Whether the binding of the substrate to the carboxyl group involves a Watson–Crick or a Hoogsteen-like arrangement (Figure 4.83) can be deduced from the stability of the complex between **4.69** and the monomethylated analog of N,N-dimethyl-9-ethyladenine, N-methyl-9-ethyladenine. For steric reasons, this compound almost exclusively adopts the

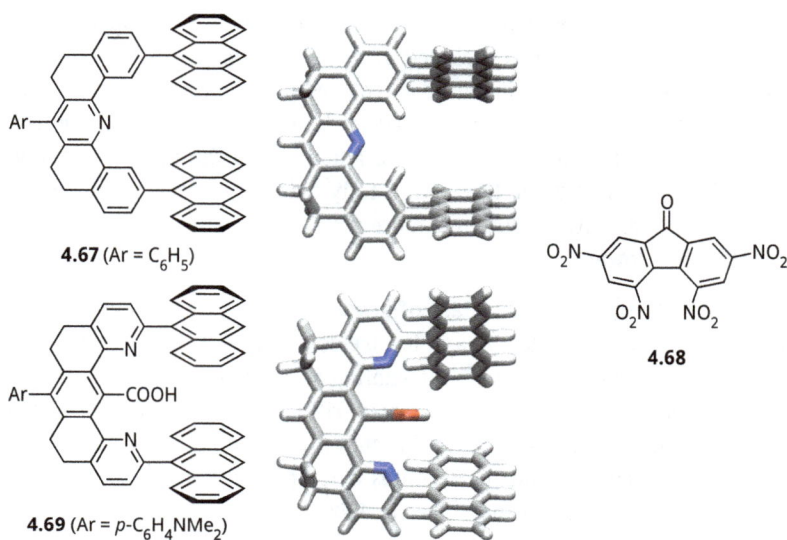

Figure 4.82: Molecular structures and calculated structures of the molecular tweezers **4.67**, **4.69**, and molecular structure of 2,4,5,7-tetranitrofluorene **4.68**. In the calculated structures, the aryl groups in the spacers are not shown for reasons of clarity.

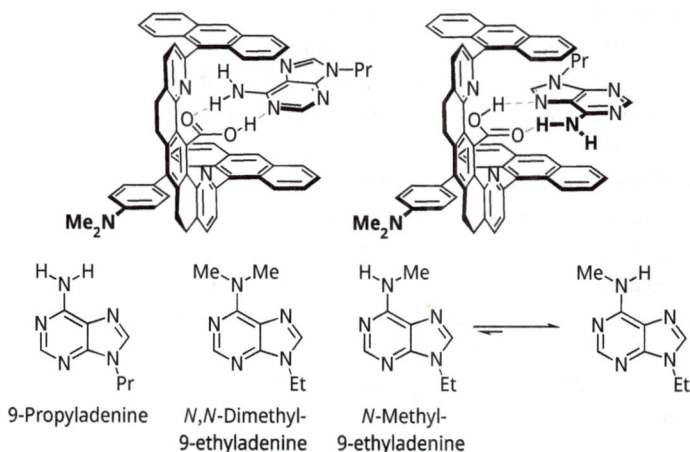

Figure 4.83: Watson–Crick- (left) and Hoogsteen-like (right) interactions of 9-propyladenine with the carboxyl group of **4.69** and structures of the adenine derivatives used for the structural assignment of the complex.

conformation with the *N*-methyl group oriented away from the imidazolyl ring (Figure 4.83). It is thus well preorganized for binding to **4.69** in the Hoogsteen arrangement, whereas it must adopt an energetically unfavorable conformation for Watson–Crick base pairing. Since the complexes of **4.69** with *N*-methyl-9-ethyladenine and 9-propyladenine have almost the same stability, it can be concluded that no energetically costly conformational reorganization of *N*-methyl-9-ethyladenine is required prior to complex formation. The Hoogsteen arrangement is therefore preferred in the complex, possibly due to a greater degree of overlap between the π-systems of adenine and the flanking anthryl groups in the complex. A derivative of **4.69** with a fully oxidized spacer has about an order of magnitude greater affinity because it is more rigid and the entropic cost of complex formation is therefore lower.

Klärner's tweezers and clips: Another class of acyclic receptors was developed in the group of Frank-Gerrit Klärner [152]. Examples are compounds **4.70**, **4.71**, and **4.72** (Figure 4.84), which are characterized by an alternating arrangement of aromatic and norbornadiene subunits. The latter provide a curved, ribbon-like arrangement of the aromatic subunits, each of which has one face directed toward the interior of the cavity. The number of the bicyclic subunits determines whether the structure is approximately cyclic, as in **4.70** and **4.71**, or U-shaped, as in **4.72**.

These receptors are all molecular tweezers in the Whitlock sense, but only the cyclic receptors are usually called tweezers, while the U-shaped receptors are called clips. These terms are preferred because they better illustrate that a substrate is included into the cavity of the tweezers and sandwiched between the aromatic sidewalls of the clips.

Figure 4.84 shows that the inner cavity surfaces of these receptors have a strong negative electrostatic potential. As a result, neutral or cationic electron-poor aromatic compounds are the preferred substrates. The cavity of **4.70** is too small to accommodate larger organic molecules, but the extended version **4.71a** forms stable complexes in chloroform with electron-poor aromatic compounds such as 1,2,4,5-tetracyanobenzene ($\log K_a > 5$) or *N*-methylpyridinium iodide ($\log K_a = 4.5$). According to the crystal structure of the 1,2,4,5-tetracyanobenzene complex, this substrate is fully included into the cavity, with the nitrile groups protruding from the cavity openings and the positively polarized protons oriented toward the receptor π-systems, indicating that aromatic interactions contribute to complex formation. In solution, the binding mode is similar, as demonstrated by the complexation-induced upfield shift of the 1,2,4,5-tetracyanobenzene signal by 3.6 ppm in the ^{1}H NMR spectrum.

1,2,4,5-Tetracyanobenzene also binds to the clip **4.72b**, but the complex has a $\log K_a$ of only 3.3. This complex is less stable than that of **4.71a** because **4.72b** is unable to fully encapsulate the guest and because complex formation is accompanied by a conformational adjustment of the receptor to reduce the distance between the aromatic sidewalls. This process is energetically costly, explaining the lower affinity of **4.72b** compared to **4.71a**. The induced fit observed for **4.72b** shows that these receptors are flexible enough to adapt to the structure of the substrate, despite the rigidity of the norbornadiene units.

The properties of water-soluble derivatives of these tweezers and clips were investigated in a collaboration between the Klärner group and the group of Thomas

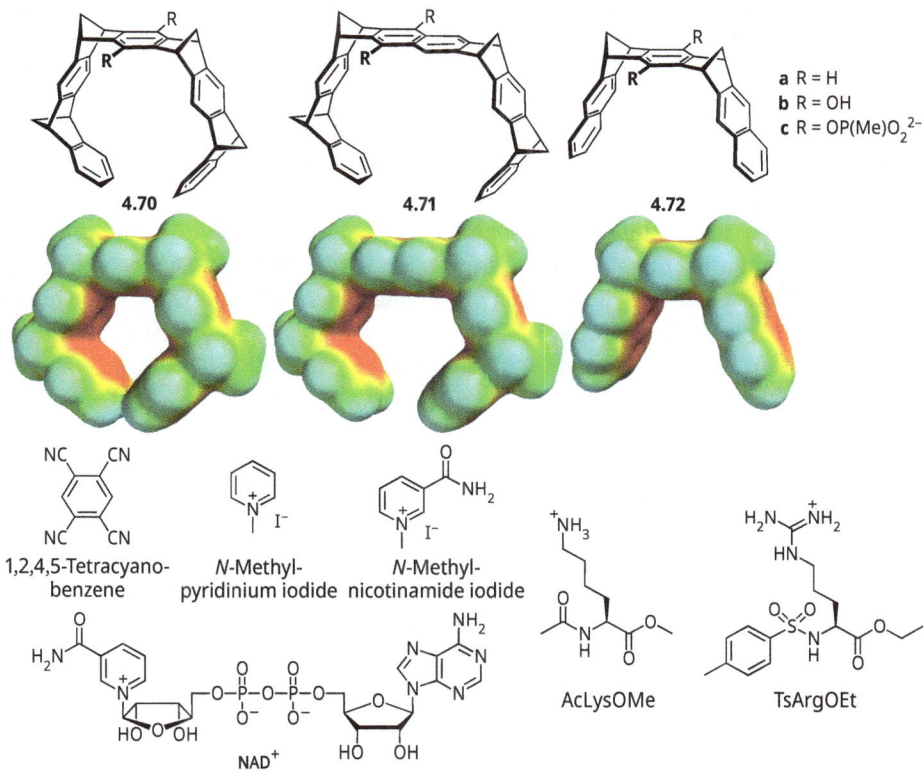

Figure 4.84: Molecular structures of the tweezers **4.70** and **4.71** and the clip **4.72** together with the corresponding electrostatic potential surfaces. The color coding covers a potential range from −100 to +100 kJ mol^{-1}, with red and blue indicating values greater than or equal to the absolute maximum in negative and positive potentials, respectively. The lower part of the figure shows structures of suitable substrates.

Schrader. Complex formation in this case benefits from the hydrophobic effect, as might be expected for receptors with a nonpolar binding site. For example, the affinity of **4.72c** for N-methylnicotinamide iodide increases when moving from methanol (log K_a = 4.2) to water (log K_a = 4.9) [153]. This clip also binds NAD$^+$ by sandwiching the nicotinamide moiety between the sidewalls, leaving the adenine moiety to interact with the receptor from the outside.

The spherical receptor **4.70c** efficiently recognizes the cationic groups in the side chains of lysine and arginine in water [154]. Lysine, in the form of the protected derivative AcLysOMe, is bound in phosphate buffer (pH 7.6) with a log K_a of 3.6, whereas the log K_a of the complex of the protected arginine derivative TsArgOEt is 3.2 (Figure 4.84). Tweezer **4.70c** and related receptors also recognize short peptides and even protein surfaces with exposed lysine and arginine side chains, thereby not only preventing the aggregation of pathological proteins involved in Alzheimer's or Parkinson's disease, but also inducing the dissolution of aggregates that have already formed [155]. See Section 12.2.1 for details.

Nolte's clips: The group of Roeland J. M. Nolte used a glycoluril unit as a core structure for the preparation of molecular clips [156]. An example is **4.73**, which contains 1,4-dimethoxybenzene-derived residues as sidewalls (Figure 4.85a). These clips interact preferentially with planar aromatic guests by sandwiching them between the aromatic residues and holding them in place by aromatic interactions. Guests containing suitably positioned hydrogen bond donors additionally interact with the carbonyl moieties of the central glycoluril moiety. To illustrate this binding mode, the calculated structure of the resorcinol complex of **4.73**, which has a log K_a of 3.4 in chloroform, is shown in Figure 4.85a [157].

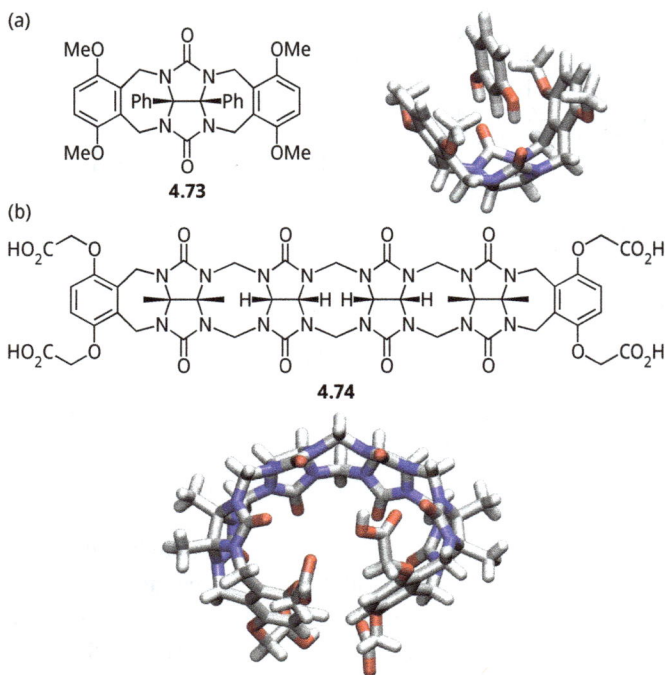

Figure 4.85: Molecular structure of the clip **4.73** and calculated structure of the complex of **4.73** with resorcinol to illustrate the hydrogen bonding interactions between the binding partners. The phenyl substituents in **4.73** are not shown for reasons of clarity. In (b), the molecular structure and crystal structure of the acyclic cucurbituril **4.74** is shown.

Isaacs's acyclic cucurbiturils: A disadvantage of both Nolte's clips and the tweezers developed by Zimmerman is that they are self-complementary. In the absence of suitable guests, these receptors therefore fill the space between the aromatic subunits by forming dimers, which must dissociate before substrate binding can take place. This tendency can be greatly reduced by giving acyclic receptors a curved shape with the end groups so close together that the cavity is virtually closed. Lyle Isaacs demonstrated this concept by replacing the single glycoluril unit in **4.73** with a C-shaped chain of four methylene-linked glycoluril units [158]. To illustrate the resulting shape, the crystal structure of a member of this receptor family, the acyclic cucurbituril **4.74**, is shown in Figure 4.85b. The four anionic substituents in this receptor ensure its solubility in water.

The binding properties of these so-called acyclic cucurbiturils are consistent with their structural relationship to cyclic cucurbiturils and cyclophanes. Accordingly, the combination of convergent carbonyl groups with aromatic residues induces a pronounced affinity for cations, which is further enhanced by the presence of the negatively charged substituents. Receptor **4.74**, for example, binds to the dihydrochloride of 1,6-diaminohexane with a $\log K_a$ of 8.2 in 20 mM NaH_2PO_4 buffer at pH 7.4. For comparison, the $\log K_a$ values of the corresponding complexes of CB[6] and CB[7] are 8.5 in 50 mM NaCl and 8.0 in 50 mM NaOAc buffer (pH 4.74), showing that the cation affinity of acyclic cucurbiturils is comparable to that of the cyclic counterparts [159]. Due to their high substrate affinity and wide structural variability, acyclic cucurbiturils have developed into a promising family of synthetic receptors for biomedical applications, as discussed in more detail in Section 12.2.1.

In conclusion, the acyclic receptors presented in this section clearly demonstrate that a cyclic structure is not always necessary to achieve efficient or selective binding. Acyclic receptors can be just as effective as cyclic ones, provided they are well preorganized. From a synthetic point of view, acyclic receptors have the advantage that they are generally easier to prepare than cyclic receptors, since the yield-limiting macrocyclization can be avoided. Therefore, such compounds should not be overlooked in receptor design.

4.1.13 Foldamers

Introduction: Although some natural receptors are macrocyclic, examples being ionophores such as valinomycin (Section 4.1.1), nature more often uses acyclic molecules with a specific sequence of subunits along the backbone for molecular recognition. These compounds fold in a characteristic way, thereby creating cavities or clefts into which the substrate is incorporated. Examples of such systems include proteins and polynucleotides. Because of their defined folding pattern, even in the absence of the substrate, these biopolymers are more or less preorganized for substrate binding, although some conformational adjustments may occur during the binding event to optimize the interactions through an induced fit (Section 3.5.1). Substrate affinity and selectivity are mediated by the shape of the binding site and the distribution of substituents along its inner surface.

Chemists have not only elucidated many of the underlying principles, but have also been able to mimic the folding and molecular recognition properties of such biosystems using synthetic oligomers. These compounds are called foldamers, a term derived from the combination of the words *folding* and *oligomer*. By definition, a foldamer is "any oligomer that folds into a conformationally ordered state in solution, the structures of which are stabilized by a collection of noncovalent interactions between nonadjacent monomer units" [160].

The design of foldamers is often inspired by nature. Amino acids, for example, are popular building blocks, but many systems also contain structural elements that have no natural counterpart. Conformations similar to those found in biopolymers, such as sheets or helices, are often targeted, but appropriately designed foldamers can also produce structures not found in nature. These conformations are stabilized intramolecularly by the same types of noncovalent interactions found in supramolecular systems. Supramolecular chemistry also comes into play when foldamers self-assemble or bind a suitable substrate within a cavity created by the folded chain. In the latter context, the term foldamer is sometimes also used for receptors that adopt a folded conformation only when binding to the substrate [161]. While the outcome is indistinguishable from a complex formed from an acyclic receptor that is folded in the absence of the binding partner, to qualify as a foldamer in the strict sense, a chain must be folded in the absence of the substrate, which is why we will limit our discussion to such systems.

Helices with cylindrical cavities: One approach to achieving substrate recognition with foldamers is to use an oligomer that folds helically around a sufficiently large cavity. Since such foldamers are often composed of sequences of identical subunits, their cavities are typically cylindrical in shape, with two openings and a constant diameter along the helix axis. The binding properties depend on how well the substrate fits into the cavity and how effectively it interacts with the functional groups lining the inner surface. These parameters are controlled by the number and structure of the foldamer subunits.

A crucial aspect of foldamer design is the appropriate choice of intramolecular interactions that stabilize the desired helical conformation. In the absence of direct interactions between the monomer units, solvophobic effects can be useful. For example, oligomer **4.75** (Figure 4.86) adopts an extended linear conformation in nonpolar media ($CDCl_3$) but a helical conformation in polar protic solvents (CD_3OD) [162]. In most cases, however, direct intramolecular interactions within the chain mediate the conformational stabilization. In this context, hydrogen bonds are particularly useful, which can be arranged in a predictable way along the foldamer backbone by the correct choice of subunits, as demonstrated by the oligohydrazide **4.76** [163]. In the case of the oligomer **4.77**, consisting of an alternating sequence of naphthyridine and pyrimidine units, the helical conformation is induced by the repulsion of the nitrogen lone pairs, which destabilize the *cisoid* conformations of the bonds between the aromatic subunits, making the alternative *transoid* arrangements the preferred ones [164].

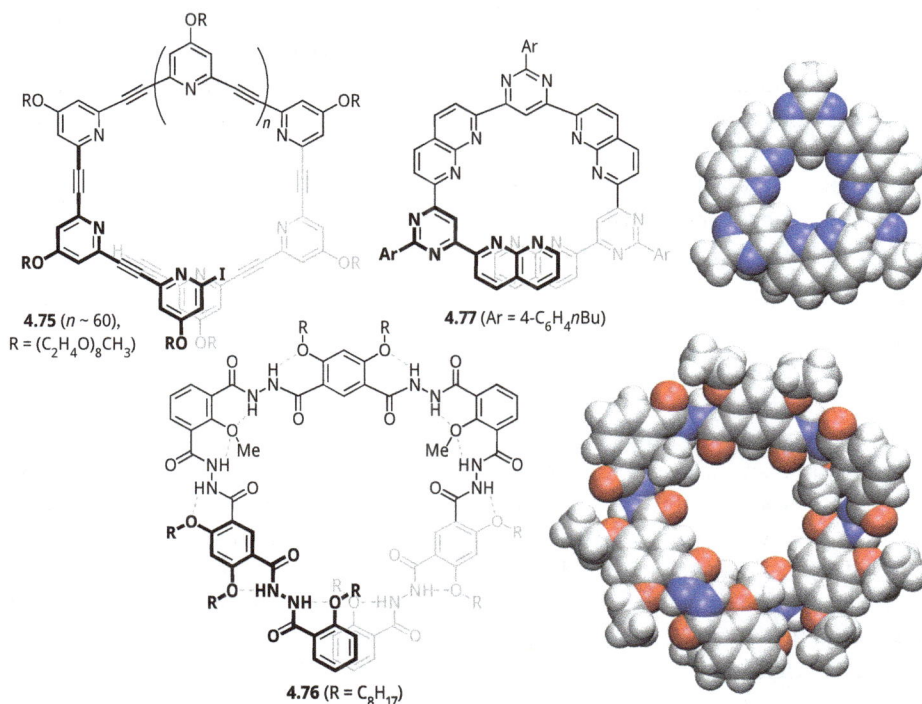

Figure 4.86: Molecular structures of foldamers **4.75**, **4.76**, and **4.77**. The calculated structures of **4.76** and **4.77** illustrate the helical arrangement of the chains surrounding cavities whose inner surfaces are lined with polar functional groups. The residues R in **4.76** and the aryl groups in **4.77** are replaced by ethyl and methyl groups, respectively, for the sake of clarity. The hydrogen bonds stabilizing **4.76** are shown as dashed lines.

All three foldamers contain negatively polarized heteroatoms along the inner cavity surface. As a result, foldamer **4.77** binds electropositive metal ions such as K^+, while the larger cavities of foldamers **4.75** and **4.76** allow them to complex monosaccharides. In the case of **4.76**, complex formation occurs in nonpolar media such as chloroform, resulting in complexes with stabilities in the mM range. The foldamer **4.75** with solubilizing polyethylene glycol residues, binds monosaccharides even in competitive protic media such as methanol or methanol/water, 10:1 (v/v), but not very efficiently ($\log K_a \sim 1$).

These achiral foldamers are dynamic, constantly reversing the sense of the two enantiomeric helix forms. When interacting with an achiral substrate, the right-handed and left-handed helices form complexes of equal stability, but when the substrate is chiral, the two complexes are diastereomeric and therefore differ in stability. Consequently, one helix sense usually predominates when the chiral substrate is enantiomerically pure or enriched. The induction of a specific helix sense by a chiral substrate can be conveniently monitored by circular dichroism (c.d.) spectroscopy, as it is accompanied by the appearance of characteristic bands in the c.d. spectrum. In contrast, no bands are visible in spectra of solutions containing two rapidly interconverting enantiomeric helices or foldamer complexes of achiral or racemic substrates where P- and M-configured helices are present in equal amounts. Since the two enantiomers of a chiral substrate induce helices of opposite sense, the c.d. spectra of both complexes are mirror images, and if the correlation between the absolute configuration of the substrate and the sign of the c.d. band is known, foldamer-based receptors can be used to probe the absolute configuration or *e.e.* of a chiral substrate.

Helices with bulged cavities: The group of Ivan Huc developed a toolbox of heterocyclic building blocks to construct helical oligoamide-based foldamers that are conformationally stabilized along the chain by strategically placed hydrogen bond donors (e.g., amide NH groups) and acceptors (e.g., ring nitrogen atoms) [165]. Each building block has a characteristic influence on the folding pattern and the dimension of the resulting helix, allowing precise conformational control by adjusting the subunit sequence.

This approach allows the design of foldamers with noncylindrical cavities. An example is **4.78** (Figure 4.87a), which contains a central sequence of three pyridine-derived units, flanked on either side by dimers of 8-substituted 2-quinolinecarboxylic acid. The terminal quinoline units cap the openings of a completely closed cavity that is wider in the central region and narrower at both ends. This structure resembles the skin of a peeled apple, which is why **4.78** has been termed the *molecular apple peel*.

While **4.78** can only bind one molecule of water, variations in the lengths of such oligoamides, the structures of the monomers, and their sequence can lead to foldamers with cavities large enough to accommodate organic molecules such as aliphatic alcohols or amines, dicarboxylic acids, or even monosaccharides. To illustrate the binding modes, the crystal structures of the water complex of **4.78** and the 1,4-butanediol and tartaric acid complexes of foldamers with larger cavities are shown in Figure 4.87b–d.

The lack of openings in the folded structures requires these oligoamides to undergo a substantial conformational reorganization during complex formation or dissociation. As a consequence, guest exchange is slow on the NMR timescale. However, in contrast to hemicarcerands, where guest exchange is also slow, the complexes of these oligoamides are thermodynamically stabilized by the interactions of the substrate with the convergent hydrogen bond donors and acceptors lining the inner cavity surface. This gain in ΔG^0 is reflected in the appreciable binding constants determined for some of these complexes. For example, the 1,4-butanediol complex of the foldamer shown in Figure 4.87c has a $\log K_a$ of 3.5 in chloroform [166]. Binding is also highly selective, as substrates that are too bulky to fit into the cavity, such as 1,3-butanediol, branched analogs of 1,4-butanediol, or 1,5-pentanediol, are not bound.

Chiral guests again induce one helix conformation over the corresponding mirror image. For example, D-tartaric acid is preferentially bound by a right-handed helix [167]. Due to of the high stability of this complex ($\log K_a > 6$ in CDCl$_3$/DMSO-d_6, 99:1 (v/v)), analogous foldamers with suitable solubilizing groups bind tartaric acid in more com-

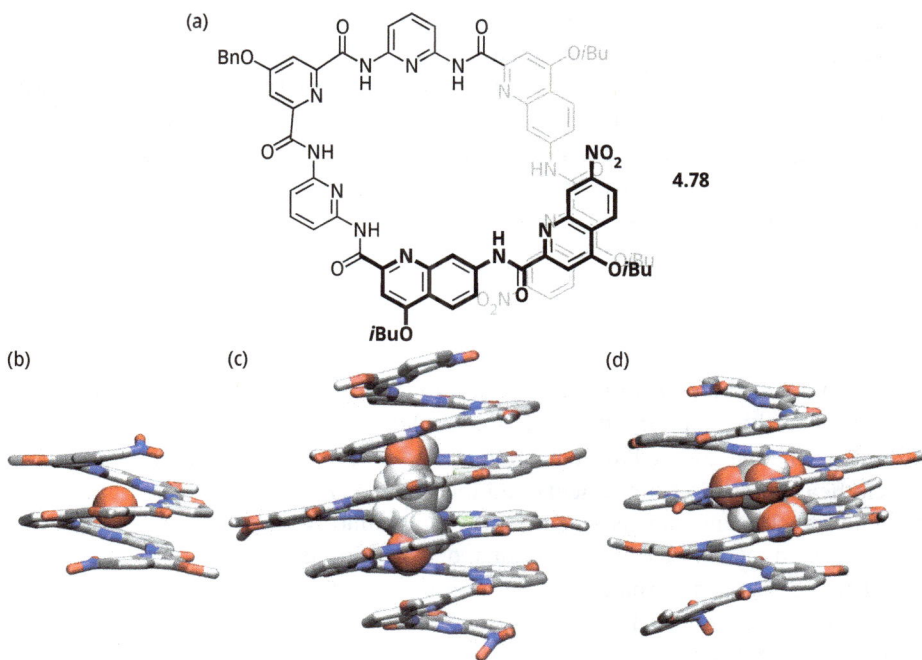

Figure 4.87: Molecular structure of foldamer **4.78**, which folds into a helix with a cavity large enough to host a water molecule (a), crystal structure of the corresponding complex (b), and crystal structures of related foldamers binding to 1,4-butanediol (c), and tartaric acid (d). Protons and side chains in the foldamers are not shown for reasons of clarity.

petitive media such as methanol or even water. Among the largest substrates that can be bound are monosaccharides, again with good selectivity [168].

Such foldamers thus combine high structural variability with the possibility to reliably control the folding pattern through the subunit sequence. In addition, they have chiral three-dimensional cavities in which substrates are shielded from the environment and engage in directed interactions with functional groups distributed along the inner surface. These attractive features are not found in many other receptors.

4.2 Substrates

Imagine that you have to design a receptor for a particular substrate. To approach this task, you would need to evaluate the properties and functional groups in the substrate and then decide which structural elements in the receptor are required to mediate binding. Sometimes there are additional aspects to consider. For example, if the receptor is to be used in water, its design must be based on water-soluble scaffolds that are known to work in a highly competitive medium. In addition, if the receptor is to be used in a biochemical context or even *in vivo*, toxicological aspects must be considered and the use of the toxicologically problematic crown ethers would not be an option. Finally, if your research group specializes in calixarenes, it may not be opportune to choose a cucurbituril as a receptor scaffold. In any case, your chances of success will benefit from a good overview of which receptors are available for which type of substrate. The purpose of this section is to provide such an overview and to help identify receptors for a given target.

4.2.1 Inorganic and organic cations

We start with cationic substrates because, after the discovery of crown ethers, cations were initially the main targets in supramolecular chemistry. Many receptors bind to positively charged guests, but their binding modes differ depending on the family to which they belong. To better understand the relationship, it is helpful to classify cations into transition metal ions, electropositive inorganic cations, and organic cations.

Transition metals have a rich coordination chemistry due to their ability to interact with Lewis-basic ligands. Some aspects of coordination chemistry are also important in supramolecular chemistry. For example, the chelate effect explains the thermodynamic stability of metal complexes with multidentate ligands and that of receptors with multiple binding sites. Similarly, the formation of kinetically labile transition metal complexes proceeds under thermodynamic control, as does the formation of supramolecular complexes held together by reversible interactions. Despite these relationships, the recognition of transition metal ions is best described by using the concepts of coordination chemistry and will therefore not be discussed here.

Electropositive metal ions from the first two groups of the periodic table have pharmacological properties (lithium), are responsible for maintaining the membrane potential of cells (sodium and potassium), are components of bone structure (calcium), or are responsible for hard water (calcium and magnesium). Salts of alkali or alkaline earth metal ions are also widely used in synthetic procedures, and the availability of receptors for these cations therefore has several practical implications (Chapter 12). Noncovalent interactions on which the complexation of these cations can be based are ion–dipole interactions with electronegative heteroatoms, especially the oxygen atoms of ethers or carbonyl compounds.

Important receptors for hard metal ions are the crown ethers (Section 4.1.1). Their cation affinity and selectivity are controlled by the ring size and by the presence of additional substituents, as in the case of the lariat ethers. Moving from monocyclic crown ethers to bicyclic cryptands usually leads to an improvement in affinity and selectivity (Section 4.1.2). Other receptors containing ethylene glycol-derived subunits also often interact with electropositive metal ions, even if they do not immediately qualify as crown ethers or cryptands. Spherands typically have an even higher cation affinity than cryptands (Section 4.1.3). However, they suffer from difficult synthetic accessibility and only bind well to the smallest cations (Li^+, Na^+). If other cations are to be targeted, hemispherands may be useful, but their cation affinity is lower than that of spherands due to their poorer preorganization. Calixarenes prefer to bind electropositive metal ions with their cyclically arranged OH groups along the narrower cavity opening (Section 4.1.7). One way to modulate substrate affinity and selectivity is to vary the ring size and thus the number of available OH groups. A more versatile approach is to modify the OH groups by introducing additional binding sites, as in the calix[4]arene derivative **4.37** (Figure 4.44). Cucurbiturils also have a cyclic arrangement of carbonyl groups along the cavity openings (Section 4.1.11). They therefore interact with metal ions, but usually not with marked selectivity. Table 4.10 summarizes the classes of receptors available for binding electropositive metal ions.

Table 4.10: Classes of receptors available for the recognition of electropositive metal ions and organic cations together with their preferred binding modes.

Substrate	Binding modes	Receptors
Alkali and alkaline earth metal ions	Ion–dipole interactions	Crown ethers, cryptands, spherands, calixarenes, cucurbiturils
Protonated amines	Salt bridges (ionic interactions combined with hydrogen bonding)	Negatively charged receptors such as appropriately substituted cryptophanes, calixarenes, deep cavitands, pillararenes, acyclic cucurbiturils, and molecular tweezers
	Hydrogen bonding	Neutral receptors such as crown ethers, cryptands, hemispherands, cucurbiturils
Protonated amines and quaternary ammonium ions	Cation–π interactions	Cryptophanes, calixarenes, deep cavitands, pillararenes
Quaternary ammonium ions	Ion–dipole interactions	Cucurbiturils

The third category of cationic guests includes protonated amines or quaternary ammonium ions. These cations differ structurally in a wide range, but they have in common that they are softer than electropositive metal ions, making them more prone to interact with a receptor by cation–π interactions. Important organic cations include the neurotransmitters acetylcholine, dopamine, noradrenaline, and adrenaline, or amino acids including side-chain methylated lysine and arginine derivatives, which play key roles in controlling gene activity and expression. These substrates are characterized by one or more positively charged head groups and additional organic residues. Suitable receptors therefore often, but not necessarily, contain two binding sites, one for the cationic group and one for the other part of the substrate.

Complexation of the cationic group can be mediated by oppositely charged residues in the receptor. This mode of binding is found, for example, in CDs with anionic substituents, water-soluble cryptophanes (e.g., **4.32**, Figure 4.35), sulfonatocalixarenes (e.g., **4.39a** or **4.39b**, Figure 4.44), deep cavitands (e.g., **4.49** or **4.50**, Figure 4.57), pillararenes (e.g., **4.56**, Figure 4.69), molecular tweezers (e.g., **4.70c**, Figure 4.84), and acyclic cucurbiturils (e.g., **4.74**, Figure 4.85). The efficiency of the underlying electrostatic interactions varies with the receptor structure. For example, the negatively charged groups in the octa acid **4.50** mainly mediate water solubility rather than substrate affinity.

All of these receptors have cavities that allow the inclusion of the uncharged part of the substrate, with the associated release of solvent molecules favorably contributing to complex stability. In some cases, the orientation of the substrate within the cavity is controlled by the arrangement of the negatively charged groups. For example, organic cations are preferentially included into the deep cavitand **4.49** with their cationic head group oriented toward the carboxyl groups lining the cavity opening.

The formation of these complexes is independent of whether the positively charged group of the substrate is a protonated amine or a quaternary ammonium group, unless the negatively charged binding sites of the receptor can contribute to substrate recognition by hydrogen bonding. In this case, protonated amines are bound by a combination of ionic interactions and hydrogen bonding, whereas the complexation of quaternary ammonium ions is mediated by ionic interactions alone.

A similar distinction can be made when the receptors are neutral. Protonated amines form hydrogen bonds with the carbonyl groups of cucurbiturils, for example, whereas the binding of the same hosts to quaternary ammonium ions is mediated by ion–dipole interactions. Hydrogen bonds between oxygen atoms and protonated amines also stabilize the complexes of crown ethers, cryptands, or hemispherands (e.g., **4.18**, Figure 4.20), whereas ion–dipole or even CH···O hydrogen bonds to quaternary ammonium ions are usually insufficient to induce binding (Section 3.1.5). Receptors with cavities lined by electron-rich aromatic subunits, such as cryptophanes, calixarenes, deep cavitands, or pillararenes, interact with cationic substrates through cation–π interactions. An overview of the receptors and the binding modes used to recognize organic cations is given in Table 4.10.

4.2.2 Inorganic and organic anions

It is of historical interest that the first receptors for anions were described by Chung Ho Park and Howard E. Simmons only one year after Pedersen reported the discovery of the cation-binding crown ethers [169]. These anion receptors were named *katapinands*, from the Greek καταπίνω (to swallow, to engulf), to illustrate that complex formation involves complete incorporation of the guests into the receptor cavity. An example of a katapinand is **4.79** (Figure 4.88a), in which two nitrogen atoms are linked by three nonamethylene chains. In its diprotonated form, **4.79** binds halides between the two bridgehead ammonium groups, where they are held by a combination of ionic interactions and hydrogen bonding. However, the interactions are not very strong, as shown by the K_a of the chloride complex, which is only 4 M^{-1} in CF$_3$COOD/D$_2$O, 1:1 (v/v). This weak binding is partly due to the poor preorganization of **4.79**, which, in the absence of anionic substrates, prefers a conformation with the NH protons outside the cavity and the protonated nitrogen atoms at the maximum distance due to charge repulsion. Anion binding therefore requires the receptor to undergo a significant conformational reorganization from the preferred *out–out* conformation to the less favorable *in–in* conformation, as shown in Figure 4.88b.

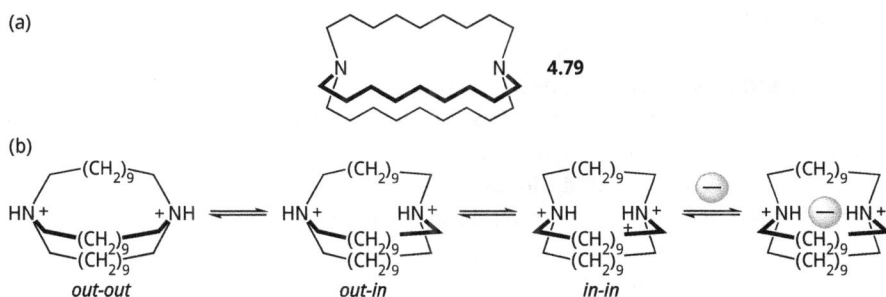

(a)

(b)

Figure 4.88: Molecular structure of katapinand **4.79** (a), and schematic representation of the conformational preorganization that the diprotonated form of this receptor must undergo during anion complexation (b).

Although the development of cation and anion receptors began almost simultaneously, the two fields progressed at very different rates. While the development of cation receptors flourished immediately after Pedersen's first paper, also due to the development of cryptands by Lehn and chiral crown ethers and spherands by Cram, the field of anion coordination chemistry took longer to develop, mainly because anions are more challenging substrates than cations for the following reasons:

– Inorganic anions are relatively large and therefore require receptors with larger cavities than cations.

– Electrostatic interactions with an anion are weaker than with an isoelectronic cation because the larger anion has a lower electron density.

– We have also seen in Section 3.1.3 that anions have higher free energies of hydration than cations of the same absolute charge and comparable size. Strong interactions are therefore required in water for a receptor to compete efficiently with the water molecules in the hydration shells of anions.

– Many oxoanions are involved in protonation equilibria in water. Phosphate buffer, for example, contains almost equal amounts of HPO$_4^{2-}$ and H$_2$PO$_4^-$ ions at physiological pH, making anion recognition pH-dependent.

– Finally, the structural diversity of anions, which in the case of inorganic anions can be spherical, linear, trigonal planar, tetrahedral, or octahedral, while (organic) polyanions have even more complex structures, must be taken into account in receptor design.

While these factors may have made the design of anion receptors more difficult than that of cation receptors in the early days of supramolecular chemistry, many of these difficulties have been overcome. As a result, anion coordination chemistry has become a mature and influential field of research, partly due to the relevance of many anions. Potential anionic substrates range from inorganic anions such as nitrate and phosphate, which are responsible for the eutrophication of water bodies, to toxic anions such as cyanide and arsenate, and biologically relevant anions such as chloride, sulfate, and especially phosphate. In addition, more than 70% of the substrates and cofactors involved in biological processes are negatively charged. The most important ones are nucleotides, polynucleotides, phosphorylated carbohydrates, phosphorylated proteins, and related biomolecules with carboxylate or sulfate groups. Accordingly, there is considerable interest in anion receptors for practical applications such as environmental monitoring or medicinal chemistry.

The main types of noncovalent interactions that mediate anion binding are ionic interactions with a positively charged receptor or hydrogen bonding with a hydrogen bond donor. More recently, the use of anion–π interactions or halogen bonding has also become popular. Many anion receptors also contain metal centers integrated into a suitable ligand framework to which an anion can coordinate. Table 4.11 provides an overview of receptors that utilize these binding modes.

Positively charged receptors that use only ionic interactions for anion binding are relatively rare, but we have encountered such receptors in Section 4.1.2 in the form of the macrotricyclic systems **4.10** and **4.11** (Figure 4.15), which contain quaternary ammonium ions in the bridgehead positions. These receptors adopt conformations with wide open cavities because charge repulsion causes the cationic groups to maximize their distance. These cavities are available for anion binding even in water, but complex stabilities are generally low.

Much higher affinities are typically observed for polyammonium-based receptors with protonated nitrogen atoms that combine ionic interactions with hydrogen bonding. Examples are the macrocyclic receptor **4.7** (Figure 4.9) or the polyazacryptands **4.12** and **4.13** (Figure 4.16). In contrast to receptors with quaternary ammonium groups, the charge state of these systems varies with pH, so that anion affinity typically increases with increasing degree of protonation. This pH dependence of anion binding is less pronounced for receptors containing amidinium or guanidinium groups, which have pK_a values between 11 and 13 and thus remain protonated over a much wider pH range than amines. The ability of amidinium and guanidinium groups to form two parallel hydrogen bonds to carboxylates, carbonates, phosphates, or sulfates (Figure 4.89a), which strengthen the ionic interactions, makes them particularly effective binding mo-

Table 4.11: Classes of receptors available for the recognition of anionic substrates and their preferred binding modes.

Binding modes	Receptors
Electrostatic interactions	Macrocyclic or macropolycyclic receptors with quaternary ammonium ions
Electrostatic interactions and hydrogen bonding	Polyammonium-based receptors or receptors containing guanidinium or amidinium groups
Hydrogen bonding	Receptors with hydrogen bond donors along a chain or within a macrocyclic or macropolycyclic framework
Anion–π interactions	Receptors with electron-deficient aromatic subunits
Halogen bonding	Receptors with halogen atoms bound to electron-deficient aromatic subunits
Metal coordination	Ligand frameworks incorporating coordinatively unsaturated metal centers or metal centers with weakly bound ligands

tifs for oxoanions. They are found, for example, in the tripodal receptor **4.62** shown in Figure 4.78b, or in the receptors shown in Figure 4.89b.

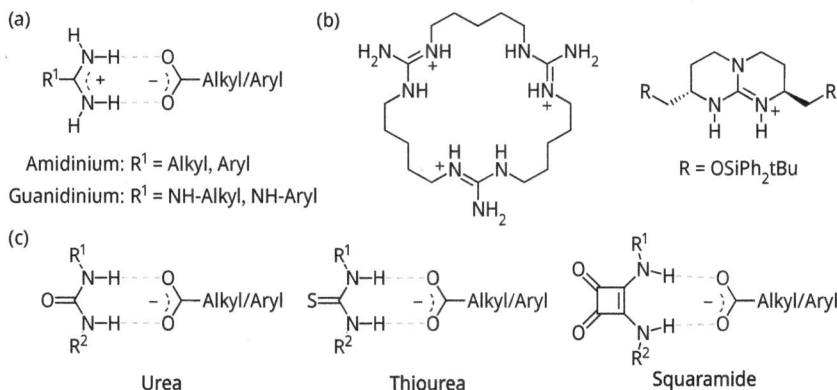

Figure 4.89: Schematic representation of the interaction of an amidinium group or a guanidinium group with a carboxylate ion (a), and examples of receptors containing guanidinium groups as binding sites (b). In (c), neutral binding motifs for oxoanions based on a urea, a thiourea, and a squaramide are shown. Phosphate or sulfate ions are bound in a similar way to the carboxylates in (a) and (c).

Neutral analogs of amidinium and guanidinium ions are ureas, thioureas, and squaramides (Figure 4.89c). These groups also form two hydrogen bonds to the oxygen atoms of oxoanions, but the corresponding complexes lack the additional stabilization provided by ionic interactions. Other receptors that form hydrogen bonds to anions contain two or more NH groups from amide or pyrrole moieties along the molecular framework. Calix[4]pyrrole **4.40** (Figure 4.46) and the receptors shown in Figure 4.90a are examples. Hydroxy groups also act as hydrogen bond donors, but are less common. Because the corresponding receptors are usually uncharged and relatively nonpolar, they are soluble only in organic media. Complex formation is therefore typically studied in solvents such as chloroform, DMSO, acetone, or acetonitrile.

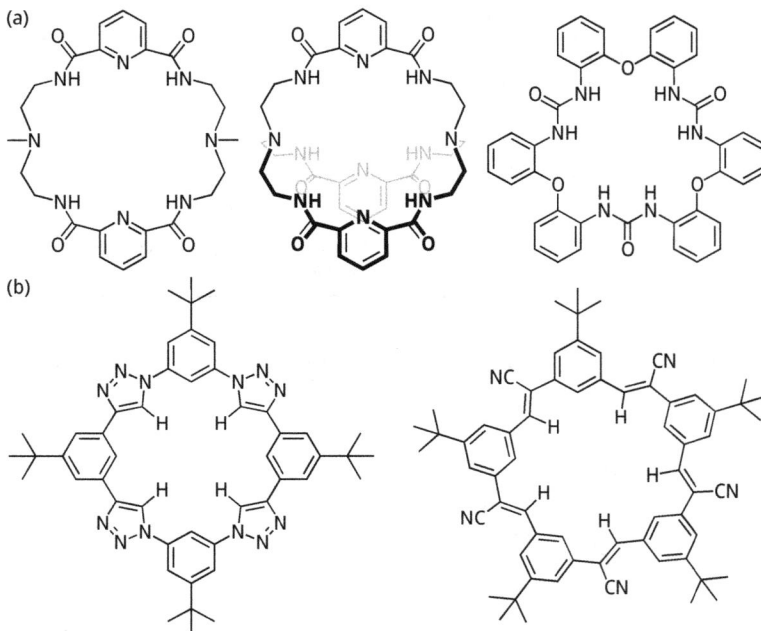

Figure 4.90: Examples of receptors containing NH (a) or CH (b) hydrogen bond donors for anion recognition.

The propensity of NH groups in ureas or amides to interact with anions can be fine-tuned by varying the substituents on the nitrogen atoms. In general, electron-withdrawing groups increase the acidity of the NH groups and thus their ability to act as hydrogen bond donors. A potential caveat is that NH groups that are too acidic will be deprotonated by basic anions, resulting in ion pairs. These ion pairs may be held together by a salt bridge, but if no structurally defined complex is formed or the ion pair even dissociates, the proton shift represents a simple acid–base reaction that should not be confused with a molecular recognition event.

Polarized CH bonds are also used for anion recognition. They are found, for example, in the two macrocyclic receptors shown in Figure 4.90b, both developed in the group of Amar Flood, and in the bambus[6]uril **4.60** (Figure 4.77) introduced in Section 4.1.11.

Anion–π interactions or halogen bonding require the receptor to contain electron-deficient aromatic moieties or halogen atoms attached to electron-withdrawing groups, respectively. We encountered such systems in Sections 3.1.9 and 3.1.6. Although the use of these interactions in anion coordination chemistry is still limited, halogen bonding, in particular, is becoming increasingly popular because of the high directionality of halogen bonds and their lower susceptibility to solvent effects compared to hydrogen bonds. Further applications of halogen bonding will be discussed in Sections 5.4, 7.7, and 9.3.1.

Metal-containing receptors are very common in anion coordination chemistry. They are often constructed by using multidentate ligands as molecular scaffolds that form chelate-type complexes with one or more transition metal ions, leaving at least one coordination site available for anion recognition. The macrocyclic and macrobicyclic polyamines **4.5**, **4.6** (Figure 4.9), and **4.14** (Figure 4.17) described in Sections 4.1.1 and 4.1.2 are examples of such ligands. One of the most frequently used metal ions in such receptors is zinc(II), which is diamagnetic and thus allows the use of NMR spectroscopy to characterize the structure and affinity of the corresponding receptors. In addition, copper(II) is popular because anion coordination to the metal center typically causes the receptor solutions to change color, which facilitates binding studies. These receptors are also efficient in water, benefiting from the directionality of coordinate bonds and their strength, which is hardly affected by solvation effects.

When targeting organic anions, complex formation usually benefits from the presence of additional binding motifs in the receptor with which the organic parts of the substrates can interact. This general principle was introduced in Section 3.5.3 on multivalent interactions. It applies not only to receptors for organic anions, such as carboxylates, nucleotides, or other phosphate esters, but also to receptors for ammonium ions, as discussed earlier. An application is the attachment of aromatic subunits to the backbone of polyammonium receptors to induce affinity for nucleotides. For example, in the adenosine triphosphate (ATP) receptor shown in Figure 4.91, the central ring binds to the triphosphate group of the substrate and the peripheral aromatic residues serve as binding sites for the nucleobase.

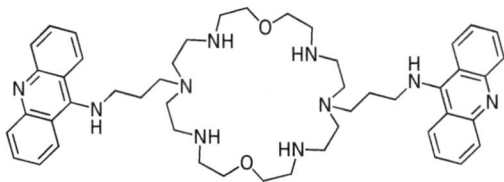

Figure 4.91: Example of a receptor that combines a polyazamacrocycle with appended aromatic residues to mediate the simultaneous recognition of the triphosphate and the nucleobase group of ATP.

4.2.3 Zwitterions and ion pairs

Zwitterion receptors contain two complementary receptor subunits, allowing them to bind simultaneously to both the cationic and anionic groups of the substrate. When designing such receptors, care must be taken to ensure that the two binding sites do not interact with each other, which would lead to either collapse of the receptor cavity if the interactions occur intramolecularly, or to noncovalent polymerization. Strategies to avoid these effects include using complementary recognition elements that have little or no affinity for each other, linking them by rigid spacers, or incorporating them into a macro(bi)cyclic framework.

Figure 4.92 shows a selection of crown ether-derived receptors for unprotected amino acids or short peptides to illustrate these concepts. The crown ether moiety in these receptors interacts with the ammonium groups of the substrates, while carboxylate binding is mediated by electrostatic interactions with either a quaternary ammonium ion, the protonated form of the macrocyclic polyamine **4.7** (Figure 4.9), the macrotricyclic cage **4.11** (Figure 4.15), or a guanidinium group.

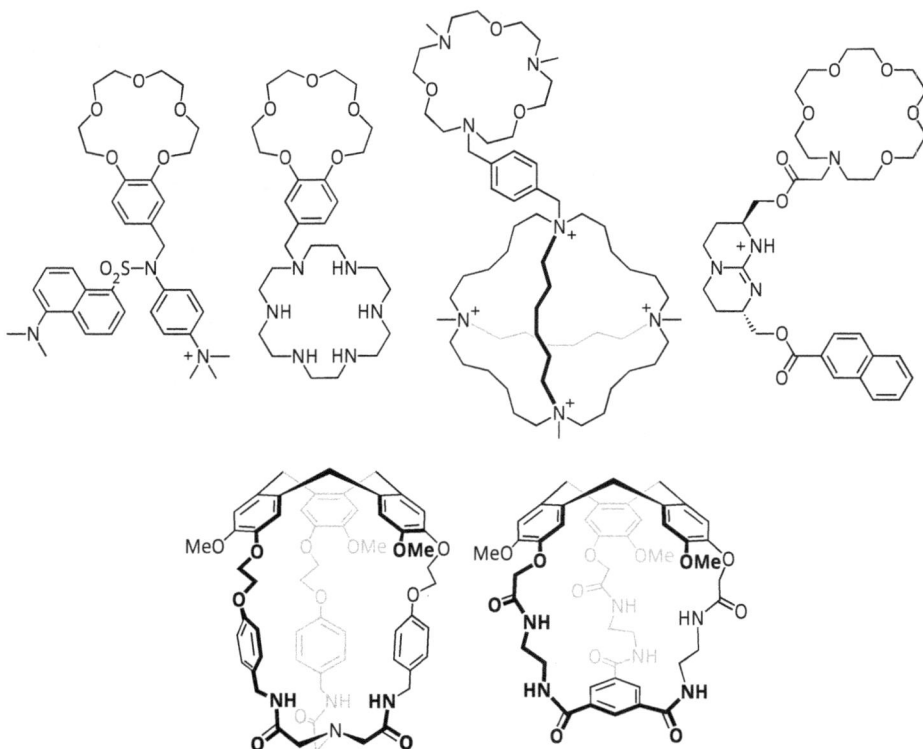

Figure 4.92: Examples of ditopic receptors for the recognition of zwitterions.

The other receptors in Figure 4.92 are hemicryptophanes in which the CTV moieties mediate cation binding while the amide NH groups at the opposite end of the cavity interact with the anionic part of the zwitterionic substrate by hydrogen bonding. These receptors bind to ω-aminocarboxylic acids, ω-aminosulfonic acids, and ω-aminophosphonic acids as we saw in Section 4.1.6.

Ion pair receptors are conceptually very similar to zwitterion receptors in that they also combine binding sites for a cation and an anion. However, unlike zwitterions, the components of an ion pair are not part of the same molecule. Nevertheless, they tend to stay together in nonpolar solvents because they are not well solvated. Three binding modes are distinguished for such receptors.

The first involves the binding of both ions as a contact ion pair. In this case, complex formation does not require the energetically unfavorable separation of cation and anion. Cascade complexes, such as that formed between the dicopper(II) complex of polyaza-cryptand **4.14** and an azide anion (Figure 4.17), are examples of this binding mode. Another example is based on receptor **4.80** (Figure 4.93a), developed by Bradley D. Smith, which contains two amide groups for chloride binding and the crown ether moiety as a binding site for the cation [170]. The cavity is small enough to accommodate KCl as a contact ion pair, as shown in the crystal structure in Figure 4.93a. That the binding of the first ion influences the binding of the counterion with positive cooperativity is demonstrated by the fact that the potassium complex of **4.80** binds chloride 13 times more strongly in DMSO than **4.80** alone. Conversely, the chloride complex of **4.80** has an approximately 40-fold higher K^+ affinity than **4.80**.

The second case involves complexation of the ions as a solvent-separated ion pair. This binding mode is observed for **4.81**, which is structurally related to **4.80** but has a larger cavity [171]. In the crystal structure of the NaCl complex, a chloroform molecule resides between the two ions (Figure 4.93b). The resulting separation of cation and anion is energetically unfavorable, which explains why the potassium complex of **4.81** has only about 9 times the affinity for Cl^- in DMSO-d_6/CD$_3$CN, 3:1 (v/v) than **4.81** alone.

Finally, ion pairs can be bound separately as host-separated ion pairs. Since complete separation of the ion pair is energetically very unfavorable, the communication between the cation and anion is rarely completely prevented in such complexes. In addition, interactions between the individual ions and the receptor partially compensate for the energy required to separate the ions, so that their binding also proceeds with positive cooperativity. An example is the CsF complex of calix[4]pyrrole **4.40** (Figure 4.93c) [79]. In this complex, the fluoride anion is hydrogen-bonded to the NH groups and the cesium cation occupies the shallow bowl-shaped cavity opposite the anion binding site, where it is held by electrostatic interactions with the anion and by cation–π interactions with the pyrrole moieties, as illustrated by the crystal structure in Figure 4.93c. Because of this binding mode, **4.40** can extract CsCl and CsBr but not CsNO$_3$ from an aqueous phase into nitrobenzene.

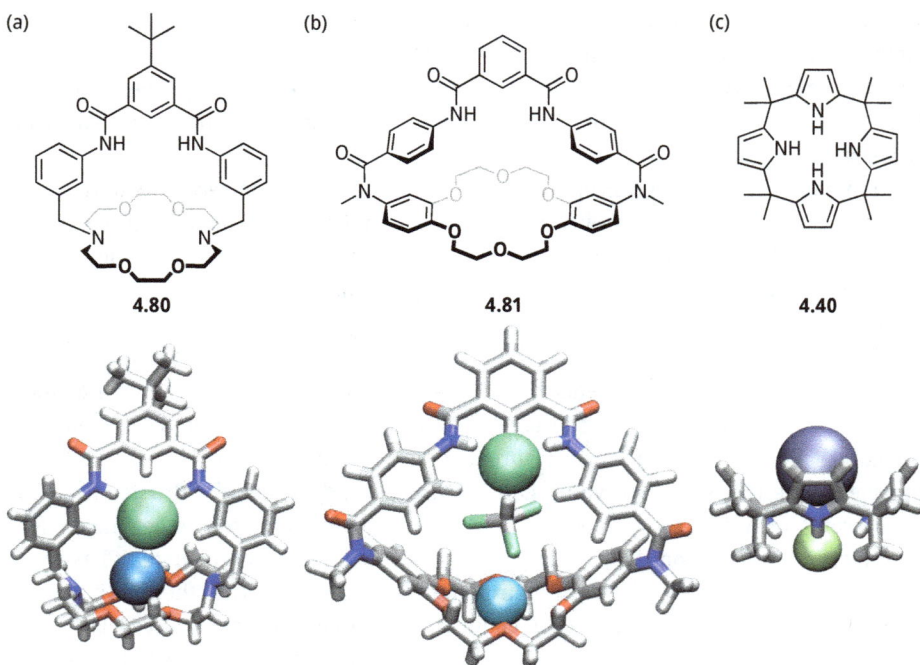

(a)   4.80   (b)   4.81   (c)   4.40

Figure 4.93: Molecular structures of the ion pair receptors **4.80** (a), **4.81** (b), and **4.40** (c) and crystal structures of their KCl (a), NaCl (b), and CsF (c) complexes.

How does the counterion influence the binding of an ion to a receptor?

While the complementary binding sites in the above receptors certainly help organize the ion pair, they are not responsible for bringing the two ions together. This is because cations and anions are strongly paired in organic solvents even without a receptor. It is therefore almost impossible to study the binding of an ion under these conditions without the interference of the counterion. Indeed, several studies have shown that the complexation of quaternary ammonium ions in chloroform by cyclophanes [172] or calixarenes [173] is influenced by the counterion in a characteristic way, with the cation affinity typically decreasing in the order picrate > I⁻ > Br⁻ > Cl⁻. This order is receptor-independent and therefore anion-specific. Analysis of the underlying equilibria using double-mutant cycles showed that the intrinsic strength of cation–π interactions between a receptor and a cation is not affected by the anion, leading the authors to conclude that the observed anion effect is likely due to the "multiple equilibria that are present and [the] competition for interaction sites between the anion and cation" [174]. Describing the binding of a cation to a receptor in organic solvents with a single equilibrium that considers only the formation of the cation complex is therefore an oversimplification. At the very least, the direct competition between ion pairing and cation binding must be considered, since the ion pair must generally dissociate to some extent to allow the cation to be complexed. Consequently, stronger interactions in the ion pair will weaken cation binding. This notion is consistent with the experimental observation that cation binding is particularly weak when the anion is small and has a high charge density. Conversely, anions with lower charge densities that do not form strong ion pairs, allow the cation to be strongly bound. In the case of organic anions, the ability of the host to position the anion close to the cation without having to separate the ion pair plays an additional role. Ion pairing therefore strongly influences complex formation in organic solvents and cannot be neglected. In water, ions of opposite charge are usually completely dissociated and strongly hydrated, making the binding of an ion to a receptor generally independent of the counterion.

4.2.4 Neutral organic molecules

The binding of polar neutral molecules to synthetic receptors benefits from dipole–dipole interactions or hydrogen bonding and becomes particularly effective when several of these interactions occur simultaneously. In the previous section, we have seen a number of receptors that use hydrogen bonding for substrate recognition, such as the Hamilton receptor **4.28** (Figure 4.30), the Kemp's triacid derivatives **4.65b–e** and **4.66** (Figure 4.81), and the foldamers with hydrogen bond acceptors and donors along their cavities. Other types of interactions that are used to mediate the binding of neutral substrates are aromatic interactions, such as the edge-to-face interactions that stabilize the durene complex of cyclophane **4.23** (Figure 4.28), the p-benzoquinone complex of cyclophane **4.27** (Figure 4.30), and the 1,2,4,5-tetracyanobenzene complex of the molecular tweezers **4.71** (Figure 4.84). Charge-transfer interactions between electron-rich and electron-deficient π-systems, which stabilize the 1,4-dimethoxybenzene complex of the blue box **4.26** (Figure 4.29) and the complexes of the molecular tweezers **4.67** (Figure 4.82), are also useful.

Substrates with regions of strong positive or negative electrostatic potential thus have anchor points with which a receptor can interact. In contrast, nonpolar substrates are much more difficult to bind in nonpolar solvents. This is because the contribution of electrostatic interactions to binding is small. In addition, nonpolar substrates are well solvated in organic media, and the situation inside the receptor cavity is often not very different from that in the bulk solvent, unless large surface areas are in contact, increasing the contributions of dispersion interactions. The release of solvent molecules from the receptor cavity can help, but usually does not lead to pronounced thermodynamic stability. A useful trick to force nonpolar substrates into a cavity is to perform binding studies in solvents that are too large to fit into the cavity, but this only works well for receptors that have shielded cavities, such as molecular cages. Examples are cryptophane-E **4.30**, cryptophane-A **4.31** (Figure 4.35), or deep cavitands. Complex formation in this case is driven by the gain in stability on the way from the thermodynamically unfavorable, potentially strained empty receptor to the state where the receptor makes contacts, however weak, with the substrate bound in the cavity.

Obviously, polar neutral substrates are relatively easy to bind in organic media, whereas the complexation of nonpolar substrates is more difficult. The situation is reversed in water, where complex formation is favored if the substrate has large nonpolar surfaces that are poorly hydrated and/or if the receptor cavity contains water molecules whose release is enthalpically or entropically favorable. We have encountered a number of receptors that benefit from these effects, not only CDs, but also sulfonatocalixarenes, an example of which is **4.39a** (Figure 4.44), water-soluble pillararenes, and cucurbiturils.

If the target is a polar substrate that is strongly solvated in water, binding is more difficult. In this context, carbohydrates are probably the most challenging substrate class because they differ from a cluster of water molecules mainly by the presence of the aliphatic CH groups. Let us first look at how carbohydrate recognition is achieved in nonpolar solvents. Under these conditions it is actually not very difficult to bind a carbohydrate whose OH groups are available for hydrogen bonding. Many carbohydrate receptors therefore contain a suitable array of hydrogen bond donors and acceptors along the binding site. The helical foldamers mentioned in Section 4.1.13 are examples. Other interactions useful for carbohydrate recognition are CH–π interactions between one or more aromatic subunits in the receptor and the aliphatic CH groups of the substrates. The acyclic tripodal receptors derived from triethylbenzene and the molecular cage shown in Figure 4.94a contain subunits that mediate these interactions. These receptors efficiently bind to carbohydrates containing long chain alcohols at their anomeric centers in organic media such as chloroform [175].

The same binding motifs are used to achieve carbohydrate binding in water. In this environment, it is necessary to effectively shield the binding site from the surrounding solvent molecules. Receptors **4.82** and **4.83** (Figure 4.94b) illustrate that this can be achieved by using cage-type receptor architectures that have aromatic residues to mediate CH–π interactions and linkers with multiple hydrogen bonding sites [176]. Both receptors contain peripheral substituents with multiple negative charges to ensure water solubility, help shield the cavities, and may even contribute to substrate recognition. These receptors were developed in the group of Anthony P. Davis, who introduced the term *temple* to describe their structures.

Receptor **4.82** binds preferentially to all-equatorial monosaccharides, such as β-D-glucose **4.84a** or the β-D-methylglucoside **4.84b** (Figure 4.94c). With a K_a of 9 M^{-1}, the affinity for **4.84a** in water is low, but **4.82** was the first receptor to demonstrate that monosaccharide binding is possible with abiotic receptors in this solvent. *N*-Acetylglucosamine **4.84c** and its β-methylglycoside **4.84d** form more stable complexes. For example, the K_a of the complex between **4.82** and **4.84d** is 630 M^{-1}, which is very similar to that of the complex between the same monosaccharide and the natural carbohydrate binder wheat germ agglutinin. With a log K_a of 4.3, **4.83** is the current record holder for glucose affinity in water [177]. Glucose is bound about 100 times more strongly than

(a)

(b)

4.82

4.83

(c)

4.84a R¹ = H, R² = OH
4.84b R¹ = CH₃, R² = OH
4.84c R¹ = H, R² = NHAc
4.84d R¹ = CH₃, R² = NHAc

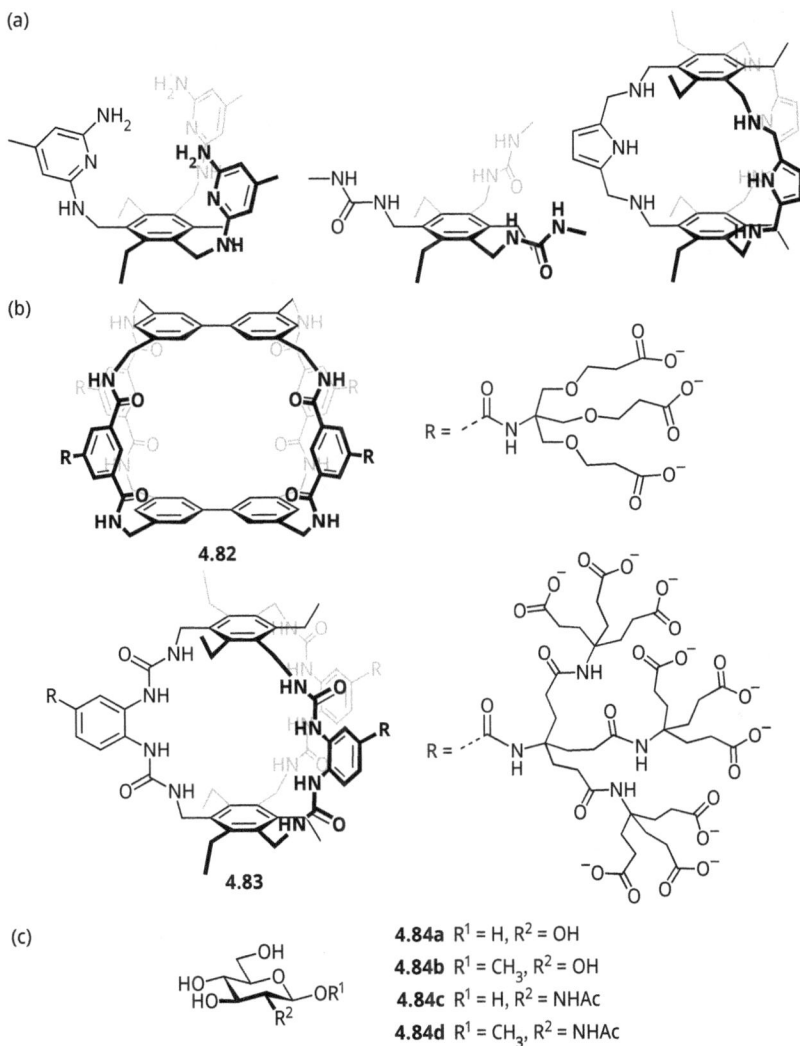

Figure 4.94: Examples of carbohydrate receptors active in nonpolar organic solvents (a) and in water (b), and structures of monosaccharides that were used as substrates (c).

galactose and about 1,000 times more strongly than other potential substrates, making this receptor useful for medicinal applications.

Carbohydrate recognition in water can also be achieved by using boronic acids, which form cyclic boronates with 1,2- and 1,3-diols [178]. Although substrate binding through covalent bonds does not strictly fall within the domain of supramolecular chemistry, the formation of boronates from carbohydrates has many features of supramolecular systems. First, boronate formation is reversible and relatively fast, even at room temperature and physiological pH. Second, the selectivity of boronate formation can be controlled by the structure of the boronic acid. The major advantage of binding carbohydrates with boronic acids is that the covalent bonds in the boronate esters are much more stable than any noncovalent interaction.

Figure 4.95a shows the equilibria underlying the reaction between a boronic acid and a diol. As can be seen, ester formation involves both trigonal and tetrahedral species. Because of the smaller O–B–O bond angle, the tetrahedral species is less strained than the ester with the trigonal boron center. Therefore, the tetrahedral product is usually favored, but its formation also depends on the pH, as indicated by the participation of hydroxide ions in the reaction.

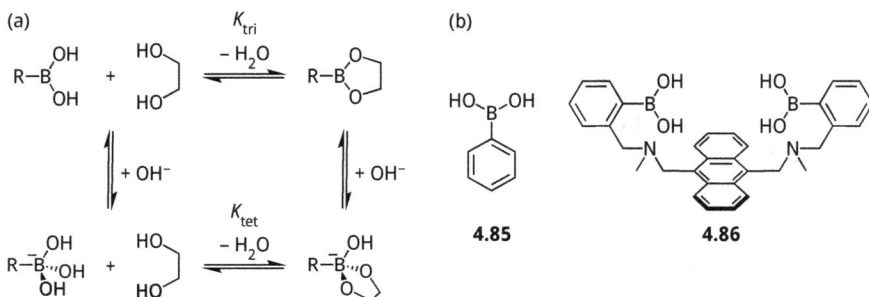

Figure 4.95: Equilibria describing the formation of boronates from a boronic acid and 1,2-ethanediol (a), and molecular structures of phenylboronic acid **4.85** and the bis(boronic acid)-derived glucose receptor **4.86** (b).

Simple boronic acids, such as phenylboronic acid **4.85** (Figure 4.95b), generally show selectivity for D-fructose. For practical applications, this intrinsic selectivity must be shifted away from D-fructose to other sugars, preferably D-glucose. A receptor with which this has been achieved is the bis(boronic acid) **4.86** developed by Tony D. James and Seiji Shinkai, which can react with two pairs of OH groups in the substrates if they are appropriately arranged [179]. The corresponding bis(boronate) of D-glucose has a high $\log K_a$ of 3.6 in methanol/water, 1:2 (v/v) buffered to pH 7.8, while D-fructose and D-galactose are inferior substrates under the same conditions, forming complexes with $\log K_a$ values of only 2.5 and 2.2, respectively. For applications of **4.86** in glucose sensing, see Section 12.2.3.

Bibliography

[1]   Pedersen CJ. Cyclic polyethers and their complexes with metal salts. J. Am. Chem. Soc. 1967, 89, 7017–36.

[2]   Pedersen CJ. The discovery of crown ethers (Nobel Lecture). Angew. Chem. Int. Ed. Engl. 1988, 27, 1021–7.

[3]   Hendrixson, RR, Mack, MP, Palmer, RA, Ottolenghi, A, Ghirardelli RG. Oral toxicity of the cyclic polyethers – 12-crown-4, 15-crown-5, and 18-crown-6 – in mice. Toxicol. Appl. Pharmacol. 1978, 33, 263–8.

[4]   Greene RN. 18-Crown-6: a strong complexing agent for alkali metal cations. Tetrahedron Lett. 1972, 13, 1793–6.

[5]   Busch DH. Structural definition of chemical templates and the prediction of new and unusual materials. J. Inclusion Phenom. Mol. Recognit. Chem. 1992, 12, 389–95.

[6]   Rose MC, Henkens RW. Stability of sodium and potassium complexes of valinomycin. Biochim. Biophys. Acta 1974, 372, 426–35.

[7]   Solov'ev VP, Strakhova NN, Raevsky OA, Rüdiger V, Schneider HJ. Solvent effects on crown ether complexations, J. Org. Chem. 1996, 61, 5221–6.

[8]   Schalley CA. Molecular recognition and supramolecular chemistry in the gas phase. Mass Spectrom. Rev. 2001, 20, 253–309.

[9]   Arnaud-Neu F, Delgado R, Chaves S. Critical evaluation of stability constants and thermodynamic functions of metal complexes of crown ethers (IUPAC Technical Report). Pure Appl. Chem. 2003, 75, 71–102.

[10]   Gokel GW, Goli DM, Minganti C, Echegoyen L. Clarification of the hole-size cation-diameter relationship in crown ethers and a new method for determining calcium cation homogeneous equilibrium binding constants. J. Am. Chem. Soc. 1983, 105, 6786–8.

[11]   Izatt RM, Bradshaw JS, Nielsen SA, Lamb JD, Christensen JJ, Sen D. Thermodynamic and kinetic data for cation-macrocycle interaction. Chem. Rev. 1985, 85, 271–339.

[12]   Bianchi A, Micheloni M, Paoletti P. Thermodynamic aspects of the polyazacycloalkane complexes with cations and anions. Coord. Chem. Rev. 1991, 110, 17–113.

[13]   Dietrich B, Lehn JM, Sauvage JP. Diaza-polyoxa-macrocycles et macrobicycles. Tetrahedron Lett. 1969, 10, 2885–8.

[14]   Dietrich B, Lehn JM, Sauvage JP. Les cryptates. Tetrahedron Lett. 1969, 10, 2889–92.

[15]   Brachvogel RC, Hampel H, von Delius M. Self-assembly of dynamic orthoester cryptates. Nat. Commun. 2015, 6, 7129.

[16]   Weber E, Vögtle F. Classification and nomenclature of coronands, cryptands, podands, and of their complexes. Inorg. Chim. Acta. 1980, 45, L65–L67.

[17]   Krakowiak KE, Bradshaw JS, Kou X, Dalley NK. One- and two-step metal ion templated syntheses of the cryptands. Tetrahedron. 1995, 51, 1599–606.

[18]   Haymore BL, Lamb JD, Izatt RM, Christensen JJ. Thermodynamic origin of the macrocyclic effect in crown ether complexes of sodium(1+), potassium(1+), and barium(2+). Inorg. Chem. 1982, 21, 1598–602.

[19]   Graf E, Kintzinger JP, Lehn JM, LeMoigne J. Molecular recognition. selective ammonium cryptates of synthetic receptor molecules possessing a tetrahedral recognition site. J. Am. Chem. Soc. 1982, 104, 1672–8.

[20]   Schmidtchen FP. Inclusion of anions in macrotricyclic quaternary ammonium salts. Angew. Chem. Int. Ed. Engl. 1977, 16, 720–1.

[21]   Dietrich B, Dilworth B, Lehn JM, Souchez JP, Cesario M, Guillhem J, Pascard C. Anion cryptates: synthesis, crystal structures, and complexation constants of fluoride and chloride inclusion complexes of polyammonium macrobicyclic ligands. Helv. Chim. Acta. 1996, 79, 569–87.

[22]   Dietrich B, Guillhem J, Lehn JM, Pascard C, Sonveaux E. Molecular recognition in anion coordination chemistry – Structure, binding constants and receptor–substrate complementarity of a series of anion cryptates of a macrobicyclic receptor molecule. Helv. Chim. Acta. 1984, 67, 91–104.

[23]   Amendola V, Fabbrizzi L, Mangano C, Pallavicini P, Poggi A, Taglietti A. Anion recognition by dimetallic cryptates. Coord. Chem. Rev. 2001, 219, 821–37.

[24]   Lehn JM. Dinuclear cryptates: dimetallic macropolycyclic inclusion complexes: concepts – design – prospects. Pure Appl. Chem. 1980, 52, 2441–59.

[25] Cram DJ. Preorganization – from solvents to spherands. Angew. Chem. Int. Ed. 1986, 25, 1039–57.

[26] Cram DJ, Lein GM. Host–guest complexation. 36. Spherand and lithium and sodium ion complexation rates and equilibria. J. Am. Chem. Soc. 1985, 107, 3657–68.

[27] Cram DJ, Dicker IB, Lauer M, Knobler CB, Trueblood KN. Host–guest complexation. 32. Spherands composed of cyclic urea and anisyl units. J. Am. Chem. Soc. 1984, 106, 7150–67.

[28] Crini G. Review: a history of cyclodextrins. Chem. Rev. 2014, 114, 10940–75.

[29] Saenger W. Cyclodextrin inclusion compounds in research and industry. Angew. Chem. Int. Ed. Engl. 1980, 19, 344–62.

[30] Manor PC, Saenger W. Topography of cyclodextrin inclusion complexes. III. Crystal and molecular structure of cyclohexaamylose hexahydrate, the water dimer inclusion complex. J. Am. Chem. Soc. 1974, 96, 3630–9.

[31] Rekharsky MV, Inoue Y. Complexation thermodynamics of cyclodextrins. Chem. Rev. 1998, 98, 1875–918.

[32] VanEtten RL, Sebastian JF, Clowes GA, Bender ML. Acceleration of phenyl ester cleavage by cycloamyloses. A model for enzymic specificity. J. Am. Chem. Soc. 1967, 89, 3242–53.

[33] Biedermann F, Nau WM, Schneider HJ. The hydrophobic effect revisited – studies with supramolecular complexes imply high-energy water as a noncovalent driving force. Angew. Chem. Int. Ed. 2014, 53, 11158–71.

[34] Assaf KI, Ural MS, Pan F, Georgiev T, Simova S, Rissanen K, Gabel D, Nau WM. Water structure recovery in chaotropic anion recognition: high-affinity binding of dodecaborate clusters to γ-cyclodextrin. Angew. Chem. Int. Ed. 2015, 54, 6852–6.

[35] Rekharsky MV, Inoue Y. Chiral recognition thermodynamics of β-cyclodextrin: the thermodynamic origin of enantioselectivity and the enthalpy–entropy compensation effect. J. Am. Chem. Soc. 2000, 122, 4418–35.

[36] Wenz G. Cyclodextrins as building blocks for supramolecular structures and functional units. Angew. Chem. Int. Ed. Engl. 1994, 33, 803–22.

[37] Wenz G. Influence of intramolecular hydrogen bonds on the binding potential of methylated β-cyclodextrin derivatives. Beilstein J. Org. Chem. 2012, 8, 1890–5.

[38] Liu J, Wang B, Przybylski C, Bistri-Aslanoff O, Ménand M, Zhang Y, Sollogoub M. Programmed synthesis of hepta-differentiated β-cyclodextrin: 1 out of 117655 arrangements. Angew. Chem. Int. Ed. 2021, 60, 12090–6.

[39] Breslow R, Greenspoon N, Guo T, Zarzycki R. Very strong binding of appropriate substrates by cyclodextrin dimers. J. Am. Chem. Soc. 1989, 111, 8296–7.

[40] Diederich F. Complexation of neutral molecules by cyclophane hosts. Angew. Chem. Int. Ed. Engl. 1988, 27, 362–86.

[41] Odashima K, Itai A, Iitaka Y, Koga K. Biomimetic studies using artificial systems. 3. Design, synthesis, and inclusion complex forming ability of a novel water-soluble paracyclophane possessing diphenylmethane skeletons. J. Org. Chem. 1985, 50, 4478–84.

[42] Smithrud DB, Sanford EM, Chao I, Ferguson SB, Carcanague DR, Evansek JD, Houk KN, Diederich F. Solvent effects in molecular recognition. Pure Appl. Chem. 1990, 62, 2227–36.

[43] Kearney PC, Mizoue LS, Kumpf RA, Forman JE, McCurdy A, Dougherty DA. Molecular recognition in aqueous media. New binding studies provide further insights into the cation–π interaction and related phenomena. J. Am. Chem. Soc. 1993, 115, 9907–19.

[44] Odell B, Reddington MV, Slawin AMZ, Spencer N, Stoddart JF, Williams DJ. Cyclobis(paraquat-*p*-phenylene). A tetracationic multipurpose receptor. Angew. Chem. Int. Ed. Engl. 1988, 27, 1547–50.

[45] Bühner M, Geuder W, Gries WK, Hünig S, Koch M. A novel type of cationic host molecules with π-acceptor properties. Angew. Chem. Int. Ed. Engl. 1988, 27, 1553–6.

[46] Chen XY, Chen H, Stoddart JF. The story of the little blue box: a tribute to Siegfried Hünig Angew. Chem. Int. Ed. 2023, 62, e202211387.

[47]   Hunter CA. Molecular recognition of *p*-benzoquinone by a macrocyclic host. J. Chem. Soc. Chem. Commun. 1991, 749–51.

[48]   Allott C, Adams H, Bernad Jr. PL, Hunter CA, Rotger C, Thomas JA. Hydrogen-bond recognition of cyclic dipeptides in water. Chem. Commun. 1998, 2449–50.

[49]   Chang SK, Hamilton AD. Molecular recognition of biologically interesting substrates: synthesis of an artificial receptor for barbiturates employing six hydrogen bonds. J. Am. Chem. Soc. 1988, 110, 1318–9.

[50]   Palmer LC, Rebek Jr. J. The ins and outs of molecular encapsulation. Org. Biomol. Chem. 2004, 2, 3051–9.

[51]   Avram L, Bar-Shir A. ^{19}F-GEST NMR: studying dynamic interactions in host–guest systems. Org. Chem. Front. 2019, 6, 1503–12.

[52]   Collet A. Cyclotriveratrylenes and cryptophanes. Tetrahedron. 1987, 43, 5725–59.

[53]   Canceill J, Collet A, Gottarelli G. Optical activity due to isotopic substitution. Synthesis, stereochemistry, and circular dichroism of (+)- and (–)-[2,7,12-^{2}H$_3$]cyclotribenzylene. J. Am. Chem. Soc. 1984, 106, 5997–6003.

[54]   Zimmermann H, Tolstoy P, Limbach HH, Poupko R, Luz Z. The saddle form of cyclotriveratrylene. J. Phys. Chem. B 2004, 108, 18772–8.

[55]   Canceill C, Lacombe L, Collet A. A new cryptophane forming unusually stable inclusion complexes with neutral guests in a lipophilic solvent. J. Am. Chem. Soc. 1986, 108, 4230–2.

[56]   Haberhauer G, Woitschetzki S, Bandmann H. Strongly underestimated dispersion energy in cryptophanes and their complexes. Nat. Commun. 2014, 5, 3542.

[57]   Canceill J, Cesario M, Collet A, Guilhem J, Lacombe L, Lozach B, Pascard C. Structure and properties of the cryptophane-E/CHCl$_3$ complex, a stable van der Waals molecule. Angew. Chem. Int. Ed. Engl. 1989, 28, 1246–8.

[58]   Collet A, Dutasta JP, Lozach B, Canceill J. Cyclotriveratrylenes and cryptophanes: Their synthesis and applications to host–guest chemistry and to the design of new materials. Top. Curr. Chem. 1993, 165, 103–29.

[59]   Canceill J, Lacombe L, Collet A. Analytical optical resolution of bromochlorofluoromethane by enantioselective inclusion into a tailor-made cryptophane and determination of its maximum rotation. J. Am. Chem. Soc. 1985, 107, 6993–6.

[60]   Garel L, Lozach B, Dutasta JP, Collet A. Remarkable effect of receptor size in the binding of acetylcholine and related ammonium ions to water-soluble cryptophanes. J. Am. Chem. Soc. 1993, 115, 11652–3.

[61]   Bartik K, Luhmer M, Dutasta JP, Collet A, Reisse J. ^{129}Xe and ^{1}H NMR Study of the reversible trapping of xenon by cryptophane-A in organic solution. J. Am. Chem. Soc. 1998, 120, 784–91.

[62]   Jacobson DR, Khan NS, Collé R, Fitzgerald R, Laureano-Pérez L, Bai Y, Dmochowski IJ. Measurement of radon and xenon binding to a cryptophane molecular host. Proc. Natl. Acad. Sci. U. S. A. 2011, 108, 10969–73.

[63]   Zhang D, Martinez A, Dutasta JP. Emergence of hemicryptophanes: from synthesis to applications for recognition, molecular machines, and supramolecular catalysis. Chem. Rev. 2017, 117, 4900–42.

[64]   Perraud O, Lefevre S, Robert V, Martinez A, Dutasta JP. Hemicryptophane host as efficient primary alkylammonium ion receptor. Org. Biomol. Chem. 2012, 10, 1056–9.

[65]   Perraud O, Robert V, Martinez A, Dutasta JP. The cooperative effect in ion-pair recognition by a ditopic hemicryptophane host. Chem. Eur. J. 2011, 17, 4177–82.

[66]   Perraud O, Robert V, Martinez A, Dutasta JP. A designed cavity for zwitterionic species: selective recognition of taurine in aqueous media. Chem. Eur. J. 2011, 17, 13405–8.

[67]   Zinke A, Ziegler E. Zur Kenntnis des Härtungsprozesses von Phenol-Formaldehyd-Harzen, X. Mitteilung. Ber. Chem. Chem. Ges. 1944, 77, 264–72.

[68] Cornforth JW, D'Arcy Hart P, Nicholls GA, Rees RJW, Stock JA. Antituberculous effects of certain surface-active polyoxyethylene ethers. Brit. J. Pharmacol. 1955, 10, 73–86.

[69] Gutsche CD, Muthukrishnan R. Calixarenes. 1. Analysis of the product mixtures produced by the base-catalyzed condensation of formaldehyde with *para*-substituted phenols. J. Org. Chem. 1978, 43, 4905–6.

[70] Original image taken by an anonymous photographer and published under the CC0 1.0 Universal (CC0 1.0) Public Domain Dedication license.

[71] Gutsche CD, Bauer LJ. Calixarenes. 13. The conformational properties of calix[4]arenes, calix[6]arenes, calix[8]arenes, and oxacalixarenes, J. Am. Chem. Soc. 1985, 107, 6052–9.

[72] Gutsche CD, Gutsche AE, Karaulov AI. Calixarenes 11. Crystal and molecular structure of *p-tert*-butylcalix[8]arene. J. Inclusion Phenom. 1985, 3, 447–51.

[73] Iwamoto K, Ikeda A, Araki K, Harada T, Shinkai S. "Cone"-"partial-cone" isomerism in tetramethoxycalix[4]arenes. influence of solvent polarity and metal ions. Tetrahedron. 1993, 49, 9937–46.

[74] Arduini A, Ghidini E, Pochini A, Ungaro R, Andreetti GD, Calestani G, Ugozzoli F. *p-t*-Butylcalix[4]arene tetra-acetamide: a new strong receptor for alkali cations. J. Inclusion Phenom. 1988, 6, 119–34.

[75] Pappalardo S, Parisi MF. Selective *endo*-calix complexation of linear alkylammonium cations by functionalized (1,3)-*p-tert*-butylcalix[5]crown ethers. J. Org. Chem. 1996, 61, 8724–5.

[76] Cui J, Uzunova VD, Guo DS, Wang K, Nau WM, Liu Y. Effect of lower-rim alkylation of p-sulfonatocalix[4]arene on the thermodynamics of host–guest complexation. Eur. J. Org. Chem. 2010, 1704–10.

[77] Gale PA, Sessler JL, Král V. Calixpyrroles. Chem. Commun. 1998, 1–8.

[78] Baeyer, A. Ueber ein Condensationsproduct von Pyrrol mit Aceton. Ber. Dtsch. Chem. Ges. 1886, 19, 2184–5.

[79] Custelcean R, Delmau LH, Moyer BA, Sessler JL, Cho WS, Gross D, Bates GW, Brooks SJ, Light ME, Gale PA. Calix[4]pyrroles: old yet new ion-pair receptor. Angew. Chem. Int. Ed. 2005, 44, 2537–42.

[80] Gross DE, Schmidtchen FP, Antonius W, Gale PA, Lynch VM, Sessler JL. Cooperative Binding of calix[4]pyrrole-anion complexes and alkylammonium cations in halogenated solvents. Chem. Eur. J. 2008, 14, 7822–7.

[81] Gale PA, Anzenbacher Jr. P, Sessler JL. Calixpyrroles II. Coord. Chem. Rev. 2001, 222, 57–102.

[82] Gale PA, Sessler JL, Král V, Lynch V. Calix[4]pyrroles: old yet new anion-binding agents. J. Am. Chem. Soc. 1996, 118, 5140–1.

[83] Sessler JL, Gross DE, Cho WS, Lynch VM, Schmidtchen FP, Bates GW, Light ME, Gale PA. Calix[4] pyrrole as a chloride anion receptor: solvent and countercation effects. J. Am. Chem. Soc. 2006, 128(37), 12281–8.

[84] Peng S, He Q, Vargas-Zúñiga GI, Qin L, Hwang I, Kim SK, Heo NJ, Lee CH, Dutta R, Sessler JL. Strapped calix[4]pyrroles: from syntheses to applications. Chem. Soc. Rev. 2020, 49, 865–907.

[85] Yoon DW, Hwang H, Lee CH. Synthesis of a strapped calix[4]pyrrole: structure and anion binding properties. Angew. Chem. Int. Ed. 2002, 41, 1757–9.

[86] Gil-Ramírez G, Escudero-Adán EC, Benet-Buchholz J, Ballester P. Quantitative evaluation of anion–π interactions in solution. Angew. Chem. Int. Ed. 2008, 47, 4114–8.

[87] Escobar L, Sun Q, Ballester P. Aryl-extended and super aryl-extended calix[4]pyrroles: design, synthesis, and applications. Acc. Chem. Res. 2023, 56, 500–13.

[88] Niederl JB, Vogel HJ. Aldehyde-resorcinol condensations. J. Am. Chem. Soc. 1940, 62, 2512–4.

[89] Högberg AGS. Cyclooligomeric phenol-aldehyde condensation products. 2. Stereoselective synthesis and DNMR study of two 1,8,15,22-tetraphenyl[14]metacyclophan-3,5,10,12,17,19,24,26-octols. J. Am. Chem. Soc. 1980, 102, 6046–50.

[90] Timmerman P, Verboom W, Reinhoudt DN. Resorcinarenes. Tetrahedron. 52, 1996, 2663–704.

[91]   Chwastek M, Szumna A. Higher analogues of resorcinarenes and pyrogallolarenes: bricks for supramolecular chemistry. Org. Lett. 2020, 22, 6838–41.

[92]   Chwastek M, Cmoch P, Szumna A. Dodecameric anion-sealed capsules based on pyrogallol[5]arenes and resorcin[5]arenes. Angew. Chem. Int. Ed. 2021, 60, 4540–4.

[93]   Abis L, Dalcanale E, Du Vosel A, Spera S. Structurally new macrocycles from the resorcinol-aldehyde condensation. Configurational and conformational analyses by means of dynamic NMR, NOE, and T1 experiments. J. Org. Chem. 1988, 53, 5475–9.

[94]   Schneider U, Schneider HJ. Synthese und Eigenschaften von Makrocyclen aus Resorcinen sowie von entsprechenden Derivaten und Wirt-Gast-Komplexen. Chem. Ber. 1994, 127, 2455–69.

[95]   Tunstad LM, Tucker JA, Dalcanale E, Weiser J, Bryant JA, Sherman JC, Helgeson RC, Knobler CB, Cram DJ. Host–guest complexation. 48. Octol building blocks for cavitands and carcerands. J. Org. Chem. 1989, 54, 1305–12.

[96]   Tanaka Y, Kato Y, Aoyama Y. Molecular recognition. 8. Two-point hydrogen-bonding interaction: a remarkable chain-length selectivity in the binding of dicarboxylic acids with resorcinol-aldehyde cyclotetramer as a multidentate host. J. Am. Chem. Soc. 1990, 112, 2807–8.

[97]   Moran JR, Karbach S, Cram DJ. Cavitands: synthetic molecular vessels. J. Am. Chem. Soc. 1982, 104, 5826–8.

[98]   Tucker JA, Knobler CB, Trueblood KN, Cram DJ. Host–guest complexation. 49. Cavitands containing two binding cavities. J. Am. Chem. Soc. 1989, 111, 3688–99.

[99]   Moran JR, Ericson JL, Dalcanale E, Bryant JA, Knobler CB, Cram DJ. Vases and kites as cavitands. J. Am. Chem. Soc. 1991, 113, 5707–14.

[100] Cram DJ, Choi HJ, Bryant JA, Knobler CB. Host–guest complexation. 62. Solvophobic and entropic driving forces for forming velcraplexes, which are 4-fold, lock-key dimers in organic media. J. Am. Chem. Soc. 1992, 114, 7748–65.

[101] Soncini P, Bonsignore S, Dalcanale E, Ugozzoli F. Cavitands as versatile molecular receptors. J. Org. Chem. 1992, 57, 4608–12.

[102] Rudkevich DM, Hilmersson G, Rebek Jr. J. Self-folding cavitands. J. Am. Chem. Soc. 1998, 120, 12216–25.

[103] Biedermann F, Schneider HJ. Experimental binding energies in supramolecular complexes. Chem. Rev. 2016, 116, 5216–300.

[104] Biros SM, Rebek Jr. J. Structure and binding properties of water-soluble cavitands and capsules. Chem. Soc. Rev. 2007, 36, 93–104.

[105] Biros SM, Ullrich EC, Hof F, Trembleau L, Rebek Jr. J. Kinetically stable complexes in water: the role of hydration and hydrophobicity. J. Am. Chem. Soc. 2004, 126, 2870–6.

[106] Laughrey Z, Gibb BC. Water-soluble, self-assembling container molecules: an update. Chem. Soc. Rev. 2011, 40, 363–86.

[107] Gibb CLD, Gibb BC. Anion binding to hydrophobic concavity is central to the salting-in effects of Hofmeister chaotropes. J. Am. Chem. Soc. 2011, 133, 7344–7.

[108] Jasat A, Sherman JC. Carceplexes and hemicarceplexes. Chem. Rev. 1999, 99, 931–68.

[109] Sherman JC, Cram DJ. Carcerand interiors provide a new phase of matter. J. Am. Chem. Soc. 1989, 111, 4527–8.

[110] Chapman RG, Chopra N, Cochien ED, Sherman JC. Carceplex formation: scope of a remarkably efficient encapsulation reaction. J. Am. Chem. Soc. 1994, 116, 369–70.

[111] Cram DJ. Molecular container compounds. Nature. 1992, 356, 29–36.

[112] Tanner ME, Knobler CB, Cram DJ. Hemicarcerands permit entrance to and egress from their inside phases with high structural recognition and activation free energies. J. Am. Chem. Soc. 1990, 112, 1659–60.

[113] Cram DJ, Blanda MT, Paek K, Knobler CB. Constrictive and intrinsic binding in a hemicarcerand containing four portals. J. Am. Chem. Soc. 1992, 114, 7765–73.

[114] Cram DJ, Tanner ME, Knobler CB. Host–guest complexation. 58. Guest release and capture by hemicarcerands introduces the phenomenon of constrictive binding. J. Am. Chem. Soc. 1991, 113, 7717–27.

[115] Houk KN, Nakamura K, Sheu C, Keating AE. Gating as a control element in constrictive binding and guest release by hemicarcerands. Science. 1996, 273, 627–9.

[116] Cram DJ, Tanner ME, Thomas R. The taming of cyclobutadiene. Angew. Chem. Int. Ed. Engl. 1991, 30, 1024–7.

[117] Warmuth R. o-Benzyne: strained alkyne or cumulene? NMR characterization in a molecular container. Angew. Chem. Int. Ed. Engl. 1997, 36, 1347–50.

[118] Warmuth R. Inner-phase stabilization of reactive intermediates. Eur. J. Org. Chem. 2001, 423–37.

[119] Ogoshi T, Kanai S, Fujinami S, Yamagishi TA, Nakamoto Y. para-Bridged symmetrical pillar[5]arenes: their Lewis acid catalyzed synthesis and host–guest property. J. Am. Chem. Soc. 2008, 130, 5022–3.

[120] Ogoshi T, Yamagishi TA, Nakamoto Y. Pillar-shaped macrocyclic hosts pillar[n]arenes: new key players for supramolecular chemistry. Chem. Rev. 2016, 116, 7937–8002.

[121] Ogoshi T, Masaki K, Shiga R, Kitajima K, Yamagishi TA. Planar-chiral macrocyclic host pillar[5]arene: no rotation of units and isolation of enantiomers by introducing bulky substituents. Org. Lett. 2011, 13, 1264–6.

[122] Cao D, Kou Y, Liang J, Chen Z, Wang L, Meier H. A facile and efficient preparation of pillararenes and a pillarquinone. Angew. Chem. Int. Ed. 2009, 48, 9721–3.

[123] Strutt NL, Zhang H, Schneebeli ST, Stoddart JF. Functionalizing pillar[n]arenes. Acc. Chem. Res. 2014, 47, 2631–42.

[124] Vincent SP, Chen W. Copillar[5]arene chemistry: synthesis and applications. Synthesis. 2023, 55, 246–62.

[125] Ogoshi T, Hashizume M, Yamagishi TA, Nakamoto Y. Synthesis, conformational and host–guest properties of water-soluble pillar[5]arene. Chem. Commun. 2010, 46, 3708–10.

[126] Yu G, Zhou X, Zhang Z, Han C, Mao Z, Gao C, Huang F. Pillar[6]arene/paraquat molecular recognition in water: high binding strength, pH-responsiveness, and application in controllable self-assembly, controlled release, and treatment of paraquat poisoning. J. Am. Chem. Soc. 2012, 134, 19489–97.

[127] Li Z, Yang J, Yu G, He J, Abliz Z, Huang F. Water-soluble pillar[7]arene: synthesis, pH-controlled complexation with paraquat, and application in constructing supramolecular vesicles. Org. Lett. 2014, 16, 2066–9.

[128] Ma Y, Ji X, Xiang F, Chi X, Han C, He J, Abliz Z, Chen W, Huang F. A cationic water-soluble pillar[5]arene: synthesis and host–guest complexation with sodium 1-octanesulfonate. Chem. Commun. 2011, 47, 12340–2.

[129] Behrend R, Meyer E, Rusche F. I. Ueber Condensationsproducte aus Glycoluril und Formaldehyd. Liebigs Ann. Chem. 1905, 339, 1–37.

[130] Freeman WA, Mock WL, Shih N.-Y. Cucurbituril. J. Am. Chem. Soc. 1981, 103, 7367–768.

[131] Lagona J, Mukhopadhyay P, Chakrabarti S, Isaacs L. The cucurbit[n]uril family. Angew. Chem. Int. Ed. 2005, 44, 4844–70.

[132] Cong H, Ni XL, Xiao X, Huang Y, Zhu QJ, Xue SF, Tao Z, Lindoy LF, Wei G. Synthesis and separation of cucurbit[n]urils and their derivatives. Org. Biomol. Chem. 2016, 14, 4335–64.

[133] Zhang S, Grimm L, Miskolczy Z, Biczók L, Biedermann F, Nau WM. Binding affinities of cucurbit[n]urils with cations. Chem. Commun. 2019, 55, 14131–4.

[134] He S, Huang B, Xiao B, Chang S, Podalko M, Nau WM. Stabilization of guest molecules inside cation-lidded cucurbiturils reveals that hydration of receptor sites can impede binding. Angew. Chem. Int. Ed. 2023, 62, e202313864.

[135] Mock WL, Shih NY. Structure and selectivity in host–guest complexes of cucurbituril. J. Org. Chem. 1986, 51, 4440–6.

[136] Meschke C, Buschmann HJ, Schollmeyer E. Complexes of cucurbituril with alkyl mono- and diammonium ions in aqueous formic acid studied by calorimetric titrations. Thermochim. Acta. 1997, 297, 43–8.

[137] Rekharsky MV, Mori T, Yang C, Ko YH, Selvapalam N, Kim H, Sobransingh D, Kaifer AE, Liu S, Isaacs L, Chen W, Moghaddam S, Gilson MK, Kim K, Inoue Y. A synthetic host–guest system achieves avidin–biotin affinity by overcoming enthalpy–entropy compensation. Proc. Natl. Acad. Sci. U. S. A. 2007, 104, 20737–42.

[138] Cao L, Šekutor M, Zavalij PY, Mlinarić-Majerski K, Glaser G, Isaacs L. Cucurbit[7]uril·guest pair with an attomolar dissociation constant. Angew. Chem. Int. Ed. 2014, 53, 988–93.

[139] Biedermann F, Uzunova V. D, Scherman O. A, Nau W. M, De Simone A. Release of high-energy water as an essential driving force for the high-affinity binding of cucurbit[n]urils. J. Am. Chem. Soc. 2012, 134, 15318–23.

[140] Vaitheeswaran S, Yin H, Rasaiah JC, Hummer G. Water clusters in nonpolar cavities. Proc. Natl. Acad. Sci. U. S. A. 2004, 101, 17002–5.

[141] Barrow SJ, Kasera S, Rowland MJ, del Barrio J, Scherman OA. Cucurbituril-based molecular recognition. Chem. Rev. 2015, 115, 12320–406.

[142] Kim HJ, Heo J, Jeon WS, Lee E, Kim J, Sakamoto S, Yamaguchi K, Kim K. Selective Inclusion of a hetero-guest pair in a molecular host: formation of stable charge-transfer complexes in cucurbit[8]uril. Angew. Chem. Int. Ed. 2001, 40, 1526–9.

[143] Svec J, Necas M, Sindelar V. Bambus[6]uril. Angew. Chem. Int. Ed. 2010, 49, 2378–81.

[144] Lizal T, Sindelar V. Bambusuril anion receptors. Isr. J. Chem. 2018, 58, 326–33.

[145] Hennrich G, Anslyn EV. 1,3,5–2,4,6-Functionalized, facially segregated benzenes – Exploitation of sterically predisposed systems in supramolecular chemistry. Chem. Eur. J. 2002, 8, 2219–24.

[146] Rebek Jr. J. Recent progress in molecular recognition. Top. Curr. Chem. 1988, 149, 189–210.

[147] Rebek Jr. J. Molecular recognition with model systems. Angew. Chem. Int. Ed. Engl. 1990, 29, 245–55.

[148] Williams K, Askew B, Ballester P, Buhr C, Jeong KS, Jones S, Rebek Jr. J. Molecular recognition with convergent functional groups. VII. Energetics of adenine binding with model receptors. J. Am. Chem. Soc. 1989, 111, 1090–4.

[149] Benzing T, Tjivikua T, Wolfe J, Rebek Jr. J. Recognition and transport of adenine derivatives with synthetic receptors. Science. 1988, 242, 266–8.

[150] Chen CW, Whitlock HW. Molecular tweezers: a simple model of bifunctional intercalation. J. Am. Chem. Soc. 1978, 100, 4921–2.

[151] Zimmerman SC. Rigid molecular tweezers as hosts for the complexation of neutral guests. Top. Curr. Chem. 1993, 165, 71–102.

[152] Klärner FG, Kahlert B. Molecular tweezers and clips as synthetic receptors. Molecular recognition and dynamics in receptor–substrate complexes. Acc. Chem. Res. 2003, 36, 919–32.

[153] Fokkens M, Jasper C, Schrader T, Koziol F, Ochsenfeld C, Polkowska J, Lobert M, Kahlert B, Klärner FG. Selective complexation of *N*-alkylpyridinium salts: binding of NAD$^+$ in water. Chem. Eur. J. 2005, 11, 477–94.

[154] Fokkens M, Schrader T, Klärner FG. A molecular tweezer for lysine and arginine. J. Am. Chem. Soc. 2005, 127, 14415–21.

[155] Schrader T, Bitan G, Klärner FG. Molecular tweezers for lysine and arginine – powerful inhibitors of pathologic protein aggregation. Chem. Commun. 2016, 52, 11318–34.

[156] Sijbesma RP, Kentgens APM, Lutz ETG, van der Maas JH, Nolte RJM. Binding features of molecular clips derived from diphenylglycoluril. J. Am. Chem. Soc. 1993, 115, 8999–9005.

[157] Reek JNH, Priem AH, Engelkamp H, Rowan AE, Elemans JAAW, Nolte RJM. Binding Features of Molecular Clips. Separation of the Effects of Hydrogen Bonding and π–π Interactions. J. Am. Chem. Soc. 1997, 119, 9956–64.

[158] Ganapati S, Isaacs L. Acyclic cucurbit[n]uril-type receptors: preparation, molecular recognition properties and biological applications. Isr. J. Chem. 2018, 58, 250–63.

[159] Ma D, Zavalij PY, Isaacs L. Acyclic cucurbit[n]uril congeners are high affinity hosts. J. Org. Chem. 2010, 75, 4786–95.

[160] Hill DJ, Mio MJ, Prince RB, Hughes TS, Moore JS. A field guide to foldamers. Chem. Rev. 2001, 101, 3893–4012.

[161] Juwarker H, Suka JM, Jeong KS. Foldamers with helical cavities for binding complementary guests. Chem. Soc. Rev. 2009, 38, 3316–25.

[162] Waki M, Abe H, Inouye M. Helix formation in synthetic polymers by hydrogen bonding with native saccharides in protic media. Chem. Eur. J. 2006, 12, 7839–47.

[163] Hou JL, Shao XB, Chen GJ, Zhou YX, Jiang XK, Li ZT. Hydrogen bonded oligohydrazide foldamers and their recognition for saccharides. J. Am. Chem. Soc. 2004, 126, 12386–94.

[164] Petitjean A, Cuccia LA, Lehn JM, Nierengarten H, Schmutz M. Cation-promoted hierarchical formation of supramolecular assemblies of self-organized helical molecular components. Angew. Chem. Int. Ed. 2002, 41, 1195–8.

[165] Ferrand Y, Huc I. Designing helical molecular capsules based on folded aromatic amide oligomers. Acc. Chem. Res. 2018, 51, 970–7.

[166] Bao C, Kauffmann B, Gan Q, Srinivas K, Jiang H, Huc I. Converting sequences of aromatic amino acid monomers into functional three-dimensional structures: second-generation helical capsules. Angew. Chem. Int. Ed. 2008, 47, 4153–6.

[167] Ferrand Y, Kendhale AM, Kauffmann B, Grélard A, Marie C, Blot V, Pipelier M, Dubreuil D, Huc I. Diastereoselective encapsulation of tartaric acid by a helical aromatic oligoamide. J. Am. Chem. Soc. 2010, 132, 7858–9.

[168] Chandramouli N, Ferrand Y, Lautrette G, Kauffmann B, Mackereth CD, Laguerre M, Dubreuil D, Huc I. Iterative design of a helically folded aromatic oligoamide sequence for the selective encapsulation of fructose. Nat. Chem. 2015, 7, 334–41.

[169] Park CH, Simmons HE. Macrobicyclic amines. III. Encapsulation of halide ions by in,in-1,(k + 2)-diazabicyclo[k.l.m.]alkane ammonium ions. J. Am. Chem. Soc. 1968, 90, 2431–2.

[170] Mahoney JM, Beatty AM, Smith BD. Selective recognition of an alkali halide contact ion-pair. J. Am. Chem. Soc. 2001, 123, 5847–8.

[171] Deetz MJ, Shang M, Smith BD. A macrobicyclic receptor with versatile recognition properties: simultaneous binding of an ion pair and selective complexation of dimethylsulfoxide. J. Am. Chem. Soc. 2000, 122, 6201–7.

[172] Bartoli S, Roelens S. Electrostatic attraction of counterion dominates the cation–π interaction of acetylcholine and tetramethylammonium with aromatics in chloroform. J. Am. Chem. Soc. 1999, 121, 11908–9.

[173] Böhmer V, Dalla Cort A, Mandolini L. Counteranion effect on complexation of quats by a neutral calix[5]arene receptor. J. Org. Chem. 2001, 66, 1900–2.

[174] Hunter CA, Low CMR, Rotger C, Vinter JG, Zonta C. The role of the counteranion in the cation–π interaction. Chem. Commun. 2003, 834–5.

[175] Mazik M. Molecular recognition of carbohydrates by acyclic receptors employing noncovalent interactions. Chem. Soc. Rev. 2009, 38, 935–56.

[176] Davis AP. Synthetic lectins. Org. Biomol. Chem. 2009, 7, 3629–38.

[177] Tromans RA, Carter TS, Chabanne L, Crump MP, Li H, Matlock JV, Orchard MG, Davis AP. A biomimetic receptor for glucose. Nat. Chem. 2019, 11, 52–6.

[178] Wu X, Li Z, Chen XX, Fossey JS, James TD, Jiang YB. Selective sensing of saccharides using simple boronic acids and their aggregates. Chem. Soc. Rev. 2013, 42, 8032–48.

[179] James TD, Sandanayake KRAS, Shinkai S. A glucose-selective molecular fluorescence sensor. Angew. Chem. Int. Ed. Engl. 1994, 33, 2207–9.

5 Assembling molecules

CONSPECTUS: The receptors introduced in the previous chapter have cavities surrounded by inwardly directed functional groups to mediate the recognition of suitable substrates. In this chapter, we look at recognition processes that occur between molecules with outwardly directed binding sites. Two or more of these molecules interact to form larger assemblies whose structure and stability are controlled by the number and properties of their components. This chapter begins with an introduction to the principles underlying such self-assembly processes. We then look at the different molecular architectures that can be accessed, the interactions that stabilize them, and the ways in which product formation can be controlled by template effects. The chapter ends with a look at how mixtures of interacting molecules behave as they become more complex.

5.1 Self-assembly and template effects

When a synthetic chemist *assembles a molecule,* he or she carries out a series of reactions to establish the connectivity of the atoms in the product. These syntheses allow the precise control over how the atoms are linked by carefully designing the nature and the order of the individual steps, taking into account selectivity issues and/or the use of protecting groups if functional groups are incompatible with the reaction conditions. This strategy ultimately provides access to virtually any molecule of reasonable size, even strained or unstable ones, although the synthetic effort is high if the product is structurally complex. The overall approach is reminiscent of manufacturing a macroscopic object from its components on an assembly line, in that the starting material is sequentially modified by adding, removing, or modifying groups at specific positions until the product is complete.

When a supramolecular chemist sets out to *assemble molecules,* the task is different. The synthetic approach in this case involves the use of carefully designed molecular building blocks (synthesized by using the covalent strategy described above) that interact in predictable ways by virtue of their built-in recognition motifs. The goal is to obtain larger aggregates with structures that are controlled by the shape and arrangement of the individual subunits. This noncovalent approach to assembling molecules extends synthetic chemistry into the supramolecular realm. It makes it possible to achieve levels of structural complexity that are difficult, if not impossible, to realize using covalent chemistry. The underlying principles differ in many respects from those of (most) covalent syntheses.

One difference is that noncovalent syntheses of even complex products proceed continuously from the starting materials to the final product without external interference, rather than in a stepwise fashion as in the case of covalent syntheses. Mechanistically, product formation involves continuous association and dissociation processes along an interconnected network of possible pathways. The final product must be thermodynamically favored under the chosen conditions, or it will not form. Kinetically

https://doi.org/10.1515/9783111315171-005

favored products are not accessible because the reversibility of noncovalent interactions ensures that unwanted byproducts are converted into the product once the thermodynamic equilibrium is reached. This continuous error correction is advantageous because it allows even structurally very complex products to be produced, ideally in quantitative yields, simply by mixing the required components in the correct ratio, as long as there is a thermodynamic driving force for product formation.

Another aspect of noncovalent syntheses is that the products remain dynamic after assembly. They may be stable under the reaction conditions, but will continue to exchange building blocks unless they have been permanently stabilized by postfunctionalization. Changing the external conditions thus leads to dissociation or rearrangement, which is a disadvantage because it compromises structural integrity, but can also be an advantage if responsiveness, that is, the ability of the product to change its structure or composition in response to external stimuli, is desired. Note that the principles of covalent and noncovalent synthesis differ only when the covalent approach involves irreversible and therefore kinetically controlled reactions. Covalent syntheses based on reversible bond formation share many of the characteristics of noncovalent syntheses, as we will see in Sections 5.6 and 5.7.

An analogy from the macroscopic world that is often used to illustrate the spontaneous assembly of molecular building blocks is the construction kit, whose components miraculously assemble in the right way to produce the finished model when mixed or otherwise agitated. Indeed, molecules in solution are constantly moving and potentially searching for a binding partner. However, they also stick together autonomously when properly arranged, which is usually not the case with the components of a construction kit. Another aspect where the above analogy fails has to do with thermodynamics, because the finished model does not represent a more stable state in the macroscopic world than the disassembled components. Accordingly, there is no thermodynamic driving force for the macroscopic components to assemble, whereas in the case of molecules, the energetic gain associated with the formation of the product drives the assembly and its outcome. Thus, molecules assemble spontaneously not only because they are moving all the time, but also because the overall system ends up in a more stable state. This state results from the conditions as well as the structure and ratio of the building blocks, but does not require additional external influences. Rather, the molecules find the way from the disassembled state to the product by *themselves*, which is why the process is called *self-assembly*. Based on these considerations, the following definition of self-assembly can be derived:

Self-assembly is the spontaneous, thermodynamically controlled association of molecules into larger aggregates with structures that are a direct consequence of the shapes, arrangements, and ratios of the building blocks.

Although this definition does not exclude the receptor–substrate complexes discussed in the previous chapter, self-assembly is a much broader concept because it does not limit the number of interacting molecules to two or only a few, and it makes no distinction between receptors and substrates. In addition, in order for molecular building blocks to assemble into aggregates larger than themselves, they must have divergent binding sites as opposed to typical receptors in which the binding sites are arranged in a convergent fashion around a cavity or cleft.

The ultimate potential of self-assembly processes can best be appreciated by looking at examples from nature. In this context, virus assembly is a particularly instructive example. Viruses consist of protein shells, so-called capsids, that surround and thereby protect DNA or RNA, the polynucleotides needed to reprogram infected cells and induce them to produce new virus particles. Although viruses can have intricate shapes, their capsids are usually composed of multiple copies of a few or even a single protein to minimize the amount of genetic information required for their biosynthesis. Capsid proteins must therefore be able to spontaneously assemble into the three-dimensional shape characteristic of the final virus, an ability that derives from their folding pattern, the resulting shape, and the distribution of amino acid side chains along their surfaces. This process is especially well understood for the tobacco mosaic virus (TMV) (Figure 5.1a).

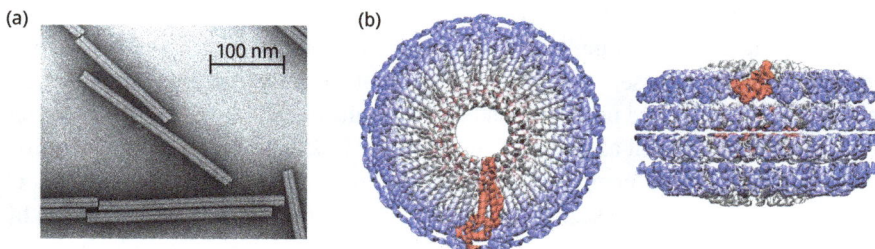

Figure 5.1: Transmission electron microscope image of a tobacco mosaic virus (a) and top and side views of the crystal structure of a four-layer disk containing 68 subunits of the TMV coat protein [2]. A protein subunit in this structure is highlighted in red. The image in (a) was taken from the International Committee on Taxonomy of Viruses (ICTV) database.

TMV infects various plants, particularly tobacco, causing characteristic discoloration patterns on the leaves. It has a rod-like shape with a length of 300 nm and a diameter of 18 nm. Its capsid consists of 2,130 identical protein subunits helically surrounding a single RNA strand with 6,395 nucleobase residues. The virus can be lysed by treatment with a detergent (sodium dodecyl sulfate), and the proteins can subsequently be separated from the polynucleotides by precipitation with ammonium sulfate. The mixing of the disassembled components in an aqueous buffer results in reassembly, ultimately yielding fully reconstituted and infectious virus particles [1]. The mechanism of TMV formation involves several steps, starting with the self-assembly of 34 protein subunits into a disk consisting of two stacked 17-membered rings. To illustrate this structure, the crystal structure of a four-layer disk containing 68 individual protein subunits is shown in Figure 5.1b [2]. This disk has a central hole into which a short segment of the single-stranded RNA is inserted. RNA recognition is highly specific because it occurs at a specific nucleation site in the single-stranded viral RNA, located approximately 1,000 nucleobases from the 3' end, while foreign RNAs that do not have this site are rejected. RNA binding then leads to a rearrangement of the disk into a helical structure, which grows by the attachment of additional protein disks and the simultaneous threading of the RNA through the central hole. Self-assembly stops when the RNA strand is completely buried [3].

As complex as it may seem, this process is completely autonomous and extremely efficient. Each step is reversible, so that errors can be corrected, and the final result is subject to thermodynamic control. It goes without saying that a structure of similar structural complexity would be impossible to obtain by covalent synthesis, demonstrating the enormous potential of self-assembly processes to generate large and ordered structures.

The construction of virus capsids is only one of the many impressive thermodynamically controlled self-assembly processes found in nature. Other systems include DNA, which is rarely found in the monomeric form in natural systems, but – at the first level of organization – as a dimer of two complementary DNA strands. In addition, DNA interacts with proteins in the nuclei of eukaryotic cells, forming compact aggregates that help protect against degradation. Proteins assemble for various reasons. The assembly of enzymes, for example, results in multienzyme complexes in which the product of one reaction step is efficiently passed on to the enzyme that mediates the next step. Protein aggregation also leads to the plaques found in the brains of patients with Parkinson's or Alzheimer's disease. The self-assembly of lipids, which is driven primarily by the hydrophobic effect, leads to cell membranes whose curvature is determined by the shapes of the individual building blocks (Section 10.1). The large vesicular structures surrounded by these membranes ensure the confinement of molecules or larger entities within the cell.

The latter example shows that structural organization in biological systems occurs at different levels, from the assembly of molecules such as lipids, proteins, or polynucleotides to the formation of single cells or multicellular living organisms (whose assembly further leads to societies). However, there are fundamental differences in the outcome of these processes. The self-assembly of molecules is thermodynamically controlled, proceeds autonomously even in the absence of the complete biochemical machinery, and leads to an equilibrium state. Living cells, on the other hand, are out-of-equilibrium systems. Thus, the term *self-assembly* applies only to the first case, while the interaction of self-assembled entities into larger functional out-of-equilibrium systems is termed *self-organization*.

The concept of self-assembly is not limited to biochemical systems, but can readily be applied to abiotic molecules. All that is required are appropriate molecular building blocks that engage in intermolecular interactions to produce larger aggregates of defined structure and size. The exact mode of self-assembly depends on several factors, including the nature and number of interacting groups in the building blocks, their shape and mutual arrangement, the strength of the underlying interactions, the solvent, and the presence of templates. The underlying principles will be derived for the example shown in Figure 5.2. The same rules apply to more complex systems.

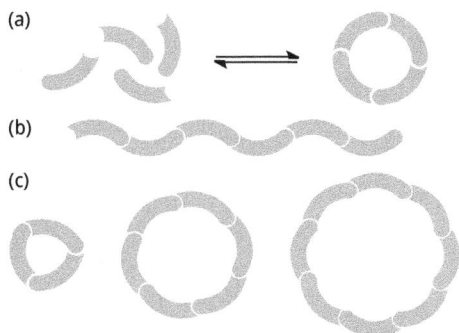

(a)

(b)

(c)

Figure 5.2: Schematic representation of a self-assembly process in which a curved building block with two complementary recognition units assembles into a macrocyclic tetramer (a). Alternative products such as an acyclic oligomer and rings of different sizes are shown in (b) and (c), respectively.

Figure 5.2a shows the arbitrarily chosen ideal situation for this self-assembly process, which involves the formation of a macrocyclic tetramer from a single ditopic building block with complementary recognition units. This tetramer is only one of many possible products, since self-assembly can alternatively lead to acyclic oligomers of different lengths (Figure 5.2b) or smaller or larger rings (Figure 5.2c). To identify the factors that control product distribution, it is helpful to divide self-assembly into oligomerization and cyclization, as shown in Figure 5.3.

According to Figure 5.3, oligomerization is a stepwise process, each step involving the addition of a monomeric building block to the chain end of a self-assembled linear oligomer. These oligomers have different lengths, but since the interactions underlying chain elongation are all identical, these equilibria are characterized by the same intermolecular equilibrium constant K_a^{inter}. The functional groups responsible for chain elongation also mediate cyclization, but the extent of cyclization depends not only on the interaction strength but also on the ease with which the corresponding ring is formed. The intramolecular binding constants K_i^{intra} therefore vary with the ring size i. The same set of interconnected equilibria also describes the ring–chain mechanism of supramolecular polymerization, which will be discussed in Section 6.1.2, but while the desired product in that case is the linear polymer, here we are interested in the formation of the cyclic product.

The first reaction in Figure 5.3, involving the cyclization of the linear dimer and the competing chain elongation to yield the trimer, is closely related to the equilibria we discussed in Section 3.5.3 on chelate cooperativity. In this context, we derived equation (3.13), which correlates the ratio of cyclized over linear species c_c/c_o with the intrinsic strength of the intermolecular interactions K_a and the effective molarity EM.

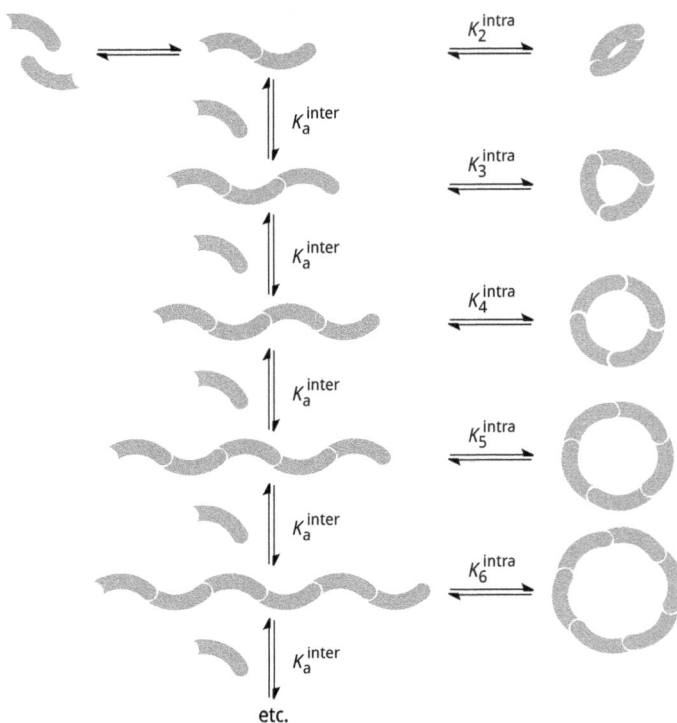

Figure 5.3: Stepwise self-assembly of a ditopic building block leading to acyclic oligomers (shown in the center), which subsequently cyclize in intramolecular processes to yield the corresponding macrocyclic products (shown on the right). Each oligomerization step is associated with an intermolecular equilibrium constant K_a^{inter}, while the ring-chain equilibria have intramolecular constants K_i^{intra} that vary with the ring size i.

By adapting equation (3.13) to the formation of the cyclic dimer in Figure 5.3, we obtain equation (5.1), in which the efficiency of the interaction underlying chain elongation is denoted by K_a^{inter}, the ratio c_c/c_o is expressed as the corresponding equilibrium constant K_2^{intra}, and in which statistical factors are absent because K_a^{inter} and K_a^{intra} refer to experimental stability constants:

$$K_a^{inter}\, EM_2 = K_2^{intra} \tag{5.1}$$

Rearrangement yields equation (5.2), which shows that the effective molarity quantitatively describes the extent to which cyclization is favored over chain elongation:

$$EM_2 = K_2^{intra} / K_a^{inter} \tag{5.2}$$

More specifically, EM_2 is the concentration at which the monomeric building block must be present for chain elongation to be as efficient as cyclization. The higher the value of EM_2, the more cyclization is favored. Since the self-assembly shown in Figure 5.3 allows the formation of rings of different sizes, we can generalize equation (5.1) by writing the following equation:

$$K_a^{inter} \, EM_i = K_i^{intra} \tag{5.3}$$

Accordingly, the formation of each ring is associated with a characteristic effective molarity, and the extent to which these EM_i values differ determines the product ratio. If different rings have similar EM_i values, they will all be present in the mixture. However, if the EM_i of one macrocycle is much larger than the values of other macrocycles, that ring will dominate the equilibrium. For example, if the macrocyclic tetramer is to be the dominant product, EM_4 must be larger than the other EM_i values. This situation is observed for the macrocyclization of **5.1** (Figure 5.4).

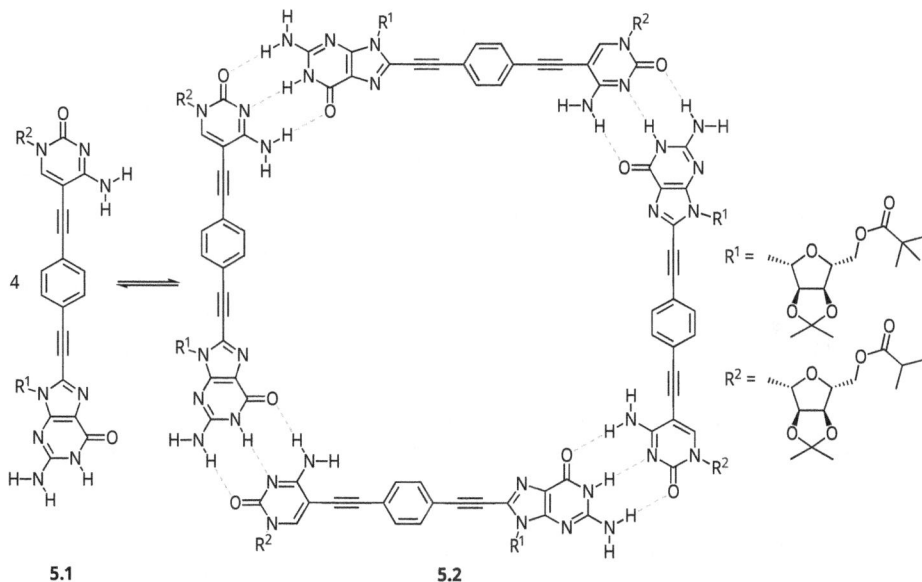

Figure 5.4: Self-assembly of the ditopic building block **5.1** containing cytidine and guanosine residues into the corresponding macrocyclic tetramer **5.2**.

This monomer assembles by Watson–Crick base pairing between the terminal cytidine and guanosine residues to yield exclusively **5.2**, which is thermodynamically very stable, even in competitive media such as DMF [4]. The actual EM_4 associated with the self-assembly can be estimated by relating the intrinsic stability of the guanosine-cytidine base pair K_a^{inter} to K_T, the overall stability of **5.2**. Since the formation of **5.2** involves three consecutive oligomerization steps followed by cyclization, K_T is given by equation (5.4) as the product of $\left(K_a^{\text{inter}}\right)^3$ (oligomerization) and $K_a^{\text{inter}} \ EM_4$ (cyclization):

$$K_T = \left(K_a^{\text{inter}}\right)^4 EM_4 \tag{5.4}$$

The value K_a^{inter} can be determined by measuring the stability of a base pair formed by two monotopic building blocks, one containing a guanosine and the other a cytidine group. The stability constant K_T is more difficult to quantify because the self-assembled complex is formed from identical subunits so that titrations, in which the degree of complex formation is assessed at different ratios of the binding partners (Section 2.2), are not possible. The corresponding law of mass action in equation (5.5) shows that K_T is given by the quotient of the concentrations of the final complex c_{Complex} and the monomeric building blocks c_{Monomer}, where i is the number of building blocks in the complex. In the example above, i is 4.

$$K_T = \frac{c_{\text{Complex}}}{\left(c_{\text{Monomer}}\right)^i} \tag{5.5}$$

Accordingly, K_T can be determined by following the effects of concentration or temperature on the degree of complex formation, as will be discussed in more detail in Chapter 6. In the example above, self-assembly is slow on the NMR timescale and separate signals are observed in the NMR spectrum for **5.1** and **5.2**. Thus, the ratio of the monomeric building block to the tetrameric complex can be calculated by integrating characteristic signals of **5.1** and **5.2**, taking into account the number of absorbing protons. From this ratio and the known initial concentration of **5.1**, the equilibrium concentrations of **5.1** and **5.2** can be determined. The stability constant K_T is then obtained from the slope of a plot of $\left(c_{\text{Monomer}}\right)^4$ *versus* c_{Complex} for a series of measurements.

With K_a^{inter} and K_T now available, EM_4 can be calculated using equation (5.4). In this case, values between 200 and 900 M were obtained, depending on the solvent, which are among the highest effective molarities observed for a synthetic system. The high fidelity of the self-assembly process is at least partly related to the hydrogen bonding pattern in the base pairs, since the macrocyclization of a structurally related building block containing 2-aminoadenosine and uridine groups instead of cytidine and guanosine residues is associated with an *EM* of only 0.1 M [5].

Having seen that the *EM* values control the product ratio in a self-assembly process, we need to look at the factors on which they depend. The general form of equation (5.2), equation (5.6), provides information in this regard when the equilibrium constants are expressed in terms of ΔH^0 and ΔS^0. Equation (5.7), after rearrangement, leads to (5.8), which shows that EM_i is a product of two terms, the first accounting for the enthalpic and the second for the entropic differences of cyclization and chain elongation:

$$EM_i = K_i^{\text{intra}} / K_a^{\text{inter}} \tag{5.6}$$

$$EM_i = e^{\left(-\frac{1}{RT}\Delta G_{\text{intra}}^0\right)} \Big/ e^{\left(-\frac{1}{RT}\Delta G_{\text{inter}}^0\right)}$$

$$= e^{\left[-\frac{1}{RT}\left(\Delta H_{\text{intra}}^0 - T\Delta S_{\text{intra}}^0\right)\right]} \Big/ e^{\left[-\frac{1}{RT}\left(\Delta H_{\text{inter}}^0 - T\Delta S_{\text{inter}}^0\right)\right]} \tag{5.7}$$

$$EM_i = e^{\left[-\frac{1}{RT}\left(\Delta H_{\text{intra}}^0 - H_{\text{inter}}^0\right)\right]} \, e^{\left[\frac{1}{R}\left(\Delta S_{\text{intra}}^0 - \Delta S_{\text{inter}}^0\right)\right]} \tag{5.8}$$

The first term relates the strength of the interactions between the monomer and an end group in the chain ($\Delta H_{\text{inter}}^0$) to that between the same groups when they form the ring ($\Delta H_{\text{intra}}^0$). If the ring is free of strain, the enthalpy values are equal because oligo-merization and cyclization are mediated by the same types of interactions. In this case, $\Delta H_{\text{intra}}^0 - \Delta H_{\text{inter}}^0 = 0$, so that the first term in equation (5.8) has the optimal value of 1. More often, however, cyclization is associated with the buildup of strain, which makes cyclization enthalpically less favorable than chain elongation. The difference $\Delta H_{\text{intra}}^0 - \Delta H_{\text{inter}}^0$ is then positive, leading to a reduction of EM_i.

These considerations are valid if the enthalpic driving force for cyclization and chain elongation comes only from end group interactions. However, if a pathway leads to a product that is stabilized by other types of interactions, such as interactions that stabilize a tightly folded conformation, additional effects on $\Delta H_{\text{intra}}^0 - \Delta H_{\text{inter}}^0$ arise that could promote either chain elongation or cyclization. In this case, EM_i can become sol-vent-dependent if the solvent affects the conformational behavior of the acyclic and cy-clic products to different degrees. Note also the temperature dependence of the enthalpy term, which causes EM_i to increase with temperature if the self-assembled rings are not free of strain.

Entropically, chain elongation has a disadvantage over cyclization because the molecule added to the chain must give up degrees of translational and rotational free-dom. The entropic component of EM_i is thus mainly due to $\Delta S_{\text{inter}}^0$ ($\Delta S_{\text{inter}}^0 < 0$), although cyclization also usually has a small entropy term. In the absence of solvent effects, this cyclization entropy reflects the restriction of torsional flexibility during ring clo-sure, and while this effect can be minimized by using rigid building blocks, it becomes increasingly unfavorable as the number of subunits in the self-assembled product in-creases. Thus, entropy drives the self-assembly toward smaller aggregates, which can also be attributed to the fact that the formation of many copies of a product contain-ing only a few building blocks is entropically more favorable than the formation of a smaller number of structurally more complex alternatives. The contribution of $\Delta S_{\text{intra}}^0$ thus dictates the size of the products but does not prevent the formation of ordered aggregates.

The effective molarity of each step in a self-assembly process thus reflects a bal-ance of enthalpic and entropic factors, with enthalpy generally favoring chain elonga-tion unless the cyclic product is unstrained, in which case there is no enthalpic difference between the intermolecular and intramolecular processes. Entropy, on the other hand, favors cyclization, and the smaller the cyclic products, the stronger the preference. Therefore, high EM_i values are expected for building blocks that are *predisposed* to yield the smallest possible unstrained ring, as well as *preorganized* to minimize the entropic penalty of cyclization.

What is the difference between preorganization and predisposition?

It is important to distinguish carefully between the terms predisposition and preorganization. We saw in Section 3.5.1 that a receptor is preorganized if it undergoes minimal structural changes when forming a complex. Analogously, building blocks are preorganized for self-assembly if they are rigid and lose little torsional flexibility as they form the product. In both cases, preorganization refers to the effects of *entropy* on the recognition process. Predisposition, on the other hand, means that a self-assembling building block has a strong conformational or structural preference to yield a particular product when incorporated into a larger structure [6]. This effect is therefore related to the *enthalpic* stability of the final product. Building blocks can be preorganized *and* predisposed if they are rigid and if their structure induces the preferential formation of one product among several possible ones.

The above principles were derived for the example shown in Figure 5.3, where self-assembly leads to cyclic products. The same rules apply to more complex systems. Accordingly, self-assembly preferentially leads to products in which all binding sites are occupied to maximize the number of interactions, a criterion that is known as the *principle of maximum site occupancy* [7]. Any alternative product with vacant binding sites should be less stable and should therefore not form to a large extent. Furthermore, unstrained products are preferred over strained ones, and smaller products composed of fewer subunits are preferred over larger ones.

Based on these rules, it is possible to identify building blocks and reaction conditions that reliably lead to self-assembled products of high structural complexity. Important design criteria are the rigidity, size, and shape of the building blocks and the spatial orientation of the interacting functional groups, all of which control the mutual orientation of the subunits in the final assembly and, hence its overall structure. The actual interactions between the building blocks should not only be efficient under the assembly conditions but also directional to allow reliable prediction of the outcome. As a result, only a few of the noncovalent interactions discussed in Section 3.1 are used for self-assembly, most notably hydrogen bonding. In addition, coordinative interactions mediate metal-directed self-assembly. It should be noted that a crucial driving force for self-assembly in water is the hydrophobic effect. The lack of directionality in this case causes the structures of the products – examples are micelles or vesicles – to be usually structurally less defined than those obtained from self-assembly processes involving directed interactions. However, there are exceptions, as we will see in Section 5.2.

In the ideal case, self-assembly proceeds with high fidelity, yielding mainly or even exclusively a single product. However, when different products have similar thermodynamic stabilities, they coexist in the equilibrium. This situation arises, for example, when enthalpic and entropic factors work in opposite directions, with enthalpy favoring a larger unstrained product and entropy favoring a smaller but slightly strained analog. In this case, it is useful to take advantage of the dynamic nature of the self-assembly process by controlling product formation through the addition of suitable templates.

How do templates work?

The concept of using templates to control the outcome of a reaction was first mentioned in Section 4.1.1, which also contains Busch's definition of a template. We saw there that crown ether syntheses benefit from the presence of alkali metal ions in the reaction mixture, which preorganize the linear precursor of the product, thus facilitating cyclization. Let us take a closer look at this reaction. The starting material is an oligoethylene glycol with a free OH group at one end of the chain and a tosylate group at the opposite end. This compound is treated with a base that induces deprotonation of the free hydroxy group. The resulting alkoxide now displaces a tosylate group on a carbon atom, yielding an ether. Figure 5.5 shows that there are two competing pathways for this step – an intermolecular one that yields an acyclic oligomer and an intramolecular one that yields the macrocycle. Assuming the ring is unstrained, the activation enthalpies of the two pathways should be similar, but their activation entropies differ because the acyclic precursor must adopt a bent conformation in which the two end groups are in close proximity. The associated unfavorable activation entropy makes cyclization slower than oligomerization, as reflected in the energy profiles of the two reaction pathways (Figure 5.5).

Since the Williamson ether synthesis underlying crown ether formation is an irreversible reaction, the ratio of oligomerization to cyclization is determined by the difference in the Gibbs free energy of activation $\Delta\Delta G^{\ddagger}$ of the two reactions. Thus, changing the product ratio requires changing the rate of one reaction with respect to the other one, which is exactly how the alkali metal ions that serve as templates work. They interact with the linear precursor by inducing a folded conformation that is well suited for ring formation. Complex formation thus reduces the unfavorable activation entropy of cyclization so that the crown ether becomes the favored product. Since the template selectively increases the rate with which the desired product is formed, it acts as a *kinetic template*. We conclude:

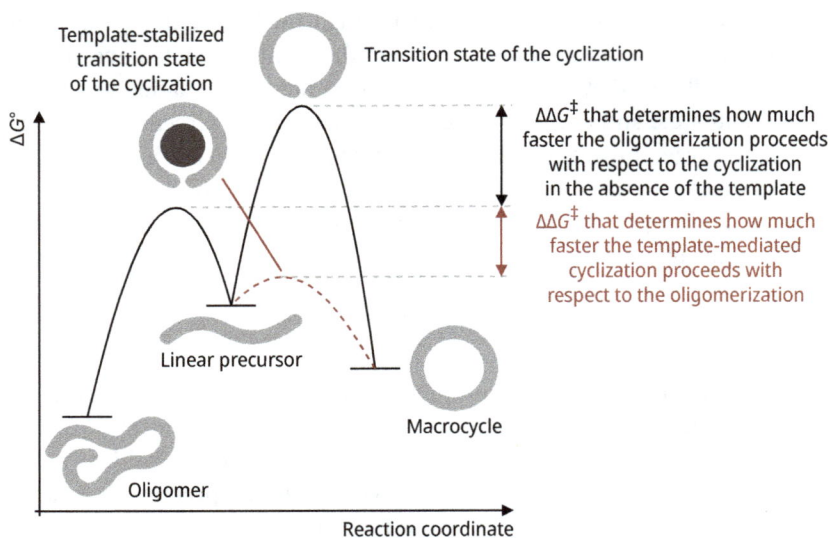

Figure 5.5: Competing pathways of an irreversible reaction between functional groups at the ends of a linear precursor with the associated energy profiles. The presence of a template in the reaction mixture preorganizes the precursor for the intramolecular reaction, reducing the associated $\Delta G^{\ddagger}$ and causing cyclization to become the preferred reaction pathway.

Irreversible reactions that proceed along **competing pathways** are controlled by **kinetic templates** when the template selectively stabilizes the **transition state** of the normally unfavorable reaction, thereby **increasing the rate** at which the corresponding product is formed.

In the case of reversible reactions, the product ratio is determined by the thermodynamic stabilities of the different products, that is, by how much their ΔG^0 values differ. This relationship is illustrated by the energy profile shown in Figure 5.6 for a reversible reaction yielding two products. An example of such a reaction could be the self-assembly process mentioned above, which yields either the entropically favored cyclic trimer or the entropically disfavored but less strained tetrameric analog. In the absence of a template, the trimer dominates in the equilibrium shown in Figure 5.6, and for the tetramer to become the preferred product, it must be thermodynamically stabilized. This can be achieved by adding a template to the reaction mixture that selectively interacts with the tetramer. The formation of the corresponding complex leads to the removal of the tetramer from the equilibrium, which must adapt by regenerating the tetramer at the expense of the cyclic trimer according to the Le Chatelier principle.

For this effect to work, the template must form a stable complex with the compound that should be selected. Since the template acts on the thermodynamics of the

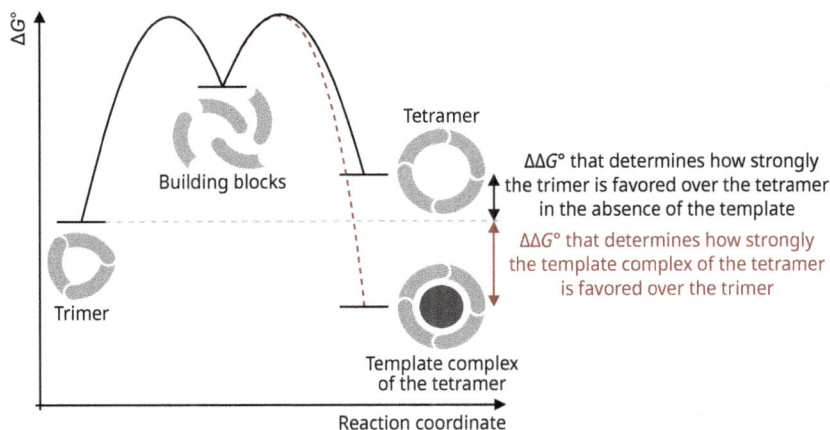

Figure 5.6: Competing pathways of a thermodynamically controlled self-assembly process, yielding a cyclic trimer and a cyclic tetramer. The trimer is favored in the absence of a template. A suitable template that binds selectively to the tetramer leads to thermodynamic stabilization of this compound and its enrichment in the equilibrium.

process, the reaction control is achieved by a *thermodynamic template* effect. Thus, we can generalize as follows:

> The product distribution of **reversible reactions** that yield different products can be shifted by using **thermodynamic templates** effect that selectively interact with one of the interconverting products, causing its **enrichment at equilibrium**.

The metal-directed self-assembly shown in Figure 5.7 is a reversible process whose preferred outcome is controlled by the addition of thermodynamic templates. In this example, the coordination of (*E*)-1,2-di(pyridin-4-yl)ethene to the square planar palladium(II) complex [Pd(en)(NO$_3$)$_2$] (en = ethylenediamine) in D$_2$O yields a mixture of the trimetallic and tetrametallic metallacycles **5.3a,b** [8]. The ratio of the two products depends on the concentration, with the trimer **5.3a** dominating at low concentrations (0.1 mM) and the tetramer **5.3b** dominating at higher concentrations (10 mM). Accordingly, **5.3b** is the enthalpically favored product, while the smaller **5.3a** is entropically favored. Addition of *p*-dimethoxybenzene to a solution containing a mixture of **5.3a** and **5.3b** (2 mM) shifts the equilibrium toward the smaller macrocycle. In contrast, **5.3b** dominates the equilibrium in the presence of 1,3-adamantanedicarboxylic acid. These templates thus lead to the selective amplification of the respective size-complementary metallacycles. Similar coordination-driven self-assembly processes are presented in Section 5.5.

While the template effect on the equilibrium shown in Figure 5.7 is easy to rationalize, with each template favoring the formation of the metallamacrocycle with which it interacts best, it is important to emphasize that dynamic equilibria respond to changes in conditions, such as variation in temperature, concentration, or the presence of an

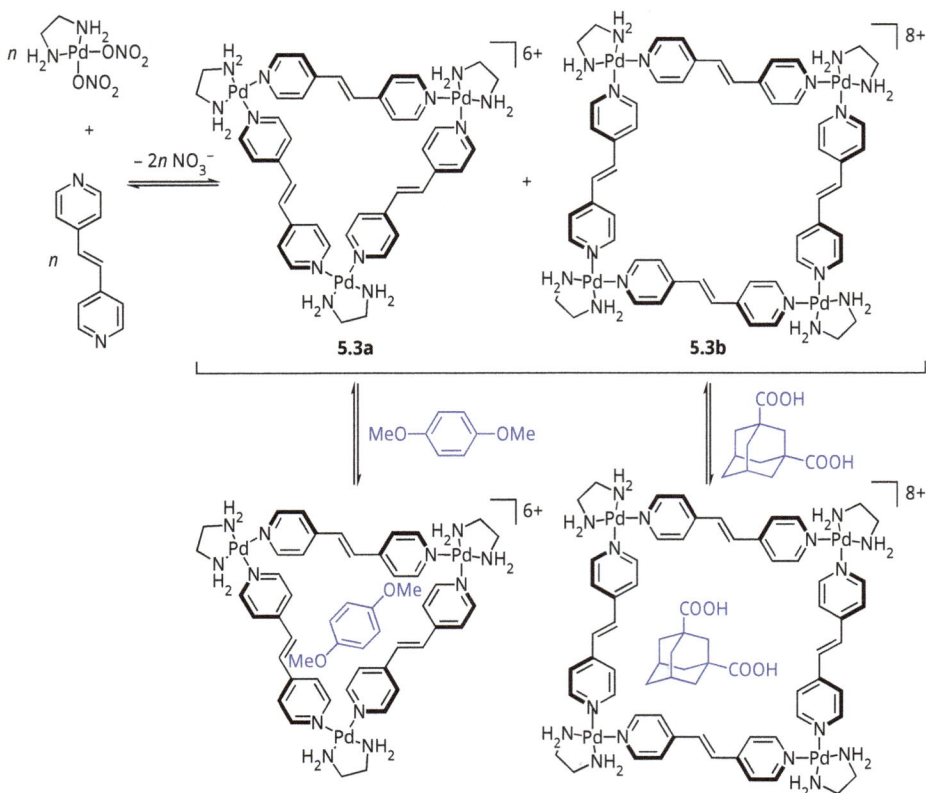

Figure 5.7: Formation of the metallacycles **5.3a** and **5.3b** by coordination of (*E*)-1,2-di(pyridin-4-yl)ethene to [Pd(en)(NO₃)₂] and effects of suitable templates on the ratio of the products.

additional binding partner by adopting a new state that represents the thermodynamic minimum of the *entire system*. This system can be quite complex because it includes the solvent molecules and all possible solute–solvent and solute–solute combinations. Templates can therefore even amplify structures that do not exist in the absence of the template, or that may not be the best binding partners. We will see examples of such cases in Section 5.7.

Can molecules be sorted?

Another way to increase the complexity of self-assembling systems is to use mixtures of building blocks that contain complementary functional groups, allowing multiple compounds to participate in product formation. Although self-assembly in this case could result in an intractable mixture of products, this outcome can often be avoided

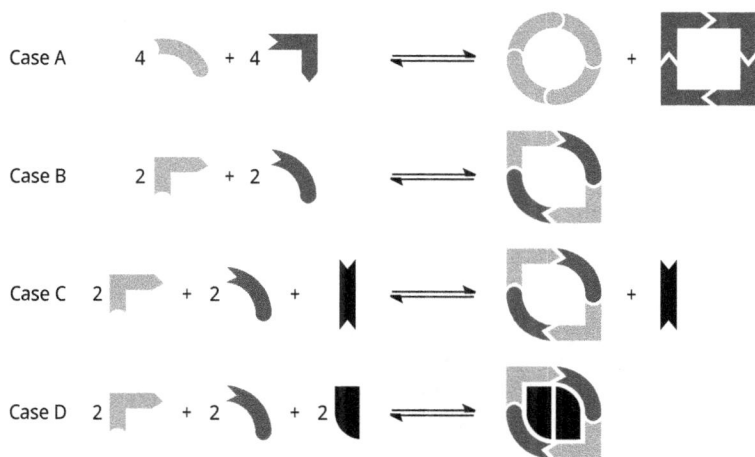

Figure 5.8: Self-assembly of mixtures of building blocks. In case A, the two products are composed of identical building blocks (narcissistic self-sorting), and in case B, the product is produced from different building blocks (social self-sorting). Both cases are completive as no unused building blocks are left over. Case C shows the situation of a social but incomplete self-assembly. In case D, all building blocks are incorporated into the product, representing integrative, social, and completive self-sorting.

by the proper choice of the building blocks. As a result, order can prevail over chaos, even in mixed systems. Possible scenarios are illustrated schematically in Figure 5.8.

Case A in Figure 5.8 involves two self-complementary compounds that interact preferentially with themselves. Since the actual interactions are the same for both compounds, involving hydrogen bonding for example, the curved building block can also bind to the angular one. However, such complexes are less stable and will eventually be outcompeted by the complexes containing the better binding partners. Thus, self-assembly should only yield products composed of identical building blocks. As a consequence, a mixture of compounds ends up autonomously in a state where no mixed products are present, which explains why the underlying process is called *self-sorting* [9].

According to the general definition, mixtures undergo self-sorting when the individual molecules "recognize their mutual counterparts selectively so that specific pairs are formed rather than a library of all possible noncovalent complexes of the compounds present in the mixture" [10]. The other cases in Figure 5.8 therefore also represent self-sorting processes, and additional terms are used to distinguish them [11]. In case A, self-sorting is *narcissistic* because only identical subunits bind to each other. It is also *completive* because no subunits are left unused. Case B is also a completive self-assembly, but it does not involve self-complementary binding partners. The two building blocks thus form crossover species, leading to a product that contains both components in a specific arrangement. This scenario is called *social* self-sorting. In case C, an additional building block is present in the mixture, which can only be integrated into structurally rather complex products. Since the formation of these products is entropically and potentially enthalpically unfavorable, the third building block remains unused, rendering this self-assembly *incomplete*. Finally, case D represents a situation in which all components are integrated into the product, and the corresponding self-sorting process is therefore social, completive, and *integrative*. The term integrative is also used for social self-sorting in case B, while narcissistic self-sorting is never integrative.

We will see several examples of self-sorting processes in the following sections, but this introduction cannot end without describing a fundamental study on self-assembly performed by Lyle Isaacs [9]. In this study, the nine compounds **5.4–5.12** shown in Figure 5.9, all of which contain hydrogen bond acceptors and donors, were mixed with barium picrate in chloroform, and the composition of the resulting mixture was investigated. Independent studies had previously shown that, individually, all of these compounds produce characteristic self-assembled structures, with **5.4** forming a macrocyclic pentamer, two of which stack in the presence of barium ions. Compound **5.5** forms a similar assembly comprising a stack of four tetrameric rings, three molecules of **5.6** form a bis(rosette) together with six barbiturates **5.7**, and all other compounds **5.8**, **5.9**, **5.10**, **5.11**, and **5.12** form dimers. We will return to some of these systems in Section 5.3. Perfect self-sorting should therefore lead to eight different products, while unspecific aggregation could lead to a very complex mixture of products.

The fact that the mixture of all nine compounds gives a ^{1}H NMR spectrum that is almost a superposition of the spectra of the eight individual self-assembled products, provided that the concentration and ratio of the individual species are chosen correctly, demonstrates that almost perfect self-sorting occurs. Moreover, the following general rules apply:

–   Increasing the temperature causes the dissociation of the self-assembled products, with the least stable aggregates dissociating first. Lowering the temperature, on the other hand, increases the amount of crossover species and thus decreases the fidelity of the self-assembly process if the exchange remains fully reversible.
–   A small difference between the stabilities of the self-sorted and crossover products will result in a significant amount of crossover products in the product mixture.
–   Similarly, the outcome of self-assembly is determined by the relative amounts of the components in the mixture. If the concentration of a component is too low, its homodimerization is less likely than its association with another species present at a higher concentration, leading to social self-sorting. Narcissistic self-sorting, on the other hand, dominates if the ratios and concentrations of the components are chosen correctly.
–   Finally, the presence of a large excess of a potential competitor in the mixture has a negative effect on the self-sorting process.

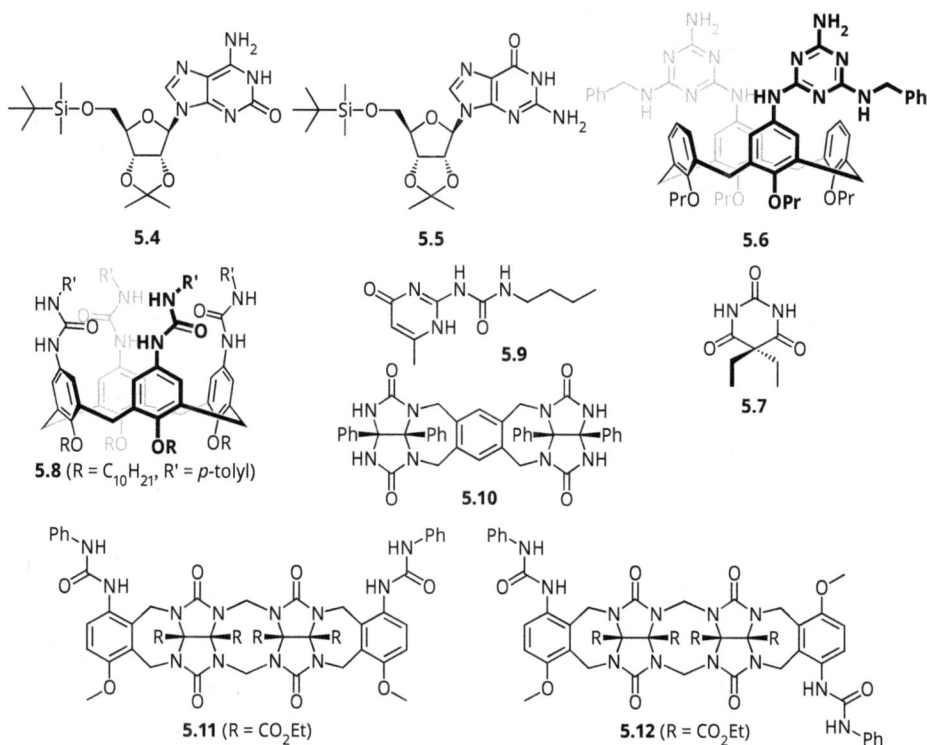

Figure 5.9: Structures of the nine compounds **5.4–5.12** investigated by Isaacs with respect to their self-sorting behavior.

The authors concluded that "self-sorting is neither the exception nor the rule" [9]. When systems become complex, their response to changing conditions depends on many parameters, making it difficult to predict their behavior. However, if the conditions and building blocks are properly chosen, even complex systems undergo self-assembly and self-sorting with high fidelity.

5.2 Self-assembly mediated by the hydrophobic effect

The hydrophobic effect is related to the tendency of water molecules to form an extended network of hydrogen bonds (Section 3.3). It results from the gain in Gibbs free energy when individually hydrated hydrophobic molecules are transferred to a larger cavity in the water matrix where they are all hydrated together. This effect is an important driving force for the self-assembly of hydrophobic molecules in water, but usually lacks a pronounced directional component, making the structure of the resulting assemblies difficult to control. Nevertheless, many self-assembled systems are formed in this way. Examples include lipid bilayers, in which the hydrophobic chains of individual lipid molecules are buried within the membranes, or many protein complexes, including the viral coat proteins mentioned earlier. Protein folding is also mediated to a significant extent by the screening of hydrophobic residues from water. However, although protein folding and self-assembly share common features in that both usually proceed autonomously and under thermodynamic control, protein folding is not strictly a self-assembly process because it is intramolecular rather than intermolecular.

Self-assembly driven by the hydrophobic effect can be used to generate micelles, vesicles, and larger multilamellar structures in water by using natural or synthetic amphiphiles. Although rules exist to correlate the structure, size, and composition of such assemblies with the structures of the underlying building blocks (Section 10.1), these large and often structurally complex systems will not be considered here. Instead, we will focus on discrete assemblies formed from building blocks such as the velcrands introduced in Section 4.1.9.

Such velcrands preferentially adopt the kite conformation, which is prone to dimerization even in organic solvents, especially when substituents such as methyl groups are present in 2-position of the resorcinol subunits. This tendency is even more pronounced for water-soluble velcrands such as **5.13** due to their large and difficult-to-hydrate hydrophobic surfaces (Figure 5.10). These compounds therefore form dimers in water to shield the hydrophobic regions from the solvent, leaving only the residues carrying the solubilizing groups exposed. This behavior is illustrated in Figure 5.10c using the calculated structure of the dimer of **5.13** as an example.

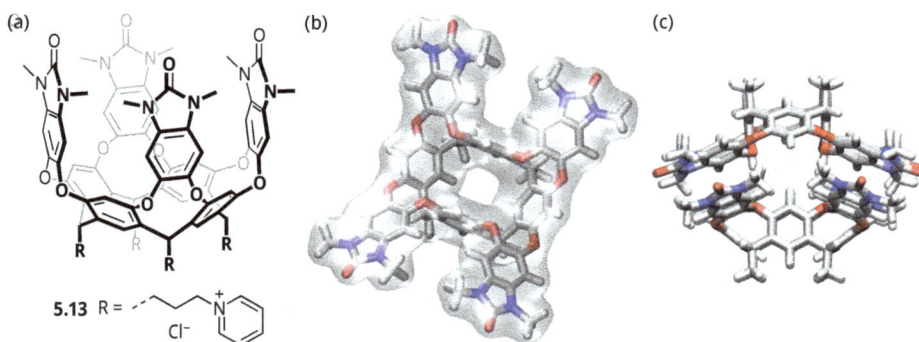

Figure 5.10: Molecular structure of velcrand **5.13** (a) and calculated structure of its kite conformation (b), illustrating the extended hydrophobic surface that mediates the self-assembly in water. The calculated structure of the dimer of **5.13** is shown in (c). The solubilizing groups are replaced by methyl groups for reasons of clarity.

Velcrand **5.13** binds hydrophobic substrates such as linear long-chain alkanes and primary alcohols in water [12]. For complex formation to occur, the self-assembled dimer must first dissociate and the resulting monomer must then adopt the vase conformation to accommodate the substrate. Velcrand **5.13** is therefore not only poorly preorganized, but its tendency to dimerize further counteracts complex formation. Gibb's conformationally rigid octa acid **5.14** (Figure 5.11a) behaves more interestingly.

This cavitand is also based on a central resorcin[4]arene unit. The four resorcinol subunits are covalently linked *via* benzal groups whose aromatic rings are further linked in their 3- and 5-positions to 3,5-dihydroxybenzoic acid moieties. Thus, cavitand **5.14** has a rigid structure with the eight aromatic subunits of the resorcin[4]arene core and the benzal groups lining a hydrophobic binding pocket. The eight carboxylate groups ensure water solubility at pH > 7, but the aromatic rings surrounding the cavity make the rim relatively hydrophobic. These solvent-exposed aromatic surfaces mediate the dimerization of **5.14** in water (Figure 5.11b). As we saw in Section 4.1.9, the propensity to self-assemble depends on the nature of the bound guest. When carboxylic acids such as adamantane-1-carboxylic acid occupy the cavity, their (at pH > 7) deprotonated carboxylate group protrudes from the cavity opening, facilitating hydration and preventing dimerization. On the other hand, filling the cavity with nonpolar substrates, such as steroids or alkanes, increases the size of the hydrophobic surface at the cavity opening, which promotes self-assembly and leads to a molecular capsule in which two molecules of **5.14** surround an encapsulated substrate [13].

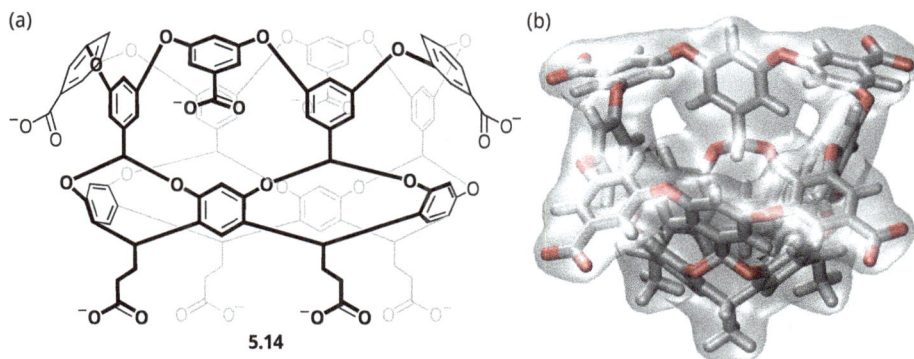

Figure 5.11: Molecular (a) and calculated (b) structure of octa acid **5.14**. The surface in the calculated structure illustrates the region along the rim of the cavity that mediates self-assembly in water.

Large substrates such as steroids have the perfect size to completely fill the cavity of the octa acid dimer. They therefore form 2:1 complexes in which one substrate molecule is surrounded by two subunits of **5.14**. Complex formation restricts the mobility of the substrate, which can only rotate along the long cavity axis. Each half of the capsule therefore hosts only part of the nonsymmetrical guest, making the two capsule halves distinguishable by NMR spectroscopy.

The complexation of linear alkanes varies with chain length. Starting with *n*-butane, short-chain alkanes are bound as 2:2 complexes with two substrate molecules occupying the cavity. This stoichiometry is maintained up to *n*-octane, which forms a mixture of 2:2 and 2:1 complexes. Longer alkanes up to *n*-heptadecane then form 2:1 complexes with only one substrate bound. These complexes differ in the conformation of the bound alkane. In the case of *n*-nonane and *n*-decane, there is enough room in the cavity for these alkanes to adopt fully extended conformations. Longer alkanes must coil into helical conformations to fit. The ability of **5.14** to discriminate *n*-butane from *n*-propane by complexing two molecules of the former substrate, while the latter is too small to efficiently fill the cavity, allows these alkanes to be separated by selectively scavenging *n*-butane from a gaseous mixture and transferring it to an aqueous receptor solution. Another application of **5.14** or structurally related derivatives is their use as nanoscale reactors to mediate photochemical reactions such as Norrish reactions or cycloadditions (Section 9.2.2) [13].

The third example of a building block that self-assembles in water is the benzene-derived amphiphile **5.15** (Figure 5.12a), developed in the group of Mitsuhiko Shionoya [14]. Among several derivatives, **5.15** with two pyridinium and one pyridine unit forms the most stable aggregate. Product formation involves the self-assembly of six subunits of **5.15**, resulting in a cube whose crystal structure is shown in Figure 5.12b. This cube forms only in water. In methanol, **5.15** is monomeric, demonstrating that the self-assembly is due to the hydrophobic effect. Because of the amphiphilic nature of **5.15**, the resulting assemblies can be regarded as structurally defined micelles.

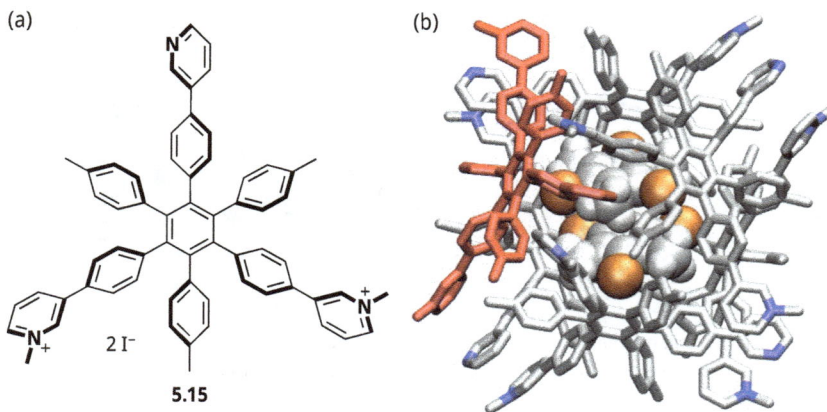

Figure 5.12: Molecular structure of the amphiphile **5.15** (a) and crystal structure of the corresponding hexameric aggregate containing two entrapped 2,4,6-tribromomesitylene molecules (b). One subunit of **5.15** is shown in red and the hydrogen atoms of the amphiphiles are omitted for reasons of clarity.

In this cube, the six subunits are arranged along the faces, with the aromatic rings interdigitating and the pyridine units intercalating between two pyridinium rings at each corner. The resulting cavity is capable of accommodating two 2,4,6-tribromomesitylene molecules. Interestingly, a derivative of **5.15** with three *N*-methylpyridinium groups assembles into a mixture of hexameric and tetrameric aggregates in water. The lack of selectivity is probably due to the fact that the interactions between three pyridinium groups at the vertices are weaker than those between pyridine and pyridinium groups. The equilibrium is shifted in the presence of suitable templates. For example, adamantane induces the exclusive formation of the tetramer [14].

These examples show that self-complementary structures with properly arranged hydrophobic surfaces assemble into ordered structures in water. For this concept to work, a good preorganization of the building blocks is advantageous. In addition, the subunits should be predisposed to assemble into defined structures, and the aromatic subunits that come into contact in the aggregates should be electronically complementary to enhance the stability of the products. When these factors are met, high thermodynamic stability is achieved.

5.3 Self-assembly mediated by hydrogen bonds

5.3.1 Introduction

Hydrogen bonds are particularly attractive interactions for mediating self-assembly [15]. The reasons for this are the ease with which hydrogen bond donors and acceptors can be introduced into molecular building blocks of varying structure and size, and the various ways in which the strength and orientation of hydrogen bonds can be controlled by using the principles outlined in Section 3.1.5. Another aspect is that hydrogen bonds are found in many natural systems where they mediate folding or assembly. Thus, nature not only serves as a blueprint for the design of hydrogen-bonded self-assembled systems, but also provides a pool of building blocks in the form of peptides, nucleobases, or larger oligonucleotides from which to choose.

Typical building blocks contain two or more subunits, each with a characteristic pattern of hydrogen bond donors and acceptors, connected by an appropriate linker. When these compounds self-assemble, as shown schematically in Figure 5.13, the structure, composition, and stability of the resulting products are controlled by several parameters, the most important of which are the number and pattern of the hydrogen bonds, the structure and flexibility of the linking units, and the solvent.

Number of hydrogen bonds: Interactions between molecular building blocks involving the formation of only a single hydrogen bond are neither strong nor directional enough to predictably control self-assembly. Therefore, the self-assembling subunits must be able to form at least two hydrogen bonds, which makes heterocyclic systems such as nucleobases or related synthetic systems attractive recognition motifs. They allow control not only of the number of hydrogen bonds but also of their arrangement.

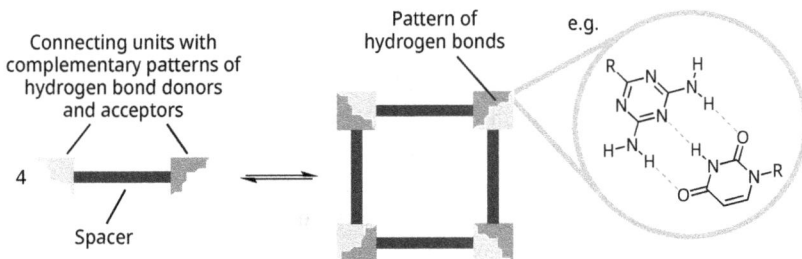

Figure 5.13: Schematic representation of a self-assembly process mediated by hydrogen bonds between the terminal groups of a ditopic building block, in this case yielding a cyclic tetramer. A specific example of such a process is shown in Figure 5.4.

Hydrogen bonding pattern: We have seen in Section 3.1.5 that the stability of hydrogen-bonded duplexes formed between two heterocyclic systems depends not only on the total number of hydrogen bonds, but also on the arrangement of the hydrogen donors and acceptors. The most stable complexes are formed when one binding partner contains only hydrogen bond acceptors and the other an equal number of donors, making all primary and secondary hydrogen bonds attractive. The inversion of one or more hydrogen bonds causes a decrease in stability because secondary interactions become repulsive. The intrinsic stability of a hydrogen-bonded complex thus depends on the exact pattern of hydrogen bonds holding the building blocks together. The chelate cooperativity operating in many self-assembling systems can still lead to high overall stability, even when repulsive secondary interactions within individual hydrogen bond patterns cannot be avoided, as in the rosettes discussed in the next section.

In addition to its influence on stability, the hydrogen bonding pattern also allows control of the mutual arrangement of the binding partners. This aspect is illustrated by using the example of the bis(pyridones) **5.16** and **5.17** shown in Figure 5.14 [16].

The bis(pyridone) **5.16** is self-complementary (when both pyridones exist in the same tautomeric form), allowing two molecules to interact to form a dimer stabilized by four hydrogen bonds. The C_{2v}-symmetric bis(pyridone) **5.17**, on the other hand, cannot form a dimer, and since its rigidity and linearity also prevent the formation of larger self-assembled macrocycles, it assembles into a polymeric structure (Section 6.2.1).

Linker structure: The length and flexibility of the subunits connecting the hydrogen bonding sites have a critical influence on the outcome of the self-assembly. If the linker is flexible, the self-assembly is associated with the restriction of conformational mobility, which reduces the stability of the final product. Thus, preorganization makes the assembly of rigid components to be entropically less costly, resulting in a more stable product. However, rigid linkers that orient the interacting groups in a way that creates

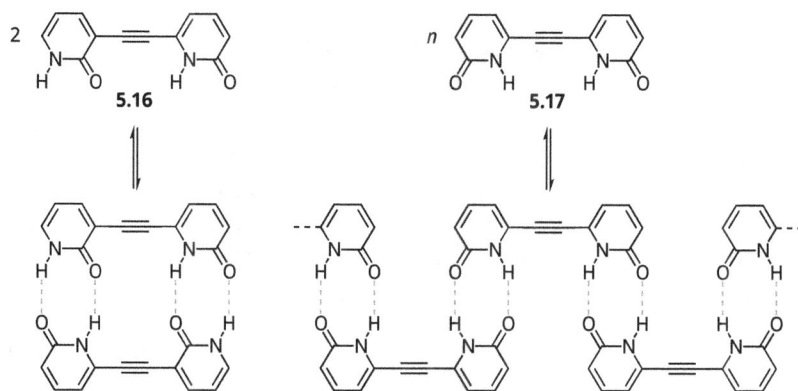

Figure 5.14: Molecular structures of the bis(pyridones) **5.16** and **5.17** and their modes of self-assembly. The hydrogen bonding patterns of the two pyridone subunits cause **5.16** to dimerize, while **5.17** polymerizes upon self-assembly.

strain in the product are disadvantageous, as illustrated by the different stabilities of the self-assembled dimers formed from the two closely related bis(nucleoside) derivatives **5.18** and **5.19** (Figure 5.15).

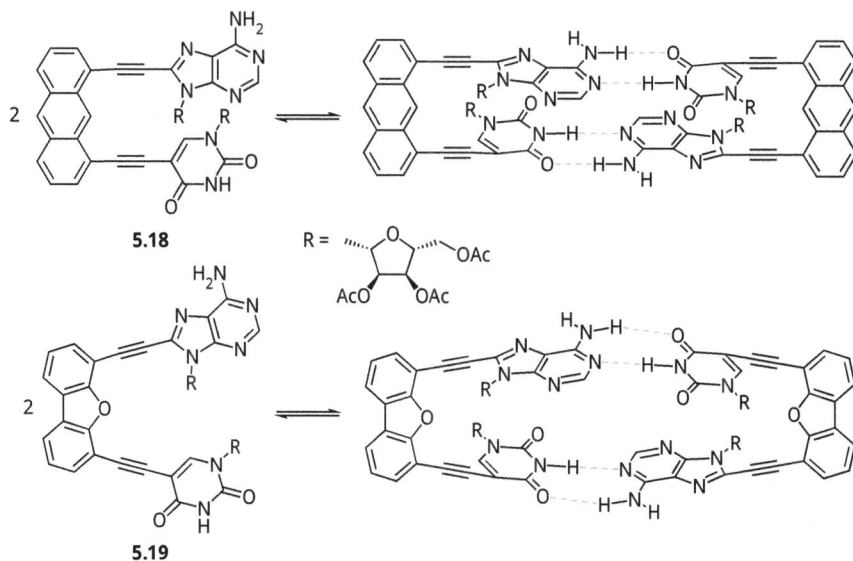

Figure 5.15: Molecular structures of the ditopic bis(nucleosides) **5.18** and **5.19** and their modes of self-assembly. Compound **5.18** dimerizes to give an unstrained product, whereas the dimer of **5.19** is strained.

Both bis(nucleosides) are fully dimerized in chloroform, but the progressive addition of DMSO affects their degree of dimerization to different extents [17]. While the dimer of **5.18** requires 60 vol% DMSO in the solvent mixture to fully dissociate, 25 vol% DMSO is sufficient to induce dissociation of the dimer of **5.19**, because the strained arrangement of the nucleobases makes this duplex less stable than that formed from **5.18**.

Finally, the structure of the linker also allows control over the structure of the assembly. Compound **5.20**, for example, arranges two complementary hydrogen bonding motifs along the two edges of the heterocyclic rings at an angle of 120°. The hydrogen bonding pattern and the angle at which the recognition motifs are oriented thus predispose **5.20** to form a cyclic hexamer, as shown in Figure 5.16 [18]. The introduction of a pyrrole ring between the two recognition sites, as in compound **5.21**, reduces the angle to 90°, and **5.21** therefore prefers to form a cyclic tetramer (Figure 5.16) [19].

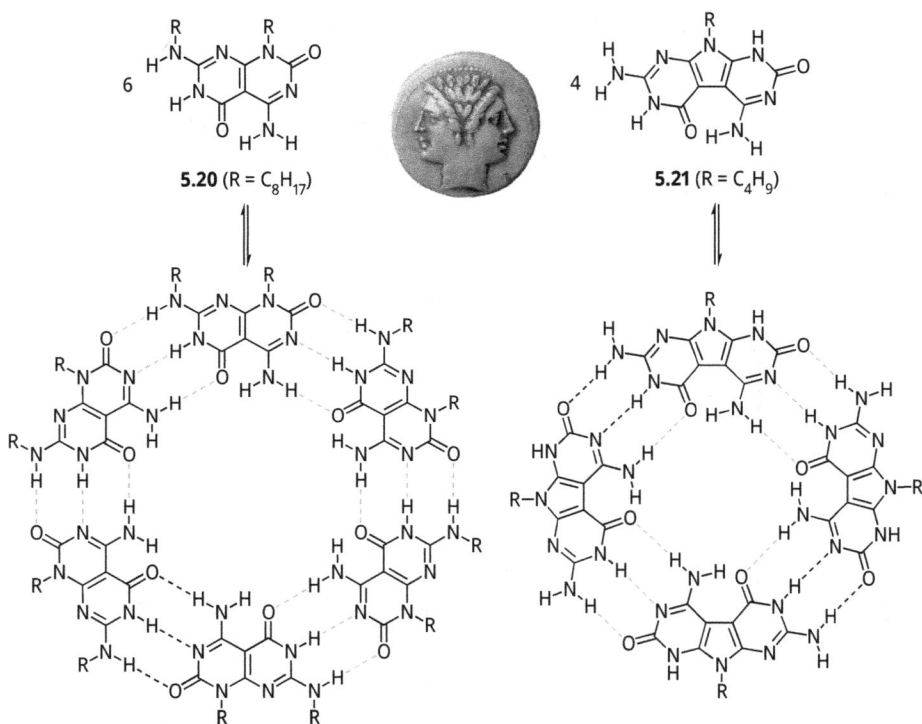

Figure 5.16: Molecular structures of the Janus molecules **5.20** and **5.21**, and structures of the hexameric assembly produced by **5.20** and of the tetrameric assembly produced by **5.21**. The image shows a coin depicting the Roman god Janus [20].

Compounds **5.20** and **5.21** are examples of so-called Janus molecules, a term coined by Jean-Marie Lehn for heterocyclic compounds with arrangements of hydrogen bond donors and acceptors on opposite sides with which complementary binding partners can interact [21]. The name refers to the Roman god Janus, who is usually depicted with two faces looking in opposite directions.

Solvent: We have seen in Section 3.1.5 that the strength of hydrogen bonding depends strongly on the polarity of the medium. Hydrogen bonds that are strong in nonpolar solvents such as chloroform become much weaker in polar aprotic or even protic solvents. This is because polar solvents not only increase the permittivity of the medium, but also directly interfere in the interactions of the binding partners. For example, DMSO is a strong hydrogen bond acceptor and its competitive binding to the NH groups of the bis(nucleosides) **5.18** and **5.19** (Figure 5.15) explains why the corresponding dimers exist in chloroform but dissociate in DMSO. Hydrogen-bonded assemblies are therefore only stable in polar solvents or even in water if the hydrogen bonds are shielded from the surrounding medium, as in the DNA double helix (whose stability in water is additionally due to the cooperativity of the multiple hydrogen bonds along the polynucleotide strands), and/or if the hydrophobic effect contributes to complex formation.

Hydrogen bond-mediated self-assembly is therefore often limited to nonpolar solvents such as chloroform, dichloromethane, benzene, or toluene. These solvents do not efficiently solvate hydrogen bond donors and acceptors, rendering polar compounds often poorly soluble because they form disordered aggregates. In contrast, discrete self-assembled products in which all binding sites are saturated and no vacant hydrogen bond donors or acceptors are exposed to the solvent dissolve more readily, especially if they contain additional solubilizing groups, such as the alkyl substituents in the aggregates formed from **5.20** and **5.21** (Figure 5.16). Thus, the dissolution of a polar molecule in a nonpolar solvent often qualitatively indicates the formation of a structurally defined self-assembled product.

These assemblies are structurally characterized using standard techniques. For example, NMR spectroscopy provides information about the symmetry of the products and their composition when more than one component is involved in product formation. Upfield shifts of signals in NMR spectra indicate which protons mediate hydrogen bonding. Colligative techniques such as vapor pressure osmometry are used to determine the average molecular weight of the assemblies. The ultimate proof that the predicted structure is correct is usually a crystal structure. In the following sections, we look at different classes of structures that are accessible using hydrogen bond-mediated self-assembly.

5.3.2 Rosettes

Macrocyclic assemblies consisting of heterocyclic subunits interconnected by an array of hydrogen bonds are called rosettes (Figure 5.16). Such rosettes are formed, for example, when a 1:1 mixture of cyanuric acid and melamine is cocrystallized, resulting in a highly stable layered structure that can be heated to 350 °C without decomposition. In these layers, each molecule of cyanuric acid interacts with three molecules of melamine through nine hydrogen bonds and *vice versa* (Figure 5.17), resulting not only in the aforementioned cyclic rosettes, but also in linear and crinkled tapes.

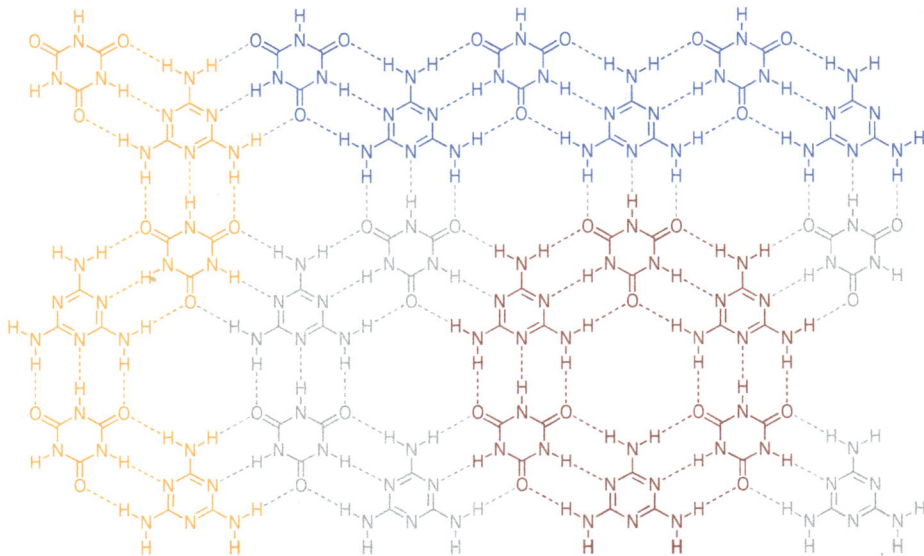

Figure 5.17: Section of an infinite layer containing equimolar amounts of cyanuric acid and melamine. The three hydrogen-bonded submotifs, the rosette, the linear, and the crinkled tape, are highlighted in red, blue, and orange, respectively.

The self-assembly of cyanuric acid and melamine into infinite structures can be prevented either by blocking certain NH groups or by using analogs of the building blocks with a reduced number of hydrogen bond donors and/or acceptors, such as barbiturates instead of cyanuric acid or pyrimidines instead of melamine. Examples of compounds that self-assemble into linear or cyclic structures and allow the isolation of a specific submotif of the cyanuric acid-melamine network are shown in Figure 5.18a.

Several strategies were developed in the group of George M. Whitesides to favor the formation of discrete rosettes over that of infinite tapes [22]. The first, called *peripheral crowding*, makes use of the fact that large substituents in the interacting binding partners are closer together in a linear self-assembled structure than in a cyclic one, as shown in Figure 5.18b. Cyanuric acid and melamine derivatives bearing sufficiently large substituents are therefore expected to preferentially form rosettes, which has been experimentally demonstrated by studying the self-assembly of diethylbarbiturate **5.7** with the *N,N*-diphenylmelamine derivatives **5.22a–c**, which differ structurally in the substituents in the 4-positions of the phenyl groups (Figure 5.18a). In this series, **5.22a** with the smallest substituent assembles into linear tapes when interacting with **5.7**, melamine **5.22b** yields the crinkled tape, and the large substituents in **5.22c** finally induce the formation of the rosette [23].

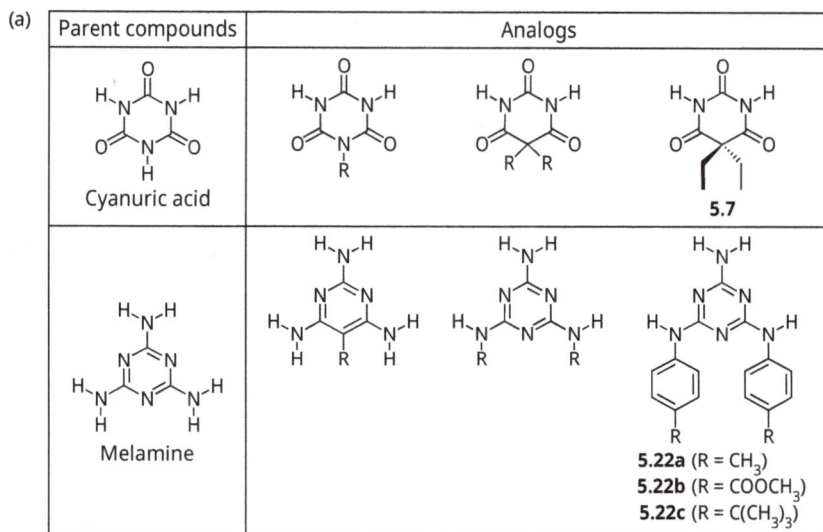

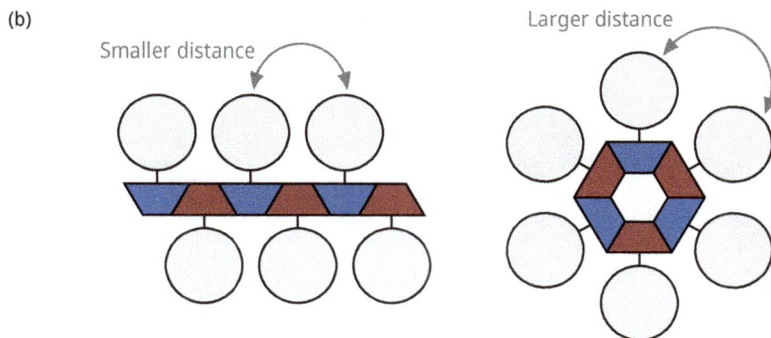

Figure 5.18: Examples of cyanuric acid and melamine analogs with reduced numbers of hydrogen bond donors and acceptors to restrict self-assembly to the formation of linear or cyclic structures (a). Compounds **5.7** and **5.22a–c** were used to demonstrate the peripheral crowding strategy shown schematically in (b).

The second strategy is based on the concept of *preorganization*. By covalently linking two or more building blocks in a suitable way, fewer molecules need to come together to form the rosette, thus reducing the entropic disadvantage of self-assembly. For example, when the tripodal hub **5.23** (Figure 5.19) interacts with mono-*N*-alkylated cyanuric acid derivatives, only four molecules are required to form a rosette, which is therefore more stable than a rosette formed from six subunits [24]. Another factor influencing the stability of such rosettes is the flexibility of the components, as demonstrated by the higher stability of the rosette containing the more rigid and thus better preorganized **5.23** compared to the analogous rosette formed from the more flexible **5.24** [25]. The combination of **5.24** with the tris(cyanurate) **5.25** leads to an even more stable rosette, since the self-assembly in this case involves only two molecules, one of which is preorganized and the other flexible enough to adapt to the required binding geometry [26]. Based on these investigations, the Whitesides group demonstrated that assemblies containing up to three rosette motifs and up to ten components can be assembled [22].

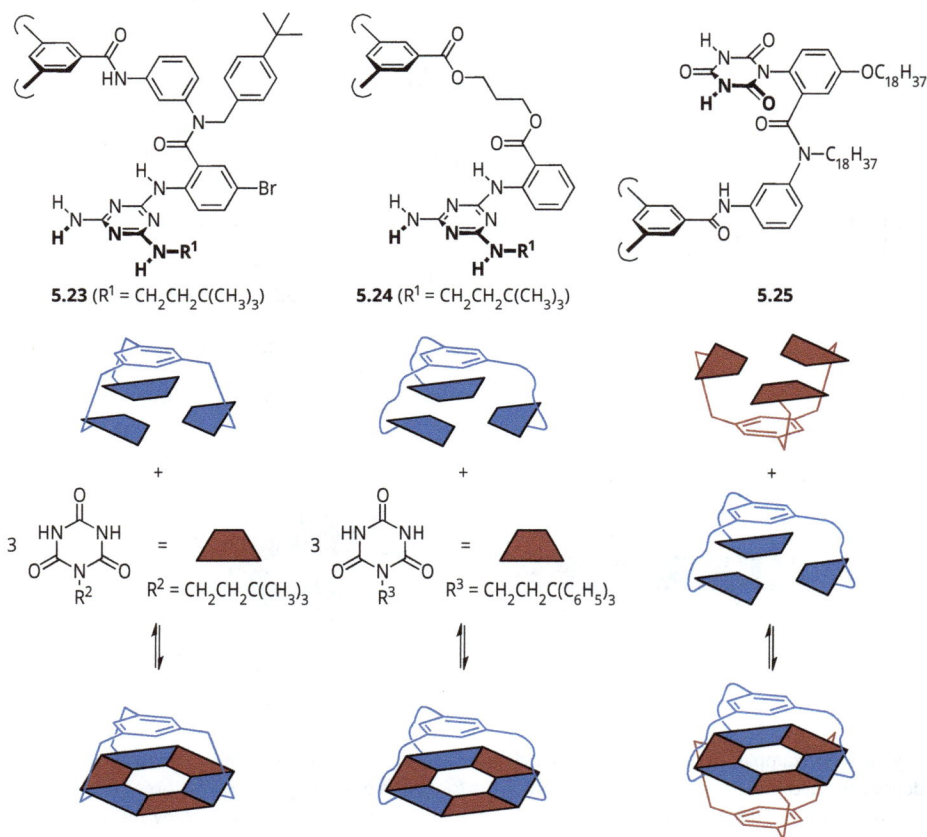

Figure 5.19: Structures of hubs **5.23**, **5.24**, and **5.25** and schematic representations of the rosettes derived therefrom.

David N. Reinhoudt and coworkers stabilized the cyanuric acid-melamine rosette motif by using **5.26** and structurally related derivatives (Figure 5.20a) [27]. This substituted calix[4]arene adopts a pinched cone conformation, induced by the propoxy groups along the narrower rim. The two melamine units along the wider rim allow three molecules of **5.26** and six molecules of diethylbarbiturate **5.7** to self-assemble, resulting in a bis(rosette) stabilized by 36 hydrogen bonds, 18 in each rosette motif (Figure 5.20b) [28]. The crystal structure of the assembly illustrates the dual role of the calix[4]arene subunits: they preorganize the melamine residues and ensure efficient peripheral crowding, thereby favoring rosette formation.

(a)   (b)

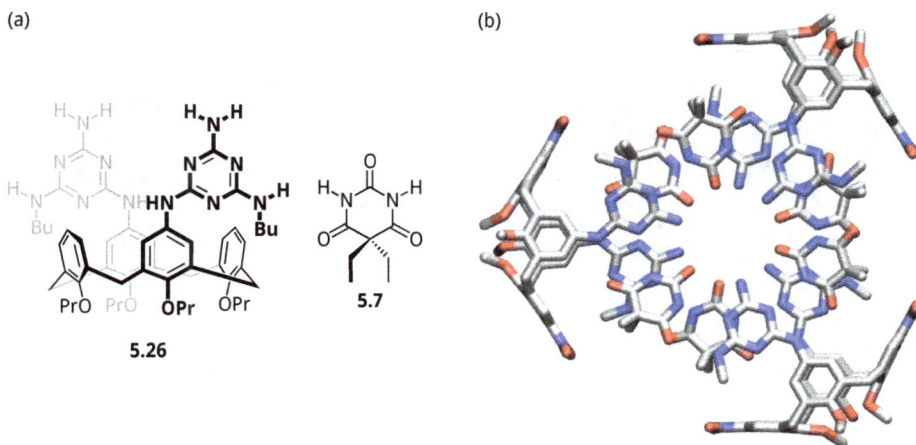

Figure 5.20: Structures of the calix[4]arene derivative **5.26** and the diethylbarbiturate **5.7** (a) and crystal structure of the bis(rosette) formed from and 6 equiv of **5.7** and 3 equiv of a derivative of **5.26** with nitro groups in the aromatic subunits flanking the melamine units (b). Protons could not be located in the crystal structure and side chains were disordered and are therefore not shown.

This bis(rosette) has a D_3 symmetry in the solid state. It is therefore chiral and is formed as a mixture of enantiomers in the absence of chiral information. The two rosette motifs are tightly stacked at a distance of about 3.3 Å, which is typical for π-stacked systems. The melamine rings are oriented away from each other, causing the two rosette motifs to assume a staggered arrangement, as shown schematically in Figure 5.21. This D_3-symmetric bis(rosette) is not the only possible product. When the calix[4]arenes adopt conformations with the melamine residues facing in the same direction, two additional assemblies are possible in which the rosette motifs are arranged in an eclipsed fashion. These assemblies differ in the orientation of the subunits. When all calix[4]arenes have the same orientation, the resulting eclipsed bis(rosette) is C_{3h}-symmetric, whereas when the orientation of one calix[4]arene unit is reversed, the bis(rosette) has C_s symmetry.

The different bis(rosette) isomers can be distinguished by NMR spectroscopy. The D_3 and C_{3h}-symmetric bis(rosettes) each give two NH signals in the region of the ^{1}H NMR spectra between 13 and 15 ppm, where the imide NH signals are located, because the two edges of the barbiturate or cyanurate subunits have different environments, but all six

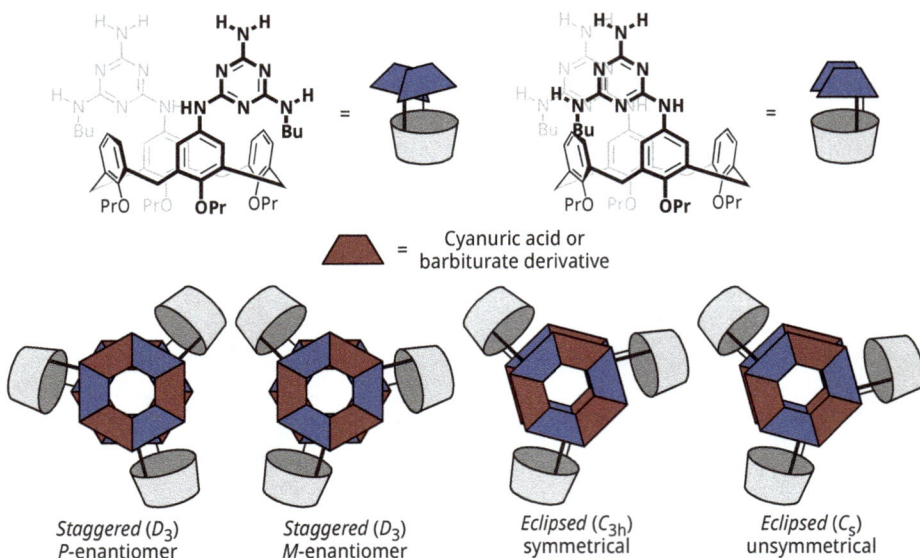

Figure 5.21: Isomers of bis(rosettes) formed from **5.26** and cyanuric acid derivatives or barbiturates.

subunits are symmetry equivalent. The C_s-symmetric bis(rosette) produces six signals in this region of the NMR spectrum due to the reduced symmetry. Consequently, the ^{1}H NMR spectrum of a mixture of all possible isomers will contain up to 10 NH signals if the exchange is slow on the NMR timescale, which is indeed observed in many cases. The integrals of the NH signals then provide information about the ratio of the different bis(rosettes). For example, **5.26** and an *N*-alkylated cyanurate with a –CH$_2$CH$_2$C(CH$_3$)$_3$ group produce the three bis(rosette) isomers in CDCl$_3$ in a ratio of 5:2:3 ($D_3/C_{3h}/C_s$) [29]. Thus, the staggered and eclipsed isomers are formed in equal amounts, and the ratio of the C_{3h} and C_s isomers is close to the expected statistical 1:3 ratio. Self-assembly is thus virtually nonselectively. In other cases, the exact correlation between the effects of the cyanuric acid substituents and the preferred structure of the product is complex, depending on both steric and electronic parameters. Barbiturates exclusively lead to the formation of staggered bis(rosettes) because the alkyl groups in their 5-position, which are oriented perpendicular to the ring plane, sterically disfavor the formation of eclipsed bis(rosettes).

In the absence of chiral information, the staggered D_3-symmetric assemblies are formed as racemic mixtures, but one of the two mirror images can also be selectively prepared using either chiral calix[4]arene derivatives or chiral barbiturates. For example, combining 3 equiv of **5.27** with two *R*-configured 1-phenylethylamine residues and 6 equiv of diethylbarbiturate **5.7** yields the *M*-form of the staggered bis(rosette) [30]. This enantiomer is also formed when the achiral calix[4]arene **5.28** and the *R*-configured barbiturate **5.29** are used (Figure 5.22) [31].

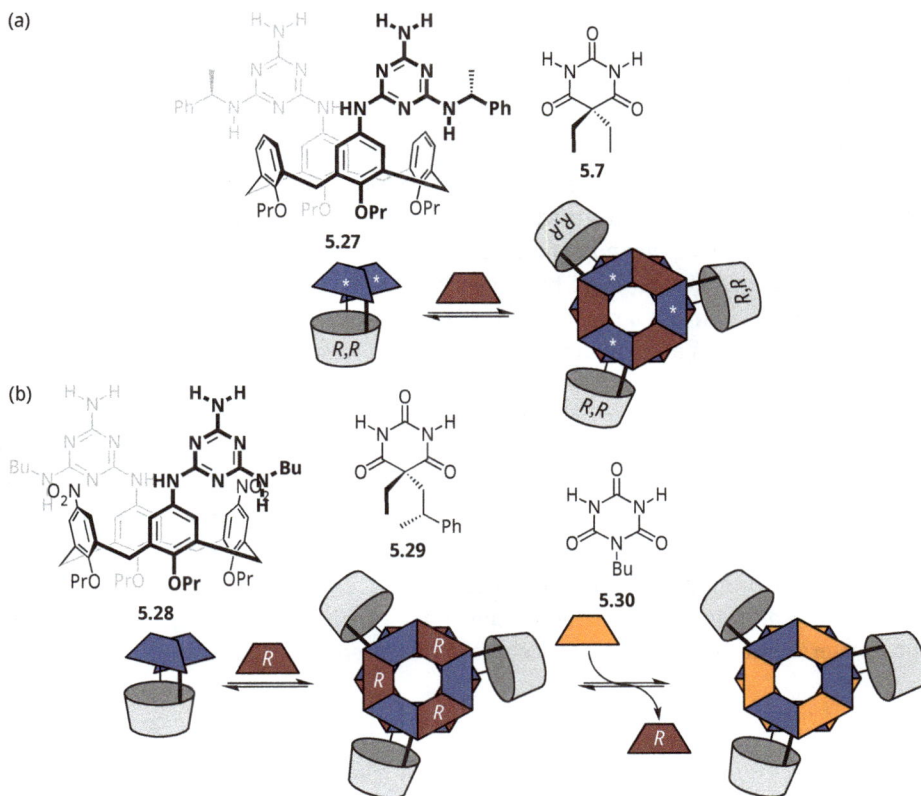

Figure 5.22: Self-assembly of a bis(rosette) from the chiral calix[4]arene derivative **5.27** and the achiral barbiturate **5.7** (a), and of a bis(rosette) from the achiral calix[4]arene derivative **5.28** and the chiral barbiturate **5.29** (b). In both cases, the corresponding *M*-configured bis(rosette) is formed. The reaction scheme in (b) shows the exchange of the barbiturate units in the chiral bis(rosette) derived from **5.28** and **5.29** with the achiral cyanurate **5.30** under the retention of the bis(rosette) configuration.

The chiral building blocks of the latter bis(rosette) can be exchanged by treating it with a slight excess (1.2 equiv per **5.29**) of cyanurate **5.30** in benzene-d_6. This exchange occurs because cyanurates form stronger hydrogen bonds than barbiturates due to their more acidic NH groups. According to circular dichroism spectroscopy, the resulting bis(rosette) still has the original configuration, although it is completely deprived of chiral building blocks. The product thus seems to *remember* the structure of the starting material, and the observed phenomenon is therefore an example of a *chiral memory* effect. The reason for the lack of isomerization during the building block exchange is that the replacement of the chiral units in the original assembly by achiral ones proceeds stepwise *via* short-lived intermediates. Only six hydrogen bonds must be broken and reformed during each exchange reaction, making this reaction faster than any alternative involving a more extensive bis(rosette) disruption. The intermediates still contain 30 intact hydrogen bonds and are therefore sufficiently stable to maintain their configuration while they exist. Although the final product is initially formed with high fidelity, it slowly isomerizes due to its dynamic nature. The interconversion half-life of the two D_3-symmetric assemblies is approximately 4.5 days at room temperature, demonstrating that the stability is significantly below that of a covalently assembled compound. Mechanistic studies suggest that the rate-limiting step of interconversion involves the dissociation of a calix[4]arene moiety from the intact bis(rosette) (requiring the dissociation of 12 hydrogen bonds), followed by rearrangement and reassociation, resulting in a mixture of the isomers.

Although the crystal structure in Figure 5.20 suggests that there is not much space between the two rosette motifs to permit the binding of additional molecules, these structures have an interesting host–guest chemistry. Alizarine **5.31** (Figure 5.23a), for example, binds to the bis(rosette) derived from **5.26** and diethylbarbiturate **5.7** in chloroform [32]. The equilibrium is slow on the NMR timescale, leading to a complex that is not fully formed until three molecules of **5.31** per bis(rosette) are present. The guest molecules intercalate between the two rosette motifs, where they form a third rosette, as shown in the crystal structure in Figure 5.23b. Complex formation not only increases the distance between the two melamine-barbiturate rosettes, but also causes the staggered bis(rosette) to rearrange into the eclipsed isomer [33].

(a)   (b)

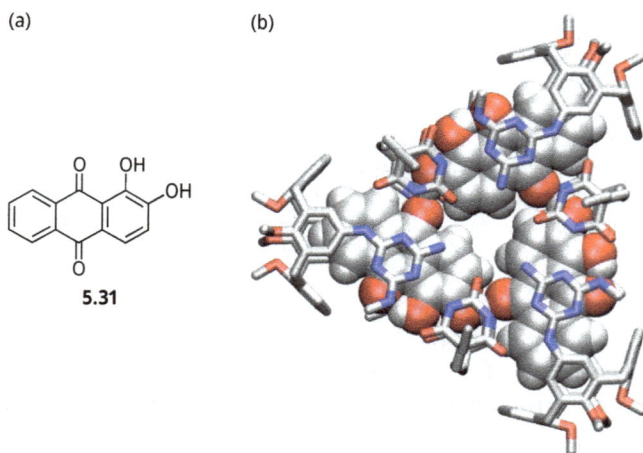

5.31

Figure 5.23: Molecular structure of alizarine **5.31** (a), and crystal structure of the complex between the bis(rosette) formed from **5.26**, **5.7**, and three molecules of **5.31** (b). Disordered side chains and protons in the bis(rosette) are not shown for reasons of clarity.

5.3.3 Capsules

Among the receptors discussed in Section 4.1, molecular cages such as cryptophanes, hemicryptophanes, carcerands, and hemicarcerands have particularly intriguing properties because they have cavities in which the substrate is surrounded on all sides and in which it is kinetically trapped. Unfortunately, the synthesis of these receptors is often challenging and low yielding, but the use of covalently assembled cages is not the only way to achieve substrate encapsulation. The alternative is self-assembly, as we have seen for viral capsid proteins. The appeal of this approach is that it proceeds under thermodynamic control and therefore potentially yields the products in high yields, provided they are the thermodynamically favored species in the equilibrium. The corresponding assemblies are called molecular capsules, as opposed to molecular cages, which are obtained by covalent synthesis [15, 34].

The assembly of capsules requires the design of suitable building blocks that surround a cavity of predictable size. An approximately spherical capsule can be obtained, for example, from two hemispherical building blocks with functional groups along their edges to mediate dimerization, but there are many other shapes that can be used to assemble spheres. The building blocks must only be curved and predisposed to assemble with their convex surfaces facing outward.

An early molecular capsule was designed in the group of Julius Rebek Jr. using the bis(glycoluril) derivative **5.10** as building block (Figure 5.24a). This C-shaped compound dimerizes by hydrogen bond formation between the glycoluril NH and C=O groups, resulting in an assembly in which the two subunits are arranged in a perpendicular fashion (Figure 5.24b). The seam between the components is similar to the way the two halves of a tennis ball are joined, which explains why this capsule is known as the molecular tennis ball [35].

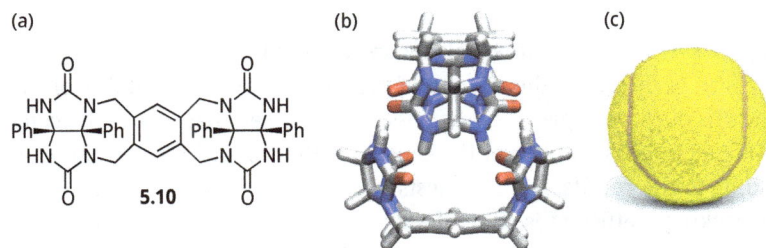

Figure 5.24: Molecular structure of bis(glycoluril) **5.10** (a), and calculated structure of the molecular tennis ball formed from two molecules of **5.10** in which the substituents in the glycoluril residues are omitted for reasons of clarity (b). The picture in (c) illustrates that a tennis ball is assembled in a similar manner.

With about 60 Å^3, the cavity of the tennis ball is large enough to accommodate guests such as methane, ethylene, chloroform, or dichloromethane. Binding of these substrates results in the shielding of their protons, causing the corresponding signals in the ^{1}H NMR spectrum to move upfield. Methane, for example, gives a signal at −0.91 ppm when bound inside the tennis ball, while free methane has a chemical shift of 0.23 ppm. Guest exchange proceeds *via* a gating mechanism, similar to that proposed for carcerands (Section 4.1.9), in which a glycoluril subunit folds away to create an opening that allows the guest to exit and enter the cavity. Only four hydrogen bonds need to be broken for this process to occur, as opposed to eight hydrogen bonds for complete tennis ball dissociation [36]. Capsule formation is restricted to nonpolar solvents such as chloroform, but can also be templated by suitable guests in solvents such as DMF, in which **5.10** alone does not dimerize.

Several other glycoluril derivatives can be used to obtain capsules with larger cavities. Examples are the extended bis(glycoluril) **5.32** and the tripodal tris(glycoluril) **5.33** (Figure 5.25a). Two molecules of **5.32** assemble in nonpolar solvents to form a capsule with a seam of hydrogen bonds similar to the tennis ball, but with a much larger cavity. This so-called molecular softball is stabilized by a combination of hydrogen bonds between the terminal glycoluril subunits and between the OH and C=O groups along the bridges, as shown in Figure 5.25b. The size of the cavity allows the incorporation of guests such as ferrocene or adamantane derivatives, or a combination of two smaller molecules [37]. Guest binding is entropically driven in CDCl$_3$, probably due to the favorable release of solvent molecules from the cavity upon guest entry. This softball can also be used as a reaction chamber, as we will see in Section 9.2.2.

The shape of the capsule derived from the tripodal building block **5.33** explains why it is called a molecular jelly doughnut (Figure 5.25c) [38]. This capsule binds disk-shaped guests such as benzene or cyclohexane, demonstrating that the size and shape of the cavities of these capsules characteristically influence their selectivity.

How much space does a substrate usually occupy in a receptor cavity?

The series of capsules available in the Rebek group was used to establish a relationship between the size of the guest, the volume of the capsule, and the stability of the complex [39]. For this purpose, a packing coefficient was defined that relates the computationally estimated internal volume of the capsule to the van der Waals volume of the guest. According to the results, neutral guests typically form the most stable complexes when the packing coefficient is about 0.55, that is, when about 55% of the available cavity volume is filled. This correlation is general for neutral guests and applies not only to complexes of self-assembled capsules, but also to those of molecular cages such as cryptophanes or carcerands.

Similar packing coefficients are found for organic solvents in their liquid state. In this case, they are determined by relating the minimum space occupied by a given number of solvent molecules, calculated by multiplying the number of molecules with their van der Waals volume, to the actual volume of the corresponding amount of solvent. Solvents therefore have a lot of empty space, which is probably required to

(a)

5.32 (Ar = 4-*n*-heptylphenyl)   **5.33**

(b)

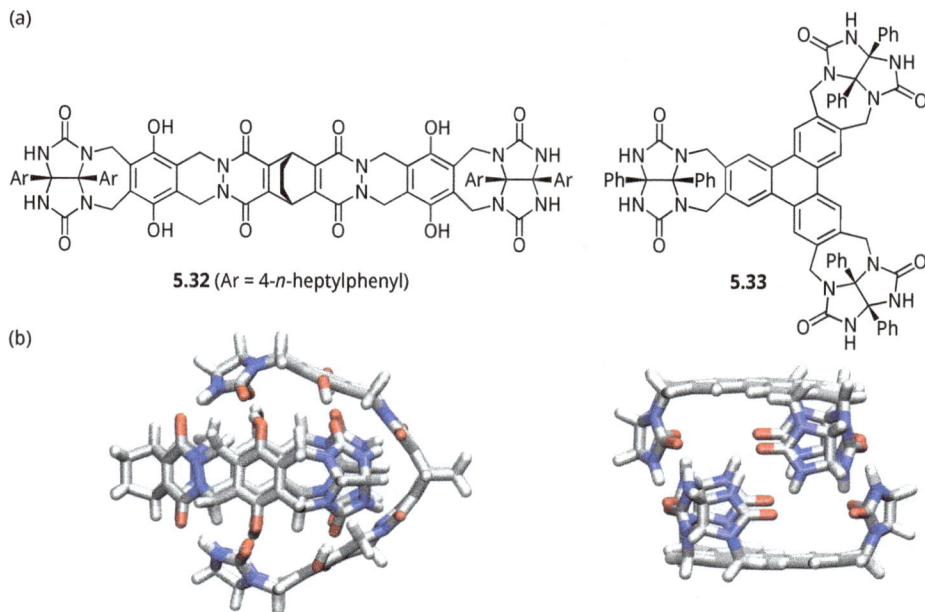

Figure 5.25: Molecular structures of bis(glycoluril) **5.32** and tris(glycoluril) **5.33** (a), and calculated structures of the molecular softball formed from two molecules of **5.32** (b), and the molecular jelly doughnut formed from two molecules of **5.33** (c). The substituents in the glycoluril residues are omitted for reasons of clarity.

allow the individual molecules to move around. Solvent packing coefficients thus reflect a trade-off between the translational freedom of molecules, which is determined by entropy, and a sufficient proximity to allow enthalpically favored intermolecular interactions. The agreement between solvent packing coefficients and those determined for receptor substrate complexes suggests that complex stability benefits when the space available for a substrate in a receptor cavity is similar to that in the bulk solvent. This conclusion, known as Rebek's 55% rule, has been confirmed for many systems and has become an important tool for assessing the binding properties of a receptor. Packing coefficients greater than 55% are possible and typically indicate attractive interactions between the binding partners.

The formation of capsules by self-assembly can also involve more than two building blocks. For example, compound **5.34** (Figure 5.26a), which contains a glycoluril and a cyclic sulfamide residue, forms a capsule in which four subunits are assembled in an up-down arrangement and held together by 16 hydrogen bonds [40]. This capsule does not form in the absence of suitable templates, as evidenced by the insolubility of **5.34** in nonpolar solvents such as CD_2Cl_2, $CDCl_3$, or benzene-d_6. Only in the presence of adamantane or adamantane derivatives, dissolution of **5.34** and capsule formation occur. The most stable complex is formed with adamantane-2,6-dione due to stabilizing interactions between the carbonyl oxygen atoms of the guest and the glycoluril NH groups. The efficient filling of this tetrameric capsule, as evidenced by the structure of the adamantane-2,6-dione complex shown in Figure 5.26b, is an important requirement for the encapsulation, since the less bulky 1,4-cyclohexanedione does not induce capsule formation.

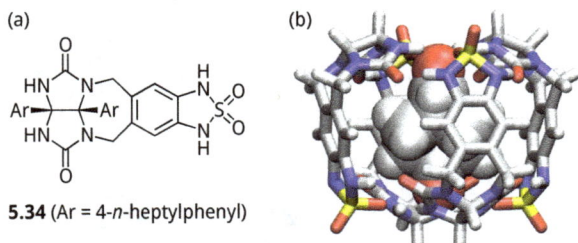

(a)

5.34 (Ar = 4-n-heptylphenyl)

(b)

Figure 5.26: Molecular structure of **5.34** (a) and calculated structure of the tetrameric capsule formed from **5.34**, containing an entrapped adamantane-2,6-dione molecule. The substituents in the glycoluril residues are omitted for reasons of clarity.

Capsules with even larger cavities are accessible by using the deep cavitand **5.35** (Section 4.1.9), whose imide groups along the cavity edge mediate hydrogen bond formation (Figure 5.27a). Two molecules of **5.35** assemble in nonpolar solvents to form a cylindrical capsule with a height of about 18 Å and a diameter of 10 Å (Figure 5.27b). The internal volume amounts to 425 Å^3, sufficient to accommodate several guest molecules. The entrapped molecules are often unable to move freely but are constrained in their mobility by the shape and size of the cavity, resulting in properties not observed outside the capsule [41].

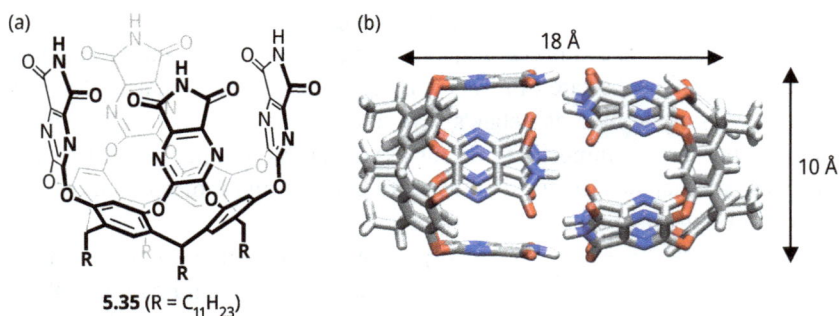

(a)

5.35 (R = C$_{11}$H$_{23}$)

(b)

Figure 5.27: Molecular structure of the self-assembling deep cavitand **5.35** (a) and calculated structure and dimensions of the corresponding cylindrical capsule. The substituents in the resorcin[4]arene ring are replaced by methyl groups for reasons of clarity.

Capsule formation occurs in $CDCl_3$, benzene-d_6, or toluene-d_8, but not in the presence of solvent molecules too large to be encapsulated, such as mesitylene-d_{12} [42]. In this solvent, the self-assembly can be mediated by suitable guests. For example, benzene and toluene induce the formation of capsules containing two molecules of either benzene or toluene. Two molecules of p-xylene are too large to be entrapped simultaneously. In the presence of a mixture of benzene and p-xylene, the capsule containing one molecule of benzene and one molecule of p-xylene forms preferentially. The same complex dominates even in the presence of a 2:1:1 mixture of toluene, benzene, and p-xylene, indicating that one molecule of benzene and one molecule of p-xylene fill the available space more efficiently than two molecules of toluene.

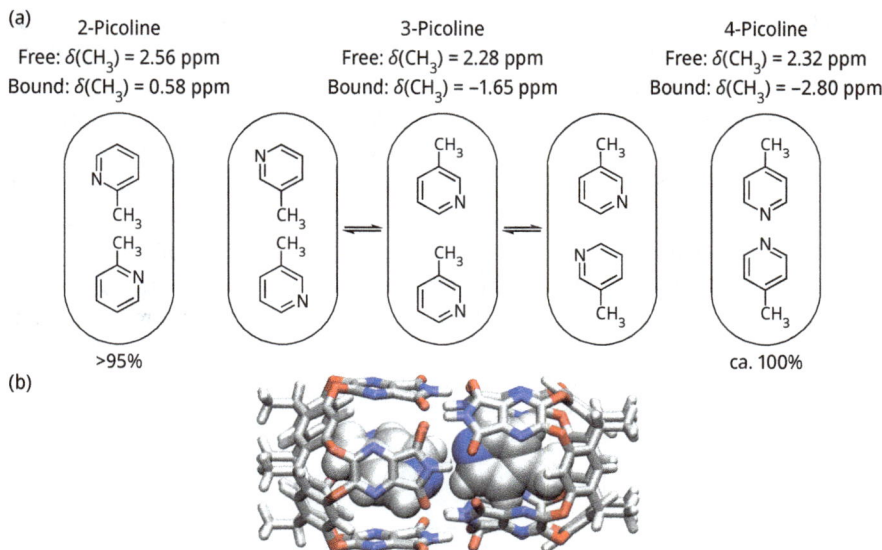

(a)

2-Picoline
Free: $\delta(CH_3)$ = 2.56 ppm
Bound: $\delta(CH_3)$ = 0.58 ppm

3-Picoline
Free: $\delta(CH_3)$ = 2.28 ppm
Bound: $\delta(CH_3)$ = −1.65 ppm

4-Picoline
Free: $\delta(CH_3)$ = 2.32 ppm
Bound: $\delta(CH_3)$ = −2.80 ppm

>95%

ca. 100%

(b)

Figure 5.28: Schematic representation of the preferred orientations of 2-, 3-, and 4-picoline inside the capsule formed from the deep cavitand **5.35** (a) and calculated structure of the 4-picoline complex of this capsule (b). The substituents in the resorcin[4]arene ring are replaced by methyl groups for reasons of clarity.

The ^{1}H NMR spectroscopic fingerprints of the complexes containing two molecules of 2-, 3-, or 4-picoline provide structural information (Figure 5.28) [43]. In the absence of **5.35**, the methyl protons of the three picolines absorb at 2.56, 2.28, and 2.32 ppm, respectively. These resonances shift to 0.58, −1.65, and −2.80 ppm upon complex formation. The different positions of these signals indicate that the picoline molecules cannot move freely within the capsule. If they could, the signal from the methyl protons should have approximately the same chemical shift in all three cases. The actual values of the chemical shifts reflect the orientation of the guests within the cavity. The large shielding of the 4-picoline methyl protons indicates that the methyl groups are almost exclusively oriented near the ends of the capsule, inside the resorcin[4] arene bowls. The two 2-picoline molecules, on the other hand, preferentially (>95%) adopt orientations with their methyl groups oriented near the equator, where the shielding is least pronounced. The arrangement of 3-picoline is less well defined, with a slight preference for the methyl groups near the ends (60–70%) rather than near the center (40–30%) of the capsule.

Complexes that differ only in the mutual orientation of the included guest molecules are isomeric. In the case of 3-picoline, the interconversion of the different arrangements inside the capsule is too fast to distinguish the corresponding isomers, but this situation changes for other guests or guest combinations. For example, the ^{1}H NMR spectrum of the capsule containing one molecule of chloroform and one molecule of 4-ethyltoluene contains two signals for the terminal protons of the ethyl group, one at about –3.8 ppm and one at –0.1 ppm, showing that the tumbling of the 4-ethyltoluene molecule inside the cavity is slow on the NMR timescale. The two signals can be assigned to stereoisomers of the complex that differ in the orientation of the aromatic guest, and their integration shows that the complex with the ethyl group near the equator (weaker shielding) is four times more abundant than the isomeric complex with the ethyl group in the bowl (stronger shielding) (Figure 5.29). Thus, the confinement of molecules within a capsule leads to an unusual type of stereoisomerism that cannot be observed in the absence of the capsule. The corresponding isomers are called *social isomers* because they differ only in the way the bound molecules approach each other [44].

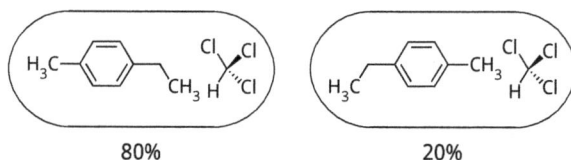

Figure 5.29: Schematic representation of the social isomerism observed in the cavity of the dimer of **5.35**, containing one molecule of chloroform and one molecule of 4-ethyltoluene.

Complex formation also influences the preferred conformation of the entrapped guest, as illustrated by the encapsulation of alkanes [45]. The table in Figure 5.30a shows how the relative stability of these complexes, all of which contain only one alkane molecule, varies with chain length. Undecane forms the most stable complex because it fits perfectly into the capsule in an energetically favorable extended formation (Figure 5.30b). Shorter n-alkanes are also bound, but fill the available space less efficiently, resulting in weaker interactions. This explains the decrease in complex stability with decreasing chain length. Longer alkanes can only be bound by adopting energetically unfavorable coiled conformations, and these alkanes therefore also form less stable complexes than undecane. Another reason for the lower stability is the weakening of the hydrogen bonds that stabilize the capsule, caused by the internal pressure generated by the coiled guests. The longest alkane complexed by this capsule is tetradecane in its fully coiled conformation (Figure 5.30c).

The pressure generated by the entrapped coiled alkanes can be released by the insertion of a ring of four glycoluril units between the capsule halves, resulting in an extended capsule with an internal volume of 620 Å^3. This capsule forms spontaneously when glycoluril is added to the tetradecane complex of **5.35** because it allows the included alkane to adopt the energetically favorable extended conformation (Figure 5.30d). In the absence of suitable guests, the extended capsule does not exist because its formation is entropically too costly. If the corresponding extended capsule contains a suitable alkane, this entropic term is compensated by the enthalpic stabilization resulting from the extensive van der Waals interactions between the guest and the cavity walls.

The corresponding extended capsule also binds alkanes longer than tetradecane, which have to adopt coiled conformations again. The internal pressure thus generated influences the rate at which the glycoluril units rearrange along the equator, that is, the rate at which this D_4-symmetric chiral capsule racemizes [46]. This rate increases as the alkane become longer because the internal pressure, which weakens the hydrogen bonds holding the assembly together, increases in the same direction. The extended capsule accommodates alkanes up to nonadecane in their coiled conformations. The next longer alkane, eicosane, templates the formation of a capsule with eight glycoluril molecules, consisting of a total of 11 components (Figure 5.30e), and suitable guest molecules even stabilize a capsule with three glycoluril rings [47].

Julius Rebek Jr. and Volker Böhmer reported independently and almost simultaneously that capsules are also accessible from calix[4]arene tetraureas with the general structure **5.8** (Figure 5.31a). These calix[4]arenes self-assemble in nonpolar solvents to form dimers held together by 18 bifurcated hydrogen bonds [48]. The crystal structure in Figure 5.31b illustrates the complementarity of the two calix[4]arene units and the interdigitation of the side chains, which allows hydrogen bonding interactions between the urea groups [49].

Such capsules can accommodate monocyclic aromatic or aliphatic guests in their 180 Å^3 large cavity, as well as structurally more complex guests such as terpenes or even cubane. Binding studies benefit from the slow exchange on the NMR timescale, which allows the simultaneous observation of different complexes. For example, the ^1H NMR spectrum of **5.36b** (Figure 5.31a) in a mixture of benzene-d_6 and toluene-d_8 contains two signal sets, one belonging to the benzene and the other to the toluene complex [50].

The slow exchange also makes it possible to study the self-assembly of mixtures of different calix[4]arene derivatives. Normally, a 1:1 ratio of two calix[4]arene tetraureas yields the two homodimeric and the heterodimeric capsules in a statistical 1:2:1 ratio [51]. However, the combination of **5.36c** and **5.36d** leads to the exclusive formation of the heterodimeric capsule. While the homodimeric capsules are S_8

(b)

(a)

Guest	K_a (rel)
$n\text{-}C_9H_{20}$	0.3
$n\text{-}C_{10}H_{22}$	16.9
$n\text{-}C_{11}H_{24}$	100
$n\text{-}C_{12}H_{26}$	24.4
$n\text{-}C_{13}H_{28}$	1.0
$n\text{-}C_{14}H_{30}$	0.008

(e)

(d)

(c)

18 Å

24 Å

30 Å

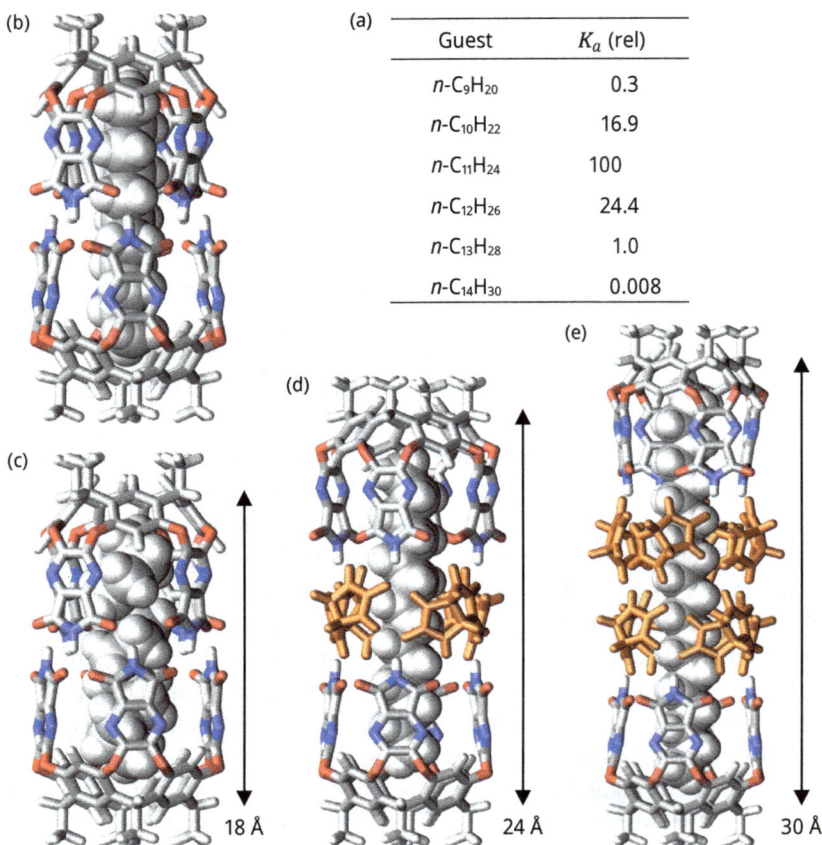

Figure 5.30: Relative stabilities of alkane complexes of the capsule formed from **5.35** (a), and calculated structures of the complexes of this capsule with undecane in the extended conformation (b) and tetradecane in the coiled conformation (c). In addition, the complexes of tetradecane in the extended conformation with a capsule containing a ring of four glycoluril units, and of eicosane ($C_{20}H_{42}$) with a capsule containing two rings of eight glycoluril subunits are shown in (d) and (e), respectively. The substituents in the resorcin[4]arene ring are replaced by methyl groups and those in the glycoluril units are omitted for reasons of clarity.

symmetric and therefore achiral, the C_4-symmetric heterodimeric capsules are chiral, with the two enantiomers differing in whether the urea units are arranged clockwise or counterclockwise. These capsules are normally formed as a racemic mixture, but if there are stereogenic centers in the substituents, one enantiomer forms preferentially [52].

The two calix[4]arene derivatives **5.37** and **5.38** (Figure 5.32) also assemble into a heterodimeric capsule, but since this capsule is stabilized by salt bridges between the amidinium and sulfonate groups, self-assembly occurs even in polar solvents such as

(a)

(b)

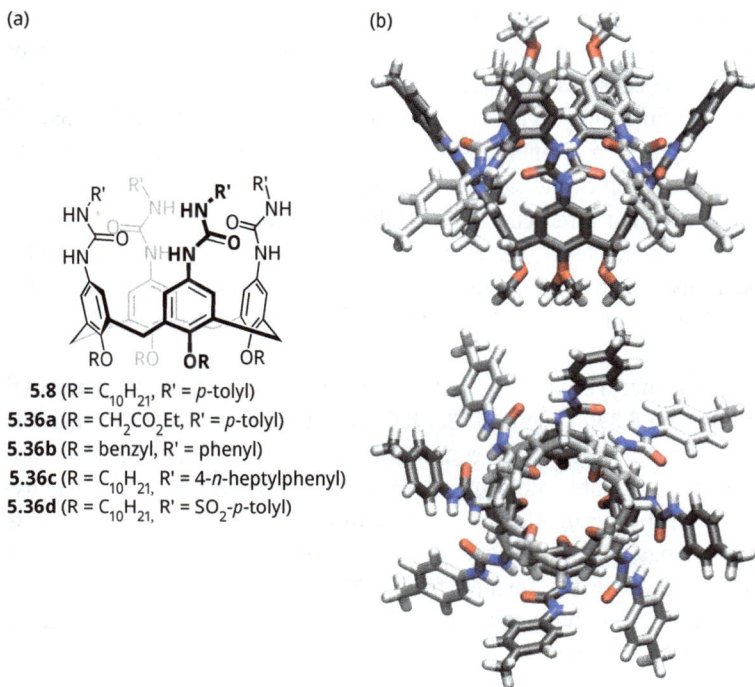

5.8 (R = C$_{10}$H$_{21}$, R' = *p*-tolyl)
5.36a (R = CH$_2$CO$_2$Et, R' = *p*-tolyl)
5.36b (R = benzyl, R' = phenyl)
5.36c (R = C$_{10}$H$_{21}$, R' = 4-*n*-heptylphenyl)
5.36d (R = C$_{10}$H$_{21}$, R' = SO$_2$-*p*-tolyl)

Figure 5.31: Molecular structures of calix[4]arene tetraureas **5.8** and **5.36a–d**, which self-assemble into dimeric capsules (a), and side and top views of the crystal structure of the capsule derived from **5.36a** (b). The substituents along the lower calix[4]arene rims are replaced by methyl groups for reasons of clarity.

methanol or methanol/water mixtures in which the calix[4]arene tetraureas do not dimerize [53]. The resulting cavity is available for the binding of quaternary ammonium ions such as acetylcholine.

5.37 (R = SO$_3$Na, R' = CH$_2$CH$_2$OC$_2$H$_5$)

5.38 (R = ⎯⎯⎯< $^{NH_2Cl}_{NHC_3H_7}$, R' = C$_3$H$_7$)

Figure 5.32: Molecular structures of calix[4]arene derivatives **5.37** and **5.38**, which self-assemble in polar solvents such as water/methanol mixtures, yielding the corresponding heterodimer.

Capsules with two hemispherical subunits can also be assembled from resorcin[4]arenes. For example, crystallization of resorcin[4]arene **5.39a** (Figure 5.33a) from aqueous methanol in the presence of tetraalkylammonium ions yields a dimeric capsule in which a cation occupies the cavity [54]. The two halves of this capsule are bridged by methanol and water molecules whose OH groups complete the hydrogen bonding pattern holding the subunits together. Much more interesting than these dimers, however, are the large resorcin[4]arene hexamers. The crystal structure of the octahydrate of **5.39a** shows that in these capsules, six resorcin[4]arene subunits are positioned along the faces of a cube, surrounding a cavity with a volume of 1,375 Å^3 (Figure 5.33b) [55]. Of the eight OH groups in each resorcin[4]arene, four serve as hydrogen bond donors to interact with a hydroxy group in the same molecule, while the remaining four donors are involved in intermolecular interactions. The eight water molecules at the corners of the cube are needed to complete the seam of hydrogen bonds. These water molecules donate additional 12 hydrogen bonds (four water molecules donate two bonds each, and the other four donate only one), bringing the total number of hydrogen bonds in this capsule to 60.

A closely related structure with a slightly larger volume of 1,510 Å^3 is formed from the pyrogallolarene **5.39b** (Figure 5.33c) [56]. Due to the presence of three OH groups in the aromatic subunits of this compound, the respective capsule does not require the participation of water molecules. It is stabilized by 48 intermolecular hydrogen bonds between the pyrogallolarene units, in addition to 24 intramolecular bonds within the six rings. Thus, this capsule contains fewer components and more hydrogen bonds than the capsule formed from **5.39a**, making it more stable. 2,6-Dihydroxypyridine also forms hexameric capsules, but product formation is more complex due to amide–iminol tautomerism [57].

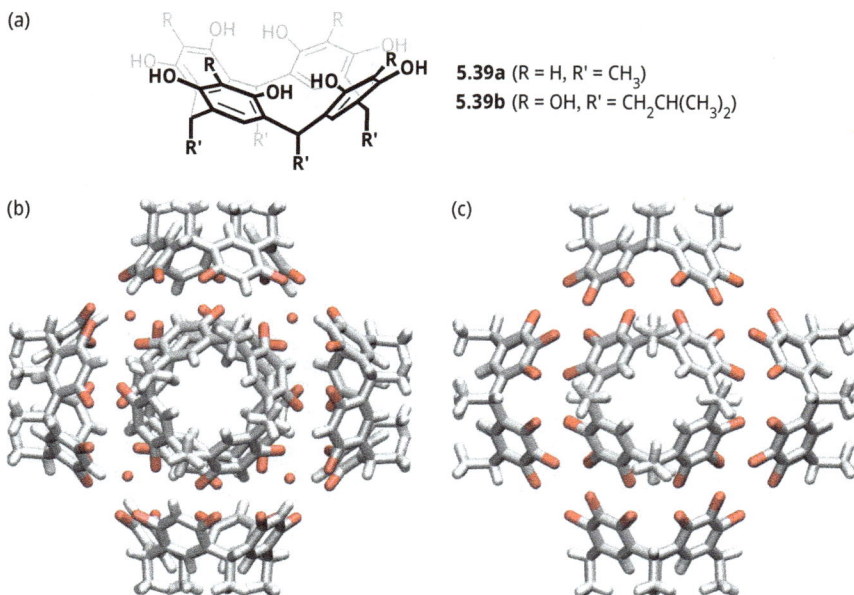

(a)

5.39a (R = H, R' = CH$_3$)
5.39b (R = OH, R' = CH$_2$CH(CH$_3$)$_2$)

(b)

(c)

Figure 5.33: Molecular structures of resorcin[4]arene **5.39a** and pyrogallolarene **5.39b** (a), and crystal structures of the hexameric capsules assembled from **5.39a** (b) and **5.39b** (c). The side chains in the pyrogallol subunits in (c) are truncated for reasons of clarity.

These capsules also exist in the gas phase and in solution. While resorcin[4]arene-derived capsules form only in nonpolar solvents such as chloroform, the pyrogallolarene-containing analogs tolerate more competitive media such as acetone/water, 1:1 (v/v). The large cavities surrounded by the macrocyclic subunits are usually filled with several solvent or guest molecules. For example, the capsule derived from **5.39a** can bind eight benzene molecules, three biphenyl molecules, different tetraalkylammonium salts, or combinations of different guests [41]. These results raised the question whether dicarboxylic acids such as glutaric acid or certain monosaccharides are indeed bound by resorcin[4]arenes in the form of simple 1:1 or 2:1 complexes, as suggested by Yasohiro Aoyama in the 1980s on the basis of NMR spectroscopic studies (Section 4.1.9). Indeed, modern NMR spectroscopic methods not available to Aoyama showed in 2006 that these guests are also bound inside hexameric capsules, with one capsule hosting either six glutaric acid molecules or three monosaccharides [58]. Thus, the host–guest ratio determined by Aoyama was correct, but the total number of components was not.

Resorcin[4]arene-based capsules are among the largest synthetic noncovalent assemblies identified to date. They not only host several guest molecules simultaneously, but also catalyze certain transformations with characteristic selectivity, as we will see in Section 9.3.2.

5.3.4 Tubes

A tube is a cylindrical object with a characteristic length and diameter. Building blocks from which tubular structures can be assembled are obtained by cutting a tube either parallel or perpendicular to its axis. In the first case, one or more sheets are obtained that must be joined at the correct edges to reassemble the tube. The second method produces rings that must be stacked to form a tube. A third option is to cut the tube in a helical fashion to produce a tape that can be folded into the original tube. Typical examples of linear tube-forming molecules are α-helical peptides, in which hydrogen bonds between amino acid residues in adjacent turns stabilize the tubular structure. Some of the foldamers discussed in Section 4.1.13 also fall into this category. However, since the tubes formed in this way are stabilized by intramolecular interactions, they are not formed by self-assembly and will therefore not be considered here.

The first strategy to assemble a tubular structure was realized by Stefan Matile using p-octiphenyl derivatives with short peptides in each aromatic subunit (Figure 5.34a) [59]. Self-assembly involves interdigitation of the peptide strands along the p-octiphenyl core, resulting in segments of antiparallel β-sheets perpendicular to the tube axis. To promote β-sheet formation, the peptides have an alternating sequence of nonpolar and polar amino acids, with the side chains of the amino acids directly attached to the p-octiphenyl core oriented outside the tube (or barrel, as the authors call the assembly). Because the peptides have an odd number of amino acids, the terminal amino acids are oriented in the same direction and the side chains in the even-numbered amino acid residues face the inside of the tube, allowing the properties of the inner surface to be controlled.

These compounds preferentially assemble into tetrameric circular structures consisting of four *p*-octiphenyl rods as staves and eight peptides as hoops (Figure 5.34b), but dimeric and hexameric assemblies are also possible. The stability depends on the length of the peptide residues and their sequence as well as on the length of the staves: biphenyl or *p*-quaterphenyl derivatives with peptide side chains do not assemble, while longer rods yield tubes whose stability increases with the number of aromatic subunits and the length of the peptides. The length of the staves also determines the height of the tubes, which is therefore well controlled, much better than the height of the tubes derived from the self-assembling cyclic peptides discussed below.

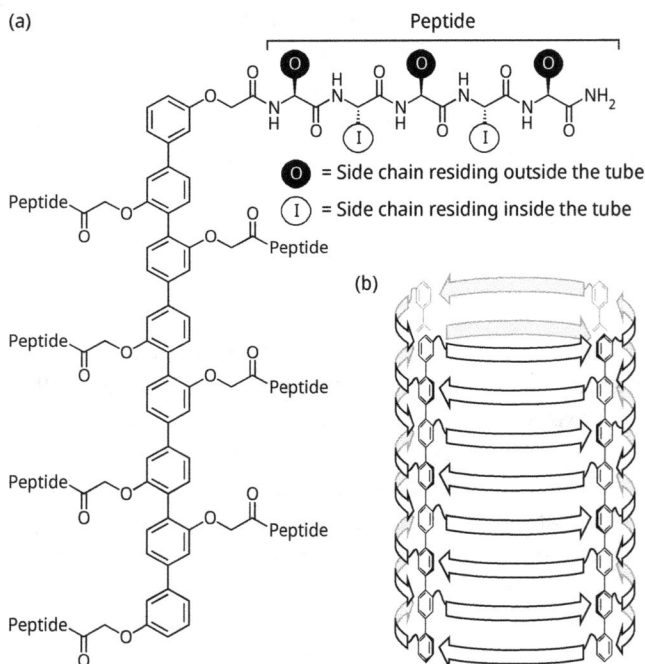

Figure 5.34: General structure of the *p*-octiphenyl-peptide conjugates that give rise to tubular structures by interdigitation and concomitant β-sheet formation of the peptide residues (a), and schematic structure of a corresponding tetrameric tube (or barrel) (b).

This strategy is very versatile and allows the diameter of the tubes to be varied to some extent by changing the length of the peptide residues. In addition, by positioning appropriate amino acids along the peptide residues, water-soluble tubes can be obtained that are polar on the outside and nonpolar on the inside, allowing them to accommodate hydrophobic substrates such as carotenoids. Conversely, tubes with a polar interior and nonpolar exterior can partition into bilayer membranes, allowing the design of pores to mediate membrane transport processes (Section 10.2.1).

The alternative strategy for assembling tubes is based on macrocyclic compounds with a suitable sequence of complementary hydrogen bond donors and acceptors to mediate stacking. Several types of macrocyclic compounds are used for this purpose, most of which are derived from cyclic peptides. Examples are those with alternating L- and D-amino acids, whose ability to stack in a β-sheet-like arrangement was predicted as early as 1974 [60]. The experimental proof had to wait until 1993, when M. Reza Ghadiri devised a way to control the assembly [61]. His strategy involves the use of cyclopeptides with acidic amino acid residues, such as the glutamic acid subunits in cyclooctapeptide **5.40a** (Figure 5.35a). These residues render the corresponding peptides soluble in basic media, where assembly does not occur due to Coulomb repulsion between the negatively charged carboxylate groups in the side chains. Careful acidification of the solutions gradually removes this repulsion, allowing the peptides to assemble in a controlled manner, ultimately leading to the precipitation of microcrystalline fibers. These fibers contain tubes of stacked peptides, as shown schematically in Figure 5.35b. Cyclopeptide **5.40b** provides insight into the structural parameters that control the assembly. In the crystal, this cyclopeptide forms a dimer stabilized by hydrogen bonds between every other peptide bond (Figure 5.35c) [62]. The methyl groups on the N-substituted peptide groups are arranged along the rims of the tubular assembly, preventing the addition of further rings. This dimer also exists in CDCl$_3$, where its K_a amounts to 80 M^{-1}.

The behavior and properties of these cyclopeptides can be varied over a wide range by changing the number, sequence, and nature of the amino acids from which they are composed. Appropriate cyclopeptides also assemble within lipid bilayer membranes to form pores whose diameter and height are determined by the size of the rings and the thickness of the membrane, respectively. These pores mediate the transport of ions or molecules of appropriate size. In the case of cyclic octapeptides, for example, the pores have an internal diameter of 7 Å, which is large enough to allow the passage of sodium or potassium ions, while the 10 Å-wide pores of cyclic decapeptides allow the passage of small organic molecules such as glutamic acid or even glucose. These cyclopeptides also have pronounced *in vitro* antibacterial activity [63].

Cyclopeptides with an alternating sequence of D-α-amino acids and (1R,3S)-3-amino-alkanecarboxylic acids (γ-amino acids), introduced by Juan R. Granja, also have conformations and arrangements of the peptide bonds that allow them to stack [64]. An example is cyclopeptide **5.41a**, which contains (1R,3S)-3-aminocyclohexanecarboxylic acid subunits (Figure 5.36a). In these cyclopeptides, one ring face contains only the α-amino acid and the other contains the γ-amino acid NH and C=O groups. Tube formation therefore involves either an antiparallel arrangement of the rings, so that they interact alternately through their α- and γ-faces, or a parallel arrangement in which the α-faces interact with the γ-faces (Figure 5.36b). The α,α- and γ,γ-interactions differ substantially

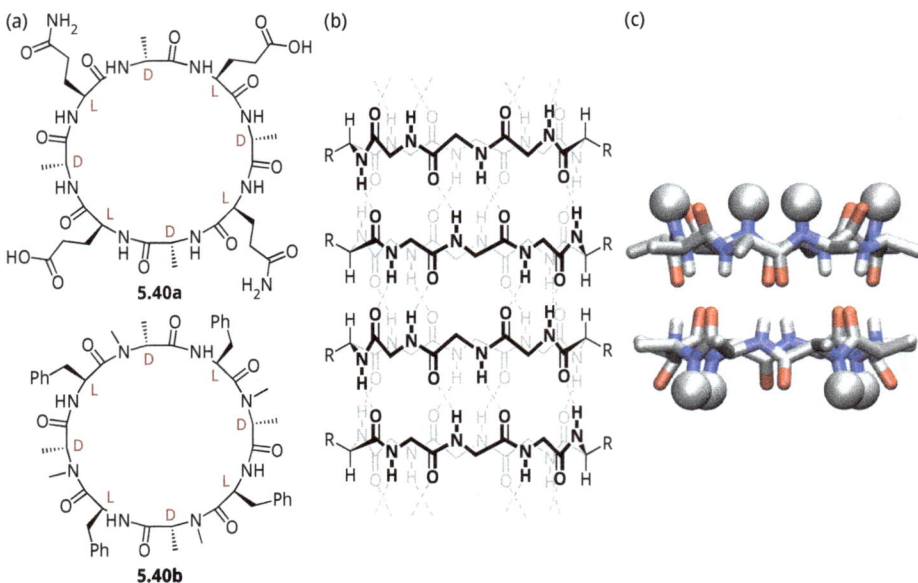

Figure 5.35: Molecular structures of cyclopeptides **5.40a** and **5.40b** (a), schematic representation of the β-sheet-like arrangement in which these peptides are stacked (b), and crystal structure of the dimer of **5.40b** (c). Phenyl rings and protons other than those on NH groups are omitted in the crystal structure for reasons of clarity. The spheres represent the *N*-methyl groups.

in strength, as demonstrated by the stability of dimers formed from the corresponding *N*-methylated cyclopeptides [65]. While **5.41b**, in which the *N*-methylation restricts the interactions to the γ-face, forms a dimer in CDCl₃ with a log K_a of 2, the α-face stacked dimer of **5.41c** has a significantly larger log K_a of 6 in chloroform. Treatment of the more stable dimer with **5.41b** yields the heterodimer, indicating that the parallel arrangement of the rings is the preferred arrangement (Figure 5.36b) [66].

Analogous tubes are formed from cyclopeptides containing cyclopentane instead of cyclohexane rings, or from cyclooctapeptides, the latter giving tubes with larger inner diameters than the hexapeptides. Corresponding cyclic tetrapeptides cannot self-assemble because they do not have the required disk-shaped conformation.

Tubes assembled from such cyclopeptides contain the β-methylene groups of the cyclic amino acid residue within their lumen. Thus, by using peptides with γ-amino acids containing functional groups in the β-position, the polarity of the inner tube surface can be controlled, which is not possible with cyclopeptides containing alternating L- and D-α-amino acids [67].

The self-assembly of **5.40** and **5.41** results in tubes in which the NH and C=O groups are oriented in both directions. A different arrangement is realized when cyclopeptides derived from β-amino acids are used as building blocks. The corresponding hydrogen bonding pattern is shown in Figure 5.37 for cyclopeptide **5.42** containing

(a)

(b)

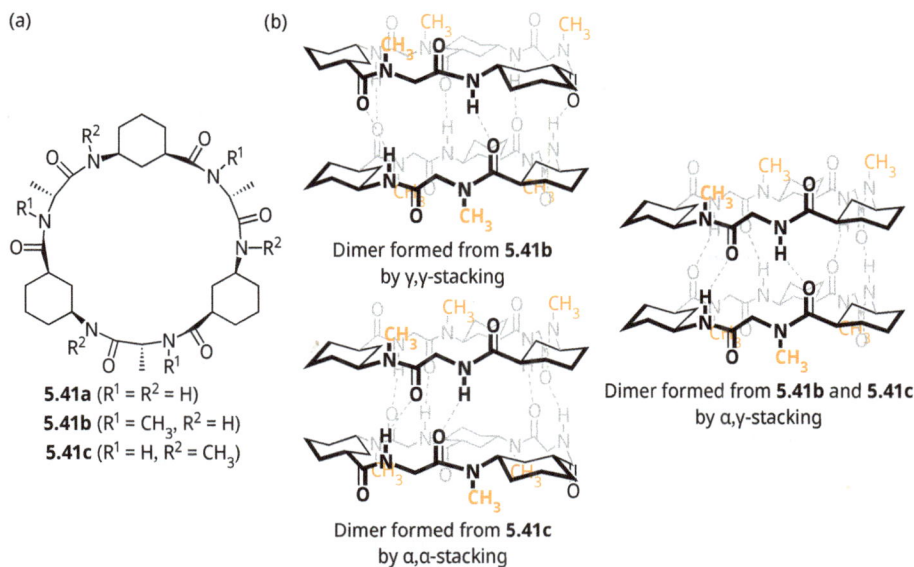

5.41a (R¹ = R² = H)
5.41b (R¹ = CH₃, R² = H)
5.41c (R¹ = H, R² = CH₃)

Dimer formed from **5.41b**
by γ,γ-stacking

Dimer formed from **5.41c**
by α,α-stacking

Dimer formed from **5.41b** and **5.41c**
by α,γ-stacking

Figure 5.36: Molecular structures of cyclopeptides **5.41a–c** (a) and schematic representation of the different modes of self-assembly of the partially methylated derivatives **5.41b** and **5.41c** (b). The interaction between two α-faces is more efficient than the interaction between two γ-faces. The most stable dimer is formed when one α- and one γ-face interact.

four S-β^3-homoleucine residues, which assembles into a tube with an inner diameter of approximately 2.6 Å [68]. In this tube, all C=O groups face in one direction and the NH groups face in the opposite direction, resulting in a dipole moment that potentially influences ion transport through the pore.

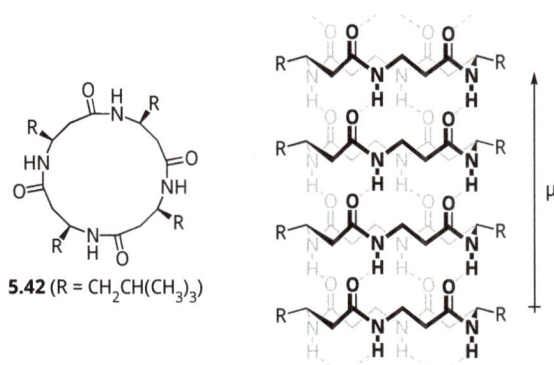

5.42 (R = CH₂CH(CH₃)₃)

Figure 5.37: Molecular structure of **5.42** and schematic representation of the mode of assembly of this cyclopeptide. The arrow indicates the dipole moment of the tube.

Another versatile approach to assembling tubes is based on macrocyclic bis(ureas) [69]. These compounds, of which **5.43** (Figure 5.38a) in an example, are rather rigid and almost planar, with the urea groups oriented perpendicular to the ring plane. The crystal structure of **5.43** (Figure 5.38b) shows that the urea groups mediate the self-assembly through similar bifurcated hydrogen bonds as found in the dimeric capsule of the calix[4]arene tetraurea **5.8** (Figure 5.31). Such macrocyclic bis(ureas) allow the fabrication of robust porous materials that have the potential to be used for gas storage, separations, or catalysis.

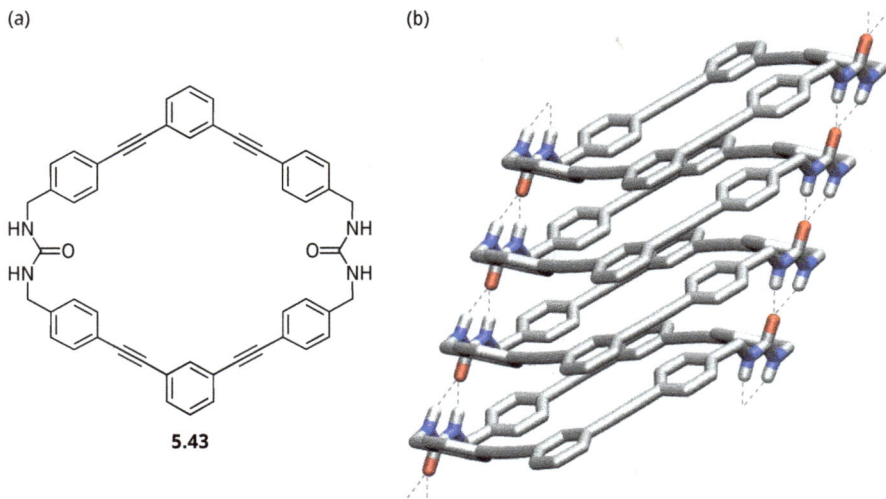

(a)   (b)

5.43

Figure 5.38: Molecular structure of bis(urea) **5.43** (a) and crystal structure of the tubular assembly of this compound (b). Protons other than those on nitrogen atoms are omitted in the crystal structure for reasons of clarity.

5.4 Self-assembly mediated by halogen and chalcogen bonds

5.4.1 Introduction

Noncovalent interactions must arrange individual building blocks in a predictable manner to ensure that self-assembly proceeds with sufficient fidelity. Properly arranged patterns of hydrogen bonds are suitable for this purpose, as discussed in the previous section, but halogen bonds are also an attractive option. The reason for this is the high directionality of halogen bonds, caused by the restriction of the σ-hole in the donor to a relatively small region at the end of the X–Hal bond (Section 3.1.6). Thus, the strongest electrostatic attraction occurs when the X–Hal bond and the halogen bond are collinear, while deviations from this optimal 180° angle cause a significant weakening of the interaction. Despite their promising properties, halogen bonds or the closely related chalcogen bonds have not been widely used to mediate self-assembly. We will only briefly look at a few examples.

5.4.2 Helices

Orion B. Berryman studied the interaction of the arylethynyl oligomer **5.44** (Figure 5.39a), which contains three 4-iodo-1-methylpyridinium moieties, with iodide ions [70]. One possible mode of interaction involves wrapping a single chain of **5.44** around an anion, but because it is not possible to arrange the iodine atoms along the chain in such a way that they simultaneously form three linear halogen bonds to a central iodide anion, this 1:1 complex does not form. Instead, self-assembly leads to the formation of a complex in which three subunits of **5.44** wrap around two iodide anions.

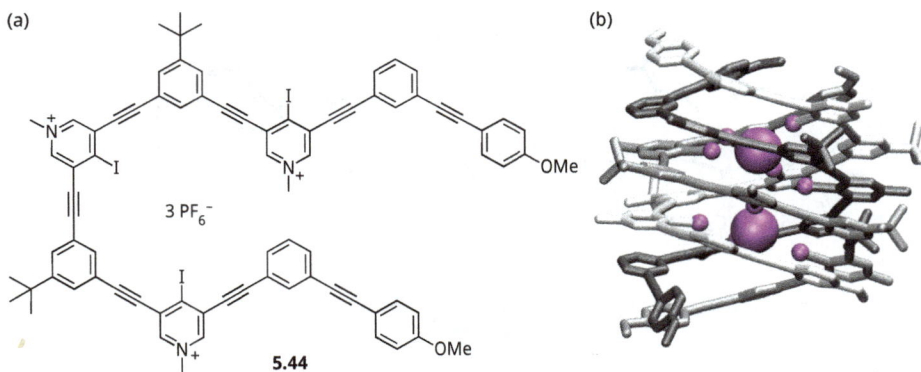

Figure 5.39: Molecular structure of oligomer **5.44** (a), and crystal structure of the triple helicate in which three of these oligomers wrap around two iodide ions (b). The protons are omitted in the crystal structure for reasons of clarity, and the backbones of the three ligands are shown in different shades of gray. The iodine atoms along the chain and the bound iodide anions are shown as small and large spheres, respectively.

To illustrate this arrangement, the corresponding crystal structure is shown in Figure 5.39b. The strong directionality of the halogen bonds thus controls the formation of a product that is structurally more complex than the entropically favored 1:1 assembly. This product also exists in DMF/acetonitrile and even in DMF/water mixtures. Although the complex shown in Figure 5.39b is structurally somewhat related to the foldamer complexes discussed in Section 4.1.13, **5.44** is not a foldamer because it adopts a helical conformation only in the presence of iodide anions.

Complexes in which directed intermolecular interactions cause an oligomeric chain with a sequence of donor atoms to wrap around one or more acceptors are called *helicates*. Depending on the number of ligands involved in helix formation, double and triple helicates are distinguished, with the complex in Figure 5.39b representing a triple helicate. While helicates stabilized by anions are rare, those between ligands and metal ions are more common. We will see examples in Section 5.5.2.

5.4.3 Capsules

François Diederich showed that the self-assembly of the two deep cavitands **5.45a** and **5.45b** (Figure 5.40a) yields a dimeric capsule held together by four linear halogen bonds between the iodine atoms and the pyridine nitrogen atoms (Figure 5.40b) [71].

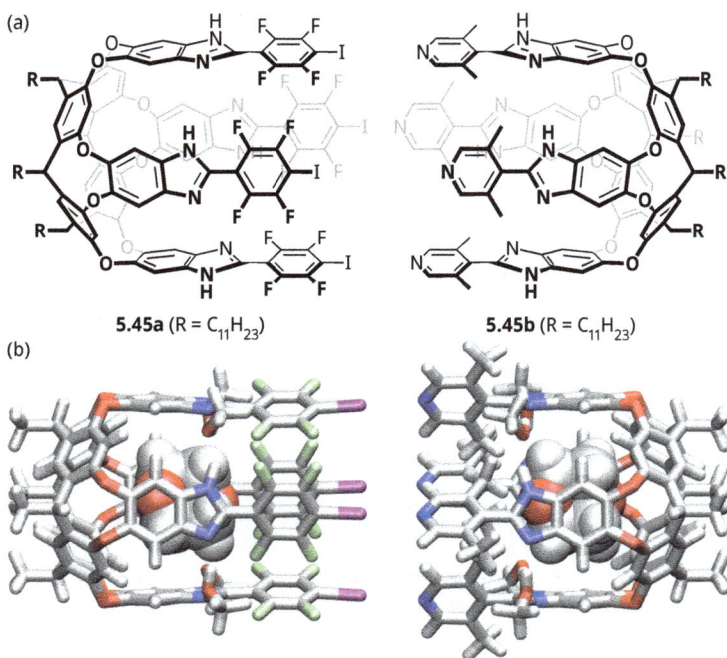

Figure 5.40: Molecular structures of cavitands **5.45a** and **5.45b** (a), and calculated structure of the capsule formed from both compounds (b). Each subunit of the capsule contains a 1,4-dioxane molecule, which is shown as space-filling model. The cavitands are further stabilized in their vase conformations by methanol molecules bridging the benzimidazole units. The alkyl residues in the resorcin[4]arene subunits are truncated for reason of clarity.

For capsule formation to occur, the solvent must contain a small amount of an alcohol that provides the OH groups necessary to stabilize the vase conformations of **5.45a** and **5.45b**. This stabilizing effect was discussed in Section 4.1.9. The capsule has a dimerization constant $\log K_a$ of 3.7 in benzene-d_6/DMSO-d_6/methanol-d_4, 70:30:1 (*v/v*) at

283 K, showing that the four halogen bonds provide substantial stabilization. In solvents too large to enter the cavity (mesitylene-d_{12} with 2 vol% 3,5-dimethylbenzyl alcohol), this capsule hosts guest molecules such as two molecules of 1,4-dioxane, one per capsule half.

Further work by the Diederich group showed that deep cavitand **5.46a**, which contains 2,1,3-benzotelluradiazole moieties along the rim, forms a dimeric capsule stabilized by Te···N chalcogen bonds [72, 73]. The crystal structure in Figure 5.41 shows that the nitrogen atoms in each half of the capsule are ideally positioned to interact with the two σ-holes of the tellurium atoms in the other half, resulting in a highly symmetric arrangement characterized by 16 short Te···N contacts. In the solid state, each cavitand contains an encapsulated benzene molecule. In solution, **5.46a** first self-assembles into a capsule with an offset arrangement of the two subunits, stabilized by only 8 chalcogen bonds. This kinetic product slowly rearranges to the symmetrical, more stable structure. The water-soluble cavitand **5.46b** with 2,1,3-benzoselenadiazole groups also dimerizes, this time in water, yielding a capsule that hosts guest molecules such as alkanes, cycloalkanes, and branched or cyclic carboxylic acids and their amides [74]. In the case of cavitand **5.46c** with 2,1,3-benzothiadiazole groups, capsule formation is only observed in solvents that stabilize the vase conformation (tetrahydrofuran or benzene). In chlorinated solvents, where **5.46c** prefers the kite conformation, the S···N chalcogen bonds are not strong enough to shift the conformational equilibrium and allow dimerization. In contrast, compound **5.46a** forms capsules in all of the above solvents due to the stronger Te···N chalcogen bonds.

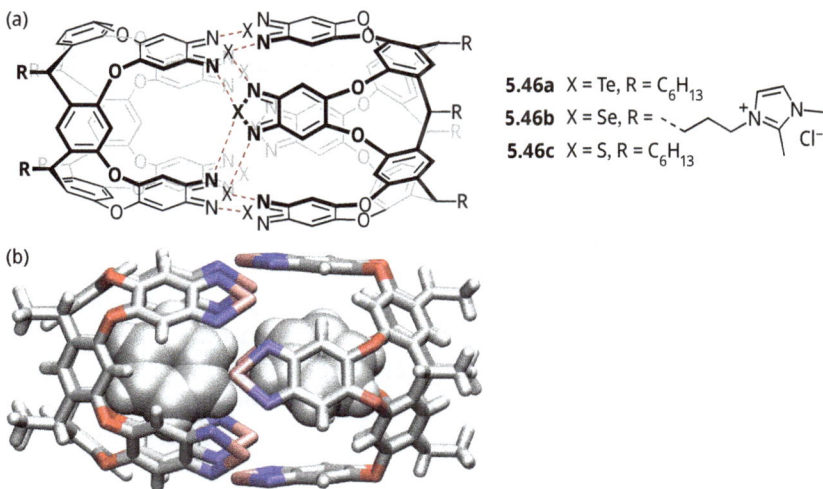

(a)

5.46a X = Te, R = C_6H_{13}
5.46b X = Se, R =
5.46c X = S, R = C_6H_{13}

(b)

Figure 5.41: Molecular structures of cavitands **5.46a–c** (a) and crystal structure of the capsule formed from **5.46a**. Each subunit of the capsule contains a benzene molecule, shown as space-filling model. The alkyl residues in the resorcin[4]arene subunits are truncated for reason of clarity.

5.5 Self-assembly mediated by coordinate bonds

5.5.1 Introduction

We now turn to self-assembly processes mediated by the coordination of Lewis base ligands to metal ions. Coordinative interactions have a pronounced covalent character, making them stronger than noncovalent interactions (60–200 kJ mol^{-1} *vs.* 1–40 kJ mol^{-1} of noncovalent interactions) (Section 3.2.1). As a consequence, self-assembly is not restricted to organic solvents, as in the case of hydrogen bonding, but also occurs in competitive media, including water. At the same time, many metal–ligand bonds are labile, which is an important prerequisite for ensuring that product formation proceeds under thermodynamic control. Frequently used metal ions are Cu^+, Zn^{2+}, Fe^{2+}, Pd^{2+}, Ga^{3+}, and some others, not only because of the kinetic lability of their complexes and the possibility to realize different coordination geometries [tetrahedral with copper (I), octahedral with iron(II) and gallium(III), square planar with palladium(II), etc.], but also because these ions are diamagnetic, which facilitates product characterization by NMR spectroscopy. A major advantage of coordinate bonds is their directionality due to their covalent nature, which makes self-assembly highly predictable. Parameters to be considered are the number and arrangement of donors in the ligands and the preferred coordination number and geometry of the connecting metal centers. The structure of the product can then be predicted using the following concepts [75].

Directional bonding approach: In this approach, the building blocks are divided into donors (the ligands) and acceptors (the metal ions). The donors are rigid organic compounds with two or more monodentate coordination sites arranged at an angle between 0° and 180°. The acceptors are metal ions, metal complexes, or organometallic compounds that serve to orient the incoming ligands in a predefined arrangement. They typically contain weakly bound ligands that are exchanged with the incoming donors during self-assembly. Additional strongly bound ligands can serve to block specific sites on the metal centers and thus control the orientation of the incoming ligands.

Predictions about the most likely structure of the product are based on geometric considerations. For example, bifunctional donors and acceptors give rise to macrocyclic compounds whose size and symmetry depend on the angles at which the newly formed bonds are arranged (Figure 5.42a). If three-dimensional assemblies are targeted, at least one of the components must be able to form three bonds. Figure 5.42b shows that a tetrahedron results from combining six linear and four tritopic precursors that arrange the newly formed bonds at 60° angles. A cube requires a 3:2 ratio of linear and trifunctional precursors with 90° bond angles. These predictions are reliable only if the precursors are sufficiently rigid. If deviations from the optimal angles occur, the fidelity of self-assembly suffers.

Paneling approach: The paneling approach is particularly useful for the construction of polyhedra, but it also provides access to tubes, barrels, or bowls [76]. The term

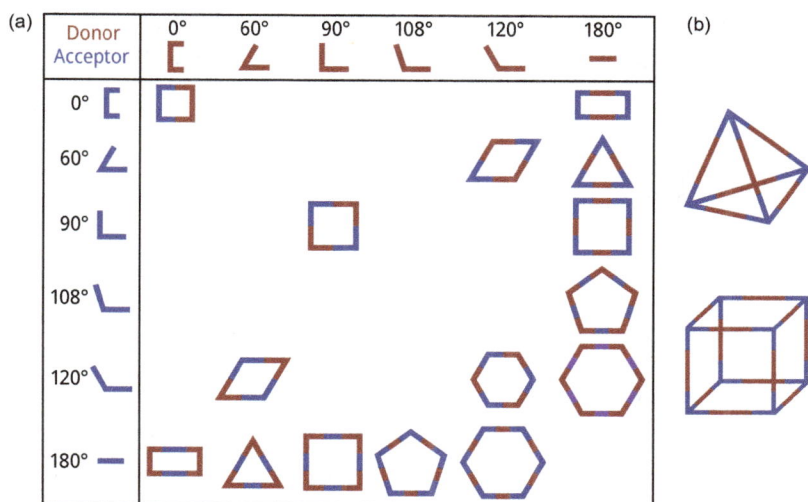

Figure 5.42: Correlation of the angles at which difunctional donors (ligands) and acceptors (metal ions) orient their binding sites with the structure of the preferred macrocyclic product (a), and assembly of a tetrahedron or a cube from suitable building blocks (b).

panel refers to flat multidentate organic ligands that are joined together to form the faces of the self-assembled product. In terms of coordination chemistry, this approach is closely related to the directional bonding approach in that the individual coordination sites in the panels are typically monodentate, and the metal centers may contain additional ligands to control the number and orientation of the donors with which they interact.

One parameter that determines the structure of the product is the shape of the panels. Four triangles make a tetrahedron, for example, or six squares make a cube. However, many panels give rise to more than one type of assembly. For example, triangular panels can produce tetrahedra, octahedra, icosahedra, or square pyramids, and the shape of the panels is therefore not the only factor that determines which product is formed. Equally important are the angles between the individual components, the number of linkages, and the arrangement of the donor atoms in the ligands.

The tripodal ligands **5.47** and **5.48** and the palladium(II) complex [Pd(en)(NO$_3$)$_2$] **5.49** should illustrate these aspects (Figure 5.43). Since **5.49** cannot accept more than two incoming ligands to replace the weakly bound nitrates, coordinatively saturated complexes derived from **5.49** and tripodal ligands must have the compositions M$_3$L$_2$, M$_6$L$_4$, M$_9$L$_6$, etc. In addition, since the ethylenediamine ligand in **5.49** directs the incoming ligands into adjacent corners of a square planar complex, the two panels at each metal center are arranged at a 90° angle. In the case of panel **5.47**, the smallest complex where this is possible is the M$_6$L$_4$ complex, and because the donors in this panel are directed to the corners, an octahedral complex is formed with four of the eight faces occupied by the ligands (Figure 5.43b). This complex forms with high fidelity.

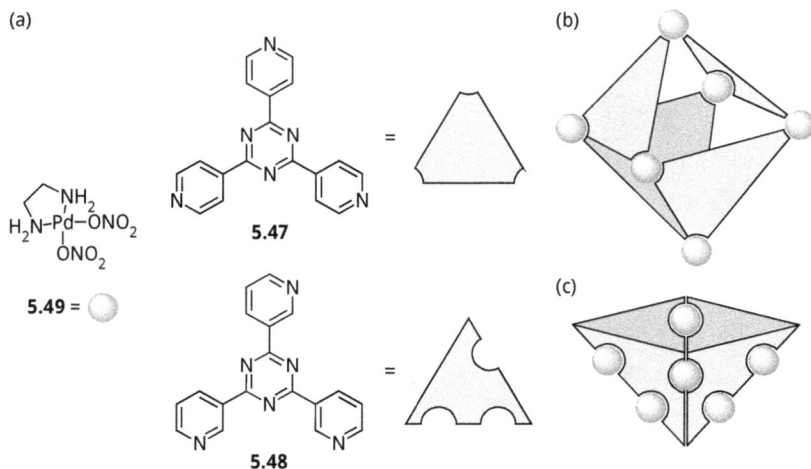

Figure 5.43: Molecular structures of the tripodal ligands **5.47** and **5.48** and the palladium(II) complex **5.49** (a), schematic representation of the octahedral complex containing four panels of **5.47** (b), and the square pyramidal complex containing four panels of **5.48** (c).

If two donor atoms point to one edge, as in panel **5.48**, four triangles form an open square pyramid (Figure 5.43c). Thus, the outcome of self-assembly depends sensitively on how the donor atoms are arranged in the panels. Many other coordination structures are accessible from triangular or rectangular panels, the latter often derived from porphyrin derivatives.

Symmetry interaction approach: This approach allows predictions to be made about the structure of coordination compounds formed between chelating ligands and metal centers [77]. The analysis is based on symmetry considerations and involves three parameters, the *coordinate vector*, the *chelate plane*, and the *approach angle*. The coordinate vector connects the ligands to the metal centers. For example, bidentate chelating groups are bisected by the coordinate vector in the direction of the metal, as shown in Figure 5.44a. The chelate plane contains all the coordinate vectors surrounding a metal, and the approach angle is a measure of the arrangement of the ligands with respect to the symmetry axis of the metal, which is orthogonal to the chelate plane (Figure 5.44b). The structure of the final assembly is determined by the orientation of the chelate planes relative to each other.

Two examples will illustrate this concept. The first involves the assembly of a triple helicate in which three ditopic ligands L are bound to two metal ions M, yielding a complex with the composition M_2L_3 (Section 5.5.2). Such a complex is characterized by

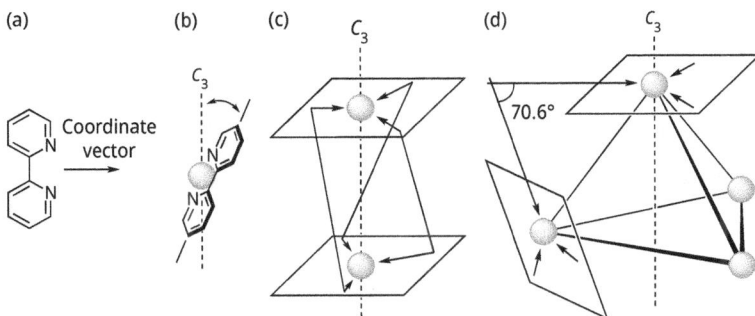

Figure 5.44: Construction of the coordinate vector (a) and approach angle (b) for a bidentate ligand, and schematic illustrations of a triple M_2L_3 helicate (c), and a tetrahedral M_4L_6 cage (d), showing the arrangement of the coordinate vectors and the chelate planes at selected metal centers.

a C_3 axis that coincides with the axis connecting the metal ions (Figure 5.44c). The threefold symmetry along this axis causes the coordinate vectors at each metal to be arranged at 120° angles, and since each ligand has two coordinate vectors, the chelate planes in a helicate must be parallel. In a tetrahedral M_4L_6 coordination cage with the metal centers at the corners and the ligands at the edges, the coordinate vectors are also arranged at 120° angles because of the threefold symmetry at each corner of the tetrahedron, but the chelate planes must be arranged at angles of 70.6° (Figure 5.44d).

5.5.2 Helices

Metal-containing helicates are formed from oligomeric ligands with suitable donor atoms along the chain. An example is the dicopper(I) complex of ligand **5.50a** (Figure 5.45), which was first described by Jean-Marie Lehn [78]. The two 2,2'-bipyridine moieties in **5.50a** cannot coordinate to a single metal ion. Product formation therefore involves the intertwining of two ligand molecules to realize (distorted) tetrahedral coordination geometries at both metal ions. In the product, the two ligands adopt helical conformations, resulting in a metal-stabilized double helix (or *helicate*) in which both strands have the same configuration. The helicate shown in Figure 5.45 is *P*(lus)-configured due to the clockwise helical screw sense of the ligands, while the equally likely *M*(inus)-form contains counterclockwise coiled ligands.

Several parameters allow structural control [79]. One is the number of coordinating subunits in the ligands, which determines how many metal centers the final helicate contains. Ligands **5.50b-d** (Figure 5.45), derived from **5.50a** by chain elongation, thus yield double helices with up to five metal centers.

According to mechanistic studies, the formation of such polynuclear complexes typically involves several intermediates. In the case of the trinuclear double helicate derived from **5.50b**, the most important ones are a complex ML in which one ligand binds

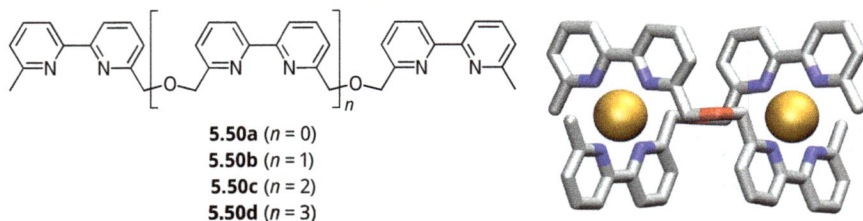

Figure 5.45: Molecular structures of ligand **5.50a–d** and calculated structure of the double helicate formed from two subunits of **5.50a** and two copper(I) centers. Protons are omitted for reasons of clarity.

5.50a ($n = 0$)
5.50b ($n = 1$)
5.50c ($n = 2$)
5.50d ($n = 3$)

to a single metal ion, presumably through its end groups, in addition to the mono- and dinuclear complexes ML_2 and M_2L_2 [80]. These incompletely formed complexes play only a minor role in the thermodynamic equilibrium because they are orders of magnitude less stable than the trinuclear final product [81]. As a consequence, M_3L_2 forms with pronounced positive cooperativity, probably due to the unstrained arrangement of the ligands in the product. However, positive cooperativity is not guaranteed. If complexation of the second or any subsequent metal ion induces strain in the ligand, negative cooperativity will result.

Factors that control the structure of helicates are the nature of the metal center and the denticity of the chelating units. Accordingly, double helicates are formed from bidentate ligands and metal ions that prefer a tetrahedral coordination geometry, as in the helicate shown in Figure 5.45, because the nearly orthogonal arrangement of the ligands at each metal center is optimal for inducing the chains to meander around them. A similar orthogonal arrangement is achieved when tridentate ligands, such as terpyridines, are coordinated to metal ions that prefer an octahedral coordination geometry (Figure 5.46). On the other hand, an octahedral complex with bidentate ligands allows the assembly of triple helicates, but only if the ligand permits the required coordination geometry. In the case of **5.50a**, the substitution pattern of the bipyridine subunits does not allow the formation of octahedral complexes, making this ligand unsuitable for the assembly of triple helicates.

Figure 5.46: Schematic illustration of the binding modes that arrange two ligands orthogonally, resulting in double helicates, and the structure of an octahedral complex that yields a triple helicate.

These helicates can be obtained from the ligands **5.51a** or **5.51b** in which the 2,2'-bipyridine units are linked *via* their 5,5'-positions (Figure 5.47) [82]. The structure of the corresponding dinickel(II) complex of **5.51a** is shown in Figure 5.47. Ligand **5.51b** forms an analogous trinuclear complex.

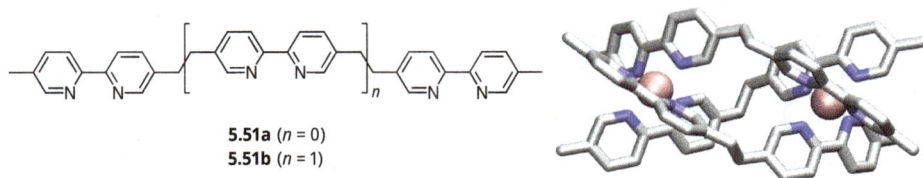

5.51a (*n* = 0)
5.51b (*n* = 1)

Figure 5.47: Molecular structures of ligands **5.51a,b** and calculated structure of the dinickel(II) triple helicate of **5.51a.**

When iron(II) salts are used instead of nickel(II) salts, the trinuclear triple helicate of **5.51b** is initially formed in a kinetically controlled reaction. This product then rearranges upon heating to yield a thermodynamically favored circular helicate [83]. The exact structure of this product depends on the counterion of the iron(II) salt. In the presence of chloride ions, the pentanuclear circular helicate $\{Cl \subset [Fe_5 \textbf{5.51b}_5]\}^{9+}$ is formed, whereas sulfate ions induce the formation of the hexanuclear helicate $\{SO_4 \subset [Fe_6 \textbf{5.51b}_6]\}^{10+}$ (Figure 5.48). The preferred formation of these products is probably due to enthalpic stabilization by the counterions, which overcompensates the unfavorable entropic term associated with bringing all the components together. In the case of the chloride complex, the crystal structure contains an anion bound in the central cavity of the circular helicate, which probably templates the assembly. Assuming that sulfate binds in a similar manner to the larger helicate, the size of the product is determined by the size of the anion occupying the cavity [84].

The structure of the ligand also controls stereochemical aspects of the products. Achiral ligands such as **5.50a** or **5.51a** form racemic chiral helicates. However, structural aspects in the ligand can prevent folding, resulting in achiral *meso*-helicates with a symmetry plane bisecting the axis connecting the metal ions. In addition, chiral helicates can be prepared enantioselectively. For example, the two stereogenic centers in the central part of **5.52a,b** (Figure 5.49) are sufficient to yield one enantiomer of the product, with the two *S* configurations in the ligand leading to the exclusive formation of the *P*-configured double helicate [85].

These helicates generally form with high fidelity even when several different ligands or metal ions are simultaneously present in solution, as the following examples illustrate. The addition of copper(I) salts to a mixture of ligands **5.50a–d** will yield exclusively the four double helicates containing identical ligands once thermodynamic equilibrium is reached [7] (Figure 5.50a). This self-assembly is an example of a completive narcissistic self-sorting process that is almost always observed in helicate chemistry (case A in Figure 5.8). Similarly, a mixture of **5.50b** and **5.51b** containing the required amounts of

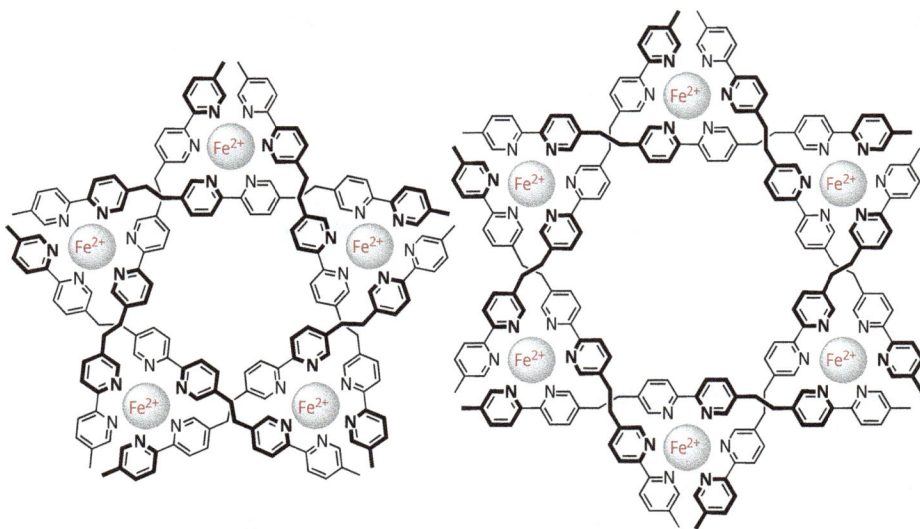

Figure 5.48: Structures of the pentanuclear and hexanuclear circular triple helicates formed from **5.51b** and FeCl$_2$ and FeSO$_4$, respectively.

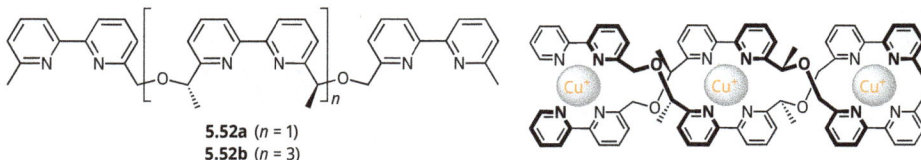

5.52a ($n = 1$)
5.52b ($n = 3$)

Figure 5.49: Molecular structure of ligands **5.52a,b** and schematic representation of the *P*-configured tricopper(I) double helicate derived from **5.52a**.

copper(I) and nickel(II) salts, will yield only the tricopper(I) double helicate of **5.50b** and the trinickel(II) triple helicate of **5.51b**, again by narcissistic self-sorting (Figure 5.50b).

Both reactions are consistent with the rules of self-assembly and self-sorting that we derived earlier (Section 5.1). In the first reaction, two ligands of different lengths would lead to a helicate in which donor atoms of the longer ligand remain unused. The formation of these complexes violates the principle of maximum site occupancy, and these complexes are therefore enthalpically disfavored. The use of the dangling donors to coordinate further metal ions is entropically unfavorable because of the large number of components involved in such products. Only the completely self-sorted system leads to the maximum number of coordinative interactions (enthalpically favorable) in combination with the maximum number of possible products (entropically favorable). Similarly, the ligands in the second reaction are structurally programmed to form either a double helicate (**5.50b**) or a triple helicate (**5.51b**). Any

(a)

(b)

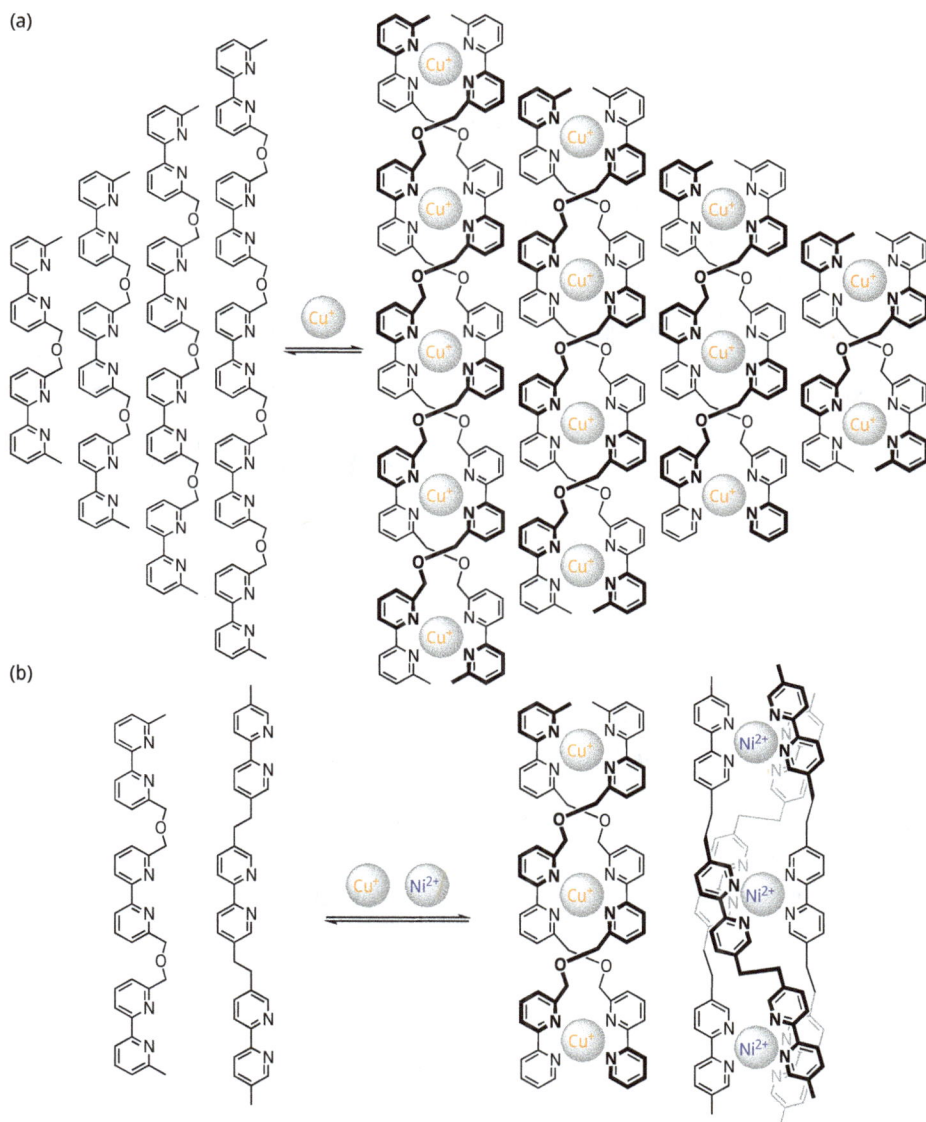

Figure 5.50: Narcissistic self-sorting process involving ligands **5.50a–d** and copper(I) ions, yielding double helicates containing pairs of identical ligands (a). The self-sorting process in (b) shows that the ligands **5.50b** and **5.51b** give a mixture of double and triple helicates when coordinating to copper(I) and nickel(II) ions.

mixed species can be assumed to be enthalpically unfavorable, as it should either be strained or contain metal ions that are coordinatively unsaturated.

Since these fundamental studies were performed by Lehn and coworkers, the chemistry of helicates has been extended to many other ligand types and metal ions

[79]. Another example will be presented, because similar ligands will again be discussed in Section 5.5.5. These complexes are formed from the biscatecholates **5.53a–c** (Figure 5.51), which coordinate to gallium(III) ions in an octahedral fashion after deprotonation, leading to the corresponding triple helicates [86].

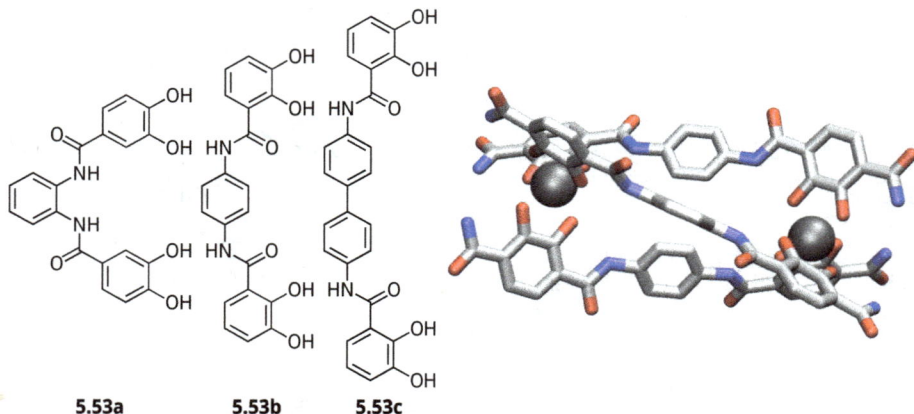

Figure 5.51: Molecular structure of the biscatecholates **5.53a–c** and crystal structure of the triple helicate formed from (deprotonated) **5.53b** and gallium(III) ions.

Unlike the bipyridine-derived ligands, the biscatecholates are negatively charged, and since these charges are not fully compensated by the gallium(III) ions, the resulting dinuclear helicates have six negative charges. When the three ligands **5.53a–c** and a corresponding number of gallium(III) ions are used simultaneously in the reaction, narcissistic completive self-sorting is observed because only ligands of the same length can be incorporated into a stable triple helicate (Figure 5.51).

5.5.3 Grids

The coordination of bidentate chelate ligands to metal ions that prefer a tetrahedral coordination geometry leads to double helicates if the ligands are sufficiently flexible. When helicate formation is prevented by the use of rigid ligands, self-assembly leads to other molecular architectures, such as the tetranuclear 2 × 2 molecular grid from the rigid rod-like ligand **5.54a** (Figure 5.52a) [87]. Similar grids are obtained from ligands with tridentate coordination sites and metal ions that prefer an octahedral coordination geometry.

The next larger ligand **5.54b** produces a 3 × 3 grid (Figure 5.52b) with silver(I) ions [88]. When a mixture of **5.54a** and **5.54b** is treated with a silver(I) salt, a rectangular 2 × 3 grid is preferentially formed (Figure 5.52c). Thus, social self-sorting predominates over narcissistic self-sorting, probably due to the destabilizing electrostatic interaction between the central and surrounding silver ions in the 3 × 3 grid [89]. For a similar reason,

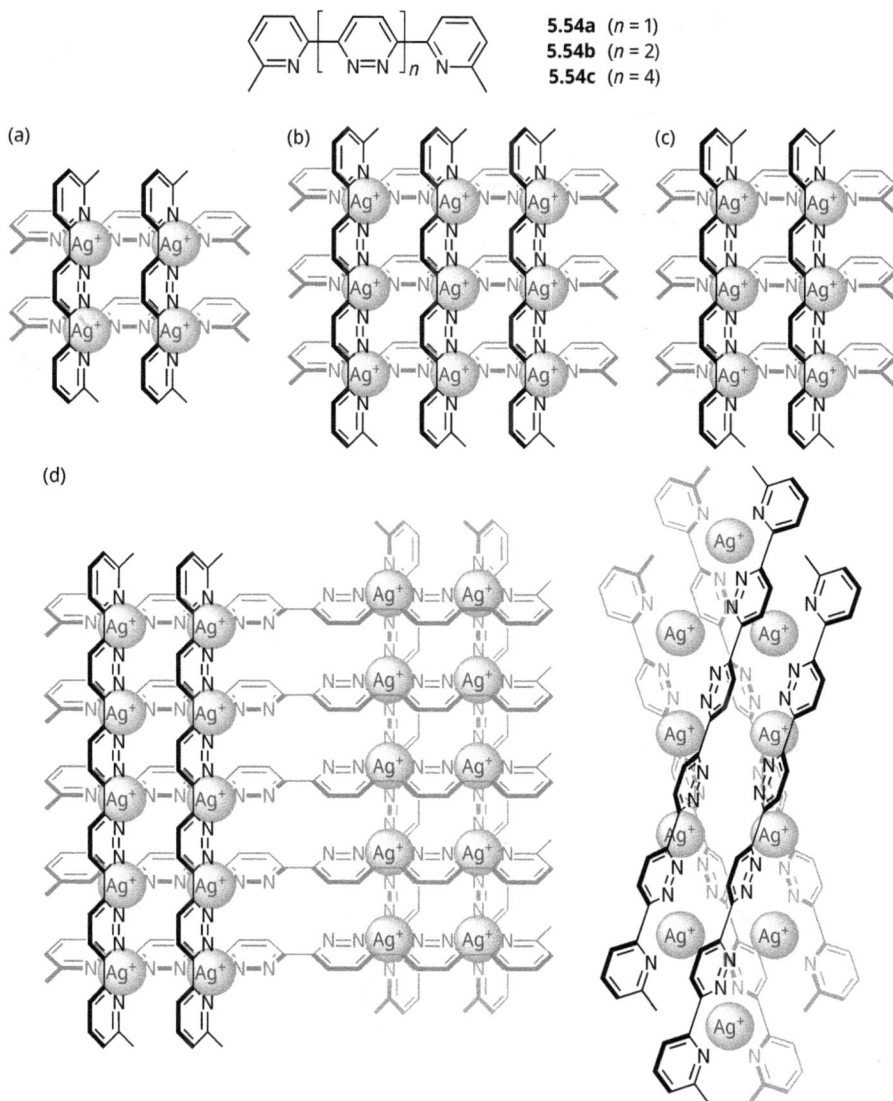

Figure 5.52: Molecular structures of ligands **5.54a–c** and schematic representation of the formation of the 2 × 2 and 3 × 3 grids from ligands **5.54a** (a) and **5.54b** (b), respectively. The 2 × 3 grid formed from a mixture of these ligands is shown in (c), and (d) shows assemblies formed from **5.54a**.

the even longer ligand **5.54c** does not form the 5 × 5 grid [90]. Instead, coordination to silver(I) ions leads to a complex in which two rectangular 2 × 5 grids are arranged on opposite sides of five parallel ligands, in addition to a decametallic quadruple helicate (Figure 5.52d). Repulsive interactions between the central silver ions are only one reason why the 5×5 grid does not form. Another is the poor structural complementarity between the ligand and the metal ions, which requires the ligand to adopt a bent conformation in the grid. In addition, all the nitrogen atoms of **5.54c** must be arranged on the same side to allow the grid to form, which leads to destabilizing interactions between their lone pairs that are not compensated by the metal coordination.

Although the concept of grid formation is mainly limited to smaller grids, the general approach of using rigid rod-like ligands as building blocks for metal-directed self-assembly is very versatile. A variety of different molecular architectures are produced in a similar manner, some of which are shown in Figure 5.53 [91]. All of these complexes form with high fidelity because of the strict orthogonal arrangement of the ligands at each metal center, and because no other ligand arrangements in which all donor atoms are used are possible. Note that these complexes contain different ligands that predictably interact to form the final product. Another strategy for controlling the formation of such heteroleptic complexes is presented in Section 5.5.4.

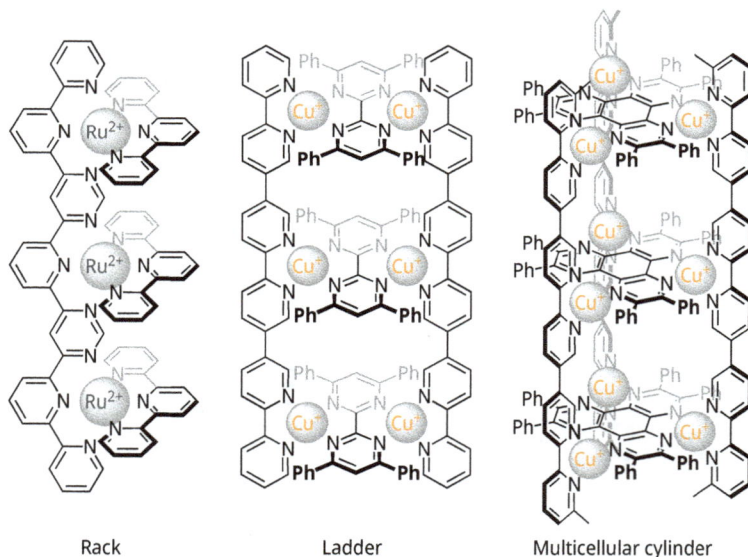

Rack   Ladder   Multicellular cylinder

Figure 5.53: Examples of molecular architectures created by metal-directed self-assembly.

5.5.4 Rings

Figure 5.42 shows that the assembly of rings requires each metal to interact with two ligands and *vice versa*. An instructive example of how to make a metallacycle in this way is the molecular box described by Makoto Fujita [92]. This box is one of many coordination compounds developed by Fujita and others using palladium(II) as acceptor. We will see more examples in the next section.

Palladium(II) ions can bind to four ligands in a square planar coordination geometry. When two adjacent coordination sites on the metal center are blocked, as in **5.49**, replacement of the remaining weakly bound counterions arranges the new ligands at a 90° angle. The *cis*-protected **5.49** is thus ideally predisposed to yield rectangular metallamacrocycles when treated with linear ligands having two divergent donor atoms. Indeed, the coordination of 4,4′-bipyridine to **5.49** yields the almost perfectly square box **5.55** in quantitative yields (Figure 5.54a). The length of the edges, estimated from the distance of the two palladium ions, is about 11 Å (Figure 5.54b). Accordingly, the cavity of **5.55** is similar in size to that of γ-CD (Table 4.4). This metallamacrocyle hosts aromatic

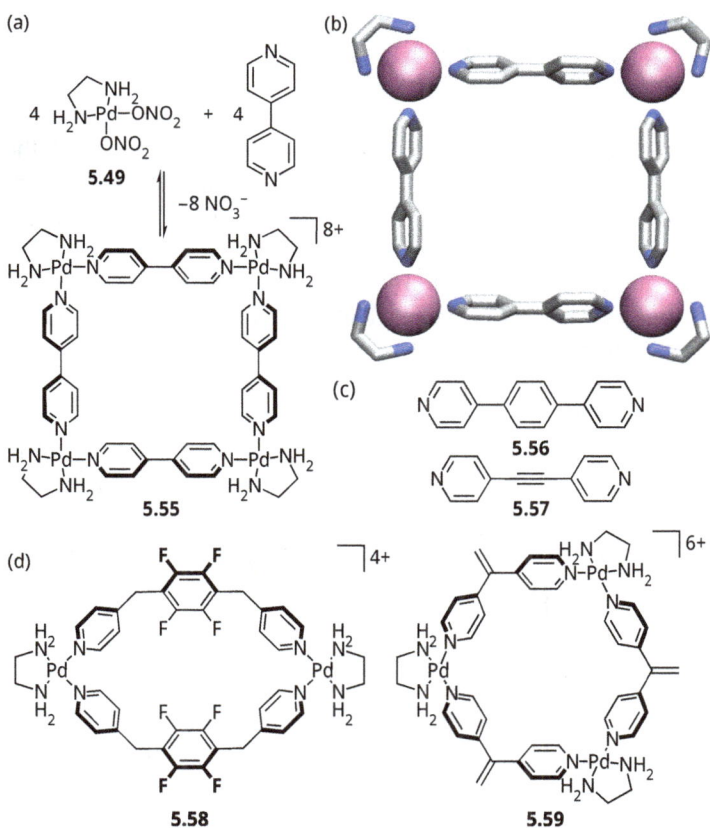

Figure 5.54: Formation of the molecular box **5.55** from **5.49** and 4,4′-bipyridine (a), and calculated structure of **5.55** (b). The structures in (b) show other linear ligands from which boxes are assembled, and those in (d) show the structures of dinuclear and trinuclear metallamacrocycles that are prepared in a similar manner.

substrates in the aqueous solution in which it is assembled, with the $\log K_a$ of the 1,3,5-trimethoxybenzene complex, for example, amounting to 2.9.

The cavity diameter can be increased by using rigid linear ligands with a larger distance between the two coordinating nitrogen atoms, such as 1,4-di(pyridin-4-yl)benzene **5.56** or 1,2-di(pyridin-4-yl)ethyne **5.57** (Figure 5.54c). More flexible ligands lead to the formation of a mixture of trinuclear and tetranuclear complexes, as we saw in Section 5.1, where template effects were discussed (Figure 5.7). Depending on the bite angle in the ligand, other macrocycles such as the dinuclear and trinuclear metallacycles **5.58** and **5.59** (Figure 5.54d) are also accessible [93].

Platinum(II) complexes are formed in a similar manner. However, their formation usually requires high temperatures and longer times to reach equilibrium because the platinum–nitrogen bond is kinetically less labile than the palladium–nitrogen bond. The advantage of platinum(II) complexes is that they do not exchange the ligands at room temperature, which allows them to be isolated as stable and inert compounds at ambient conditions.

The approach of assembling macrocycles by using metal–ligand interactions is certainly attractive from a synthetic point of view because of the efficiency with which the product is obtained without having to initially synthesize a linear precursor in a stepwise fashion that is ultimately cyclized. A potential disadvantage is the lack of control over the sequence of the subunits along the ring, which leads to a mixture of products that vary in composition and subunit sequence when a mixture of ligands is used. This limitation can be overcome by choosing metal ions and ligands whose interactions are restricted to a certain subset of combinations, as shown by Michael Schmittel [94].

This approach is based on the steric and electronic effects of the residues in 2,9-disubstituted phenanthroline derivatives on the structure of the corresponding tetrahedral metal complexes. Ligand **5.60** (Figure 5.55a), for example, is unable to form homoleptic complexes with copper(I) or zinc(II) ions because steric effects of the *ortho*-methyl groups in the mesityl substituents prevent two orthogonally arranged ligands from coming close enough. Ligand **5.60** therefore requires a less crowded ligand, such as an unsubstituted phenanthroline, to form a complex. This heteroleptic complex is further stabilized by aromatic interactions between the two mesityl groups in **5.60** and the intercalating ligand, rendering it more stable than a homoleptic complex containing only unsubstituted phenanthrolines.

The level of complexity can be further increased by using an equimolar mixture of the ligands **5.60**, **5.61**, phenanthroline, terpyridine, and both copper(I) and zinc(II) salts (Figure 5.55b). Again, only heteroleptic complexes are formed, with terpyridine coordinating preferentially to zinc(II). Copper(I) is unable to extend the coordination number beyond four and its interaction with terpyridine would therefore not benefit from the additional donor atom. Of the two possible heteroleptic terpyridine-zinc(II) complexes, the one with **5.61** is more stable than the one with **5.60** because the oxygen atoms in **5.61** also interact with the metal center. Accordingly, this mixture undergoes social self-sorting, yielding only the two complexes highlighted in Figure 5.55b.

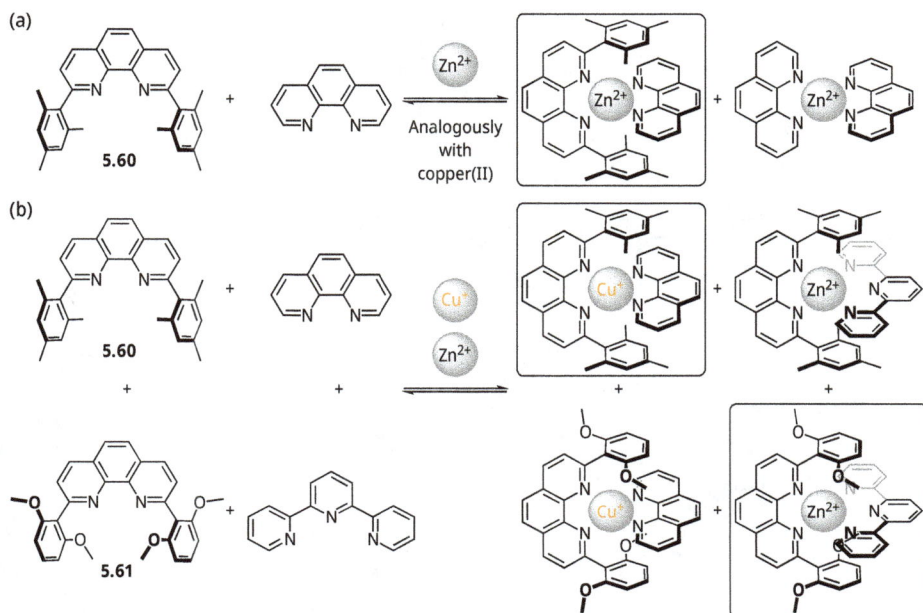

Figure 5.55: Self-sorting processes involving a 1:1:1 mixture of **5.60**, phenanthroline, and copper(I) or zinc(II) ions (a), or a 1:1:1:1:1:1 mixture of the ligands **5.60**, **5.61**, phenanthroline, terpyridine, and both copper(I) and zinc(II) (b). The complexes in the boxes are preferentially formed.

Based on these self-sorting principles, ditopic ligands can be designed to assemble in a predictable manner, which in turn allows control over the structure of the product. For example, the three ligands shown in Figure 5.56a assemble in the presence of copper(I) and zinc(II) ions to form a triangle in which the positions of the ligands and the metal ions are controlled by the rules derived above [95]. Another set of ditopic ligands yields the trapezoid shown in Figure 5.56b [96], and by adding an additional ligand to the system, which forms a homoleptic complex, a five-membered ring can also be obtained [97].

5.5.5 Cages

Coordination cages are hollow three-dimensional objects with metal centers distributed along the surface. The construction of such cages requires one type of building block, either the metal or the ligand, to interact with three or more binding partners, as shown in Figure 5.42b. Such cages are classified according to the structures of their building blocks, with Figure 5.57 illustrating the principles [98].

In the case of cages constructed from macrocyclic ligands, the number of metal centers depends on the symmetry of the macrocycle. For example, cyclotriveratry-

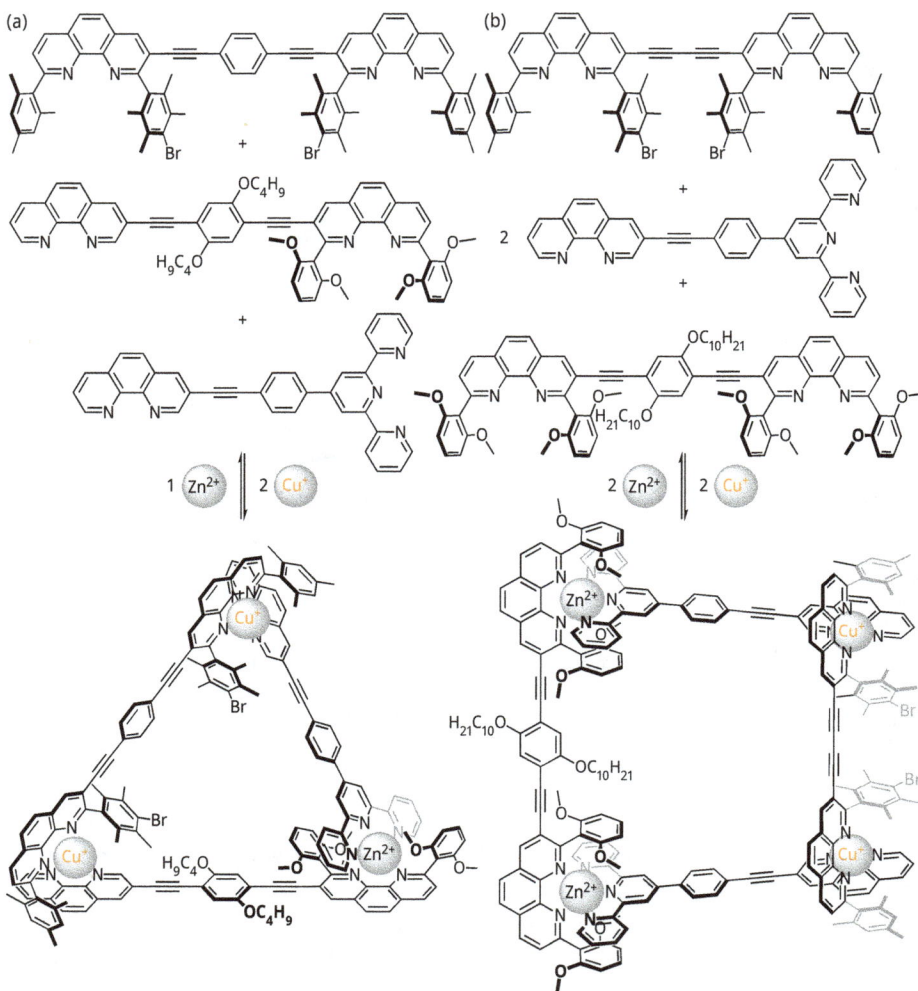

Figure 5.56: Selective formation of a triangle (a) and a trapezoid (b) from appropriate ditopic ligands whose coordination to copper(I) and zinc(II) ions is limited to the combinations in the products shown, while other metal–ligand combinations are much less stable and therefore do not contribute to product formation.

lenes are trifunctional ligands, while calix[4]arenes and resorcin[4]arenes are tetra-functional. The nature of the donor groups determines which metal ion must be used for the assembly. In the case of monodentate donors, metal centers with two vacant sites, such as the palladium(II) complex **5.49**, can serve as linking units. Linking two bidentate ligands requires metal ions that form tetrahedral complexes, while two tri-dentate ligands can be linked with metal ions that prefer an octahedral coordination geometry (Figure 5.46).

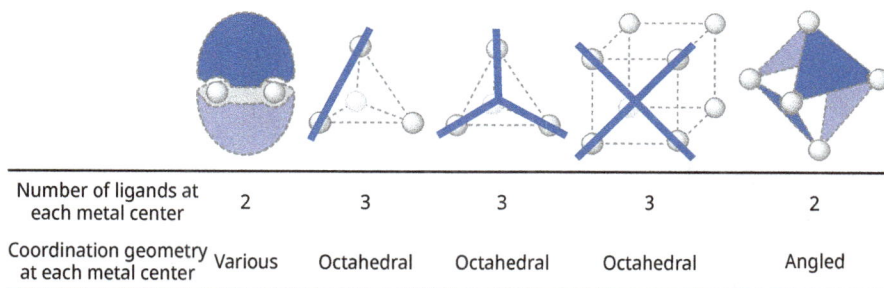

Number of ligands at each metal center	2	3	3	3	2
Coordination geometry at each metal center	Various	Octahedral	Octahedral	Octahedral	Angled

Figure 5.57: General concepts for the construction of coordination cages of different shapes, where the blue objects and the spheres represent the ligands and the metal ions, respectively.

Tetrahedral coordination cages, on the other hand, usually contain the metal ions at the corners. These ions interact either with difunctional ligands located along the edges or with trifunctional ligands capping the faces of the tetrahedron. In both cases, octahedral complexes are formed at the metal centers. The combination of linear ligands with octahedral complexes at the vertices could alternatively yield cubes or even larger cages, but the tetrahedron is usually favored entropically because it contains the smallest number of building blocks. Cubes are formed from tetrafunctional ligands arranged on the six faces and connected at the corners by octahedral metal complexes. The assembly of an octahedron requires four ligands to be connected at the corners, which is difficult to achieve by metal coordination. However, if triangular ligands are connected at the corners, octahedra are obtained with only half of the faces occupied. The following examples of coordination cages illustrate these concepts.

The linking of macrocyclic ligands through metal centers is somewhat related to the formation of hydrogen-bonded capsules from cavitands or calix[4]arenes (Section 5.3.3). For example, the cavitand **5.62**, which contains four nitrile groups along the rim, can serve as a tetrafunctionalized ligand. Bridging the nitrile groups *via* metal centers by treating **5.62** with the *cis*-protected palladium(II) complex **5.63** yields a tetranuclear cage with a cavity large enough to accommodate anions, such as one of the triflate anions released during cage formation (Figure 5.58) [99].

Such *cis*-protected palladium(II) complexes also serve to link planar ligands that end up as panels in a coordination polyhedron. For example, the ligands **5.47** and **5.48** form an octahedral and a square pyramidal complex, respectively, when combined with the palladium(II) complex **5.49**, as discussed in Section 5.5.1 (Figure 5.59a,b) [100]. Other ligands yield products with different structures. In the case of **5.64**, two products are possible, both having the composition M_8L_4, but differing in the relative orientation of the four panels (Figure 5.59c). If the panels are arranged in a parallel fashion, the complex has an open structure in the form of a square pyramid with the apex cut off, while the antiparallel orientation results in a cage with a small cavity. Product formation is controlled by the addition of appropriate templates that occupy

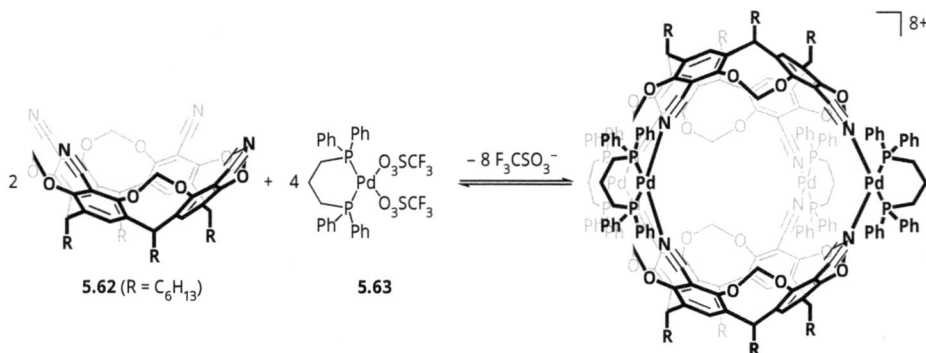

Figure 5.58: Formation of a coordination cage from the tetrafunctionalized cavitand **5.62** and the palladium(II) complex **5.63**.

the cavities of the products: large substrates, such as dibenzoyl, induce the formation of the pyramidal complex, while smaller ones, such as CBr$_4$, shift the equilibrium toward the closed cage.

Ligand **5.65** with three pyrimidine groups leads to the formation of an M$_{18}$L$_6$ trigonal-bipyramidal structure upon treatment with **5.49**, which has a volume of 900 Å^3 and a closed surface so that trapped molecules cannot escape (Figure 5.59d). The tetrafunctional ligand **5.66** gives rise to a rectangular tubular complex with two panels missing to close the cavity (Figure 5.59e).

Two different types of ligands can also be combined in this approach, although some tricks are required. One is the use of the platinum(II) analog of **5.49** to obtain inert complexes. The second is the use of templates to drive product formation. For example, a mixture of ligand **5.47**, pyrazine, and the platinum(II) complex in the presence of hexamethoxytriphenylene yields a cage in which the two tripodal panels are arranged in a parallel fashion and connected by three pyrazine pillars (Figure 5.59f). Due to the inert nature of the platinum–nitrogen bonds, the template can be removed without destroying the cage.

These cages often interact with anionic guests due to their multiple positive charges. In addition, hydrophobic compounds are also bound in the aqueous environment in which the cages are formed, with the driving force for complex formation coming from the hydrophobic effect. In this context, the platinum complex shown in Figure 5.59f binds 1,3-diketones and selectively stabilizes the enol form of these guests [101], while the platinum(II) analog of the pyramidal complex shown in Figure 5.59b mediates the folding of a designed nonapeptide with the amino acid sequence Ac-Trp-Ala-Glu-Ala-Ala-Ala-Glu-Ala-Trp-NH$_2$ [102]. This peptide has a random coil conformation in the absence of the cage. In its presence, the hydrophobic effect mediates the incorporation of the hydrophobic tryptophan side chains into the cavity. To allow both residues to bind simultaneously, a conformational reorganization of the peptide occurs, resulting in an α-

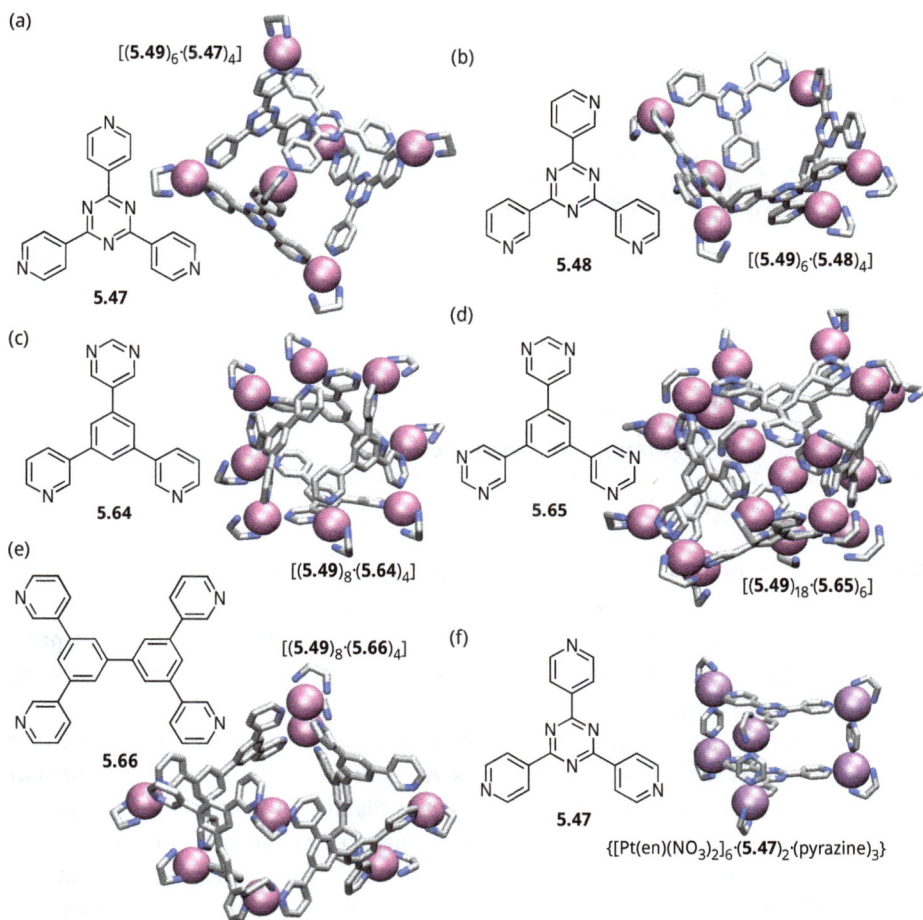

Figure 5.59: Molecular structure of ligands **5.47**, **5.48**, and **5.64–5.66**, and crystal structures of the multinuclear complexes formed from these ligands upon coordination to **5.49** or the corresponding platinum(II) analog. In the case of **5.64**, the crystal structure of the cage has been reported, but not that of the alternative pyramidal complex. Hydrogen atoms are omitted for reasons of clarity.

helix with the tryptophan side chains on one side and the glutamic side chains on the opposite side (Figure 5.60). These polar groups are exposed to the solvent in the complex and aid in its hydration. In Section 9.2.2, we will see that these palladium cages can also serve as reaction vessels to mediate Diels–Alder reactions or photochemical cycloadditions [103].

In all of the above examples, the metal complex **5.49** serves as a difunctional linking unit to orient two ligands at a 90° angle, while the ligands provide three or more donor atoms to extend the assembly into the third dimension. These roles are reversed when ligands with only two donor atoms are arranged around palladium(II)

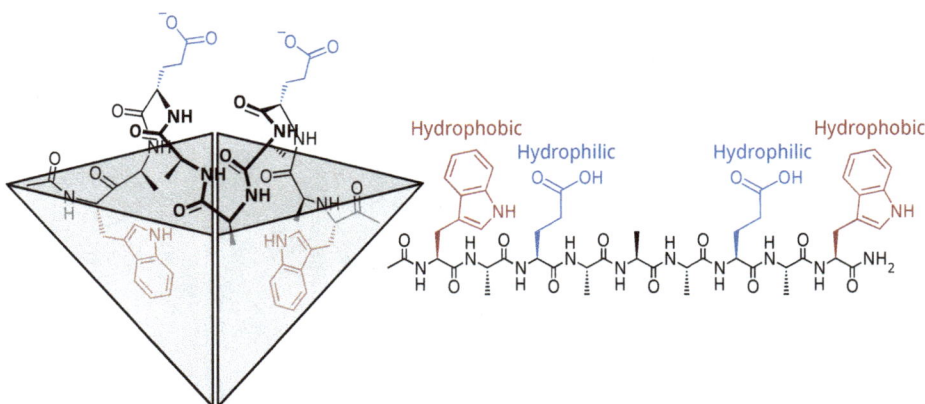

Figure 5.60: Schematic representation of the binding mode of the α-helical nonapeptide Ac-Trp-Ala-Glu-Ala-Ala-Ala-Glu-Ala-Trp-NH$_2$ inside the cavity of the M$_6$L$_4$ complex formed from ligand **5.48** and the platinum(II) analog of **5.49**.

ions as tetrafunctional linking units. Due to the planar arrangement of the ligands at the metal centers, this strategy yields discrete three-dimensional objects with a convex outer surface only when curved ligands are used, with the structure of the products depending sensitively on the shape of the ligands, that is, the angle at which the donor atoms are arranged.

Curved ditopic ligands with parallel donor atoms lead to dimetallic cages in which the two metal ions are linked by four ligands. An example is the cage formed from four banana-shaped bis(pyridyl) ligands **5.67** and two palladium(II) ions (Figure 5.61a) [104]. This compound is rigid and has a cavity large enough to accommodate aromatic disulfonates, which are bound primarily by ionic interactions. The assembly of the dinuclear palladium(II) complex of **5.68** (Figure 5.61b) follows similar principles [105]. In this case, the cavity is lined and almost completely closed by the anthracene panels, which strongly influence the guest exchange. This cage hosts hydrophobic organic molecules whose binding in water is largely driven by the hydrophobic effect.

If rigid ligands with two donor atoms oriented at an angle >0° are used for self-assembly, the planes of individual palladium complexes are tilted against each other and the formation of dimetallic complexes is no longer possible. The outcome of the self-assembly then depends on the exact structure of the ligand [106]. Ligands such as **5.69**, in which the angle between the two coordinating pyridyl units is 90°, form an M$_6$L$_{12}$ cube. Extending the angle to 127°, as in **5.70**, leads to an M$_{12}$L$_{24}$ cuboctahedron, and further to 149°, as in **5.71**, leads to an M$_{24}$L$_{48}$ rhombicuboctahedron (Figure 5.62a).

The selectivity with which these cages form is very sensitive to the bite angle of the ligands. If an 8:2 mixture of **5.70** and **5.71** is used for self-assembly, corresponding to an average bite angle of 131°, only the cuboctahedron is formed, while a 7:3 mixture with an average bite angle of 134° leads exclusively to the rhombicuboctahedron. This

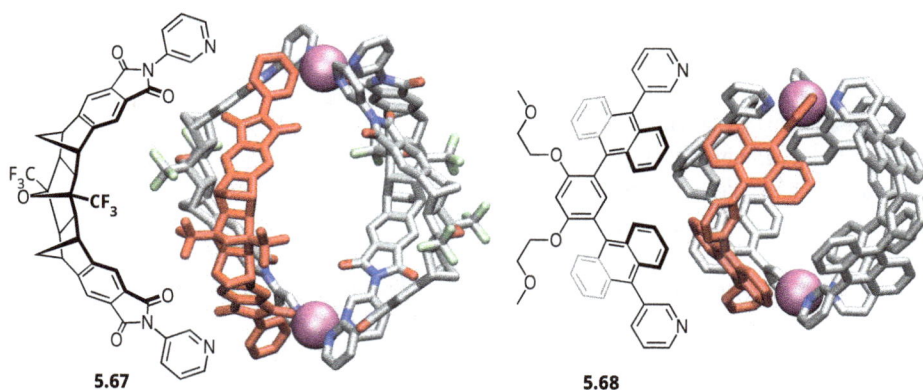

Figure 5.61: Molecular structures of ligands **5.67** and **5.68** and crystal structures of their dinuclear palladium(II) complexes. The protons in the crystal structures are omitted for reasons of clarity. One ligand in each structure is shown in red.

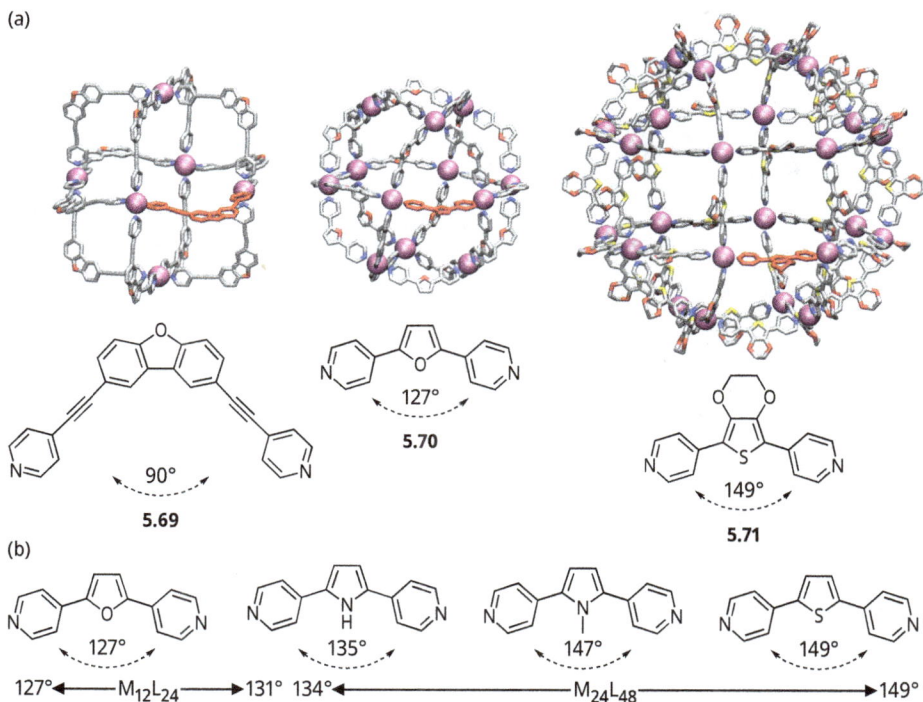

(a)

(b)

Figure 5.62: Molecular structures of ligands **5.69–5.71** and crystal structures of the polyhedra from coordination to palladium(II) ions (a). The correlation of the bite angle in such difunctional ligands with the structure of the product is shown in (b). The protons in the crystal structures are omitted for reasons of clarity. One ligand in each structure is shown in red.

correlation allows a reliable prediction of which of the ligands shown in Figure 5.62b will form the cuboctahedron and which will form the rhombicuboctahedron when interacting with palladium(II) ions.

The properties of these cages can be easily varied by changing the structure of the building blocks, with substituents in the convex region of the ligands ending up on the outside of the cages and those in the concave region ending up inside the cavities (Figure 5.63a). Since the nature of the substituents can also be varied in a wide range (alkanes, fluorinated alkanes, oligoethylene glycol residues, oligosaccharides, peptides, etc.), it is possible to tailor the properties of the cages for different applications. Coordination cages have even been prepared that are large enough to host globular proteins. The assembly of an $M_{12}L_{24}$ ubiquitin-containing cage is shown in (Figure 5.63b) [107]. This cage is composed of 23 subunits of **5.72a** and one subunit of **5.72b** to which the ubiquitin molecule is covalently attached. The characterization of another such cage showed that the confinement of the protein in the cavity leads to a pronounced stabilization of the folded structure at elevated temperatures or in aqueous solvent mixtures containing up to 90 vol% of acetonitrile [108].

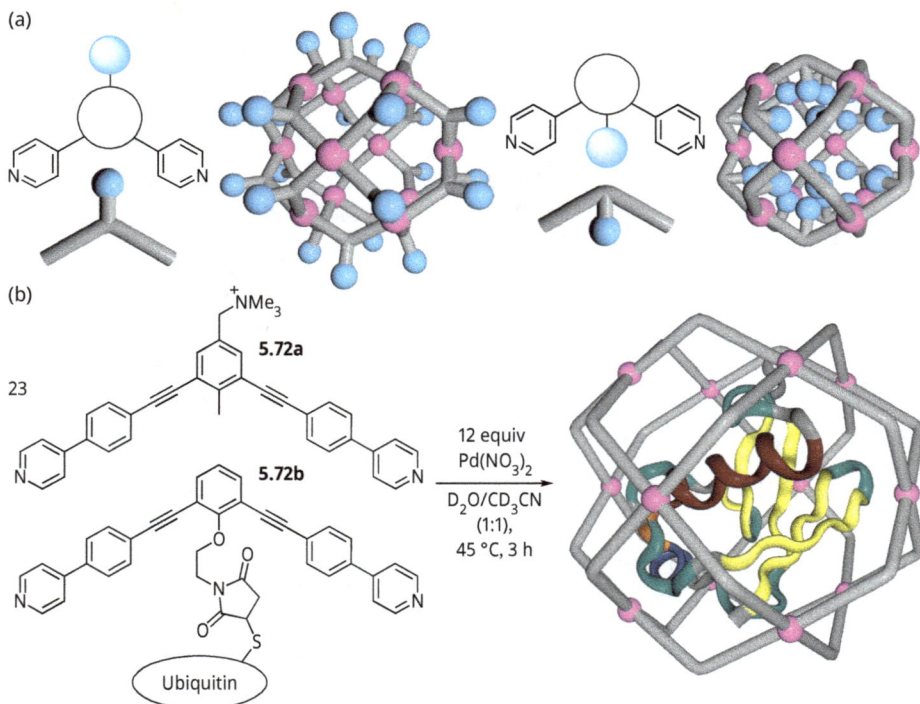

Figure 5.63: General structure of difunctional ligands that give rise to $M_{12}L_{24}$ cages with additional substituents on the outside and the inside (a), and molecular structures of ligands **5.72a,b** that give rise to a cage containing an entrapped ubiquitin molecule (b).

Another class of cages is characterized by octahedral metal complexes at the vertices. An example is the coordination cage developed by Kenneth N. Raymond [109], which is formed from the deprotonated form of bis(catechol) **5.73** (Figure 5.64a) and gallium(III) ions. The formation of this cage involves a similar coordination mode as the formation of the helicates from ligands **5.53a–c** (Figure 5.51), except that **5.73** is predisposed to form a cage rather than a helicate, according to the symmetry interaction approach.

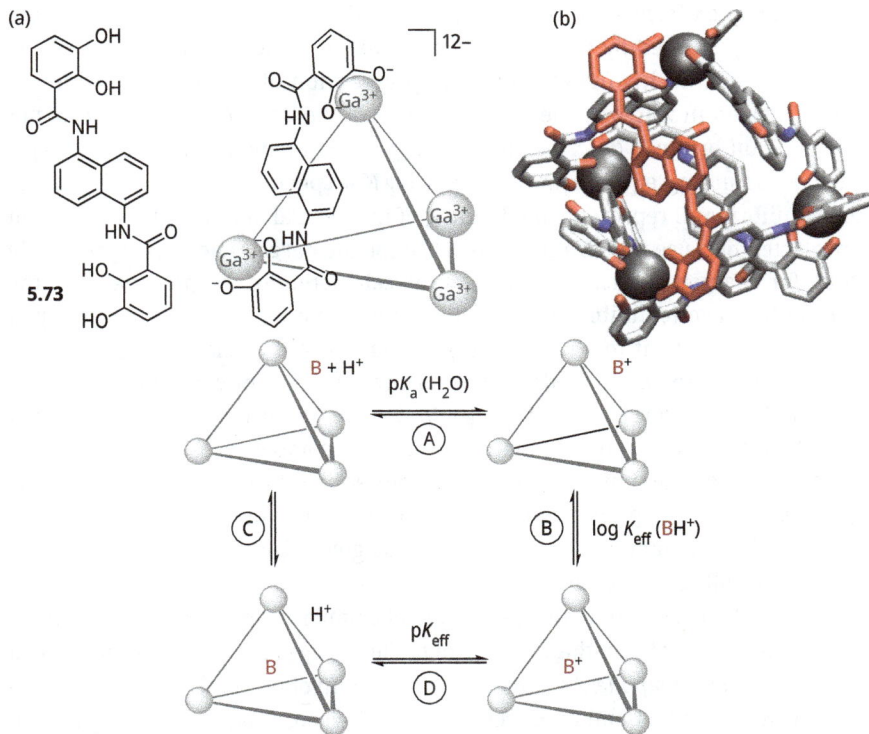

Figure 5.64: Molecular structure of ligand **5.73** and the corresponding M_4L_6 coordination cage in which six deprotonated ligands coordinate to gallium(III) ions (a). The crystal structure of this cage is shown in (b). In (c) a thermodynamic cycle is shown, which allows to estimate the shift of the effective basicity of amines resulting from their incorporation into the cage. The protons in the crystal structure are omitted for reasons of clarity. One ligand is shown in red.

In this cage, the six gallium(III) centers are located at the vertices of a tetrahedron with the ligands arranged along the edges. Each metal center has octahedral coordination geometry, with all four centers having either the Δ or Λ configuration (Figure 5.64b). Cage formation is thus diastereoselective, yielding the racemate of the $(\Delta,\Delta,\Delta,\Delta)$-configured and $(\Lambda,\Lambda,\Lambda,\Lambda)$-configured cages in the absence of chiral induction. The volume of the cavity is between 350 and 500 Å^3, which is large enough for the incorporation of small guest molecules. Since the cage is overall 12-fold negatively charged, it binds preferentially to cationic substrates such as protonated amines or quaternary ammonium ions. For example, the $\log K_a$ of the tetramethylammonium ion complex is 4.4 in water, with the driving force for complex formation coming mainly from the release of water molecules from the cavity and the associated entropy gain.

Cation binding is so efficient that protonated amines are stabilized within the cage even when the pH value of the surrounding medium would normally lead to deprotonation. Complex formation thus results in a shift in the effective basicity of the amine, which is quantified by using the thermodynamic cycle shown in Figure 5.64c [110]. Equilibrium A in this figure shows the protonation of the amine outside the cage, which is given in water by the pK_a value (where the letter *a* stands for *acid*, not *association*). Equilibrium B describes the binding of the ammonium ion to the cage, with the corresponding equilibrium constant $\log K_{eff}$ specifying the stability of the complex. Equilibrium C represents the binding of the neutral amine. This equilibrium can be neglected because the affinity of the cage for uncharged guests is significantly lower than for charged guests, which allows the calculation of the pK_{eff} value of the amine inside the cage (equilibrium D) by adding pK_a and $\log K_{eff}$ (note that the pK_a value quantifies the extent to which the *deprotonation* of the acidic ammonium ion occurs, but since it is the *negative* decadic logarithm of the acidity constant, the addition of pK_a and $\log K_{eff}$ correctly gives pK_{eff}). In the case of diisopropylamine, the pK_a is 10.8 and the $\log K_{eff}$ of the corresponding ammonium complex is 3.4, giving an effective pK_{eff} of 14.2. The cage thus shifts the pK_a by more than three orders of magnitude, leaving 50% of the amine protonated even at pH 14. Similar effects are observed for other amines. This efficient stabilization of cations gives rise to interesting catalytic properties as we will see in Section 9.3.2.

Another family of cages with octahedral metal complexes at the vertices has been developed in the group of Jonathan R. Nitschke. These cages are prepared by using a strategy known as *subcomponent self-assembly*, which involves combining two reversible reactions in the self-assembly process, namely the coordination of the ligands to the metal center and the generation of the actual ligands from smaller building blocks through a reversible covalent reaction [111]. This concept is illustrated in Figure 5.65 using a mononuclear complex as an example.

In this reaction, 2-aminoethanesulfonic acid (taurine), 2-formylpyridine, and a copper(I) salt yield a tetrahedral complex in which two chelating ligands, formed by imine formation from the amine and the aldehyde, coordinate to the metal center. This complex represents the thermodynamically most stable species of all possible products, and therefore dominates the equilibrium.

Figure 5.65 shows that in this approach, the structure of the ligands can be easily varied by selecting different subcomponents without having to synthesize and isolate the ligand prior to forming the metal complex. The ligand can even be modified after the complex has been assembled, since the imine bonds remain dynamic despite the stabilization by metal coordination. Accordingly, it is possible to exchange parts of the

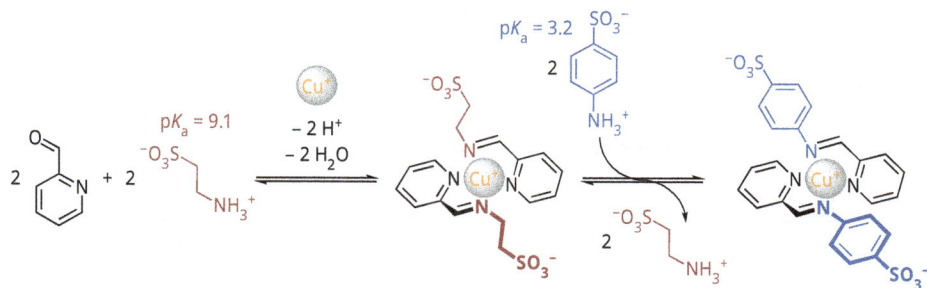

Figure 5.65: Subcomponent self-assembly of a copper(I) complex from 2-aminosulfonic acid and 2-formylpyridine and subsequent replacement of the aliphatic amine by an aromatic amine.

ligands at any time by adding another subcomponent if a thermodynamically more stable state can thus be reached. For example, in the reaction shown in Figure 5.65, the aliphatic amine is quantitatively replaced by more acidic 4-aminobenzenesulfonic acid residues (pK_a of the ammonium group 3.2 vs. 9.1 of the ammonium group of the aliphatic amine). The equilibrium thus favors a state in which the neutral form of the aromatic amine is incorporated into the complex, while the aliphatic amine binds the proton. The ability of such complexes to respond to a change in conditions or the presence of a new component thus allows the design of stimuli-responsive systems.

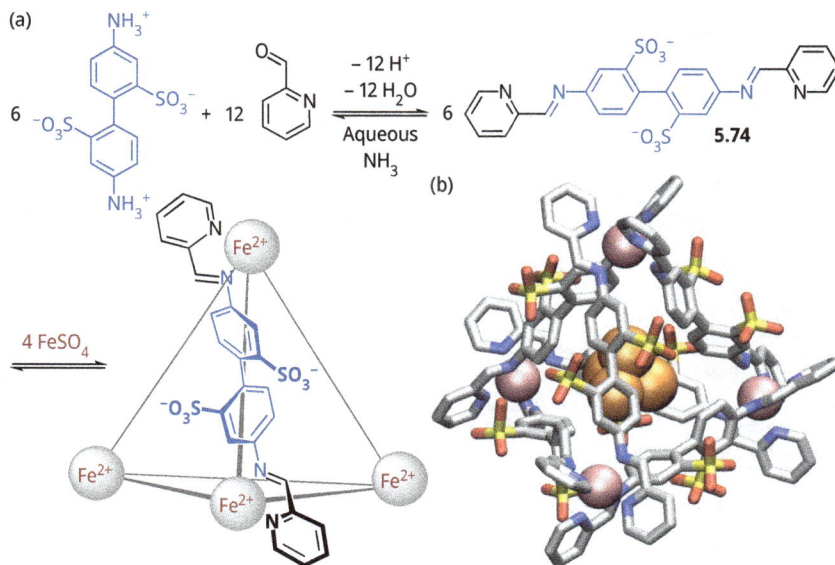

Figure 5.66: Subcomponent self-assembly of a coordination cage from 2-formylpyridine, 4,4'-diaminobiphenyl-2,2'-disulfonic acid and an iron(II) salt (a) and crystal structure of the P_4 complex of the cage (b). Counterions and hydrogen atoms are omitted for reasons of clarity.

Subcomponent self-assembly is useful for preparing helicates or cages, but we will focus primarily on cages here [112]. A prototypical example is shown in Figure 5.66a. This cage contains the diimine **5.74** as a chelating ligand, derived from 2-formylpyridine and 4,4'-diaminobiphenyl-2,2'-disulfonic acid. Six ligands are arranged at the edges of a tetrahedron and are held together by iron(II) centers occupying the vertices, where they form octahedral complexes with the nitrogen atoms at the ends of the ligands.

The twelve sulfonate groups render this cage soluble in water, where it accommodates neutral, nonpolar guest molecules such as cyclohexane. These guests can be released by different stimuli. One is to lower the pH, which causes hydrolysis of the imine bonds holding the cage together and protonation of the amines thus released (Figure 5.67a). This process is reversed by another change in pH, so that the system can be switched back and forth from the cage hosting a guest to the fully disassembled state by repeated pH changes. Alternatively, the cage can be disassembled by adding tris (2-ethylamino)amine to the solution (Figure 5.67b). This ligand scavenges the metal ions, thus inducing the cage to release the guest. The shift in the equilibrium is partly due to entropy, because the disassembly of the cage and the concomitant formation of the mono-

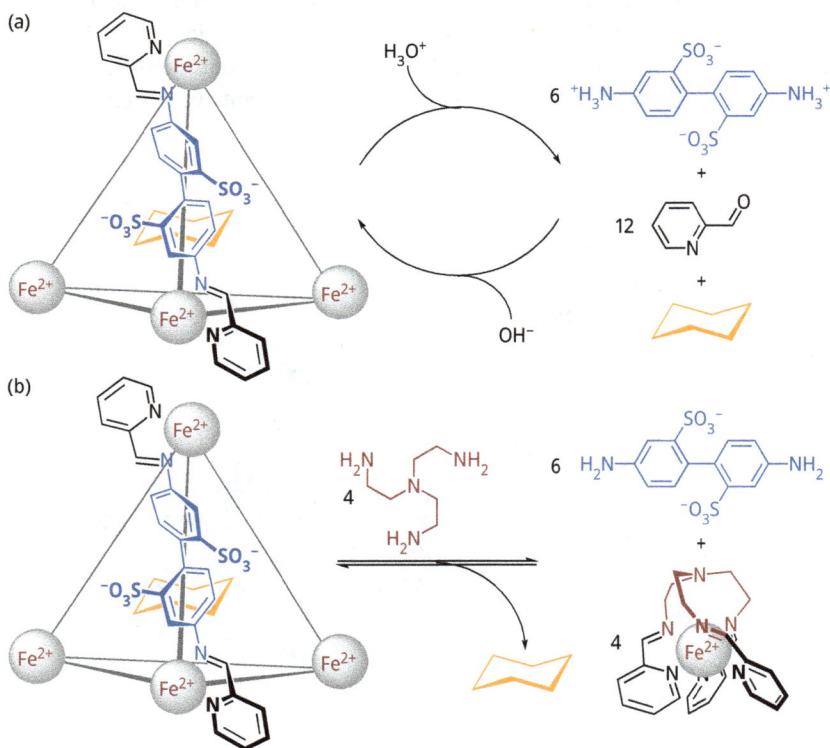

Figure 5.67: Disassembly of the cage formed from ligand **5.74** and subsequent release of its cargo by changing the pH (a) or adding a scavenging ligand (b).

nuclear iron(II) complex is associated with an increase in the number of molecules from 5 (1 filled cage and 4 tris(2-ethylamino)amine molecules) to 11 (6 difunctional ligands, 4 iron(II) complexes, and 1 cyclohexane molecule). Both of the above stimuli rely on the reversibility of the imine bonds, similar to the equilibrium shown in Figure 5.65.

This cage hosts several types of guest molecules. One is pyrophoric white phosphorus, which is stable in the cage for months (Figure 5.66b) [113]. Addition of benzene to the solution causes the P_4 molecule to be expelled, thus initiating its oxidative decomposition to phosphoric acid. Another possible guest is furan, which is unreactive when trapped in the cage, even if the external solution contains a dienophile (Figure 5.68). When furan is released from the cavity by the addition of benzene, it undergoes the expected Diels–Alder reaction.

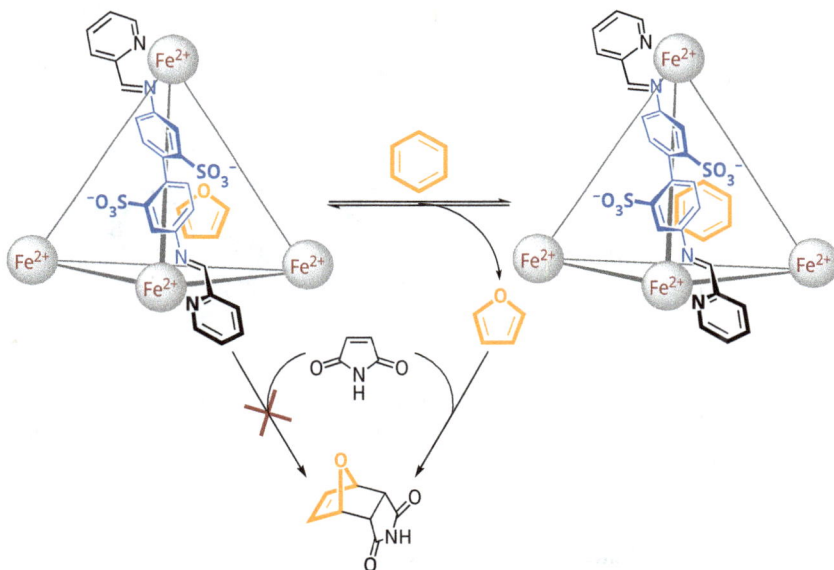

Figure 5.68: Schematic illustration of the initiation of a Diels–Alder reaction between furan and maleimide by replacing the bound and thereby protected furan molecule from the cage with benzene (b).

The concept of constructing cages through subcomponent self-assembly is very versatile, providing access to coordination cages of widely varying structure and size [112]. Figure 5.69 shows additional examples to illustrate the scope. Tripodal ligands derived from 2-formylpyridine and a triamine, when coordinated to iron(II), yield tetrahedral cages in which the ligands cover the four faces (Figure 5.69a). By changing the structure of the triamine, cages with volumes ranging from 31 to 823 $Å^3$ are accessible. Using the porphyrin-derived tetraamine and 2-formylpyridine results in a cube with iron(II) ions in the corners and nickel(II) ions bound to the six porphyrin subunits capping the faces (Figure 5.69b). These cubes have cavities large enough to accommodate buckminsterfullerenes.

(a)

(b)

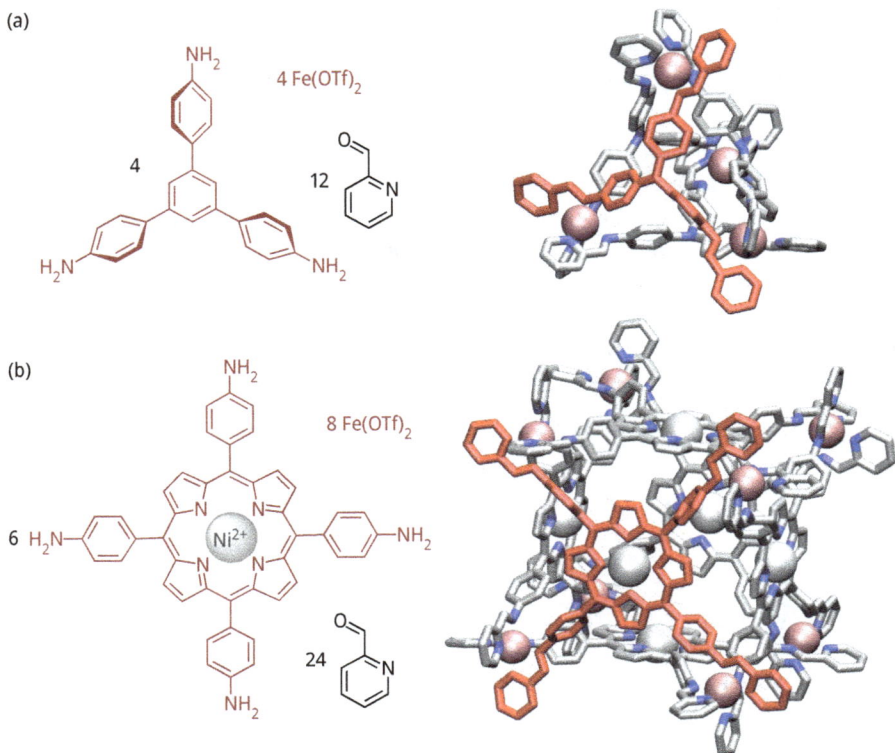

Figure 5.69: Subcomponents required to form a face-capped tetrahedral cage (a) and a face-capped cubic cage (b), along with crystal structures of the corresponding assemblies. Counterions and hydrogen atoms are omitted for reasons of clarity. One ligand in each structure is shown in red.

5.6 Self-assembly mediated by covalent bonds

5.6.1 Introduction

The principles of self-assembly outlined at the beginning of this chapter are not limited to noncovalent or metal–ligand interactions, but apply to any fully reversible process that causes one or more building blocks to hold together, including the formation of covalent bonds. One might question whether the term self-assembly is appropriate in this context; indeed, more appropriate terms are dynamic covalent chemistry [114] or dynamic covalent synthesis [115]. Another question is how such reactions relate to supramolecular chemistry, since the formation of a covalent bond is hardly a molecular recognition process. However, syntheses involving reversible reactions have so many characteristics of genuine self-assembly processes that it is justified to discuss them here. For example, as in self-assembly processes, reversible reactions permit the synthesis of structurally complex compounds in one-pot reactions without the need to join the building blocks in a stepwise reaction sequence. The reversibility of bond formation, which allows incorrectly formed bonds to dissociate during the course of the reaction, ensures that the thermodynamic product is formed with similar fidelity as in a noncovalent synthesis. Finally, the nature of the product that dominates the equilibrium is determined by factors analogous to those that control self-assembly. Accordingly, an enthalpically favorable situation arises when all functional groups available for product formation are used (principle of maximum site occupancy). In addition, thermodynamics favors the formation of the smallest possible unstrained product, and product distribution is influenced by stabilizing intramolecular interactions that occur in one product but not in the starting materials or competing products.

Such syntheses are most effective and high yielding when only a single product remains in the equilibrium. However, when several products have similar thermodynamic stability, a mixture of products is formed. In this case, there are still ways to shift the equilibrium to a single species. One is the use of templates, an attractive strategy discussed in the next section. Alternatively, it is sometimes possible to make use of kinetic traps. For example, if the product is insoluble in the solvent used for the synthesis, it will precipitate once formed, and the equilibrium will continuously adjust to produce more of that compound.

A major difference between noncovalent and covalent synthesis is that the latter typically yields robust products that can be isolated, characterized, and further processed once the exchange is stopped. Figure 5.70 lists a selection of reversible reactions where this is possible.

Reactions (a) to (f) involve the coupling of two different functional groups with the release of an equivalent of water. This water is required to mediate the corresponding back reaction, thus ensuring reversibility. The ester formation in reaction (a) is catalyzed by acids but usually requires long reaction times and elevated temperatures. An alternative is the transesterification between two esters that is mediated by catalytic amounts of alkoxides. Cyclic boronic acid esters form rapidly, even at neutral pH, as we have seen in Section 4.2.4. However, these esters are usually much less stable than the esters of carboxylic acids. Acetals (reaction (c)) and imines (reaction (d)) are readily formed from aldehydes under relatively mild acidic conditions upon treatment with alcohols or primary amines, respectively. Both functional groups are stable in the absence of acids, allowing the reaction to be stopped by changing the pH. Since imines are susceptible to hydrolysis in the presence of water, their isolation is often only possible after postfunctionalization, such as reduction to the corresponding amines. Hydrazones (reaction (e)) and oximes (reaction (f)) do not suffer from this disadvantage. Their formation typically requires more strongly acidic conditions than that of imines, but in the absence of acids, they are stable even in water.

Disulfide formation (reaction (g)) and olefin metathesis (reaction (h)) are examples of reversible reactions between two identical functional groups. In these reactions, two different thiols or olefins give rise to three products, namely the heterocoupling product

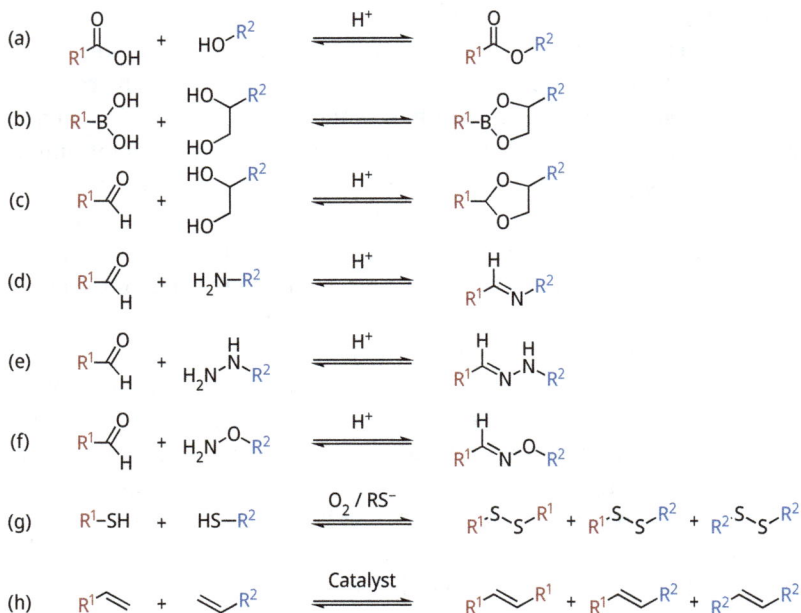

Figure 5.70: Selection of important reversible reactions used in dynamic covalent chemistry.

and the two homocoupling products. Disulfides are formed from thiols by oxidation, with the subsequent shuffling of the substituents requiring the presence of small amounts of a thiolate. The reaction therefore occurs in basic media and is stopped by lowering the pH. A major advantage of disulfide exchange is that it works in aqueous solvent mixtures and in water. Olefin metathesis requires a transition metal catalyst to mediate the coupling of two terminal olefins with the simultaneous release of ethylene. Since the catalyst also reacts with internal olefins, the initially released ethylene is not required for the exchange to work. The scope of olefin metathesis is often limited by the lifetime of the catalyst, which must remain active until the reaction reaches thermodynamic equilibrium.

In the remainder of this section, we will see how these reactions allow the preparation of structurally complex products from simple building blocks. We will focus on syntheses of rings and cages to demonstrate that supramolecular and dynamic covalent chemistry have more in common than just conceptual links. Many of the compounds produced by reversible reactions are directly relevant to molecular recognition and self-assembly. Dynamic covalent chemistry under the influence of templates even allows the identification of new receptors, as discussed in Section 5.7, and the synthesis of interlocked molecules, which is the subject of Chapter 7. This strategy is therefore extremely useful for addressing many synthetic challenges in supramolecular chemistry.

5.6.2 Rings

Before discussing specific syntheses, it should be noted that several of the receptors discussed in Section 4.1 are prepared using reversible reactions. Examples are cyclo-triveratrylenes (Section 4.1.6), calixarenes (Section 4.1.7), calixpyrroles (Section 4.1.8), resorcinarenes (Section 4.1.9), pillararenes (Section 4.1.10), and cucurbiturils (Section 4.1.11), that is, all macrocycles prepared by the condensation of suitable building blocks with carbonyl compounds, mostly formaldehyde. The reversibility of these reactions allows the use of templates, as in the synthesis of calixarenes. In addition, the reaction mixtures usually evolve over time, eventually converging on the thermodynamically favored product, as observed in resorcinarene and cucurbituril syntheses. Thus, some of the most important receptors in supramolecular chemistry owe the efficiency with which they are prepared to dynamic covalent chemistry. Moreover, since these syntheses require relatively harsh acidic or basic conditions, the respective products are inert after isolation and can thus be conveniently used for binding studies or further transformations.

An instructive example of a macrocyclization involving a transesterification reaction is the cyclotrimerization of the cinchona alkaloid derivatives **5.75a** and **5.75b** (Figure 5.71). These esters give the corresponding cyclic trimers in >90% yield when heated in toluene together with catalytic amounts of KOMe/18-crown-6 (the crown ether serves to complex the potassium ions, thus mediating the dissolution of KOMe in toluene) (Figure 5.71a) [116].

The reaction conditions suggest that the selectivity and efficiency of this reaction is due to thermodynamic control, but it cannot be excluded that the formation of the cyclic trimer is a kinetic effect and that no exchange of the building blocks occurs once the product is formed. Reference experiments are therefore required to demonstrate the reversibility of the transformation.

An important aspect that must be demonstrated is that kinetic control leads to a different product distribution. Indeed, the cyclization of the acid chloride **5.75d** at room temperature in the presence of 4-(dimethylamino)pyridine (DMAP) produces a mixture of products in which the cyclic trimer still dominates (37%), but the tetramer and larger macrocycles are also present in significant amounts (Figure 5.71b). Thus, the preferential formation of the trimer by transesterification cannot be a kinetic effect, since products other than the trimer are present when the reaction is carried out under kinetic control. The observation that the product distribution is independent of whether the monomer **5.75b** or the dimer **5.75c** is used as starting material (Figure 5.71a) also indicates that the reaction involves continuous assembly and disassembly reactions. Finally, experiments with mixtures of the starting materials using a 1:1 ratio of the two monomers **5.75a** and **5.75b**, a 1:1 ratio of the two cyclic homotrimers, or a 2:1 ratio of **5.75a** and **5.75c**, all yield mixtures containing the four possible macrocycles in the expected statistical ratio of 1:3:3:1 (Figure 5.71c). Accordingly, even the macrocycles can shuffle their components to yield the products expected for a reaction under thermodynamic

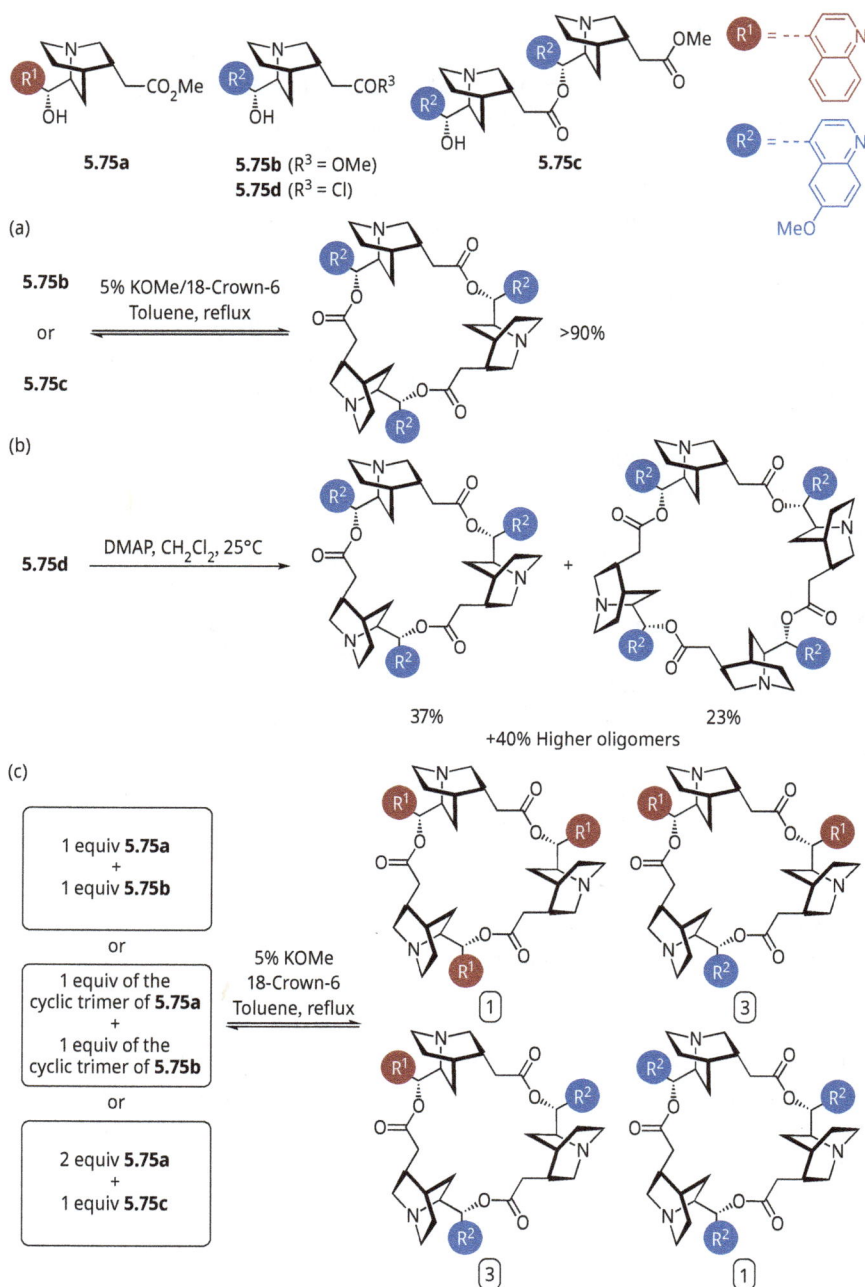

Figure 5.71: Molecular structures of building blocks **5.75a–d** and cyclization of **5.75b** and **5.75c** under thermodynamic control (a) and of **5.75d** under kinetic control (b). The results of reactions with mixtures of starting materials are shown in (c) (DMAP = 4-(dimethylamino)pyridine).

control. The exclusive formation of the observed macrocycles indicates that the monomers are strongly predisposed to form three-membered rings.

Another useful reversible reaction for the preparation of macrocycles is imine formation. In this context, it is often observed that macrocyclic imines undergo exchange reactions to form mixtures of rings of different sizes. For example, the dissolution of the tetraimine **5.76c** (Figure 5.72) in chloroform yields a mixture of the three macrocycles **5.76a–c** in a reaction mediated by traces of HCl. The presence of the three products suggests that they have comparable thermodynamic stability [117]. The corresponding oxygen analogs **5.77a–c** undergo a similar exchange reaction [118]. The thermodynamic control is reflected in the concentration dependence of the equilibrium, which shifts from the preferred formation of the dimer **5.77a** at low concentrations (0.006 M: **5.77a:5.77b:5.77c** = 72:23:5) to the formation of the larger macrocycles at higher concentrations (0.2 M: **5.77a:5.77b:5.77c** = 44:31:25) due to entropic factors.

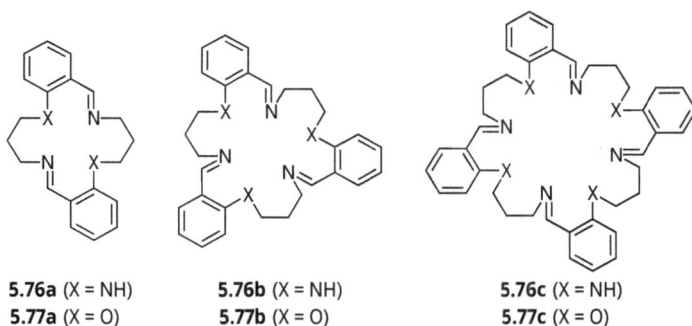

5.76a (X = NH)   **5.76b** (X = NH)   **5.76c** (X = NH)
5.77a (X = O)   **5.77b** (X = O)   **5.77c** (X = O)

Figure 5.72: Molecular structures of the macrocyclic imines **5.76a–c** and **5.77a–c**.

The fidelity of such macrocyclizations depends on the structure of the subunits, as shown by the strong preference of (R,R)-1,2-diaminocyclohexane and terephthalaldehyde to form the cyclic hexaimine **5.78** (Figure 5.73a) with E configurations at all imine bonds. This compound, which is the prototype of a larger family of macrocyclic hexaimines known as trianglimines, is surprisingly robust, given the hydrolytic lability of imines, and can be structurally varied over a wide range [119]. For example, replacing terephthalaldehyde with biphenyl-4,4'-dicarbaldehyde results in a macrocycle with a larger diameter, and additional substituents in the aromatic subunits can serve either for intramolecular conformational control or as binding sites for potential substrates.

When isophthalaldehyde is used in the macrocyclization reaction, the imine formation is somewhat less selective, with the hexamine being only the kinetically controlled product that slowly reequilibrates to yield the thermodynamically favored smaller ring. Pyridine-2,6-dicarbaldehydes yield a mixture of rings of different sizes. Some of these compounds have interesting binding and/or self-assembly properties, but the general disadvantage of trianglimines is that they hydrolyze under acidic

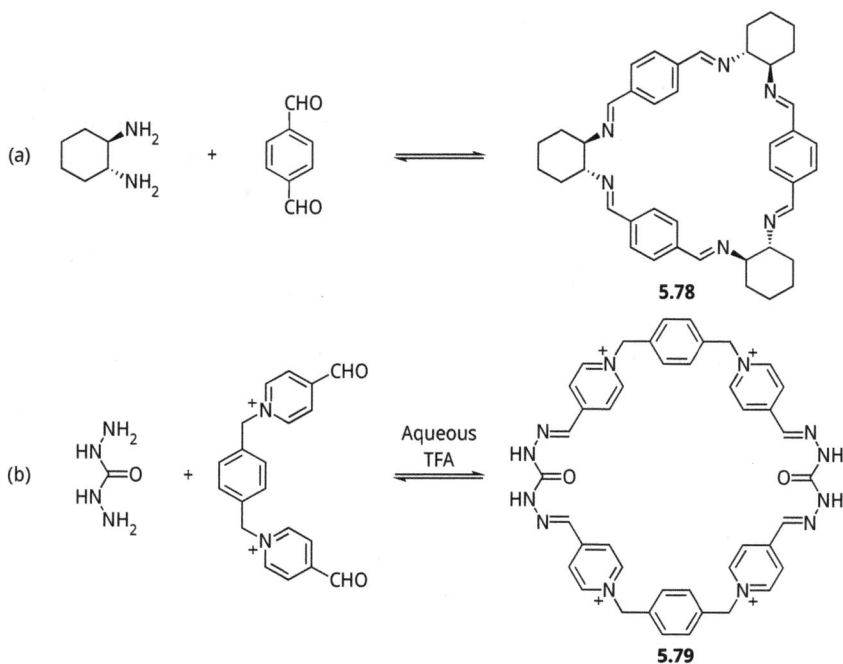

Figure 5.73: Synthesis of the macrocyclic hexaamine **5.78**, the prototype of a trianglimine (a), and of the macrocyclic hydrazone **5.79** (b).

aqueous conditions. Hydrazones do not suffer from this drawback, as demonstrated by the synthesis of macrocycle **5.79** (Figure 5.73b) from the corresponding dialdehyde and carbonohydrazide in acidic aqueous solution [120].

Figure 5.74 shows that it is also possible to combine different reversible reactions in a macrocyclization, in this case imine and boronate formation [121]. Since the bonds formed originate from different functional groups – an aldehyde and an amine in one case, and a boronic acid and a diol in the other – the corresponding reactions proceed independently of each other without giving rise to crossover products. When the condensation reaction is performed in a Dean–Stark trap, significant amounts of polymeric side products are observed, indicating that the removal of water could prevent the reaction from reaching the thermodynamic equilibrium. Mixing the required components in the presence of a small amount of solvent in a ball mill usually gives better yields [122].

An impressive example of a macrocycle obtained by disulfide exchange, described by Sijbren Otto, is shown in Figure 5.75. This compound forms spontaneously within about 16 days when **5.80** is equilibrated at a concentration of 0.5 mM in borate buffer [123]. Under similar conditions, 1,3-benzenethiols usually give cyclic trimers and tetramers. These small rings also form preferentially when the concentration of **5.80** is decreased to 0.05 mM, indicating that the enthalpic stabilization of the large 15-mer at

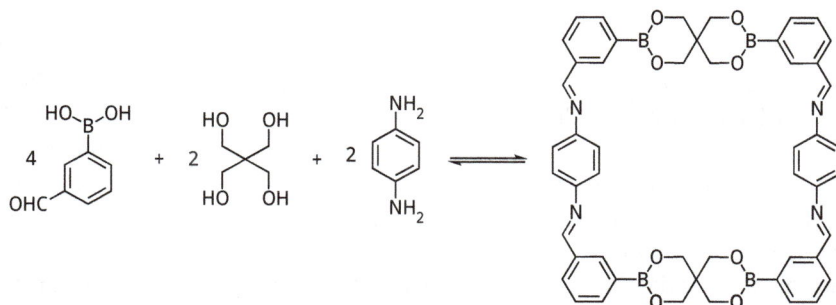

Figure 5.74: Formation of a macrocycle by the simultaneous formation of imine and boronic ester bonds.

higher concentrations overcompensates the entropic advantage of forming smaller rings. This macrocycle is tightly folded and stabilized by an intricate pattern of intramolecular hydrogen bonds, as shown in the crystal structure in Figure 5.75. The burial of hydrophobic residues in this structure further suggests that the hydrophobic effect contributes to the folding pattern. Indeed, performing the disulfide exchange in the presence of salts that enhance the hydrophobic effect favors the formation of the 15-mer, while this compound does not form in solvent mixtures in which a substantial fraction of an organic component suppresses the thermodynamic gain associated with the desolvation of hydrophobic residues. The formation of these intricately folded macrocycles is not limited to **5.80**, as other 1,3-benzenethiol derivatives give rise to related structures, some of which contain an unusual prime number of subunits (13, 17, and 23) [124].

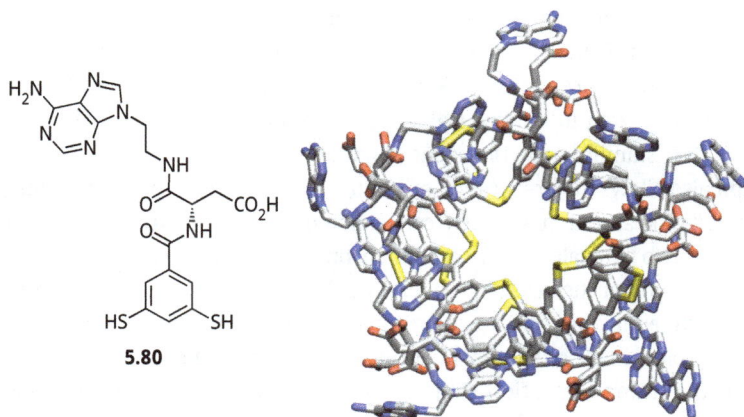

5.80

Figure 5.75: Molecular structure of building block **5.80** and crystal structure of the macrocyclic 15-mer derived therefrom.

5.6.3 Cages

The most commonly used reversible reaction for the preparation of cages is imine formation [125], which is why we will mainly focus on this strategy in this section. Again,
an important class of receptors, namely the polyazacryptands discussed in Section 4.1.2,
are synthesized in this way. For example, the reaction of isophthalaldehyde and tris(2-
aminoethyl)amine in a 3:2 ratio in methanol gives the corresponding hexaimine in
yields up to 90% (Figure 5.76) [126]. Precipitation of the product from the reaction mixture, which allows isolation by simple filtration, brings this reaction almost to completion. Treatment with sodium borohydride then provides the corresponding cryptand.

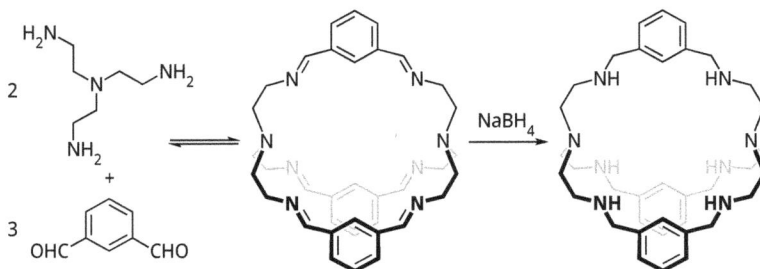

Figure 5.76: Example of the synthesis of a polyazacryptand *via* the corresponding hexaimine.

Other polyazacryptands are synthesized in a similar manner from tris(2-aminoethyl)
amine or other triamines and a variety of dialdehydes [127]. The results of the following
experiments show that this reaction is under thermodynamic control: the two triamines
and two aldehydes shown in Figure 5.77a give only the two self-sorted products, which
are the thermodynamically favored species. Similarly, the hexaimine shown in Figure
5.77b is quantitatively converted to a more stable cage by replacing the terminal tris(2-
aminoethyl)amine with tris(aminomethyl)benzene residues.

Imine formation also allows the preparation of analogs of Cram's carcerands and hemicarcerands, as shown by the almost quantitative conversion of the resorcin[4]arene-derived
tetraaldehyde **5.81a** and two equivalents of 1,3-diaminobenzene into the corresponding cage
upon treatment with trifluoroacetic acid (Figure 5.78) [128]. Dissolving this cage in chloroform together with four equivalents of a 5-substituted 1,3-diaminobenzene derivative
causes the stepwise replacement of the unsubstituted linkers by the substituted ones,
demonstrating that the imine formation remains reversible even after the cage has
been formed. Since the replacement of the linkers must proceed through short-lived intermediates in which a bridge is opened or completely removed, the acid-induced
linker exchange also facilitates the release of a guest molecule trapped inside the cage.

(a)

(b)

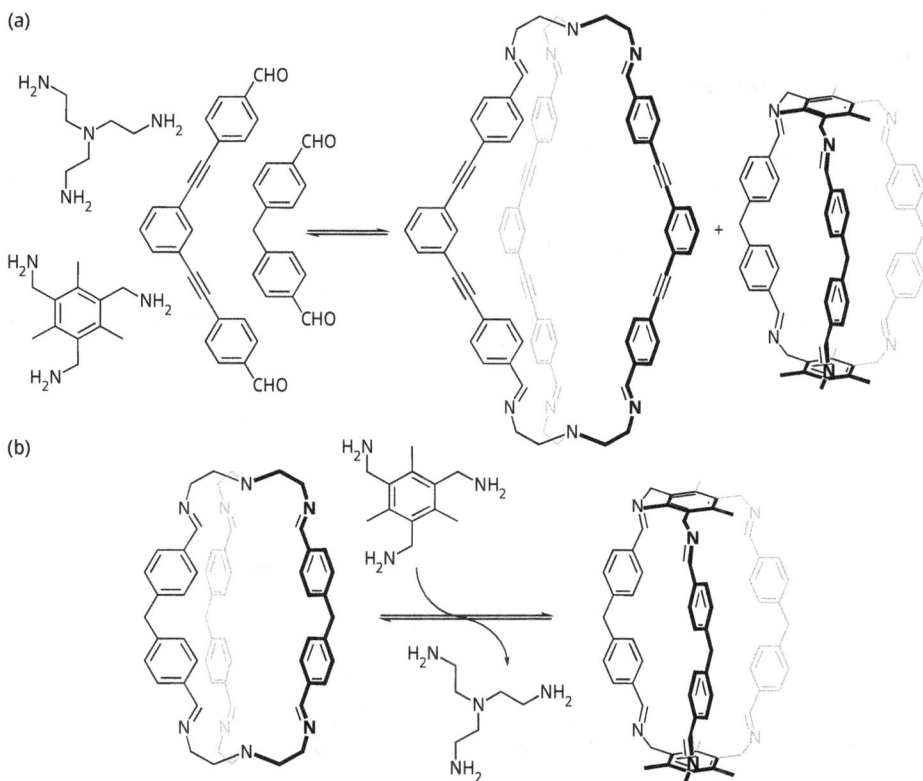

Figure 5.77: Self-sorting leads to the exclusive formation of two cages from a mixture of two triamines and two dialdehydes (a), and the exchange of the terminal triamine residues converts the original hexaimine into a less strained and therefore thermodynamically more stable analog (b).

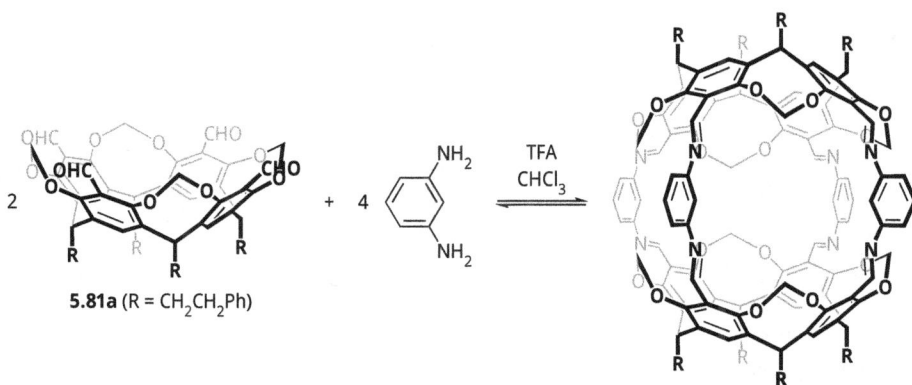

5.81a (R = CH₂CH₂Ph)

Figure 5.78: Formation of a carcerand-type octaimime from the resorcin[4]arene-derived tetraaldehyde **5.81a** and 1,3-diaminobenzene.

The outcome of the reaction shown in Figure 5.78 is largely determined by the structure of the diamine linker. 1,3-Diaminobenzene arranges two cavitands in a converging fashion during imine formation, resulting in the entropically favored octaimine. Short aliphatic diamines, such as 1,2-diaminoethane, cannot be incorporated into such a structure without strain and therefore induce the formation of larger cages. In chloroform, the trifluoroacetic acid-catalyzed condensation of the tetraaldehyde **5.81b** with 1,2-diaminoethane yields an octahedral cage containing six macrocyclic units connected by 12 linkers (Figure 5.79) [129]. Product formation is highly dependent on the solvent, as this cage does not form in tetrahydrofuran, where the same reactants yield a tetrahedral cage containing four cavitands. In dichloromethane, a square antiprismatic cage is preferentially formed with the cavitands occupying the eight corners.

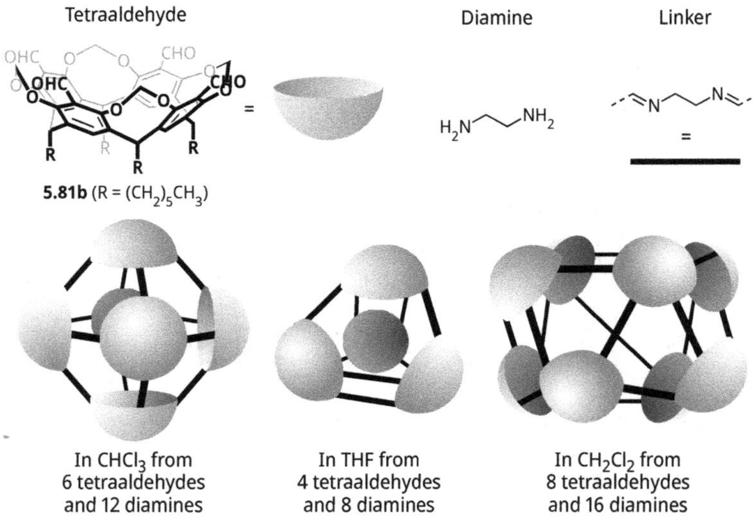

Tetraaldehyde   Diamine   Linker

5.81b (R = (CH$_2$)$_5$CH$_3$)

In CHCl$_3$ from 6 tetraaldehydes and 12 diamines

In THF from 4 tetraaldehydes and 8 diamines

In CH$_2$Cl$_2$ from 8 tetraaldehydes and 16 diamines

Figure 5.79: Schematic representation of the cages formed from the resorcin[4]arene-derived tetraaldehyde **5.81b** and 1,2-diaminoethane in chloroform, tetrahydrofuran, and dichloromethane.

The benzene-derived hexaamine **5.82**, when reacted with three equivalents of isophthalaldehyde, forms a cage in which two benzene units are connected by six linkers. The first example of these so-called superphanes, compound **5.83**, has a small cavity that can accommodate only two water molecules [130]. This cage was developed by Qing He, who later showed that superphanes can also be obtained from other dialdehydes and stabilized by reducing the imine groups to amines [131]. In this way, stable receptors with larger cavities are accessible that can also host neutral guest molecules or oxoanions [132, 133].

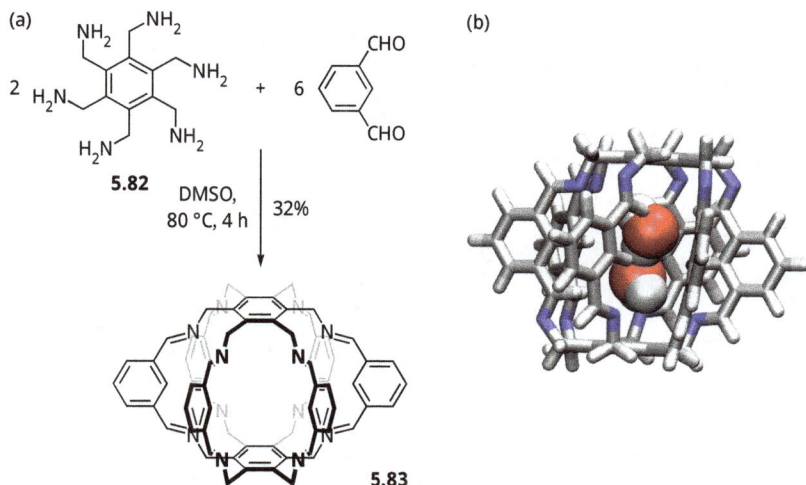

Figure 5.80: Synthesis of superphane **5.83** from the hexaamine **5.82** and isophthalaldehyde (a) and crystal structure of **5.83** with a water dimer in the cavity (b). The water molecules in the crystal structure are shown as space-filling models.

Other examples of cages obtained by imine chemistry are shown in Figure 5.81. Ralf Warmuth used the trifluoroacetic acid-mediated reaction between the cyclotriveratrylene-derived trialdehyde and 1,4-phenylenediamine to obtain a homochiral cube in which the cyclotriveratrylene units occupy the corners and the linkers the edges (Figure 5.81a) [134]. Tetrahedral cages were prepared in the group of Andrew I. Cooper from 1,3,5-triformylbenzene and 1,2-ethylenediamine or other vicinal diamines (Figure 5.81b) [135], and an *endo*-functionalized adamantoid cage was obtained in the group of Michael Mastalerz by reacting a triptycene-derived triamine with salicylic dialdehyde (Figure 5.81c) [136].

Can a liquid be porous?

In the solid state, the latter two cages yield porous microcrystalline materials with high surface areas and useful gas adsorption properties. This porosity can be transferred to the liquid state, as shown by Cooper using a derivative of the small cage [137]. Two factors are critical to ensure that the cage molecules remain empty in solution. First, they must contain solubilizing residues on the outer surface that do not enter and block the cavity. This is achieved by introducing crown ether residues into the ligands, as shown in Figure 5.81b. Second, the solvent molecules must also be too large to enter the cavity, with 15-crown-5 having an appropriate size. Accordingly, a solution of the crown ether-decorated cage in 15-crown-5 has empty pores, resulting in a high capacity for absorbing gases such as methane. Bound gas molecules are expelled from the cages by the addition of a better binding guest, causing the solution to

(a)

(b)

(c)

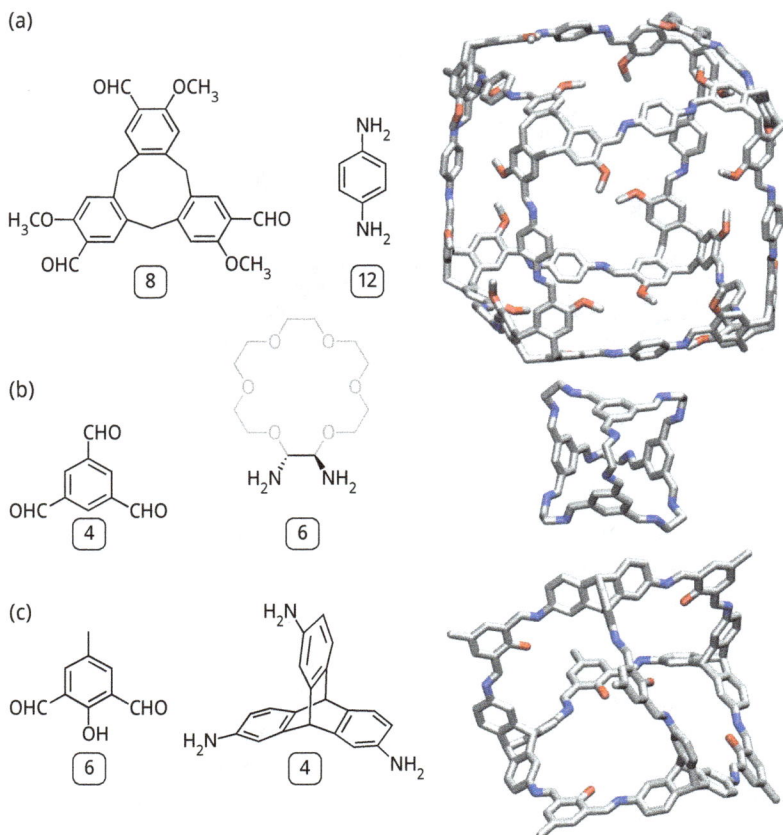

Figure 5.81: Building blocks and structures of polyimine cages formed from a cyclotriveratrylene-derived trialdehyde and 1,4-phenylenediamine (a), 1,3,5-triformylbenzene and 1,2-ethylenediamine (b), and salicylic dialdehyde and a triptycene-derived triamine (c). The crown ether in (b) allows the preparation of a soluble empty cage for a porous liquid. The structure in (a) was calculated, while the structures in (b) and (c) are crystal structures. Hydrogen atoms are omitted for reasons of clarity.

visibly bubble. The reliable and facile preparation of such cages by dynamic covalent chemistry thus makes it possible to explore their use in novel applications [138].

Despite the prevalence of imine chemistry, there are alternative strategies for the preparation of covalent cages. The formation of boronates from boronic acids and diols is one such strategy. In this context, Florian Beuerle used design principles closely related to those underlying the formation of the cube shown in Figure 5.81a to prepare an analogous cube from the tribenzotriquinacene derivative **5.84** and bis(boronic) acid **5.85** (Figure 5.82a) [139]. The synthesis is very efficient, giving the product in a yield of 94% by simply mixing **5.84** and 1.5 equivalents of **5.85** in tetrahydrofuran in the presence of molecular sieves. The reason for this high fidelity is the rigidity of **5.84** and the arrangement of the aromatic residues at angles of almost exactly 90°. When eight of these building blocks are placed at the vertices of a cube, the aromatic residues point toward the twelve edges, allowing them to be connected by the linear components through boronate formation. The calculated structure of the resulting product is shown in Figure 5.82b. Various other cage architectures can be realized by combining other bis(boronic) acids and diols [140].

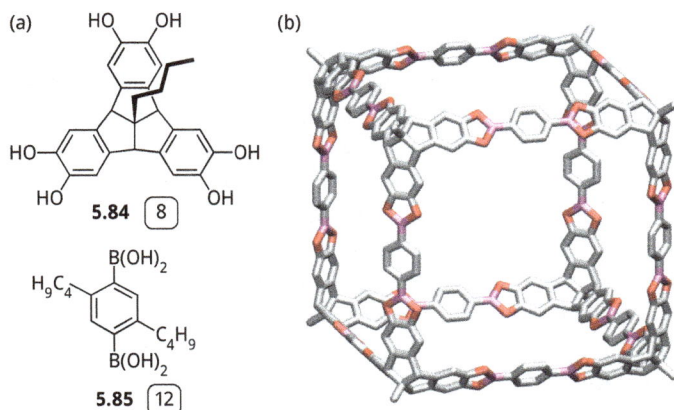

Figure 5.82: Molecular structures of the tribenzotriquinacene **5.84** and the bis(boronic acid) **5.85** (a) and calculated structure of the boronate-derived cage formed from both components (b). In the calculated structure, the protons are not shown, the side chains in the tribenzotriquinacene moieties are truncated, and those in the bis(boronic acid) moieties are omitted for reasons of clarity.

Further reading

Although we have limited the discussion of receptor architectures accessible by dynamic covalent chemistry to macrocycles and cages, there are other types of hollow molecules that can be prepared using this synthetic strategy. Examples are molecular barrels whose cylindrical shape results from a cyclic arrangement of rigid building blocks, which often contain extended π-systems around a deep cavity open at both ends. These receptors are also accessible by metal-directed self-assembly and not only form stable complexes with guest molecules threaded through the cavity, but can also be used in applications such as catalysis and molecular separation. For details, see:

Banerjee R, Chakraborty D, Mukherjee PS. *Molecular barrels as potential hosts: from synthesis to applications.* J. Am. Chem. Soc. 2023, 145, 7,692 – 711.

5.7 Dynamic combinatorial chemistry

5.7.1 Introduction

The motivation for using dynamic covalent chemistry is often to efficiently prepare a compound that would be much more difficult, if not impossible, to obtain using other strategies. Since more than one covalent bond is usually formed in these syntheses, the desired product is never the only possible one. Sometimes one product is so much more stable than potential others that it dominates the equilibrium. In this case, the other products are conceivable but not actually observed and can therefore be called virtual [141]. Sometimes the result of a thermodynamically controlled synthesis is a mixture, as we saw in Section 5.6.2. However, unlike conventional syntheses based on kinetically controlled reactions, the components that make up this mixture are not structurally fixed, but are constantly exchanging their components, a feature that has interesting implications, as will be shown in this section.

Let us first look at the possible strategies for receptor development. The traditional approach would be to design a receptor prototype, perhaps with the aid of computational methods, then prepare the compound, and finally determine its binding properties. Since the first receptor is unlikely to be the optimal one, the initial results typically lead to a redesign and another cycle of receptor optimization. After several rounds of iteration, the receptor may or may not be available. This approach is time-consuming and not necessarily successful. An alternative is to use combinatorial chemistry for receptor identification. This approach is faster than the traditional one because many receptors are synthesized and screened more or less simultaneously, but it is limited to the receptors actually present in the library under investigation. The chances of identifying a very good receptor are high, but there is always the possibility that better receptors may exist that were absent in the library screened. The design of receptors by conventional combinatorial chemistry therefore also has disadvantages.

A third strategy for receptor development is based on mixtures of interconverting molecules generated using reversible reactions. This strategy also involves libraries, but it differs from conventional combinatorial chemistry in several ways. First, the receptors are not synthesized as stable compounds, but as dynamically interconverting species. The structural diversity of the library is therefore not limited to the compounds that *have* been synthesized but to those that *can* be produced. For example, even a single building block can lead to acyclic and cyclic oligomers of different sizes, and the structural diversity becomes even greater when mixtures of building blocks are used for library generation. Some control over the properties of the target compounds can be exerted by selecting building blocks with binding sites that are expected to induce receptor properties, but it is not necessary to have a clear idea of what the product should look like before setting up the libraries. The possibility to access all combinations of building blocks, even those that are not physically present but virtual, makes it very likely that the best receptor will be identified.

The second difference between conventional and dynamic combinatorial chemistry is that receptor assembly and screening are performed simultaneously by adding the substrate to the dynamic library, which acts as a thermodynamic template. This template will shift the product distribution, inducing the formation of complementary receptors at the expense of other compounds that do not form complexes. Another way of looking at this effect is that the template changes the thermodynamic landscape of the system of interconverting species, causing the stabilization of certain receptors and their amplification in the mixture (Figure 5.83). Under ideal conditions, the degree of stabilization correlates with the stability of the corresponding complexes, with the best receptors being amplified the most. Experimentally, the effect of the template on product distribution can be estimated using modern HPLC-MS techniques, even for libraries containing many different members.

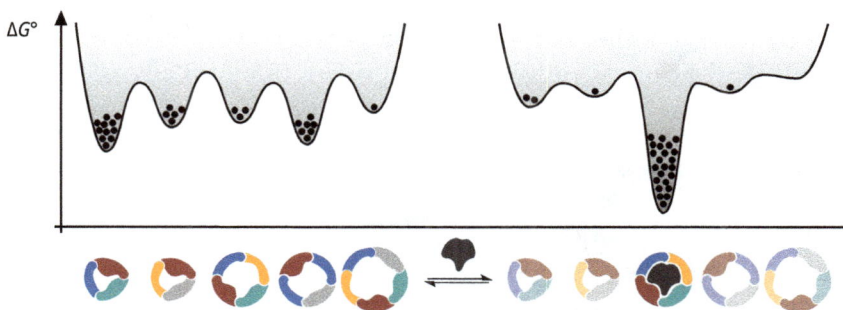

Figure 5.83: Schematic representation of the energetic landscape of a dynamic library of interconverting molecules before and after the addition of a template. The template causes the thermodynamic stabilization of the receptor complementary to the template, with a concomitant shift in the relative proportions of the library members, as symbolized by the black circles located in the minima of the curve.

A final advantage of dynamic covalent chemistry over conventional combinatorial chemistry is the ease with which the receptor, once identified, can be synthesized. This synthesis is performed under the same conditions used to generate the dynamic library, again exploiting the template effect of the substrate, but restricting the building blocks to only those that end up in the product. To isolate this product as a stable compound, it is only necessary to stop the interconversion, for example by changing the reaction conditions.

The concept of dynamic combinatorial chemistry, developed simultaneously and independently in the groups of Jean-Marie Lehn and Jeremy K. M. Sanders, is not limited to the identification of a receptor for a substrate but can also be used for other purposes [142]. The possible applications differ in the roles of the template and the library members, but the underlying principles are the same in all cases.

If the participating compounds can be classified as receptors (convergent binding sites) and substrates (divergent binding sites), two strategies can be distinguished –

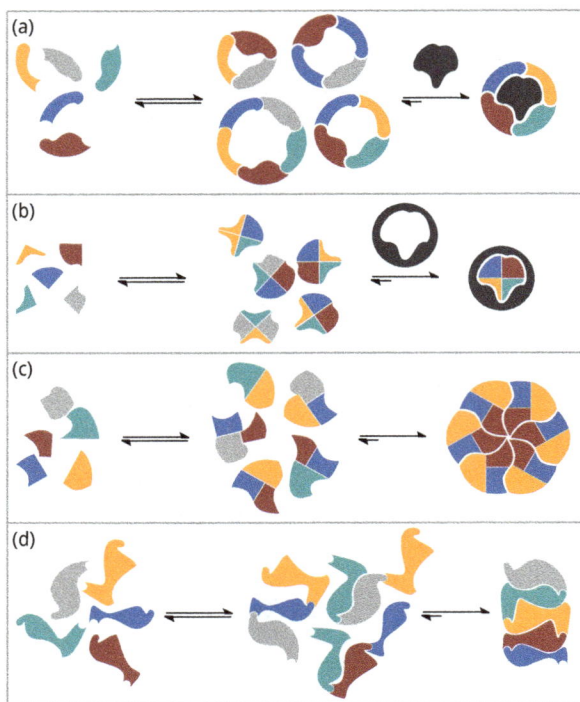

Figure 5.84: Possible applications of dynamic combinatorial chemistry include the identification of a receptor for a given substrate (a), the identification of a substrate for a receptor (b), the selection of binding partners in self-assembling systems (c), and the control of the sequence and length of oligomers of different subunits by intramolecular interactions (d).

molding and casting. Molding involves forming the receptor like a mold around the shape of the substrate (Figure 5.84a). The reverse strategy is called casting, because a cast is made of the receptor cavity (Figure 5.84b). If all library members have divergent binding sites, they can interact by self-assembly. In this case, one compound in the dynamic library can act as a template to produce more of the compounds with which it interacts (Figure 5.84c). If intramolecular recognition processes are included in this classification, a fourth possibility is the selection of the preferred sequence and length of oligomers by intramolecular interactions (Figure 5.84d).

At this point, the question may arise as to the difference between dynamic covalent chemistry and dynamic combinatorial chemistry. Focusing on product formation, is it correct to say that any reaction that leads to a single product belongs to dynamic covalent chemistry, while processes that lead to mixtures belong to dynamic combinatorial chemistry? Probably not, because whether a reaction yields a single product or a mixture can change depending on the reaction conditions. Moreover, if we consider the formation of virtual products, there is no reaction that does not have a combinatorial element. Another criterion for differentiation could be that the combinatorial aspect of dynamic combinatorial chemistry is reflected in the use of mixtures of building blocks. If only the building blocks needed to access the desired product are used, the approach belongs to dynamic covalent chemistry, while the use of mixtures of products, some of which may not end up in the product, makes the approach combinatorial. But how do we then classify reactions in which a single building block leads to mixtures (libraries) of different products? Does not this reaction also have a combinatorial aspect? A convenient approach would be to make the distinction based on the use of templates, with only dynamic combinatorial chemistry requiring templates, but this distinction makes it difficult to classify the process shown in Figure 5.84d. Can an intramolecular interaction really be considered a template effect, which would make the approach a dynamic combinatorial one, or would it not be equally true to say that the product selected by the intramolecular interactions is simply the most stable one, which would make the term dynamic covalent chemistry more appropriate? Thus, there is no easy way to make a clear distinction, and it is perhaps best to conclude that dynamic combinatorial chemistry is the overarching concept that uses dynamic covalent chemistry as a tool.

As in dynamic covalent chemistry, the key to dynamic combinatorial chemistry is the reversible reaction that mediates the interconversion of the library members. This reaction must meet several requirements for the concept to work:

- The reaction must be fast enough to be able to generate a dynamic library and follow the effects of templates on library composition over a reasonable timescale, typically within hours or days.
- There should be no side reactions between the functional groups involved in the reversible reaction and those that mediate the interactions between the template and the library members.
- Since the equilibration and templation occur simultaneously, the conditions required for the exchange to work should be mild enough not to interfere with the interactions between the template and the library members.
- All library members and their complexes should remain in solution during the equilibration to avoid kinetic traps.
- Ideally, all library members should have a similar thermodynamic stability because it is energetically costly to shift the equilibrium if the library is strongly biased toward the formation of a few or even a single product in the absence of the template.
- Finally, it must be possible to freeze the reaction, thereby converting the library members into stable products that can be isolated and characterized.

Given these requirements, some of the reversible reactions shown in Figure 5.70 are not ideal. For example, ester formation or transesterification usually requires relatively harsh conditions that are incompatible with the molecular recognition processes required for the templates to work. Boronic esters, on the other hand, are often too labile to allow further processing. The most commonly used reversible reactions in dynamic combinatorial chemistry are imine formation (usually followed by reduction and isolation of the corresponding amines), hydrazone formation, and disulfide formation, which are shown in Figure 5.85 in the form of the corresponding exchange processes. A few other reactions are possible, but have been used less frequently [142].

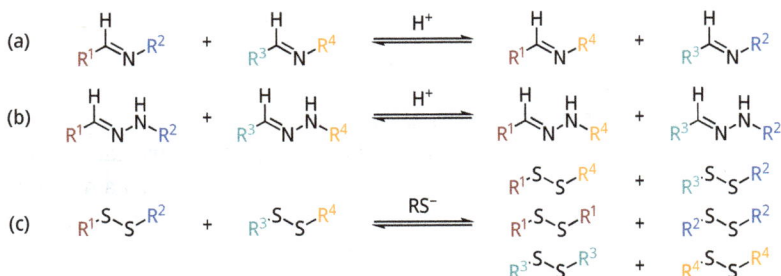

Figure 5.85: Commonly used exchange reactions in dynamic combinatorial chemistry.

Does a template always amplify the best binder?

A key question is whether the best binder is always chosen by the template. Perhaps surprisingly, the answer is no. The reason is that the dynamic library responds to the presence of the template by adopting the thermodynamically most favorable state of *the entire system* [143]. As a result, situations may arise where a template leads to the amplification of a good, but not necessarily the best binding partner, as illustrated by the following examples.

Consider a dynamic library containing macrocyclic receptors formed from a single building block (Figure 5.86a). In this library, the cyclic hexamer should form the most stable complex, but the cyclic trimers should also have weak affinity for the template. One cyclic hexamer but two cyclic trimers are formed from the same number of building blocks, and which of these macrocycles is most strongly amplified depends on whether the formation of one complex of the hexamer or two complexes of the trimer leads to an overall more favorable thermodynamic state. If the $\Delta G^0_{hexamer}$ associated with the formation of the hexamer complex is significantly more negative than twice the ΔG^0_{trimer}, the better receptor wins. However, if $2 \times \Delta G^0_{trimer} < \Delta G^0_{hexamer}$, the trimer is preferentially amplified and not the receptor that forms the more stable complex.

A similar situation arises in libraries where different building blocks form macrocyclic receptors of the same size. For example, two building blocks yield four macrocyclic trimers in a statistical ratio of 1:3:3:1 (Figure 5.86b). One of the homotrimers should be the best receptor, but the heterotrimer containing two building blocks of the best receptor also interacts with the template, albeit with a weaker affinity. Which macrocycle will be selected by the template again depends on whether the homotrimer or the statistically favored three equivalents of the mixed receptor bind the template with an overall more negative ΔG^0. These considerations thus show that if there are compounds

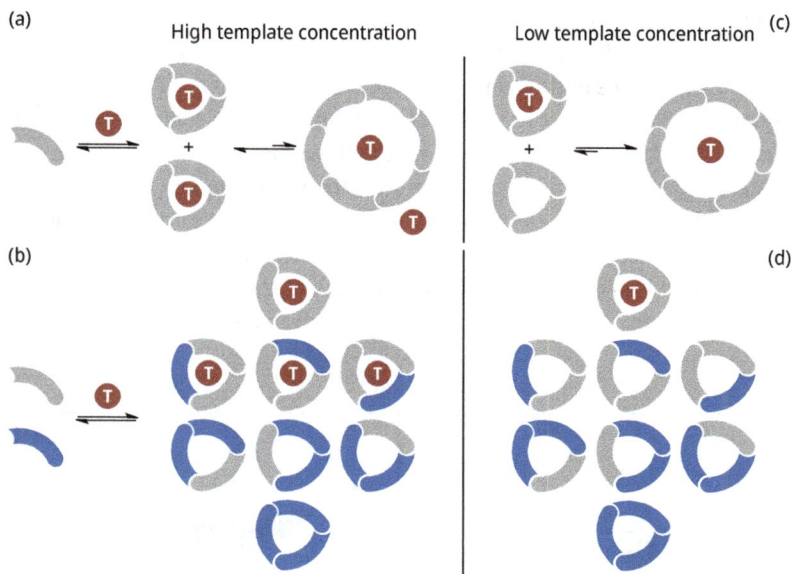

Figure 5.86: Scenarios illustrating the competition of different receptors in a dynamic library that differ in structure and affinity for the binding to the template. The receptors in (a) are macrocycles of different sizes, of which the hexamer should have a higher affinity than the trimer. In (b), cyclic trimers containing different subunits interconvert, with the homotrimer having a higher affinity than one of the heterotrimers. The equilibria in (c) and (d) show how these situations change when the template concentration is reduced.

in a dynamic library that bind to the template and have a statistical advantage over the best receptor, these compounds can be amplified the most.

There are experimental ways to avoid this unwanted result. The examples above show that the weaker binders contribute to the total ΔG^0 of the reaction only if there are enough template molecules present. If the number of template molecules in the solution is too small to include all the weaker binders in the selection process, as shown in Figures 5.86c and d, direct competition occurs between receptors with different affinities. In this case, the best receptor wins. Therefore, the lower the concentration of the template, the more likely it is that the best receptor will be identified. That this assumption is indeed correct has been verified computationally by predicting the responses of simulated dynamic combinatorial libraries to the presence of templates [144]. In this approach, a 322-member library of interconverting compounds was generated from a set of virtual building blocks, with each library member present in amounts that depend on the equilibrium constants of the exchange reactions, statistical factors, and building block concentrations. The binding constants for all possible complexes were randomly assigned and the change in library composition, specified in terms of amplification factors that denote the ratios of the concentrations of each library member before and after template addition, was then estimated. Figure 5.87a shows the result of such a calculation for a template concentration of 10 mM. In this case, the library responds strongly to the presence of the template by enriching a number of receptors and depleting many others. Notably, the receptor whose binding to the template is associated with the most negative ΔG^0 is not amplified at all, while the receptor that forms the third most stable complex is amplified the most. When the template concentration is reduced to 1 mM, the distribution of amplification factors for the same library looks completely different (Figure 5.87b). In this case, the concentrations of most library members are not or only weakly affected by the presence of the template, but the best receptor is amplified the most, confirming the above considerations.

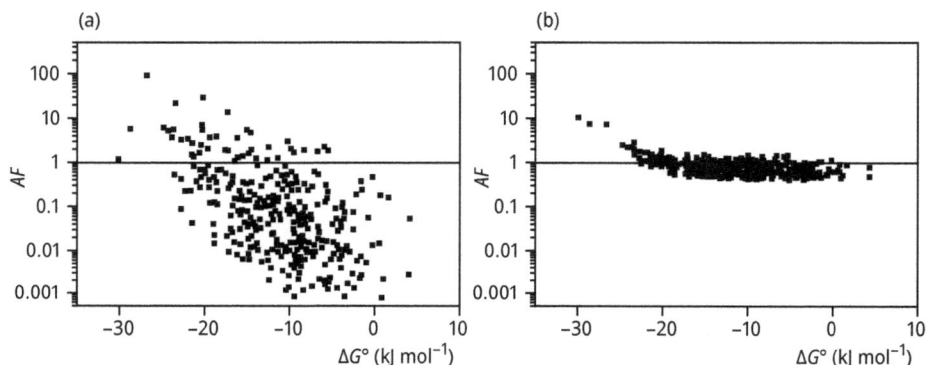

Figure 5.87: Relationship between the amplification factor *AF*, which describes the change in the concentration of a library member induced by the presence of the template, and the Gibbs free energy ΔG^0 of the complex between the template and the corresponding library member in simulated dynamic libraries differing only in the concentration of the template. The template concentration is 10 mM in (a) and 1 mM in (b).

Limiting the template concentration is not without negative consequences, as it also affects the absolute amount to which a receptor can be amplified. Thus, a good receptor may not be detected at low template concentrations because its concentration is below the detection limit. Therefore, it is usually advisable to perform a series of experiments with different template concentrations to increase the chances of identifying the best receptor for a given template.

The following examples show how dynamic combinatorial chemistry is used to select substrates for a given receptor (casting), receptors for a given substrate (molding), and complementary components in a self-assembly process. A selection process involving internal templation is the formation of the macrocyclic 15-mer discussed at the end of Section 5.6.2 (Figure 5.75).

5.7.2 Casting

Casting usually involves identifying a binder for an enzyme whose active site serves as a negative from which a cast is made. Of the range of reversible reactions available, relatively few are suitable for this purpose because the exchange must proceed under conditions where the enzyme is stable. The acidic conditions required for hydrazone exchange, for example, are often incompatible with sensitive biomolecules. The most commonly used reactions are disulfide exchange and imine exchange, the latter followed by reduction of the imines to amines.

In an early study, the Lehn group used dynamic combinatorial chemistry to identify an inhibitor for the enzyme carbonic anhydrase, which converts carbon dioxide to carbonic acid [145]. In this work, three aldehydes and four amines were combined to form twelve imines, which were then reduced to amines with sodium cyanoborohydride (Figure 5.88). When this reaction was performed in the presence of the enzyme, an amine with a structure similar to hexyl 4-sulfamoylbenzoate, a known inhibitor of carbonic anhydrase, was amplified. This amplification was less pronounced when the inhibitor was present during equilibration, indicating that the selection process involved the active site of the enzyme.

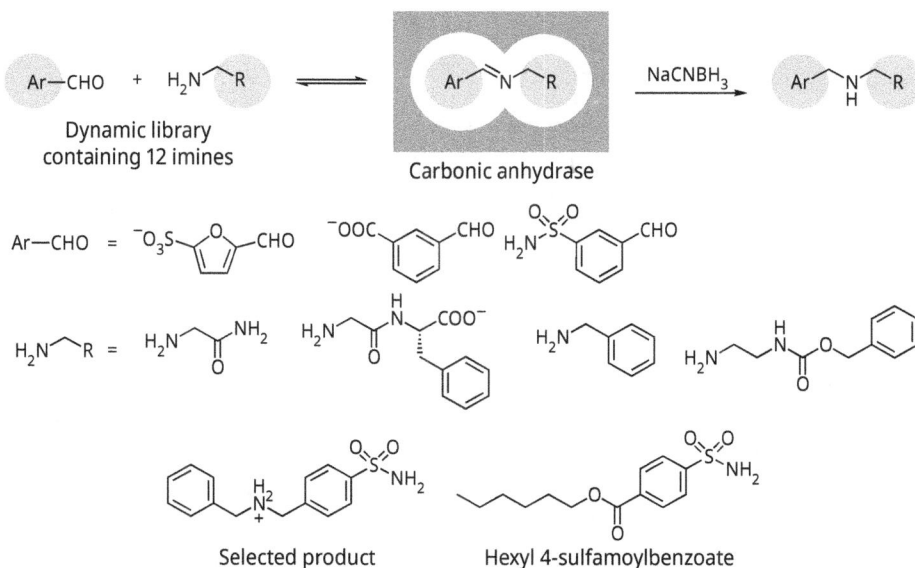

Figure 5.88: Selection of an inhibitor for carbonic anhydrase from a dynamic library of twelve imines, which are reduced to give the corresponding amines. Performing the equilibration in the presence of hexyl 4-sulfamoylbenzoate reduces the amplification of the selected amine.

A second example is shown in Figure 5.89, involving a dynamic library of 21 interconverting monosaccharide-derived disulfides [146]. A bis(mannoside) disulfide is amplified when the carbohydrate-binding protein concanavalin A is present during this exchange reaction, consistent with the selectivity of concanavalin A for oligosaccharides with terminal α-D-mannosyl residues. Scrambling is relatively fast at physiological pH, but stops when the pH is lowered to 4.

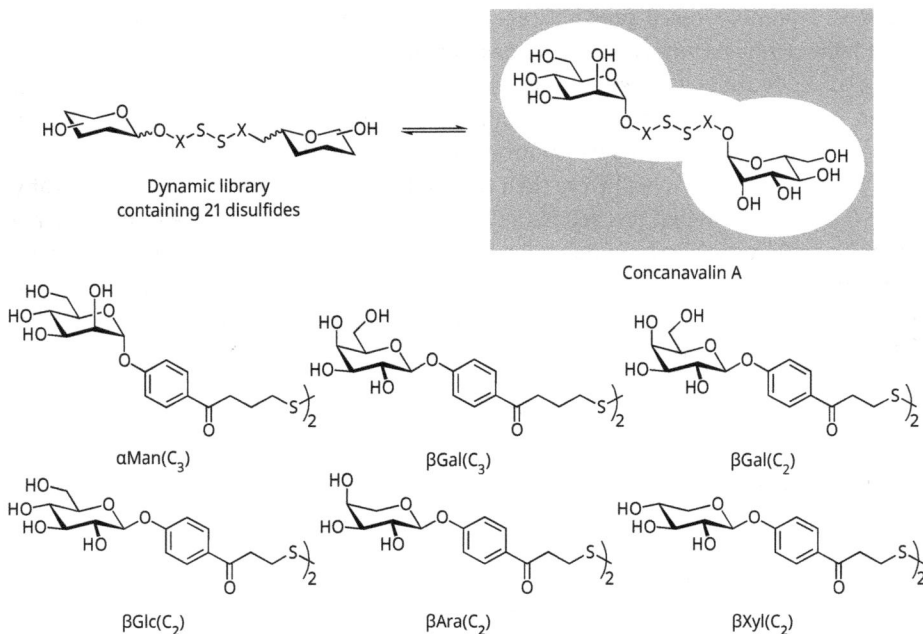

Dynamic library
containing 21 disulfides

Concanavalin A

αMan(C₃)   βGal(C₃)   βGal(C₂)

βGlc(C₂)   βAra(C₂)   βXyl(C₂)

Figure 5.89: Selection of a binder for concanavalin A from a dynamic library of 21 monosaccharide-derived disulfides.

In these examples, substrate selection and freezing of the exchange reaction occur independently. Both processes can be combined if the enzyme not only serves as a template, but also mediates the irreversible stabilization of the selected library components. An application of this concept is the reaction shown in Figure 5.90. Here, different aldehydes are equilibrated with a nitroalkane in the presence of a lipase and an acyl donor [147]. Under these conditions, chiral acetylated nitroaldols are formed, with the structure of the preferred products reflecting not only which substitution pattern of the aldehyde is optimal for the interactions with the enzyme, but also which product enantiomer is acetylated most rapidly. This strategy thus adds a combinatorial element to the separation of enantiomers by dynamic kinetic resolution and has therefore been termed dynamic combinatorial resolution.

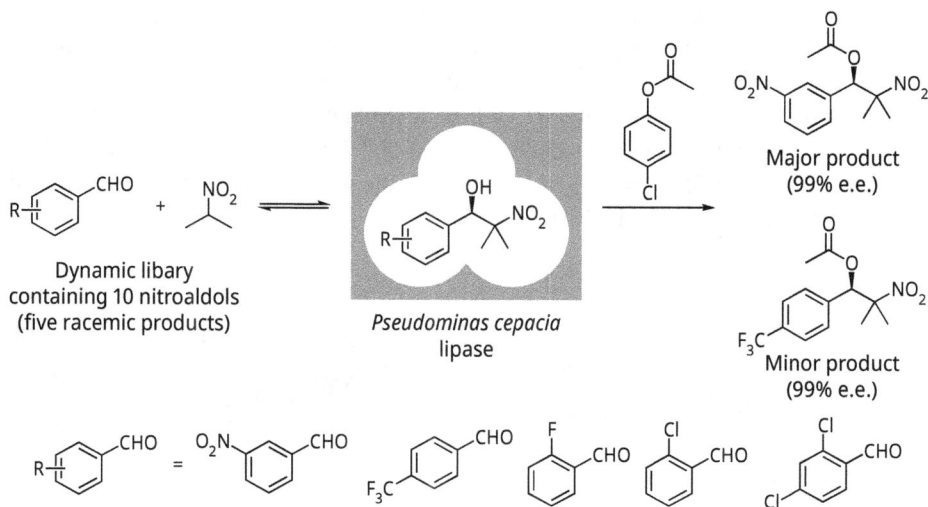

Figure 5.90: Selection of enantiopure acetylated nitroaldols by equilibration of a dynamic library generated from five aromatic aldehydes and one nitroalkane in the presence of a lipase and an acyl donor.

5.7.3 Molding

Most receptors in supramolecular chemistry are macrocyclic (Section 4.1), and the identification of receptors by dynamic combinatorial chemistry is therefore often based on building blocks with functional groups that allow oligomerization followed by cyclization once the chain is long enough to form an unstrained ring. In addition, the building blocks usually contain structural elements of known receptors to increase the chances that the products will interact with the substrate. Groundbreaking studies in this regard have been carried out in the group of Jeremy K. M. Sanders. An example is compound **5.86**, which contains an acetal group on the aromatic subunit and a hydrazide group at the opposite end of the chain [148]. After cleavage of the acetal, the released aldehyde can react with hydrazide groups of other molecules to form rings of various sizes whose cyclically arranged aromatic subunits and carbonyl groups are expected to induce an affinity for cations.

Initially, treatment of **5.86** with trifluoroacetic acid yields a series of cyclic *N*-acyl hydrazones from the dimer to the 15-mer. This product distribution changes with time until thermodynamic equilibrium is reached after two days, when only the cyclic dimer (88%) and the cyclic trimer (11%) are still present. The entropically favored formation of the smaller ring is consistent with the principles of self-assembly described earlier. The subsequent addition of quaternary ammonium ions to this dynamic library induces a pronounced change in the product distribution. *N*-Methylquinuclidinium iodide and acetylcholine chloride both shift the equilibrium in favor of the cyclic trimer, suggesting that these cations template the formation of the macrocycle to which they bind. Performing the same reaction on a larger scale and interrupting the exchange by neutralizing the reaction mixture after equilibrium is reached allows isolation of the selected cyclic trimer. This trimer does indeed interact with both templates, but forms the more stable complex with acetylcholine (K_a = 230 M^{-1} vs. 150 M^{-1} for the N-methylquinuclidinium complex in CDCl$_3$/CD$_3$OD, 95:5 (v/v)), consistent with the stronger effect of acetylcholine on the product distribution (Figure 5.91).

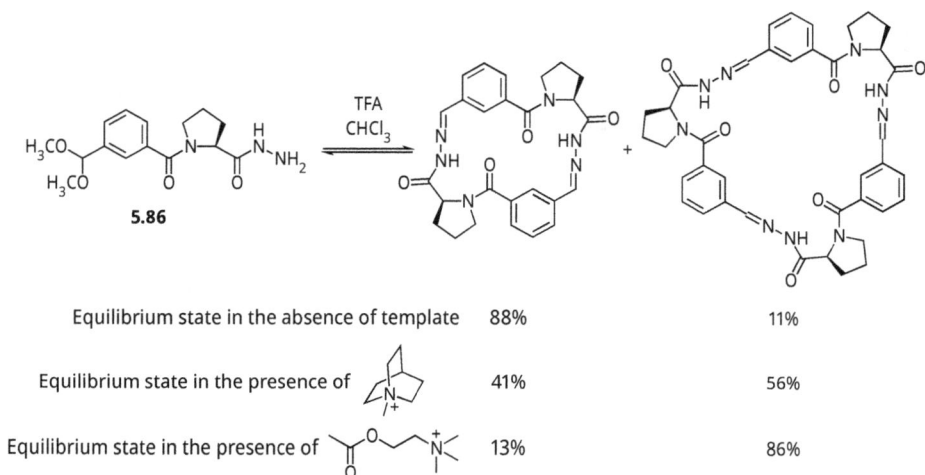

Equilibrium state in the absence of template	88%	11%
Equilibrium state in the presence of	41%	56%
Equilibrium state in the presence of	13%	86%

Figure 5.91: Molecular structure of building block **5.86** and structures of the corresponding cyclic dimer and trimer. The changes in the library composition induced by N-methylquinuclidinium iodide and acetylcholine chloride are also specified.

A significantly more efficient acetylcholine receptor can be obtained from building block **5.87** whose macrocyclization proceeds in a similar manner to that of **5.86** [149]. In this case, the library composition is complex, consisting mainly of the cyclic dimer, trimer, and tetramer, with small amounts of the pentamer, hexamer, and higher oligomers also present. The addition of acetylcholine results in the appearance of a product absent in the original library, which eventually makes up approximately 70% of the mixture. This receptor consists of six subunits, which are, however, not part of a single ring, but of two interlocked cyclic trimers. The best receptor in this library, which is only virtually present in the absence of the template, is therefore a catenane that binds to acetylcholine in CHCl$_3$/DMSO, 95:5 (v/v) with an impressive log K_a of 7.1. Due to the complexity of its structure, the exact mode of acetylcholine binding is not entirely clear, although it is likely that the binding pocket is located between the two rings (Figure 5.92).

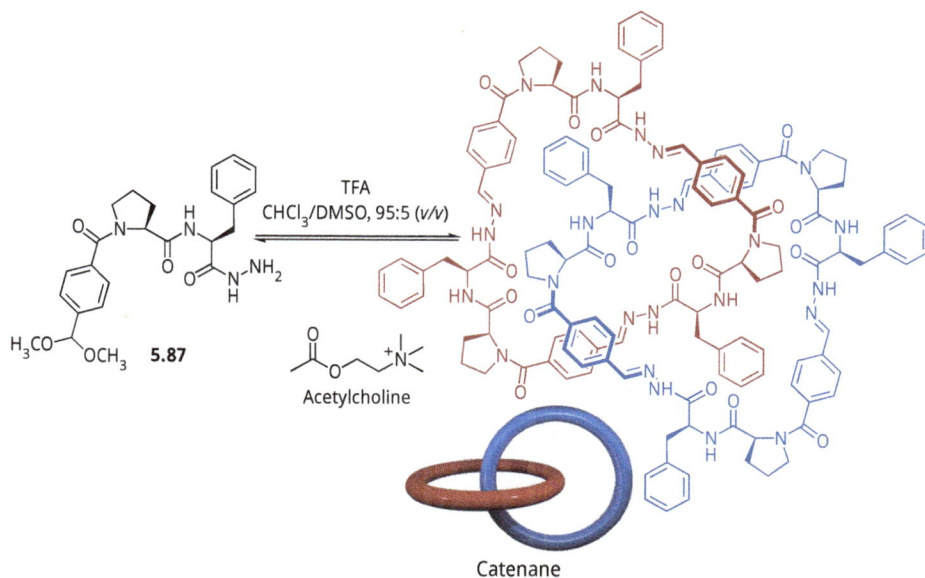

Figure 5.92: Molecular structure of building block **5.87** and schematic structure of the catenane consisting of two interlocked cyclic trimers amplified by acetylcholine chloride.

Catenanes and related structures are the subject of Chapter 7. They are often synthesized efficiently using the concepts of supramolecular chemistry. However, when faced with the task of designing a receptor for acetylcholine, it is unlikely that the first choice of a supramolecular chemist would be to prepare a catenane. The above example thus demonstrates the potential of dynamic combinatorial chemistry to deliver completely unexpected receptors if they are accessible from the chosen building blocks and if their interaction with the template leads to a thermodynamically favored state [150].

In addition to *de novo* receptor development, dynamic combinatorial chemistry also allows for the evaluation and potential improvement of existing receptors. In Section 4.1.5 we saw that cyclophane **4.25** has a high cation affinity in 10 mM borate buffer [151]. Could a similar receptor be obtained from a dynamic library of interconverting macrocycles? A dynamic combinatorial library generated from the three thiols **5.88**, **5.89**, and **5.90** does indeed contain **5.91**, the closest analog of **4.25**, but only in small amounts (Figure 5.93) [152]. The overall composition of the library is complex, comprising more than 45 compounds, partly because **5.90** is chiral and many products are therefore formed as mixtures of stereoisomers.

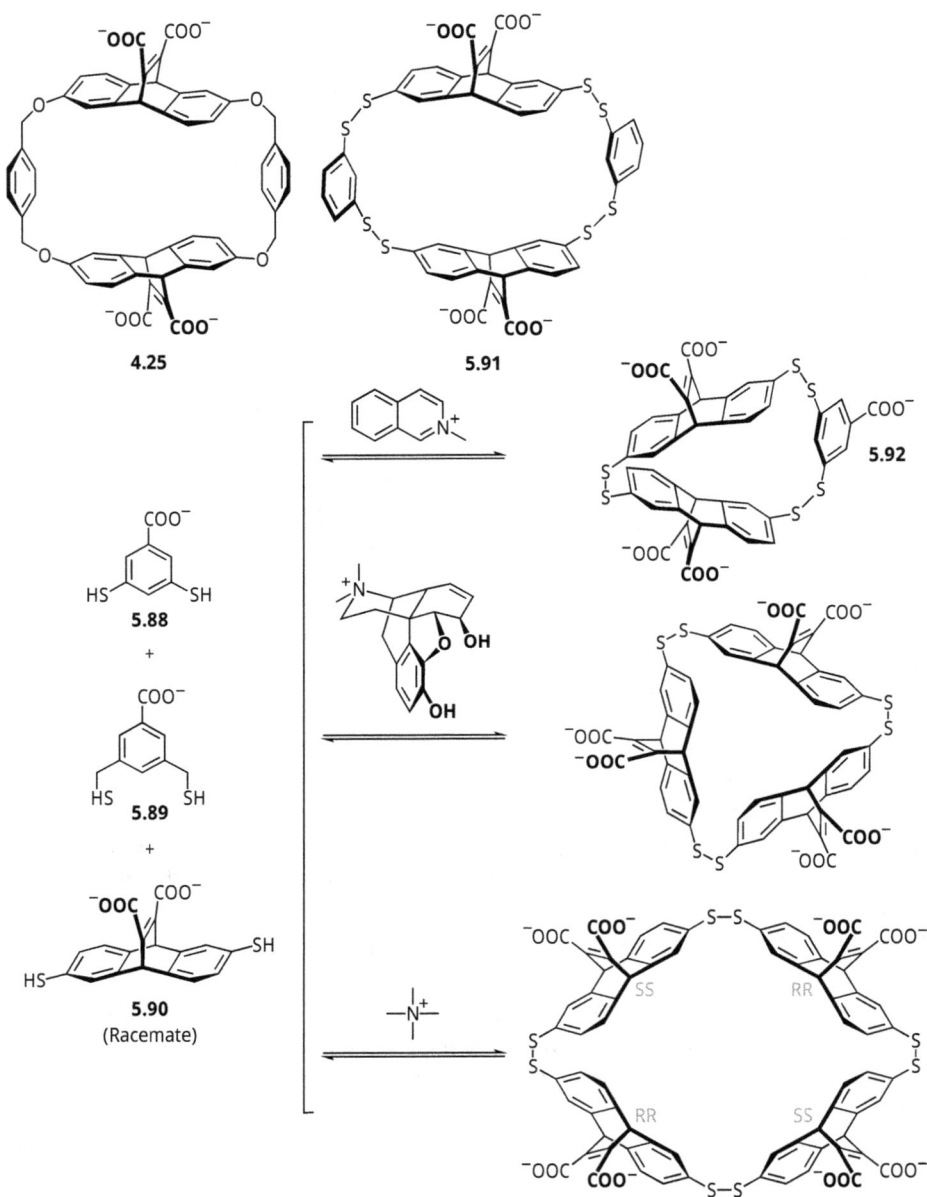

Figure 5.93: Molecular structure of building blocks **5.88**, **5.89**, and **5.90** and of the disulfide analog of receptor **4.25**, compound **5.91**. The receptor selected by *N*-methylisoquinolinium iodide is **5.92**. The structures of macrocyclic receptors amplified by other cationic templates are also shown.

N-Methylisoquinolinium iodide induces the amplification of one of these macrocycles (also as a mixture of stereoisomers) whose structure differs from that of **5.91** by the number of subunits. This amplified receptor **5.92** binds the *N*-methylisoquinolinium cation in 10 mM borate buffer with a $\log K_a$ of 5.4, which is practically the same affinity as determined for the *N*-methylisoquinolinium complex of **4.25** under the same conditions ($\log K_a$ = 5.3) [151]. Thus, although the receptor properties of **4.25** could not be improved, dynamic combinatorial chemistry demonstrated that a smaller analog of this cyclophane is also an efficient receptor.

The amplification of **5.92** by *N*-methylisoquinolinium iodide is highly specific. The use of other cationic templates, such as a morphine derivative or tetramethylammonium iodide, results in the amplification of other macrocycles from the same building blocks. In the case of the tetramethylammonium cation, the tetrameric receptor forms stereoselectively because only the selected diastereomer can fold into a compact structure with a cavity suitable to accommodate the small ammonium ion [153].

Other examples of receptors obtained by the molding strategy are the orthoester-derived cryptands introduced by Max von Delius (Section 4.1.2). The synthesis of these compounds also proceeds under thermodynamic control and involves the use of template effects. The cryptand **5.93a** is obtained, for example, by treating 1,1,1-trimethoxyethane with three equivalents of diethylene glycol in chloroform under the influence of catalytic amounts of trifluoroacetic acid and one equivalent of a sodium salt containing a large lipophilic anion (Figure 5.94) [154]. Under these conditions, the three alkoxy groups in the orthoester are shuffled, resulting in a library of different products. Of these products, **5.93a** forms predominantly because this cryptand is thermodynamically stabilized by the sodium ion.

Allowing the orthoester exchange to proceed in the presence of a mixture of diols adds a combinatorial component to this synthesis [155]. In this case, cryptands of different sizes are formed, and control of product formation is achieved by using the template effects of different metal ions. When 1,1,1-trimethoxyethane is reacted with two equivalents of diethylene glycol and one equivalent of triethylene glycol in the presence of a potassium salt, cryptand **5.93b** with a slightly larger cavity than **5.93a** is preferentially formed (Figure 5.94). Inverting the ratio of diethylene glycol to triethylene glycol and using a rubidium salt as a template yields **5.93c** as the main product, and the largest cryptand **5.93d** is obtained by reacting 1,1,1-trimethoxyethane with three equivalents of triethylene glycol in the presence of a cesium salt. Accordingly, product formation correlates with the ionic radius of the templating metal ion.

5.7.4 Self-assembly

A remarkable system in which self-assembly is responsible for the amplification of members of a dynamic combinatorial library has been reported by Sijbren Otto. This system is based on building block **5.94**, which consists of a 1,3-benzenedithiol core con-

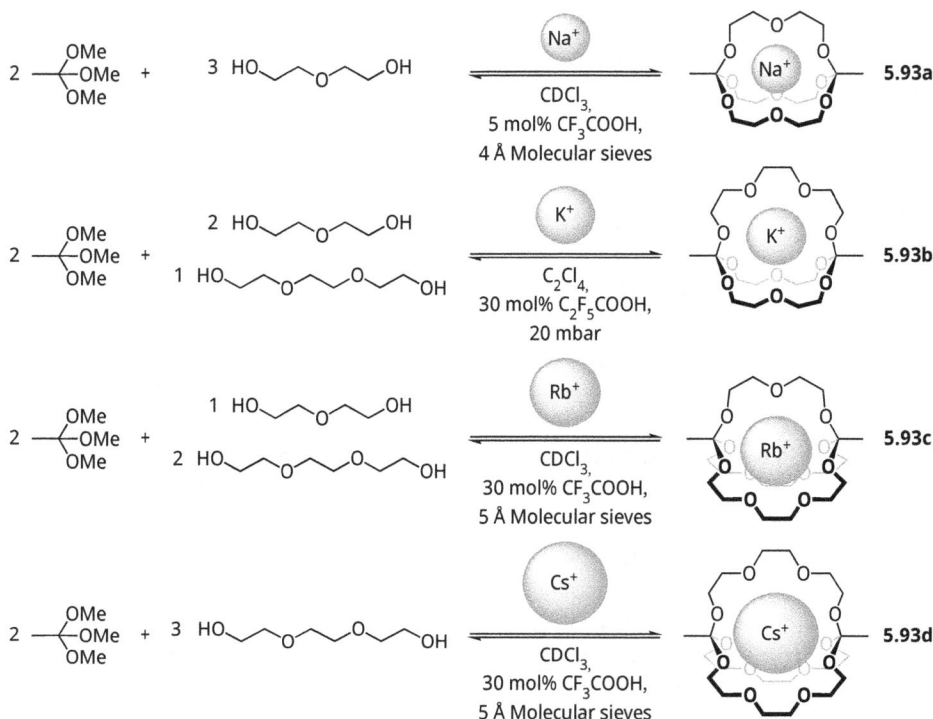

Figure 5.94: Formation of orthoester-derived cryptands **5.93a–d** from 1,1,1-trimethoxyethane and various diols in the presence of suitable metal ions as templates. Product formation correlates with the size of the cation when diethylene glycol and triethylene glycol are present in the required ratios.

taining a short pentapeptide chain with the sequence Gly-Leu-Lys-Leu-Lys in the 5-position (Figure 5.95) [156]. The alternating sequence of polar and nonpolar side chains predisposes this peptide to aggregate into amyloid-like β-sheets. Therefore, macrocycles formed from **5.94** by disulfide formation should self-assemble, resulting in stacks of rings. The ring size of the product preferentially formed when **5.94** is equilibrated in borate buffer at pH 8 is sensitively dependent on the conditions. If a solution is allowed to stand, cyclic trimers and tetramers dominate at equilibrium, which are the macrocycles normally formed when 3-benzenedithiol derivatives are cyclized under thermodynamic control. When the solution is shaken, the cyclic hexamer appears with time and is the major product after about 20 days. When the solution is stirred, the cyclic heptamer eventually dominates the equilibrium. The concentrations of both products increase exponentially with time, indicating that they autocatalytically mediate their own formation.

1,3-Benzenedithiol derivatives, without the peripheral substituents or with substituents unable to form β-sheets do not exhibit this behavior, showing that the ability to self-assemble into tubular structures is crucial for the formation of the larger rings.

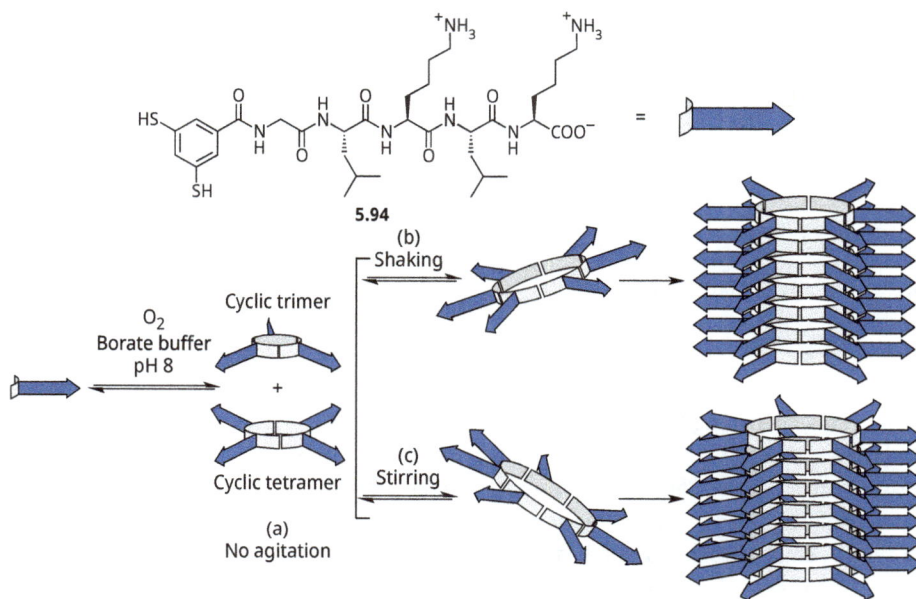

Figure 5.95: Molecular structure of building block **5.94** and schematic representation of the products formed in solutions that are left standing (a), or are shaken (b), or stirred (c).

The stability of the self-assembled tubes correlates with the number of peptide residues. Three or four peptides in the cyclic trimer and tetramer are insufficient to form stable aggregates, but the larger macrocycles with more side chains stack under the conditions of the experiments, as evidenced by the presence of fibrillar structures in cryo-transmission electron microscopy images of the solutions. Thus, if these large rings are formed by a random event, they are trapped by self-assembly. However, the formation of stacks alone does not explain the observed autocatalysis, since tubes can only grow at both ends, which should cause the concentrations of the larger macrocycles to increase linearly, not exponentially. Exponential self-replication occurs only if the stacks are continuously broken to create new ends to which more rings can attach. This breaking of aggregates requires agitation. If left standing, stacks that might form from larger macrocycles will not break and therefore grow too slowly to precipitate before the building blocks are consumed in the exchange reaction. Mechanical agitation, however, causes the stacks to break, explaining autocatalysis. The force generated by shaking is just enough to break the stacks of the cyclic hexamers, whose concentration therefore increases exponentially. Increasing the mechanical stress by stirring also breaks the stacks of the cyclic heptamers, which are now preferentially produced because they self-assemble more efficiently.

It is important to emphasize that the amplification of the large macrocycles is not a thermodynamic effect. Intrinsically, these macrocycles are less stable than the smaller ones, as evidenced by their absence in the nonagitated solutions. However, by kinetically trapping the large rings within the self-assembled stacks, they are prevented from further participation in the exchange reaction, demonstrating that kinetic products can result from a process in which all steps are reversible. The relevance of this work thus extends beyond dynamic combinatorial chemistry, showing that out-of-equilibrium states, which must have played a role in the evolution of life, can emerge spontaneously.

5.8 Systems chemistry

Systems theory is an interdisciplinary field of science that studies the behavior of complex systems whose many components do not behave independently, but interact in many ways, according to certain rules. Our world is full of such systems. Examples include the climate, the economy, the World Wide Web, and biological ecosystems. These systems are adaptive. For example, removing a food source from an ecosystem or adding a predator to it affects the network of the interacting species at many different levels. Moreover, small changes at one level can have massive effects at another level. This feature is often associated with the term *butterfly effect*, which implies that a drastic effect in one part of a complex system, such as a tornado in one part of the world, could have been triggered weeks earlier and in a completely different place by something as innocuous as the flapping of the wings of a butterfly. Complex systems are studied intensively in many scientific disciplines, including biology, physics, mathematics, and computer science, but have long been ignored in chemistry. Why is that so?

Are mixtures of molecules always messy?

One reason for this may be that the synthetic targets of synthetic chemistry generally need to be pure to be used, for example, as pharmaceuticals. Reactions that lead to mixtures of products are therefore usually avoided. Another factor is that the analysis of complex mixtures is technically challenging and has only recently become possible with the development of sufficiently sensitive analytical techniques. These developments had a strong impact on the rapid development of dynamic combinatorial chemistry, and work in this context soon showed that networks of interconverting compounds can respond unexpectedly to an external stimulus. Recall how widely the amplification factors vary when a template is added to a dynamic library at a certain concentration, and that it is not necessarily the best receptor that is amplified (Figure 5.87a). Another example is shown in Figure 5.96 [157].

Like the simulated dynamic library described in Section 5.7.1, the library shown in Figure 5.96a contains 322 different interconverting oligomeric species, ranging from dimers to tetramers. The randomly assigned template affinities of these library members vary, with most library members having weak affinities, while a few members bind strongly. In many simulations, the template most strongly amplifies the best

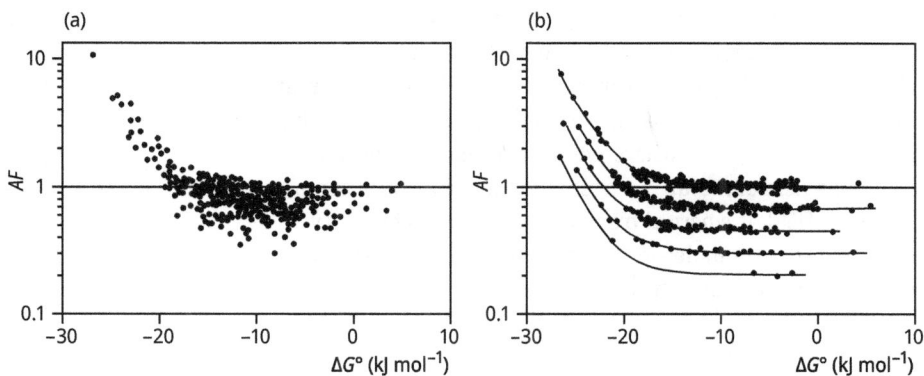

Figure 5.96: Correlation between the amplification factor *AF* and the affinity for the template in terms of ΔG^0 of the components of a simulated dynamic combinatorial library, each member of which interacts with the template with a randomly assigned binding constant. The graph in (b) shows the emergence of patterns under certain conditions. The template concentration is 1 mM in both cases.

binder, affects the concentration of other strong binders to a lesser extent, and does not strongly affect the concentrations of most library members. However, certain distributions of template affinities cause the simulation result to be drastically different. An example is shown in Figure 5.96b, where the amplification factors are clustered in lines. Thus, patterns emerge in this library, which is a hallmark of complex systems. The reason for this pattern is that the incorporation of certain building blocks into the library members penalizes their amplification in the presence of the template by a relatively large and constant factor. The top line in Figure 5.96b describes the amplification of the library members that do not contain any of these building blocks, the members on the next line contain one of them, then two, and so on.

Another and perhaps more intuitive example is the behavior of a dynamic library made from five different building blocks A through E that can only form dimeric species. Compound AB should bind strongly to template T1 and CD should bind with the same affinity to template T2. Adding both templates to the dynamic library will therefore lead to the amplification of AB and CD, and since there are no building blocks left for E to react with, it must form the homodimer EE. As a result, the concentration of EE increases greatly with the addition of T1 and T2, even though this compound does not bind to either template.

The behavior of dynamic libraries could therefore hold many surprises. One of the most fascinating aspects of complex systems is the emergence of properties that transcend those of the individual components, with the paramount example in this context undoubtedly being the emergence of life from simple molecules. We will never know exactly how life originated on Earth because the exact conditions are unknown and therefore cannot be reproduced. However, it is reasonable to assume that external conditions and the presence of networks of interacting molecules on the early planet favored the emergence of complex chemical systems that exhibited some of the characteristics of life, such as organization, replication, metabolism, and evolution. By studying how mixtures of molecules behave under various stimuli, we can try to mimic such systems and learn about the fundamental mechanisms that control their behavior. In this way, we may one day reach a point, where we have a much better understanding of how life began. The young field of research that studies the behavior of complex systems is called systems chemistry [158]. It *emerged* from supramolecular chemistry and dynamic covalent chemistry and promises to deliver fascinating insights into the behavior and possible functions of complex molecular networks. Stay tuned!

Bibliography

[1]   Fraenkel-Conrat H, Williams RC. Reconstitution of active tobacco mosaic virus from its inactive protein and nucleic acid components. Proc. Natl. Acad. Sci. U. S. A. 1955, 41, 690–8.

[2]   Bhyravbhatla B, Watowich SJ, Caspar DL. Refined atomic model of the four-layer aggregate of the tobacco mosaic virus coat protein at 2.4-Å resolution. Biophys. J. 1998, 74, 604–15.

[3]   Klug A. The tobacco mosaic virus particle: structure and assembly. Philos. Trans. R. Soc., B. 1999, 354, 531–5.

[4]   Montoro-García C, Camacho-García J, López-Pérez AM, Bilbao N, Romero-Pérez S, Mayoral MJ, González-Rodríguez D. High-fidelity noncovalent synthesis of hydrogen-bonded macrocyclic assemblies. Angew. Chem. Int. Ed. 2015, 54, 6780–4.

[5]   Montoro-García C, Camacho-García J, López-Pérez AM, Mayoral MJ, Bilbao N, González-Rodríguez D. Role of the symmetry of multipoint hydrogen bonding on chelate cooperativity in supramolecular macrocyclization processes. Angew. Chem. Int. Ed. 2016, 55, 223–7.

[6]   Rowan SJ, Hamilton DG, Brady PA, Sanders JKM. Automated recognition, sorting, and covalent self-assembly by predisposed building blocks in a mixture. J. Am. Chem. Soc. 1997, 119, 2578–9.

[7]   Krämer R, Lehn JM, Marquis-Rigault A. Self-recognition in helicate self-assembly: spontaneous formation of helical metal complexes from mixtures of ligands and metal ions. Proc. Natl. Acad. Sci. U. S. A. 1993, 90, 5394–8.

[8]   Lee SB, Hwang S, Chung DS, Yun H, Hong JI. Guest-induced reorganization of a self-assembled Pd(II) complex. Tetrahedron Lett. 1998, 39, 873–6.

[9]   Wu A, Isaacs L. Self-sorting: the exception or the rule? J. Am. Chem. Soc. 2003, 125, 4831–5.

[10] He Z, Jiang W, Schalley CA. Integrative self-sorting: a versatile strategy for the construction of complex supramolecular architecture. Chem. Soc. Rev. 2015, 44, 779–89.

[11] Safont-Sempere MM, Fernández G, Würthner F. Self-sorting phenomena in complex supramolecular systems. Chem. Rev. 2011, 111, 5784–814.

[12] Zhang KD, Ajami D, Gavette JV, Rebek Jr. J. Alkyl groups fold to fit within a water-soluble cavitand. J. Am. Chem. Soc. 2014, 136, 5264–6.

[13] Jordan JH, Gibb BC. Molecular containers assembled through the hydrophobic effect. Chem. Soc. Rev. 2015, 44, 547–85.

[14] Hiraoka S, Nakamura T, Shiro M, Shionoya M. In-water truly monodisperse aggregation of gear-shaped amphiphiles based on hydrophobic surface engineering. J. Am. Chem. Soc. 2010, 132, 13223–5.

[15] Prins LJ, Reinhoudt DN, Timmerman P. Noncovalent synthesis using hydrogen bonding. Angew. Chem. Int. Ed. 2001, 40, 2382–426.

[16] Ducharme Y, Wuest JD. Use of hydrogen bonds to control molecular aggregation. Extensive, self-complementary arrays of donors and acceptors. J. Org. Chem. 1988, 53, 5787–9.

[17] Sessler JL, Wang R. Design, synthesis, and self-assembly of "artificial dinucleotide duplexes". J. Org. Chem. 1998, 63, 4079–91.

[18] Marsh A, Silvestri M, Lehn JM. Self-complementary hydrogen bonding heterocycles designed for the enforced self-assembly into supramolecular macrocycles. Chem. Commun. 1996, 1527–8.

[19] Asadi A, Patrick BO, Perrin DM. G^C Quartet – a DNA-inspired Janus-GC heterocycle: synthesis, structural analysis, and self-organization. J. Am. Chem. Soc. 2008, 130, 12860–1.

[20] Original image taken by an anonymous photographer and published under the Public Domain Mark 1.0 license.

[21] Marsh A, Nolen EG, Gardinier KM, Lehn JM. Janus molecules: synthesis of double-headed heterocycles containing two identical hydrogen bonding arrays. Tetrahedron Lett. 1994, 35, 397–400.

[22] Whitesides GM, Simanek EE, Mathias JP, Seto CT, Chin DN, Mammen M, Gordon DM. Noncovalent synthesis: using physical-organic chemistry to make aggregates. Acc. Chem. Res. 1995, 28, 37–44.

[23] Zerkowski JA, Seto CT, Whitesides GM. Solid-state structures of rosette and crinkled tape motifs derived from the cyanuric acid melamine lattice. J. Am. Chem. Soc. 1992, 114, 5473–5.

[24] Seto CT, Whitesides GM. Self-assembly based on the cyanuric acid·melamine lattice. J. Am. Chem. Soc. 1990, 112, 6409–11.

[25] Seto CT, Whitesides GM. Molecular self-assembly through hydrogen bonding: supramolecular aggregates based on the cyanuric acid-melamine lattice. J. Am. Chem. Soc. 1993, 115, 905–16.

[26] Seto CT, Whitesides GM. Synthesis characterization and thermodynamic analysis of a 1 + 1 self-assembling structure based on the cyanuric acid·melamine lattice. J. Am. Chem. Soc. 1993, 115, 1330–40.

[27] Prins LJ, Timmerman P, Reinhoudt DN. Non-covalent synthesis of organic nanostructures. Pure Appl. Chem. 1998, 70, 1459–68.

[28] Timmerman P, Vreekamp RH, Hulst R, Verboom W, Reinhoudt DN, Rissanen K, Udachin KA, Ripmeester J. Noncovalent assembly of functional groups on calix[4]arene molecular boxes. Chem. Eur. J. 1997, 3, 1823–32.

[29] Prins LJ, Jolliffe KA, Hulst R, Timmerman P, Reinhoudt DN. Control of structural isomerism in noncovalent hydrogen-bonded assemblies using peripheral chiral information. J. Am. Chem. Soc. 2000, 122, 3617–27.

[30] Prins LJ, Huskens J, de Jong F, Timmerman P, Reinhoudt DN. Complete asymmetric induction of supramolecular chirality in a hydrogen-bonded assembly. Nature. 1999, 398, 498–502.

[31] Prins LJ, Huskens J, de Jong F, Timmerman P, Reinhoudt DN. An enantiomerically pure hydrogen-bonded assembly. Nature. 2000, 408, 181–4.

[32] Kerckhoffs JMCA, van Leeuwen FWB, Spek AL, Kooijman H, Crego-Calama M, Reinhoudt DN. Regulatory strategies in the complexation and release of a noncovalent guest trimer by a self-assembled molecular cage. Angew. Chem. Int. Ed. 2003, 46, 5717–22.

[33] Kerckhoffs JMCA, Mateos-Timoneda MA, Reinhoudt DN, Crego-Calama M. Dynamic combinatorial libraries based on hydrogen-bonded molecular boxes. Chem. Eur. J. 2006, 13, 2377–85.

[34] Conn MM, Rebek Jr. J. Self-assembling capsules. Chem. Rev. 1997, 97, 1647–68.

[35] Wyler R, de Mendoza J, Rebek Jr. J. A synthetic cavity assembles through self-complementary hydrogen bonds. Angew. Chem. Int. Ed. Engl. 1993, 32, 1699–701.

[36] Szabo T, Hilmersson G, Rebek Jr. J. Dynamics of assembly and guest exchange in the tennis ball. J. Am. Chem. Soc. 1998, 120, 6193–4.

[37] Kang J, Rebek Jr. J. Entropically driven binding in a self-assembling molecular capsule. Nature. 1996, 382, 239–41.

[38]   Grotzfeld RM, Branda N, Rebek Jr. J. Reversible Encapsulation of disc-shaped guests by a synthetic, self-assembled host. Science. 1996, 271, 487–9.

[39]   Mecozzi S, Rebek Jr. J. The 55% solution: a formula for molecular recognition in the liquid state. Chem. Eur. J. 1998, 4, 1016–22.

[40]   Martín T, Obst U, Rebek Jr. J. Molecular assembly and encapsulation directed by hydrogen-bonding preferences and the filling of space. Science. 1998, 281, 1842–5.

[41]   Rebek Jr. J. Simultaneous encapsulation: molecules held at close range. Angew. Chem. Int. Ed. 2005, 44, 2068–78.

[42]   Heinz T, Rudkevich DM, Rebek Jr. J. Pairwise selection of guests in a cylindrical molecular capsule of nanometre dimensions. Nature. 1998, 394, 764–6.

[43]   Ajami D, Theodorakopoulos G, Petsalakis ID, Rebek Jr. J. Social Isomers of picolines in a small space. Chem. Eur. J. 2013, 19, 17092–6.

[44]   Shivanyuk A, Rebek Jr. J. Social isomers in encapsulation complexes. J. Am. Chem. Soc. 2002, 124, 12074–5.

[45]   Ajami D, Liu L, Rebek Jr. J. Soft templates in encapsulation complexes. Chem. Soc. Rev. 2015, 44, 490–9.

[46]   Ajami D, Rebek Jr. J. Compressed alkanes in reversible encapsulation complexes. Nat. Chem. 2009, 1, 87–90.

[47]   Ajami D, Rebek Jr. J. Longer guests drive the reversible assembly of hyperextended capsules. Angew. Chem. Int. Ed. 2007, 46, 9283–6.

[48]   Rebek Jr. J. Host–guest chemistry of calixarene capsules. Chem. Commun. 2000, 637–43.

[49]   Mogck O, Paulus EF, Böhmer W, Thondorf I, Vogt W. Hydrogen-bonded dimers of tetraurea calix[4]arenes: unambiguous proof by single crystal X-ray analysis. Chem. Commun. 1996, 2533–4.

[50]   Shimuzu KD, Rebek Jr. J. Synthesis and assembly of self-complementary calix[4]arenes. Proc. Natl. Acad. Sci. U. S. A. 1995, 92, 12403–7.

[51]   Mogck O, Böhmer V, Vogt W. Hydrogen bonded homo- and heterodimers of tetra urea derivatives of calix[4]arenes. Tetrahedron. 1996, 52, 8489–96.

[52]   Castellano RK, Kim BH, Rebek Jr. J. Chiral capsules: asymmetric binding in calixarene-based dimers. J. Am. Chem. Soc. 1997, 119, 12671–2.

[53]   Corbellini F, Fiammengo R, Timmerman P, Crego-Calama M, Versluis K, Heck AJR, Luyten I, Reinhoudt DN. Guest encapsulation and self-assembly of molecular capsules in polar solvents via multiple ionic interactions. J. Am. Chem. Soc. 2002, 124, 6569–75.

[54]   Mansikkamäki H, Nissinen M, Schalley CA, Rissanen K. Self-assembling resorcinarene capsules: solid and gas phase studies on encapsulation of small alkyl ammonium cations. New J. Chem. 2003, 27, 88–97.

[55]   MacGillivray LR, Atwood JL. A chiral spherical molecular assembly held together by 60 hydrogen bonds. Nature. 1997, 389, 469–72.

[56]   Atwood JL, Barbour LJ, Jerga A. Hydrogen-bonded molecular capsules are stable in polar media. Chem. Commun. 2001, 2376–7.

[57]   Kiesilä A, Beyeh NK, Moilanen JO, Puttreddy R, Götz S, Rissanen K, Barran P, Lützen A, Kalenius E. Thermodynamically driven self-assembly of pyridinearene to hexameric capsules. Org. Biomol. Chem. 2019, 17, 6980–4.

[58]   Evan-Salem T, Baruch I, Avram L, Cohen Y, Palmer LC, Rebek Jr. J. Resorcinarenes are hexameric capsules in solution. Proc. Natl. Acad. Sci. U. S. A. 2006, 103, 12296–300.

[59]   Sakai N, Matile S. Synthetic multifunctional pores: lessons from rigid-rod β-barrels. Chem. Commun. 2003, 2514–23.

[60]   De Santis P, Morosetti S, Rizzo R. Conformational analysis of regular enantiomeric sequences. Macromolecules. 1974, 7, 52–8.

[61] Bong DT, Clark TD, Granja JR, Ghadiri MR. Self-assembling organic nanotubes. Angew. Chem. Int. Ed. 2001, 40, 988–1011.

[62] Ghadiri MR, Kobayashi K, Granja JR, Chadha RK, McRee DE. The structural and thermodynamic basis for the formation of self-assembled peptide nanotubes. Angew. Chem. Int. Ed. Engl. 1995, 34, 93–5.

[63] Brea RJ, Reiriz C, Granja JR. Towards functional bionanomaterials based on self-assembling cyclic peptide nanotubes. Chem. Soc. Rev. 2010, 39, 1448–56.

[64] Montenegro J, Ghadiri MR, Granja JR. Ion channel models based on self-assembling cyclic peptide nanotubes. Acc. Chem. Res. 2013, 46, 2955–65.

[65] Amorín M, Castedo L, Granja JR. New cyclic peptide assemblies with hydrophobic cavities: the structural and thermodynamic basis of a new class of peptide nanotubes. J. Am. Chem. Soc. 2003, 125, 2844–5.

[66] Calvelo M, Lamas A, Guerra A, Amorín M, Garcia-Fandino R, Granja JR. Parallel versus antiparallel β-sheet structure in cyclic peptide hybrids containing γ- or δ-cyclic amino acids. Chem. Eur. J. 2020, 26, 5846–58.

[67] Rodríguez-Vázquez N, Amorín M, Granja JR. Recent advances in controlling the internal and external properties of self-assembling cyclic peptide nanotubes and dimers. Org. Biomol. Chem. 2017, 15, 4490–505.

[68] Clark TD, Buehler LK, Ghadiri MR. Self-assembling cyclic β^3-peptide nanotubes as artificial transmembrane ion channels. J. Am. Chem. Soc. 1998, 120, 651–6.

[69] Shimizu LS, Salpage SR, Korous AA. Functional materials from self-assembled bis-urea macrocycles. Acc. Chem. Res. 2014, 47, 2116–27.

[70] Massena CJ, Wageling NB, Decato DA, Rodriguez EM, Rose AM, Berryman OB. A halogen-bond-induced triple helicate encapsulates iodide. Angew. Chem. Int. Ed. 2016, 55, 12398–402.

[71] Dumele O, Trapp N, Diederich F. Halogen bonding molecular capsules. Angew. Chem. Int. Ed. 2015, 54, 12339–44.

[72] Riwar LJ, Trapp N, Root K, Zenobi R, Diederich F. Supramolecular capsules: strong versus weak chalcogen bonding. Angew. Chem. Int. Ed. 2018, 57, 17259–64.

[73] Zhu YJ, Gao Y, Tang MM, Rebek Jr. J, Yu Y. Dimeric capsules self-assembled through halogen and chalcogen bonding. Chem. Commun. 2021, 57, 1543–9.

[74] Rahman FU, Tzeli D, Petsalakis ID, Theodorakopoulos G, Ballester P, Rebek Jr. J, Yu Y. Chalcogen bonding and hydrophobic effects force molecules into small spaces. J. Am. Chem. Soc. 2020, 142, 5876–83.

[75] Chakrabarty R, Mukherjee PS, Stang PJ. Supramolecular coordination: self-assembly of finite two- and three-dimensional ensembles. Chem. Rev. 2011, 111, 6810–918.

[76] Fujita M, Umemoto K, Yoshizawa M, Fujita N, Kusukawaa T, Biradha K. Molecular paneling via coordination. Chem. Commun. 2001, 509–18.

[77] Caulder DL, Raymond KN. Supermolecules by design. Acc. Chem. Res. 1999, 32, 975–82.

[78] Lehn JM, Rigault A, Siegel J, Harrowfield J, Chevrier B, Moras D. Spontaneous assembly of double-stranded helicates from oligobipyridine ligands and copper(I) cations: structure of an inorganic double helix. Proc. Natl. Acad. Sci. U. S. A. 1997, 84, 2565–9.

[79] Albrecht M. "Let's twist again" – Double-stranded, triple-stranded, and circular helicates. Chem. Rev. 2001, 101, 3457–98.

[80] Fatin-Rouge N, Blanc S, Pfeil A, Rigault A, Albrecht-Gary AM, Lehn JM. Self-assembly of tricuprous double helicates: thermodynamics, kinetics, and mechanism. Helv. Chim. Acta. 2001, 84, 1694–711.

[81] Hamacek J, Borkovec M, Piguet C. Simple thermodynamics for unravelling sophisticated self-assembly processes. Dalton Trans. 2006, 1473–90.

[82] Krämer R, Lehn JM, De Cian A, Fischer J. Self-assembly, structure, and spontaneous resolution of a trinuclear triple helix from an oligobipyridine ligand and Ni^{II} ions. Angew. Chem. Int. Ed. Engl. 1993, 32, 703–6.

[83] Hasenknopf B, Lehn JM, Boumediene N, Leize E, Van Dorsselaer A. Kinetic and thermodynamic control in self-assembly: sequential formation of linear and circular helicates. Angew. Chem. Int. Ed. 1998, 37, 3265–8.

[84] Hasenknopf B, Lehn JM, Boumediene N, Dupont-Gervais A, Van Dorsselaer A, Kneisel B, Fenske D. Self-assembly of tetra- and hexanuclear circular helicates. J. Am. Chem. Soc. 1997, 119, 10956–62.

[85] Zarges W, Hall J, Lehn JM, Bolm C. Helicity Induction in helicate self-organisation from chiral tris(bipyridine) ligand strands. Helv. Chim. Acta. 1991, 74, 1843–52.

[86] Caulder DL, Raymond KN. Supramolecular self-recognition and self-assembly in gallium(III) catecholamide triple helices. Angew. Chem. Int. Ed. Engl. 1997, 36, 1440–2.

[87] Youinou T, Rahmouni N, Fischer F, Osborn JA. Self-Assembly of a Cu_4 complex with coplanar copper(I) ions: synthesis, structure, and electrochemical properties. Angew. Chem. Int. Ed. Engl. 1992, 31, 733–5.

[88] Baxter PNW, Lehn JM, Fischer J, Youinou MT. Self-assembly and structure of a 3 × 3 inorganic grid from nine silver ions and six ligand components. Angew. Chem. Int. Ed. Engl. 1994, 33, 2284–7.

[89] Baxter PNW, Lehn JM, Kneisel BO, Fenske D. Multicomponent self-assembly: preferential generation of a rectangular [2×3]G grid by mixed-ligand recognition. Angew. Chem. Int. Ed. Engl. 1997, 36, 1978–81.

[90] Baxter PNW, Lehn JM, Baum G, Fenske D. Self-assembly and structure of interconverting multinuclear inorganic arrays: a [4×5]-Ag^I_{20} grid and an Ag^I_{10} quadruple helicate. Chem. Eur. J. 2000, 6, 4510–7.

[91] Swiegers GF, Malefetse TJ. New self-assembled structural motifs in coordination chemistry. Chem. Rev. 2000, 100, 3483–538.

[92] Fujita M, Yazaki J, Ogura K. Preparation of a macrocyclic polynuclear complex, [(en)Pd(4,4'-bpy)]$_4$(NO$_3$)$_8$, which recognizes an organic molecule in aqueous media. J. Am. Chem. Soc. 1990, 112, 5645–7.

[93] Fujita M. Metal-directed self-assembly of two- and three-dimensional synthetic receptors. Chem. Soc. Rev. 1998, 27, 417–25.

[94] Saha ML, Neogi S, Schmittel M. Dynamic heteroleptic metal-phenanthroline complexes: from structure to function. Dalton Trans. 2014, 43, 3815–34.

[95] Schmittel M, Mahata K. A fully dynamic five-component triangle via self-sorting. Chem. Commun. 2010, 46, 4163–5.

[96] Mahata K, Schmittel M. From 2-fold completive to integrative self-sorting: a five-component supramolecular trapezoid. J. Am. Chem. Soc. 2009, 131, 16544–54.

[97] Saha ML, Mittal N, Bats JW, Schmittel M. A six-component metallosupramolecular pentagon via self-sorting. Chem. Commun. 2014, 50, 12189–92.

[98] MacGillivray LR, Atwood JL. Structural classification and general principles for the design of spherical molecular hosts. Angew. Chem. Int. Ed. 1999, 38, 1018–33.

[99] Fochi F, Jacopozzi P, Wegelius E, Rissanen K, Cozzini P, Marastoni E, Fisicaro E, Manini P, Fokkens R, Dalcanale E. Self-assembly and anion encapsulation properties of cavitand-based coordination cages. J. Am. Chem. Soc. 2001, 123, 7539–52.

[100] Fujita M, Tominaga M, Hori A, Therrien B. Coordination assemblies from a Pd(II)-cornered square complex. Acc. Chem. Res. 2005, 38, 371–80.

[101] Tashiro S, Tominaga M, Yamaguchi Y, Kato K, Fujita M. Folding a de novo designed peptide into an α-helix through hydrophobic binding by a bowl-shaped host. Angew. Chem. Int. Ed. 2006, 45, 241–4.

[102] Kumazawa K, Biradha K, Kusukawa T, Okano T, Fujita M. Multicomponent assembly of a pyrazine-pillared coordination cage that selectively binds planar guests by intercalation. Angew. Chem. Int. Ed. 2009, 48, 3418–38.

[103] Yoshizawa M, Klosterman JK, Fujita M. Functional molecular flasks: new properties and reactions within discrete, self-assembled hosts. Angew. Chem. Int. Ed. 2003, 42, 3909–13.

[104] Han M, Engelhard DM, Clever GH. Self-assembled coordination cages based on banana-shaped ligands. Chem. Soc. Rev. 2014, 43, 1848–60.

[105] Yoshizawa M, Catti L. Bent anthracene dimers as versatile building blocks for supramolecular capsules. Acc. Chem. Res. 2019, 52, 2392–404.

[106] Harris K, Fujita D, Fujita M. Giant hollow M_nL_{2n} spherical complexes: structure, functionalisation and applications. Chem. Commun. 2013, 49, 6703–12.

[107] Fujita D, Suzuki K, Sato S, Yagi-Utsumi M, Yamaguchi Y, Mizuno N, Kumasaka T, Takata M, Noda M, Uchiyama S, Kato K, Fujita M. Protein encapsulation within synthetic molecular hosts. Nat. Commun. 2012, 3, 1093.

[108] Fujita D, Suzuki R, Fujii Y, Yamada M, Nakama T, Matsugami A, Hayashi F, Weng JK, Yagi-Utsumi M, Fujita M. Protein stabilization and refolding in a gigantic self-assembled cage. Chem. 2021, 7, 2672–83.

[109] Pluth MD, Bergman RG, Raymond KN. Proton-mediated chemistry and catalysis in a self-assembled supramolecular host. Acc. Chem. Res. 2009, 42, 1650–9.

[110] Pluth MD, Bergman RG, Raymond KN. Making amines strong bases: thermodynamic stabilization of protonated guests in a highly-charged supramolecular host. J. Am. Chem. Soc. 2007, 129, 11459–67.

[111] Nitschke JR. Construction substitution and sorting of metallo-organic structures via subcomponent self-assembly. Acc. Chem. Res. 2007, 40, 103–12.

[112] Ronson TK, Zarra S, Black SP, Nitschke JR. Metal-organic container molecules through subcomponent self-assembly. Chem. Commun. 2013, 49, 2476–90.

[113] Mal P, Breiner B, Rissanen K, Nitschke JR. White phosphorus is air-stable within a self-assembled tetrahedral capsule. Science. 2009, 324, 1697–9.

[114] Rowan SJ, Cantrill SJ, Cousins GRL, Sanders JKM, Stoddart JF. Dynamic covalent chemistry. Angew. Chem. Int. Ed. 2002, 41, 898–952.

[115] Cougnon FBL, Stefankiewicz AR, Ulrich S. Dynamic covalent synthesis. Chem. Sci. 2024, 15, 879–95.

[116] Rowan SJ, Sanders JKM. Macrocycles derived from cinchona alkaloids: a thermodynamic vs kinetic study. J. Org. Chem. 1998, 63, 1536–46.

[117] Gugger PA, Hockless DCR, Swiegers GF, Wild SB. Acid-catalyzed rearrangements between cyclic oligomers of 3,4-dihydro-2H-1,5-benzodiazocine. Crystal and molecular structures of mono- and dinuclear copper(II) complexes of tetraaza (16-membered), hexaaza (24-membered), and octaaza (32-membered) macrocycles. Inorg. Chem. 1994, 33, 5671–7.

[118] Hockless DCR, Lindoy LF, Swiegers GF, Wild SB. Acid-catalysed rearrangements between 3,4-dihydro-2H-1,5-benzooxazocine and -benzothiazocine and their macrocyclic (16-membered, 24-membered and 32-membered) oligomers. J. Chem. Soc., Perkin Trans. 1 1998, 117–22.

[119] Kwit M, Grajewski J, Skowronek P, Zgorzelak M, Gawroński J. One-step construction of the shape persistent, chiral but symmetrical polyimine macrocycles. Chem. Rec. 2019, 19, 213–37.

[120] Zhang Y, Zheng X, Cao N, Yang C, Li H. A kinetically stable macrocycle self-assembled in water. Org. Lett. 2018, 20, 2356–9.

[121] Christinat N, Scopelliti R, Severin K. Multicomponent assembly of boronic acid based macrocycles and cages. Angew. Chem. Int. Ed. 2008, 47, 1848–52.

[122] Içli B, Christinat N, Tönnemann J, Schüttler C, Scopelliti R, Severin K. Synthesis of molecular nanostructures by multicomponent condensation reactions in a ball mill. J. Am. Chem. Soc. 2009, 131, 3154–5.

[123] Liu B, Pappas CG, Zangrando E, Demitri N, Chmielewski PJ, Otto S. Complex molecules that fold like proteins can emerge spontaneously. J. Am. Chem. Soc. 2019, 141, 1685–9.

[124] Pappas CG, Mandal PK, Liu B, Kauffmann B, Miao X, Komáromy D, Hoffmann W, Manz C, Chang R, Liu K, Pagel K, Huc I, Otto S. Emergence of low-symmetry foldamers from single monomers. Nat. Chem. 2020, 12, 1180–6.

[125] Montà-González G, Sancenón F, Martínez-Máñez R, Martí-Centelles V. Purely covalent molecular cages and containers for guest encapsulation. Chem. Rev. 2022, 122, 13636–708.

[126] Harding CJ, Lu Q, Malone JF, Marrs DJ, Martin N, McKee V, Nelson J. Hydrolytically-sensitive hexaimino and hydrolytically-inert octaamino-cryptand hosts for dicopper. J. Chem. Soc., Dalton Trans. 1995, 1739–47.

[127] Acharyya K, Mukherjee PS. Organic imine cages: molecular marriage and applications. Angew. Chem. Int. Ed. 2019, 58, 8640–53.

[128] Stuart SR, Rowan J, Pease AR, Cram DJ, Stoddart JF. Dynamic hemicarcerands and hemicarceplexes. Org. Lett. 2000, 21, 2411–4.

[129] Liu X, Warmuth R. Solvent effects in thermodynamically controlled multicomponent nanocage syntheses. J. Am. Chem. Soc. 2006, 128, 14120–7.

[130] Li A, Xiong S, Zhou W, Zhai H, Liu Y, He Q. Superphane: a new lantern-like receptor for encapsulation of a water dimer. Chem. Commun. 2021, 57, 4496–9.

[131] Li A, Liu Y, Zhou W, Jiang Y, He Q. Superphanes: facile and efficient preparation, functionalization and unique properties. Tetrahedron Chem. 2022, 1, 100006.

[132] Xie H, Finnegan TJ, Gunawardana VWL, Pavlović RZ, Moore, CE, Badjić JD. A hexapodal capsule for the recognition of anions. J. Am. Chem. Soc. 2021, 143, 3874–80.

[133] Zhou W, Li A, Gale PA, He Q. A highly selective superphane for ReO_4^- recognition and extraction. Cell Rep. Phys. Sci. 2022, 3, 100875.

[134] Xu D, Warmuth R. Edge-directed dynamic covalent synthesis of a chiral nanocube. J. Am. Chem. Soc. 2008, 130, 7520–1.

[135] Tozawa T, Jones JTA, Swamy SI, Jiang S, Adams DJ, Shakespeare S, Clowes R, Bradshaw D, Hasell T, Chong SY, Tang C, Thompson S, Parker J, Trewin A, Bacsa J, Slawin AMZ, Steiner A, Cooper AI. Porous organic cages. Nat. Mat. 2009, 8, 973–8.

[136] Mastalerz M. One-pot synthesis of a shape-persistent endo-functionalised nano-sized adamantoid compound. Chem. Commun. 2008, 4756–8.

[137] Giri N, Del Pópolo MG, Melaugh G, Greenaway RL, Rätzke K, Koschine T, Pison L, Costa Gomes MF, Cooper AI, James SL. Liquids with permanent porosity. Nature. 2015, 527, 216–20.

[138] Yang X, Ullah Z, Stoddart JF, Yavuz CT. Porous organic cages. Chem. Rev. 2023, 123, 4602–34.

[139] Klotzbach S, Scherpf T, Beuerle F. Dynamic covalent assembly of tribenzotriquinacenes into molecular cubes. Chem. Commun. 2014, 50, 12454–7.

[140] Beuerle F, Gole B. Covalent organic frameworks and cage compounds: design and applications of polymeric and discrete organic scaffolds. Angew. Chem. Int. Ed. 2018, 57, 4850–78.

[141] Lehn JM. Dynamic combinatorial chemistry and virtual combinatorial libraries. Chem. Eur. J. 1999, 5, 2455–63.

[142] Corbett PT, Leclaire J, Vial L, West KR, Wietor JL, Sanders JKM, Otto S. Dynamic combinatorial chemistry. Chem. Rev. 2006, 106, 3652–711.

[143] Severin K. The advantage of being virtual – target-induced adaption and selection in dynamic combinatorial libraries. Chem. Eur. J. 2004, 10, 2565–80.

[144] Corbett PT, Otto S, Sanders JKM. Correlation between host–guest binding and host amplification in simulated dynamic combinatorial libraries. Chem. Eur. J. 2004, 10, 3139–43.

[145] Huc I, Lehn JM. Virtual combinatorial libraries: dynamic generation of molecular and supramolecular diversity by self-assembly. Proc. Natl. Acad. Sci. U. S. A. 1997, 94, 2106–10.

[146] Ramström O, Lehn JM. In situ generation and screening of a dynamic combinatorial carbohydrate library against concanavalin A. ChemBioChem. 2000, 1, 41–8.

[147] Vongvilai P, Angelin M, Larsson R, Ramström O. Dynamic Combinatorial resolution: direct asymmetric lipase-mediated screening of a dynamic nitroaldol library. Angew. Chem. Int. Ed. 2007, 46, 948–50.

[148] Cousins GRL, Furlan RLE, Ng YF, Redman JE, Sanders JKM. Identification and isolation of a receptor for *N*-methyl alkylammonium salts: molecular amplification in a pseudo-peptide dynamic combinatorial library. Angew. Chem. Int. Ed. 2001, 40, 423–8.

[149] Lam RTS, Belenguer A, Roberts SL, Naumann C, Jarrosson T, Otto S, Sanders JKM. Amplification of acetylcholine-binding catenanes from dynamic combinatorial libraries. Science. 2005, 308, 667–9.

[150] Cougnon FBL, Sanders JKM. Evolution of dynamic combinatorial chemistry. Acc. Chem. Res. 2012, 45, 2211–21.

[151] Kearney PC, Mizoue LS, Kumpf RA, Forman JE, McCurdy A, Dougherty DA. Molecular recognition in aqueous media. New binding studies provide further insights into the cation–π interaction and related phenomena. J. Am. Chem. Soc. 1993, 115, 9907–19.

[152] Otto S, Furlan RLE, Sanders JKM. Selection and amplification of hosts from dynamic combinatorial libraries of macrocyclic disulfides. Science. 2002, 297, 590–3.

[153] Corbett PT, Tong LH, Sanders JKM, Otto S. Diastereoselective amplification of an induced-fit receptor from a dynamic combinatorial library. J. Am. Chem. Soc. 2005, 127, 8902–3.

[154] Brachvogel RC, Hampel H, von Delius M. Self-assembly of dynamic orthoester cryptates. Nat. Commun. 2015, 6, 7129.

[155] Shyshov O, Brachvogel RC, Bachmann T, Srikantharajah R, Segets D, Hampel F, Puchta R, von Delius M. Adaptive Behavior of dynamic orthoester cryptands. Angew. Chem. Int. Ed. 2017, 56, 776–81.

[156] Carnall JMA, Waudby CA, Belenguer AM, Stuart MCA, Peyralans JJP, Otto S. Mechanosensitive self-replication driven by self-organization. Science. 2010, 327, 1502–6.

[157] Corbett PT, Sanders JKM, Otto S. Systems chemistry: pattern formation in random dynamic combinatorial libraries. Angew. Chem. Int. Ed. 2007, 46, 8858–61.

[158] Ashkenasy G, Hermans TM, Otto S, Taylor AF. Systems chemistry. Chem. Soc. Rev. 2017, 46, 2543–54.

6 Making polymers

CONSPECTUS: *Polymeric materials are ubiquitous today and are used for many different purposes because their properties can be fine-tuned by varying structural parameters such as the type of monomer, the molecular weight, the degree of cross-linking, and others. In addition, noncovalent interactions within or between polymer chains sometimes contribute to the material properties. In polyamides, for example, interchain hydrogen bonding influences melting behavior and flexibility. However, the mere presence of reversible interactions is not sufficient for a polymer to be considered supramolecular. Rather, supramolecular polymers require the monomers to be linked by reversible (often noncovalent) interactions. Supramolecular polymerization is thus a self-assembly process that results in chains rather than discrete objects. In this chapter, we will look at the thermodynamic and kinetic aspects of supramolecular polymerization and the mechanism by which the polymers are formed. We will then discuss a selection of reversible interactions that are commonly used to prepare supramolecular polymers. The last part of the chapter is devoted to polymers whose behavior is controlled by reversible interactions between the chains.*

6.1 Supramolecular polymerization

In the last chapter, we focused on rosettes, helicates, capsules, cages and other self-assembled systems with discrete structures, a limited number of building blocks, and no vacant binding sites. Only in a few cases have we seen larger and less defined products, such as those formed by the cyclic peptides and cyclic ureas discussed in Section 5.3.4, which can in principle stack indefinitely, or the macrocycles presented in Section 5.7.4, which form microcrystalline fibers. Recall also the bis(pyridone) derivative **5.17** (Figure 5.14), which must form chains because the mutual orientation of the two pyridone units prevents dimerization. Thus, when building blocks are unable to form discrete products, an ordered assembly is still possible, but leads to polymers.

In this chapter we will take a closer look at such supramolecular polymers, their formation, and their properties. With this topic, we enter an area of supramolecular chemistry that uses molecular recognition to design materials with useful and even novel properties. The interest in this branch of supramolecular chemistry is due to the enormous impact that polymeric materials have on our daily lives, with applications ranging from simple packaging materials and consumer products to sophisticated high-strength, conducting, or light-emitting polymers. Modern life would be inconceivable without these materials, although their current overuse and negative impact on the environment in many cases calls for less problematic alternatives.

The versatility of (conventional) polymeric materials is a direct result of the large number of monomers available, the different ways in which they can be linked and distributed along the polymer backbone, the relative configuration of substituents (tacticity), the degree of cross-linking, and so on. In addition, the way a polymer is synthesized (radical, anionic, or cationic polymerization, polyaddition, polycondensa-

https://doi.org/10.1515/9783111315171-006

tion) and processed, and whether the final material is a single component or a blend or composite, will affect the performance. In any case, an important reason for the robustness of conventional polymers is their covalent nature, which keeps their structure intact even under mechanical stress or at elevated temperatures.

What makes a polymer supramolecular?

In stark contrast, the monomers in supramolecular polymers are held together by reversible interactions according to the definition of supramolecular polymers proposed by Egbert (Bert) W. Meijer: "Supramolecular polymers are defined as polymeric arrays of monomeric units that are brought together by reversible and highly directional secondary interactions, resulting in polymeric properties in dilute and concentrated solution as well as in the bulk" [1]. This definition concludes by stating that supramolecular polymers "[...] behave according to well-established theories of polymer physics." Thus, the behavior of supramolecular polymers is expected to mirror that of conventional polymers. Still, the bonds between the monomers are reversible, allowing them to exchange their subunits under appropriate conditions, which is impossible for covalently stabilized materials. This ability to constantly break and reform the bonds along the chain results in unique properties that are useful in many applications.

For example, supramolecular polymers can be repeatedly depolymerized and polymerized by simply changing the temperature. In this way, they can be switched back and forth from a processable, free-flowing state to a viscous or solid material. The actual polymerization is spontaneous, requires no reagents or catalysts, and is virtually quantitative. Because of the reversibility of the interactions between the monomers, cracks and fractures in the material caused by mechanical stress can heal by reestablishing the reversible interactions, a behavior called *self-healing*. The result is a material with nearly the same mechanical stability as before [2]. Finally, the ease of depolymerization also makes supramolecular polymers much easier to degrade than conventional polymers. These properties can be important for tissue engineering, drug delivery, and the development of adhesives or personal care products [3], especially if the polymers are stable under aqueous conditions [4]. Some of these applications are discussed in Chapter 12. Here we will look at the different mechanisms of supramolecular polymerization and the types of reversible interactions that induce self-assembly. Intermolecular cross-linking of polymers using reversible interactions will be the subject of Section 6.3.

Monomeric building blocks that can be used to assemble supramolecular polymers must have two recognition sites to allow chain growth in both directions. These groups may be connected by flexible or rigid linkers, an aspect that affects the polymerization mechanism. They may be self-complementary so that polymerization yields a homopolymer (Figure 6.1a), or they may require an additional component, such as a metal ion, to mediate the interactions (Figure 6.1b). In addition, monomers

with different but mutually complementary binding sites will yield homopolymers
(Figure 6.1c), while polymerization of two different complementary monomers will
yield alternating copolymers (Figure 6.1d).

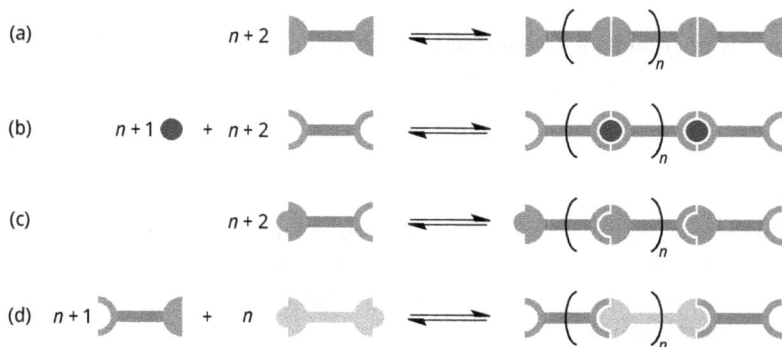

Figure 6.1: Schematic representation of supramolecular polymerizations, classified according to the
binding motifs in the monomer(s). Homopolymers are formed from monomers with two identical
recognition subunits that bind to each other in the absence (a) or the presence (b) of an auxiliary
component, and from monomers with two different complementary recognition units (c). The copolymer
in (d) is formed from two different but complementary monomers.

Each interaction between the monomer units is fully reversible, so mixing monomers
under appropriate conditions leads to equilibria in which the ratio of the individual
self-assembled products and the degree of polymerization depend on external parameters, such as concentrations, solvent, temperature, and so on. The general formalism
for treating these equilibria is similar to what we saw for receptor–substrate interactions in Chapter 2 and for self-assembly in Chapter 5, except that the interactions do
not lead to discrete systems, but can in principle continue indefinitely, yielding high
molecular weight products. To understand the principles, it is helpful to start by looking at the simplest system, the self-assembly of two identical building blocks M that
can only form a dimer D (Figure 6.2).

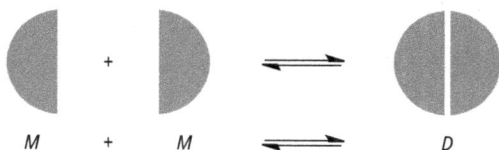

Figure 6.2: Schematic representation of the dimerization of two self-complementary building blocks M to
yield a dimer D and the corresponding reaction scheme.

The law of mass action for the equilibrium shown in Figure 6.2 is given by equation (6.1), with the association constant K_a indicating the stability of the dimer:

$$K_a = \frac{c_D}{c_M^2} \qquad (6.1)$$

The concentrations of c_M and c_D are related to the initial concentration of M (c_M^0) by the mass balance in the following equation:

$$c_M^0 = c_M + 2\,c_D \qquad (6.2)$$

Rearranging (6.2) leads to (6.3), and using this expression for c_D in equation (6.1) gives a quadratic equation. Of the two possible solutions, equation (6.5) correctly describes how the concentration of the monomer varies with the initial concentration c_M^0 for any given K_a. An analogous relationship can be obtained for c_D by combining equation (6.5) with the mass balance (6.3):

$$c_D = 0.5\left(c_M^0 - c_M\right) \qquad (6.3)$$

$$K_a = \frac{0.5\left(c_M^0 - c_M\right)}{c_M^2} \qquad (6.4)$$

$$c_M = \frac{-1 + \sqrt{1 + 8K_a c_M^0}}{4K_a} \qquad (6.5)$$

Since increasing c_M^0 also causes c_D and c_M to increase continuously, equation (6.5) does not directly show how K_a affects the extent of dimerization. A better way to estimate this effect is to look at the degree of aggregation α, which is defined by equation (6.6). Under conditions where no dimer is formed, $c_M = c_M^0$ and the degree of dimerization is zero. Conversely, when all the monomers are bound in the dimer, $c_M = 0$ and $\alpha = 1$.

$$\alpha = 1 - \frac{c_M}{c_M^0} \qquad (6.6)$$

Combining equations (6.5) and (6.6) gives equation (6.7), which describes the degree of dimerization as a function of K_a and c_M^0. The dependence of α on c_M^0 for a series of complexes with different K_a values is shown in Figure 6.3.

$$\alpha = 1 - \frac{-1 + \sqrt{1 + 8K_a c_M^0}}{4K_a c_M^0} \qquad (6.7)$$

When c_M^0 is plotted on a linear scale (Figure 6.3a), saturation curves similar to those in Section 2.1 are obtained. These curves become steeper and reach saturation earlier as the K_a values increase. The complex with a K_a of 10^6 M^{-1} is almost fully formed at low

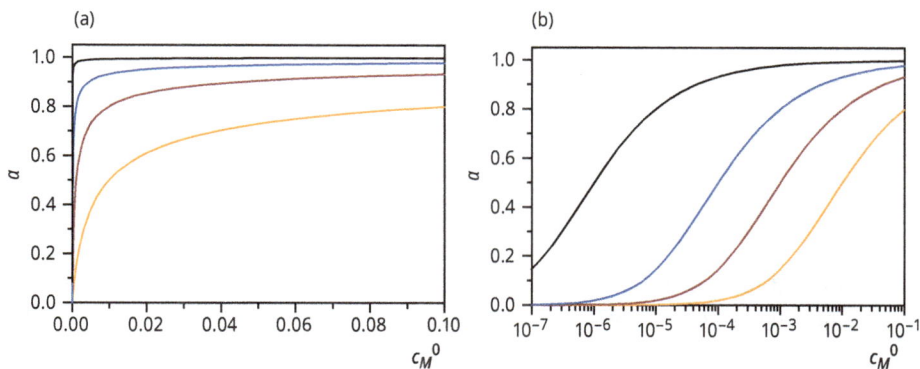

Figure 6.3: Graphs showing how the degree of aggregation α changes with c_M^0 for association constants K_a of 10^2 M^{-1} (orange), 10^3 M^{-1} (red), 10^4 M^{-1} (blue), and 10^6 M^{-1} (black). In (a), the concentration c_M^0 is shown on a linear and in (b) on a logarithmic scale.

values of c_M^0, for example, while the extent of dimerization is only 80% even at a concentration of 0.1 M when K_a is 100 M^{-1}. Association constants can be estimated using NMR spectroscopy, UV–vis spectroscopy, or any other analytical technique that provides information on the degree of dimerization. The results are plotted, and the resulting isotherms are fitted to the mathematical model derived above. Since binding isotherms do not have a suitable curvature when the complexes are very stable, the nonlinear regressions give more reliable results when c_M^0 is plotted on a logarithmic scale. The corresponding binding isotherms have a sigmoidal shape, independent of K_a, with an inflection point that shifts progressively to lower concentrations as complex stability increases (Figure 6.3b). They can be evaluated most reliably if α values are available for concentrations that vary over several orders of magnitude. Such measurements have been used to determine the stabilities of some of the hydrogen-bonded or halogen-bonded capsules presented in Sections 5.3.3 and 5.4.3, or other dimeric systems.

An alternative approach to estimating complex stability is to measure the temperature dependence of the dimerization. In this case, the experimentally determined measure of α is plotted against the temperature T and the resulting curve is fitted to equation (6.8), which is derived from equation (6.7) by expressing K_a in terms of ΔH^0 and ΔS^0. Since both ΔH^0 and ΔS^0 must be fitted during the nonlinear regression, this method provides information about the enthalpic and entropic contributions to self-assembly.

$$\alpha = 1 - \frac{-1 + \sqrt{1 + 8e^{-X}c_M^0}}{4e^{-X}c_M^0} \quad \text{with} \quad X = \frac{\Delta H^0 - T\Delta S^0}{RT} \qquad (6.8)$$

Finally, K_a can also be determined by following the degree of dimerization as the solvent composition is gradually changed from a solvent mixture that promotes dimerization to a solvent mixture in which the interactions are weak. A comparison of the above three methods for determining the dimerization constant of a merocyanine dye has been reported [5].

How do supramolecular polymers form?

The dimerization shown in Figure 6.2 can be regarded as the first step in a supramolecular polymerization. If the self-assembly does not stop after the dimerization, and if each further addition of monomers is associated with the same K_a, a chain is formed by an isodesmic supramolecular polymerization mechanism, which is the subject of the next section.

6.1.1 Isodesmic supramolecular polymerization

Having mathematically described a dimerization process, we can take the next step and adapt this treatment to equilibria where self-assembly proceeds indefinitely, leading to extended chains. To simplify the model, we assume that no byproducts are formed and that all species in equilibrium continuously exchange building blocks without encountering kinetic traps, such as precipitation, that would prevent intermediates from further participating in product formation. If we additionally assume an isodesmic (from the Greek words *isos* = equal and *desmos* = bond) supramolecular polymerization, also called multistage open association model or free association model, in which each monomer addition is associated with the same K_a, we can write the following series of equilibria and the associated laws of mass action [6, 7]:

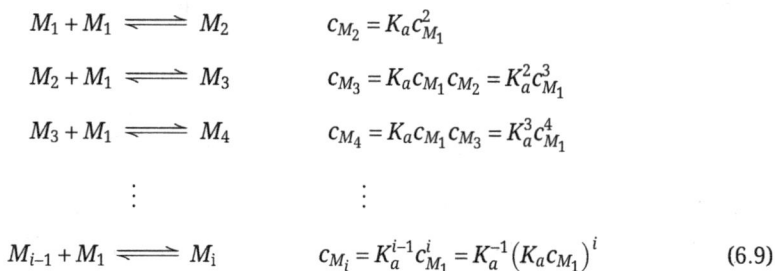

$$M_1 + M_1 \rightleftharpoons M_2 \qquad c_{M_2} = K_a c_{M_1}^2$$

$$M_2 + M_1 \rightleftharpoons M_3 \qquad c_{M_3} = K_a c_{M_1} c_{M_2} = K_a^2 c_{M_1}^3$$

$$M_3 + M_1 \rightleftharpoons M_4 \qquad c_{M_4} = K_a c_{M_1} c_{M_3} = K_a^3 c_{M_1}^4$$

$$\vdots \qquad\qquad \vdots$$

$$M_{i-1} + M_1 \rightleftharpoons M_i \qquad c_{M_i} = K_a^{i-1} c_{M_1}^i = K_a^{-1}\left(K_a c_{M_1}\right)^i \qquad (6.9)$$

The total concentration of the monomer $c_{M_1}^0$ is given by the following equation:

$$c_{M_1}^0 = \sum_{i=1}^{\infty} i\, c_{M_i} \qquad (6.10)$$

Combining equations (6.9) and (6.10) leads to equation (6.11), in which the infinite series can be expressed with the quadratic term $(1 - K_a c_{M_1})^{-2}$ if $K_a c_{M_1} < 1$. This condition is always satisfied because, as we will see in Figure 6.4, $K_a c_{M_1}$ approaches but never equals 1.

$$c_{M_1}^0 = \sum_{i=1}^{\infty} i\, K_a^{-1} \left(K_a c_{M_1} \right)^i = c_{M_1} \sum_{i=1}^{\infty} i\, \left(K_a c_{M_1} \right)^{i-1} = \frac{c_{M_1}}{\left(1 - K_a c_{M_1}\right)^2} \tag{6.11}$$

$$c_{M_1} = \frac{1 + 2K_a c_{M_1}^0 - \sqrt{4K_a c_{M_1}^0 + 1}}{2K_a^2 c_{M_1}^0} \tag{6.12}$$

Equation (6.11) can be solved, yielding the expression (6.12), which, like equation (6.5), describes the dependence of the equilibrium monomer concentration c_{M_1} on the total concentration $c_{M_1}^0$. Based on c_{M_1}, the degree of aggregation or polymerization α can be calculated by combining equations (6.6) and (6.12). The resulting equation (6.13) describes the dependence of α on $K_a c_{M_1}^0$:

$$\alpha = 1 - \frac{c_{M_1}}{c_{M_1}^0} = 1 - \frac{1 + 2K_a c_{M_1}^0 - \sqrt{4K_a c_{M_1}^0 + 1}}{2\left(K_a c_{M_1}^0\right)^2} \tag{6.13}$$

The advantage of plotting α versus $K_a c_{M_1}^0$ (rather than versus $c_{M_1}^0$ for different values of K_a as in Figure 6.3b) is that it is easier to estimate the concentration $c_{M_1}^0$ required to achieve a given degree of aggregation for a system with a known K_a. In other words, the same degree of polymerization that results from a particular combination of K_a and $c_{M_1}^0$ can also be obtained for a system in which the interaction between the monomers is weaker (smaller K_a), only at higher values of $c_{M_1}^0$. The graph in Figure 6.4a shows that the resulting binding isotherm is similar to that of a dimerization (Figure 6.3b). Note also that – regardless of the values of the equilibrium constant and the total concentration – the number of monomers at equilibrium is always greater than the number of polymers in an isodesmic supramolecular polymerization.

Based on equation (6.13), the number and weight average degrees of polymerization, $\langle dp \rangle_n$ and $\langle dp \rangle_w$, respectively, can be calculated. These averages are closely related to the corresponding average molecular weights in polymer chemistry, except that the molecular weight is replaced by the degree of polymerization i in the respective equations. Accordingly, $\langle dp \rangle_n$ and $\langle dp \rangle_w$ are given as follows:

$$\langle dp \rangle_n = \frac{\sum\limits_{i=1}^{\infty} i c_{M_i}}{\sum\limits_{i=1}^{\infty} c_{M_i}} \qquad\qquad \langle dp \rangle_w = \frac{\sum\limits_{i=1}^{\infty} i^2 c_{M_i}}{\sum\limits_{i=1}^{\infty} i c_{M_i}} \tag{6.14a, b}$$

The term in the numerator of equation (6.14a) is the total number of monomers before self-assembly $c_{M_1}^0$, as expressed in equation (6.11), and the term in the denominator de-

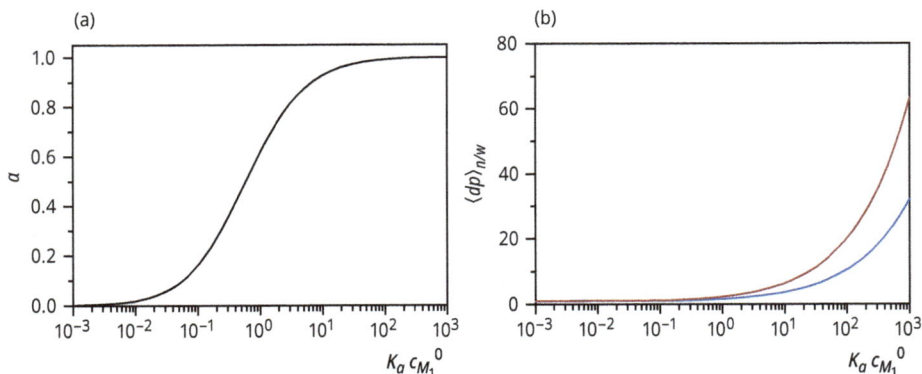

Figure 6.4: Graphs showing how the degree of aggregation (or polymerization) a in an isodesmic supramolecular polymerization changes with $K_a c_{M_1}^0$ (a) and dependence of the number and weight average degrees of polymerization, $\langle dp \rangle_n$ (blue line) and $\langle dp \rangle_w$ (red line), respectively, on $K_a c_{M_1}^0$ (b).

scribes the concentration of the products c_P. This concentration is given by equation (6.15), in which the infinite series can again be simplified using expression (6.16). The rearrangement of the resulting equation gives equation (6.17), which can be used to calculate $\langle dp \rangle_n$. The corresponding equation (6.18) together with the expression for $K_a c_{M_1}$ in equation (6.12) allows the calculation of $\langle dp \rangle_n$ as a function of $K_a c_{M_1}^0$:

$$c_P = \sum_{i=1}^{\infty} c_{M_i} = \sum_{i=1}^{\infty} K_a^{-1} \left(K_a c_{M_1} \right)^i \tag{6.15}$$

$$\sum_{i=0}^{\infty} (x)^i = \frac{1}{1-x} \tag{6.16}$$

$$c_P = \sum_{i=1}^{\infty} K_a^{-1} \left(K_a c_{M_1} \right)^i = -K_a^{-1} + K_a^{-1} \sum_{i=0}^{\infty} \left(K_a c_{M_1} \right)^i = -K_a^{-1} + K_a^{-1} \frac{1}{1 - K_a c_{M_1}}$$

$$c_P = \frac{c_{M_1}}{1 - K_a c_{M_1}} \tag{6.17}$$

$$\langle dp \rangle_n = \frac{c_{M_1}^0}{c_P} = \frac{c_{M_1}}{\left(1 - K_a c_{M_1}\right)^2} \bigg/ \frac{c_{M_1}}{1 - K_a c_{M_1}} = \frac{1}{1 - K_a c_{M_1}} \tag{6.18}$$

The weight average degree of polymerization $\langle dp \rangle_w$ is calculated in a similar way. Starting from equation (6.19), the numerator is rearranged using the infinite series expression (6.20). The denominator is equal to $c_{M_1}^0$, which is given by equation (6.11). The rearrangement of (6.19) gives equation (6.21), which allows the estimation of $\langle dp \rangle_w$:

$$\langle dp \rangle_w = \frac{\sum_{i=1}^{\infty} i^2 K_a^{-1} \left(K_a c_{M_1} \right)^i}{\sum_{i=1}^{\infty} i K_a^{-1} \left(K_a c_{M_1} \right)^i} \tag{6.19}$$

$$\sum_{i=1}^{\infty} i^2 (x)^{i-1} = \frac{1+x}{(1-x)^3} \tag{6.20}$$

$$\langle dp \rangle_w = \frac{c_{M_1} \left(1 + K_a c_{M_1} \right)}{\left(1 - K_a c_{M_1} \right)^3} \Big/ \frac{c_{M_1}}{\left(1 - K_a c_{M_1} \right)^2} = \frac{1 + K_a c_{M_1}}{1 - K_a c_{M_1}} \tag{6.21}$$

The plots in Figure 6.4b, calculated using equations (6.12), (6.18), and (6.21), illustrate the dependence of $\langle dp \rangle_n$ and $\langle dp \rangle_w$ on $K_a c_{M_1}^0$. These graphs show that in isodesmic supramolecular polymerizations, the polymerization is a continuous process that lacks a critical concentration (and a critical temperature). In addition, high degrees of polymerization can only be achieved at high values of $K_a c_{M_1}^0$. Since the achievable monomer concentration is limited, the interaction between the monomers must be very effective ($K_a > 10^6$ M^{-1}) to obtain high molecular weight polymers. The actual value of K_a can be determined using concentration-dependent UV–vis or NMR spectroscopic measurements and fitting the resulting binding isotherm to the appropriate model, as discussed earlier for dimerization.

Another important parameter characterizing the self-assembly process is the polydispersity of the products, which is given by the ratio $\langle dp \rangle_n / \langle dp \rangle_w$. This ratio is equal to $1 + K_a c_{M_1}$ according to equations (6.18) and (6.21), which means that the polydispersity approaches a value of 2 at high values of $K_a c_{M_1}^0$, where $K_a c_{M_1}$ is close to 1. Thus, supramolecular polymers obtained by an isodesmic supramolecular polymerization have a relatively broad molecular weight distribution. Related statistical models have been developed to describe the more complex formation of supramolecular copolymers from more than one building block [8].

Regarding the thermodynamics of chain elongation, it is important to note that the temperature dependence of the entropy has a characteristic effect on the polymerization. When both the enthalpy and entropy of chain propagation are positive, polymerization is driven by entropy alone. Accordingly, chain elongation will only occur at a temperature high enough for entropy to overcompensate the unfavorable enthalpy term. In this case, the system has a floor temperature below which it is in the unassembled state. Conversely, when both enthalpy and entropy are negative, there is a ceiling temperature above which entropy favors the dissociated state.

6.1.2 Ring–chain supramolecular polymerization

Ring–chain supramolecular polymerizations are related to isodesmic supramolecular polymerization with the major difference that the monomers also form rings at each step because their end groups are tethered by a flexible linker. Accordingly, the polymerization mechanism involves not only the stepwise growth of the linear chains, but also the formation of cyclic products. This mechanism is shown schematically in Figure 6.5.

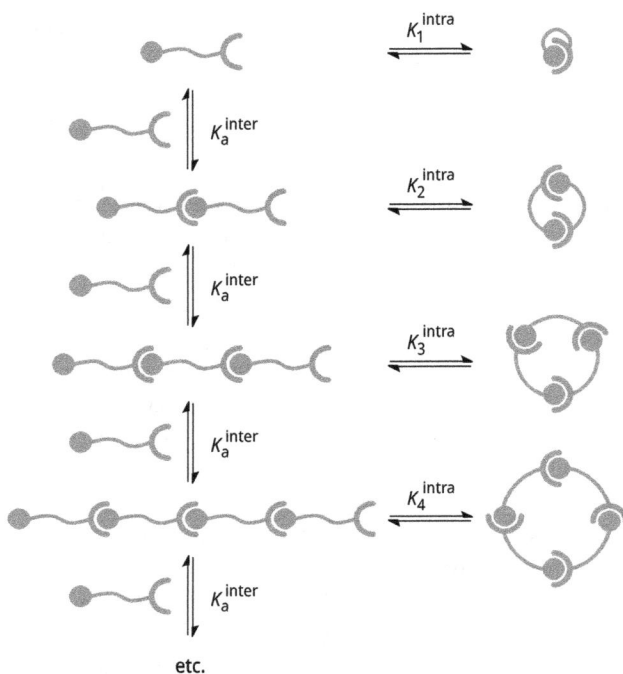

etc.

Figure 6.5: Stepwise self-assembly of a self-complementary monomer leading to acyclic oligomers. Each building block, including the monomer itself, can cyclize to yield the corresponding self-assembled macrocycle containing the same number of monomeric units. Each oligomerization step is associated with an intermolecular equilibrium constant K_a^{inter}, while the ring–chain equilibria have intramolecular constants K_i^{intra} that vary with the ring size i. Additional equilibrium arrows between the macrocycles, which can also directly interconvert, are omitted for reasons of clarity.

Not coincidentally, this figure looks very similar (except for the very first equilibrium – the cyclization of the monomer) to Figure 5.3, which was used to illustrate the formation of cyclic self-assembled products from acyclic building blocks. In Chapter 5, we were interested in the factors that determine whether a particular cyclic product is formed preferentially over other cyclic or linear compounds. The crucial parameter turned out to be the effective molarity EM, that is, the ratio of the intramolecular equilibrium constants of ring formation and the equilibrium constant associated with the intrinsic strength of the interactions (K_i^{intra}/K_a^{inter}). Cyclization reactions with high EM_i values will cause the corresponding macrocycles to dominate at equilibrium. Recall that the effective molarity can be viewed as the concentration at which a binding partner must be present for the intermolecular reaction to outcompete the intramolecular one. In this section, we will look at how the interplay between the intermolecular and intramolecular binding constants and the individual EM_i values affect the formation of the linear products.

The actual chain elongation in Figure 6.5 is treated in the same way as discussed in the previous section since each step is associated with the same K_a. The cyclization is accounted for by a series of equilibria similar to those used to derive equation (6.9), describing the formation of the ring C_1 from a single monomer M_1 and the subsequent formation of the larger rings C_i from the corresponding linear oligomers M_i [9]. All reactions are intramolecular reactions, so we can use the product of the intermolecular binding constant K_a and the corresponding EM_i value as the equilibrium constants. The monomer concentration c_{M_i} is expressed using the laws of mass balances derived for the polymerization (equation (6.9)):

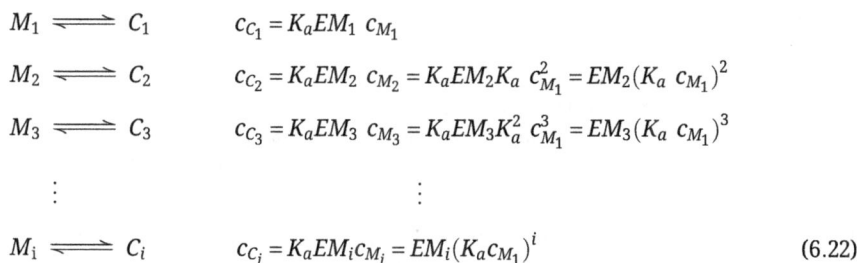

$$M_1 \rightleftharpoons C_1 \qquad c_{C_1} = K_a EM_1 \, c_{M_1}$$

$$M_2 \rightleftharpoons C_2 \qquad c_{C_2} = K_a EM_2 \, c_{M_2} = K_a EM_2 K_a \, c_{M_1}^2 = EM_2 (K_a \, c_{M_1})^2$$

$$M_3 \rightleftharpoons C_3 \qquad c_{C_3} = K_a EM_3 \, c_{M_3} = K_a EM_3 K_a^2 \, c_{M_1}^3 = EM_3 (K_a \, c_{M_1})^3$$

$$\vdots \qquad\qquad \vdots$$

$$M_i \rightleftharpoons C_i \qquad c_{C_i} = K_a EM_i c_{M_i} = EM_i (K_a c_{M_1})^i \qquad (6.22)$$

We now use equations (6.9) and (6.22) to specify the concentrations of chains and rings in the mass balance, which is the sum of the total concentration of linear oligomers and rings according to equation (6.23). In the resulting equation (6.24), the series describing the chain elongation is the same as in equation (6.11) and can therefore be simplified as described above. Since it has been shown that EM_i varies inversely with the 5/2 power of the degree of polymerization i multiplied by EM_1 ($EM_i = EM_1 \, i^{-5/2}$) when all rings are strainless [10], the second series can also be rewritten. If some rings are strained, a third term must be included in equation (6.25) [9], which will not be discussed here to keep the discussion simple:

$$c_{M_1}^0 = \sum_{i=1}^{\infty} i \, c_{M_i} + \sum_{i=1}^{\infty} i \, c_{C_i} \qquad (6.23)$$

$$c_{M_1}^0 = \sum_{i=1}^{\infty} i \, K_a^{-1} (K_a c_{M_1})^i + \sum_{i=1}^{\infty} i \, EM_i \, (K_a c_{M_1})^i \qquad (6.24)$$

$$c_{M_1}^0 = \frac{c_{M_1}}{(1 - K_a c_{M_1})^2} + EM_1 \sum_{i=1}^{\infty} i^{-3/2} \, (K_a c_{M_1})^i \qquad (6.25)$$

Equation (6.25) cannot be solved algebraically, but it is possible to find solutions using Newton's iterative method, as explained for the mathematical treatment of 1:2 complexes in Section 13.1. With the $K_a c_{M_1}$ values thus obtained, the dependence of the degree of polymerization on $K_a c_{M_1}^0$ is estimated using strategies similar to those described in the previous section on isodesmic supramolecular polymerizations. The results are shown in Figure 6.6.

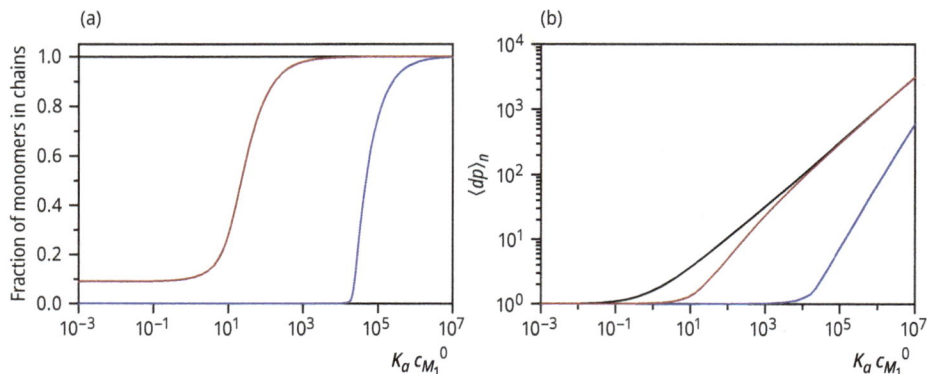

Figure 6.6: Fraction of the monomers incorporated into the chain during a ring–chain supramolecular polymerization (a) and number average degree of polymerization $\langle dp \rangle_n$ (b) as a function of $K_a c^0_{M_1}$. The curves correlate to EM_1 values of 0 M (black), 10^{-5} M (red), and 10^{-2} M (blue). The value of K_a is 10^6 M^{-1}.

If the monomer has no tendency to cyclize ($EM_1 = 0$ M), the second term in equation (6.25) can be neglected, and self-assembly follows a simple isodesmic supramolecular polymerization mechanism. Plotting the degree of aggregation α versus $K_a c^0_{M_1}$ yields a graph identical to that in Figure 6.4a, and an alternative way to illustrate this behavior is the straight line in Figure 6.6a, which indicates that regardless of $K_a c^0_{M_1}$, the ratio of consumed monomers to monomers incorporated into the chains is one, meaning that no cyclic products are formed and all monomers end up in the polymers.

The red and blue curves in Figure 6.6a show that the ability of the monomers to cyclize reduces the fraction of the monomers incorporated into the chains. Depending on EM_1, the extent of polymerization is low or even negligible at low values of $K_a c^0_{M_1}$ until a critical concentration is reached at which polymerization abruptly sets in. The exact value of this concentration coincides with EM_1, as shown by the fact that the red curve, which correlates with an EM_1 value of 10^{-5} M, begins to rise visibly only when $K_a c^0_{M_1} > 10$. Since K_a is 10^6 M^{-1}, the critical $c^0_{M_1}$ value is 10^{-5} M, which is equal to EM_1. Similarly, the abrupt rise of the blue curve is observed at $K_a c^0_{M_1} = 10^4$, which correlates with a $c^0_{M_1}$ value of 10^{-2} M or EM_1. Thus, only at concentrations higher than EM_1 does intermolecular polymerization outcompete intramolecular cyclization, but once this concentration is exceeded, polymerization becomes the dominant process. This critical concentration is a characteristic feature of ring–chain supramolecular polymerizations, which is not observed for isodesmic supramolecular polymerizations where chain growth proceeds continuously.

A similar trend can be seen in Figure 6.6b: as soon as $c^0_{M_1} > EM_1$ polymerization starts, yielding chains with high degrees of polymerization at high $K_a c^0_{M_1}$ values. At concentrations below EM_1, no significant amounts of linear products are present in the equilibria, which explains why we neglected polymer formation in the systems discussed in Chapter 5, which were mostly studied in dilute solutions.

As in the case of isodesmic supramolecular polymerizations, self-assembly is temperature dependent, and depending on whether it is primarily controlled by enthalpy or entropy, there is a ceiling temperature above which or a floor temperature below which cyclic monomers are thermodynamically favored over polymers.

6.1.3 Cooperative supramolecular polymerization

So far, we have focused on polymerization mechanisms in which each addition of the monomer to the chain is associated with the same binding constant K_a. Although there are many examples for such systems, there are cases where the chain elongation proceeds in two steps, the first being an isodesmic supramolecular polymerization to yield an oligomer of a specific length. Once this oligomer is present, the addition of further monomers proceeds with an association constant that is either higher or lower than that associated with the formation of the nucleus. Higher association constants lead to a cooperative supramolecular polymerization, also known as the *nucleation–elongation mechanism* [1]. They can be caused by electronic effects when the interactions between the initial monomers enhance subsequent interactions, as in the polarization-enhanced hydrogen bonding discussed in Section 3.1.5. Alternatively, the addition of monomers to the growing chain can benefit from stabilizing interactions with monomer units further away from the chain end. In the formation of helical polymers, for example, these interactions could involve monomers in a previous helix turn. Finally, solvophobic effects can contribute to the addition of monomers, but they only become effective once the chain has reached a certain size.

To estimate the effect of the nucleation on the polymerization equilibrium, it is helpful (and sometimes sufficient) to consider the simplest case, where two monomers initially form a dimer with the association constant K_a^D [7]. This dimer serves as the nucleus for the polymerization, whose individual steps proceed in the form of an isodesmic supramolecular polymerization, characterized by the association constant K_a^P. The ratio $\sigma = K_a^D / K_a^P$ is called the *nucleation factor*. It is smaller than unity in the case of a cooperative supramolecular polymerization and describes the extent to which the polymerization is more favorable than the dimerization. The individual equilibria are similar to those in Section 6.1.1, differing only in the first step:

$$M_1 + M_1 \rightleftharpoons M_2 \qquad c_{M_2} = K_a^D c_{M_1}^2$$

$$M_2 + M_1 \rightleftharpoons M_3 \qquad c_{M_3} = K_a c_{M_1} c_{M_2} = K_a^P K_a^D c_{M_1}^3$$

$$M_3 + M_1 \rightleftharpoons M_4 \qquad c_{M_4} = K_a c_{M_1} c_{M_3} = (K_a^P)^2 K_a^D c_{M_1}^4$$

$$\vdots \qquad\qquad \vdots$$

$$M_{i-1} + M_1 \rightleftharpoons M_i \qquad c_{M_i} = (K_a^P)^{i-2} K_a^D c_{M_1}^i = \sigma (K_a^P)^{-1} (K_a^P c_{M_1})^i \tag{6.26}$$

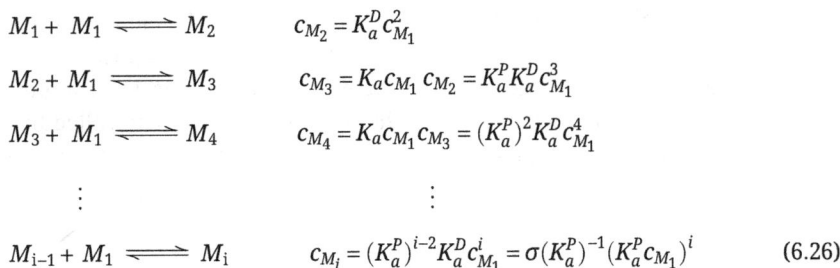

The series in equation (6.26) describes the polymerization for $i > 1$, and to consider all species in the equilibrium, the mass balance in equation (6.10) must be rewritten to give the following expression:

$$c_{M_1}^0 = \sum_{i=1}^{\infty} i \, c_{M_i} = c_{M_1} + \sum_{i=2}^{\infty} i \, \sigma \, (K_a^P)^{-1} (K_a^P \, c_{M_1})^i \tag{6.27}$$

This equation is transformed in a similar way as described in Section 6.1.1. The rearrangement of the resulting equation (6.28) results in a cubic equation for which solutions are found using Newton's iterative method. With the values of c_{M_1} thus obtained, the degree of aggregation α can be estimated as a function of $K_a^P c_{M_1}^0$:

$$c_{M_1}^0 = c_{M_1} - \sigma c_{M_1} + \sigma \sum_{i=1}^{\infty} i \, (K_a^P)^{-1} (K_a^P \, c_{M_1})^i = (1-\sigma) c_{M_1} + \frac{\sigma \, c_{M_1}}{(1 - K_a^P c_{M_1})^2} \tag{6.28}$$

Plots calculated for different values of σ are shown in Figure 6.7. As can be seen, the equilibrium polymer and monomer concentrations in cooperative supramolecular polymerizations do not rise simultaneously as in isodesmic supramolecular polymerizations (black curve in Figure 6.7a). Instead, no substantial self-assembly occurs until $K_a^P c_{M_1}^0 = 1$ (i.e., as long as $c_{M_1}^0 < 1/K_a^P$), and polymerization only takes place at concentrations higher than the critical concentration $1/K_a^P$, from which point on the concentration of free monomers remains constant and all additional monomers are incorporated into polymers. The graphs in Figure 6.7a also show that polymerization starts at higher $K_a^P c_{M_1}^0$ values in a nucleation–elongation polymerization than in an isodesmic supramolecular polymerization associated with the same K_a^P.

The abrupt appearance of polymers at $K_a^P c_{M_1}^0 = 1$ can also be seen in Figure 6.7b, which shows the dependence of the number average degree of polymerization on $K_a^P c_{M_1}^0$. The definition of $\langle dp \rangle_n$ is identical to that in an isodesmic supramolecular polymerization (equation (6.14a)), but the infinite series must be rewritten as in equation (6.29). Solving this equation gives equation (6.30), which is the basis for the graphs in Figure 6.7b:

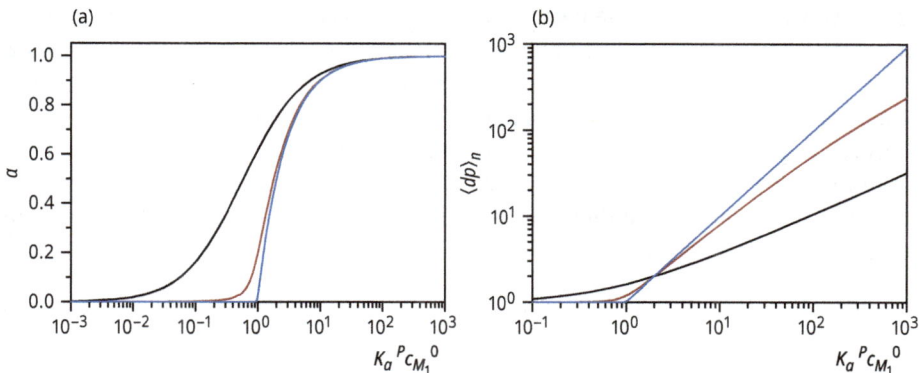

Figure 6.7: Graphs showing how the degree of polymerization a changes in a cooperative supramolecular polymerization with $K_a^P c_{M_1}^0$ (a) and dependence of the number average degrees of polymerization, $\langle dp \rangle_n$ on $K_a^P c_{M_1}^0$ (b). The curves describe the polymerization for σ values of 1 (isodesmic supramolecular polymerization) (black), 10^{-2} M (red), and 10^{-5} M (blue).

$$\langle dp \rangle_n = \frac{\sum\limits_{i=1}^{\infty} i\, c_{M_i}}{\sum\limits_{i=1}^{\infty} c_{M_i}} = \frac{c_{M_1} + \sum\limits_{i=2}^{\infty} i\sigma \left(K_a^P\right)^{-1}\left(K_a^P c_{M_1}\right)^i}{c_{M_1} + \sum\limits_{i=2}^{\infty} \sigma \left(K_a^P\right)^{-1}\left(K_a^P c_{M_1}\right)^i}$$

$$\langle dp \rangle_n = \left((1-\sigma)c_{M_1} + \frac{\sigma\, c_{M_1}}{\left(1-K_a^P c_{M_1}\right)^2}\right) \Big/ \left(c_{M_1} + \frac{\sigma}{K_a^P}\left(\frac{1}{1-K_a^P c_{M_1}} - 1 - K_a^P c_{M_1}\right)\right)$$

$$= \left(\frac{c_{M_1}\left[(1-\sigma)\left(1-K_a^P c_{M_1}\right)^2 + \sigma\right]}{\left(1-K_a^P c_{M_1}\right)^2}\right) \Big/ \left(\frac{c_{M_1}\left[1-(1-\sigma)K_a^P c_{M_1}\right]}{\left(1-K_a^P c_{M_1}\right)}\right) \qquad (6.29)$$

$$= \frac{(1-\sigma)\left(1-K_a^P c_{M_1}\right)^2 + \sigma}{\left(1-(1-\sigma)K_a^P c_{M_1}\right)\left(1-K_a^P c_{M_1}\right)} \qquad (6.30)$$

The dependence of $\langle dp \rangle_n$ on $K_a^P c_{M_1}^0$ in Figure 6.7b is similar to that of a ring–chain supramolecular polymerization, in that in both cases polymerization only starts at a certain $K_a^P c_{M_1}^0$ value. The reason for this is that both polymerization mechanisms are characterized by two equilibrium constants. While the onset of polymerization in a ring–chain supramolecular polymerization depends on EM_1, polymer formation in a cooperative supramolecular polymerization starts at $K_a^P c_{M_1}^0 = 1$, independent of K_a^D. The smaller σ, that is, the stronger polymerization is favored over nucleation, the steeper the increase in the degree of polymerization.

At this point we have discussed the three most important mechanisms of supra-molecular polymerization, but by far not all possibilities. For example, in the case of cooperative supramolecular polymerization, we have limited the discussion to the simplest case where nucleation involves the dimer of two monomers. Larger nuclei are possible, and models have been developed to describe these cases [1]. In addition, anticooperative supramolecular polymerizations exist, where chain growth is disfavored once a certain chain length is reached. In this case, polymers of defined length and a narrow size distribution are obtained under appropriate conditions. Other polymerization pathways include the formation of byproducts or of copolymers from more than one building block. Obviously, supramolecular polymerizations are complex, which also means that they are versatile, allowing the properties of the products to be varied over a wide range.

6.1.4 Kinetic aspects

In accordance with our definition of supramolecular chemistry, which states that "supramolecular chemistry deals with recognition phenomena mostly under thermodynamic control" (Chapter 1), we have so far ignored kinetic aspects of supramolecular polymerizations and treated chain elongation as a fully reversible process. However, once noncovalently assembled systems become large, with many individual building blocks held together by strong interactions (to achieve high degrees of polymerization), intermediates may be trapped in a metastable state. In this case, kinetics becomes critical to the outcome of the self-assembly. We saw an example of such a process in Section 5.7.4, where we discussed the self-assembled hexameric and heptameric macrocycles that autocatalytically mediate their own formation. Individually, these macrocycles are thermodynamically less disfavored over smaller three- or four-membered analogs. However, the entrapment of the large macrocycles in fibers prevents reequilibration and reaching the thermodynamic minimum. One way to illustrate this influence is the Gibbs free energy landscape of a polymerization mechanism shown in Figure 6.8 [11].

The energy curve in Figure 6.8 has several minima, of which the global minimum represents the thermodynamically most stable state. This state is still dynamic, with the monomers and polymers continuously exchanging building blocks, but the average composition of the polymers and the monomer to polymer ratio remain constant. The local minimum to the left of the global minimum represents a metastable state. The product in this state is not a transient species but will remain in solution for some time. If the energy barrier to reach the thermodynamic minimum is low enough, on the order of $k_{B}T$ (where k_{B} is the Boltzmann constant and T is the temperature), the system will gradually relax to the stable structure. However, if the energy barrier is too high to reach the most stable state, as in the third local minimum, the corresponding structure remains kinetically trapped and requires an energy input or a catalyst to reach the thermodynamic minimum. Finally, a continuous supply of energy can keep the system in a dissipative nonequilibrium state, and spontaneous relaxation to more stable states is only possible when the energy input is removed. Two examples will illustrate the impact of kinetics on supramolecular polymerizations.

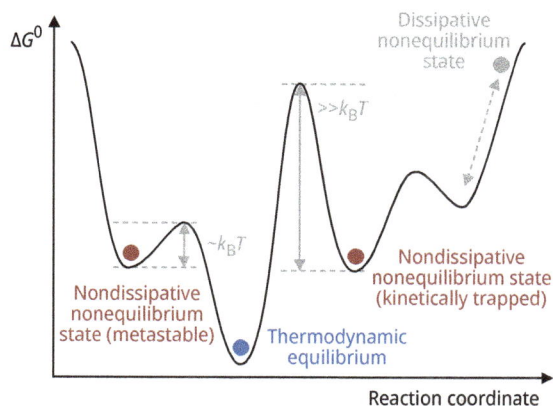

Figure 6.8: Schematic representation of a possible Gibbs free energy landscape of a supramolecular polymerization. State (a) is the global thermodynamic minimum, while states (b) and (c) are metastable and kinetically trapped states, respectively.

The first example comes from the group of Frank Würthner and involves the merocyanine derivative **6.1** (Figure 6.9) [12]. This monomer polymerizes by intermolecular dipole–dipole interactions between merocyanine groups from different monomers. Polymerization is induced by adding low-polarity solvents such as methylcyclohexane (MCH) to a solution of **6.1** in tetrahydrofuran (THF), where the monomers are dissociated. Linear chains are formed that helically intertwine to form hexameric nanorods. In the case of the chiral **6.1**, only single-handed helices are observed, as shown by circular dichroism (c.d.) spectroscopy and atomic force microscopy (AFM). Interestingly, the bands observed in the c.d. spectrum at the beginning of self-assembly change sign with time, indicating that polymers formed in a kinetically controlled first step are slowly rearranging into thermodynamically more stable products. Since the nanorods observed by AFM at different stages of polymerization are all right-handed, the change in sign of the c.d. bands must be attributed to a change in the twist of the antiparallel merocyanine dimers that stabilize the polymers. These dimers initially form with a left-handed twist between the closest neighbors, which subsequently changes to a more tightly packed, thermodynamically more stable right-handed arrangement.

The oligo(p-phenylenevinylene) derivative **6.2**, developed in the group of Bert Meijer, forms helical stacks in nonpolar solvents [13]. The self-assembly proceeds by a nucleated growth mechanism, involving the initial formation of hydrogen-bonded dimers that serve as nuclei for the chain elongation, which is mediated by aromatic interactions. Due to the chirality of **6.2**, single-handed helices are preferentially formed, which are exclusively *M*-configured when a solution of **6.2** in MCH is slowly cooled from the disassembled state at 70 °C to room temperature. In contrast, rapid cooling leads to a mixture of *M*- and *P*-configured helices. Thus, there are two competing self-assembly pathways, leading to assemblies with opposite helicity. Short oligomers that grow into *P*-configured helices are formed more rapidly, allowing them to grow under kinetic control. When chain growth proceeds under thermodynamic control, the more stable *M*-configured helices dominate at equilibrium.

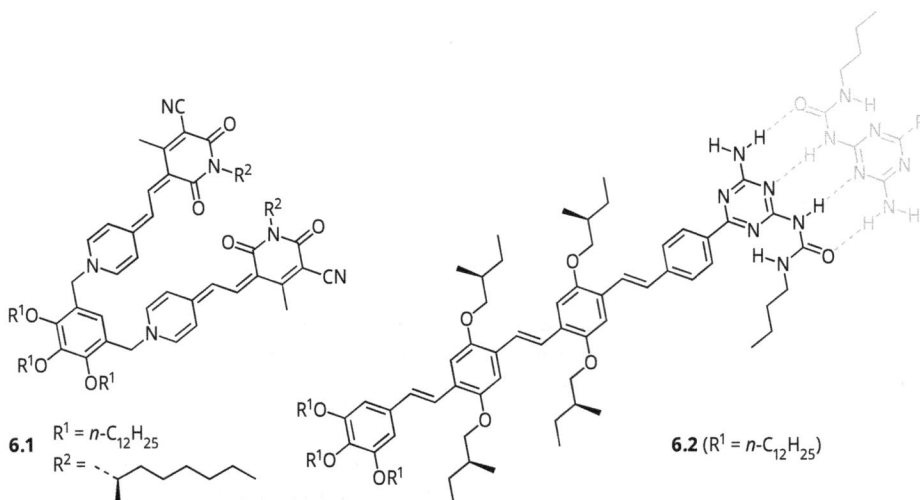

Figure 6.9: Structures of the merocyanine derivative **6.1** and the oligo(*p*-phenylenevinylene) derivative **6.2** along with the hydrogen bond pattern that mediates dimerization.

Kinetics can therefore play an important role in supramolecular polymerizations, providing a means of structural control that is rarely available in other self-assembly processes. Moreover, kinetic control can be achieved in many ways, including by changing the solvent composition or the rate at which the temperature is varied, but also by the use of templates. In this way, the structure of supramolecular polymers can be tuned much more precisely than by simple equilibrium to reach the thermodynamic minimum, which explains the importance of supramolecular polymerization in the development of functional materials [14].

6.2 Reversible interactions within polymer chains

6.2.1 Supramolecular polymerization mediated by hydrogen bonds

How can monomers be held together by reversible interactions?

Early examples of building blocks for supramolecular polymers, developed in the group of Jean-Marie Lehn, are the ditopic monomers **6.3** and **6.4** (Figure 6.10), of which **6.3** contains two terminal Hamilton receptor motifs and **6.4** represents the complementary bis(cyanuric acid) [15]. Monotopic analogs of these monomers interact in chloroform with a $\log K_a$ of 4.8, which is strong enough to mediate polymerization of the ditopic analogs. In toluene, where the interaction is even stronger, a 1:1 mixture of both monomers produces a viscous solution, which is indicative of polymer formation. The downfield shifts and broadening of the NH signals in the ^{1}H NMR spectrum and the observation of fibers in electron microscopic images further support the presence of high molecular weight products. The mechanism of polymerization can be described by the equations derived for an isodesmic supramolecular polymerization, taking into account that alternating copolymers are formed (case (d) in Figure 6.1). The length of the polymers can be controlled by adding monotopic building blocks to the mixture that act as stoppers.

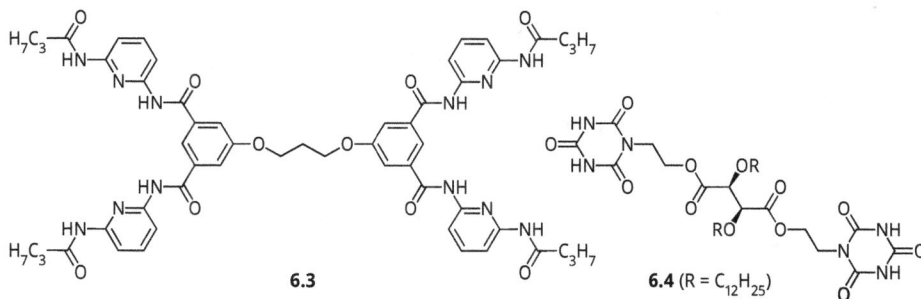

Figure 6.10: Structures of the monomers **6.3** and **6.4** containing terminal Hamilton receptors and cyanurates, respectively.

Another potent binding motif for supramolecular polymers is the 2-ureido-4[1*H*]-pyrimidinone (UPy) unit **6.5**, introduced by Bert Meijer and Rint P. Sijbesma (Figure 6.11a) [16]. Its 4[1*H*]-pyrimidinone tautomer is conformationally stabilized by an intramolecular hydrogen bond and exhibits a self-complementary DDAA pattern that mediates the formation of stable dimers in nonpolar solvents with dimerization constants $\log K_a$ of 7.3 in chloroform and 8.0 in toluene. Depending on the substituent at the 6-position, UPy derivatives also exist in the pyrimidin-4-ol tautomeric form, which dimerizes with a similar efficiency but with a DADA pattern [17]. Both hydrogen bonding patterns are observed for the 6-phenyl derivative, as shown by the crystal structures in Figure 6.11a. 6-Trifluoromethyl-2-butylureidopyrimidone, in contrast, exists almost exclusively in the pyrimidine-4-ol form. In the third 6[1*H*]-pyrimidinone form, pairing with 2,7-diamido-1,8-naphthyridine derivatives is possible *via* an ADDA–DAAD hydrogen bonding pattern [18]. Thus, UPy units provide a versatile basis for supramolecular polymerization.

Polymer assembly requires the use of ditopic monomers such as **6.6a** or a heteroditopic monomer such as **6.7**, containing a UPy and a 2,7-diamido-1,8-naphthyridine unit (Figure 6.11b). In the first case, homopolymerization occurs spontaneously when a small amount of **6.6a** is dissolved in chloroform, leading to polymers with a number average

(a)

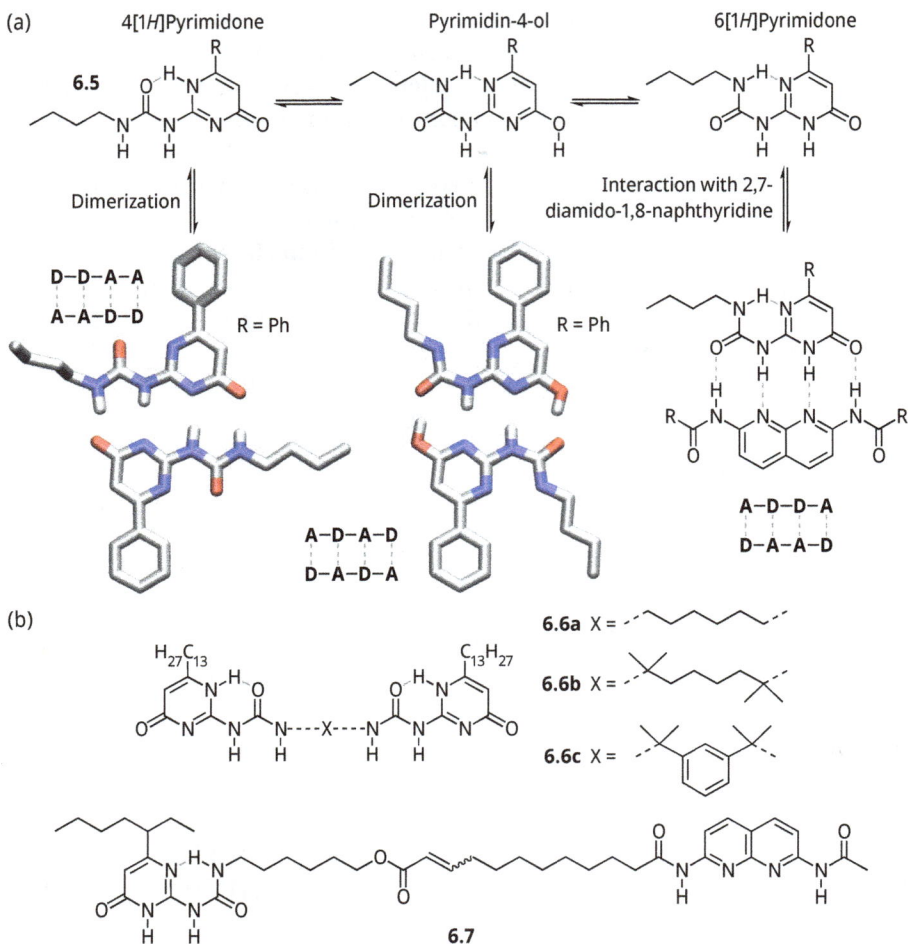

4[1*H*]Pyrimidone   Pyrimidin-4-ol   6[1*H*]Pyrimidone

6.5

Dimerization   Dimerization   Interaction with 2,7-diamido-1,8-naphthyridine

D–D–A–A
A–A–D–D   R = Ph   R = Ph

A–D–A–D
D–A–D–A

A–D–D–A
D–A–A–D

(b)

6.6a X =

6.6b X =

6.6c X =

6.7

Figure 6.11: Dimerization equilibria of the monotopic UPy derivative **6.5**, with the 4[1*H*]-pyrimidinone tautomer forming a self-complementary DDAA dimer, the pyrimidin-4-ol tautomer forming a self-complementary DADA dimer, and the 6[1*H*]-pyrimidinone forming an ADDA–DAAD complex with a 2,7-diamido-1,8-naphthyridine derivative (a). The DDAA and DADA dimers are shown as crystal structures of 6-phenyl-2-ureido-4[1*H*]-pyrimidinone and the corresponding pyrimidin-4-ol form. The structures of the ditopic monomers **6.6a–c** and **6.7** are shown in (b). Nonacidic protons are omitted in the crystal structures for reasons of clarity.

degree of polymerization $\langle dp \rangle_n$ of 700 even at a concentration as low as 40 mM [16]. The addition of the monofunctional UPy derivative **6.5** causes a shift in the degree of polymerization to shorter chains, with a concomitant decrease in viscosity. The solution of **6.6a** can be processed into an elastic, flexible material that resembles a conventional polymer, but can be returned to a processable state simply by heating. Heating can also be used to repair damage such as scratches or cracks, and a control over the material properties to tailor the product for different applications can be achieved by structurally varying the linking unit between the UPy units. For example, UPy units that are shielded from the surrounding medium self-assemble even in water, allowing the development of biocompatible materials for medical applications [19].

The dependence of the specific viscosity of a solution of **6.6a** in chloroform on the concentration of the monotopic stopper **6.5** indicates that an isodesmic supramolecular polymerization mechanism is operative [16]. However, solutions of ditopic UPy monomers in chloroform always also contain cyclic species in equilibrium together with high molecular weight chains so that it is more accurate to describe polymer formation with the formalism of a ring–chain supramolecular polymerization [20]. Since the tendency of the monomers to cyclize can be controlled by the structure of the linker between the UPy moieties, this linker, although having no effect on the actual strength of the interactions, has a pronounced influence on the degree of polymerization. For example, the viscosity of a solution of **6.6b** (Figure 6.11c) in chloroform is significantly higher than that of a solution of **6.6a** under the same conditions because the methyl groups in the α-positions of the linker in **6.6b** increase the tendency to form cyclic dimers. In the case of **6.6c**, no polymerization is observed because this monomer forms cyclic dimers at all concentrations [21].

A cooperative supramolecular polymerization mechanism is observed for the trialkyl-1,3,5-tricarboxamide **6.8a** (Figure 6.12a) that forms helical columnar stacks in nonpolar solvents [22]. For this polymerization to occur, the amide groups of **6.8a**, which are preferentially coplanar with the central benzene ring in the thermodynamically most stable conformer, must rotate, thereby allowing the formation of intermolecular hydrogen bonds. The crystal structure of a stack of three molecules of **6.8b**, an achiral analog of **6.8a**, illustrates this arrangement (Figure 6.12b) [23]. Monomer **6.8a** starts to polymerize abruptly at a critical concentration, which is typical of a cooperative mechanism. This cooperativity is likely due to the preorganization and polarization of the amide groups once the dimer is formed, which strengthens further intermolecular hydrogen bonding interactions.

The Rebek group used the capsules formed from calix[4]arene tetraureas as linking units in supramolecular polymers. The bis(calix[4]arene) **6.9** (Figure 6.12c), in which two calix[4]arene tetraurea derivatives are linked at their narrower opening by a spacer unit, is an example of such a monomer [24]. These calixarene units dimerize by forming hydrogen bonds between the urea groups as discussed in Section 5.3.3, resulting in supramolecular polymers called *polycaps*. In chloroform, the capsule units along the chain contain solvent molecules that can be exchanged for more strongly bound guests such as 1,4-difluorobenzene without affecting the structural integrity of the polymers. The addition of monotopic calix[4]arene tetraureas as stoppers causes a decrease in the viscosity of the solution, indicating a reduction in the chain length. For examples of supramolecular polymers stabilized by the interactions of receptors with their substrates, see Section 6.2.4.

(a)

(b)

6.8a R =

6.8b R =

(c)

6.9

R =

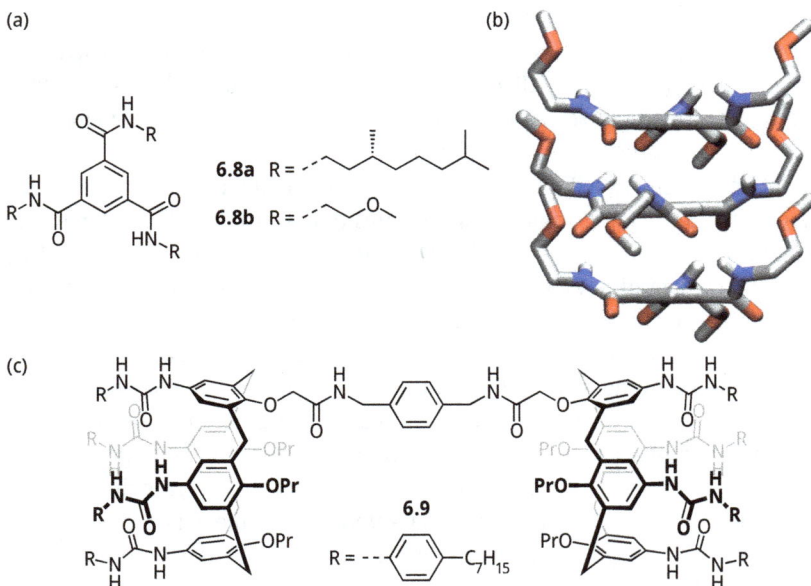

Figure 6.12: Structure of the trialkyl-1,3,5-tricarboxamides **6.8a** and **6.8b** (a) and stack of three molecules of **6.8b** in the solid state to illustrate the intermolecular hydrogen bonding pattern (b). The nonacidic protons are omitted in the crystal structure for reasons of clarity. The structure of the ditopic calix[4] arene-derived monomer **6.9** that self-assembles into polycaps is shown in (c).

6.2.2 Supramolecular polymerization mediated by coordinate bonds

Supramolecular polymers in which coordinate bonds link the subunits are formed from metal ions and monomers with two (or more) divergent donor sites [25]. The interactions lead to coordination polymers in which the ligands and the metal ions are arranged in an alternating manner (case (b) in Figure 6.1). Note that the term coordination polymer is also often used for repeated one-dimensional arrangements of ligands and metal ions in solid state structures, but while these structures usually exist only in the crystal and disintegrate then the crystal is dissolved, the polymers we are interested in persist in solution and give rise to typical properties such as increasing viscosity with increasing degree of polymerization. There are many different types of coordination polymers, but we will focus on just a few general principles.

An important aspect of coordination polymers is whether the coordinate bonds are kinetically inert or labile. The former case is typically observed for heavy transition metal ions such as platinum(II) or ruthenium(II) and gives rise to polymers that are robust and have properties similar to those of conventional covalent polymers. In contrast, polymers containing labile metal complexes of first-row transition metal ions in low oxidation states are dynamic and exhibit the stimuli-responsive and self-healing properties typical of supramolecular polymers.

The overall charge state of the polymer depends on whether the positive charge of the metal ion is compensated by the ligand or not. For example, porphyrins or salicylal-dimines form neutral complexes with dicationic metal ions (Figure 6.13), while neutral ligands form charged complexes whose charge must be compensated by additional anions. These counterions influence the properties of the resulting polymer in a characteristic way, for example, by controlling whether the material is soluble in water or not.

As with hydrogen-bonded supramolecular polymers, there is a correlation between the degree of polymerization, the concentration of the building blocks, and the strength of the metal–ligand interactions, although the exact dependence is different from that discussed earlier because coordination polymers are alternating copolymers composed of two components. The strength of the interaction depends on the metal ion and the structure of the ligand, especially its denticity. The latter aspect is illustrated by the zinc(II) complexes shown in Figure 6.13, whose stability increases with increasing number of donor atoms from a $\log K_a$ of about 3 for a single pyridine–Zn^{2+} interaction to a $\log K_a > 8$ for the terpyridine complex. The 1,10-phenanthroline complex is about an order of magnitude more stable than the 2,2'-bipyridine complex because the phenanthroline is preorganized, whereas the bipyridine must rearrange from the preferred conformation with divergent nitrogen atoms to the one in the complex prior to metal coordination. Based on the above stability constants, it can be estimated that at a 1 mM concentration of a metal salt and the same concentration of a ditopic ligand, virtually no polymerization takes place for complexes with a $\log K_a$ of 3, whereas chains of approximately 100 repeat units are formed under the same conditions when the $\log K_a$ is 7. For a given ligand and different metal ions, the complex stability usually decreases in the order $Fe^{2+} \sim Ni^{2+} > Co^{2+} > Cu^{2+} > Zn^{2+}$. For example, the bis(terpyridine) complexes of Fe^{2+}, Ni^{2+}, Co^{2+} have $\log K_a$ values of 20.9, 21.8, and 18.3, respectively [26].

Finally, the degree of polymerization depends on the metal–ligand ratio, as shown by the group of Ulrich S. Schubert, who followed the dependence of the viscosity of a solution of monomer **6.10** in methanol on the metal ion concentration [27]. Figure 6.14 shows that the viscosity increases with increasing concentration of a metal salt until a maximum is reached at one equivalent, after which the viscosity drops again. Thus, the maximum viscosity is observed at an equimolar monomer–ligand ratio, where the degree of polymerization is highest. If less than one equivalent of metal ions is present, not all ligands can participate in the polymerization and the degree of polymerization remains low. Conversely, if the metal ions are present in excess, they act as stoppers, which also results in a low degree of polymerization.

The different stabilities of the metal complexes are reflected in the extent to which the viscosity changes when the metal salts are varied, with Fe^{2+} and Ni^{2+} having greater effects than Co^{2+} or Cu^{2+}. Accordingly, metal ions that form more stable complexes lead to higher degrees of polymerization, as mentioned earlier.

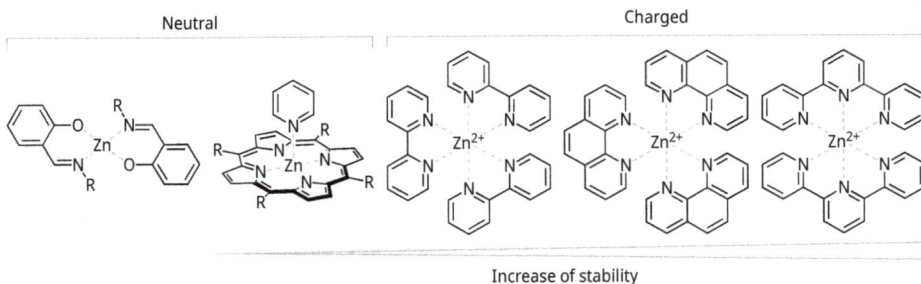

Figure 6.13: Examples of neutral metal–ligand complexes and charged complexes that can serve as linking units in coordination polymers. The stability of the neutral zinc(II) complexes increases from left to right.

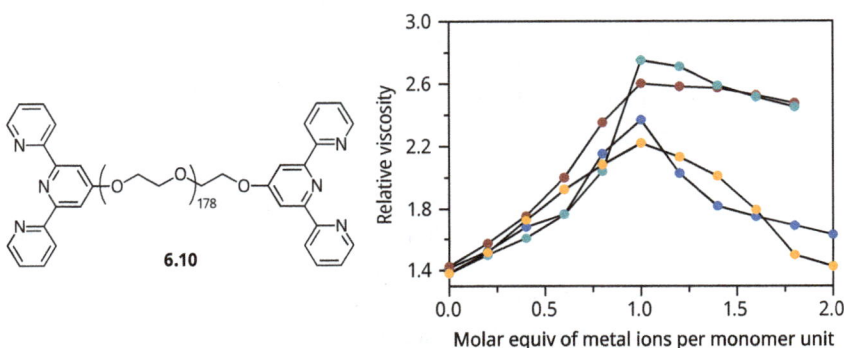

Figure 6.14: Structure of monomer **6.10** and dependence of the viscosity of a solution of **6.10** in methanol on the amount of different metal salts. The metal salt concentrations are specified as equivalents per monomer unit. The curves show the effect of Cu(OAc)$_2$ (orange), Co(OAc)$_2$ (blue), Ni(OAc)$_2$ (red), and FeSO$_4$ (green).

6.2.3 Supramolecular polymerization mediated by aromatic interactions

Molecules with extended π-systems tend to stack under appropriate conditions, which makes them attractive building blocks for supramolecular polymers. Extensive work on these systems has been carried out in the group of Frank Würthner, partly with the aim of correlating the effects of π-stacking on the electronic and optical properties of the polymeric products [28]. One of the monomers studied in this context is the ditopic mer-ocyanine derivative **6.1**, discussed in Section 6.1.4. Like other building blocks, **6.1** contains two divergent binding sites that can interact with merocyanine groups from other monomers. The interaction is due to the tendency of merocyanines to form dimers in which the two subunits are arranged in an antiparallel fashion so that the electrostatic interactions between the two ends of the merocyanine dipole become maximally attractive (Figure 6.15a). This self-assembly is quite efficient: the derivative **6.11**, for example, has a dimerization constant $\log K_a$ of 5.4 in 1,4-dioxane at 298 K.

(a)
(b)

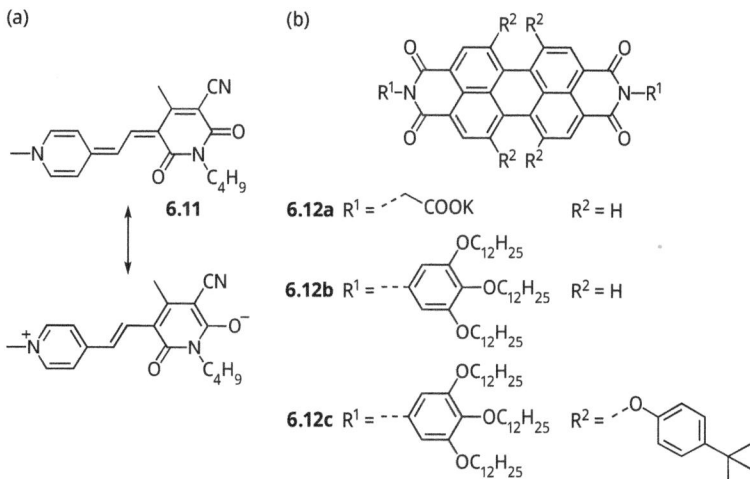

Figure 6.15: Resonance structures of **6.11**, illustrating the large dipole moment of merocyanines (a), and structures of the perylene bisimides **6.12a–c** (b).

Other classes of polycyclic aromatics that stack efficiently are perylene or naphthalene bisimides. The polymerization of water-soluble derivatives such as the planar dianionic perylene bisimide **6.12a** (Figure 6.15b) is mainly driven by the hydrophobic effect. The concentration dependence of the degree of aggregation is consistent with the isodesmic supramolecular polymerization model and a $\log K_a$ of 7.0. Almost the same $\log K_a$ underlies the polymerization of the neutral analog **6.12b** in methylcyclohexane, although the interactions between the monomer units are different. In this case, electrostatic interactions exist between the regions with the highest electron density near the carbonyl oxygen atoms and the lowest electron density near the bay region of the monomers, inducing an offset arrangement of the monomers and resulting in helical columnar stacks. In addition, dispersion interactions between the large molecular surface areas contribute to the efficiency of the interactions.

Perylene bisimides with substituents in the bay area have nonplanar, twisted structures. Their stacking is therefore less efficient as evidenced by the aggregation constant $\log K_a$ of only 5.1 associated with the polymerization of **6.12c**. However, by increasing the size of the peripheral hydrophobic substituents, the dispersion interactions between the side chains in the columnar stacks can be enhanced to the extent that nonplanar monomers polymerize as efficiently as the planar analogs.

6.2.4 Supramolecular polymerization mediated by host–guest interactions

A very versatile approach to the formation of supramolecular polymers is based on monomers equipped with receptor moieties that mediate chain elongation by binding to a suitable substrate. Although, in principle, almost any receptor can be used for this purpose, provided the interaction with the corresponding substrate is strong enough, work to date has mainly focused on a few receptor classes, most notably crown ethers, cyclodextrins, calixarenes, pillararenes, and cucurbiturils. Xi Zhang has proposed a method for classifying the various systems (Figure 6.16) [3].

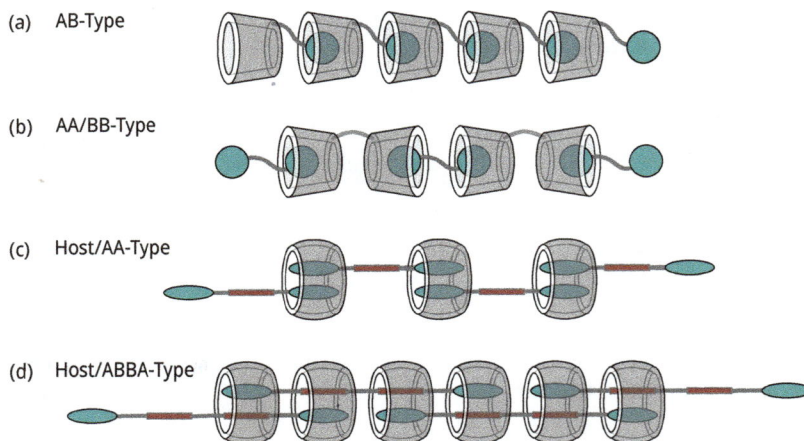

(a) AB-Type

(b) AA/BB-Type

(c) Host/AA-Type

(d) Host/ABBA-Type

Figure 6.16: Types of supramolecular polymers formed from monomers with receptor units and the corresponding binding partners.

In polymers of the AB-type, self-complementary monomers containing a receptor and a substrate moiety in the same molecule self-assemble. Care must be taken that the two subunits do not form an intramolecular complex that would prevent polymerization. An example of a monomer capable of polymerization is the α-CD derivative **6.13** described by Akira Harada with an *N*-Boc-4-aminocinnamoyl group in the 3-position (Figure 6.17a) [29]. This compound forms supramolecular polymers in water, even at low concentrations, with degrees of polymerization up to 15 [30]. In contrast, the cavity of the corresponding β-CD **6.14** is blocked by the attached substituent so that no polymerization takes place. In the case of **6.15**, the adamantane group is too large to enter the α-CD cavity, so when **6.15** is added to an aqueous solution of **6.13**, the free adamantane residues enter the β-CD cavities, simultaneously releasing the cinnamoyl groups. These groups can now be bound by the empty α-CD rings, leading to the formation of an alternating copolymer (Figure 6.17b) [31].

AA/BB-type polymers are obtained by mixing a ditopic monomer with two receptor units with a monomer containing end groups that bind to the receptor. Examples of

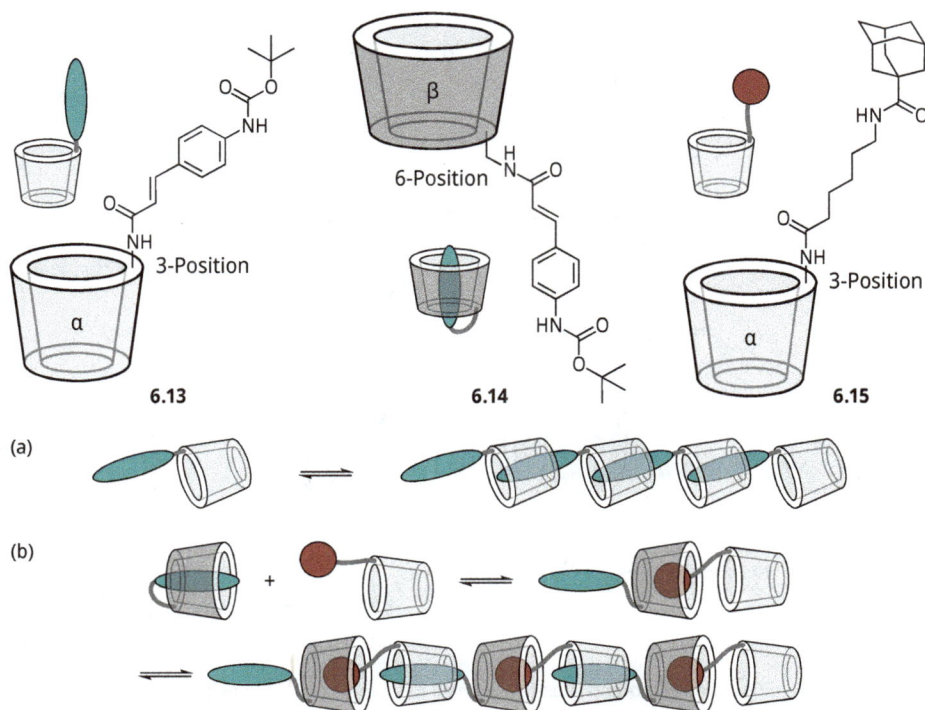

Figure 6.17: Structures of the cyclodextrin-derived monomers **6.13**, **6.14**, and **6.15**, and supramolecular homopolymer formed from **6.13** (a) and copolymer formed from **6.14** and **6.15** (b).

receptor–substrate combinations that can be used as linking units are the complexes between β-CD and adamantane, 24-crown-8 and secondary ammonium ions, sulfonato-calix[4]arenes and paraquat derivatives, and calix[5]arenes and primary ammonium ions. Figure 6.18 shows a selection of monomer combinations whose interactions lead to alternating copolymers (case (d) in Figure 6.1).

In the two remaining types of supramolecular polymers, a macrocyclic receptor serves to connect two monomer units. Since only receptors are suitable in this context with large enough cavities, this approach is mainly restricted to the use of γ-CD and, most importantly, CB[8]. A host/AA-type polymer can be obtained, for example, from the bis(anthracene) monomer **6.16** and CB[8] (Figure 6.19) [32]. The structurally related monomer **6.17** with two paraquat groups prefers to form a host/ABBA-type polymer because of the favorable interactions between the electron-rich anthracene and electron-poor paraquat units in the cucurbit[8]uril cavity [33].

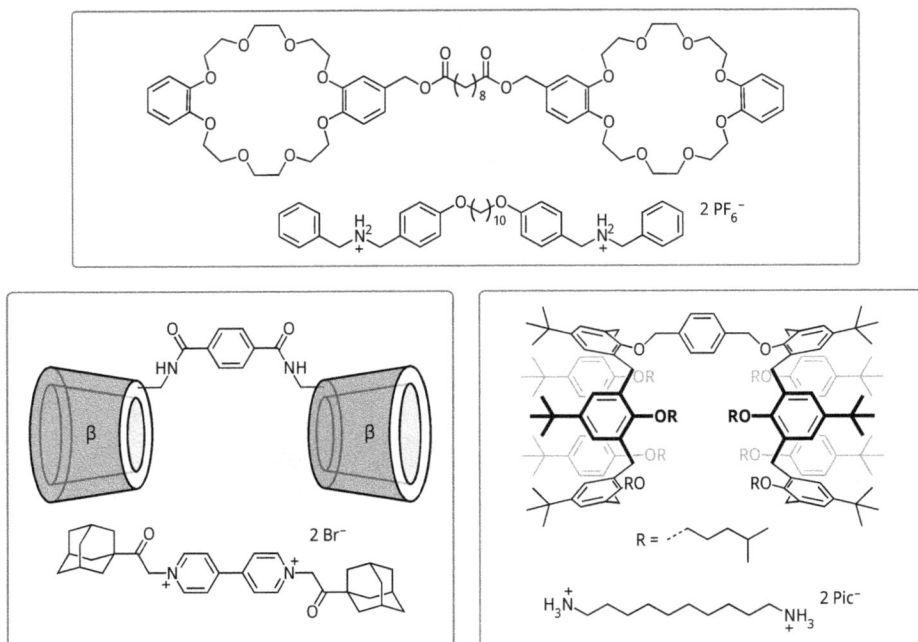

Figure 6.18: Examples of receptor- and substrate-derived monomer combinations that give rise to AA/BB-type supramolecular polymers (Pic⁻ = picrate).

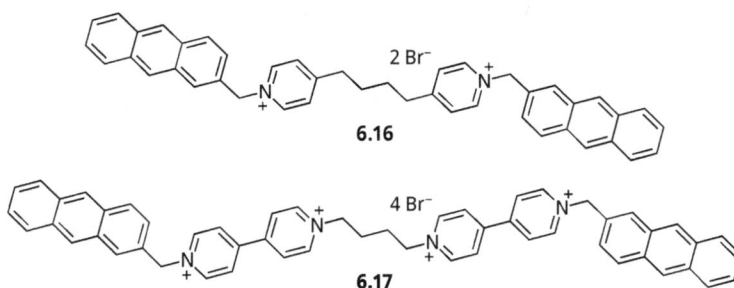

Figure 6.19: Structures of monomers **6.16** and **6.17** that form host/AA- and host/ABBA-type supramolecular polymers with CB[8].

Further reading

In Section 5.6 we have seen that self-assembly can also be mediated by reversible covalent bonds, and that the underlying concepts are closely related to self-assembly involving noncovalent interactions. Similarly, the interactions responsible for linking the monomers in supramolecular polymer chemistry must, by definition, be reversible, but not necessarily noncovalent. Indeed, the incorporation of covalent bonds into a conventional polymer that can be reversibly opened and closed under certain stimuli leads to materials with properties reminiscent of supramolecular polymers. For example, polymerization based on repeated Diels–Alder reactions leads to products that can be depolymerized at higher temperatures, polymers containing imine or hydrazone bonds can exchange building blocks under certain conditions, and such polymers can also exhibit self-healing properties. These dynamic materials, called *dynamers* by Jean-Marie Lehn, bridge supramolecular and conventional supramolecular chemistry. For more information, see:

Roy N, Bruchmann B, Lehn JM. DYNAMERS: dynamic polymers as self-healing materials. Chem. Soc. Rev. 2015, 44, 3786–807. Roy N, Schädler V, Lehn JM. Supramolecular polymers: inherently dynamic materials. Acc. Chem. Soc. 2024, 57, 349–61.

6.3 Reversible interactions between polymer chains

Polymers in which the building blocks are held together by reversible interactions have unique properties, such as the ability to self-heal or to repeatedly disassemble and reassemble under certain conditions. An alternative approach to these materials is to introduce reversible cross-links into conventional polymers by distributing receptor and substrate residues along the chains, either by grafting suitable precursors onto a preformed polymer or by copolymerization. As in covalently cross-linked polymers, the interaction between the binding partners stabilizes the polymer network, reduces solubility, and affects mechanical properties, but the reversibility of the interactions has the added advantage of allowing the material properties to be controlled by temperature, light, pH, solvent, the presence of competing binding partners, and other stimuli (Section 12.4) [34]. Accordingly, applications can be envisioned for which covalent polymers are less suitable.

How can receptor–substrate interactions be used to control material properties?

An example of a polymer containing both receptor and substrate units is the polyacrylamide described by Akira Harada, which contains additional comonomers with adamantyl and β-CD residues (Figure 6.20) [35]. Since adamantane is a good guest for β-CD (Section 4.1.4), interactions between these groups lead to cross-links in the material [36]. In addition, the presence of uncomplexed binding partners throughout the polymer network ensures that uncomplexed adamantyl and cyclodextrin residues are exposed on the surfaces when the polymer is cut. These residues bind to each other when the pieces are pressed back together, inducing self-healing and yielding a material with nearly the same mechanical stability as before.

Cross-linked polymeric materials usually behave like gels that swell by adsorbing large amounts of solvent molecules without disintegrating because the chains cannot separate. If the cross-links are noncovalent in nature, it is often possible to reversibly switch between the sol and gel states, which can be demonstrated experimentally using the tube inversion test: while the solution of a polymer that is not cross-linked

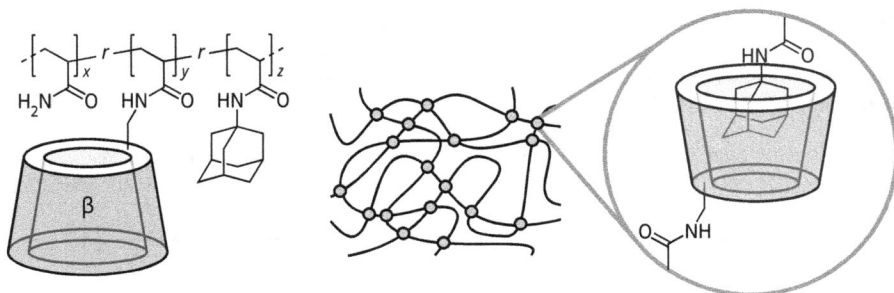

Figure 6.20: Chemical structure of the copolymer containing adamantyl and β-CD residues and schematic representation of the corresponding hydrogel.

flows freely in a vial, a gel remains at the bottom when the vial is inverted. This behavior is illustrated in Figure 6.21.

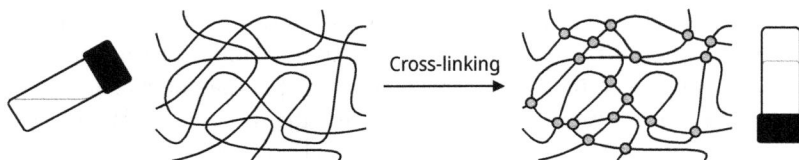

Figure 6.21: Schematic representations of polymer networks without (left) and with (right) cross-links and a method for distinguishing between the two situations using a tube inversion experiment. While a solution of the polymer without cross-links flows freely, the cross-linked version forms a gel that remains at the bottom of the inverted vial.

An example of a switchable hydrogel is the blend of poly(acrylic acids) described by Harada in which β-CD–ferrocene complexes act as cross-links (Figure 6.22) [37]. Electrochemical or chemical oxidation of the ferrocene to ferrocenium units causes the complexes to dissociate and the material to change from the gel to the sol state. Reduction of the ferrocenium groups leads back to the gel. Similarly, a hydrogel containing azobenzene and β-CD residues can be reversibly switched between the sol and gel states by light [38].

Harada also showed that receptor–substrate interactions can mediate the assembly of small cubes of a covalently cross-linked polyacrylamide hydrogel containing either β-CD or adamantyl moieties (Figure 6.23) [39]. These cubes stick together when immersed in a small amount of water and agitated in a petri dish because the surface-exposed residues bind to each other (Figure 6.23a) [40]. When four different hydrogels are used, one containing *tert*-butyl, one *n*-butyl, one α-, and one β-CD residues, self-sorting is observed (Figure 6.23b), with pairwise interactions occurring only between the *tert*-butyl and β-CD -containing hydrogels and between those containing the *n*-butyl and α-CD residues. The reason for this is that α-CD binds only to *n*-butyl groups but not to bulky *tert*-butyl

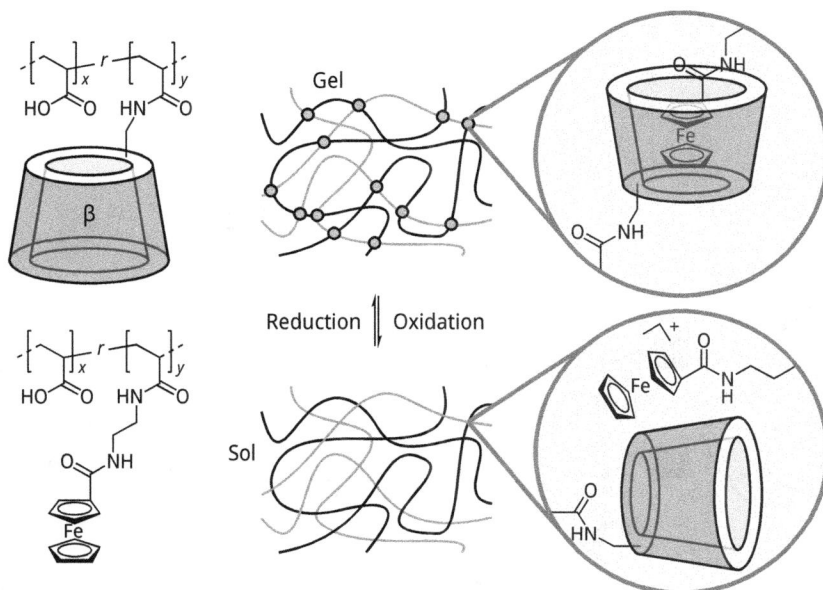

Figure 6.22: Chemical structures of the host copolymer with β-CD moieties and the guest copolymer with ferrocene units, and schematic representation of the gel-to-sol transition upon oxidation and reduction of the ferrocene moieties.

groups, while β-CD interacts with both alkyl residues, but forms more stable complexes with *tert*-butyl groups. The latter complexes are therefore formed preferentially, leaving the hydrogel containing α-CD to interact with the gel containing the *n*-butyl residues.

Self-assembly can also be controlled by external stimuli. For example, when two hydrogels, one containing α-CD residues and the other β-CD residues, are mixed with a third gel containing *E*-configured azobenzene units, the azobenzene-containing gel binds preferentially to the gel containing α-CD residues (Figure 6.23c) [41]. When the azobenzene groups are converted to the *Z* state by irradiation, the gel containing the azobenzene units dissociates from the α-CD-containing gel and binds to the β-CD-containing gel.

Crown ethers, pillararenes, or cucurbiturils are also useful recognition elements to control polymer properties [36]. An example of the use of cucurbiturils to achieve adhesion between two macroscopic objects is the molecular velcro described by Kimoon Kim, which consists of two silicon surfaces, one modified with CB[7] and one with ferrocene units (Figure 6.24) [42]. These surfaces interact so strongly underwater that a 1×1 cm^2 contact area is sufficient to support a 2 kg weight. This system thus demonstrates that many weak interactions lead to high binding strengths when they work cooperatively. Since the stability of the complex between CB[7] and ferrocene is largely mediated by the release of cavity water (Section 4.1.11), the adhesion is lost in the dry state. In water, it

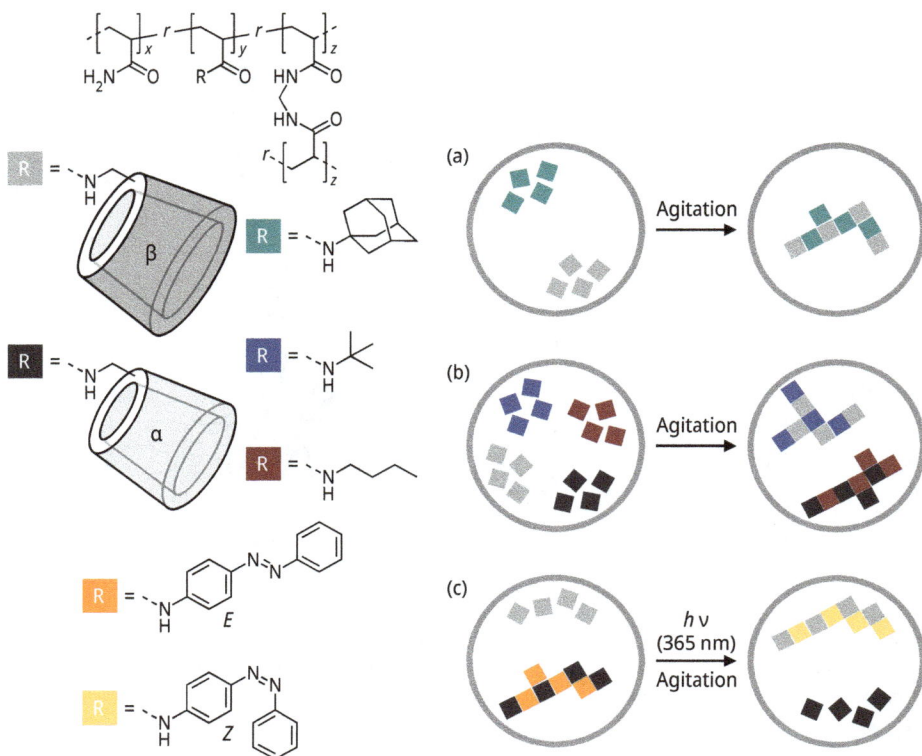

Figure 6.23: Schematic structures of hydrogels containing cyclodextrins as hosts and different types of guest units and their self-assembly behavior. The interaction of the hydrogels containing β-CD and adamantyl groups in an agitated petri dish is shown in (a). Four different hydrogels with α-CD, β-CD, n-butyl, and *tert*-butyl residues self-sort as shown in (b), and (c) illustrates that the interactions of hydrogels containing α-CD, β-CD, and azobenzene units can be controlled by light.

can be controlled by appropriate stimuli. For example, an agent that oxidizes the ferrocene to weaker binding ferrocenium units will cause the surfaces to separate.

Another approach to controlling the mechanical properties of polymers using concepts from supramolecular chemistry is to use mechanical bonds as cross-links, such as those created by threading a polymer through a ring (Chapter 7) [43]. An example is the polyethylene glycol network developed by Kohzo Ito (Figure 6.25) [44]. The cross-links in this so-called slide-ring gel consist of pairs of α-CD rings, which facilitate the dissipation of mechanical strain by allowing the polymer chains to slide past each other, which is not possible in a conventional cross-linked material [45].

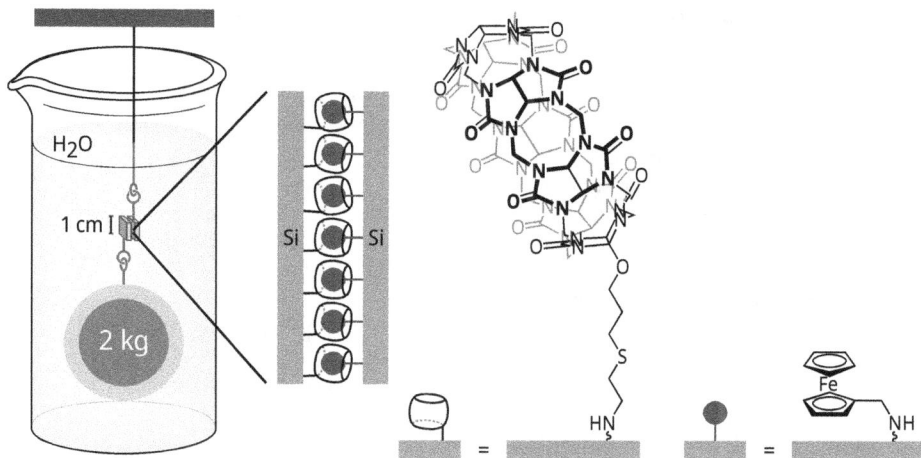

Figure 6.24: Schematic illustration of the molecular components and the mode of assembly of the molecular velcro, which consists of two 1×1 cm^2 large silicon wafers, one with CB[7] residues and the other with ferrocene residues on the surface. Both surfaces interact strongly enough in water to suspend a 2 kg weight.

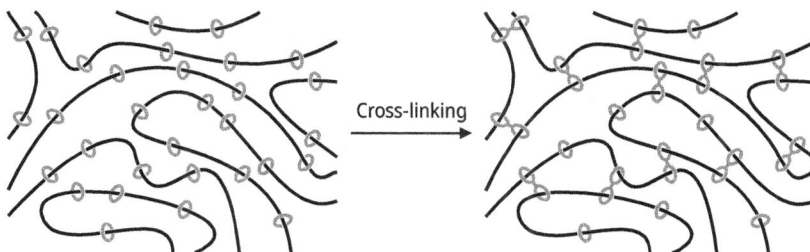

Figure 6.25: Schematic illustration of the formation and structure of a slide-ring gel.

Bibliography

[1]   De Greef TFA, Smulders MMJ, Wolffs M, Schenning APHJ, Sijbesma RP, Meijer EW. Supramolecular polymerization. Chem. Rev. 2009, 109, 5687–754.

[2]   Hart, LR, Harries JL, Greenland BW, Colquhoun HM, Hayes W. Healable supramolecular polymers. Polym. Chem. 2013, 4, 4860–70.

[3]   Yang L, Tan X, Wang Z, Zhang X. Supramolecular polymers: historical development, preparation, characterization, and functions. Chem. Rev. 2015, 115, 7196–239.

[4]   Krieg E, Bastings MMC, Besenius P, Rybtchinski B. Supramolecular polymers in aqueous media. Chem. Rev. 2016, 116, 2414–77.

[5]   Vonhausen Y, Würthner F. Concentration-, temperature- and solvent-dependent self-assembly: merocyanine dimerization as a showcase example for obtaining reliable thermodynamic data. Chem. Eur. J. 2023, 29, e202300359.

[6]   Martin RB. Comparisons of indefinite self-association models. Chem. Rev. 1996, 96, 3043–64.

[7]   Zhao D, Moore JS. Nucleation–elongation: a mechanism for cooperative supramolecular polymerization. Org. Biomol. Chem. 2003, 1, 3471–91.

[8]   Odille FGJ, Jónsson S, Stjernqvist S, Rydén T, Wärnmark K. On the characterization of dynamic supramolecular systems: a general mathematical association model for linear supramolecular copolymers and application on a complex two-component hydrogen-bonding system. Chem. Eur. J. 2007, 13, 9617–36.

[9]   Ercolani G, Mandolini L, Mencarelli P, Roelens S. Macrocyclization under thermodynamic control. A theoretical study and its application to the equilibrium cyclooligomerization of β-propiolactone. J. Am. Chem. Soc. 1993, 115, 3901–8.

[10]   Jacobson H, Stockmayer WH. Intramolecular reaction in polycondensations. I. The theory of linear systems. J. Chem. Phys. 1950, 18, 1600–6.

[11]   Sorrenti A, Leira-Iglesias J, Markvoort AJ, De Greef TFA, Hermans TM. Non-equilibrium supramolecular polymerization. Chem. Soc. Rev. 2017, 46, 5476–90.

[12]   Lohr A, Lysetska M, Würthner F. Supramolecular stereomutation in kinetic and thermodynamic self-assembly of helical merocyanine dye nanorods. Angew. Chem. Int. Ed. 2005, 44, 5071–4.

[13]   Korevaar PA, George SJ, Markvoort AJ, Smulders MMJ, Hilbers PAJ, Schenning APHJ, De Greef TFA, Meijer EW. Pathway complexity in supramolecular polymerization. Nature. 2012, 481, 492–6.

[14]   Hashim PK, Bergueiro J, Meijer EW, Aida T. Supramolecular polymerization: a conceptual expansion for innovative materials. Prog. Polym. Sci. 2020, 105, 101250.

[15]   Berl V, Schmutz M, Krische MJ, Khoury RG, Lehn JM. Supramolecular polymers generated from heterocomplementary monomers linked through multiple hydrogen-bonding arrays – formation, characterization, and properties. Chem. Eur. J. 2002, 8, 1227–44.

[16]   Sijbesma RP, Beijer FH, Brunsveld L, Folmer BJB, Hirschberg JHKK, Lange RFM, Lowe JKL, Meijer EW. Reversible polymers formed from self-complementary monomers using quadruple hydrogen bonding. Science. 1997, 278, 1601–4.

[17]   Beijer FH, Sijbesma RP, Kooijman H, Spek AL, Meijer EW. Strong dimerization of ureidopyrimidones via quadruple hydrogen bonding. J. Am. Chem. Soc. 1998, 120, 6761–9.

[18]   Scherman OA, Ligthart GBWL, Sijbesma RP, Meijer EQ. A selectivity-driven supramolecular polymerization of an AB monomer. Angew. Chem. Int. Ed. 2006, 45, 2072–6.

[19]   Goor OJGM, Hendrikse SIS, Dankers PYW, Meijer EW. From supramolecular polymers to multi-component biomaterials. Chem. Soc. Rev. 2017, 46, 6621–37.

[20]   ten Cate AT, Kooijman H, Spek AL, Sijbesma RP, Meijer EW. Conformational control in the cyclization of hydrogen-bonded supramolecular polymers. J. Am. Chem. Soc. 2004, 126, 3801–8.

[21]   Folmer BJB, Sijbesma RP, Kooijman H, Spek AL, Meijer EW. Cooperative dynamics in duplexes of stacked hydrogen-bonded moieties. J. Am. Chem. Soc. 1999, 121, 9001–7.

[22]   Smulders MMJ, Schenning APHJ, Meijer EW. Insight into the mechanisms of cooperative self-assembly: the "sergeants-and-soldiers" principle of chiral and achiral C_3-symmetrical discotic triamides. J. Am. Chem. Soc. 2008, 130, 606–11.

[23]   Lightfoot MP, Mair FS, Pritchard RG, Warren JE. New supramolecular packing motifs: π-stacked rods encased in triply-helical hydrogen bonded amide strands. Chem. Commun. 1999, 1945–6.

[24]   Castellano RK, Rudkevich DM, Rebek Jr. J. Polycaps: reversibly formed polymeric capsules. Proc. Natl. Acad. Sci. U. S. A. 1997, 94, 7132–7.

[25]   Dobrawa R, Würthner F. Metallosupramolecular approach toward functional coordination polymers. J. Polym. Sci. A. 2005, 43, 4981–95.

[26] Holyer RH, Hubbard CD, Kettle SFA, Wilkins RG. The kinetics of replacement reactions of complexes of the transition metals with 2,2′,2″-terpyridine. Inorg. Chem. 1966, 5, 622–5.

[27] Schmatloch S, van den Berg AMJ, Alexeev AS, Hofmeier H, Schubert US. Soluble high-molecular-mass poly(ethylene oxide)s via self-organization. Macromolecules. 2003, 36, 9943–9.

[28] Chen Z, Lohr A, Saha-Möller CR, Würthner F. Self-assembled p-stacks of functional dyes in solution: structural and thermodynamic features. Chem. Soc.Rev. 2009, 38, 564–84.

[29] Harada A, Takashima Y, Yamaguchi H. Cyclodextrin-based supramolecular polymers. Chem. Soc. Rev. 2009, 38, 875–82.

[30] Miyauchi M, Takashima Y, Yamaguchi H, Harada A. Chiral supramolecular polymers formed by host−guest interactions. J. Am. Chem. Soc. 2005, 127, 2984–9.

[31] Miyauchi M, Harada A. Construction of supramolecular polymers with alternating α-, β-cyclodextrin units using conformational change induced by competitive guests. J. Am. Chem. Soc. 2004, 126, 11418–9.

[32] Liu Y, Liu K, Wang Z, Zhang X. Host-enhanced π–π interaction for water-soluble supramolecular polymerization. Chem. Eur. J. 2011, 17, 9930–5.

[33] Liu Y, Yu Y, Gao J, Wang Z, Zhang X. Water-soluble supramolecular polymerization driven by multiple host-stabilized charge-transfer interactions. Angew. Chem. Int. Ed. 2010, 49, 6576–9.

[34] Xia D, Wang P, Ji X, Khashab NM, Sessler JL, Huang F. Functional supramolecular polymeric networks: the marriage of covalent polymers and macrocycle-based host−guest interactions. Chem. Rev. 2020, 120, 6070−123.

[35] Kakuta T, Takashima Y, Nakahata M, Otsubo M, Yamaguchi H, Harada A. Preorganized hydrogel: self-healing properties of supramolecular hydrogels formed by polymerization of host−guest−monomers that contain cyclodextrins and hydrophobic guest groups. Adv. Mater. 2013, 25, 2849–53.

[36] Sinawang G, Osaki M, Takashima Y, Yamaguchi H, Harada A. Supramolecular self-healing materials from non-covalent cross-linking host−guest interactions. Chem. Commun. 2020, 56, 4381–95.

[37] Nakahata M, Takashima Y, Yamaguchi H, Harada A. Redox-responsive self-healing materials formed from host−guest polymers. Nat. Commun. 2011, 2, 511.

[38] Takashima Y, Nakayama T, Miyauchi M, Kawaguchi Y, Yamaguchi H, Harada A. Complex formation and gelation between copolymers containing pendant azobenzene groups and cyclodextrin polymers. Chem. Lett. 2004, 3, 890–1.

[39] Harada A, Takashima Y, Nakahata M. Supramolecular polymeric materials via cyclodextrin−guest interactions. Acc. Chem. Res. 2014, 47, 2128–40.

[40] Harada, A, Kobayashi, R, Takashima Y, Hashidzume A, Yamaguchi H. Macroscopic self-assembly through molecular recognition. Nat. Chem. 2011, 3, 34–7.

[41] Yamaguchi H, Kobayashi Y, Kobayashi R, Takashima Y, Hashidzume A, Harada A. Photoswitchable gel assembly based on molecular recognition. Nat. Commun. 2012, 3, 603.

[42] Ahn Y, Jang Y, Selvapalam N, Yun G, Kim K. Supramolecular velcro for reversible underwater adhesion. Angew. Chem. Int. Ed. 2013, 52, 3140–4.

[43] Chen L, Sheng X, Li G, Huang F. Mechanically interlocked polymers based on rotaxanes. Chem. Soc. Rev. 2022, 51, 7046–65.

[44] Okumura Y, Ito K. The polyrotaxane gel: a topological gel by figure-of-eight cross-links. Adv. Mater. 2001, 13, 485–7.

[45] Ito K. Novel entropic elasticity of polymeric materials: why is slide-ring gel so soft? Polym. J. 2012, 44, 38–41.

7 Threading molecules

CONSPECTUS: *Have you ever wondered if it is possible to make a knot in a molecule, to connect cyclic molecules like links in a chain, or to thread them onto a linear molecule like pearls on a necklace? In this chapter, we will see that all this can be done quite efficiently by using the concepts of supramolecular chemistry. In the first part of the chapter, we focus on the structural and stereochemical aspects of knots and molecules containing two or more interlocked components. We then look at general strategies for the preparation of such entangled molecules, followed by a series of examples classified according to the noncovalent interactions that mediate their formation. The topics discussed in these contexts also serve as an introduction to the next chapter, which will show how mechanically interlocked molecules are used to create machines.*

7.1 Molecular topology

Molecules are structurally characterized on several levels. The most fundamental parameter is composition, which describes the number and types of atoms in a molecule. Constitution then specifies how the atoms are connected, and configuration refers to their orientation in space. Closely related to these terms is the concept of isomerism: two molecules are structural (or constitutional) isomers if they have the same composition but differ in the connectivity of the atoms. Configurational isomers, on the other hand, have the same composition and connectivity but differ in the spatial arrangement of the atoms. While these concepts are frequently used, it is less common to structurally distinguish molecules based on their topology.

> What is the difference between topology and structure?

Topology is a mathematical concept that describes the geometric properties of three-dimensional objects. A key idea is that topologically equivalent objects can be transformed into each other by continuous deformations such as stretching, twisting, crumpling, or bending. Accordingly, an empty, crumpled balloon is topologically equivalent to a filled one, but even objects that appear to have different shapes can be topologically equivalent. An example is the crown on the cover of this book and a doughnut.

When we apply the concept of topology to molecules, we must allow them to be completely flexible, as if the atoms were connected by infinitely stretchable rubber bands. The molecules can thus be deformed into any shape by changing the distance between the atoms or their arrangement in space, as long as the connectivity remains the same. Whether two molecules are topologically equivalent is then determined by constructing graphs with all atoms in the same plane. If no bonds cross in these graphs, they are planar, while graphs containing crossings are not. The number of

https://doi.org/10.1515/9783111315171-007

crossings determines the topological properties of the molecule, with molecules having the same number of crossings being topologically equivalent.

Let us perform this analysis for the molecules shown in Figure 7.1. The chair conformation of cyclohexane can be easily drawn as a regular hexagon in a two-dimensional projection with no crossed bonds. Bicyclo[2.2.2]octane can be similarly flattened, while buckminsterfullerene is more difficult to draw in a two-dimensional manner. One way to do this is to pull a five-membered ring wide apart, flattening the entire molecule in the process. The resulting graph also contains no crossed bonds, rendering all three molecules, as different as they may seem, topologically equivalent.

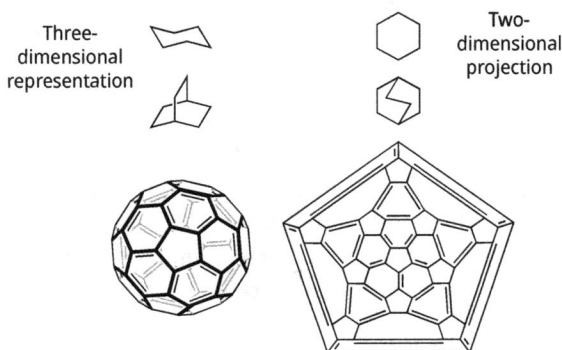

Figure 7.1: Analysis of the topological properties of cyclohexane, bicyclo[2.2.2]octane, and buckminsterfullerene.

Bond crossing is inevitable in some of the interlocked molecules discussed in this chapter. An example is the catenane shown in Figure 7.2a, which consists of two interlocked macrocycles that are inseparable, even though they are not linked by an actual bond. The schematic illustration shows that the two interlocked rings cross twice, preventing the construction of a planar graph. The structure in Figure 7.2b is a single macrocycle whose folding into a trefoil knot results in three crossings, which can also only be represented by a nonplanar graph. The rotaxane in Figure 7.2c consists of a ring through which a linear component, the so-called axle, is threaded. Bulky end groups at the ends of the axle prevent the ring from escaping. Despite the interlocking of the two components, this compound is topologically equivalent to those in Figure 7.1 because the concepts of topology allow the ring to be infinitely flexible so that it can slide over the end groups of the axle. A rotaxane can thus be drawn as a planar graph of the two separate components. All three molecules shown in Figure 7.2 are therefore topologically distinct, with the rotaxane having the same topological properties as the compounds in Figure 7.1, while the catenane and the trefoil knot have two and three crossings, respectively.

(a)

(b)

(c)

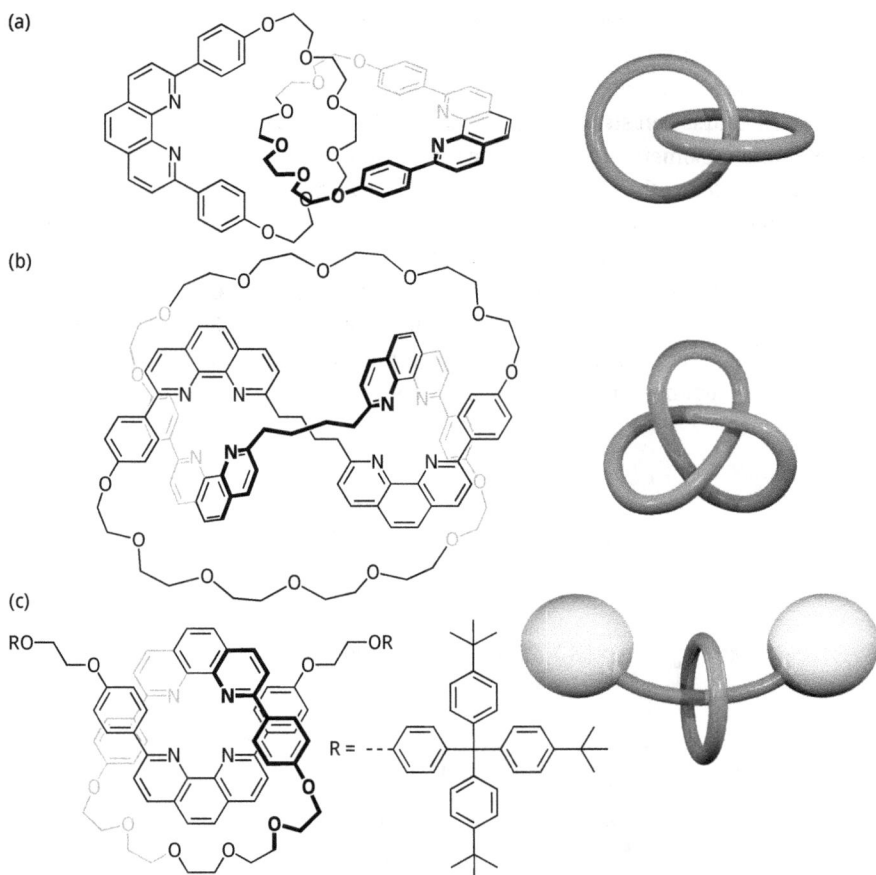

Figure 7.2: Examples of a catenane (a), a knot (b), and a rotaxane (c) along with the corresponding schematic representation of these mechanically interlocked molecules.

Let us now integrate the concept of topology into the structural classification of molecules with which this chapter began [1, 2]. At the first level of structural classification, the topology of a molecule is defined by the connectivity of the atoms. If two molecules have a different composition or constitution, they are never topologically identical, because there is no way to transform them into each other by deformation. However, different compounds can be topologically equivalent.

Consider a cycloalkane with the composition $(CH_2)_{40}$, a ring with only half the number of carbon atoms $(CH_2)_{20}$ and a catenane containing two of the smaller macrocycles (Figure 7.3a). The large and small rings have different topologies because they differ in composition, but they are topologically equivalent because they can both be represented as planar graphs. The large macrocycle and the catenane are constitutional isomers whose topologies differ not only because of the different connectivity of the atoms but also because one molecule can be represented as a planar graph while the other cannot.

Turning to stereoisomers, we find that most of the stereoisomers we have learned to distinguish in courses on stereochemistry are in fact topologically equivalent. For example, the *E*- and *Z*-isomers of an olefin or the two enantiomers of a chiral compound are generally topologically equivalent because they can be represented as planar graphs. Topological isomerism occurs only when two isomers cannot be transformed into each other by bond deformation. An example is the macrocycle in Figure 7.3b and the corresponding trefoil knot, which are topological diastereomers. The chirality of the knot also causes it to exist in two enantiomeric forms (Figure 7.3c), which have the same number of crossings in their nonplanar graphs but cannot be interconverted by bond deformation, making them topological isomers.

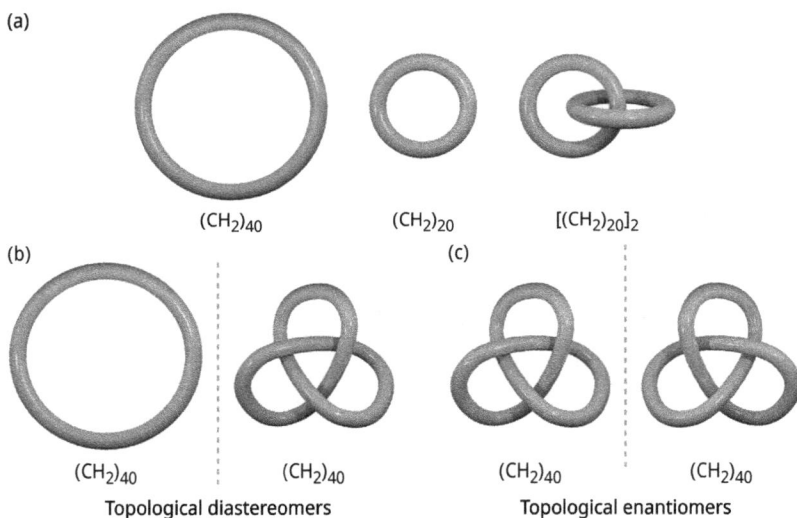

(a)

$(CH_2)_{40}$ $(CH_2)_{20}$ $[(CH_2)_{20}]_2$

(b) (c)

$(CH_2)_{40}$ $(CH_2)_{40}$ $(CH_2)_{40}$ $(CH_2)_{40}$

Topological diastereomers Topological enantiomers

Figure 7.3: Classification of topological isomers. The structural isomers in (a) have different topologies, regardless of whether they are topologically equivalent or not. The ring and the knot in (b) are topological diastereomers, while the two knots in (c) are topological enantiomers.

Describing molecules in terms of topology obviously requires careful structural analysis. Importantly, the term topology is not synonymous with structure, and its use in this sense can lead to incorrect conclusions. Furthermore, it is not wrong to say that two different molecules have different topologies, but when topological aspects are irrelevant to the discussion, using the term topology may not be necessary and even confusing.

What is a mechanical bond?

Disentangling the interlocked structures shown in Figure 7.2 requires cleavage of bonds. In the case of the catenane, opening one ring yields a pseudorotaxane in which an acyclic component is threaded through an intact ring. Since the acyclic subunit does not contain bulky end groups, it is easily dethreaded, yielding the separate components. To separate the components of the rotaxane, one could cleave either the ring or the axle, and the opening of the knot is achieved by cleaving the chain, followed by conformational reorganization of the resulting acyclic product. The structural integrity of these molecules is therefore closely linked to the structural integrity of their components. If the latter is lost, entanglement is no longer guaranteed. Thus, the stability of the interlocked structure is merely a mechanical effect that does not necessarily require special interactions in the regions of the molecules where the crossings occur. Nevertheless, the entanglement holds the individual components tightly together and therefore qualifies as a bond, a so-called mechanical bond. Such a mechanical bond does not involve any orbital or electrostatic interactions, but still leads to a considerable degree of stabilization that is lost only when covalent bonds are broken [3, 4].

The structures in Figure 7.2 are examples of large and structurally diverse families of mechanically interlocked molecules. All three subgroups come in different flavors. Catenanes, for example, can consist of two or more rings of different structures and sizes. Rotaxanes can contain more than one axle and/or ring. How many components are actually connected is usually specified in square brackets in front of the family name. For example, a [3]catenane consists of three interlocked rings.

Knots are characterized not only by the number of components but also by the number of crossings. The variation of these structural parameters leads to a large number of theoretically possible knots, which are treated mathematically in an overarching theory. A number of these knots have been prepared as molecules, not only the trefoil knot shown in Figure 7.2 but also more complex knots, as we will see later.

The first ideas to synthesize mechanically interlocked molecules date back to the first half of the twentieth century [5]. The initial work was mainly driven by curiosity, probably motivated by questions similar to those posed in the outline of this chapter. Only later did it become clear that, as so often the case, nature had solved the problem already [6, 7]. For example, there are a number of biological processes that involve catenated or knotted single or double-stranded DNA, including DNA recombination and replication. A selection of structures demonstrating the structural versatility of interlocked DNA is shown in Figure 7.4. Similar knotted structures have also been identified in proteins, where they serve to improve (thermal) stability or to precisely arrange multiple subunits in space. In some cases, entanglements in biomolecules are the basis of complex functions, such as in ATP synthase, an archetypal natural molecular machine. Mechanically interlocked molecules are therefore more than just laboratory curiosities, and the efficient synthetic strategies available today have even made it possible to make them functional.

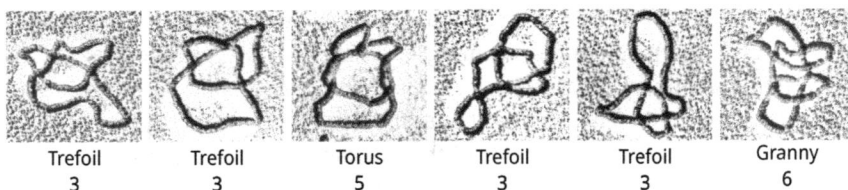

Trefoil	Trefoil	Torus	Trefoil	Trefoil	Granny
3	3	5	3	3	6

Figure 7.4: Examples of knotted DNA structures. The images are electron micrographs of knots formed by treating circular DNA with 13S condensin, a protein involved in the structural maintenance of chromosomes, and a topoisomerase. The number of crossings and the name of the corresponding knot are indicated below each image. Images adapted with permission from [8]. Copyright Elsevier, 1999.

7.2 Stereochemical aspects

Mechanical bonds have stereochemical consequences when they stabilize the way in which parts of an interlocked molecule are oriented, folded, or twisted [9]. While the macrocycle in Figure 7.3b is achiral, for example, the corresponding trefoil knot is a chiral structure, even though none of the carbon atoms along the chain is a stereogenic center. More precisely, the knot has topological chirality because interconversion of the enantiomers requires opening of the chain, which simultaneously leads to the disappearance of the mechanical bond. Rotaxanes and catenanes become chiral when the constituents have a reduced symmetry, as shown in Figure 7.5a.

In the case of the [2]rotaxane, the chirality results from the combination of a C_{nh} (or C_s) symmetric ring with a C_{nv}-symmetric axle, both of which have functional groups or subunits arranged in a specific orientation or sequence. Because a rotaxane is a topologically trivial structure that can be represented as a planar graph after the components have been separated, racemization does not require the cleavage of one component. Instead, the ring can simply slide off the axle, change its orientation, and slide back on, which is allowed by the concept of topology. The chirality of rotaxanes is therefore not a topological effect, but a special case of planar chirality, where the ring acts as a stereogenic element whose orientation with respect to the axle is fixed by the mechanical bond. Accordingly, the term mechanically planar chiral rotaxane is used [10].

The [2]catenane in Figure 7.5a also owes its chirality to the mechanical bond, which fixes the orientation of the two achiral but oriented rings. When more than two rings are interlocked, several different types of chiral catenanes exist, all of which have in common that at least two rings must be C_{nh}- or C_s-symmetric [10]. As in the case of knots, racemization requires the cleavage of one ring and the concomitant loss of the mechanical bond. Most catenanes are therefore topologically chiral, but there are special systems whose chirality results from the orientation of a substituent externally attached to an otherwise symmetric ring. In this case, reversing the orientation of this substituent, which can be achieved without ring opening, also leads to

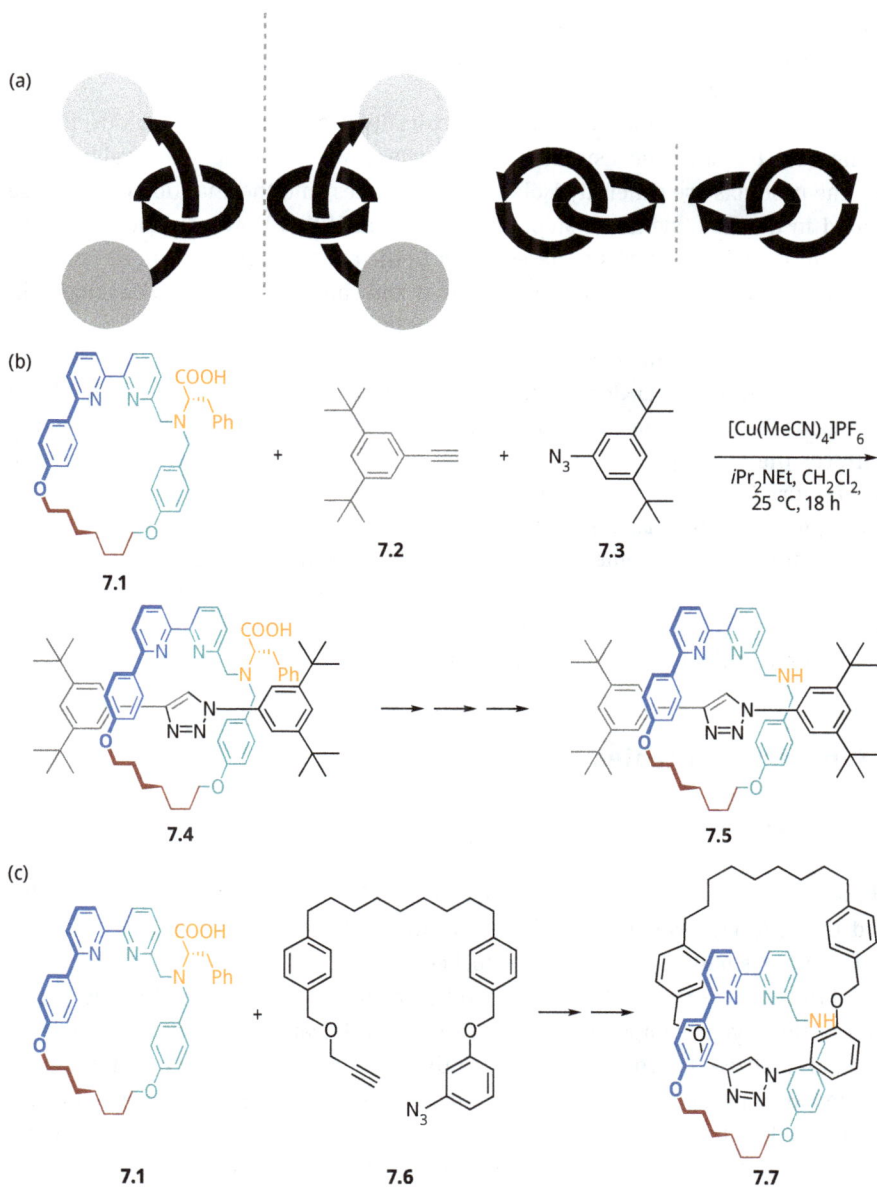

Figure 7.5: Schematic representation of the enantiomers of a mechanically planar chiral [2]rotaxane and a topologically chiral [2]catenane (a). The arrows indicate the orientation of the constituents in both compounds. The enantioselective synthesis of the chiral [2]rotaxane **7.5** and [2]catenane **7.7** is shown in (b) and (c), respectively.

racemization [11]. Thus, the stereogenic unit of catenanes is not necessarily topological, and it has therefore been suggested that chiral catenanes should also be called mechanically planar chiral.

As we will see in the following sections, chirality is quite common for interlocked molecules [10], but since most synthetic approaches involve the use of achiral building blocks, the products are generally isolated as racemates, and the enantiomers must be separated in the final synthetic step, often by chiral HPLC. However, there are also examples of enantioselective syntheses. A very efficient strategy that can be used for the enantioselective synthesis of both rotaxanes and catenanes has been developed in the group of Stephen M. Goldup [12]. The central building block in this synthesis is the chiral cyclophane **7.1**, which contains a 2,2′-bipyridine subunit and an attached substituent with a stereocenter derived from an amino acid. Treatment of **7.1** with alkyne **7.2** and azide **7.3** under the conditions of copper(I)-catalyzed azide–alkyne cycloaddition affords the chiral [2]rotaxane **7.4** in 74% yield, essentially as a single diastereoisomer (Figure 7.5b). The high yield is due to the dual role of the copper(I) ions in mediating not only the triazole formation but also the preorganization of the components in a threaded arrangement, as we will see in Section 7.4.3. The enantiomerically pure rotaxane **7.5** is obtained by cleavage of the chiral auxiliary group. Similarly, the reaction of **7.1** with **7.6** in the presence of a copper(I) catalyst gives the chiral [2]catenane **7.7** with excellent stereoselectivity (Figure 7.5c).

7.3 Synthetic strategies

7.3.1 Molecular strategies

The first catenane synthesis was published by Edel Wasserman in 1960 [13]. The idea behind the approach was that a mixture of an acyclic and a macrocyclic molecule should contain small amounts of interlocked species, with the acyclic component threaded through the cyclic one. If this pseudorotaxane could be trapped by macrocyclization of the acyclic component, the corresponding [2]catenane should be obtained. Wasserman realized this concept by first synthesizing the 34-membered deuterated cycloalkane **7.8** from the long-chain α,ω-dicarboxylic acid ester **7.9** by acyloin condensation followed by Clemmensen reduction with Zn/DCl (Figure 7.6). Compounds **7.8** and **7.9** were then mixed and subjected to the conditions of a further acyloin condensation. After chromatographic removal of the alkane **7.8**, the macrocyclic acyloin **7.10** was characterized by IR spectroscopy, with the spectrum obtained providing evidence for the presence of C–D bonds in the isolated material. Thus, the product appeared to contain not only **7.10** but also the catenane **7.11**. The successful formation of **7.11** was supported by the isolation of a small amount (about 1%) of **7.8** after cleavage of the acyloin-containing ring in **7.11** with hydrogen peroxide followed by dethreading of **7.12**. Wasserman later reported the isolation of a few milligrams of **7.11** in a yield of 0.0001%

[14]. Criticism of the validity of Wasserman's claims [15] prompted the Leigh group to repeat the reported synthesis and analyze the products using modern analytical techniques [16]. The results provided clear evidence that **7.11** is indeed formed under the reported conditions. Furthermore, **7.11** is not the only interlocked product, since a mixture of macrocyclic alkanes and alkenes is formed in the first step of the reaction. Thus, Wasserman's approach led to a mixture of catenanes containing saturated and unsaturated deuterated rings, with 0.07% of the diester ending up in this mixture during the second acyloin condensation under the conditions reported by Leigh.

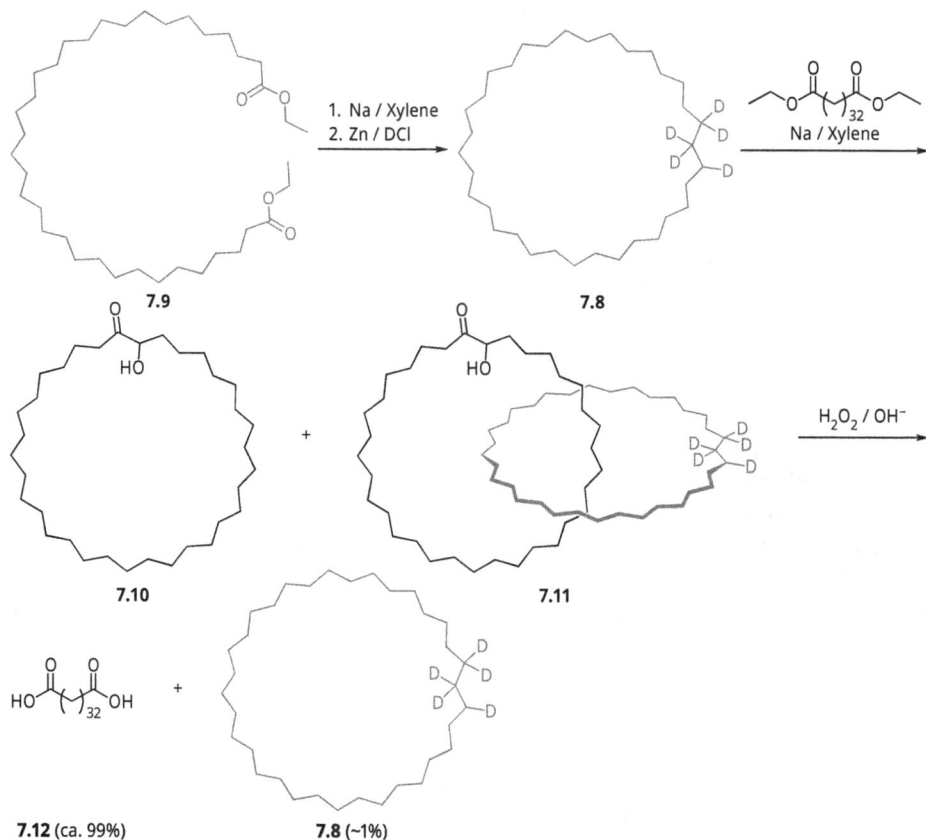

Figure 7.6: Synthesis of catenane **7.11** using a statistical approach involving the acyloin condensation of the diester **7.9** in the presence of the deuterated macrocycle **7.8**. The synthesis begins with the preparation of **7.8**. Proof of the presence of the catenane was based on the release of **7.8** upon treatment of the product mixture with hydrogen peroxide.

Although Wasserman's approach demonstrated that catenanes are accessible from preformed macrocycles by threading an acyclic molecule through the ring and cyclizing it, the results also showed that statistical approaches based on the probability (or hope) of capturing pseudorotaxanes in solution are clearly inefficient. A much more promising strategy would be to somehow hold the building blocks together before the mechanical bond is formed. On the basis of this idea, Gottfried Schill and Arthur Lüttringhaus developed a series of template-directed syntheses of mechanically interlocked molecules in which the preorganization of the subunits was achieved by covalent bonds that eventually had to be cleaved to release the free interlocked components [15]. Figure 7.7 shows the key steps of their first [2]catenane synthesis, described in 1964 [17]. This synthesis starts with the preparation of the *ansa* catechol derivative **7.13** in a linear sequence of 10 steps. This catechol is condensed with an aliphatic dichloroketone to obtain the ketal **7.14**, in which the sp^3 hybridization of the ketal carbon atom causes the two side chains to be oriented orthogonal to the plane of the aromatic ring. Consequently, the double *ansa* compound **7.15**, formed from **7.14** by nitration, reduction, and two intramolecular nucleophilic substitutions, already has the topology of the final catenane with the rings still covalently linked. Separation of the two rings involves cleavage of the ketal group, oxidation of the resulting catechol, and hydrolysis of the enamine to give **7.16**.

Further transformation of **7.16** involves the ozonolysis of the hydroxyquinone intermediate followed by reduction to yield catenane **7.17**, in which only the acetylated amino group provides evidence for the covalent template that served to direct the interlocked topology [18]. The complete reaction sequence, for which the total yield has not been reported, consists of 26 steps.

Using conceptually similar approaches, often based on the use of ketals as covalent templates, Schill and Lüttringhaus later independently pursued the synthesis of mechanically interlocked molecules with remarkable success [15]. They were the first to prepare a [2]catenane containing two cycloalkanes, a [3]catenane, a rotaxane, and a precursor for a knot. With these elegant syntheses, they laid much of the conceptual groundwork in the field and inspired modern approaches to the use of covalent templates for the synthesis of mechanically interlocked molecules [19]. Despite these impressive achievements, mechanically interlocked molecules remained laboratory curiosities until supramolecular syntheses became available. The advantages of supramolecular syntheses are that they are typically much shorter than the methods described by Schill and Lüttringhaus, yield the products in larger quantities, and allow for easier structural variation. In addition, the products additionally contain built-in binding sites that facilitate the design of functional systems, such as those described in Chapter 8.

Before discussing these syntheses, we will look at another and conceptually different approach to accessing interlocked molecules. This approach is based on the peculiar properties of Möbius strips. Such strips are easily obtained from a rectangular strip of paper by gluing the two narrow ends together to form a loop. Depending on how many times the strip is twisted before the loop is fixed, systems with different topologies are obtained. No twist results in a simple ribbon that has an outer and an inner surface (Figure 7.8a). Cutting this ribbon in half produces two rings with the

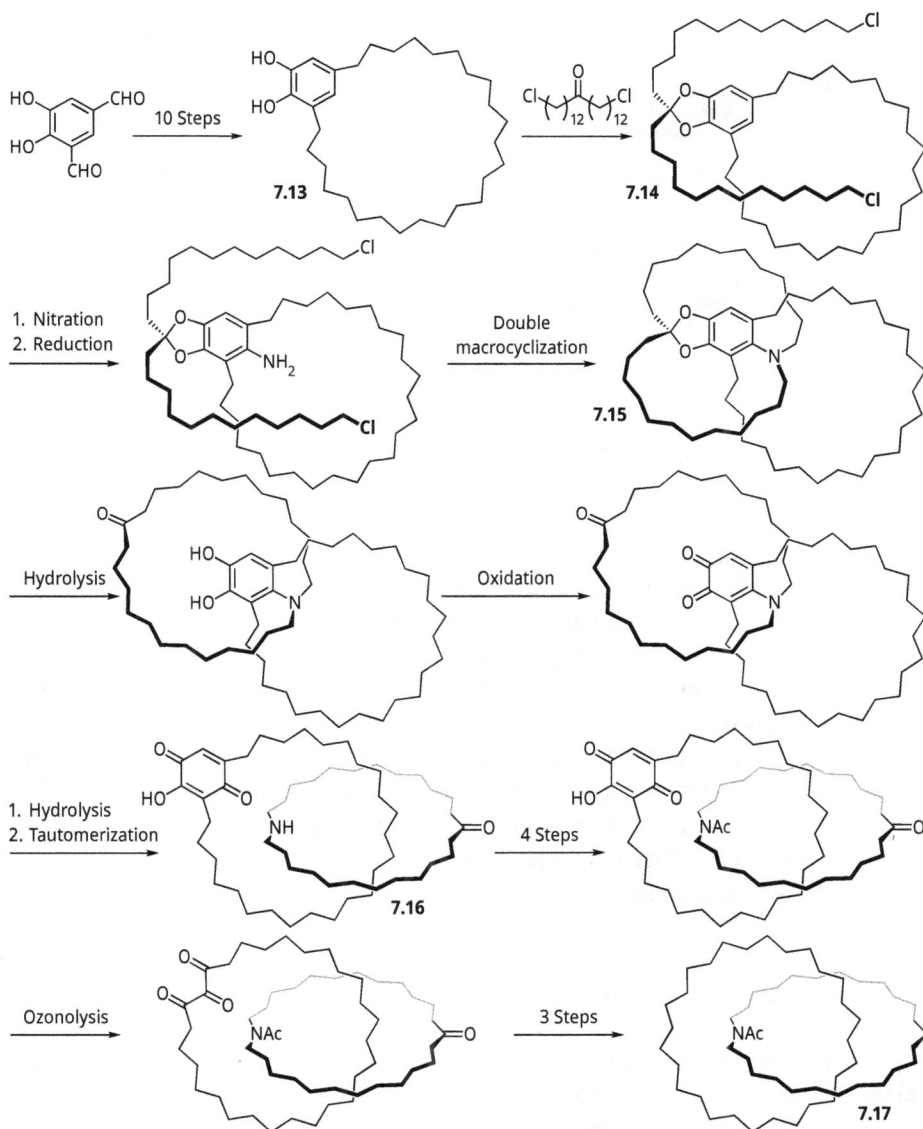

Figure 7.7: First catenane synthesis described by Lüttringhaus and Schill in which the ketal **7.14** serves as a covalent template to direct the topology of the precatenane **7.15**.

same diameter as the original. A Möbius strip is obtained by twisting one end by 180° before joining the ends (Figure 7.8b). In this system, any point on the surface can be reached without crossing the edges and a bisection yields a single ring with twice the

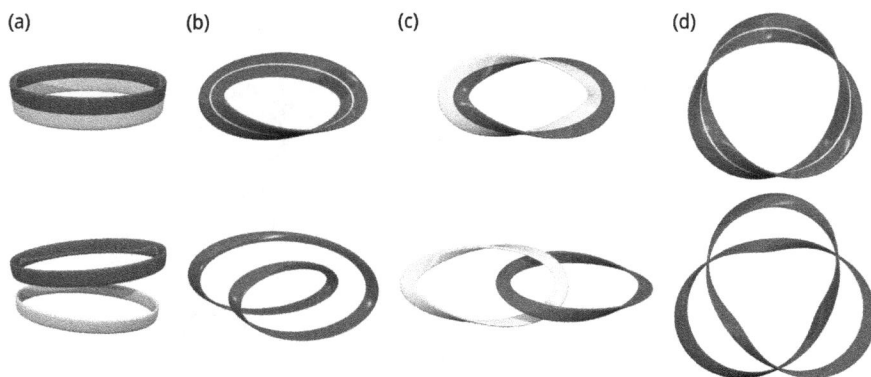

Figure 7.8: Structures of ribbons differing in the number of half-twists and products obtained by bisecting these ribbons. The ribbons have no (a), one (b), two (c), and three (d) half-twists. The ribbon in (b) is a Möbius strip.

diameter of the untwisted ring. A strip with two half-twists gives a catenane after bisection (Figure 7.8c), and one with three half-twists gives a trefoil knot (Figure 7.8d).

Molecules with similar topologies are obviously attractive precursors for accessing interlocked systems. This concept was pursued by David M. Walba, who based his approach on the polycyclic crown ether **7.18** (Figure 7.9) [20]. Macrocyclization of this compound yields two products in an almost 1:1 ratio, one representing the nontwisted ribbon **7.19** and the other the Möbius ladder **7.20** as a pair of enantiomers. Cleavage of the double bonds by ozonolysis yields two equivalents of the small macrocycle **7.21** in the case of **7.19** and one equivalent of **7.22** in the case of **7.20**. Unfortunately, the doubly twisted ribbon that should give rise to the catenane was not detected in the reaction mixture, presumably because it would be too strained.

7.3.2 Supramolecular strategies

Supramolecular syntheses of mechanically interlocked molecules are based on noncovalent or reversible coordinative interactions that hold the precursors together or fix them in an interlocked conformation. Accordingly, such syntheses require the presence of suitable subunits in the building blocks that interact either directly or through an external template. If these interactions cause the final interlocked product to be thermodynamically more stable than potential other products, the synthesis can be carried out under thermodynamic control using reversible reactions such as those discussed in Sections 5.6 and 5.7. An example is the acetylcholine-mediated catenane synthesis described in Section 5.7.3. Similar thermodynamically controlled syntheses also allow the preparation of rotaxanes and have proven to be particularly useful for accessing complex knotted molecular architectures, as we will see.

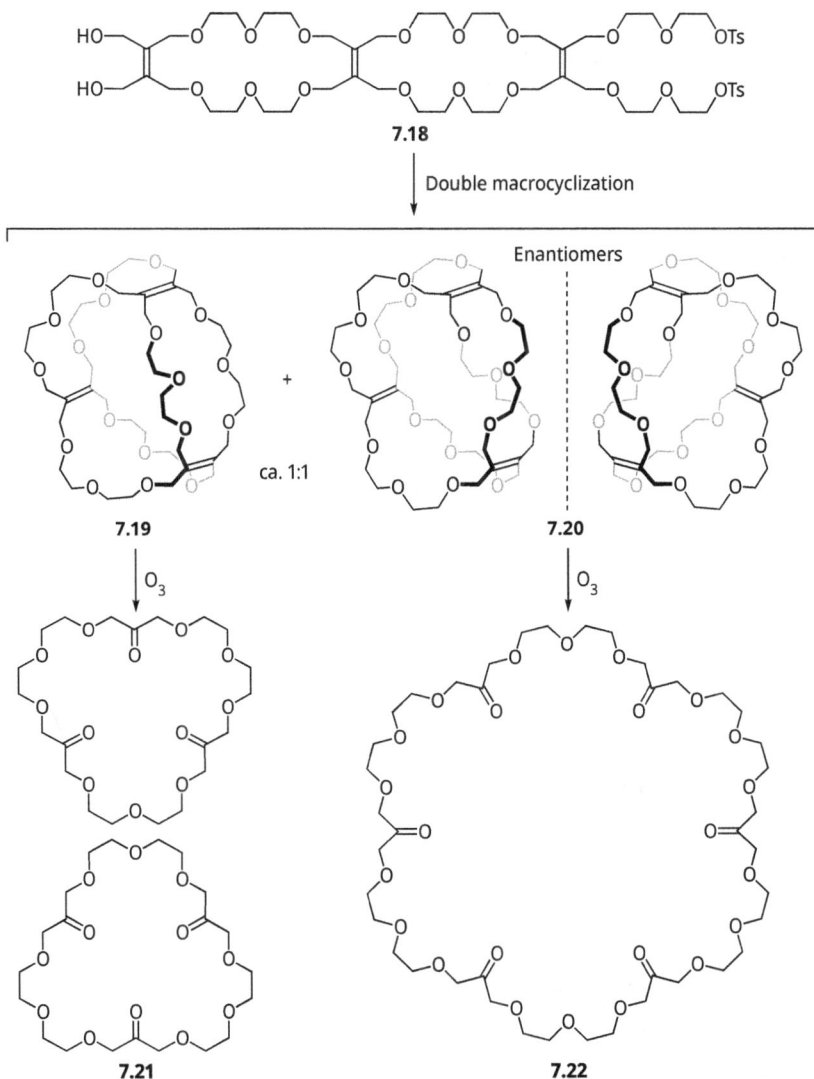

Figure 7.9: Attempted synthesis of a catenane using the Möbius approach. The ribbon **7.19** and the racemate of the Möbius ladder **7.20** were formed in almost equal amounts and could be converted into the corresponding rings by ozonolysis. The doubly twisted product was not observed.

Alternatively, suitable interactions can facilitate the formation of the correct bonds on the way from the precursor to the mechanically interlocked product by preorganizing the reacting groups. The reaction leading to the desired product thus becomes faster than other competing reaction pathways. Conceptually, this approach is closely related to template-directed crown ether syntheses (Sections 4.1.1 and 5.1) or the catenane synthesis shown in Figure 7.7. In the latter case, the arrangement of the two side chains of intermediate **7.14** is responsible for the formation of a product with the topology of a [2]catenane. The covalent template in **7.14** thus kinetically favors the formation of the desired product and disfavors the formation of unwanted side products. Supramolecular catenane syntheses are based on similar strategies, but use reversible interactions to preorganize the precursors rather than covalent bonds. The actual bond-forming reactions are irreversible, and preorganization therefore lowers the Gibbs free energy of activation associated with the formation of the desired product. Possible approaches are illustrated in Figure 7.10.

The route (a) in Figure 7.10 begins with the orthogonal arrangement of two acyclic precursors, which can be achieved, for example, by coordinating divalent ligands to a metal ion that favors a tetrahedral coordination geometry (Section 7.4.1). Cyclization of the two ligands and, if necessary, subsequent removal of the template then leads to the [2]catenane. Alternatively, the catenane synthesis can start from a cyclic and an acyclic precursor, both forming a pseudorotaxane (route (b)). Intermolecular interactions between the two pseudorotaxane components stabilize the complex and thus favor catenane formation. Finally, it is also possible that the first ring mediates the formation of the second interlocked ring (route (c)). This strategy underlies the enantioselective catenane synthesis discussed in the previous section and is also useful for the synthesis of rotaxanes, as discussed in Section 7.4.3.

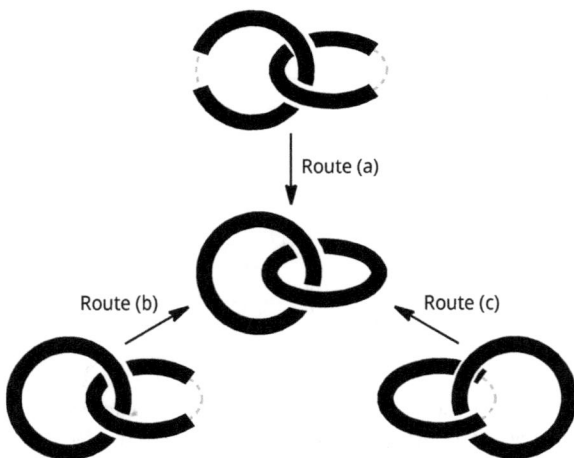

Figure 7.10: Schematic illustration of synthetic strategies leading to [2]catenanes, all of which rely on the preorganization of suitable precursors through reversible interactions.

The synthesis of knots requires the stabilization of a knotted conformation of the precursors, which is ultimately fixed by closing the ring (Figure 7.11) [21]. The formation of a trefoil knot, for example, requires a linear oligomer with subunits along the chain whose intramolecular interactions, or interactions with appropriate external templates, stabilize a conformation in which the chain ends pass through two loops. Joining these ends leads to the knot (route (a)). Alternatively, a trefoil knot is obtained by pairwise joining the correct chain ends in a double helicate in which the two precursors cross three times (route (b)). Structurally more complex knots are prepared using similar strategies. Thermodynamically controlled reactions are useful when the products contain several intricately interlocked components.

Figure 7.11: Possible routes for the synthesis of a trefoil knot, starting from precursors whose conformations and mutual orientations are stabilized by appropriate interactions. That the two central structures have the same topology is illustrated in Figure 7.20.

Rotaxanes can be synthesized in several ways. The simplest approach is to slip the ring over the fully formed axle that already contains the two stoppers (*slipping*) (route (a) in Figure 7.12). This strategy requires a finely tuned size match between the ring and the stopper groups so that the activation energy required for the ring to bypass the stoppers can just be overcome under certain conditions, for example at elevated temperatures. At lower temperatures, the product must be kinetically inert. Since no bonds are formed in this synthesis, it proceeds under thermodynamic control and the yield correlates with the extent to which the rotaxane is more stable than the free components under the reaction conditions. A variation of this approach is the use of a large ring whose diameter allows it to slide easily over the end groups of the axle. A chemical reaction that reduces the effective diameter of the ring then leads to the stabilization of the rotaxane (*shrinking*). Conversely, a rotaxane with small stoppers that do not prevent the ring from sliding over them is stabilized by enlarging the end groups (*swelling*).

The most common approaches to the preparation of rotaxanes are based on the covalent synthesis of one component in the presence of the preformed other component. The formation of the ring in the presence of the axle is called *clipping* (route (b) in Figure 7.12). In this case, the axle templates the macrocyclization of the acyclic precursor by stabilizing a conformation in which the end groups are in close proximity. The other two synthetic strategies involve the assembly of the axle, which is achieved either by attaching the bulky end groups to the chain ends of the acyclic component in a pseudorotaxane (*stoppering*) (route (c) in Figure 7.12) or by joining the two dumbbell halves of the axle (*trapping*) (route (d) in Figure 7.12). In the latter approach, the ring usually plays an active role in the bond formation and thus remains trapped in

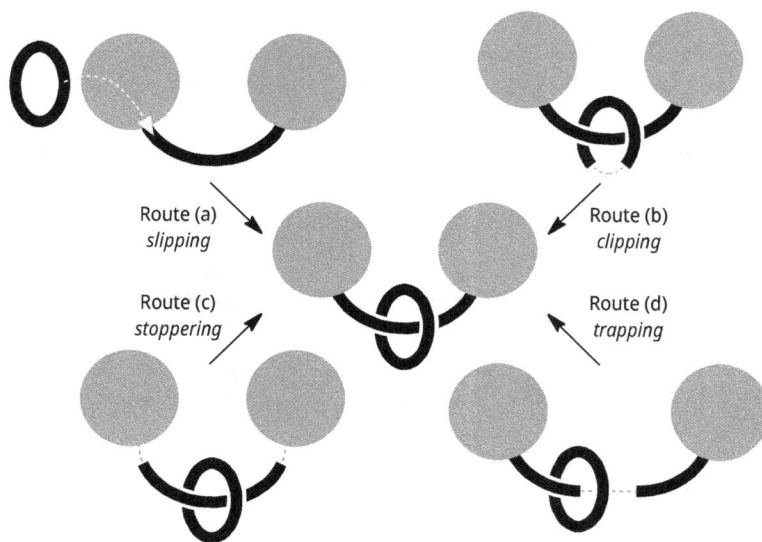

Figure 7.12: Rotaxane syntheses involving the slipping of the ring over the stoppers (a), the clipping of a ring around a preformed axle (b), the stoppering of the axle of a pseudorotaxane (c), and the capture (or trapping) of the two components of the axle by the ring (d).

the product. All of these kinetically controlled approaches rely on irreversible bond formation reactions.

7.4 Syntheses using metal coordination

7.4.1 Catenanes

Metal–ligand interactions were used to preorganize the building blocks in the first successful supramolecular synthesis of a [2]catenane, and numerous other syntheses of mechanically interlocked molecules have since been based on related concepts. Due to their covalent nature, coordinate bonds hold the precursors tightly together in a well-defined arrangement, similar to the covalent bonds in the syntheses of Schill and Lüttringhaus (Section 7.3.1). However, while covalent templates require additional synthetic steps to separate the components, the reversibility of coordinative interactions makes template removal relatively easy.

In 1980, Jean-Pierre Sauvage reported the first metal-directed synthesis of a catenane in a short communication in *Tetrahedron Letters* [22]. The pioneering approach described in this article demonstrated for the first time how easy and effective the synthesis of mechanically interlocked molecules can be when using the concepts of supramolecular chemistry. Sauvage went on to develop many other interlocked molecules of increasing structural complexity, establishing his reputation as a "master of chemical topology" [23] and laying the groundwork for the Nobel Prize in Chemistry that he received in 2016.

Conceptually, Sauvage's catenane synthesis is based on a tetrahedral metal complex with two orthogonally arranged ligands (Figure 7.13). By covalently bridging the end groups of each ligand *via* linkers of appropriate length, a metal complex with the topology of a [2]catenane is created. To indicate that the transition metal is still part of the structure, Sauvage proposed the term catenate for such a complex. Removal of the templating metal ion yields the corresponding catenand. The same catenand is also obtained by starting with a preformed ring through which an acyclic precursor is threaded. Since four bonds are formed in the former so-called *entwining* approach, and only two in the alternative *threading* strategy, the latter is usually more efficient.

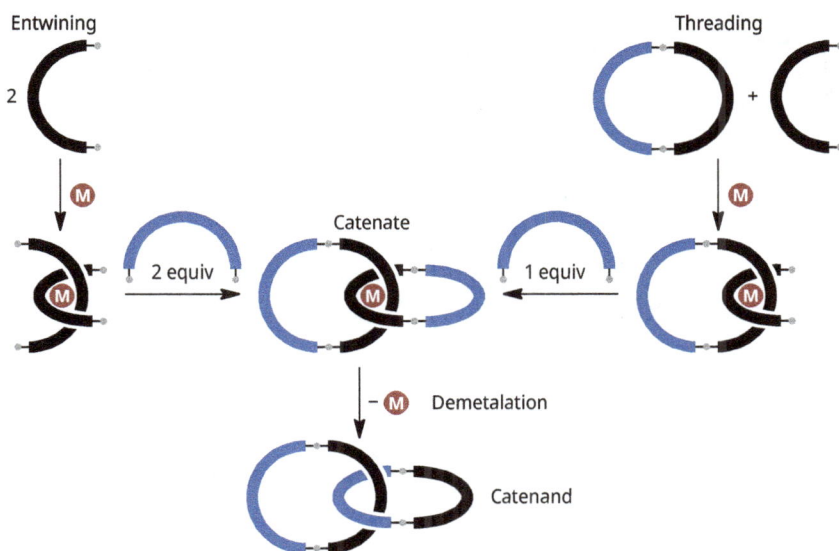

Figure 7.13: Catenane syntheses using metal templates. In the entwining strategy, two ligands are arranged orthogonally by the metal ion. Bridging their end groups yields a catenate, which is transferred to the corresponding catenand by removal of the metal template. The alternative threading strategy proceeds *via* a metal complex in which a linear ligand is threaded through a cyclic one. In this strategy, only two covalent bonds are formed in the second step instead of four in the entwining strategy.

The actual catenane synthesis described by Sauvage, shown in Figure 7.14, involves the use of Williamson ether syntheses between the ligand **7.23** and the ethylene glycol-derived diiodide **7.24**. The corresponding catenate **7.25** is isolated in 27% yield using the entwining approach and in a remarkable yield of 42% using the alternative and clearly more efficient threading strategy [24]. Note that the addition of 1 equiv of a copper(I) source to a 1:1 mixture of **7.23** and the corresponding crown ether in the threading

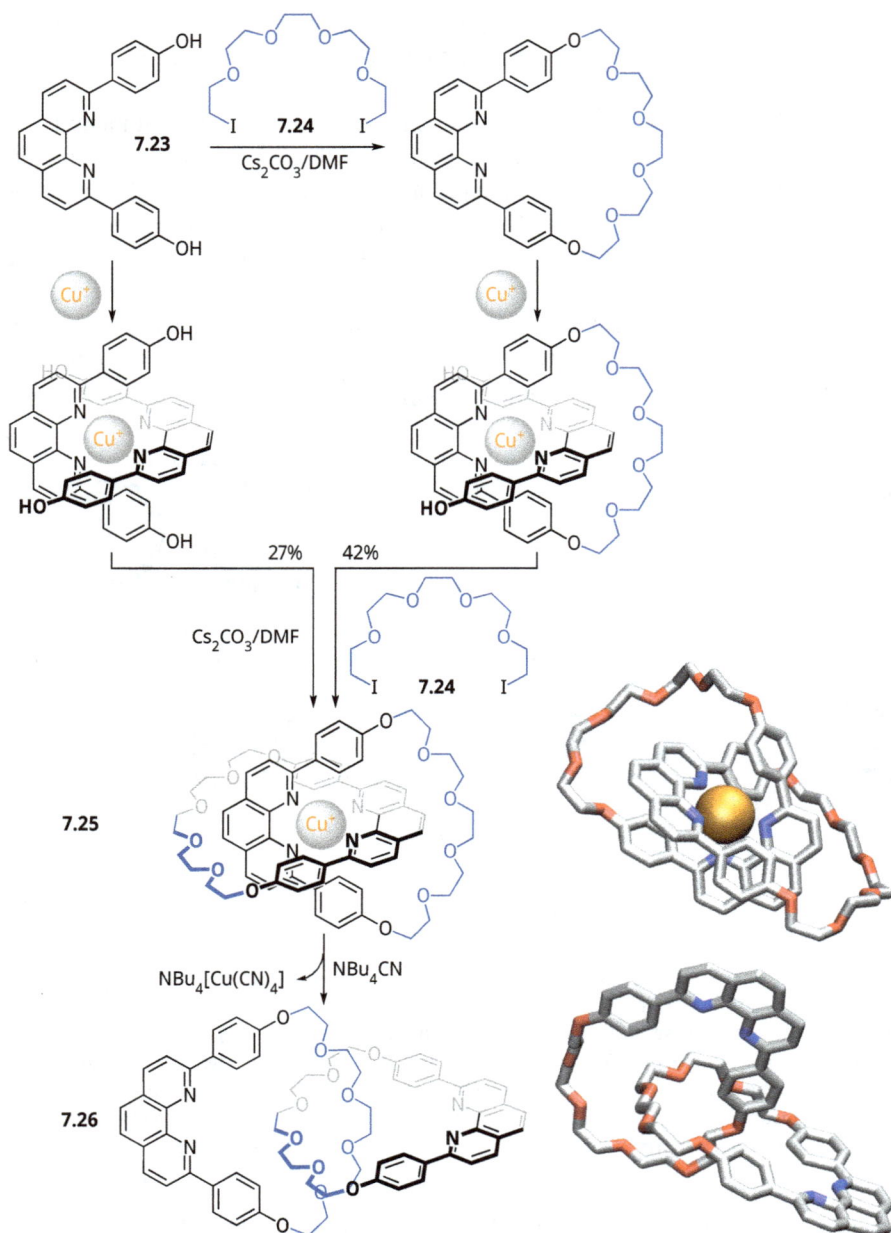

Figure 7.14: [2]Catenane syntheses developed by Sauvage, involving the entwining of the ligands in the copper(I) complex of **7.23** upon reaction with the diiodide **7.24** (a), or the macrocyclization of the acyclic ligand in the copper(I) complex containing a preformed ring (b). The yields of the corresponding Williamson ether syntheses show that the threading strategy is more efficient. The crystal structures show the interlocked topologies of **7.25** and **7.26**. Hydrogen atoms are omitted in the crystal structures for reasons of clarity.

strategy leads to the exclusive formation of the heteroleptic complex. The reason for this is that the macrocyclic component alone is unable to form a complex in which two ligand molecules are arranged orthogonally. Phenanthroline **7.23** can form a homoleptic complex,, but the formation of this complex would leave copper(I) ions unused. They can bind to the phenanthroline units in the macrocycle but remain coordinatively unsaturated. The formation of the homoleptic complex thus violates the principle of maximum site occupancy. Consequently, the formation of the heteroleptic complex leads to the thermodynamically most favorable situation. We have seen other strategies to induce the formation of heteroleptic complexes in Section 5.5.4.

Treatment of **7.25** with an excess of cyanide salts, which compete in the copper(I)-phenanthroline coordination by inducing the formation of the stable $[Cu(CN)_4]^{3-}$ complex, causes demetalation and release of the interlocked rings. Thus, the synthesis of catenand **7.26** involves either a three-step (entwining) or a four-step (threading) reaction sequence, both of which yield the product in substantial absolute amounts. Instead of using complexes between copper(I) ions and bidentate ligands, also other metal complexes can be used in such syntheses as long as the metals induce the appropriate arrangement of the ligands. For example, an orthogonal arrangement of two ligands also results when tridentate ligands are coordinated to metals that prefer an octahedral coordination geometry (Section 5.5.1). Alternatively, suitable linear or square planar metal complexes can be used for the construction of catenanes.

The crystal structures in Figure 7.14 confirm the interlocked arrangement of the rings in **7.25** and **7.26**. While the orientation of the two phenanthroline moieties in **7.25** is fixed by coordination to the metal ion, **7.26** exhibits a co-conformation in the solid state in which the phenanthroline moieties are oriented away from each other for steric and electronic reasons (repulsion of the nitrogen lone pairs).

What is a co-conformation?

The term co-conformation used in the previous paragraph was introduced by J. Fraser Stoddart to describe the mutual arrangement of different components in an interlocked molecule [4]. Co-conformations are interconvertible by translation, pirouette, or rocking motions. In the case of **7.26**, for example, the conversion of the co-conformation found in the crystal structure to the less favorable co-conformation with two facing phenanthroline units involves changing the relative orientations of the two rings.

How can mechanical bonds be detected?

Besides crystal structures, NMR spectroscopy and mass spectrometry are important methods for the structural characterization of mechanically interlocked molecules. The spatial proximity of the components of an interlocked molecule, enforced by the mechanical bond, produces characteristic effects on the proton resonances, for example, and ^{1}H NMR spectroscopy can thus be used to assign the preferred co-conformation in solution. Temperature-dependent NMR measurements are useful to characterize dynamic processes, such as the rate of interconversion of different co-conformations.

Interlocked and noninterlocked structures also give rise to distinctly different fragmentation patterns in mass spectra. In the case of macrocyclic compounds, these spectra typically contain a series of peaks with m/z ratios smaller than that of the molecular ion because the ring, once opened, becomes progressively smaller as it fragments. In contrast, when one ring of a catenane is opened during ionization or subsequent collisions in the gas phase, the catenane components separate. As a result, the spectrum contains the peak of the molecular ion and, as the next smaller fragment, the peak with the m/z ratio of a single ring, followed by smaller fragments, while ions with m/z ratios between those of the intact catenane and the single ring are missing. Thus, mass spectrometry is a convenient analytical technique to distinguish catenanes from noninterlocked rings.

By adapting the strategy shown in Figure 7.13 to other metal complexes, a wide range of different catenanes can be obtained. For example, cyclodimerization of an appropriate metal complex yields a [3]catenane after demetalation (Figure 7.15a). This concept has been realized using the copper(I) complex **7.27**, in which a 1,10-diphenylphenanthroline derivative containing two terminal alkyne groups is threaded through the same macrocyclic ligand used in the synthesis of **7.26** [25]. Treatment of **7.27** under Glaser coupling conditions (CuCl, CuCl$_2$, air) leads to the formation of the catenate **7.28** containing three interlocked rings (Figure 7.15b). The remarkable efficiency of this reaction, resulting in an overall yield of 58%, allows the isolation of **7.28** in gram quantities. As a byproduct, the [4]catenane is formed by cyclotrimerization in a yield of 22%. Catenate **7.28** is converted to the corresponding catenand by treatment with a cyanide salt.

A [2]catenane like **7.26** is called a Hopf link in knot theory. It represents the simplest link with two interlocking rings. More complex links are the Solomon link featuring four crossings between two rings, the Star of David catenane with six crossings between two rings, and the Borromean rings with six crossings between three rings (Figure 7.16). These interlocked structures are also accessible using appropriate coordination complexes. We will look at the synthesis of a Solomon link in the next chapter because it is closely related to the synthesis of a trefoil knot, which will be discussed there. Synthetic strategies to access the other catenanes are outlined below.

A Star of David catenane consists of two rings intertwined in a manner similar to the two triangles in the flag of the State of Israel. David A. Leigh showed that the synthesis of such a catenane can be based on the circular helicate discussed in Section 5.5.2 using tris(bipyridine) **7.29** and iron(II) sulfate (Figure 5.48) [26]. The helicate obtained from these components contains pairs of terminal alkene groups at the six corners (Figure 7.17) that are close enough to be covalently linked by ring-closing metathesis, giving the corresponding catenate in up to 92% yield. The crystal structure of

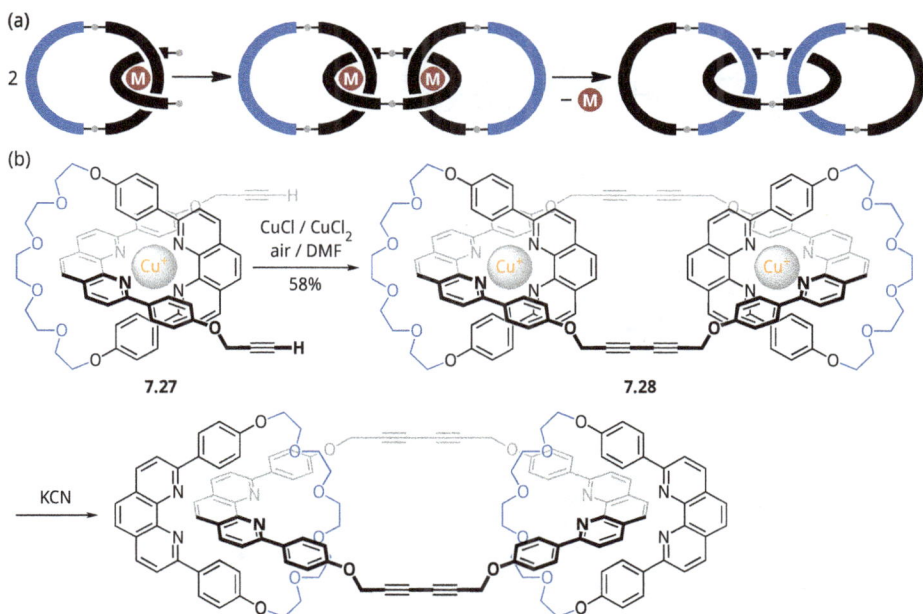

Figure 7.15: Schematic representation of a [3]catenane synthesis by cyclodimerization of a metal complex comprising a cyclic and an acyclic ligand (a) and actual synthetic approach using two equiv of copper(I) complex **7.27** by means of a Glaser coupling (b).

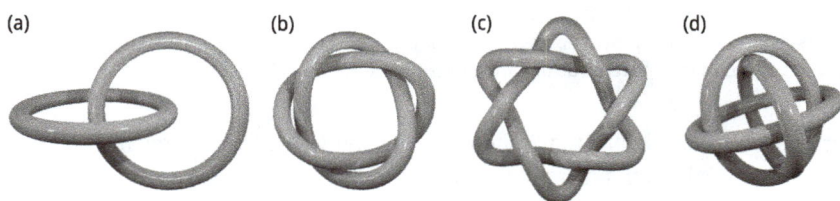

Figure 7.16: Schematic illustrations of a Hopf link (a), a Solomon link (b), a Star of David catenane (c), and a [3]catenane with Borromean ring topology (d).

the product confirms the expected topology. Removal of the metal ions with ethylene-diaminetetraacetate (EDTA) yields the metal-free form of this catenane.

Borromean rings are links consisting of three rings that are interlocked without one ring penetrating another. As a result, when one ring is opened, the remaining two rings are no longer connected. Historically, Borromean rings have been used as symbols in various contexts, with the name deriving from the Borromeo family, a family of merchants and bankers in Renaissance Italy, whose coat of arms features three similarly interlocked rings. Several attempts to synthesize their molecular counter-

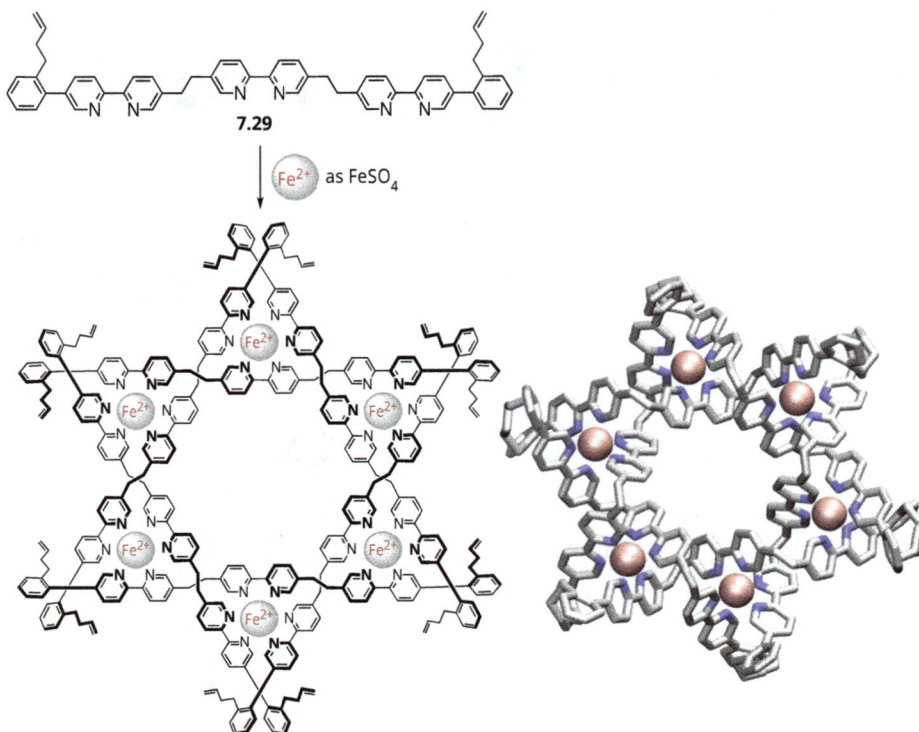

Figure 7.17: Molecular structures of ligand **7.29** and of the circular triple iron(II) helicate derived from **7.29**, and crystal structure of the product obtained by treating **7.29** under ring-closing metathesis conditions. The protons in the crystal structure are omitted for reasons of clarity.

parts were unsuccessful before the Stoddart group showed that Borromean rings are synthetically accessible in a surprisingly simple one-step procedure from equimolar amounts of the 2,2'-bipyridine-derived diamine **7.30**, 2,6-diformylpyridine, and zinc(II) acetate by imine formation and zinc(II) coordination (Figure 7.18) [27]. Since both reactions are reversible, the product is formed under thermodynamic control, unlike in the other syntheses discussed so far. The rules of self-assembly derived in Section 5.1 thus control the formation of the thermodynamically favored product, which in this case happens to be a hexazinc(II) complex containing three interlocked macrocyclic ligands, as confirmed by a crystal structure. This synthesis, which gives **7.31** in almost quantitative yield, is so reliable that it can even be carried out in undergraduate teaching laboratories [28]. Reduction of the imine groups and removal of the zinc(II) ions in the form of their EDTA complexes yields the metal-free product.

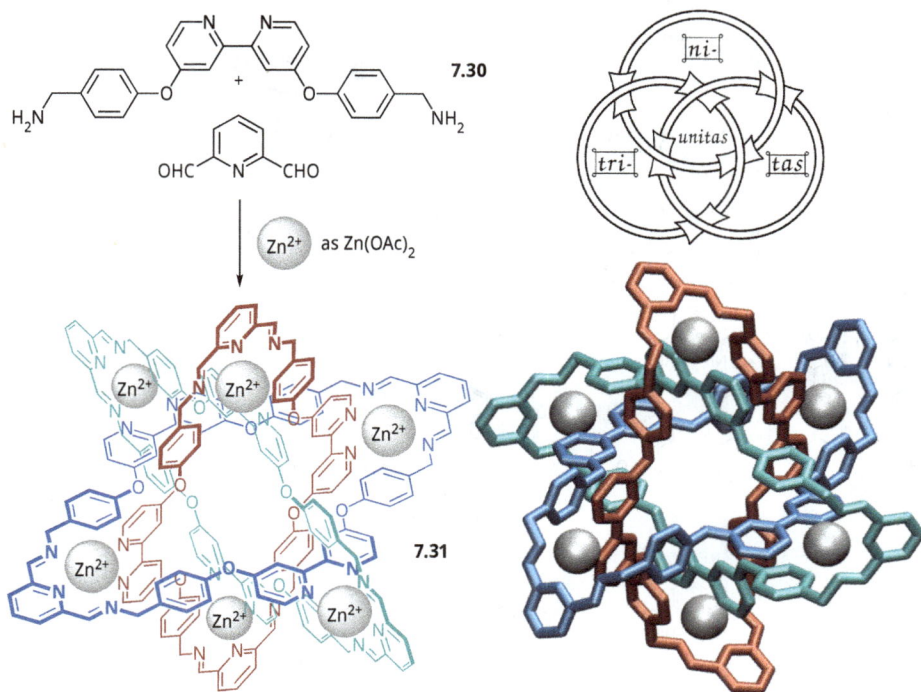

Figure 7.18: Self-assembly of ligand **7.30**, 2,6-diformylpyridine, and zinc(II) acetate to give the Borromean rings **7.31** and the crystal structure of **7.31**. The protons in the crystal structure are omitted for reasons of clarity, and the three rings are shown in different colors to illustrate their arrangement. The icon shows the symbol of the Christian Trinity, which also has the topology of Borromean rings [29].

7.4.2 Knots

Trefoil knots are obtained by inducing an acyclic oligomeric ligand to adopt a knotted conformation through metal coordination. This concept was pioneered by Christopher A. Hunter, who showed that ligand **7.32** is predisposed to form an open trefoil knot when coordinated to a zinc(II) ion (Figure 7.19a) [30]. This structure is stabilized by linking the chain ends [31].

Certain tris(2,6-pyridinedicarboxamides) form similar open knot complexes with lanthanide ions [32]. One of these complexes was used by David A. Leigh to tie a knot in the axle of a rotaxane (Figure 7.19b) [33]. The corresponding axle component, compound **7.33**, contains a central dibenzylamine unit flanked by a bulky tetraphenyl moiety and a chain of three 2,6-pyridinedicarboxamide subunits. It forms a pseudorotaxane with dibenzo-24-crown-8 in which the crown ether is preferentially located near the ammonium group of the dibenzylammonium moiety with which it interacts by hydrogen bonding. Since the tetraphenyl moiety is too large to slip through the

(a)

7.32

(b)

7.33

Lu³⁺ as Lu(CH₃SO₃)₃

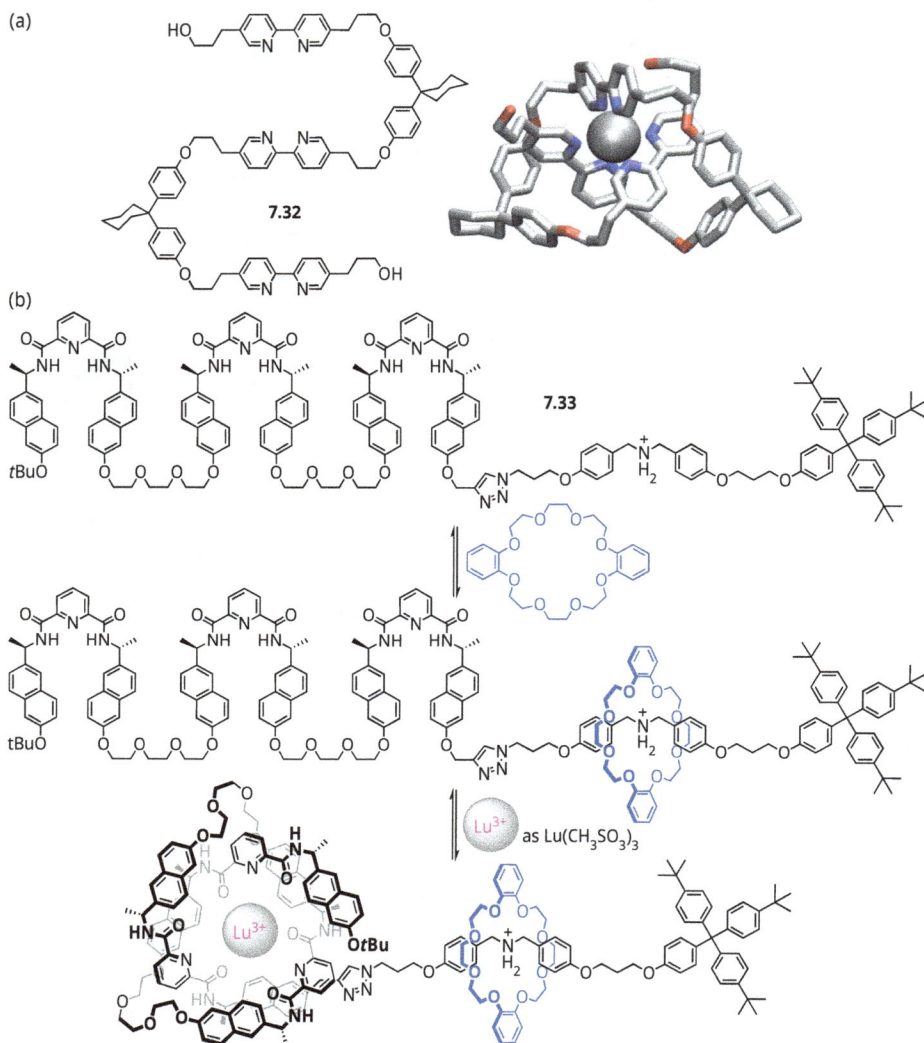

Figure 7.19: Molecular structure of the oligomeric ligand **7.32** containing three 2,2′-bipyridine units folded around a zinc(II) ion in the form of an open trefoil knot (a), and the formation of a rotaxane by stoppering the axle at one end and tying a knot at the other end (b). The axle **7.33** of this rotaxane contains a tris(2,6-pyridinedicarboxamide) moiety at one end, a dibenzylamine moiety in the middle, and a tetraphenylmethane moiety at the other end of the chain. The rotaxane is formed by threading the tris (2,6-pyridinedicarboxamide) moiety through the dibenzo-24-crown-8 ring, and then knotting this part of the axle by metal coordination.

ring, the crown ether must be threaded through the three 2,6-pyridinedicarboxamide units during pseudorotaxane formation. The subsequent addition of Lu(CH₃SO₃)₃ causes the 2,6-pyridinedicarboxamide units to fold into a knot whose steric bulk prevents the ring from leaving the axle. Untangling the knot by removing the metal ion results in the release of the ring. This synthesis impressively illustrates the extent of structural control that can be achieved by cleverly combining appropriate structural motifs and interactions.

The Sauvage group has also made seminal contributions to the metal-templated synthesis of molecular knots. Their approach is based on pairwise linking of the terminal groups in a double helicate, yielding a product with the topology of a trefoil knot. The structural relationship between the double helicate and the knot is shown in Figure 7.20a. The corresponding synthesis follows route (b) in Figure 7.11, using metal ions to stabilize the double helical intermediate (Figure 7.20b). Specifically, the dicopper(I) double helicate derived from the bis(phenanthroline) ligand **7.34** is reacted under high dilution conditions with an ethylene glycol-derived diiodide of appropriate length under Williamson ether synthesis conditions (Figure 7.20c) [34]. A number of products are formed in this step, with the desired double helicate **7.35** with linkages between the correct pairs of OH groups (a–c and b–d) being produced in only 3% yield. Major byproducts are macrocycles derived from the monocopper(I) complex of **7.34**, in which both phenanthroline units coordinate to a copper ion, or from incorrectly formed linkages in the double helicate, with the linking of end groups a–d and b–c leading to the formation of a macrocycle, while the linking of a–b and c–d leads to a catenane. The crystal structure confirms the knotted topology of **7.35** (Figure 7.20d), and demetalation yields the metal-free analog. The chirality of the product can be demonstrated by NMR spectroscopy using a chiral shift reagent.

The efficiency of knot formation is significantly improved by changing the ligand structure and the reaction with which the end groups are joined. For example, the dicopper(I) complex of a ligand with a 1,3-phenylene moiety instead of the tetramethylene bridge yields the corresponding trefoil knot in an isolated yield of 74% when the linkages are formed by ruthenium(II)-catalyzed ring-closing metathesis. The synthesis of the product from commercially available 1,10-phenanthroline requires only seven steps with an overall yield of 35% [35].

The analogous stabilization of the trinuclear double helicate formed from the tris(phenanthroline) ligand **7.36** yields a link rather than a knot because the newly formed linkages are part of two different macrocycles (Figure 7.21) [36]. These rings cross four times, making them topologically equivalent to a Solomon link.

The covalent stabilization of helicates thus yields knots or links, depending on whether the number of metal centers is even or odd. However, this approach becomes increasingly difficult as the distance between the end groups increases, which is why complex knots are often more efficiently obtained by using circular helicates, with helicates containing an even number of metal ions yielding links, as we have seen in Figure 7.17. Conversely, those with an odd number yield knots. An example of the latter strategy is the conversion of a pentanuclear circular helicate into a pentafoil knot. The corresponding synthesis involves the thermodynamically controlled subcomponent self-assembly of the 2,2′-bipyridine derivative **7.37**, the diamine **7.38**, and iron(II)

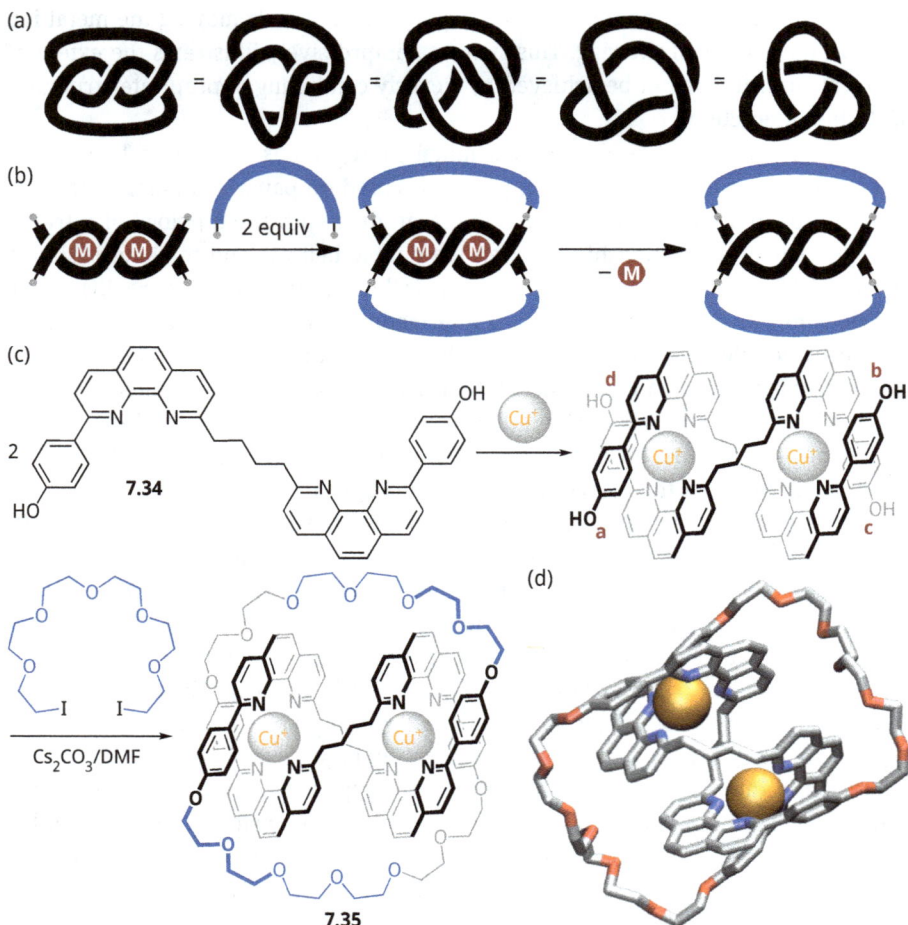

Figure 7.20: Structural relationship of a macrocyclic double helicate and a trefoil knot (a) and general strategy for synthesizing such a knot from a suitable precursor (b). An actual synthesis is shown in (c). The crystal structure of **7.35** in (d) confirms the knot topology. The protons in the crystal structure are omitted for reasons of clarity.

chloride (Figure 7.22) [37]. Recall that iron(II) chloride templates the assembly of pentanuclear circular helicates, while the corresponding sulfate salt leads to hexanuclear helicates. Suitably functionalized ligands can therefore be used to prepare a Star of David catenane, as shown in Section 7.4.1, or the pentafoil knot in Figure 7.22, depending on the choice of counterion in the metal salt. Enantioselective strategies exist for the preparation of both the pentafoil knot and the Star of David catenane [38, 39].

One of the largest endless knots prepared to date has seven crossings [40]. It is obtained by connecting pairs of ligand end groups in a 3 × 3 grid. After demetalation,

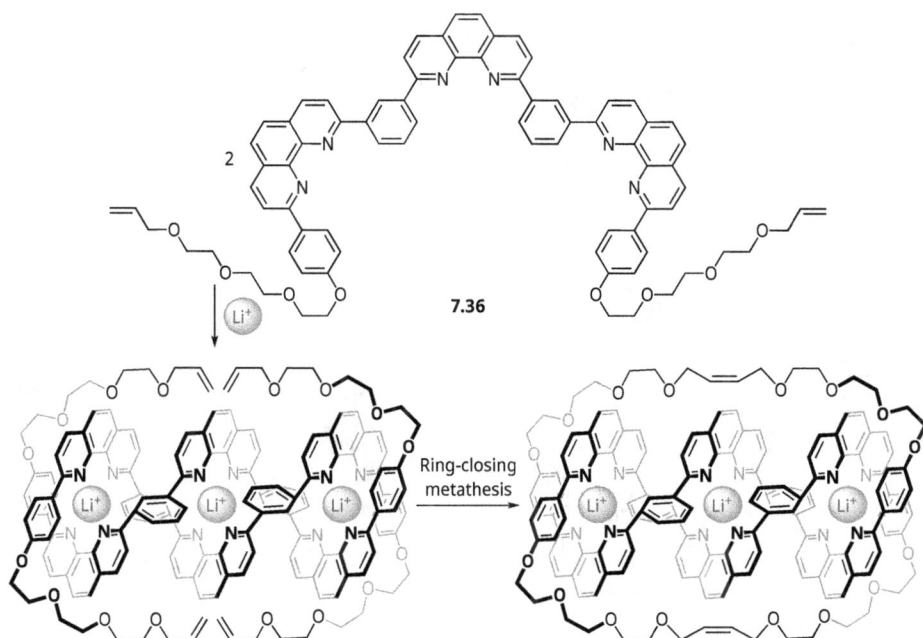

Figure 7.21: Synthesis of a Solomon link involving covalently joining the end groups in the trilithium double helicate of ligand **7.36**.

two of the six loops can unfold, reducing the number of crossings from nine in the grid to seven in the knot.

Further reading

The molecular grids discussed in Section 5.5.3 can be viewed as tiny two-dimensional interwoven structures in which the individual ligands are arranged like threads in a fabric. Similarly, the interdigitated loops formed by the ligands in helicates and molecular knots also have macroscopic counterparts. All of these structures are still discrete objects, raising the question of whether it is possible to make (formally indefinite) polymeric materials that contain periodic entanglements reminiscent of those in fabrics or nets. Omar Yaghi, a pioneer in this field, has shown that molecular weaving is indeed possible, by combining the concepts of metal-directed self-assembly of helicates or interlocked molecules with dynamic covalent chemistry to produce interwoven structures, some as complex as chicken wire. Similar approaches have been described by other groups. For more information, see:

Zhang ZH, Andreassen BJ, August DP, Leigh DA, Zhang L. Molecular weaving. Nat. Mater. 2022, 21, 275–83. Han X, Ma T, Nannenga BL, Yao X, Neumann SE, Kumar P, Kwon J, Rong Z, Wang K, Zhang Y, Navarro JAR, Ritchie RO, Cui Y, Yaghi OM. Molecular weaving of chicken-wire covalent organic frameworks. Chem. 2023, 9, 2509–17.

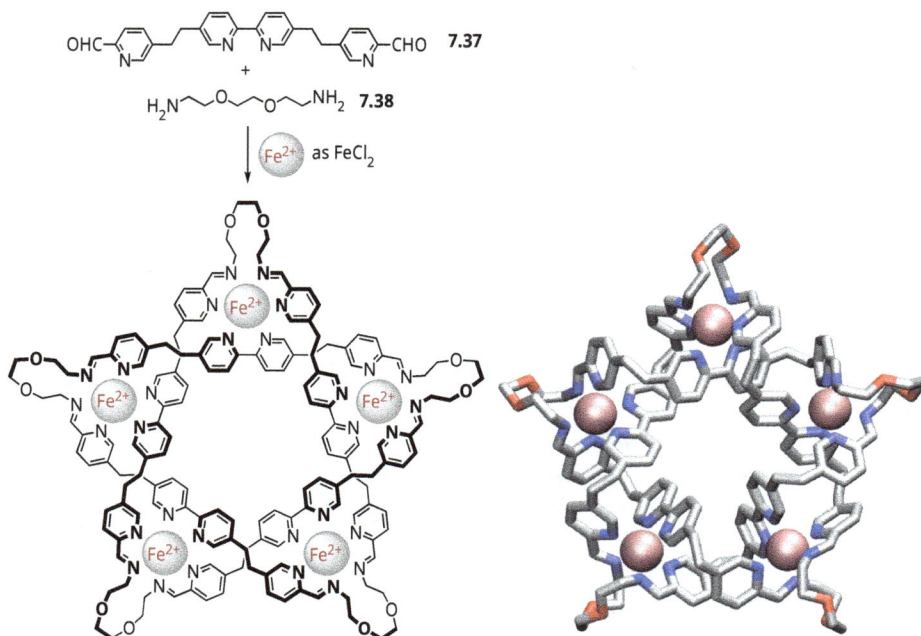

Figure 7.22: Synthesis of a pentafoil knot by subcomponent self-assembly of dialdehyde **7.37**, diamine **7.38**, and iron(II) chloride. The crystal structure of the product confirms the knot topology. The protons in the crystal structure are omitted for reasons of clarity.

7.4.3 Rotaxanes

Rotaxanes can also be synthesized using suitable metal templates as demonstrated by Harry W. Gibson using the stoppering approach shown in Figure 7.14 [41]. This synthesis is based on the copper(I) complex of **7.23** and a macrocyclic phenanthroline derivative with free hydroxy groups that are used to attach the stopper (Figure 7.23). The subsequent cyanide-induced removal of the template yields the metal-free rotaxane **7.39** in 42% yield over both steps.

> When is a template passive and when is it active?

In all of the syntheses discussed so far, the metal templates serve structural purposes. They either preorganize the precursors into an interlocked arrangement or contribute to product formation by thermodynamically stabilizing the product without otherwise participating in the actual bond formation. However, transition metals can also actively mediate coupling reactions. This concept of combining the structural and catalytic properties of metal centers to produce mechanically interlocked molecules was

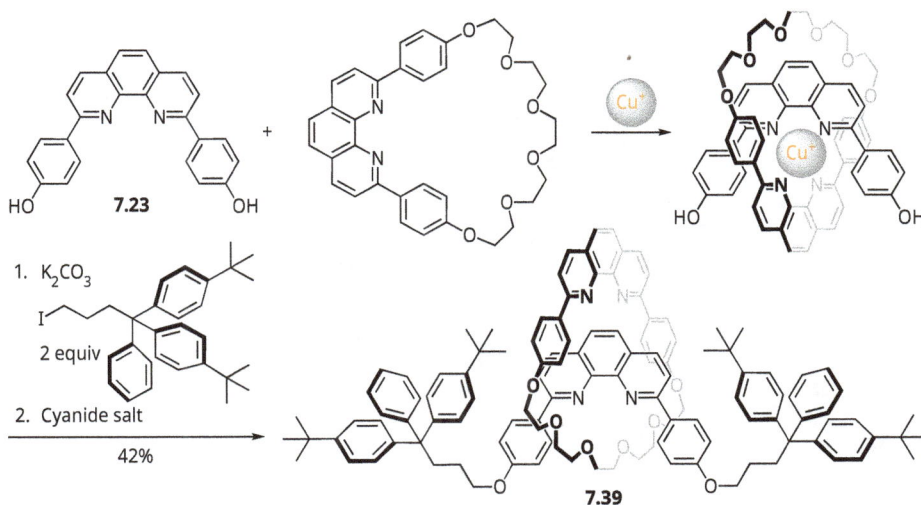

Figure 7.23: Synthesis of rotaxane **7.39**, using the copper(I) complex of a macrocyclic ligand and **7.23**, in which free OH groups serve to introduce the bulky end groups.

pioneered by David A. Leigh, who introduced the term active metal template synthesis for the underlying strategy, as opposed to the passive metal template syntheses discussed so far [42, 43]. Active metal template syntheses can be used to prepare catenanes and rotaxanes, but we will only focus on rotaxane syntheses here. For an example of a catenane synthesis, see Section 7.2.

The general strategy is illustrated in Figure 7.24. It involves the use of a macrocycle with a coordination site directed toward the center of the ring. This site serves to bind the metal template and arrange it within the macrocyclic cavity. The interactions with the individual components of the axle then preorganize the reaction partners in the manner required to produce the interlocked product. In addition to this structural stabilization, the template actively induces the coupling of the axle components, initially yielding a rotaxane that still contains the metal template. If this template is only weakly bound, it can now dissociate to find another empty ring and mediate further transformations.

This approach has several attractive features, the most important of which is the possibility to use the template in catalytic amounts. The product is then produced in the metal-free form without the need to remove stoichiometric amounts of metal ions in a separate step. Another advantage is that the axle components need to contain only those functional groups involved in the coupling reaction, while additional chelating ligands that hold the rotaxane precursors together in passive metal template syntheses are not required.

Figure 7.25 shows a rotaxane synthesis using an active metal template [44]. In this case, a copper(I) ion introduced as the acetonitrile complex [Cu(CH₃CN)₄]PF₆ catalyzes

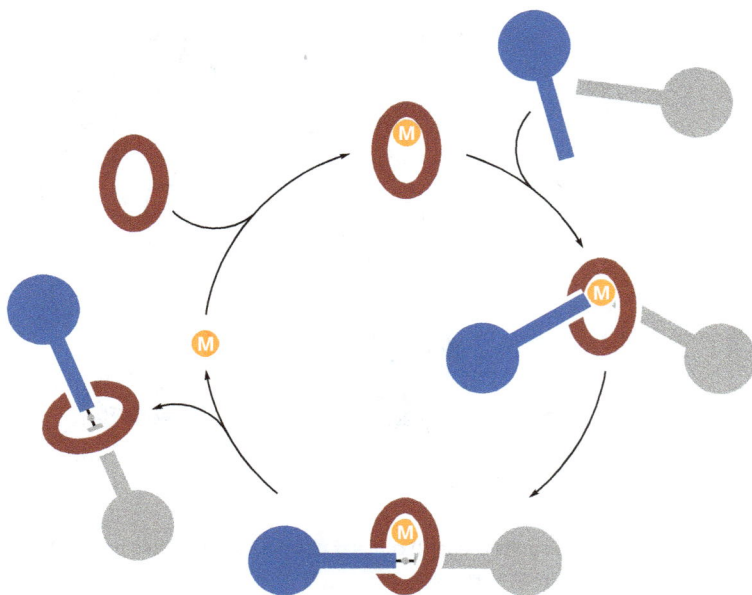

Figure 7.24: General strategy of a rotaxane synthesis using an active metal template. This metal template initially binds to a coordination site oriented toward the interior of a macrocycle. The subsequent interaction with the individual axle components leads to a rotaxane-like arrangement. When the reaction is complete, the product releases the template, which can then induce further reactions.

the azide–alkyne cycloaddition between the axle components. The other components are the macrocycle **7.40**, which contains an inwardly directed nitrogen donor in the pyridine moiety, as well as the azide **7.41** and terminal alkyne **7.42,** both of which contain bulky end groups. Mixing all four components in equimolar amounts yields the expected [2]rotaxane in 57% yield after demetalation along with 41% of the free axle. Using an excess of **7.41** and **7.42** (5 equiv with respect to **7.40**) results in the incorporation of 94% of the originally present macrocycle into the rotaxane. Substoichiometric amounts of the copper(I) source cause a decrease in yield, probably because the tight binding of the copper(I) ion to the product prevents turnover. This problem is circumvented by adding 3 equiv of pyridine as a competing ligand. In this case, the reaction of 1 equiv of **7.40** and 5 equiv each of **7.41** and **7.42** in the presence of 0.2 equiv of the copper(I) catalyst gives the product in 82% yield.

The central intermediate of this transformation is a dicopper(I) complex, consistent with the generally accepted mechanism of the copper(I)-catalyzed azide–alkyne cycloaddition [45]. If this bimetallic intermediate contains one macrocyclic ligand, the expected [2]rotaxane is formed, while coordination of two macrocycles should yield a [3]rotaxane. Indeed, the [3]rotaxane becomes a prominent byproduct when the concentration of the macrocyclic ligand is increased, supporting the proposed mechanism.

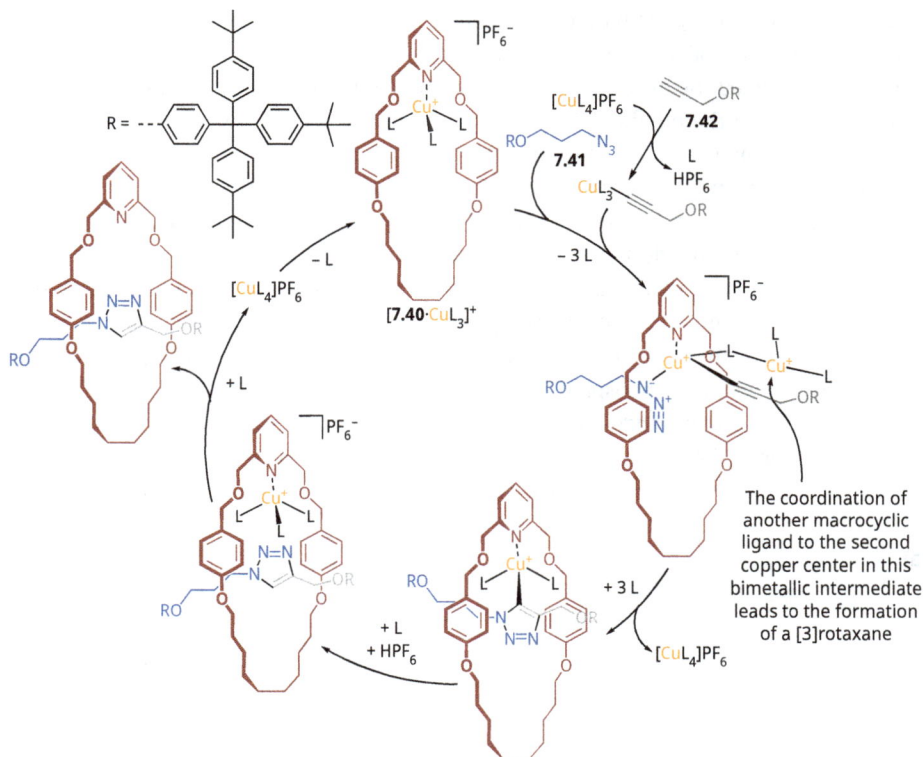

Figure 7.25: Active metal template synthesis of a [2]rotaxane by copper(I)-catalyzed azide–alkyne cycloaddition. The macrocycle **7.40** serves as a ligand for the copper(I) center, which in turn mediates the coupling of the azide **7.41** and the alkyne **7.42**. The central intermediate is a bimetallic complex in which one copper ion is bound to **7.40**.

Active metal template syntheses of rotaxanes or catenanes can be based on several other transition metal-catalyzed transformations. Examples include alkyne–alkyne heterocouplings (Cadiot–Chodkiewicz reaction), oxidative Heck reactions, and palladium(II)-mediated Michael additions [42, 43]. This versatility and the possibility of using the active metal species in catalytic amounts make the approach an attractive tool for accessing interlocked molecules.

7.5 Syntheses using charge-transfer interactions

7.5.1 Catenanes

A large family of interlocked molecules in which charge-transfer interactions facilitate synthesis and control co-conformational behavior is based on the so-called blue box **7.43**

(Section 4.1.5) or analogous macrocycles (Figure 7.26a) [46]. These systems were introduced by J. Fraser Stoddart, who not only developed reliable methods for their preparation but also studied the dynamic processes occurring in them, such as the motion of one
subunit relative to the others [47]. This work led to switchable systems and, somewhat
later, to molecular machines, achievements for which he shared the 2016 Nobel Prize in
Chemistry with Jean-Pierre Sauvage and Bernard (Ben) L. Feringa. In this chapter, we
will focus on the synthetic strategies and look at the other aspects in Chapter 8.

Interlocked systems derived from **7.43** are often prepared by *clipping* the blue box
around an appropriate template with an electron-rich subunit. For example, the reaction
of equimolar amounts of the bis(pyridylpyridinium) precursor **7.44** and 1,4-bis(bromomethyl)benzene in the presence of 2.5 equiv of the hydroquinone-containing crown ether
7.45 yields the hexafluorophosphate salt of the [2]catenane **7.46** in a yield of 70% after
purification and ion exchange (Figure 7.26b) [48]. Product formation involves the initial
chain elongation of **7.44** to yield a tricationic intermediate. This compound folds around a
hydroquinone moiety of **7.45** to maximize charge-transfer interactions between the elec-

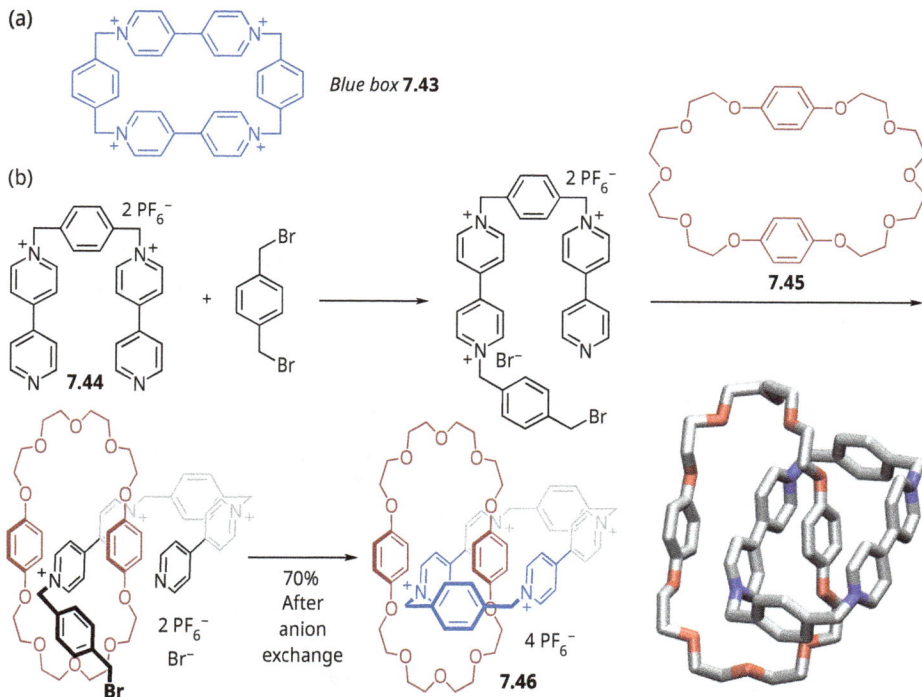

Figure 7.26: Molecular structure of blue box **7.43** (a), formation of a [2]catenane by clipping the blue box
around one of the hydroquinone units of crown ether **7.45** (b), and crystal structure of the product **7.46**
(c). Protons and counterions in the crystal structure are omitted for reasons of clarity.

tron-rich and electron-poor aromatic subunits. In the corresponding complex, the end groups of the linear component are in close proximity, facilitating ring closure.

Based on this general strategy, a variety of structurally more complex catenanes are accessible by using other macrocycles or blue box derivatives. For example, the use of a large crown ether with four hydroquinone moieties yields a [3]catenane when two blue boxes are clipped to the ring (Figure 7.27a). Conversely, clipping a blue box with biphenyl subunits to **7.45** leads to a [3]catenane in which two crown ethers are threaded through an enlarged ring (Figure 7.27b).

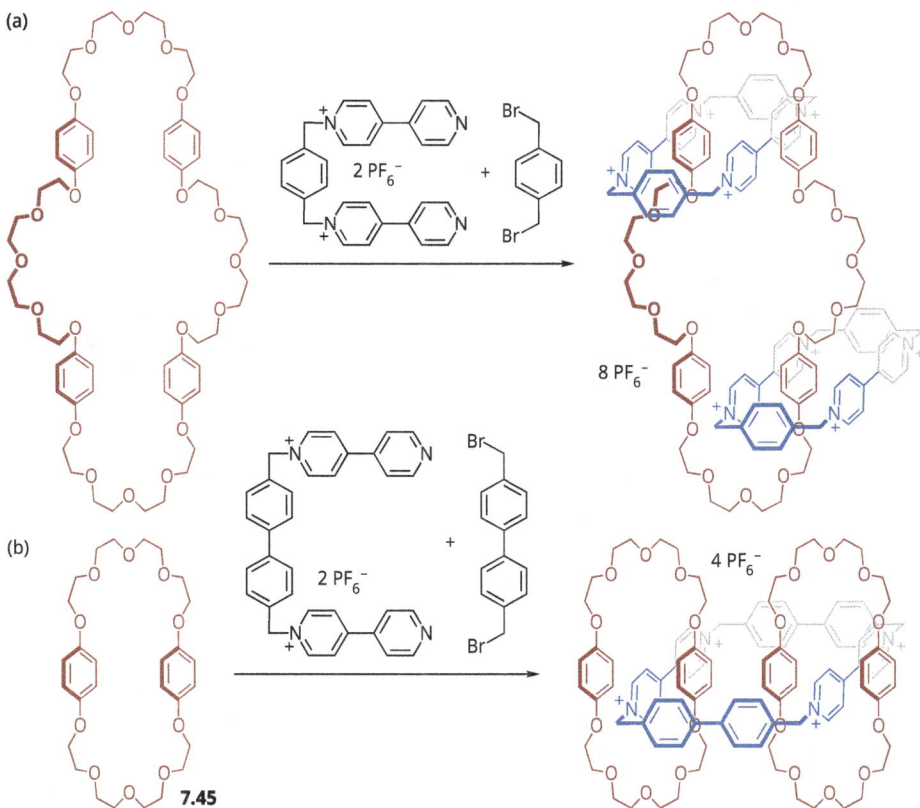

Figure 7.27: Formation of [3]catenanes by clipping two blue boxes to a crown ether (a), or by joining two crown ethers with an enlarged blue box (b).

This approach also allows two rings of crown ether **7.47** and the enlarged blue box to be combined into a [3]catenane (Figure 7.28). The two crown ethers in this product are large enough to allow an additional blue box to be clipped to each ring, resulting in the [5]catenane **7.48**. This catenane has the topology of the five interlocking rings that form the symbol of the Olympic Games, hence the name olympiadane [49].

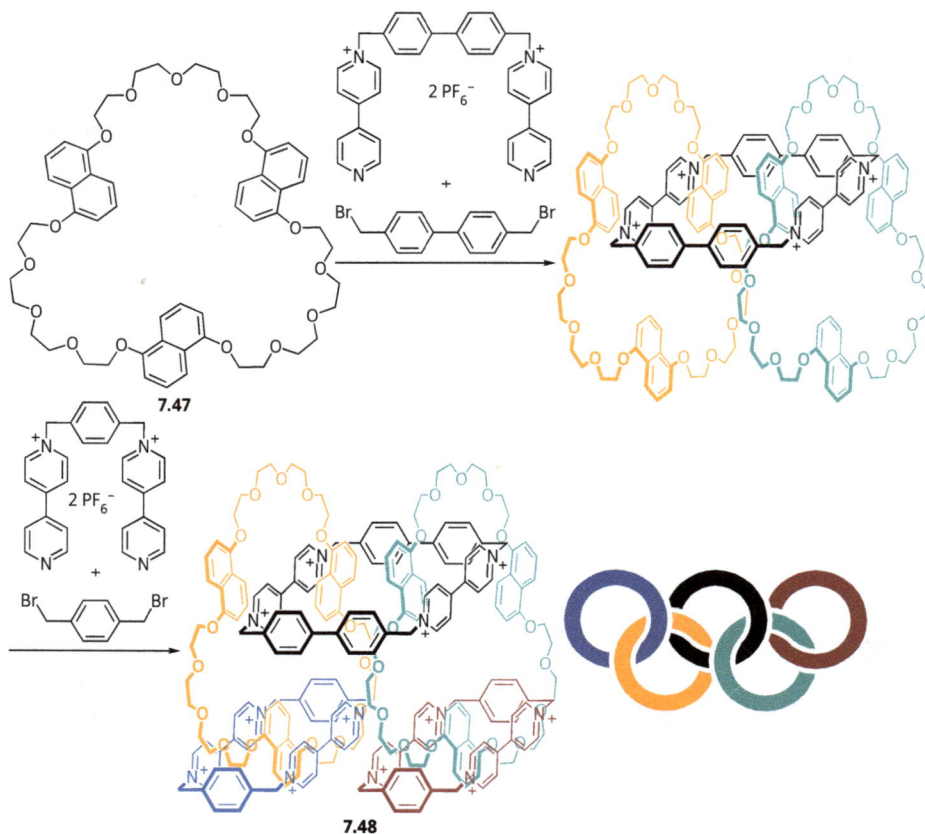

Figure 7.28: Synthetic strategy to obtain olympiadane **7.48**. A [3]catenane is first prepared from the large crown ether **7.47** and the enlarged blue box, to which two further blue boxes are clipped in the next step.

[2]Catenanes combining electron-rich and electron-poor aromatic moieties are also obtained by clipping a macrocycle with 1,4,5,8-naphthalenetetracarboxylic diimide (NDI) moieties around a 1,5-dioxynaphthalene-derived crown ether. The corresponding synthesis, developed in the group of Jeremy K. M. Sanders, involves the macrocyclization of the dialkyne **7.49** under Glaser coupling conditions in the presence of **7.50** to give the [2]catenane **7.51** in 52% yield (Figure 7.29) [50]. This catenane produces a strong band in the UV–vis spectrum between 480 and 650 nm that is not observed for either building block alone, reflecting the charge-transfer interactions between the NDI (electron-poor, acceptor, A) and 1,5-dioxynaphthalene (electron-rich, donor, D) units. Catenane **7.51** differs structurally from Stoddart's catenanes by the alternating A-D-A-D arrangement of the aromatic subunits and by the fact that it is neutral, whereas catenanes derived from **7.43** always carry the positive charges of the paraquat units.

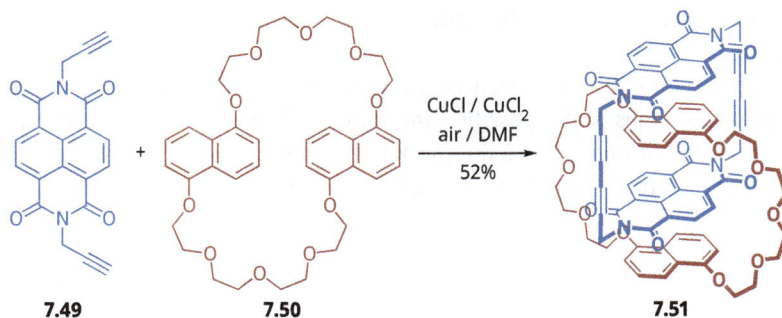

Figure 7.29: Synthesis of the [2]catenane **7.51** containing an alternating arrangement of electron-poor 1,4,5,8-naphthalenetetracarboxylic diimide and electron-rich 1,5-dioxynaphthalene residues by clipping the NDI-containing ring around the crown ether.

The alternative thermodynamically controlled preparation of catenanes from naphthalene and NDI derivatives involves equilibrating the cysteine-derived building blocks **7.52** and **7.53** in water under conditions that mediate disulfide formation and exchange (Figure 7.30) [51]. Although many different products could potentially result from this reaction, only a few macrocycles are formed along with the [2]catenane **7.54**. This catenane does not have the alternating arrangement of electron-rich and electron-poor residues, where charge-transfer interactions would be strongest, but a D-A-A-D stacking sequence. The strong dependence of the concentration of **7.54** at equilibrium on salt concentration, with 1 M NaNO$_3$ leading to a sixfold increase in the amount of **7.54** compared to salt-free conditions, strongly suggests that a major contribution to catenane formation comes from the hydrophobic effect. Charge-transfer interactions thus appear to play a minor role in the selection of **7.54**, whose preferential formation is likely due to a trade-off between optimizing the number of charge-

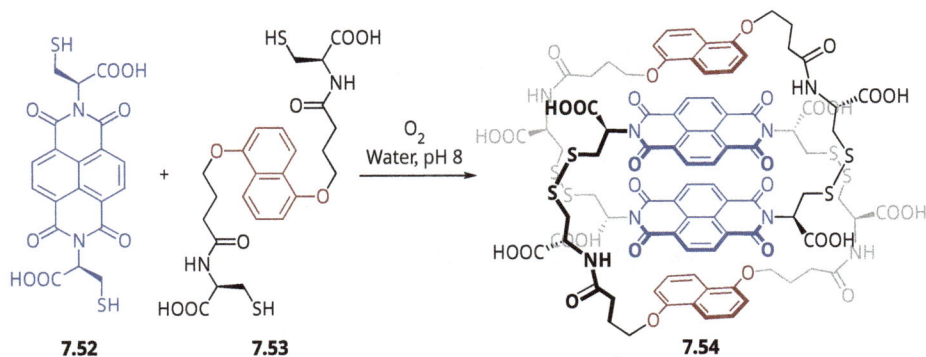

Figure 7.30: Synthesis of the [2]catenane **7.54** from dithiols **7.52** and **7.53** under thermodynamic control. The product is shown in the preferred co-conformation with a D-A-A-D stacking sequence of the aromatic moieties.

transfer interactions and minimizing the strain in the product as well as the size of the solvent-exposed surface.

Depending on the building block structure and the equilibration conditions, the formation of other catenanes is also possible, and some building blocks even give rise to more complex links and knots. Since the formation of these interlocked systems is mainly driven by the hydrophobic effect, this work is presented in Section 7.8.1.

7.5.2 Rotaxanes

The clipping approach used by Stoddart to prepare catenanes also allows the synthesis of rotaxanes. An example is shown in Figure 7.31a. In this reaction, the axle **7.55**,

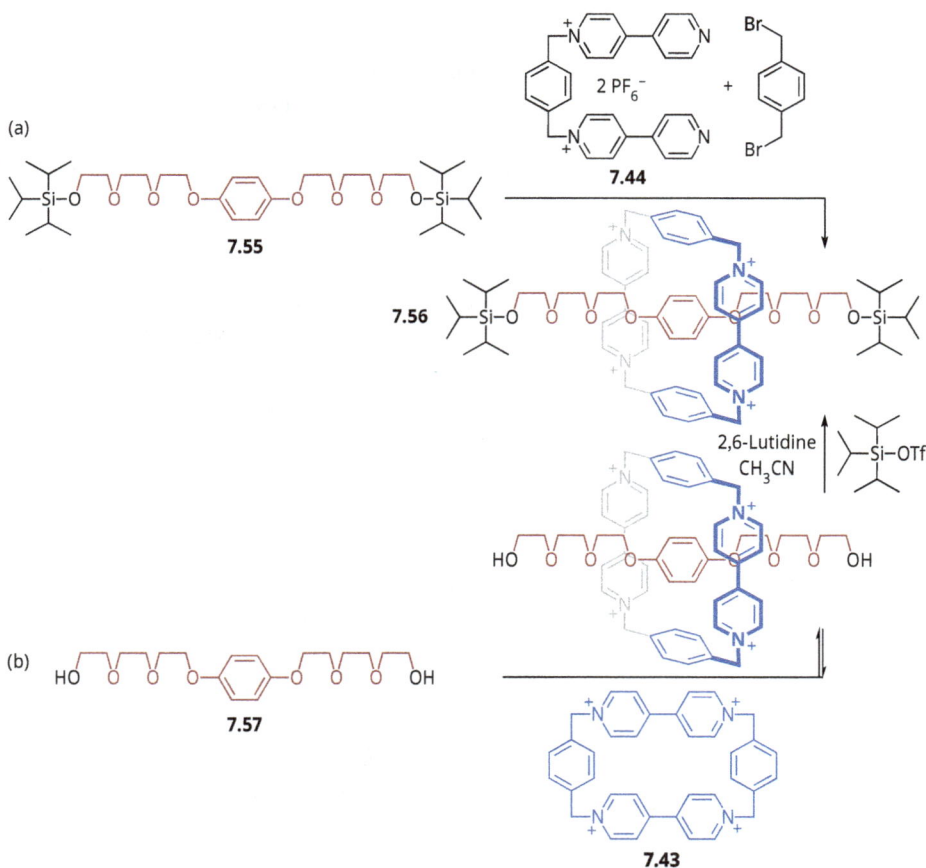

Figure 7.31: Synthesis of the [2]rotaxane **7.56** by clipping **7.43** around the axle **7.55** (a), or by stoppering the pseudorotaxane formed from **7.57** and **7.43** (b).

which contains a central hydroquinone moiety and terminal triisopropylsilyloxy groups, is treated with **7.44** and 1,4-bis(bromomethyl)benzene to afford the corresponding [2]rotaxane **7.56** in 14% yield [52]. The same [2]rotaxane is obtained by stoppering the end groups in the pseudorotaxane in which the diol **7.57** is threaded through the blue box **7.43** (Figure 7.31b) [52]. This stoppering strategy gives the product in a yield of 22%, slightly better than the yield associated with clipping.

An example of a [2]rotaxane containing an electron-poor aromatic unit in the axle and electron-rich ones in the ring is **7.58**, which is obtained by slipping the crown ether **7.45** over the end groups of the acyclic component (Figure 7.32) [53].

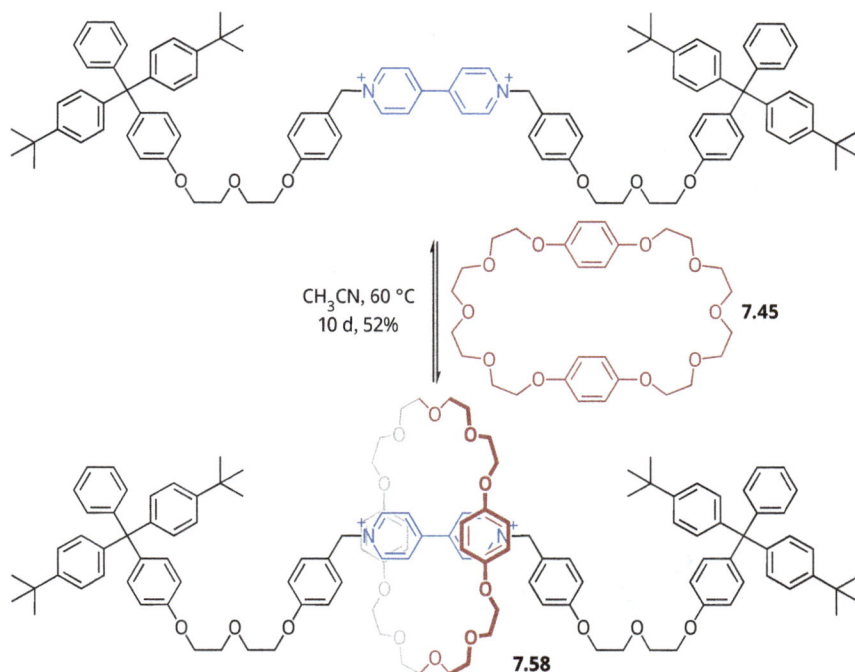

Figure 7.32: Synthesis of the [2]rotaxane **7.58** by slipping the crown ether **7.45** with two hydroquinone moieties over the end groups of an axle containing a paraquat unit.

7.6 Syntheses using hydrogen bonds

7.6.1 Catenanes

Good starting points for the hydrogen bond-templated synthesis of catenanes and rotaxanes are pseudorotaxanes in which functional groups in the axle can be used for cyclization or the attachment of bulky end groups. Suitable precursors are complexes between crown ethers and protonated dialkylamines in which the ammonium NH groups are hydrogen-bonded to the crown ether oxygen atoms. We saw such a complex in Section 7.4.2 in the form of the dibenzo-24-crown-8-containing rotaxane with a knot at one end of the threaded chain (Figure 7.19a). 21-Crown-7 is also suitable for the formation of such pseudorotaxanes, but 18-crown-6 is too small to thread ammonium ions, which are instead bound in a perched arrangement, as shown in Section 4.1.1.

Alternative precursors for the synthesis of interlocked molecules are complexes of macrocyclic lactams whose NH and C=O groups serve as hydrogen bond donors and acceptors, respectively. Two such catenanes are shown in Figure 7.33, both prepared from simple starting materials in a surprisingly efficient manner. Catenane **7.59** precipitates spontaneously when dilute solutions of isophthaloyl dichloride and 1,4-bis(aminomethyl)benzene are added at room temperature to a solution of triethylamine in chloroform. The product can be isolated in 20% yield by simple filtration (Figure 7.33a) [54]. Product formation likely involves the initial oligomerization of the starting materials, resulting in products that dimerize by hydrogen bond formation. The subsequent macrocyclization then traps these complexes in the interlocked arrangement.

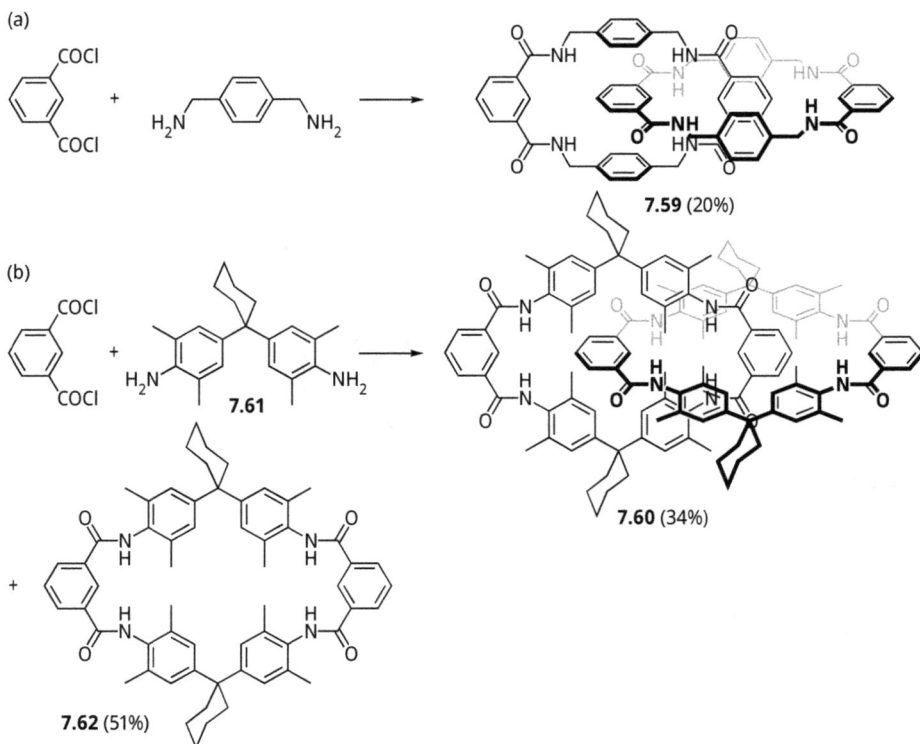

Figure 7.33: Synthesis of [2]catenanes **7.59** and **7.60** by reaction of isophthaloyl dichloride with 1,4-bis(aminomethyl)benzene (a) or the diamine **7.61** (b).

The [2]catenane **7.60** is prepared in a similar manner from isophthaloyl dichloride and the diamine **7.61** in 34% yield along with 51% of the free macrocycle **7.62** (Figure 7.33b). The synthesis was carried out by Christopher A. Hunter with the aim of preparing a benzoquinone receptor, as discussed in Section 4.1.5. However, the synthetic conditions yielded not only the desired macrocycle but also the corresponding [2]catenane [55]. Following this serendipitous discovery, Fritz Vögtle systematically investigated structural aspects of such catenanes and the mechanism of their formation [56]. This work benefited from the rigidity of **7.60**, imposed by the bulky cyclohexyl residues, which are too large to pass through the cavity openings and therefore fix the two rings in a defined relative orientation. As a result, derivatives of **7.60** in which one isophthaloyl moiety in each ring contains an additional substituent exist in three noninterconverting co-conformations, *in,in*, *in,out*, and *out,out* (Figure 7.34). The selectivity with which these isomers are formed provides insight into the mechanism of catenane formation.

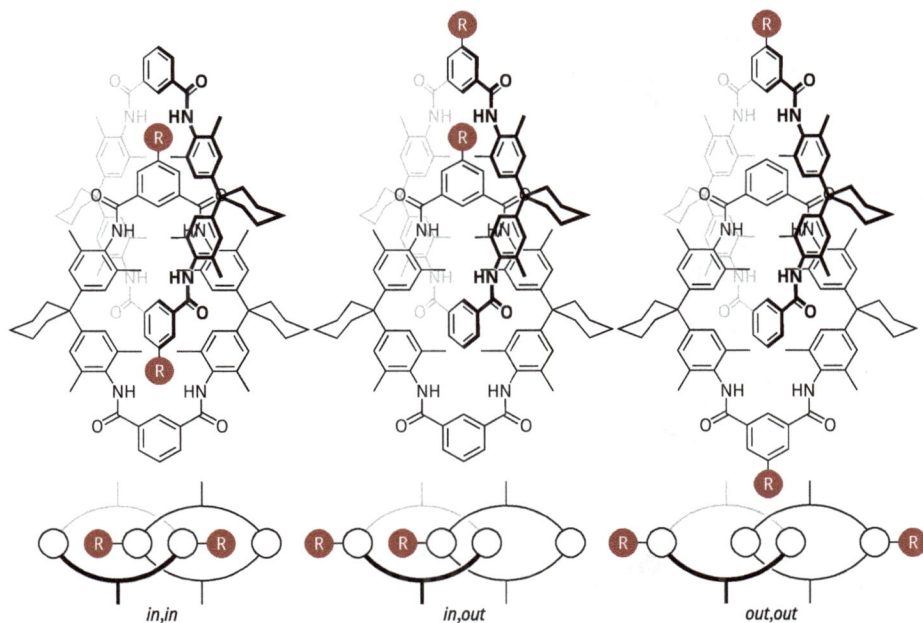

Figure 7.34: Possible co-conformations of a derivative of **7.60** in which each ring contains a substituent in one isophthaloyl subunit.

When isophthaloyl dichloride **7.63a** with an additional methoxy group is reacted with diamine **7.64a**, only the *in,out*- and *in,in*-isomers are obtained (Figure 7.35a, b). On the other hand, when the additional methoxy group is located in the central ring of diamine **7.64b** and the isophthaloyl dichloride is unsubstituted, the *in,out*- and *out,out*-isomers are formed (Figure 7.35c). This selectivity indicates that catenane formation begins with the formation of a macrocycle. In the next step, a pseudorotaxane is formed, but since the diamine containing the cyclohexyl residues is too large to thread through the ring, only the isophthaloyl dichloride can act as the axle in this step. The resulting complex is stabilized by hydrogen bonding between the isophthaloyl dichloride C=O groups and the NH groups along the ring. This complex contains the guest either near the unsubstituted or substituted subunit. Trapping the two orientations in the final step of catenane formation thus leads to two products whose substitution patterns depend on the choice of starting materials, with the substituted isophthaloyl dichloride leading to the *in,out*- and *in,in*-isomers and the substituted diamine leading to the *in,out*- and *out,out*-isomers, as shown in Figure 7.35.

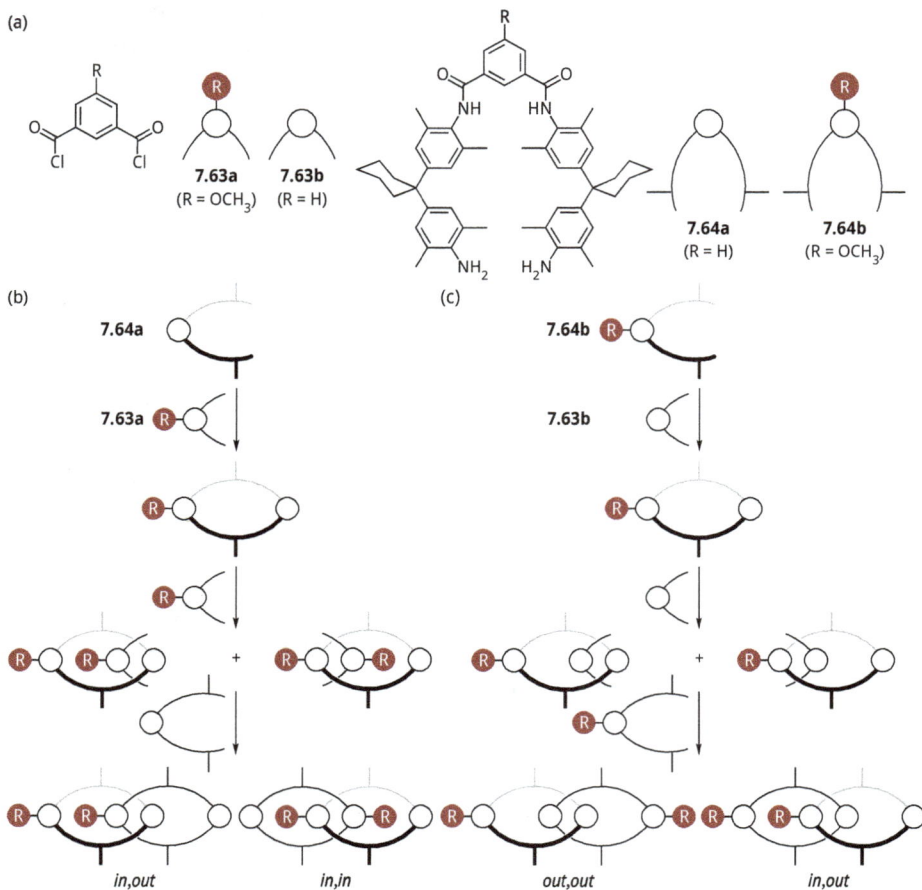

Figure 7.35: Structures of the unsubstituted and substituted precursors used for the synthesis of the dimethoxy analogs of **7.60** (a), and mechanism of catenane formation explaining the selectivity with which the different co-conformations are formed. The scheme in (b) shows the reaction with the methoxy group in the isophthaloyl dichloride **7.63a** and that in (c) the case where the substituent is in the diamine **7.64b**.

A further level of selectivity is achieved by using an analog of **7.64** in which one amide group is replaced by a sulfonamide. Since sulfonamide NH groups are better hydrogen bond donors than amide NH groups, the isophthaloyl dichloride prefers to be bound to the sulfonamide group in the pseudorotaxane. As a consequence, the second macrocyclization leads exclusively to the *in,out* catenane (Figure 7.36) [57]. This catenane is formed as a mixture of two enantiomers because the orientation with which the diamine reacts with the pseudorotaxane is not controlled. The topological enantiomerism is a consequence of the directionality of the groups in the rings, as discussed in Section 7.2.

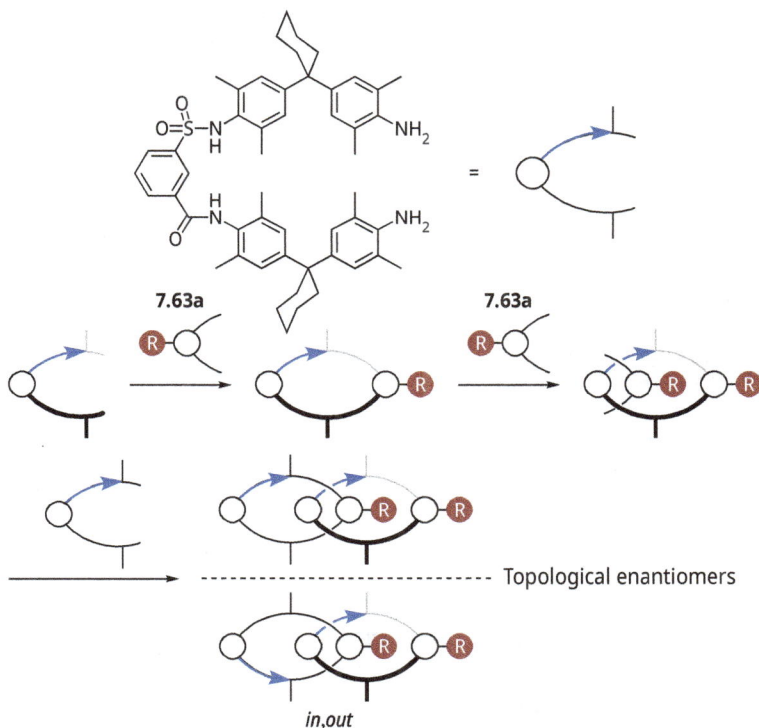

Figure 7.36: Catenane formation between **7.63a** and an analog of **7.64a** in which one amide group is replaced by a sulfonamide group. Only the corresponding *in,out*-isomer of the catenane is formed, in this case as a mixture of two topological enantiomers.

The higher acidity of the sulfonamide NH groups compared to the amide NH groups makes it possible to selectively connect the two sulfonamide groups in such a catenane using appropriate linkers. The resulting product has the topology of a pretzel and is therefore called a pretzelane (Figure 7.37) [58].

Another class of catenanes, whose formation is templated by hydrogen bonds between NH and C=O groups, derives from the calix[4]arene tetraureas introduced in

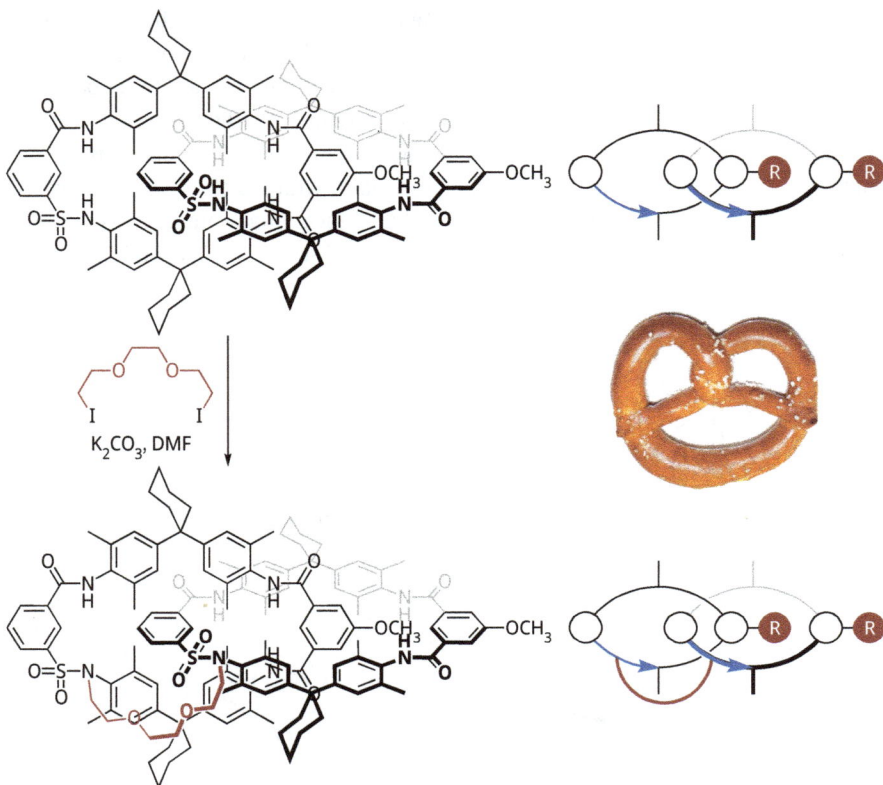

Figure 7.37: Formation of a pretzelane by covalent linkage of the two sulfonamide NH groups in a [2] catenane. The image of the pretzel serves to illustrate the structural relationship [59].

Section 5.3.3. The interdigitation of these calixarenes in the self-assembled capsule causes the urea groups to be alternately oriented along its seam. Pairwise linking of residues oriented in the same direction thus results in interlocked rings. This concept was realized by Volker Böhmer using the calix[4]arene derivative **7.65** (Figure 7.38a) with terminal double bonds [60]. Assembling the corresponding dimeric capsule in dichloromethane/benzene, 95:5 (v/v), treating it with Grubbs catalyst under high dilution conditions, and hydrogenating the double bonds formed during the ring-closing metathesis yields three isomeric products with different topologies. Figure 7.38b shows that linking adjacent substituents (a–b, c–d, e–f, and g–h) leads to a topologically trivial product that can be represented as a planar graph, consisting of the two calixarene moieties connected by four linkers. The other extreme is a product in which every other substituent is linked (a–c, b–d, e–g, f–h). In this product, adjacent substituents facing into the same direction form macrocycles through which substituents facing in the opposite direction are threaded, resulting in an interlocked structure with the to-

(a)

R' =

R = C_5H_{11}

7.65

(b)

| a–c / b–d / e–g / f–h | a–b / c–d / e–g / f–h | a–b / c–d / e–f / g–h |

Bis([2]catenane)
5–12%

Doubly-bridged
[2]catenane
26–32%

Quadruply-bridged
dimer
10–15%

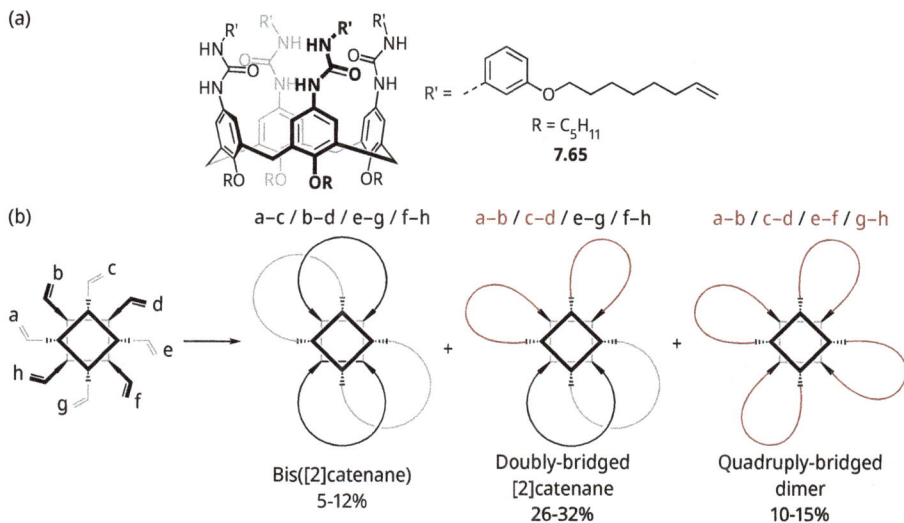

Figure 7.38: Molecular structure of the substituted calix[4]arene tetraurea **7.65** (a), and schematic structures of the linked products formed from **7.65** upon ring-closing metathesis between the marked double bonds and subsequent hydrogenation (b).

pology of a bis([2]catenane). A third product contains two interlocked rings and two direct linkages.

Due to the low yield of this statistical bis([2]catenane) synthesis, the Böhmer group also developed a more efficient approach using the calixarene derivatives **7.66** and **7.67** (Figure 7.39) [61]. Calix[4]arene **7.66** is an analog of **7.65** with shorter side chains, while **7.67** is derived from **7.66** but has both rings already closed. The latter calixarene cannot homodimerize because the rings prevent interdigitation. However, **7.67** forms a capsule with **7.66** in which two substituents of **7.66** are threaded through the rings. This capsule is perfectly preorganized to provide the corresponding bis([2]catenane) after ring-closing metathesis and hydrogenation. The 65% yield with which this product is isolated demonstrates the superiority of this approach over the statistical strategy. Note that unlike **7.65**, **7.66** cannot form a bis([2]catenane) on its own because the linkers are too short. Moreover, since the ring-closing metathesis in the dimer formed from **7.66** and **7.67** proceeds in two different ways, two topologically equivalent but enantiomeric bis ([2]catenanes) are obtained, which can be separated chromatographically on a chiral stationary phase. A whole family of related mechanically interlocked molecules has been developed using similar approaches [62].

The final examples of catenane syntheses involving hydrogen bonding interactions should illustrate that the preorganization of the precursors does not necessarily require direct interactions between the building blocks, but can also be mediated by an additional anionic binding partner. Important work in this regard came from the

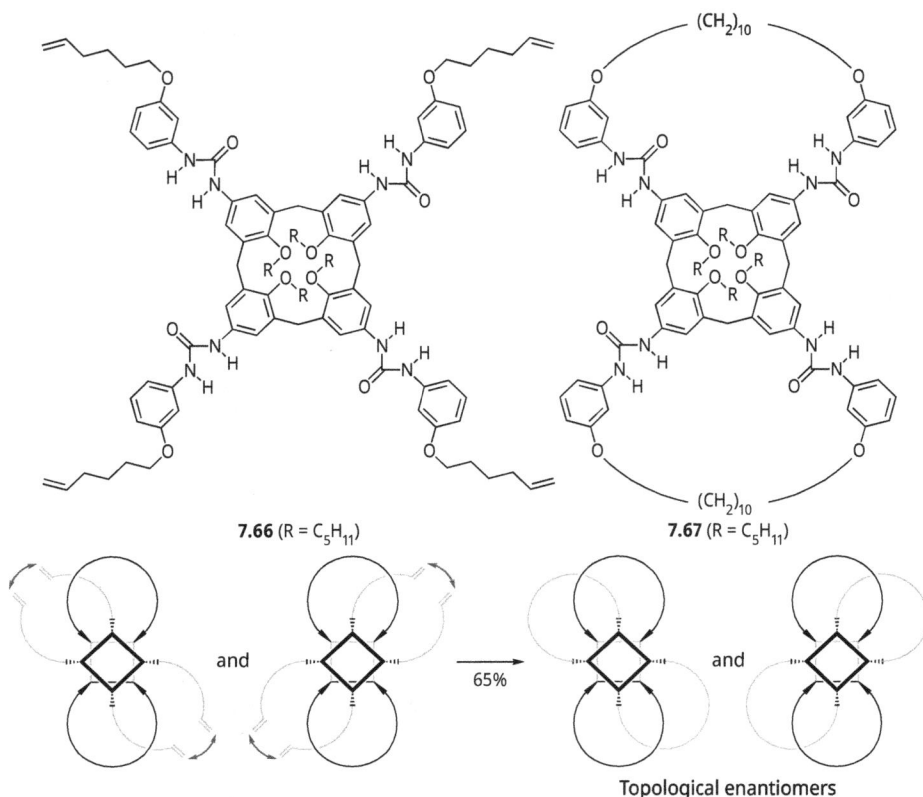

Figure 7.39: Rational approach to calix[4]arene-derived bis([2]catenanes) based on the self-assembly of **7.66** and **7.67** followed by ring-closing metathesis and hydrogenation.

group of Paul D. Beer [63]. Note that this strategy is complementary to the use of cationic metal templates described in Section 7.4.

Beer's synthetic approach is based on the interlocked chloride complexes between *N,N*-dialkylated isophthalamides and corresponding pyridinium derivatives. For example, the isophthalamide **7.68** and the pyridinium ion **7.69** (Figure 7.40) bind chloride with a K_a of 260 M^{-1} in CD$_2$Cl$_2$. This complex is stabilized by hydrogen bonding between the NH groups and the anion, charge-transfer interactions between the coplanar electron-poor and electron-rich aromatic moieties, C–H$\cdots$O hydrogen bonds between the oxygen atoms and the protons of the pyridinium methyl group, and electrostatic interactions [64]. The two binding partners are arranged orthogonally, like bidentate ligands in a copper(I) complex, making this complex a suitable precursor for the preparation of mechanically interlocked molecules.

The synthesis of a [2]catenane is accomplished by either a threading or an entwining approach. An example of the first strategy is the chloride-mediated threading

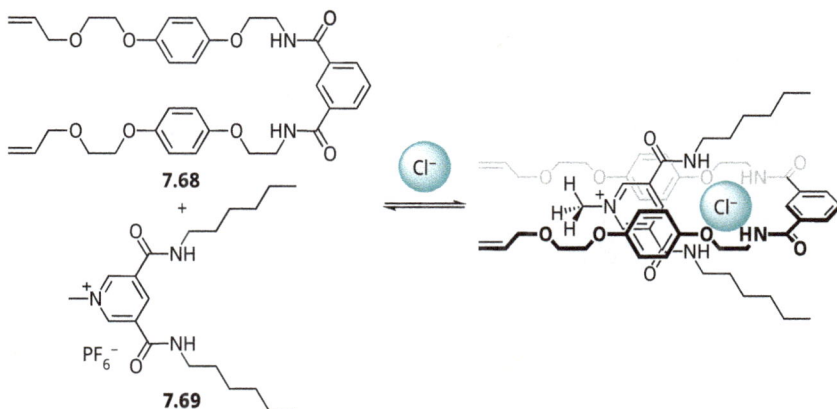

Figure 7.40: Molecular structures of **7.68** and **7.69** and schematic representation of the arrangement of these compounds in their chloride complex.

of the disubstituted *N*-methylpyridinium precursor **7.70** with two terminal double bonds through the cyclophane **7.71** followed by ring-closing metathesis. In this way, the [2]catenane **7.72** is obtained in 45% yield (Figure 7.41a). The pyridinium derivative **7.73** allows the preparation of the [2]catenane **7.74** by entwining (Figure 7.41b). The yield in this case is strongly dependent on the anion, with the hexafluorophosphate salt of **7.73** giving a yield of 16%, while equimolar amounts of a chloride and a hexafluorophosphate salt give a yield of 78%, clearly demonstrating the template effect of the chloride anion [63].

Since these catenanes contain a convergent arrangement of NH donors, they can serve as receptors for appropriate guests, especially for the anionic templates present during their synthesis [63]. For example, catenane **7.72** as hexafluorophosphate salt interacts in CDCl$_3$/CD$_3$OD, 1:1 (*v/v*) with chloride, dihydrogen phosphate, and acetate, showing the highest affinity among these anions for chloride (K_a = 730 M^{-1}). Compound **7.70** alone also binds to these anions, but with a significantly different selectivity, strongly favoring acetate over chloride. Thus, the interlocking of the two macrocycles leads to characteristic binding properties that differ from those of the individual components.

7.6.2 Knots

In the course of their work on tetralactam-derived [2]catenanes, the Vögtle group unexpectedly discovered that the reaction between the diamine **7.64a** and pyridine-2,6-dicarbonyl dichloride yields not only the expected [1 + 1] and [2 + 2] macrocycles, but also a [3 + 3] condensation product with the topology of a trefoil knot (Figure 7.42) [65]. Conclusive evidence for knot formation came from a crystal structure.

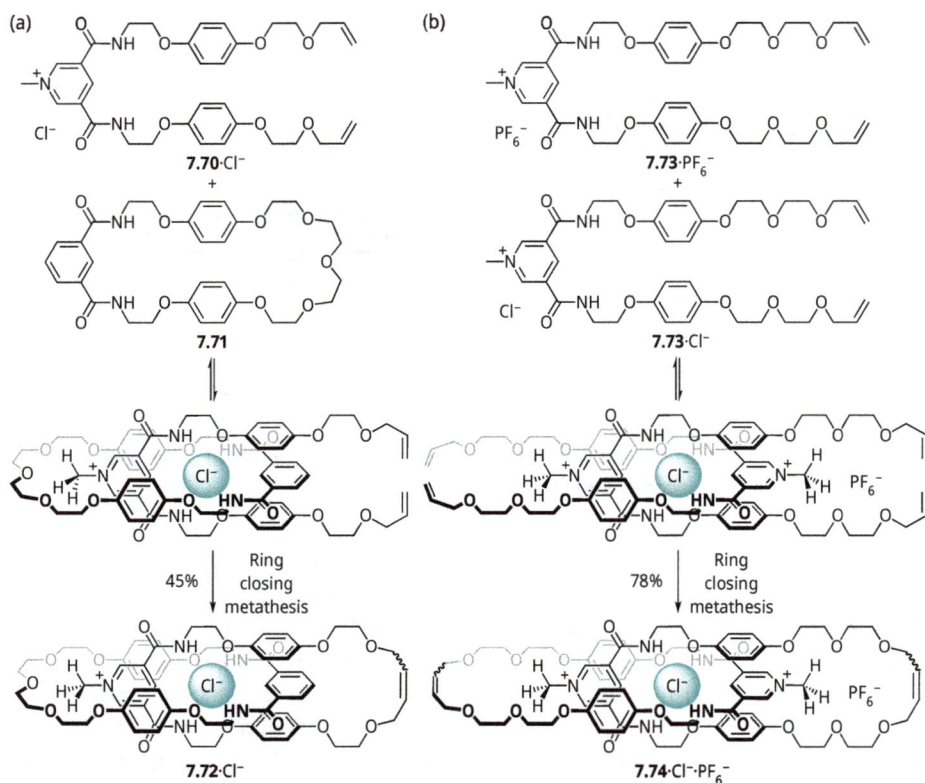

Figure 7.41: Anion-templated syntheses of [2]catenanes by the threading (a) and the entwining (b) approach.

The two enantiomers of the product can be separated chromatographically using a chiral stationary phase and their absolute configurations assigned by circular dichroism spectroscopy [66]. Knot formation is likely due to the propensity of oligomers of the starting materials to adopt tightly folded, interlocked conformations that are eventually trapped in the final macrocyclization step [67].

A similar knot is obtained from **7.75**, which contains an L-valine and aminodeoxycholanic acid subunit. The synthesis in this case involves stepwise chain elongation to the trimer, whose fully deprotected form is then cyclized to give the knotted cyclic hexamer in 21% yield (Figure 7.43) [68]. In contrast to Vögtle's synthesis, the formation of the knot is enantioselective due to the chirality of the subunits. Mechanistically, knot formation again involves the tight folding of the acyclic precursor into an interlocked conformation. Certain foldamers, such as oligomeric amides or peptides, are therefore potentially predisposed to adopt knotted conformations, suggesting a possible pathway of knot formation in nature.

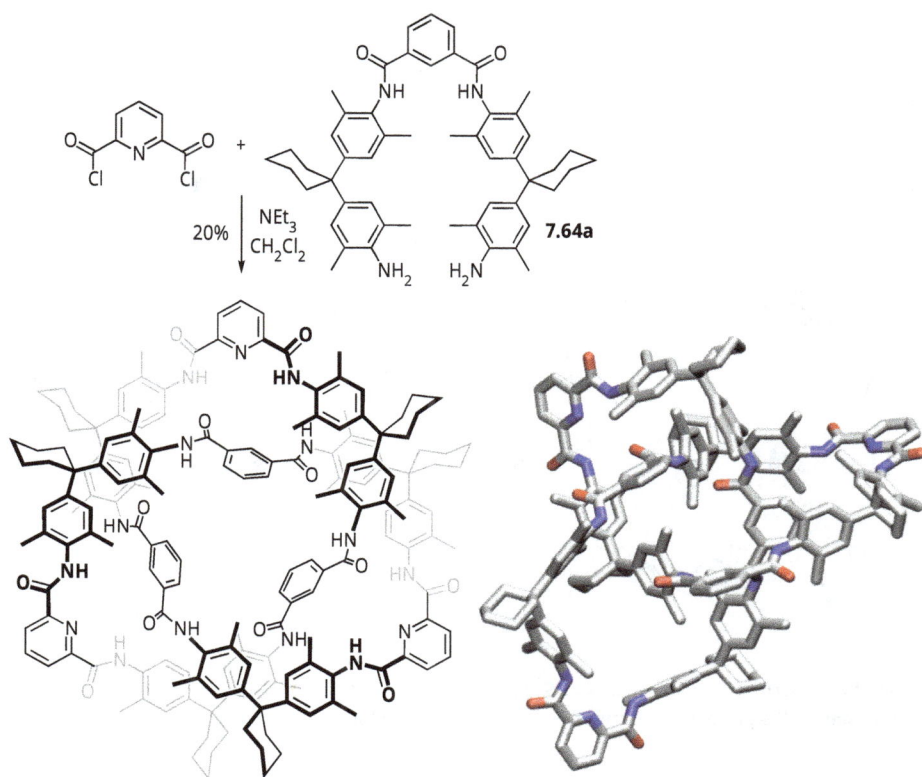

Figure 7.42: Reaction between diamine **7.64a** and pyridine-2,6-dicarbonyl dichloride, resulting in a trefoil knot, as shown in the corresponding crystal structure. The protons in the crystal structure are omitted for reasons of clarity.

7.6.3 Rotaxanes

Pseudorotaxanes used in the synthesis of catenanes are converted to rotaxanes by introducing bulky end groups into the threaded acyclic subunits. How the anion-templated strategy introduced by Beer is used in this context will be discussed in Section 7.7.1. Here we focus on rotaxanes derived from the tetralactams used by Hunter and Vögtle. These macrocycles form complexes with dicarboxylic acid dichlorides, which are trapped as the axles of rotaxanes upon treatment with bulky amines. An example is the [2]rotaxane synthesis from isophthaloyl dichloride and **7.62**, shown in Figure 7.44 [56].

Vögtle also developed a conceptually different approach to rotaxanes based on macrocyclic tetralactams. This strategy is based on the affinity of **7.62** for anions such as phenolates. In the corresponding complexes, the amide NH groups form hydrogen bonds to the phenolate oxygen atom, resulting in steric shielding of this nucleophile

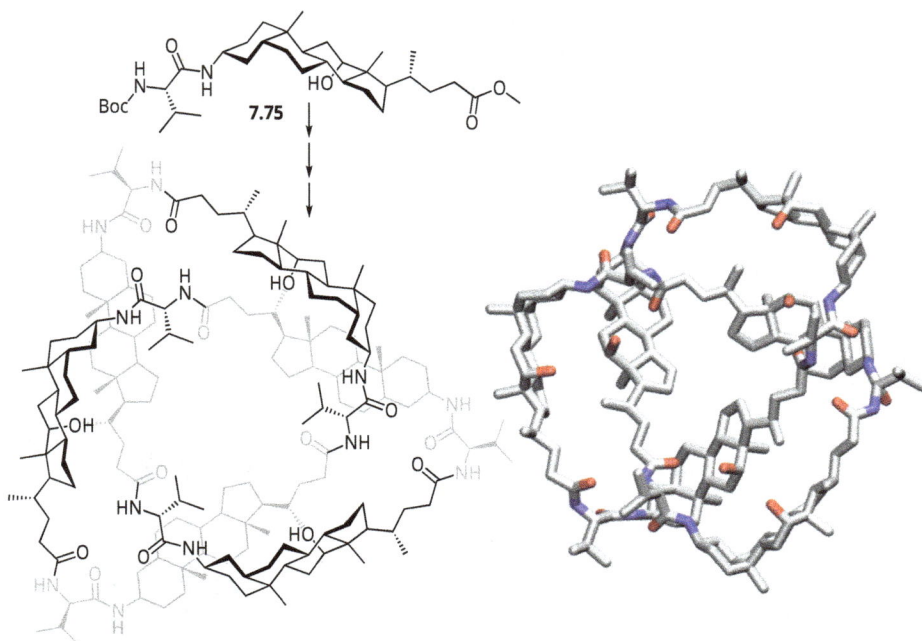

Figure 7.43: Structure of building block **7.75** and schematic structure and crystal structure of the knotted cyclic hexamer. The protons in the crystal structure are omitted for reasons of clarity.

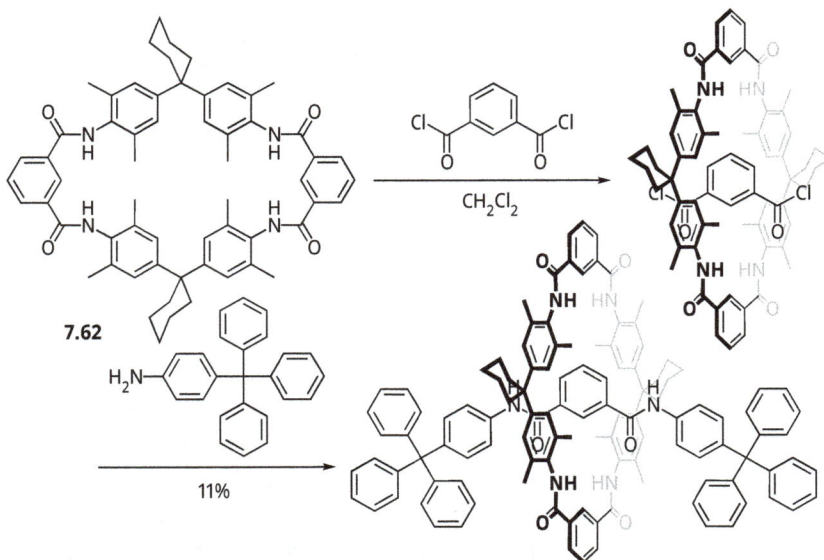

Figure 7.44: Formation of a [2]rotaxane by stoppering the pseudorotaxane formed from cyclophane **7.62** and isophthaloyl dichloride.

from almost all directions except the ring opening on the opposite side (Figure 7.45). As a result, an electrophile must approach from this side, causing the newly formed bond in the Williamson ether synthesis to pass through the cavity.

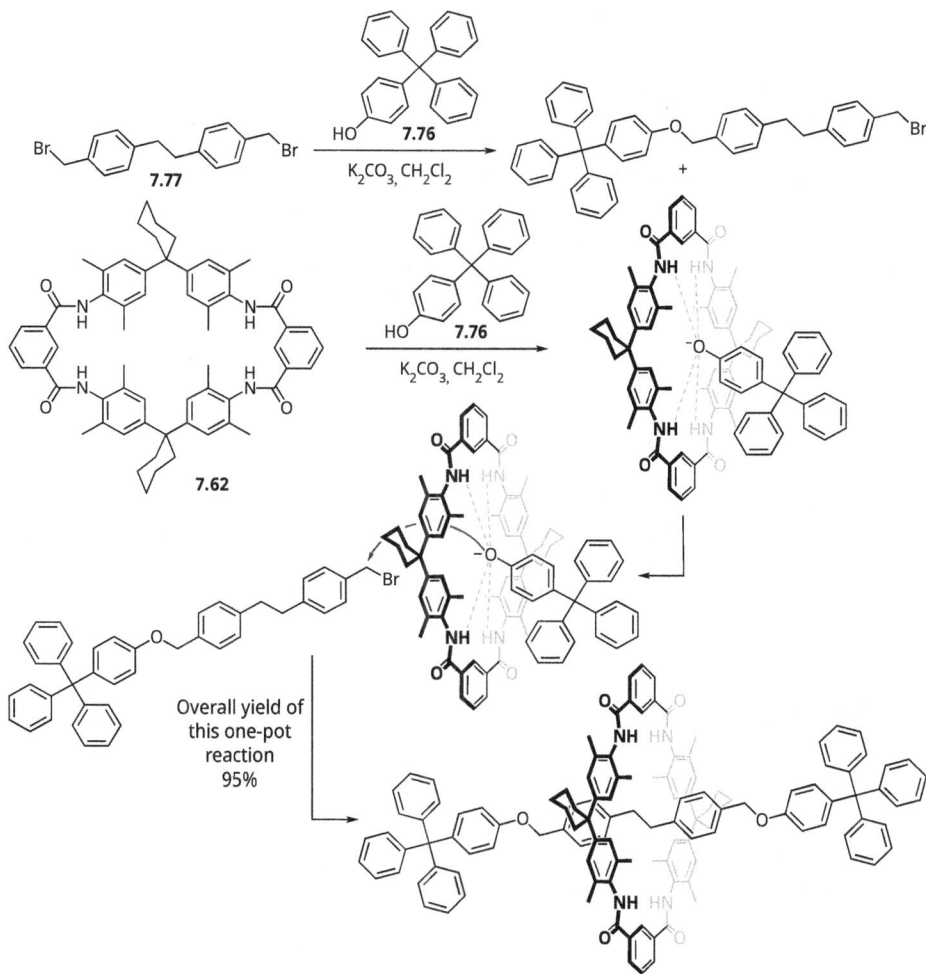

Figure 7.45: Example of a rotaxane synthesis using the trapping strategy. In the first step, the phenol **7.76** and dibromide **7.77** react under the influence of a base to form one half of the axle. In the second S$_N$2 reaction, this product reacts with the phenolate complex of **7.62** to give the [2]rotaxane.

An application of this concept is shown in Figure 7.45. It uses **7.76** as the phenolic component and **7.77** as the electrophile. Both starting materials are reacted with **7.62** in dichloromethane in the presence of solid potassium carbonate. Under these conditions, the corresponding rotaxane is efficiently formed in a remarkable yield of 95% [69]. Rotaxane formation involves the initial S_N2 reaction between **7.76** and **7.77** to produce one half of the axle. The subsequent reaction of this intermediate with the complex between **7.62** and the deprotonated form of **7.76** then leads to the formation of the [2]rotaxane. Since the macrocycle is trapped during the second S_N2 reaction, this synthetic strategy is associated with the term *trapping* [70]. It is somewhat related to active metal template syntheses in that a reactive intermediate bound to the ring is involved in rotaxane formation. However, this reaction is stoichiometric rather than catalytic.

Further reading

A rotaxane is an interlocked molecule in which a ring is threaded onto an axle from which it cannot slip due to the presence of bulky end groups. We usually assume that opening the ring will cause the rotaxane to fall apart, but there are stable rotaxane-like structures in which the component threaded onto the axle is acyclic. These compounds are based on the oligoamide-derived foldamers developed by Ivan Huc, whose diameter varies along the foldamer axis and which have a cavity large enough to accommodate guests (see Section 4.1.13). If properly designed, such foldamers can fold around a dumbell-shaped molecule to form so-called *foldaxanes*, in which the foldamer chain is trapped due to the substantial energy barrier associated with unfolding. Many aspects of foldaxanes mirror those associated with the formation and behavior of conventional rotaxanes. For example, foldaxanes can be prepared with more than one foldamer chain on the axle, and the position of the folded chain on the axle can be controlled. Stereochemical aspects of rotaxanes arising from the directionality of the ring find their counterpart in foldaxane chemistry in the handedness of the foldamer helix. For further details see:

Koehler V, Roy A, Huc I, Ferrand Y. Foldaxanes: rotaxane-like architectures from foldamers. Acc. Chem. Res. 2022, 55, 1074–85.

7.7 Syntheses using halogen bonds

7.7.1 Rotaxanes

The anion-mediated strategy developed by Beer to prepare mechanically interlocked molecules also allows the use of halogen bonds to preorganize the components. We focus here on rotaxanes, although catenanes can be prepared in a similar manner. The synthetic approach is closely related to that introduced in Section 7.6.1, except that the isophthalamide moiety in one component, typically the ring, is combined with a pyridinium derivative containing two flanking 1,4-disubstituted 5-iodo-1,2,3-triazole moieties. An example is bis(triazole) **7.78a**, which contains permethylated β-CD stoppers to impart water solubility. A ring is clipped around the chloride salt of **7.78a** by treating it with **7.79**, pyridine-3,5-dicarbonyl dichloride, and triethylamine in dichloromethane (Figure 7.46) [71]. Methylation of the pyridine moiety followed by ion exchange gives the corresponding [2]rotaxane **7.80a** in 33% yield over the three steps. The formation of the product is mediated by the interaction of the two components in their complex with

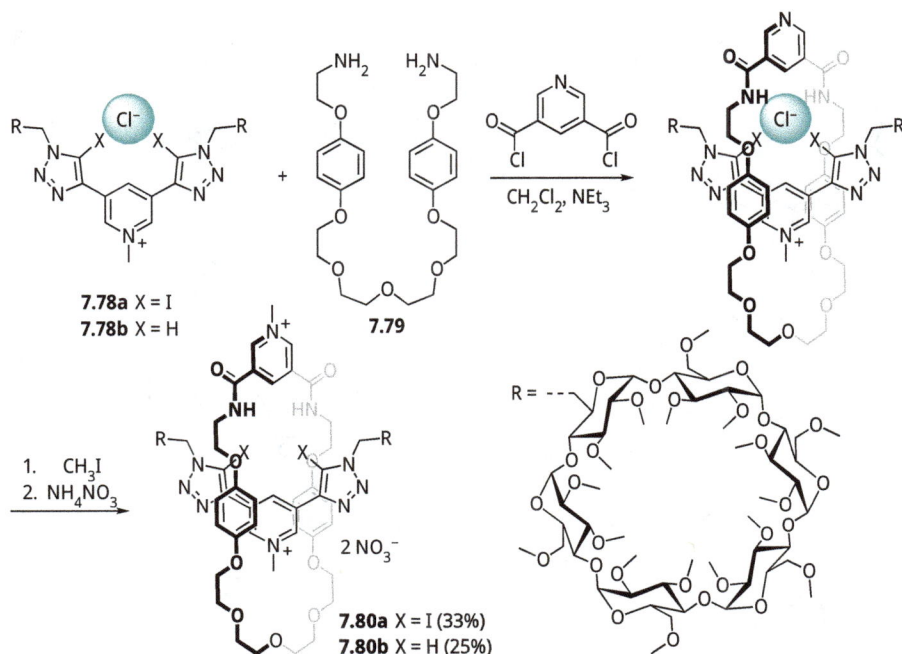

Figure 7.46: Formation of the [2]rotaxanes **7.80a, b** by clipping a ring around an axle containing either 1,4-disubstituted 5-iodo-1,2,3-triazole groups (**7.78a**) or 1,2,3-triazole groups (**7.78b**). In the first case, rotaxane formation is mediated by a combination of halogen and hydrogen bonding, and in the second case, only by hydrogen bonding.

the chloride anion through a combination of halogen and hydrogen bonding interactions with the C–I groups of **7.78a** and the NH groups of **7.79**, respectively.

Rotaxane **7.80a** has a pronounced iodide affinity in water (K_a = 2,200 M^{-1}), significantly higher than the corresponding prototriazole derivative **7.80b**, which is obtained in 25% yield by clipping the ring around **7.78b**. The lower yield of **7.80b** and its lower iodide affinity compared to **7.80a** demonstrate the beneficial effects of halogen bonding in these systems.

Adding electron-withdrawing substituents to the triazole moieties enhances the anion affinity by increasing the positive potential of the iodide σ-holes. Accordingly, clipping a ring around **7.81a**, in which two tetrafluorinated 1,4-phenylene moieties link the triazole rings to the cyclodextrin residues, yields the corresponding rotaxane **7.82a** in 91% yield, whereas the analogous reaction with the nonfluorinated **7.81b** yields the rotaxane **7.82b** in only 45% yield (Figure 7.47) [72].

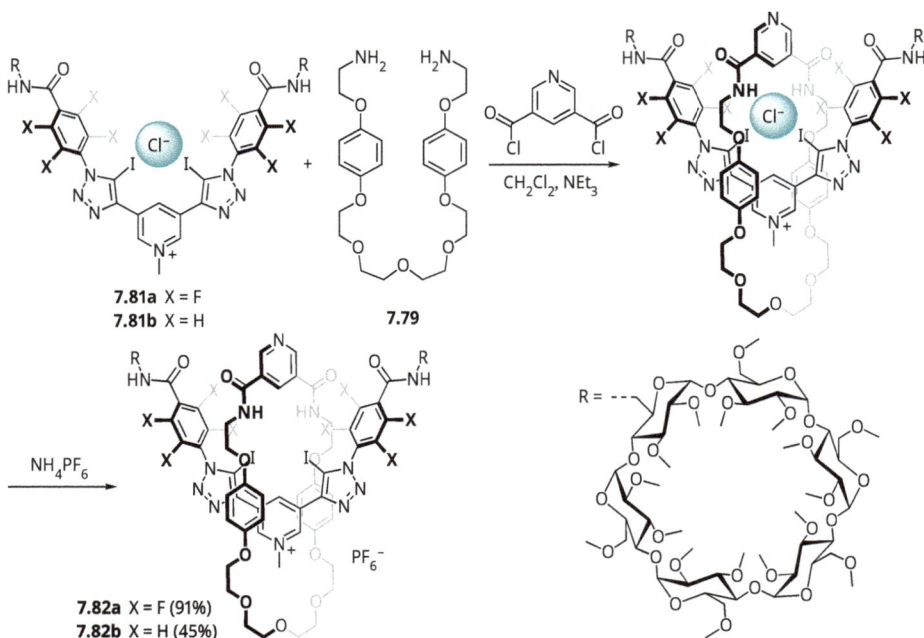

Figure 7.47: Influence of electron-withdrawing substituents on the efficiency of rotaxane formation using an axle containing two 1,4-disubstituted 5-iodo-1,2,3-triazole moieties. Clipping the ring around **7.81a**, in which the electron-withdrawing nature of the tetrafluorinated 1,4-phenylene substituents enhances the positive electrostatic potential of the iodine σ-holes, is significantly more efficient than when the nonfluorinated axle **7.81b** is used.

7.8 Syntheses using the hydrophobic effect

7.8.1 Knots

In the course of their work on the formation of catenanes from cysteine derivatives with electron-rich 1,5-dioxynaphthalene or electron-poor 1,4,5,8-naphthalenetetracar-boxylic diimide (NDI) groups, the Sanders group found that knotted structures are formed when the NDI derivatives alone are equilibrated under disulfide exchange conditions. For example, **7.83** yields not only a macrocyclic monomer and dimer in this reaction but also a trimer with the topology of a trefoil knot (Figure 7.48) [73]. When the initial concentration of **7.83** is 5 mM, the three products coexist in almost equal amounts at equilibrium, but the additional presence of NaNO₃ (1 M) leads to the almost exclusive formation of the knot, indicating that a major driving force for knot formation comes from the hydrophobic effect. Among the three observed products, the knot is indeed the one with the smallest solvent-accessible external surface due to the

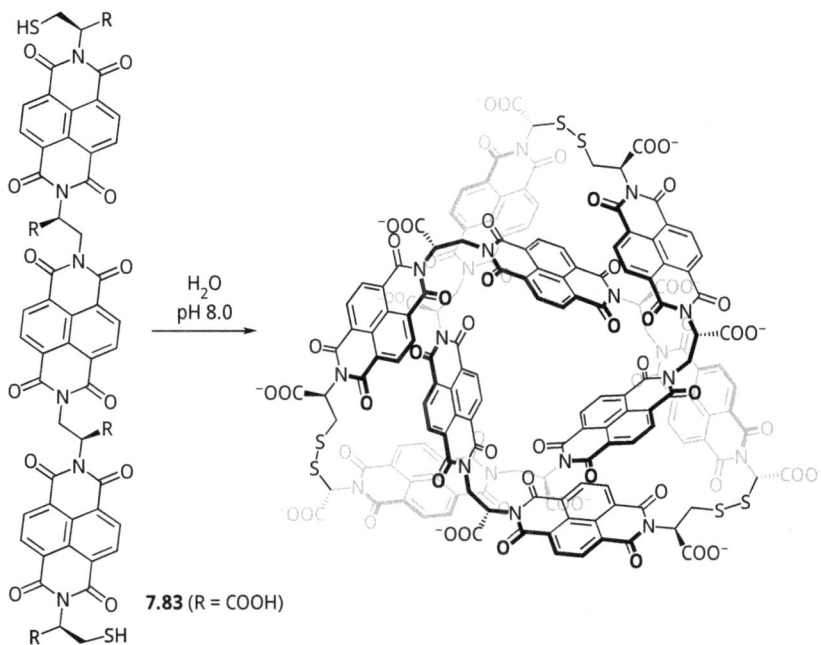

Figure 7.48: Molecular structure of dithiol **7.83** and structure of the trefoil knot resulting from disulfide formation.

screening of hydrophobic NDI subunits from the solvent. The efficiency with which this product is formed is likely due to the folding of the acyclic precursor into a knotted structure. Knot formation is enantioselective because the starting material is chiral.

The equilibration of dithiol **7.84** with two NDI residues in water at pH 8.0 yields several different mechanically interlocked structures, the most abundant of which are a Solomon link (60%) and a figure eight knot (18%) (Figure 7.49a) [74]. The formation of these products is again primarily driven by the hydrophobic effect, that is, the shielding of hydrophobic surfaces from the solvent. When the racemate of **7.84** is used instead of the homochiral compound, the figure eight knot is virtually the only product, indicating that the knot containing cysteine units of opposite absolute configuration is thermodynamically more stable than that formed from a single enantiomer of **7.84**.

The three products formed from homochiral and racemic **7.84** give rise to interesting cases of stereoisomerism. The homochiral products formed from only one enantiomer of **7.84** are stereochemically chiral because they contain only cysteine units with the same absolute configuration. The Solomon link is also topologically chiral and exists in two enantiomeric forms. However, the figure eight knot is topologically achiral because it is S_4-symmetric. One way to confirm this is to compare the struc-

(a)

7.84 (R = COOH)

H₂O
pH 8.0

60%   +   18%

(b)

Topologically chiral
Solomon link

90° Rotation

Homochiral figure eight knots tied into
one and into the opposite direction

Both knots are
interconverted by a
90° rotation, showing
that they are identical
and therefore
topologically achiral

90° Rotation

Meso figure eight knots tied into
one and into the opposite direction

The 90° rotation of one
knot illustrates that
both knots are mirror
images and therefore
topologically chiral

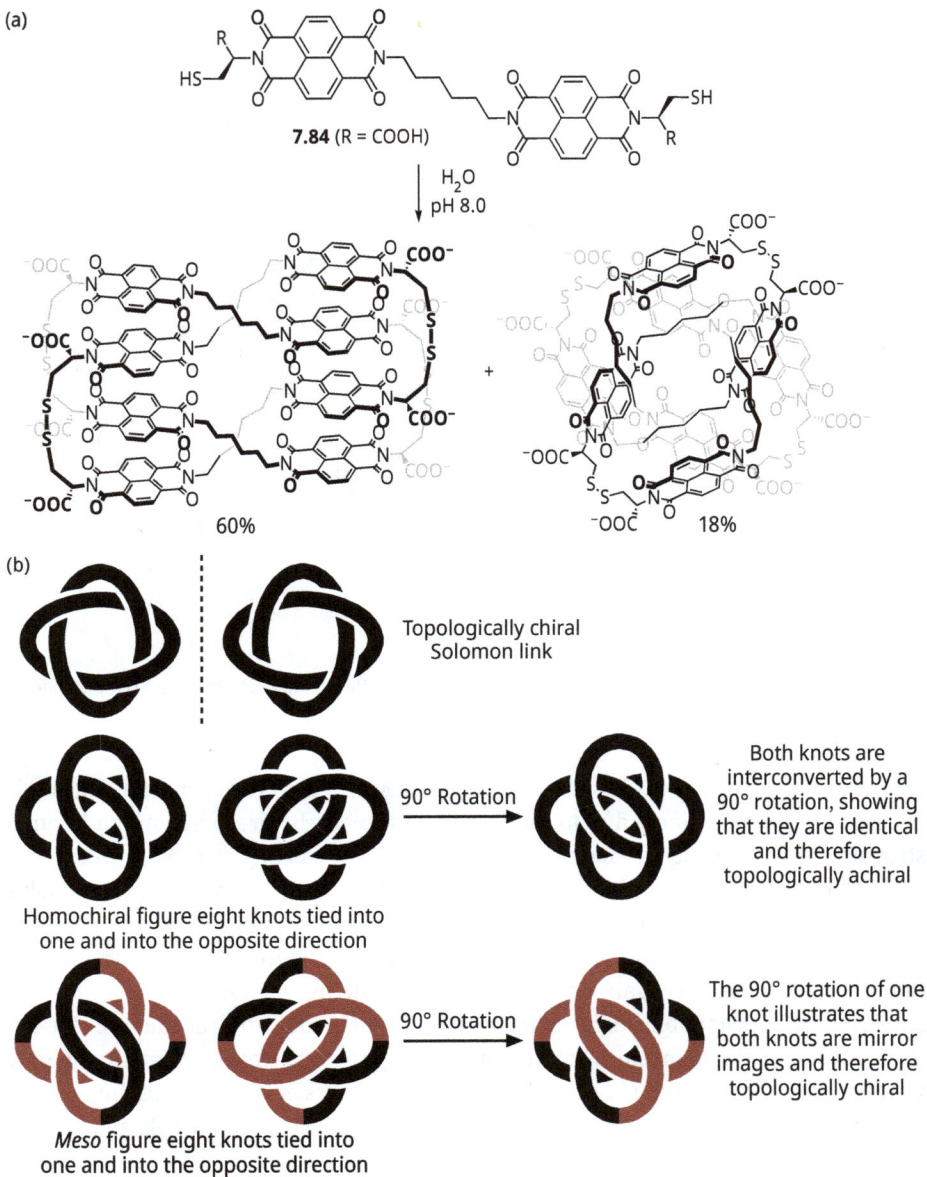

Figure 7.49: Molecular structure of dithiol **7.84** and structures of the Solomon link (left) and the figure eight knot (right) formed from this building block (a). Topologically, the Solomon link containing only L-cysteine is chiral, the corresponding figure eight knot is achiral, and the knot derived from racemic **7.84** is again chiral, as shown in (b). The red and black lines in the *meso*-knot symbolize that the respective subunits contain cysteine residues with opposite absolute configurations.

tures of two knots linked in opposite directions, which can be interconverted by simple rotation (Figure 7.49b).

In terms of stereochemistry, the figure eight knot formed from racemic **7.84** is achiral because it contains an equal number of *R* and *S* residues, making it a *meso*-form. Topologically, however, it is chiral because the combination of two different building blocks causes this knot to have a lower symmetry (C_2) than if all the cysteine residues had the same absolute configuration. This chirality is again illustrated by comparing the two knots tied in opposite directions. In this case, rotating one form by 90° results in a knot that is the mirror image of the other form. Note, however, that the two chiral forms can be interconverted by conformational rearrangement.

7.8.2 Rotaxanes

Many examples in the previous chapters illustrate that macrocyclic compounds through which acyclic molecules are threaded to yield pseudorotaxanes are convenient starting points for the preparation of mechanically interlocked molecules. Suitable rings should have a roughly cylindrical shape with two openings of similar size, making crown ethers, cyclodextrins, diphenylmethane or paraquat-derived cyclophanes, pillararenes, or cucurbiturils useful building blocks for the synthesis of such interlocked molecules. The cone-shaped cyclotriveratrylenes, calix[4]arenes, calix[5]arenes, or resorcin[4]arenes are not suitable because one of their cavity openings is too narrow to allow threading. In the case of calixarenes, both openings become wide enough only when the ring contains six or more aromatic subunits.

Interlocked molecules derived from receptors that bind their substrates through direct interactions have already been presented. Examples include the cyclophanes in Section 7.6 and the paraquat-derived blue box in Section 7.5. Pillararenes also belong to this category because they bind positively charged guests through a combination of charge-transfer interactions, hydrogen bonding, and cation–π interactions [75]. In the case of water-soluble receptors, substrate binding in the aqueous medium additionally benefits from the hydrophobic effect. Rotaxanes are thus accessible by threading suitable α,ω-difunctionalized substrates through the ring and stoppering the chain ends with large groups. Figure 7.50 shows examples of such rotaxanes prepared from α-CD, CB[6], or a cyclophane. Since the synthesis of these rotaxanes does not involve any aspects beyond those already discussed, we will not go into detail here.

Of the vast number of mechanically interlocked molecules known today, only a fraction could be presented in this chapter. The interested reader is referred to the monograph by Carson J. Bruns and J. Fraser Stoddart entitled "The Nature of the Mechanical Bond" [4], which gives an excellent and much broader overview. At this point, we will move away from the structural aspects of mechanically interlocked molecules and the concepts of their preparation to discuss in the next chapter one of the most

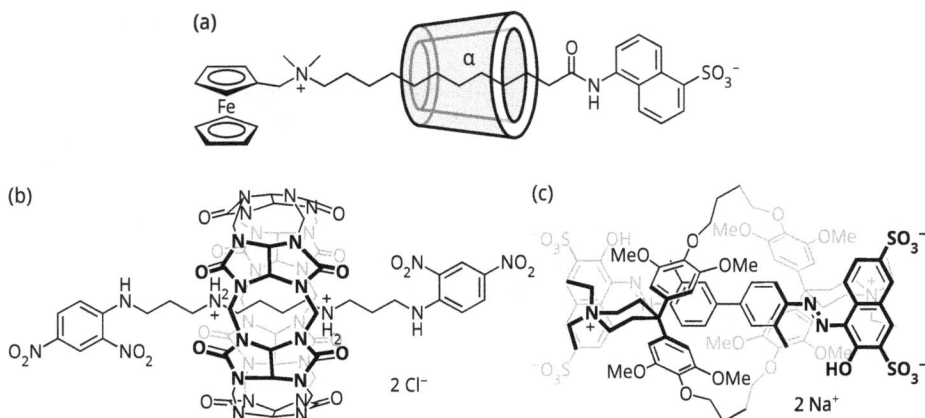

Figure 7.50: Examples of rotaxanes containing an α-CD (a), a CB[6] (b), and a positively charged cyclophane (c).

fascinating applications of these compounds, namely their use to control molecular motion and thus gain access to molecular machines.

Bibliography

[1]   Walba DM. Topological stereochemistry. Tetrahedron 1985, 41, 3161–212.
[2]   Mitchell DK, Chambron JC. Chemical topology. J. Chem. Edu. 1995, 72, 1059–64.
[3]   Stoddart JF. The chemistry of the mechanical bond. Chem. Soc. Rev. 2009, 38, 1802–20.
[4]   Bruns CJ, Stoddart JF. The Nature of the Chemical Bond. John Wiley & Sons: Hoboken, 2017.
[5]   Frisch HL, Wasserman E. Chemical topology. J. Am. Chem. Soc. 1961, 83, 3789–95.
[6]   Forgan RS, Sauvage JP, Stoddart JF. Chemical topology: complex molecular knots, links, and entanglements. Chem. Rev. 2011, 111, 5434–64.
[7]   Beeren SR, McTernan CT, Schaufelberger F. The mechanical bond in biological systems. Chem 2023, 9, 1378–412.
[8]   Kimura K, Rybenkov VV, Crisona NJ, Hirano T, Cozzarelli NR. 13S Condensin actively reconfigures DNA by introducing global positive writhe: implications for chromosome condensation. Cell 1999, 98, 239–48.
[9]   Walba, DM. Topological stereochemistry. Tetrahedron 1985, 41, 2161–212.
[10]   Jamieson EMG, Modicom F, Goldup SM. Chirality in rotaxanes and catenanes. Chem. Soc. Rev. 2018, 47, 5266–311.
[11]   Pairault N, Rizzi F, Lozano D, Jamieson EMG, Tizzard GJ, Goldup SM. A catenane that is topologically achiral despite being composed of oriented rings. Nat. Chem. 2023, 15, 781–6.
[12]   Zhang S, Rodríguez-Rubio A, Saady A, Tizzard GJ, Goldup SM. A chiral macrocycle for the stereoselective synthesis of mechanically planar chiral rotaxanes and catenanes. Chem 2023, 9, 1195–207.
[13]   Wasserman E. The preparation of interlocking rings: a catenane. J. Am. Chem. Soc. 1960, 82, 4433–4.
[14]   Wasserman E. Chemical topology. Sci. Am. 1962, 207, 94–102.

[15] Brückner R. Pioneering work on catenanes, rotaxanes, and a knotane in the University of Freiburg 1558–1988. Eur. J. Org. Chem. 2019, 3289–319.

[16] Baluna AS, Galan A, Leigh DA, Smith GD, Spence JDJ, Tetlow DJ, Vitorica-Yrezabal IJ, Zhang M. In search of Wasserman's catenane. J. Am. Chem. Soc. 2023, 145, 9825–33.

[17] Schill G, Lüttringhaus A. The preparation of catena compounds by directed synthesis. Angew. Chem. Int. Ed. Engl. 1964, 3, 546–7.

[18] Schill G, Logemann E, Vetter W. Ozonolytic degradation of a catenane. Angew. Chem. Int. Ed. Engl. 1972, 11, 1089–90.

[19] Cornelissen MD, Pilon S, van Maarseveen JH. Covalently templated syntheses of mechanically interlocked molecules. Synthesis 2021, 53, 4527–48.

[20] Walba DM, Homan TC, Richards RM, Haltiwanger RC. Synthesis and cutting "in half" of a molecular Möbius strip – applications of low dimensional topology in chemistry. New J. Chem. 1993, 17, 661–81.

[21] Ashbridge Z, Fielden SDP, Leigh DA, Pirvu L, Schaufelberger F, Zhang L. Knotting matters: orderly molecular entanglements. Chem. Soc. Rev. 2022, 51, 7779–809.

[22] Dietrich-Buchecker CO, Sauvage JP, Kintzinger JP. Une nouvelle famille de molecules: les metallo-catenanes. Tetrahedron Lett. 1983, 24, 5095–8.

[23] Stoddart JF. The master of chemical topology. Chem. Soc. Rev. 2009, 38, 1521–9.

[24] Dietrich-Buchecker CO, Sauvage JP. Interlocking of molecular threads: from the statistical approach to the templated synthesis of catenands. Chem. Rev. 1987, 87, 795–810.

[25] Dietrich-Buchecker CO, Khemiss A, Sauvage JP. High-yield synthesis of multiring copper(I) catenates by acetylenic oxidative coupling. J. Chem. Soc., Chem. Commun. 1986, 1376–8.

[26] Leigh DA, Pritchard RG, Stephens AJ. A Star of David catenane. Nat. Chem. 2014, 6, 978–82.

[27] Chichak KS, Cantrill SJ, Pease AR, Chiu SH, Cave GWV, Atwood JL, Stoddart JF. Molecular Borromean rings. Science 2004, 304, 1308–12.

[28] Pentecost CD, Tangchaivang N, Cantrill SJ, Chichak KS, Peters AJ, Stoddart JF. Making molecular Borromean Rings. A gram-scale synthetic procedure for the undergraduate organic lab. J. Chem. Educ. 2007, 84, 855–9.

[29] Original image created by Anon Moos and published under the Public Domain Mark 1.0 license.

[30] Adams H, Ashworth E, Breault GA, Guo J, Hunter CA, Mayers PC. Knot tied around an octahedral metal centre. Nature 2001, 411, 763.

[31] Guo J, Mayers PC, Breault GA, Hunter CA. Synthesis of a molecular trefoil knot by folding and closing on an octahedral coordination template. Nat. Chem. 2010, 2, 218–22.

[32] Gil-Ramírez G, Hoekman S, Kitching MO, Leigh DA, Vitorica-Yrezabal IJ, Zhang G. Tying a molecular overhand knot of single handedness and asymmetric catalysis with the corresponding pseudo-D_3-symmetric trefoil knot. J. Am. Chem. Soc. 2016, 138, 13159–62.

[33] Leigh DA, Pirvu L, Schaufelberger F, Tetlow DJ, Zhang L. Securing a supramolecular architecture by tying a stopper knot. Angew. Chem. Int. Ed. 2018, 57, 10484–8.

[34] Dietrich-Buchecker CO, Sauvage JP. A synthetic molecular trefoil knot. Angew. Chem. Int. Ed. Engl. 1989, 28, 189–92.

[35] Dietrich-Buchecker CO, Rapenne G, Sauvage JP. Efficient synthesis of a molecular knot by copper(I)-induced formation of the precursor followed by ruthenium(II)-catalysed ring-closing metathesis. Chem. Commun. 1997, 2053–4.

[36] Dietrich-Buchecker CO, Sauvage JP. Lithium templated synthesis of catenanes: efficient synthesis of doubly interlocked [2]-catenanes. Chem. Commun. 1999, 615–6.

[37] Ayme JF, Beves JE, Leigh DA, McBurney RT, Rissanen K, Schultz D. A synthetic molecular pentafoil knot. Nat. Chem. 2012, 4, 15–20.

[38] Zhang ZH, Zhou Q, Li Z, Zhang N, Zhang L. Completely stereospecific synthesis of a molecular cinquefoil (5_1) knot. Chem 2023, 9, 847–58.

[39] Feng HN, Sun Z, Chen S, Zhang ZH, Li Z, Zhong Z, Sun T, Ma Y, Zhang L. A Star of David [2]catenane of single handedness. Chem 2023, 9, 859–68.

[40] Leigh DA, Danon JJ, Fielden SDP, Lemonnier JF, Whitehead GFS, Woltering SL. A molecular endless (7_4) knot. Nat. Chem. 2021, 13, 117–22.

[41] Wu C, Lecavalier PR, Shen YX, Gibson HW. Synthesis of a rotaxane via the template method. Chem. Mater. 1991, 3, 569–72.

[42] Crowley JD, Goldup SM, Lee AL, Leigh DA, McBurney RT. Active metal template synthesis of rotaxanes, catenanes and molecular shuttles. Chem. Soc. Rev. 2009, 38, 1530–41.

[43] Denis M, Goldup SM. The active template approach to interlocked molecules. Nat. Chem. Rev. 2017, 1, 0061.

[44] Aucagne V, Hänni KD, Leigh DA, Lusby PJ, Walker DB. Catalytic "click" rotaxanes: a substoichiometric metal-template pathway to mechanically interlocked architectures. J. Am. Chem. Soc. 2006, 128, 2186–7.

[45] Rodionov VO, Fokin VV, Finn MG. Mechanism of the ligand-free Cu^I-catalyzed azide–alkyne cycloaddition reaction. Angew. Chem. Int. Ed. 2005, 44, 2210–5.

[46] Chen XY, Chen H, Stoddart JF. The story of the little blue box: a tribute to Siegfried Hünig Angew. Chem. Int. Ed. 2023, 62, e202211387.

[47] Philp D, Stoddart JF. Self-assembly in natural and unnatural systems. Angew. Chem. Int. Ed. Engl. 1996, 35, 1154–96.

[48] Ashton PR, Goodnow TT, Kaifer AE, Reddington MV, Slawin AMZ, Spencer N, Stoddart JF, Vicent C, Williams DJ. A [2]catenane made to order. Angew. Chem. Int. Ed. Engl. 1989, 28, 1396–9.

[49] Amabilino DB, Ashton PR, Reder AS, Spencer N, Stoddart JF. Olympiadane. Angew. Chem. Int. Ed. Engl. 1994, 33, 1286–90.

[50] Hamilton DG, Davies JE, Prodi L, Sanders JKM. Synthesis, structure and photophysics of neutral π-associated [2]catenanes. Chem. Eur. J. 1998, 4, 608–20.

[51] Au-Yeung HY, Dan Pantos G, Sanders JKM. Dynamic combinatorial synthesis of a catenane based on donor–acceptor interactions in water. Proc. Natl. Acad. Sci. U. S. A. 2009, 106, 10466–70.

[52] Amabilino DB, Ashton PR, Brown CL, Cordova E, Godinez LA, Goodnow TT, Kaifer AE, Newton SP, Pietraszkiewicz M, Prodi L, Reddington MV, Slawin AMZ, Spencer N, Stoddart JF, Vicent C, Williams DJ. Molecular meccano. 2. Self-assembly of [n]catenanes. J. Am. Chem. Soc. 1995, 117, 1271–93.

[53] Ashton PR, Bělohradský M, Philp D, Stoddart JF. Slippage – an alternative method for assembling [2] rotaxanes. J. Chem. Soc., Chem. Commun. 1993, 1269–74.

[54] Johnston AG, Leigh DA, Pritchard RJ, Deegan MD. Facile synthesis and solid-state structure of a benzylic amide [2]catenane. Angew. Chem. Int. Ed. Engl. 1995, 34, 1209–12.

[55] Hunter CA. Synthesis and structure elucidation of a new [2]-catenane. J. Am. Chem. Soc. 1992, 114, 5303–11.

[56] Jäger R, Vögtle F. A new synthetic strategy towards molecules with mechanical bonds: nonionic template synthesis of amide-linked catenanes and rotaxanes. Angew. Chem. Int. Ed. Engl. 1997, 36, 930–44.

[57] Ottens-Hildebrandt S, Schmidt T, Harren J, Vögtle F. Sulfonamide-based catenanes – regioselective template synthesis. Liebigs Ann. 1995, 1855–60.

[58] Jäger R, Schmidt T, Karbach D, Vögtle F. The first pretzel-shaped molecules – via catenane precursors. Synlett 1996, 723–5.

[59] Original image taken by an anonymous photographer and published under the Pixabay license that allows free use.

[60] Vysotsky MO, Bolte M, Thondorf I, Böhmer V. New molecular topologies by fourfold metathesis reactions within a hydrogen-bonded calix[4]arene dimer. Chem. Eur. J. 2003, 9, 3375–82.

[61] Bogdan A, Vysotsky MO, Ikai T, Okamoto Y, Böhmer V. Rational synthesis of multicyclic bis[2]catenanes. Chem. Eur. J. 2004, 10, 3324–30.

[62]   Böhmer V, Rudzevich Y, Rudzevich V. Rotaxanes and catenanes derived from tetra-urea calix[4]arenes. Macromol. Symp. 2010, 287, 42–50.

[63]   Lankshear MD, Beer PD. Interweaving anion templation. Acc. Chem. Res. 2007, 40, 657–68.

[64]   Wisner JA, Beer PD, Drew MGB, Sambrook MR. Anion-templated rotaxane formation. J. Am. Chem. Soc. 2002, 124, 12469–76.

[65]   Safarowsky O, Nieger M, Fröhlich R, Vögtle F. A molecular knot with twelve amide groups – one-step synthesis, crystal structure, chirality. Angew. Chem. Int. Ed. 2000, 39, 1616–8.

[66]   Vögtle F, Hünten A, Vogel E, Buschbeck S, Safarowsky O, Recker J, Parham AH, Knott M, Müller WM, Müller U, Okamoto Y, Kubota T, Lindner W, Francotte E, Grimme S. Novel amide-based molecular knots: complete enantiomeric separation, chiroptical properties, and absolute configuration. Angew. Chem. Int. Ed. 2001, 40, 2468–71.

[67]   Brüggemann J, Bitter S, Müller S, Müller WM, Müller U, Maier NM, Lindner W, Vögtle F. Spontaneous knotting – from oligoamide threads to trefoil knots. Angew. Chem. Int. Ed. 2007, 46, 254–9.

[68]   Feigel M, Ladberg R, Engels S, Herbst-Irmer R, Fröhlich R. A trefoil knot made of amino acids and steroids. Angew. Chem. Int. Ed. 2006, 45, 5698–702.

[69]   Hübner GM, Gläser J, Seel C, Vögtle F. High-yielding rotaxane synthesis with an anion template. Angew. Chem. Int. Ed. 1999, 38, 383–6.

[70]   Seel C, Vögtle F. Templates, "wheeled reagents", and a new route to rotaxanes by anion complexation: the trapping method. Chem. Eur. J. 2000, 6, 21–4.

[71]   Langton MJ, Robinson SW, Marques I, Félix V, Beer PD. Halogen bonding in water results in enhanced anion recognition in acyclic and rotaxane hosts. Nat. Chem. 2014, 6, 1039–43.

[72]   Bunchuay T, Docker A, Martinez-Martinez AJ, Beer PD. A potent halogen-bonding donor motif for anion recognition and anion template mechanical bond synthesis. Angew. Chem. Int. Ed. 2019, 58, 13823–7.

[73]   Ponnuswamy N, Cougnon FBL, Clough JM, Pantoş GD, Sanders JKM. Discovery of an organic trefoil knot. Science 2012, 338, 783–5.

[74]   Ponnuswamy N, Cougnon FBL, Pantoş D, Sanders JKM. Homochiral and *meso* figure eight knots and a Solomon link. J. Am. Chem. Soc. 2014, 136, 8243–51.

[75]   Ogoshi T, Yamagishi TA, Nakamoto Y. Pillar-shaped macrocyclic hosts pillar[n]arenes: new key players for supramolecular chemistry. Chem. Rev. 2016, 116, 7937–8002.

8 Controlling molecular motion

CONSPECTUS: A machine is a device that performs a task by the controlled motion of its components. In this chapter, we will see that molecules can behave in a similar way when the permanent and unavoidable Brownian motion they and their subunits undergo is overcompensated by a directional component. The mechanically interlocked molecules discussed in the previous chapter are particularly attractive (but not the only) starting points for the development of such machines. The chapter begins with a look at the dynamic behavior of molecules and then introduces methods to control it. In the following two sections, we look at examples of interlocked molecules, in which the components perform either a lateral back-and-forth motion or a circular motion relative to each other. The latter systems ultimately give rise to molecular motors, and since not only molecules with mechanical bonds can be used for this purpose, other systems will also be discussed.

8.1 Introduction

Molecules are always in motion. Not only are bond lengths and angles constantly changing, but molecules in solution are also moving around all the time. Triggered by external stimuli such as light or the protonation of a functional group, these random events can be superimposed by a directional component. An example is the photochemical E to Z isomerization of alkenes, which leads to a substantial structural change, bringing substituents that are far apart in one state close together in the other. These processes are reminiscent of motions in mechanical devices, suggesting that molecules that change their structure in response to a particular stimulus may be able to perform tasks at the molecular level like machines in the macroscopic world. Although molecular and macroscopic machines can indeed behave in surprisingly similar ways, it must be remembered that molecular machines never really rest, even in the absence of an energy supply, because of the inevitable Brownian motion. Moreover, the environment in which they operate is also extremely dynamic.

What is a molecular machine?

In this chapter, we will discuss various strategies for controlling the relative motion of two molecules or two subunits in a single molecule. Perhaps not surprisingly, nature has come up with a number of ingenious solutions for controlling motion and performing work at the molecular level, which is why begin our discussion with a biological system, namely the protein complex that allows certain bacteria to actively move. These bacteria carry on their outer surface rotating filaments, hollow tubular structures composed of many copies of the protein flagellin, which propel them forward (Figure 8.1a). A curved subunit just outside the cell wall, the hook, connects the filament to a group of proteins that span the cell membrane and terminate in the rings of the basal protein body. These rings can rotate and are surrounded inside the

https://doi.org/10.1515/9783111315171-008

cell membrane by static proteins, the stator. The overall assembly thus resembles the structure of a rotaxane, with the stator representing the ring through which a rod-shaped protein is threaded. This axle component contains the filament at one end and the basal protein body at the opposite end. The unidirectional rotation of the whole assembly is driven by an inward flux of protons across the cell membrane, which causes the basal protein body to move relative to the stator.

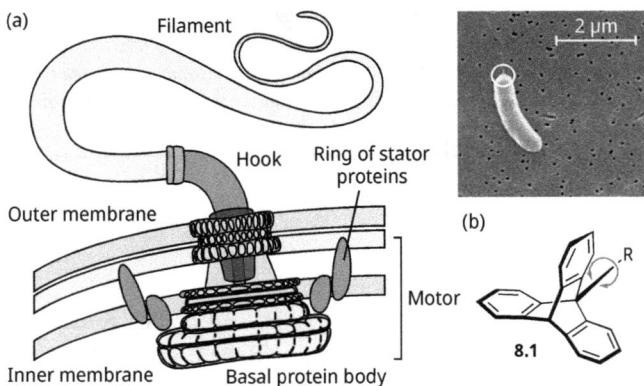

Figure 8.1: Schematic representation of the proteins that make up a bacterial flagellum (a) and the molecular structure of triptycene derivative **8.1** (b). The image shows a flagellated *Vibrio vulnificus* bacterium with the site of the attached filament marked [1].

Another important motor protein in which protons drive the rotation of a rotaxane-like assembly is ATP synthase. Natural machines in which components move laterally are myosin, which is responsible for muscle contraction, or kinesin, which transports cargo within cells. There are fascinating movies on YouTube that illustrate how these machines work [2].

Although small in absolute dimensions, natural molecular machines are still large enough to behave almost like conventional machines. This situation changes when molecules become smaller and are no longer attached to a support. In a solution of **8.1** (Figure 8.1b), for example, many copies of the same molecule tumble around freely, with the propeller-like triptycene group constantly rotating around the residue R. These rotations are largely random, and since no larger structure, such as a membrane or other type of support, can be used as an external reference point, internal motions and motions of the molecule as a whole can only be discussed in relative terms. However, even in this small molecule, they can be controlled to some extent, as shown by the work of T. Ross Kelly [3].

The triptycene derivative **8.2** (Figure 8.2), for example, acts as a molecular brake because metal coordination to the bipyridine moiety almost completely stops the rotation around the central bond [4]. In its metal-free form, **8.2** is dynamic according to the ^{1}H NMR spectrum, which contains a single set of four sharp signals for the protons in the triptycene group. These signals are broadened in the spectrum of the Hg^{2+} complex, indicating that the metal-induced conformational rigidification of the bipyridyl unit has a strong effect on the rate at which the two subunits rotate around each other. The ^{1}H NMR spectrum recorded at –30 °C contains two sharp signal sets for the protons of the triptycene group, consistent with a rigid structure in which the two aromatic rings flanking the bipyridyl residue have a different environment than the third ring. The brake is released by the addition of ethylenediaminetetraacetic acid (EDTA), which binds to Hg^{2+} more strongly than **8.2**.

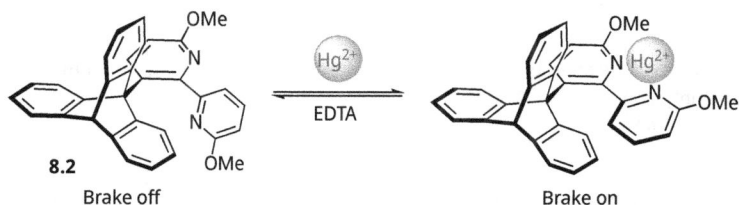

Figure 8.2: Mode of action of the molecular brake **8.2**. In the metal-free form, the triptycene group rotates freely, whereas the coordination of a mercury ion results in a conformation in which the terminal pyridyl group is intercalated between two aromatic rings of the triptycene group, thus stopping its rotation.

The Kelly group also introduced the molecular ratchet **8.3** (Figure 8.3a), in which the chiral helicene moiety makes the potential energy curve of conformational interconversion asymmetric and steeper when the triptycene group rotates in one direction than when it rotates in the opposite direction (Figure 8.3b) [5]. ^{1}H NMR spectroscopy showed that the three aromatic rings of the triptycene group are indeed in different environments as expected for a ratchet, but also that the rotation occurs at the same rate in both directions. This behavior is consistent with the principle of microscopic reversibility, which states that reaction rates are governed only by the energy distance between the transition state and the ground state, and not by how the transition state is reached. Therefore, the asymmetry of the energy profile in Figure 8.3b does not cause the rotation to be unidirectional.

A famous thought experiment described by Richard Feynman led to the same conclusion [6]. Feynman proposed a device consisting of an axle with a ratchet and pawl system at one end, located in a gas-filled container at temperature T_1 (Figure 8.3c). The other end of the axle is connected to a paddle wheel in a second container at temperature T_2. Feynman then asked whether the ratchet would be able to convert the random motion of the paddlewheel caused by colliding gas molecules into unidirectional rotation that could eventually be used to do work. Unidirectional rotation is indeed possible for $T_1 < T_2$. However, in the absence of a temperature difference between the two reservoirs, the paddlewheel moves in a random manner, as in Kelly's ratchet **8.3**. This is be-

(a)   (b)   (c)

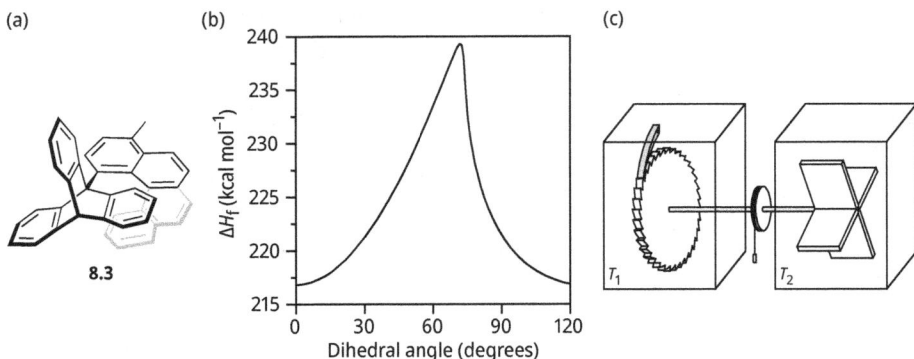

Figure 8.3: Molecular structure of the molecular ratchet (a), calculated potential energy profile of the clockwise rotation of the triptycene residue in **8.3** around the central bond (b), and schematic illustration of Feynman's ratchet and pawl (c).

cause it takes a relatively large number of gas molecules hitting the paddlewheel from one side to move the wheel in the correct direction. On the other hand, Brownian motion causes the pawl to bounce off the wheel from time to time, and the number of gas molecules needed to move the wheel a short distance backward is small. In the end, both processes have the same probability, and the motion is therefore not directional.

Thus, Feynman's device and Kelly's ratchet cannot produce work simply by absorbing heat from a single thermal reservoir, in accordance with the second law of thermodynamics. Unidirectional rotation requires an additional input of energy to bring the system into a state where rotation in one direction is easier (or faster) than rotation in the other. This idea led Kelly to develop the triptycene derivative **8.4** (Figure 8.4) [7], which is activated by converting the aromatic amino group to an isocyanate group with phosgene. Random rotations of the triptycene propeller bring this group and the hydroxy group in the substituent close enough to undergo carbamate formation. From the high-energy conformation stabilized in this way, back rotation is not possible because the strain would be too high. Moreover, completing the 120° clockwise rotation is associated with a lower activation energy than returning from the resulting conformation. Accordingly, arresting **8.4** in a high-energy conformation creates an asymmetry in the energy potential curve that causes the propeller to rotate in only one direction. Unfortunately, further control of the dynamic behavior of **8.4** is not possible.

These triptycene derivatives demonstrate that the individual subunits that move relative to each other in a molecular machine must be connected and able to communicate, either sterically or electronically, so that the motion of one component can be transmitted to the other. Separate molecules can also perform machine-like motions: for example, the insertion of a substrate into the cavity of a macrocyclic receptor resembles the movement of a piston in a cylinder. However, such systems fall apart

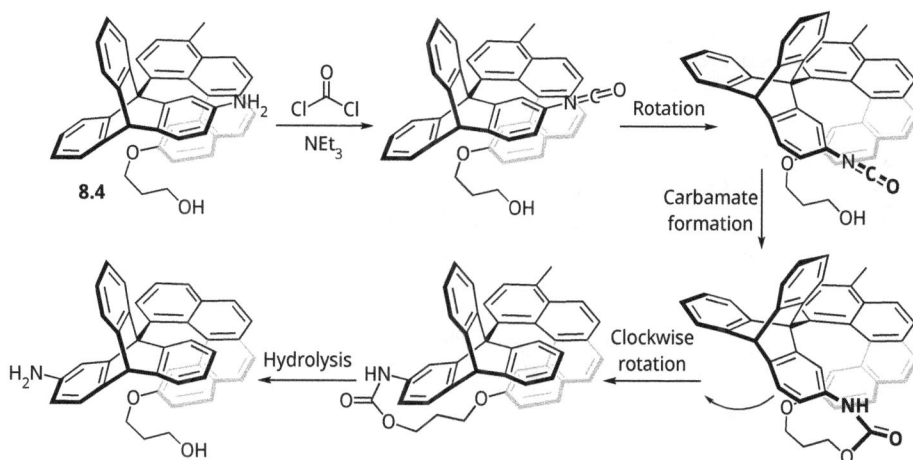

Figure 8.4: Reaction scheme illustrating the steps that mediate the unidirectional rotation of the triptycene group in the molecular motor **8.4**. The free amine in the initial atropisomer reacts with phosgene to form an isocyanate. Intramolecular carbamate formation then results in an energetically unfavorable conformation, which can only reach a more stable state by clockwise rotation of the triptycene group. Subsequent hydrolysis of the carbamate yields a new atropisomer, with the entire process involving a unidirectional rotation around the central bond.

when the components dissociate, which is generally not desired in a machine. In addition, covalent bonds are only one way to connect the moving parts in a molecular machine. The use of mechanical bonds is another attractive option because the motions in interlocked molecules follow well-defined trajectories, such as the lateral motions of the ring along the axle of a rotaxane (Figure 8.5a), or the migration of one ring along the perimeter of the other ring in a catenane (Figure 8.5b) [8]. Less important in the context of molecular machines, because they do not lead to substantial structural changes, is the rotation of an axle along its longitudinal axis or of a ring around its center (Figure 8.5d).

Fundamental work on the use of mechanically interlocked molecules for the development of molecular machines began in the 1980s in the group of Jean-Pierre Sauvage. Soon after, J. Fraser Stoddart joined the field. In 2016, both researchers, along with Bernard (Ben) L. Feringa, who invented a light-driven molecular motor (Section 8.4), were awarded the Nobel Prize in Chemistry in recognition of their pioneering achievements. This decision was based in part on the expectation that molecular machines have the potential to lead to "new materials, sensors and energy storage systems" [9]. Indeed, we will see in Section 12.6 that a rotaxane-based data storage medium has already been realized, but before we come to applications, it is instructive to take a closer look at the dynamics of these systems and see how their behavior is controlled.

A system whose dynamic behavior has been particularly well studied is Stoddart's [2]catenane **8.5** (Figure 8.6a) [10]. In this compound, three dynamic processes occur

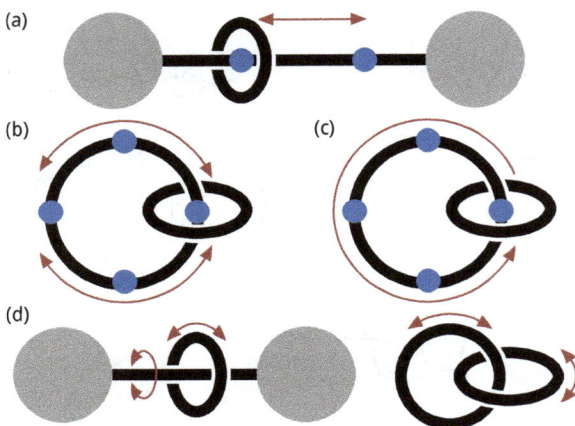

Figure 8.5: Typical motions in mechanically interlocked molecules are the lateral back-and-forth motion of the ring along the axle of a rotaxane (a) and the change in position of one ring of a catenane with respect to another (b). The unidirectional rotation of one ring of a catenane around the perimeter of the other ring gives rise to a molecular motor that potentially performs work (c). The rotation of a ring around its center or the rotation of an axle in a rotaxane around its main axis is not of primary importance in the context of molecular machines (d).

on different timescales. The fastest is a rocking motion that involves changes in the angle at which the two rings are arranged without significantly affecting the interactions between them (Figure 8.6b). This process can be slowed by lowering the temperature, but occurs rapidly at 25 °C, where it has a rate constant of 1.6×10^6 s^{-1}. The pirouetting rotation of the outer hydroquinone ring around the blue box is much slower, occurring at 7,000 times per second because it involves the loss of one charge-transfer interaction in the transition state (Figure 8.6c). The slowest process is the rotation of the blue box around its center (Figure 8.6d), which takes place only 22 times per second at 25 °C because it involves the loss of two charge-transfer interactions.

Similar dynamic processes occur in the [2]catenane **8.6,** in which the blue box is threaded onto a crown ether with four hydroquinone moieties (Figure 8.7a) [11]. In this case, the pirouetting motion occurs 2.8×10^4 times per second, significantly faster than in **8.5** due to the greater conformational flexibility of the crown ether. Another process in this catenane is the shuttling of the blue box from one hydroquinone unit to the next. With a rate constant of 300 s^{-1}, this process is slow because the blue box must completely give up the interactions with a hydroquinone ring to reach the next. Co-conformations with the blue box located in the region of the oligo(ethylene glycol) chains are populated only transiently.

In rotaxane **8.7,** the blue box shuttles between the two hydroquinone stations 1,800 times per second at 25 °C (Figure 8.7b) [12]. Shuttling stops at −50 °C, allowing the complexed and uncomplexed hydroquinone moieties to be distinguished by ^{1}H NMR spectroscopy.

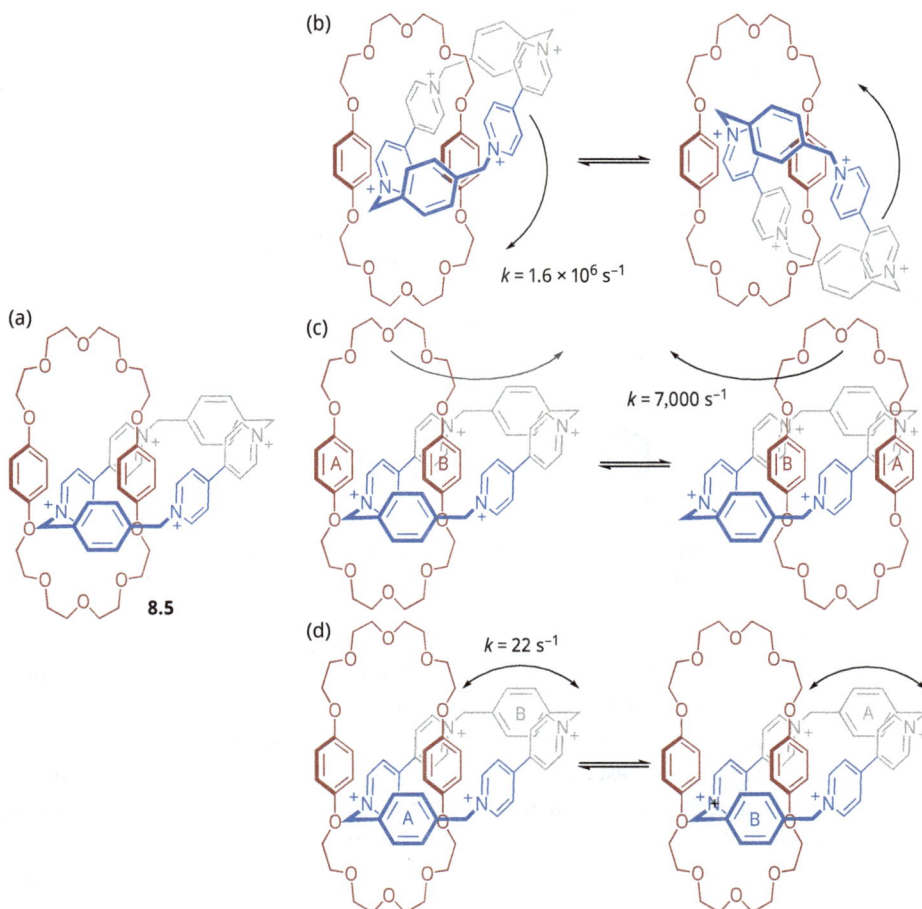

Figure 8.6: Molecular structure (a) and dynamic processes in the [2]catenane **8.5**. The fastest is the rocking motion of the rings (b). The pirouetting motion of the crown ether is much slower (c), and the slowest process is the rotation of the blue box around its center (d).

What is the difference between a shuttle and a switch?

The dynamic processes that occur in these mechanically interlocked molecules resemble the lateral motions in simple machines, but because they lack a directional component and instead proceed randomly between thermodynamically degenerate states (Figure 8.8a), neither the molecular train **8.6** nor the molecular shuttle **8.7** exhibit true machine-like behavior. However, by breaking the symmetry of the track along which the blue box moves and thermodynamically favoring one co-conformation over another, control can be achieved (Figure 8.8b).

(a)

(b)

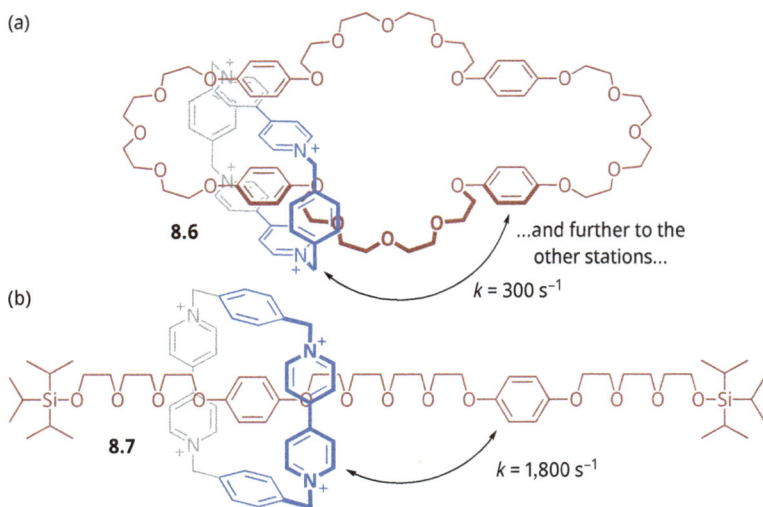

Figure 8.7: Molecular structures of the catenane-derived molecular train **8.6** (a) and the rotaxane-derived molecular shuttle **8.7** (b). The motions in these systems are shown along with the corresponding rate constants at 25 °C.

An example is rotaxane **8.8** with two different stations on the axle, of which the blue box prefers the benzidine group, which is a better π-donor than the 4,4′-biphenol moiety (Figure 8.8c) [13]. Once this imbalance is established, all that is needed for the ring to move is a stimulus that destabilizes the initial co-conformation (Figure 8.8b). In rotaxane **8.8**, this destabilization is achieved by protonation or oxidation of the benzidine group. In both cases, positive charges develop near the blue box, which moves away to avoid electrostatic repulsion. Both stimuli thus thermodynamically favor the co-conformation in which the blue box binds to the 4,4′-biphenol moiety. This bias is reversed by removing the charges on the benzidine subunits by deprotonation or reduction.

Rotaxane **8.8** thus behaves like a molecular switch whose states are controlled by changing the electronic properties of the subunits in the axle. Applying the stimulus results in a directional movement of the ring from one side to the other due to a change in the relative thermodynamic stabilities of the two co-conformations. Note that unlike a real switch, which stays on or off after being pressed, such molecular switches must be permanently activated or deactivated to remain in one position.

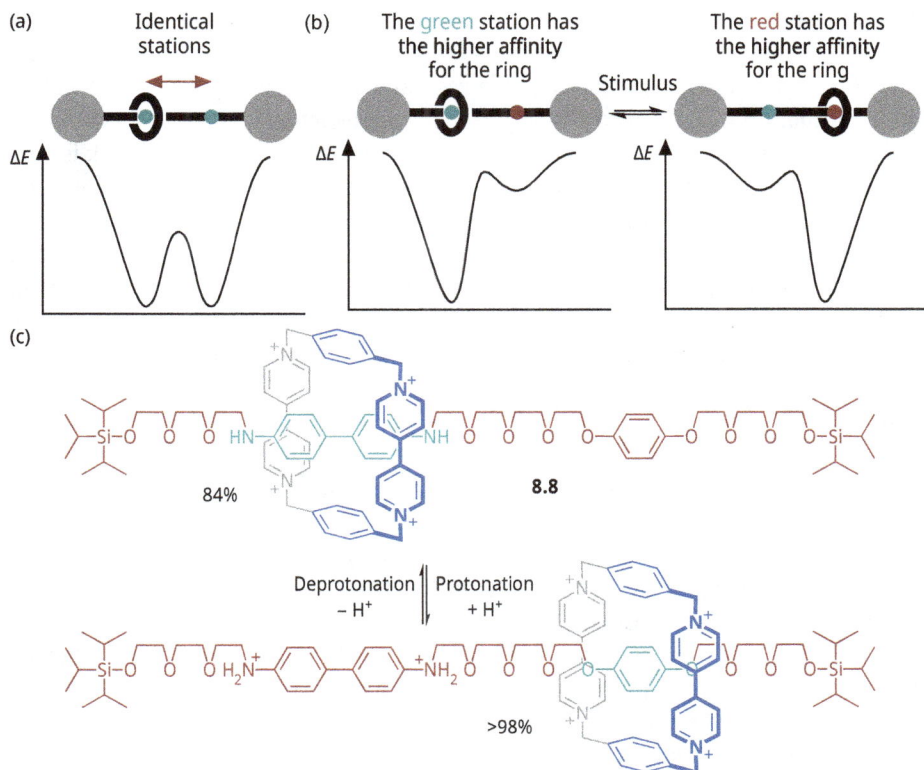

Figure 8.8: Energy profiles associated with random and directed motion of a ring along the axle of a rotaxane. The rotaxane in (a) contains two identical subunits in the axle, while that in (b) contains two different subunits whose affinity to the ring can be controlled. In (c), the switching between the co-conformations of rotaxane **8.8** is shown, triggered by protonation and deprotonation of the benzidine subunit. Switching by oxidation and reduction of the same group proceeds analogously.

Further reading

Control of the co-conformation of a molecular switch, such as the one shown in Figure 8.8b, is usually achieved by applying a stimulus that persists until the next stimulus shifts the system to a different state. In the case of rotaxane **8.8**, for example, the first stimulus is the addition of an acid, the effect of which is then reversed by the addition of a base. However, there are strategies to apply a stimulus that disappears on its own over time. In this way, a system, not necessarily just an interlocked molecule, can be induced to adopt a nonequilibrium state that dissipates autonomously and returns to the thermodynamic ground state. Such dissipative systems can be used to mimic fundamental behaviors of living systems that also operate out-of-equilibrium and may ultimately lead to artificial living systems. For more information, see:

Rieß B, Grötsch RK, Boekhoven J. The design of dissipative molecular assemblies driven by chemical reaction cycles. Chem 2020, 6, 552–78. Del Giudice D, Di Stefano S. Dissipative systems driven by the decarboxylation of activated carboxylic acids. Acc. Chem. Res. 2023, 56, 889–99.

Molecular switches cannot be used to perform work because any mechanical effect associated with motion in one direction is lost when the direction is reversed (unless the back-and-forth motion of the ring in a rotaxane drives the unidirectional rotation of an attached ring like a piston in an internal combustion engine, which would be a really cool molecular machine that has, however, not yet been realized). Work is only performed when there is continuous motion along a given trajectory, such as when a ring in a catenane returns to its original position through a 360° unidirectional rotation (Figure 8.5c). Such a process requires that the activation energies for forward and backward rotation are different, which is the operating principle of a ratchet in the macroscopic world. In the case of molecules, there are two ways to achieve directional motion, either through energy ratchets or information ratchets [14].

How do molecular ratchets work?

An example of an energy ratchet is the molecular pump shown in Figure 8.9. In this system, a sequence of stimuli causes the blue box to thread first onto the paraquat moiety and then onto the alkyl chain of the axle component **8.9** [15, 16]. When mixed in acetonitrile, **8.9** and the blue box initially do not interact because both the ring and the paraquat moiety are positively charged. The addition of zinc induces reduction of both compounds to their respective radical cations, which form a complex stabilized by radical–radical interactions. Complex formation requires the blue box to squeeze past the two methyl groups of the terminal pyridinium ring. The subsequent oxidation of the radical cations, which is faster than the dissociation of the complex, restores the original positively charged species and forces the blue box to move away from the paraquat subunit. The blue box does so by sliding over the isopropyl group in the central ring to reach the alkyl chain, because movement in the opposite direction is associated with a higher activation energy due to the positive charge of the pyridinium moiety. This reduction–oxidation sequence thus pumps the blue box in one direction to yield a [2]rotaxane. Free rings remaining in solution after the first stroke of the pump can be threaded by repeating the stimulus sequence because the isopropyl group in the central subunit prevents the blue box already threaded onto the alkyl chain from moving back to the paraquat moiety. As a result, threading in the reduced state occurs exclusively *via* the pyridinium moiety. Pump **8.9** is thus able to perform work repeatedly for two cycles, driving the rings from a low concentration in solution to a higher local concentration on the chain. The different stimuli modulate the energy profile of the system, making the position of the blue box on the paraquat subunit alternately energetically unfavorable and favorable, which is the operating principle of an energy ratchet. This process, combined with the finely tuned activation energies associated with the movements of the ring in different directions, ultimately causes the threading to be unidirectional.

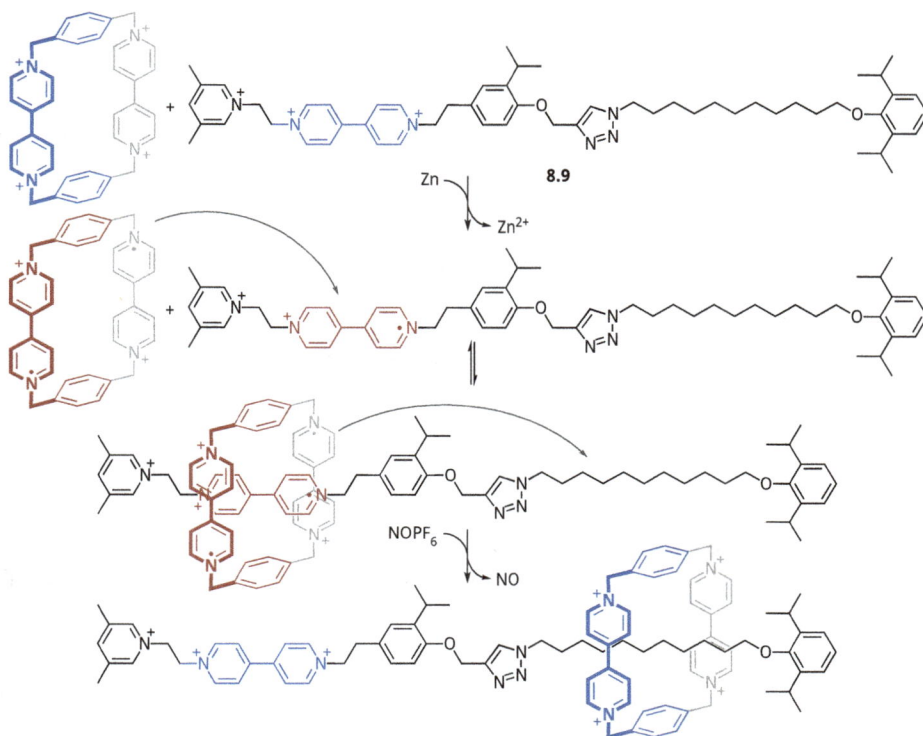

Figure 8.9: Structure of molecular pump **8.9**, which acts as an energy ratchet, using a reduction–oxidation sequence to pump a blue box across the paraquat subunit onto the alkyl chain.

Unlike an energy ratchet, the stimuli that drive an information ratchet do not affect the strength with which subunits on a track interact, but rather the activation energies associated with movement in different directions. A system that behaves like an information ratchet is the triptycene derivative **8.4** because the covalent tether holds the propeller in a position from which rotation in one direction has a lower Gibbs free energy of activation than rotation in the other direction. Another example is shown in Figure 8.10 [17].

In this system, up to three molecules of 24-crown-8 are pumped to the catching region of axle **8.10** by treating a solution of **8.10** in toluene containing N,N-diisopropylethylamine and an excess of the crown ether with the 9-fluorenylmethyl 4-nitrophenyl carbonate **8.11**. The crown ether and **8.10** alone remain separate because the interactions between the ring and the triazole moieties are not associated with a pronounced thermodynamic gain. In the presence of **8.11**, however, rapid threading occurs because the crown ether acts as a catalyst, mediating the reaction between **8.11** and the amino group of **8.10**. Rotaxane formation is thus a kinetic effect and not caused by the thermodynamic stabilization of a specific intermediate. As long as the crown ether is close to

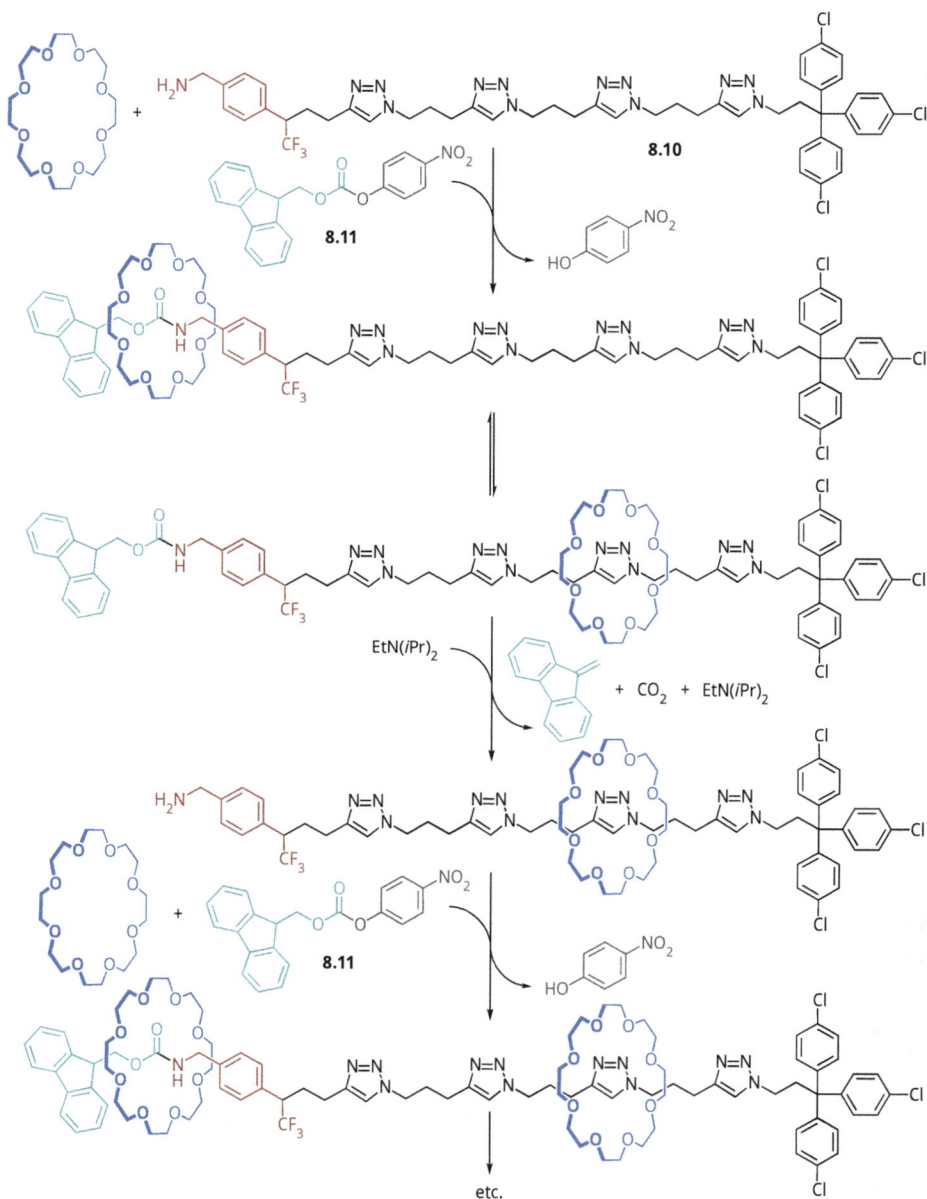

Figure 8.10: Structure of molecular pump **8.10** in which the threading of up to three molecules of 24-crown-8 onto the catching region of the axle is mediated by the catalytic effect of the crown ether on the reaction between the amino group of **8.10** and the 9-fluorenylmethyl 4-nitrophenyl carbonate **8.11**. N,N-Diisopropylethylamine serves to cleave the carbamate once the ring has left it.

the carbamate group in the initially formed [2]rotaxane, the *N,N*-diisopropylethylamine -mediated cleavage of the fluorenylmethyl group is slow. The bulky end group also prevents dethreading, so that the crown ether passes slowly over the trifluoromethyl group to the catching region of the axle. Base-mediated cleavage of the carbamate then restores the free amino group, leading to a pseudorotaxane that dissociates very slowly because the strategically placed trifluoromethyl stopper inhibits the passage of the ring. As a result, the next threading reaction occurs, yielding the [3]rotaxane. After another round of pumping, the axle is full. Thus, at each pumping step, the directionality of motion in this information ratchet is controlled by energy barriers, after the ring is trapped by a barrier that prevents dethreading, and then by a barrier that prevents the ring from moving backward from the catching region over the stopper.

Based on these general concepts we will look at other examples of molecular machines in the next sections, focusing on the different functions that have been realized with them.

8.2 Rotaxane-derived machines

Molecular switches: The translocation of the ring in a rotaxane can be induced not only by changing the properties of individual subunits along the axle, as we have seen for **8.8**, but also by an external binding partner. A classic example is the rotaxane **8.12** developed by the Sauvage group (Figure 8.11) [18]. This rotaxane, which contains a phenanthroline moiety in the ring and both a phenanthroline and a terpyridine subunit in the axle, binds to copper ions, with the position of the ring depending on the oxidation state of the metal ion. Copper(I) prefers to be tetracoordinated and therefore links the two phenanthroline units. When the metal center is oxidized to copper(II), the ring and the metal ion move to the terpyridine unit because the pentacoordinated complex geometry is now preferred. Rotaxane **8.12** can thus be electrochemically switched between two co-conformations.

In **8.13**, metal coordination and the motion of the ring are also linked, although the metal center does not participate in the interactions of the rotaxane components. In its metal-free form, **8.13** preferentially contains the ring at the succinamide moiety, where hydrogen bonding to the carbonyl groups is stronger than to the amide and ester carbonyl groups at the opposite end of the axle (Figure 8.12a) [19]. When cadmium(II) binds to the dipicolylamine moiety, steric effects destabilize the hydrogen bonding interactions, resulting in the translocation of the ring.

A number of structurally related switches have been developed by the Leigh group, containing the same tetralactam as in **8.13**, but differing in the subunits along the axle [8]. In **8.14**, the axle contains two amide groups on one side and a cinnamic acid-derived residue on the other side (Figure 8.12b). Switching is achieved by changing the protonation state of the phenolic OH group, with the ring preferring to be near the amides when the phenol is protonated and on the other side of the axle

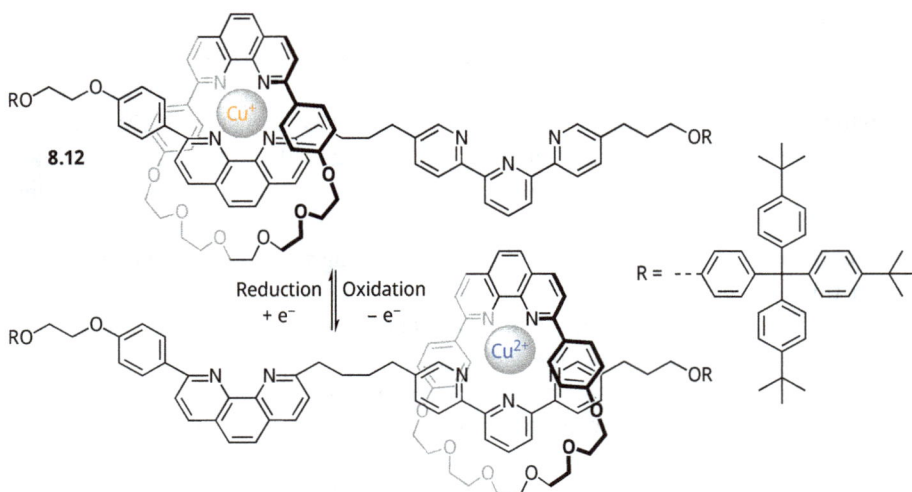

Figure 8.11: Switching between two co-conformations of **8.12** by electrochemically changing the oxidation state of the copper ions. Copper(I) prefers the tetrahedral coordination geometry in the complex with two phenanthroline units, while copper(II) binds preferentially to a phenanthroline and a terpyridine ligand.

when it can bind to the deprotonated phenolate group by charge-assisted hydrogen bonding [20]. Rotaxane **8.15** is switched photochemically by isomerization of the double bond [21]. The fumaramide group with the *E*-configured double bond has two hydrogen bond acceptors and is therefore the preferred binding site for the tetralactam. Photochemical isomerization of the double bond leads to a maleamide subunit in which one carbonyl group participates in an intramolecular hydrogen bond. As a result, the ring moves to the other side of the axle where it finds more accessible hydrogen bond acceptors (Figure 8.12c).

The additional introduction of a blocking group into these rotaxanes allows the ring to be kinetically trapped in a thermodynamically unfavorable state. An example is **8.16**, which after synthesis contains the ring exclusively on one side of the axle (Figure 8.13a) [22].

Because of the silyl ether, the ring cannot move to the other side of the axle where it would find an equally suitable binding partner. When the silyl group is cleaved, this movement becomes possible, resulting in a 1:1 ratio of the two co-conformers. Since the two stations are identical, it is not possible to further influence the final state of the co-conformational equilibrium. This is different in **8.17**, where the combination of a fumaramide and a succinamide group provides an additional level of control (Figure 8.13b) [22]. In the absence of the silyl group, **8.17** behaves like **8.15**, with the ring favoring the succinamide subunit only when the double bond has the *Z* configuration. When the silyl group is present, the translocation of the ring induced by the isomerization of the double bond is no longer possible. Accordingly, the

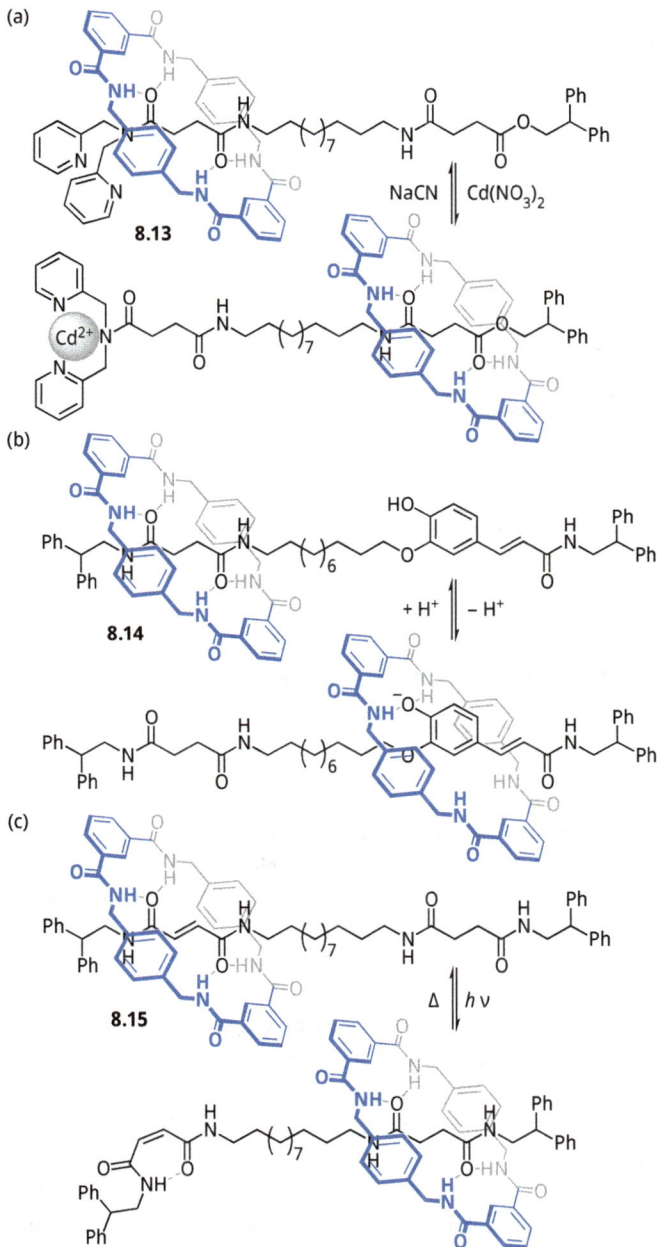

Figure 8.12: Control of the ring position in rotaxanes containing a macrocyclic tetralactam by metal coordination (a), deprotonation/protonation (b), and *E,Z* isomerization (c).

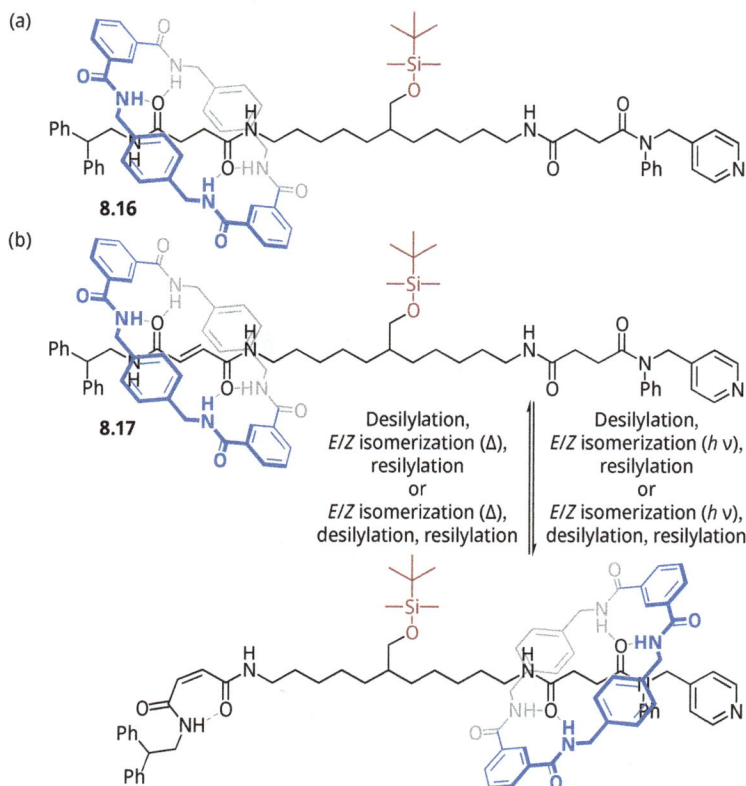

Figure 8.13: Effect of blocking groups in the axle of a molecular switch on the co-conformational equilibrium. After cleavage of the silyl ether in **8.16**, the rotaxane adopts both co-conformations with equal probability (a). In **8.17**, the silyl group prevents the ring from reaching the preferred binding site after *E,Z* isomerization (b). Thus, the redistribution of the co-conformations occurs only after cleavage of the silyl group.

ring is trapped on one side of the axle even if the other side contains the better binding site. For example, if the tetralactam is located near the unsaturated residue and this residue is switched to the *Z* form, the ring can only move if the silyl ether is cleaved before or after isomerization. After the change in position, the ring can be prevented from moving back by reintroducing the silyl group. This rotaxane can thus be structurally fixed in certain states even after the stimulus affecting the position of the ring has been removed. It therefore behaves more like a conventional switch than the other rotaxane-based switches described so far.

Molecular muscles: The expansion and contraction of a muscle is based on the lateral sliding of actin and myosin filaments past each other. Sauvage proposed that molecular daisy chains, doubly interlocked rotaxanes whose subunits contain a macrocycle at one

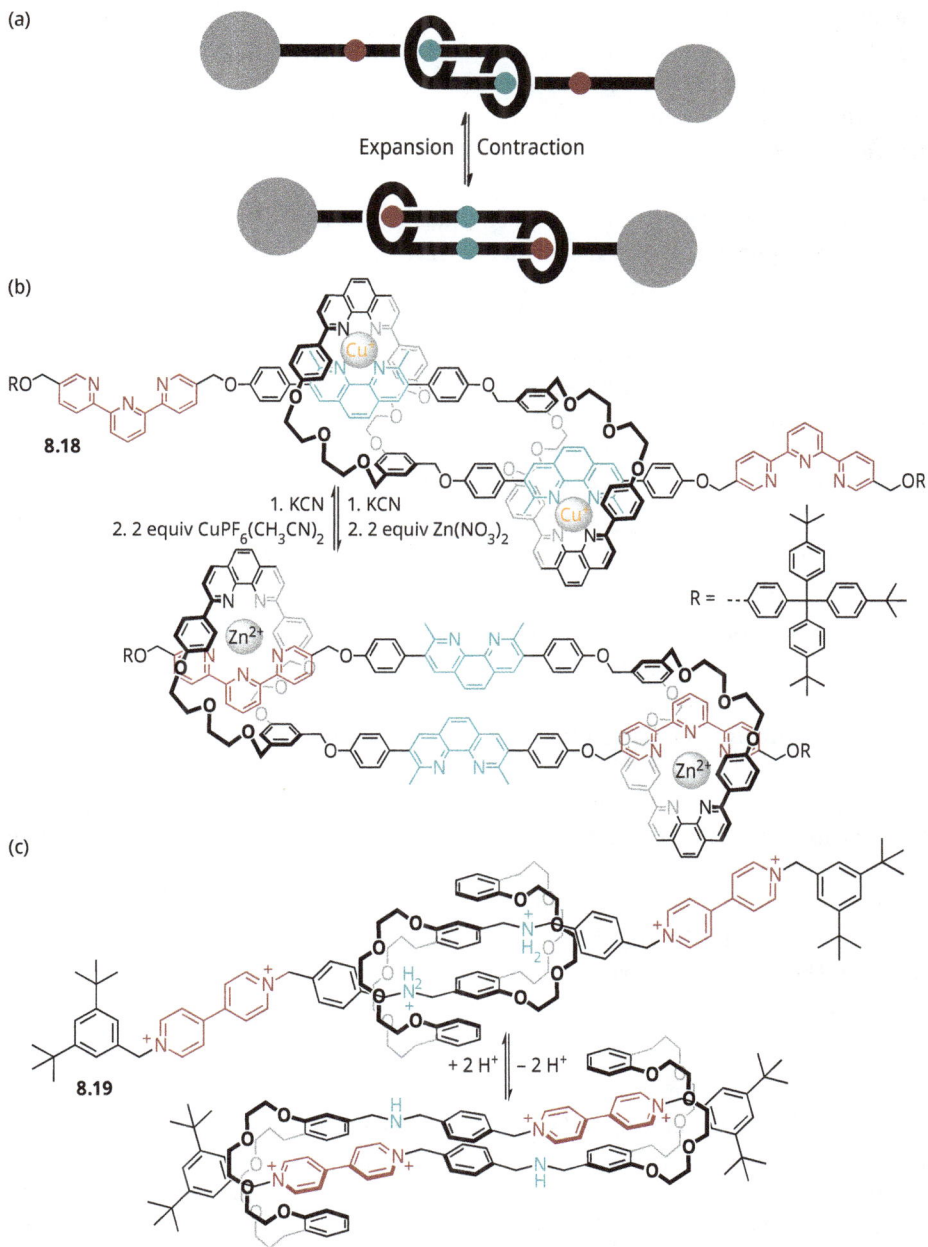

Figure 8.14: General structure and function of a molecular muscle (a), and molecular muscle **8.18**, whose expansion and contraction is based on Cu$^+$/Zn^{2+} exchange (b), and **8.19**, which is switched by changing the degree of protonation.

end of a chain and a bulky residue at the other end, can perform a similar lateral motion (Figure 8.14a) [23]. When the two rings in these systems move in opposite directions, the expansion of the molecules changes, similar to how a muscle contracts and expands. This sliding motion can be controlled by incorporating stations in the axles between which the rings can be switched [24].

Sauvage's molecular muscle **8.18** contains a combination of a phenanthroline and a terpyridine unit in each chain segment, making its switching mode somewhat related to the behavior of **8.12** (Figure 8.14b) [25]. The dicopper(I) complex of **8.18**, in which the metal ions connect the phenanthroline units, represents the expanded state. This state is contracted by demetalation followed by introduction of zinc(II) ions, which connect the phenanthroline units to the terpyridine groups in the axles. This sequence causes the bulky end groups to move together. The reverse movement back to the expanded state is induced by another metal exchange.

The expansion and contraction of such daisy chain molecules can also be induced by other stimuli. In **8.19**, for example, the crown ethers bind preferentially to the paraquat subunits when they are the only charged groups in the system, but prefer the protonated amino groups (Figure 8.14c) [26]. Thus, the muscle adopts the expanded state in acidic media and contracts upon addition of a base.

Molecular elevators: Molecular elevators are switches in which large platforms with several macrocyclic subunits move on molecular frameworks with a corresponding number of axles. An example is the tripodal system **8.20** introduced by Stoddart, in which the switching is based on the protonation states of the three amino groups (Figure 8.15) [27].

When all three amino groups are protonated, the platform is on the *upper floor*. The downward motion is induced by deprotonation of the amino groups, which in their neutral form bind less strongly to the crown ethers than the positively charged paraquat units. This movement ends when the deprotonation is complete, but since the crown ethers in **8.20** are flexible, each deprotonation step causes only one subunit to move. The platform thus reaches the *ground floor* in a stepwise fashion, leading the authors to conclude that "molecular elevators are more reminiscent of a legged animal" and less suitable for transporting cargo [27].

Molecular valves: Immobilized switchable rotaxanes act as valves when they prevent molecules from entering or leaving the pores of a solid support in one state and allow diffusion when switched to the other state. These valves are typically obtained by covalently anchoring rotaxanes to mesoporous silica nanoparticles *via* end groups in the axles. The position of the ring then determines whether the valve is open or closed, as shown schematically in Figure 8.16a.

In the open state, the rings in the rotaxanes are far enough away from the surface to allow guest molecules to be loaded into the pores. In the case of **8.21**, for example, rhodamine B can be incorporated into the porous support when the blue boxes bind to the electron-poor tetrathiafulvalene moieties (Figure 8.16b) [28]. The valves are

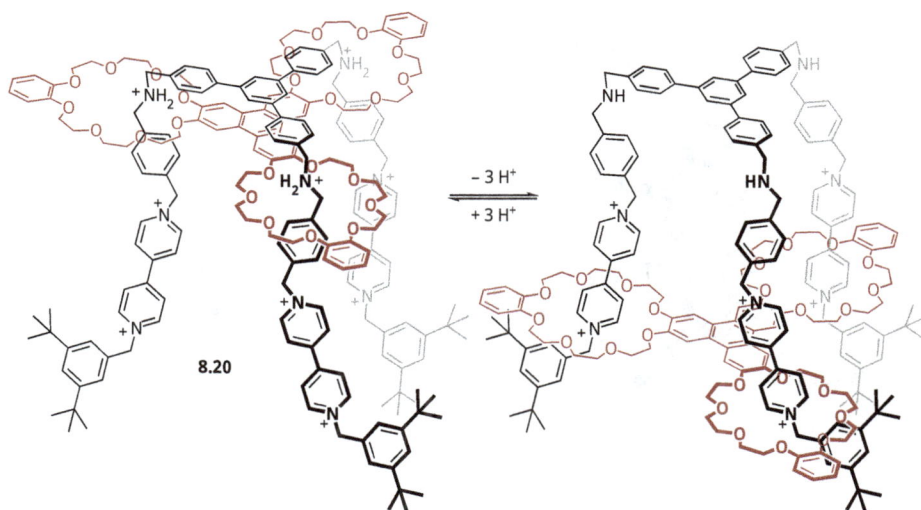

Figure 8.15: Structure and states of the molecular elevator **8.20** in which the movement of the platform between the upper floor and the ground floor is controlled by changing the protonation states of the amino groups in the axles.

closed by bringing the rings closer to the surface, which is achieved by iron(III)-mediated oxidation of the tetrathiafulvalenes, so that they block the pores and trap the bound dye molecules. Once the valves are closed, the nanoparticles can be isolated and washed to remove any remaining dye. Switching the rotaxanes back to the open position, which in the case of **8.21** is achieved by adding ascorbic acid, is accompanied by dye release.

Another example of such a valve is the pseudorotaxane **8.22** with a dibenzylammonium group in the axle (Figure 8.16c) [29]. Porous silica particles containing this axle on their surface can be loaded with coumarin 460 molecules, and the dye-loaded particles can then be capped with dibenzo-24-crown-8. The captured molecules are released by deprotonation of the ammonium group. Although the crown ether is lost when the valve is opened, the silica particles containing the immobilized axle components can be reused.

Peptide synthetase: Another example of a rotaxane in which the ring moves in a specific direction while performing a task is **8.23**. This rotaxane mediates the sequence-specific synthesis of an oligopeptide, with each component having a critical function (Figure 8.17) [30]. The axle serves as the donor of the amino acids that are sequentially transferred to the growing product, with the rigid subunits ensuring that this transfer proceeds in the correct sequence. The ring is the acceptor, containing a tripeptide fragment with a cysteine residue whose thiol group mediates the transfer reaction.

(a)

Valve closed   Valve opened

(b)

8.21

(c)

8.22

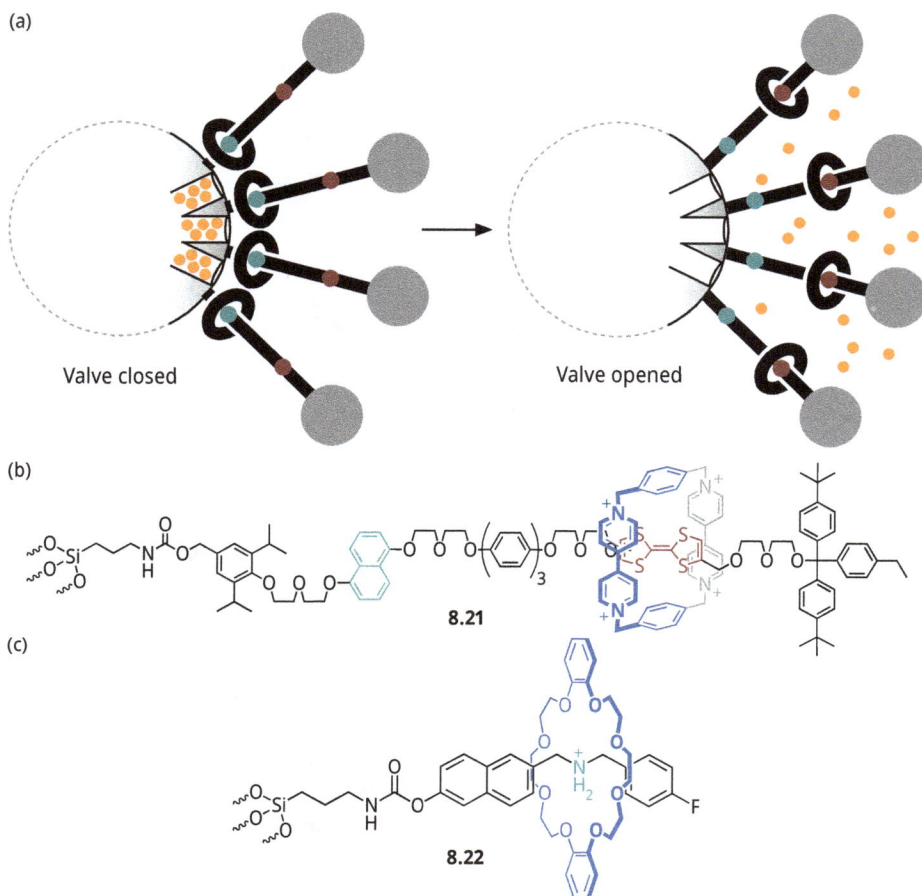

Figure 8.16: Schematic illustration of the switching process of a rotaxane-based molecular valve immobilized on a mesoporous silica nanoparticle (a), and structures of the redox-activated bistable rotaxane **8.21** (b), and the pH-driven rotaxane **8.22** (c). The rings of these rotaxanes block the pores of the support on which they are immobilized when close to the particle surface. Switching the rotaxanes to states where the rings are away from the surface induces the release of the molecules trapped in the pores.

Compound **8.23** is obtained in three steps, the first of which is an active metal template synthesis to assemble the rotaxane through copper(I)-catalyzed azide–alkyne cycloaddition (Figure 8.17). In the next step, the aldehyde group in the ring is used to anchor a tripeptide hydrazide containing an *S*-trityl protected cysteine residue and a *tert*-butyloxycarbonyl (Boc) group at the *N*-terminus. The final step is the cleavage of the protecting groups, yielding **8.23** after deprotonation, which performs the peptide synthesis autonomously.

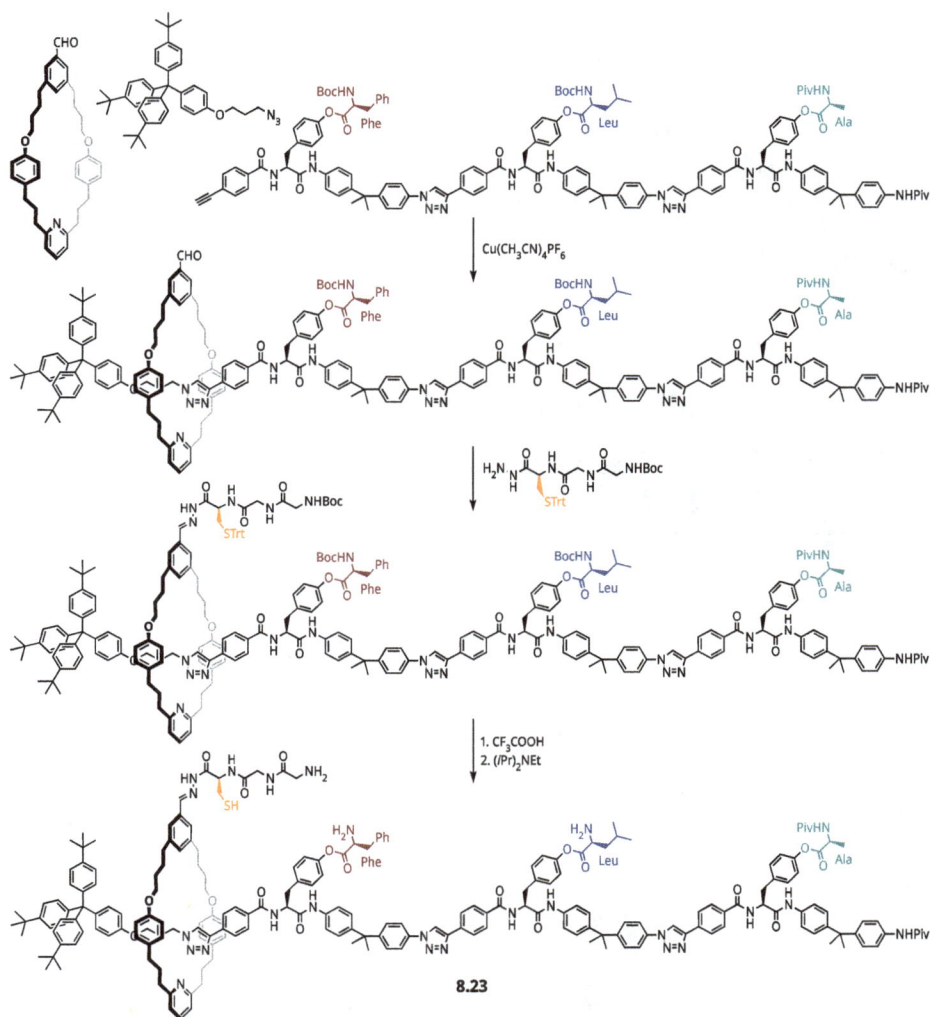

Figure 8.17: Synthesis of **8.23**, involving the assembly of the rotaxane by active metal template synthesis, followed by the introduction of the accepting oligopeptide and deprotection (Trt, triphenylmethyl, trityl; Boc, *tert*-butyloxycarbonyl; Piv, pivaloyl, trimethylmethylcarbonyl).

In this rotaxane, the ring moves freely between the blocking groups, but cannot pass over the first amino acid. The thiol group therefore reacts with the only amino acid it can reach, forming a thioester (Figure 8.18). A rearrangement then transfers the amino acid to the *N*-terminus of the peptide and simultaneously regenerates the thiol. With the first amino acid removed, the ring moves on to undergo the second and then the third chain elongation. Once the last amino acid has been transferred to the product, there is no blocking group left to prevent the ring from leaving the axle, allowing

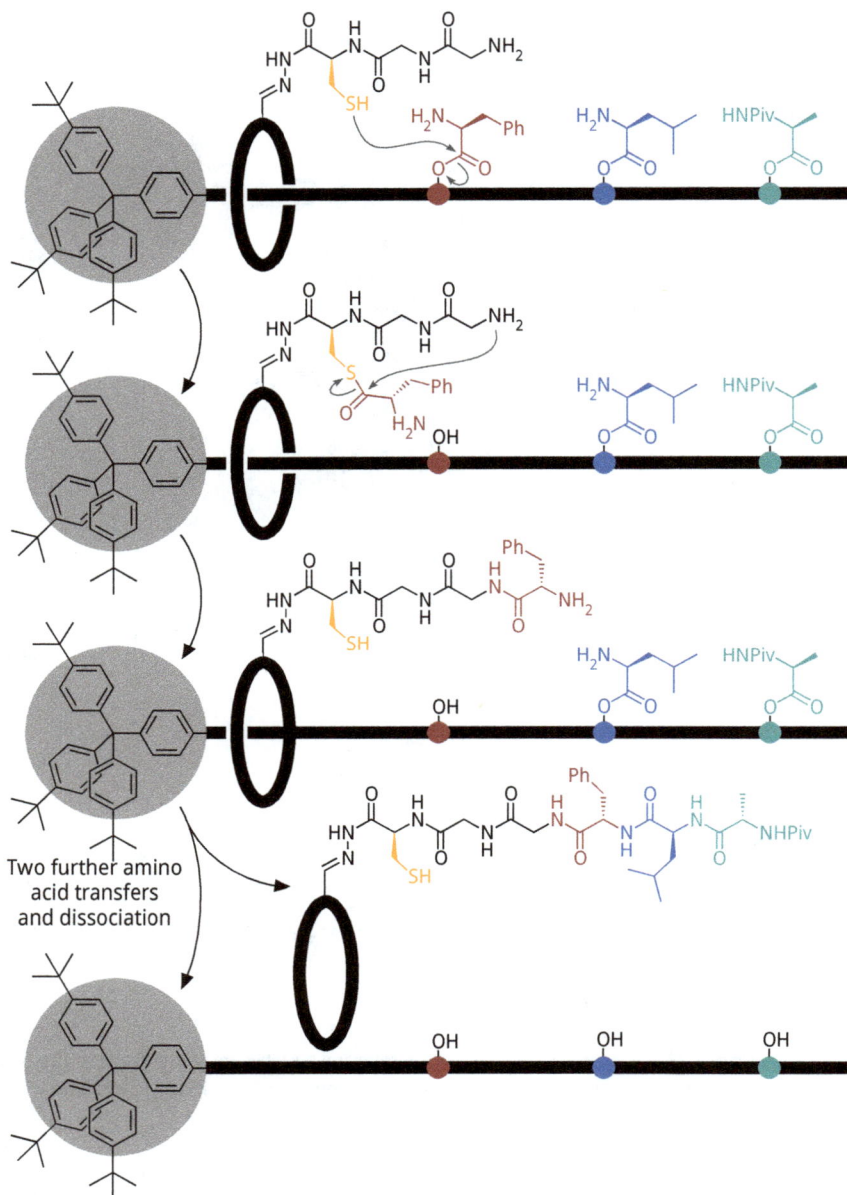

Figure 8.18: Operation of the synthetic peptide synthetase **8.23**. The thiol group undergoes transacylation followed by an *S–N*–acyl transfer reaction, resulting in the transfer of the amino acid to the *N*-terminus of the accepting peptide. After each step, the ring moves to the next amino acid where it reacts in a similar manner. When all three amino acids have been transferred, the rotaxane dissociates. Cleavage of the peptide from the ring yields the product.

the rotaxane components to dissociate. Since many molecules (~1,018 in the actual experiment) work in parallel in this reaction, the peptide is isolated in milligram quantities after cleavage from the ring. Rotaxane **8.23** thus demonstrates that molecular machines can perform iterative synthetic tasks that are closely related to those observed in some natural systems.

8.3 Catenane-derived machines

Catenanes allow the design of molecular machines in which one ring performs a continuous directed motion along the perimeter of another ring, potentially giving rise to molecular motors. However, many catenanes operate merely like switches. The two catenanes **8.24** and **8.25** (Figure 8.19), for example, find their direct counterparts in the rotaxanes **8.8** and **8.12**. In **8.24**, the arrangement of the rings depends on whether the amino groups are protonated or not [31], while the preferred structure of **8.25** is controlled electrochemically by changing the oxidation state of the copper ion [32]. Although the switching between the possible co-conformations in both cases involves the movement of one ring along the perimeter of the other, the direction is not controlled and occurs clockwise or counterclockwise. Possible strategies for achieving directionality are outlined in this section.

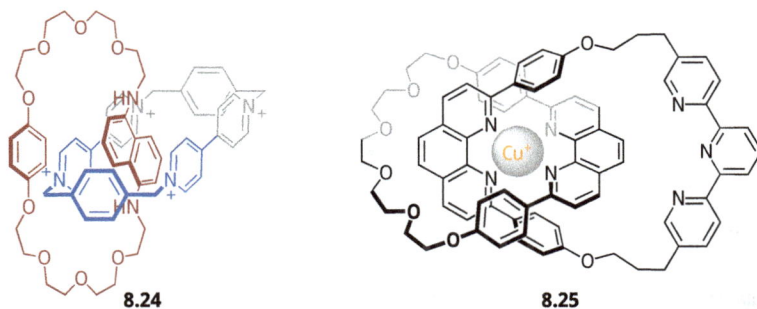

Figure 8.19: Examples of switchable catenanes. The arrangement of the rings in catenane **8.24** depends on whether the 1,5-diaminonaphthalene subunit is protonated or not, and that in **8.25** on the oxidation state of the copper ion.

One approach is to use catenanes that exist in at least three different co-conformations. If one ring in a [2]catenane has three stations, A, B, and C, the other ring will make a complete circumrotation by visiting the stations either in the order A-B-C-A or A-C-B-A. By controlling this sequence, it is possible to control the directionality of motion. An example is catenane **8.26**, in which the large ring has four stations A, B, C, and D, to which the tetralactam binds with different affinities (Figure 8.20) [33]. Binding to the fumaramides A and B is preferred, with the tetralactam binding less strongly to B than to A because the *N*-methyl groups in B destabilize the interactions for steric reasons. The least preferred station is the succinic amide ester C, and the amide D does not play a role in this case. Station A is located next to a benzophenone group, which allows selective isomerization to a maleamide by irradiation at 350 nm. The isomerization of B is achieved by irradiation at 254 nm, while station C is not photoactive.

If all double bonds of **8.26** are *E*-configured, the tetralactam is preferentially located at station A. Isomerization of this station forces the ring to move to B and, if this subunit is also isomerized, to C. Thermal regeneration of the *E*-configured double bonds finally induces the ring to move back to A. Thus, the tetralactam moves in a defined sequence from A to B to C and back to A if a suitable sequence of stimuli is applied. However, none of these steps is directional, since the stations can always be reached from either side.

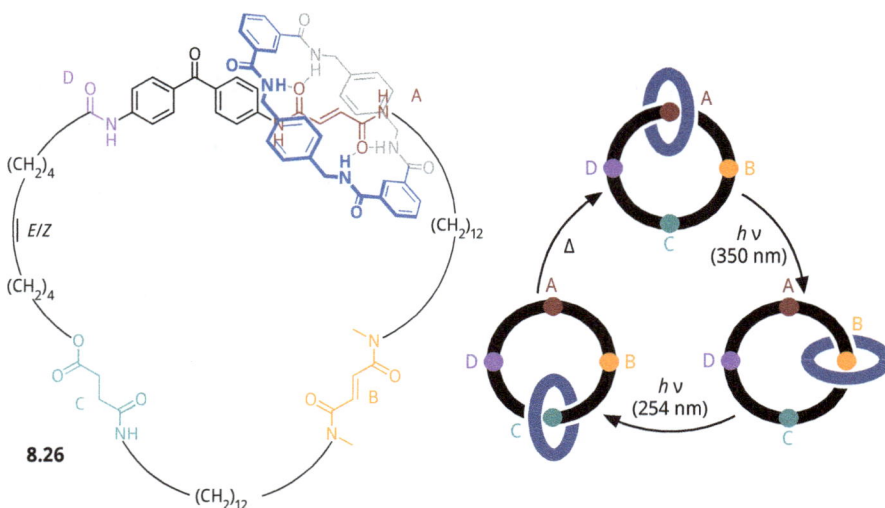

Figure 8.20: Molecular structure of **8.26** and sequential switching of the position of the tetralactam in this [2]catenane by irradiation first at 350 nm, then at 254 nm, and final thermal isomerization of the *Z,Z*-form back to the original *E,E*-form.

Directional movement requires an additional component that blocks movement in the wrong direction, such as the second ring in [3]catenane **8.27**. In this compound, the two rings are preferentially located at the two fumaramide stations (A and B) (Figure 8.21) [33]. Irradiation at 350 nm causes ring 1 to move counterclockwise (as shown) toward the succinic amide ester C, because movement in the opposite direction is prevented by ring 2. The isomerization of B now forces the other ring to move, and since there are no good stations left, it moves counterclockwise to the isolated amide D. The thermal isomerization of both double bonds back to the *E* configurations causes ring 1 to move to station B and ring 2 to move to station A. At this point, both

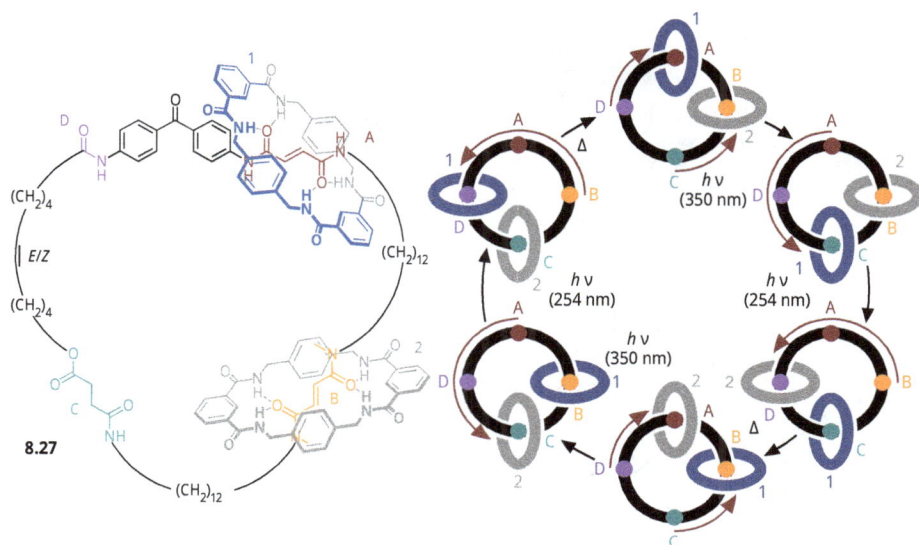

Figure 8.21: Molecular structure of **8.27** and counterclockwise circumrotation of the two tetralactams in this [3]catenane along the perimeter of the large ring by repeating the stimuli irradiation at 350 nm, irradiation at 254 nm, and thermal treatment twice.

rings have swapped their positions with respect to the original state. Repeating the cycle one more time results in another exchange, with both rings having completed a full counterclockwise circumrotation of the large ring by the time the original state is restored.

If the larger ring of a catenane contains only two stations, it is usually impossible to tell from which direction the smaller ring arrives, unless the larger ring contains blocking groups that prevent the smaller ring from moving in a particular direction. This concept is realized in **8.28** (Figure 8.22) [34]. Starting from the co-conformation in which the ring is at the fumaramide station, the tetralactam can move counterclockwise or clockwise after photochemical isomerization of the double bond. However, both paths are blocked by bulky protecting groups. The directionality of the movement is therefore controlled by which protecting group is cleaved after isomerization. For example, cleavage of the silyl ether (with tetrabutylammonium fluoride) induces clockwise movement to the succinamide. The complete clockwise circumrotation along the larger ring is completed by reintroduction of the silyl group, thermal isomerization of the maleamide, and cleavage of the trityl group (with boron trichloride). After reintroduction of the trityl group, the process can be restarted. The corresponding counterclockwise circumrotation is realized by reversing the order in which the protecting groups are cleaved.

Although unidirectional rotation can be realized with **8.28**, each rotation requires intervention by performing synthetic steps. In contrast, the unidirectional rotation of

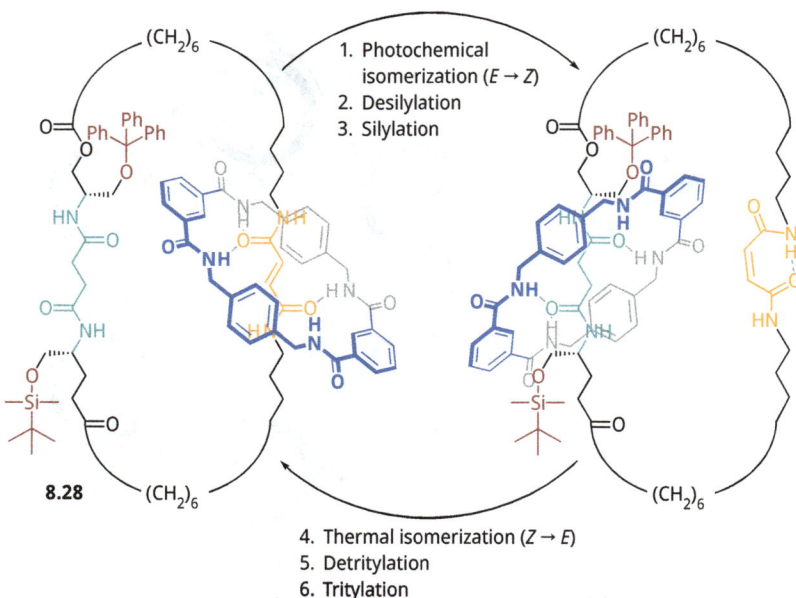

Figure 8.22: Molecular structure of **8.28** and sequence that induces the clockwise (as drawn) circumrotation of the tetralactam along the perimeter of the large ring.

the structurally related [2]catenane **8.29** proceeds autonomously. In this compound, the macrocyclic tetralactam moves along a track containing two fumaramide stations, one of which is deuterated to make the two stations distinguishable by NMR spectroscopy (Figure 8.23) [35]. The movement of the ring is prevented by two bulky 9-fluorenylmethyloxy-carbonyl (FMOC) groups. These groups are cleaved under basic conditions, allowing the ring to move. The rate of cleavage is independent of the position of the ring, but the reaction of the resulting hydroxy group with FMOC-Cl is about five times faster when the ring is at the other station. This kinetic asymmetry leads to a preference for the ring to move in one direction. Starting from state A, the base can cleave either FMOC group 1 or 2 with equal probability. If 2 is cleaved, the ring can move counterclockwise, bringing it close to the deprotected OH group, which will therefore react more slowly with FMOC-Cl than if the ring had just stayed where it was. However, if the FMOC group 1 is cleaved, the clockwise movement of the ring will cause the free OH group to react more rapidly with FMOC-Cl than if the ring had remained in its position. Therefore, clockwise movement is preferred. The same arguments apply if one of the FMOC groups is cleaved in state B. If both FMOC groups are cleaved at some point, the accuracy of the movement will suffer, but unidirectionality will be restored as the reaction proceeds. Movement stops when the FMOC-Cl is consumed [36].

Since the speed of the tetralactam in **8.29** is determined by the kinetics of FMOC cleavage and reintroduction and not by the size of the other ring, it takes about

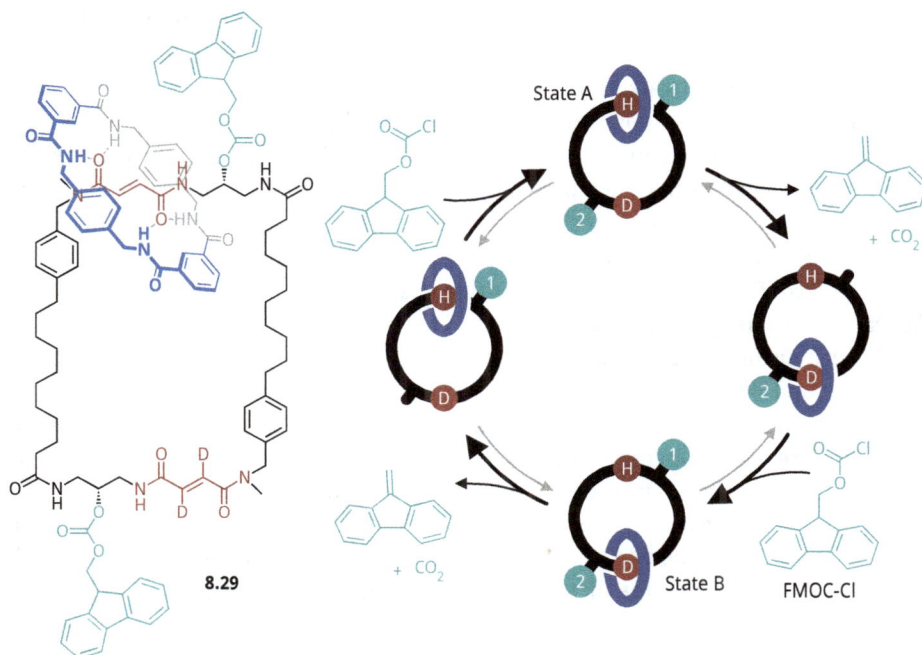

Figure 8.23: Molecular structure of **8.29** and sequence that induces the clockwise (as drawn) circumrotation of the tetralactam along the perimeter of the large ring. The reaction involves continuous cleavage and introduction of FMOC groups. Since the rate of cleavage is independent of the position of the ring, but the introduction is faster when the ring is far away, there is a kinetic asymmetry that favors one direction of ring movement over the other.

twelve hours for the tetralactam to complete the entire track. Both catenanes **8.28** and **8.29** are examples of information ratchets.

8.4 Machines without mechanical bonds

Although the examples in the previous chapter show that rings in catenanes can be made to perform directional movements, the modes of operation of these interlocked systems are complex and often require a sequence of stimuli, making it difficult to perform work. However, not only molecules with mechanical bonds but also structurally simpler compounds with moving parts whose spatial arrangement is changed in a controlled manner exhibit typical features of machines. Examples include Kelly's triptycene **8.4**, discussed in Section 8.1, although this molecule cannot perform a full unidirectional rotation. Other examples of noninterlocked molecules that behave like machines will be discussed in this section, focusing on systems in which the relative motion of the subunits is directional.

Molecular walkers: In a molecular walker, one subunit performs a directional movement along the track of another subunit by reversibly breaking and rejoining covalent bonds [37]. Molecular walkers thus mimic the behavior of proteins such as dynein and kinesin, which transport various cargoes, including cell organelles, along cytoskeletal polymers. An example is shown in Figure 8.24 [38]. In this molecule, the red subunit is attached to the track by two footholds, a disulfide and a hydrazone bond. The track also contains a photoswitchable stilbene moiety, so that movement along the track can be controlled by both chemical and photochemical stimuli. Starting from the *E*-configured **8.30**, the central stilbene subunit is first switched to the *Z* configuration, resulting in an *E*/*Z* ratio of 12:88 in the photostationary state. Under basic conditions, disulfide exchange is initiated, causing the disulfide foot to move to the other position because the resulting ring is less strained than the original one. The residue originally occupying the thiol group serves to keep the molecule fully oxidized and therefore also changes position. Another light-induced isomerization now leads to 75% of the strained *E*-isomer, allowing the hydrazone foot to complete the final step under the subsequent acidic conditions. After the reaction sequence, 48% of the walkers are found at the end of the path. Since the different stimuli affect the stabilities of the intermediates, the directionality of the movement is controlled by thermodynamics, characteristic of an energy ratchet.

Chemically driven molecular motors: In addition to Kelly's triptycene derivative **8.4**, various other systems have been described in which a sequence of chemical reactions induces unidirectional rotation [39]. In the biaryl derivative **8.31** described by Ben L. Feringa, the *ortho* substituents prevent random rotation of the two rings (Figure 8.25) [40]. The stereogenic center in the side chain causes the two atropisomers of **8.31** to be diastereomeric and therefore to have different stability. Starting from the (*S,S*)-diastereomer, the first rotation step involves the cleavage of the methyl ester and the methoxymethyl group under acidic conditions. The subsequent carbodiimide-mediated lactone formation, which initially leads to the thermodynamically less stable diastereoisomer, significantly reduces the energy barrier to rotation. As a result, the molecule relaxes to the more stable structure with the inverted configuration of the chirality axis. The next steps involve cleavage of the lactone, esterification, and introduction of the methoxymethyl group. After cleavage of the methyl ester and the benzyl ether, another lactone is formed, again initially as a strained system, which upon thermal equilibration yields the more stable structure. Several additional synthetic steps lead back to the starting compound. A 360° rotation therefore requires six synthetic steps and more than seven days to return to the starting material with an overall yield of 57%.

The group of David A. Leigh described an autonomous unidirectional molecular motor based on the arylpyrrole-2,2'-dicarboxylic acid **8.32** (Figure 8.26) [41]. This compound exists in two atropisomeric forms that are interconverted by rotation around the central C–N bond. When this compound is treated with a chiral carbodiimide, one of the atropisomers reacts more rapidly than the other to give the corresponding an-

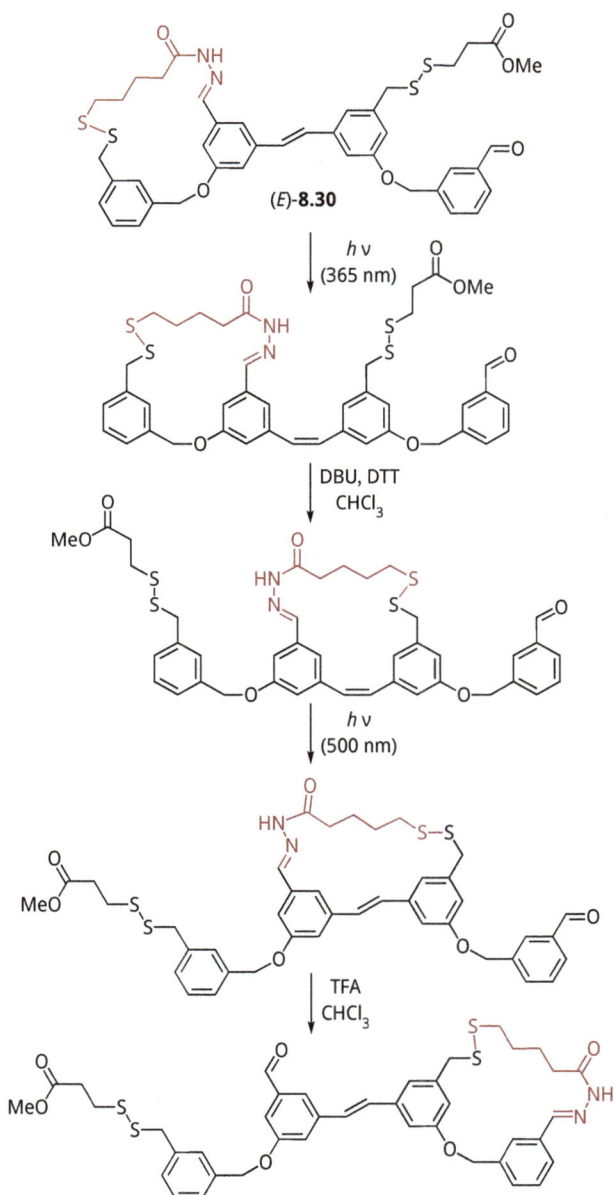

Figure 8.24: Structure of molecular walker **8.30**. The movement of the red subunit along the track from left to right (as shown) is mediated by a sequence of stimuli involving the initial light-induced *E,Z* isomerization, followed by disulfide exchange under basic conditions, *Z,E* isomerization, and hydrazone exchange under acidic conditions (DBU = diazabicycloundecene, DTT = dithiothreitol, TFA = trifluoroacetic acid).

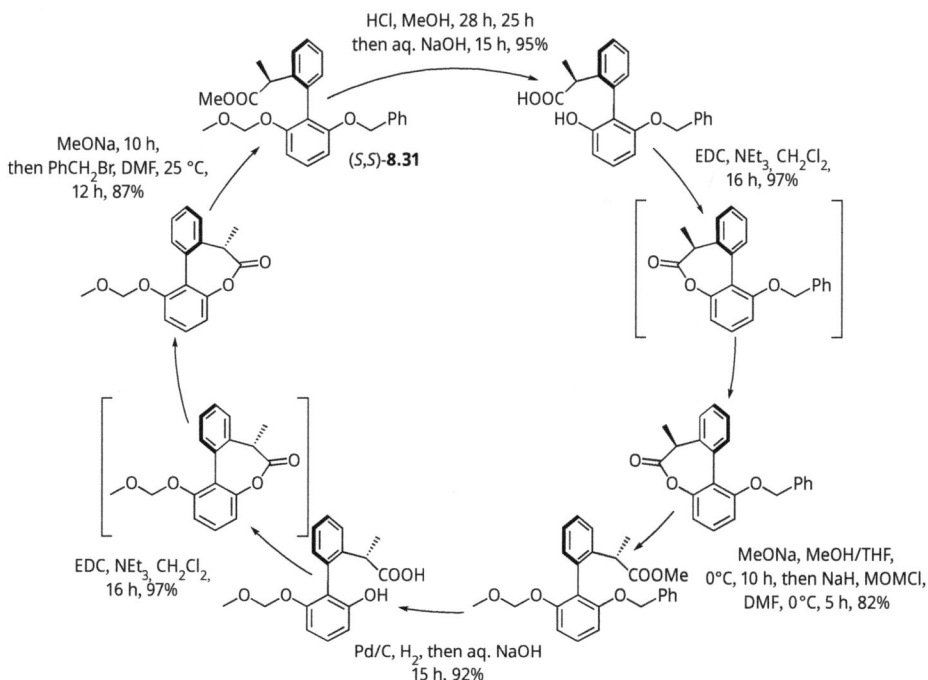

Figure 8.25: Chemically driven molecular motor **8.31** described by Feringa, in which a 360° rotation is achieved by a sequence of six synthetic steps (EDC = 1-ethyl-3-(3-dimethylaminopropyl)carbodiimide hydrochloride).

hydride, inducing directionality in the first step. The product undergoes rapid isomerization, but the ring opening of one enantiomeric form, mediated by a chiral organocatalyst, is again faster than the opening of the other enantiomer. This results in a directed 180° rotation of the pyrrole moiety relative to the other ring. The pyrrole ring continues to rotate in the same direction to restore the atropisomer that preferentially reacts with the carbodiimide. This motor completes one revolution approximately every three hours and rotates in the wrong direction every three or four turns.

Light-driven molecular motors: Efficient light-driven molecular motors are based on sterically congested alkenes in which the two substituents at the double bond perform continuous unidirectional rotation relative to each other [42, 43]. These motors were developed by Ben L. Feringa who shared the 2016 Nobel Prize in Chemistry with Jean-Pierre Sauvage and J. Fraser Stoddart for his seminal contributions to the development of such molecular machines. An example is **8.33**, in which a thioxanthene group is linked *via* a double bond to a group of three condensed six-membered rings, one of which contains an *R*-configured stereogenic center (Figure 8.27a) [44]. Due to

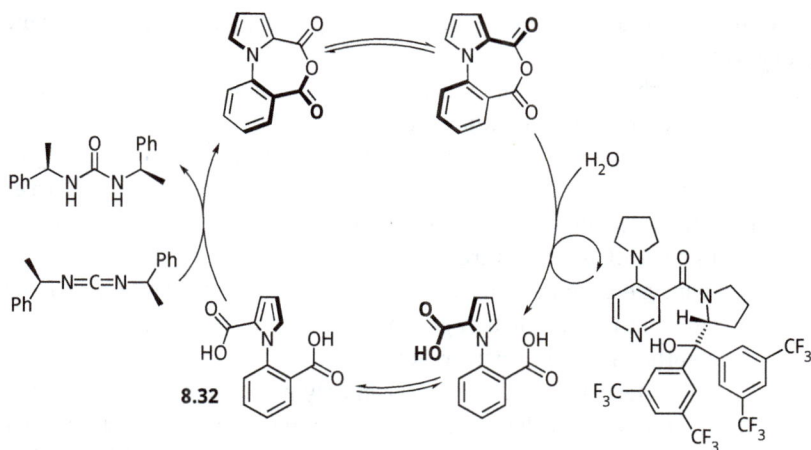

Figure 8.26: Autonomously rotating chemically driven molecular motor **8.32** described by Leigh in which unidirectional rotation is achieved by enantioselective lactone formation mediated by a chiral carbodiimide in the first step, and enantioselective ring opening mediated by a chiral organocatalyst in the second step.

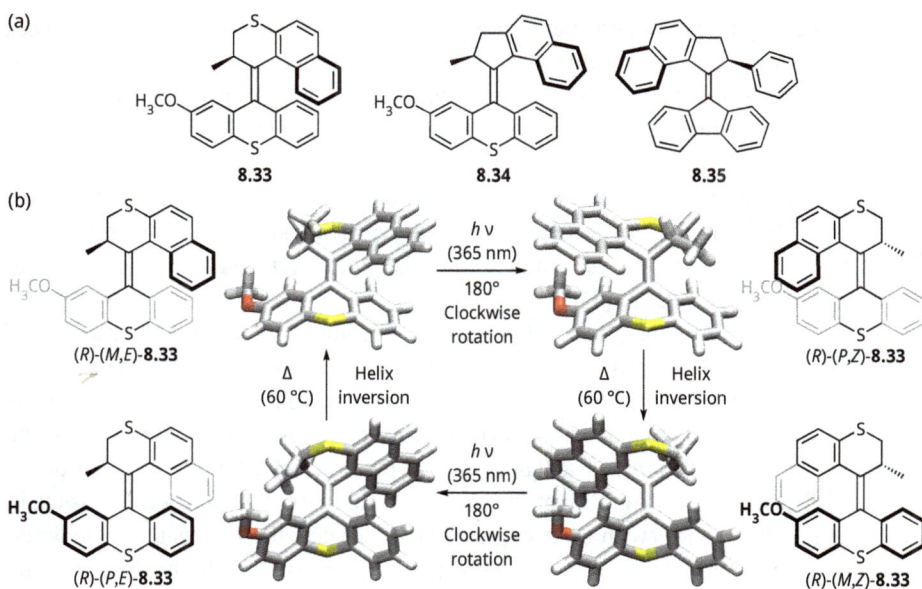

(a)

8.33   **8.34**   **8.35**

(b)

(R)-(M,E)-**8.33** hv (365 nm) 180° Clockwise rotation (R)-(P,Z)-**8.33**

Δ (60 °C) | Helix inversion   Δ (60 °C) | Helix inversion

(R)-(P,E)-**8.33** hv (365 nm) 180° Clockwise rotation (R)-(M,Z)-**8.33**

Figure 8.27: Structures of molecular motors **8.33**, **8.34**, and **8.35** (a) and mode of operation of **8.33**, which involves four steps, two photochemical E,Z isomerizations and two thermal helix inversions (b). The respective states are also shown as stick models.

the bulky substituents, the double bond is forced out of planarity and one substituent must adopt a helical arrangement.

To understand how this motor works, it is helpful to think of the thioxanthene group as the stator around which the other half of the molecule, the rotor, rotates. The full rotation involves four steps, two photochemical E,Z isomerizations leading to thermodynamically unfavorable states, and two thermal relaxations of the corresponding products. The complete cycle is shown schematically in Figure 8.27b, along with the corresponding calculated structures.

The cycle starts with the E-configured stereoisomer, which prefers an M helical arrangement of the rotor and has the methyl group in a pseudoaxial position, pointing away from the thioxanthene group. Upon irradiation at 365 nm, this isomer is converted to the Z form. For steric reasons, the rotor can only move clockwise during this process (viewed from the side of the double bond carrying the rotor) and must undergo a transition from the M helix to the P helix to prevent the subunits from colliding. This helix inversion involves the flipping of the six-membered ring in the rotor, bringing the methyl group into the unfavorable pseudoequatorial position. The molecule then relaxes in a thermal process that does not change the configuration at the double bond but pushes the two stacked aromatic rings past each other. At this point, another helix conversion has occurred at the rotor, resulting in an M-configured arrangement with the methyl group in the pseudoaxial position. Repeating the two steps initially produces a thermodynamically unfavorable intermediate with a P-configured rotor, which relaxes into the thermodynamically stable structure from which the cycle began.

The speed at which this motor operates depends mainly on the thermal steps that convert the thermodynamically unfavorable structures into stable ones. When these conformational rearrangements are associated with large energy barriers, full rotation is slow. This rate of rotation can be varied over several orders of magnitude by structural variation. The fastest motor to date is compound **8.34**, which operates in the MHz range (Figure 8.27a) [45].

These molecular motors, like their macroscopic counterparts, drive the rotation of structures much larger than themselves [46]. For example, compound **8.35** (Figure 8.27a) can induce rotation of a 5×28 μm glass rod [47]. The actual experiment involves the use of small amounts of **8.35** as a dopant in a cholesteric liquid crystalline phase in which the mesogens are arranged in either a P or M helical pattern. The actual orientation of the mesogens depends on the chirality of the dopant, in this case the helicity of the rotor in **8.35**. This helicity changes with each step of rotation, forcing the mesogens in the liquid crystalline phase to follow. The corresponding motion is transmitted to the surface of the liquid crystalline phase, causing a glass rod placed on top of the film to rotate, with the light-induced isomerization from the E to the Z configuration resulting in a clockwise rotation. This rotation stops after about ten minutes when the reorientation of the cholesteric phase is complete. When the light is turned off, the rotation begins again, but in the opposite direction, as the rotor returns to its original configuration during thermal relaxation. The motion of the glass rod therefore reflects the changes in helicity of the rotor during the switching, rather than its full rotation.

Can a molecular motor power a car?

Another fascinating application of such systems is their use to induce the gliding of molecules on surfaces, such as the nanocars developed in the group of James M. Tour [48]. An example is **8.36** (Figure 8.28), which has a Z-shaped chassis derived from oligo(phenylene ethynylene) with four fullerene wheels rotating around the alkynyl axes. Once deposited on a gold surface, these molecules can be imaged by scanning tunneling microscopy (STM) as four bright spots representing the fullerene wheels [49]. Due to the relatively strong adhesive force between the fullerenes and the underlying gold, **8.36** remains stationary on the surface up to 170 °C. As the temperature is further increased, **8.36** begins to move through a combination of translation and pivoting. The translation is perpendicular to the axes, indicating that it involves rotation of the wheels. Pivoting is probably caused by the rotation of pairs of wheels in opposite directions.

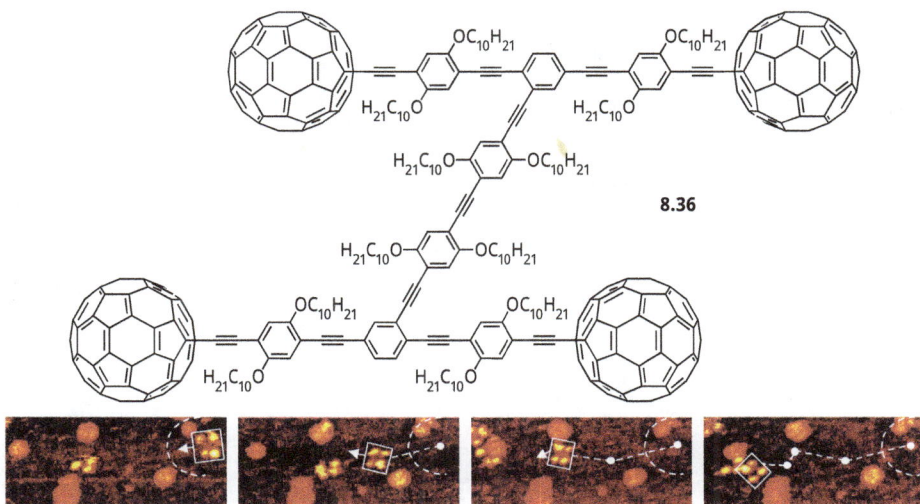

Figure 8.28: Structure of nanocar **8.36** and series of STM images showing the movement of **8.36** on a gold surface at about 200 °C. The orientation of the nanocar is determined by the separation of the fullerene wheels, which makes it possible to distinguish whether the motion is perpendicular to the axis or not. The images also show that the translation of the nanocar on the surface is accompanied by a pivoting motion. Images adapted with permission from [49]. Copyright American Chemical Society, 2005.

Several attempts have been made to combine these nanocars with Feringa's molecular motors, with the goal of converting their normally uncontrolled thermal motion into a directional light-controlled motion. One example is **8.37**, which integrates the motor **8.34** into a chassis containing four *p*-carborane wheels (Figure 8.29) [50]. These wheels were chosen because, unlike strongly absorbing fullerenes, they do not interfere with photoisomerization. It was expected that the motor would push the molecule forward each time it contacted the surface during rotation. Although **8.37** was successfully deposited on a copper surface, it did not perform any lateral motion upon light irradiation, perhaps because a single motor is insufficient to propel a car of this size. A more successful design is **8.34**, which contains a single axle with two adamanyl-derived wheels at the ends [51]. The corresponding nanoroadster **8.38** diffuses at temperatures above 150 K on a copper surface and becomes significantly faster upon irradiation, indicating that the built-in motor contributes to the motion

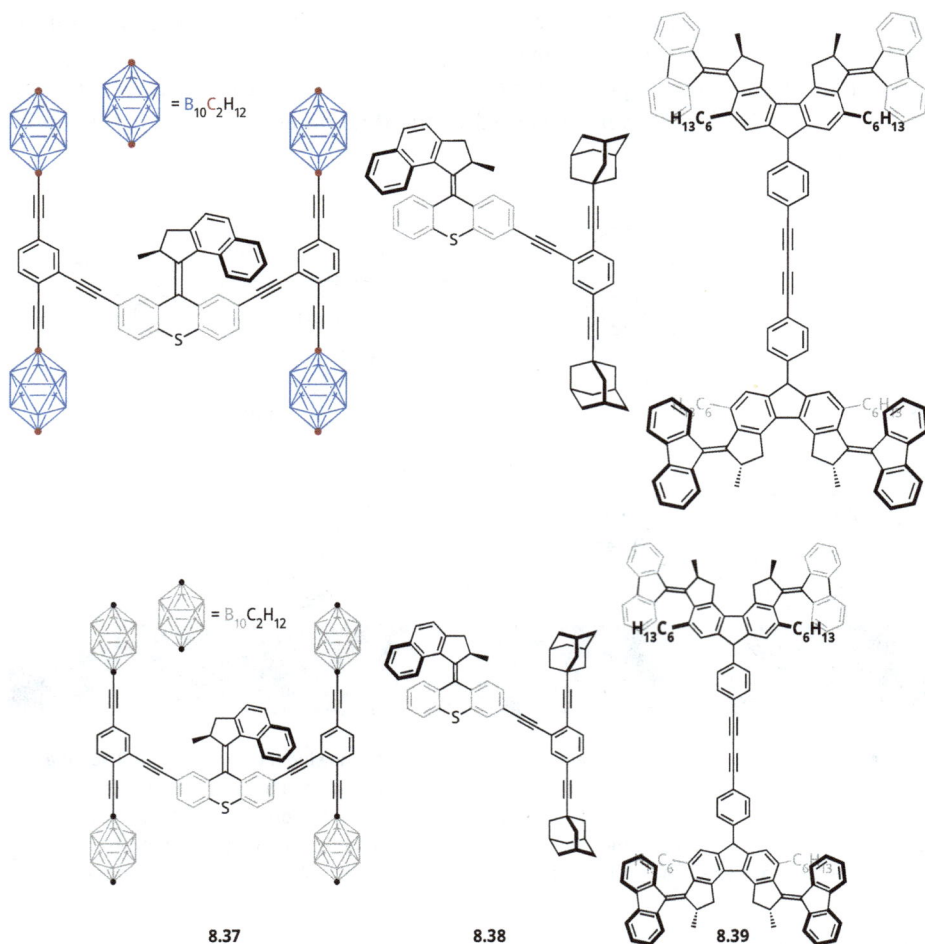

Figure 8.29: Molecular structures of nanocars **8.37**, **8.38**, and **8.39** with built-in motors.

when it begins to rotate. The Feringa group showed that the nanocar **8.39** with four motors as wheels also moves in a directional manner [52].

With these developments, a competition was launched to see which molecular design would lead to the fastest nanocar. The first race, in which six teams from around the world competed for the championship, was held in Toulouse in 2015. The track covered a distance of 100 nm on a gold surface and had two bends. Four STMs were used to follow the race. The car that set the speed record drove at a maximum speed of 95 nm per hour, which means that it would take more than 1,000 years to travel the distance of one meter. However, only this car and one other reached the finish line, while others returned after a few nanometers, broke down, or were lost on the surface. The second international nanocar race, which took place in 2022, ended in a tie between two teams. Videos of both races are available on YouTube.

These achievements may not seem remarkable to Formula 1 enthusiasts, but they demonstrate how well matter can now be manipulated at the molecular level. It is therefore appropriate to return at the end of this chapter to the physicist Richard Feynman and his famous lecture entitled "There's Plenty of Room at the Bottom" [53]. In this lecture, given in 1959 at the annual meeting of the American Physical Society, Feynman speculated about the many potential benefits of miniaturization, and also referred to tiny machines and their potential applications. Since 1959, many of Feynman's ideas, which may have been purely speculative at the time, have become reality, and we have seen in this chapter that controlling motion at the molecular level is no longer just a vision. The future, then, awaits the next step, which is to translate these developments into applications.

Bibliography

[1]   Original image taken by Janice Haney Carr and published under the Public Domain license (CDC PHIL ID# 6937).
[2]   Videos on YouTube that give interesting insight into cellular processes can be found under the titles "Inner Life of a Cell" or "Your Body's Molecular Machines".
[3]   Kelly RT. Progress toward a rationally designed molecular motor. Acc. Chem. Res. 2001, 34, 514–22.
[4]   Kelly TR, Bowyer MC, Bhaskar KV, Bebbington D, Garcia A, Lang F, Kim MH, Jette MP. A molecular brake. J. Am. Chem. Soc. 1994, 116, 3657–8.
[5]   Kelly TR, Sestelo JP, Tellitu I. New molecular devices: in search of a molecular ratchet. J. Org. Chem. 1998, 63, 3655–65.
[6]   Feynman RP, Leighton RB, Sands M. The Feynman Lectures on Physics, Vol. 1. Addison-Wesley: Reading, MA, 1963.
[7]   Kelly TR, Silva RA, De Silva H, Jasmin S, Zhao Y. A rationally designed prototype of a molecular motor. J. Am. Chem. Soc. 2000, 122, 6935–49.
[8]   Kay ER, Leigh DA, Zerbetto F. Synthetic molecular motors and mechanical machines. Angew. Chem. Int. Ed. 2006, 46, 72–191.
[9]   The Nobel Prize in Chemistry 2016 (Accessed February 01, 2024, https://www.nobelprize.org/nobel_ prizes/chemistry/laureates/2016/press.html).

[10] Anelli PL, Ashton PR, Ballardini R, Balzani V, Delgado M, Gandolfi MT, Goodnow TT, Kaifer AE, Philp D, Pietraszkiewicz M, Prodi L, Reddington MV, Slawin AMZ, Spencer N, Stoddart JF, Vicent C, Williams DJ. Molecular meccano. 1. [2]Rotaxanes and a [2]catenane made to order. J. Am. Chem. Soc. 1992, 114, 193–218.

[11] Ashton PR, Brown CL, Chrystal EJT, Parry KP, Pietraszkiewicz M, Spencer N, Stoddart JF. Molecular trains: the self-assembly and dynamic properties of two new catenaries. Angew. Chem. Int. Ed. Engl. 1991, 30, 1042–5.

[12] Anelli PL, Spencer N, Stoddart JF. A molecular shuttle. J. Am. Chem. Soc. 1991, 113, 5131–3.

[13] Bissell RA, Córdova E, Kaifer AE, Stoddart JF. A chemically and electrochemically switchable molecular shuttle. Nature 1994, 369, 133–7.

[14] Sangchai T, Shehimy SA, Penocchio E, Ragazzon G. Artificial molecular ratchets: tools enabling endergonic processes. Angew. Chem. Int. Ed. 2023, 62, e202309501.

[15] Cheng C, McGonigal PR, Schneebeli ST, Li H, Vermeulen NA, Ke C, Stoddart JF. An artificial molecular pump. Nat. Nanotechnol. 2015, 10, 547–53.

[16] Feng Y, Ovalle M, Seale JSW, Lee CK, Kim DJ, Astumian RD, Stoddart JF. Molecular pumps and motors. J. Am. Chem. Soc. 2021, 143, 5569–91.

[17] Amano S, Fielden SDP, Leigh DA. A catalysis-driven artificial molecular pump. Nature 2021, 594, 529–35.

[18] Armaroli N, Balzani V, Collin JP, Gaviña P, Sauvage JP, Ventura B. Rotaxanes incorporating two different coordinating units in their thread: synthesis and electrochemically and photochemically induced molecular motions. J. Am. Chem. Soc. 1999, 121, 4397–408.

[19] Marlin DS, González Cabrera D, Leigh DA, Slawin AMZ. An allosterically regulated molecular shuttle. Angew. Chem. Int. Ed. 2006, 45, 1385–90.

[20] Keaveney CM, Leigh DA. Shuttling through anion recognition. Angew. Chem. Int. Ed. 2004, 43, 1222–4.

[21] Altieri A, Bottari G, Dehez F, Leigh DA, Wong JKY, Zerbetto F. Remarkable positional discrimination in bistable light- and heat-switchable hydrogen-bonded molecular shuttles. Angew. Chem. Int. Ed. 2003, 42, 2296–300.

[22] Chatterjee MN, Kay ER, Leigh DA. Beyond switches: ratcheting a particle energetically uphill with a compartmentalized molecular machine. J. Am. Chem. Soc. 2006, 128, 4058–73.

[23] Moulin E, Carmona-Vargas CC, Giuseppone N. Daisy chain architectures: from discrete molecular entities to polymer materials. Chem. Soc. Rev. 2023, 52, 7333–58.

[24] Bruns CJ, Stoddart JF. Rotaxane-based molecular muscles. Acc. Chem. Res. 2014, 47, 2186–99.

[25] Jiménez MC, Dietrich-Buchecker C, Sauvage JP. Towards synthetic molecular muscles: contraction and stretching of a linear rotaxane dimer. Angew. Chem. Int. Ed. 2000, 39, 3284–7.

[26] Wu J, Leung KCF, Benítez D, Han JY, Cantrill SJ, Fang L, Stoddart JF. An acid-base-controllable [c2] daisy chain. Angew. Chem. Int. Ed. 2008, 47, 7470–4.

[27] Badjic JD, Ronconi CM, Stoddart JF, Balzani V, Silvi S, Credi A. Operating molecular elevators. J. Am. Chem. Soc. 2006, 128, 1489–99.

[28] Nguyen TD, Tseng HR, Celestre PC, Flood AH, Liu Y, Stoddart JF, Zink JI. A reversible molecular valve. Proc. Natl. Acad. Sci. U. S. A. 2005, 102, 10029–34.

[29] Nguyen TD, Leung KCF, Liong M, Pentecost CD, Stoddart JF, Zink JI. Construction of a pH-driven supramolecular nanovalve. Org. Lett. 2006, 8, 3363–6.

[30] Lewandowski B, De Bo G, Ward JW, Papmeyer M, Kuschel S, Aldegunde MJ, Gramlich PME, Heckmann D, Goldup SM, D'Souza DM, Fernandes AE, Leigh DA. Sequence-specific peptide synthesis by an artificial small-molecule machine. Science 2013, 339, 189–93.

[31] Livoreil A, Dietrich-Buchecker CO, Sauvage JP. Electrochemically triggered swinging of a [2]-catenate. J. Am. Chem. Soc. 1994, 116, 9399–400.

[32] Grunder S, McGrier PL, Whalley AC, Boyle MM, Stern C, Stoddart JF. A water-soluble pH-triggered molecular switch. J. Am. Chem. Soc. 2013, 135, 17691–4.

[33] Leigh DA, Wong JKY, Dehez F, Zerbetto F. Unidirectional rotation in a mechanically interlocked molecular rotor. Nature 2003, 424, 174–9.

[34] Hernández JV, Kay ER, Leigh DA. A reversible synthetic rotary molecular motor. Science 2004, 306, 1532–7.

[35] Wilson MR, Solà S, Carlone A, Goldup SM, Lebrasseur N, Leigh DA. An autonomous chemically fuelled small-molecule motor. Nature, 2016, 534, 235–40.

[36] Benny R, Sahoo D, George A, De S. Recent advances in fuel-driven molecular switches and machines. ChemistryOpen 2022, 11, e202200128.

[37] von Delius M, Leigh DA. Walking molecules. Chem. Soc. Rev. 2011, 40, 3656–76.

[38] Barrell MJ, Campaña AG, von Delius M, Geertsema EM, Leigh DA. Light-driven transport of a molecular walker in either direction along a molecular track. Angew. Chem. Int. Ed. 2010, 50, 285–90.

[39] Mondal A, Toyoda R, Costil R, Feringa BL. Chemically driven rotatory molecular machines. Angew. Chem. Int. Ed. 2022, 61, e202206631.

[40] Zhang Y, Chang Z, Zhao H, Crespi S, Feringa BL, Zhao D. A chemically driven rotary molecular motor based on reversible lactone formation with perfect unidirectionality. Chem 2020, 6, 2420–9.

[41] Borsley S, Kreidt E, Leigh DA, Roberts BMW. Autonomous fuelled directional rotation about a covalent single bond. Nature 2022, 604, 80–5.

[42] Kassem S, van Leeuwen T, Lubbe AS, Wilson MR, Feringa BL, Leigh DA. Artificial molecular motors. Chem. Soc. Rev. 2017, 46, 2592–621.

[43] Pooler DRS, Lubbe AS, Crespi S, Feringa BL. Designing light-driven rotary molecular motors. Chem. Sci. 2021, 12, 14964–86.

[44] Koumura N, Geertsema EM, Meetsma A, Feringa BL. Light-driven molecular rotor: unidirectional rotation controlled by a single stereogenic center. J. Am. Chem. Soc. 2000, 122, 12005–6.

[45] Klok M, Boyle N, Pryce MT, Meetsma A, Browne WR, Feringa BL. MHz Unidirectional rotation of molecular rotary motors. J. Am. Chem. Soc. 2008, 130, 10484–5.

[46] García-López V, Liu D, Tour JM. Light-activated organic molecular motors and their applications. Chem. Rev. 2020, 120, 79–124.

[47] Eelkema R, Pollard MM, Vicario J, Katsonis N, Serrano Ramon B, Bastiaansen CWM, Broer DJ, Feringa BL. Nanomotor rotates microscale objects. Nature 2006, 440, 163.

[48] Vives G, Tour JM. Synthesis of single-molecule nanocars. Acc. Chem. Res. 2009, 42, 473–87.

[49] Shirai Y, Osgood AJ, Zhao Y, Kelly KF, Tour JM. Directional control in thermally driven single-molecule nanocars. Nano Lett. 2005, 5, 2330–4.

[50] Chiang PT, Mielke J, Godoy J, Guerrero JM, Alemany LB, Villagómez CJ, Saywell A, Grill L, Tour JM. Toward a light-driven motorized nanocar: synthesis and initial imaging of single molecules. ACS Nano 2012, 6, 592–7.

[51] Saywell A, Bakker A, Mielke J, Kumagai T, Wolf M, García-López V, Chiang PT, Tour JM, Grill L. Light-induced translation of motorized molecules on a surface. ACS Nano 2016, 10, 10945–52.

[52] Kudernac T, Ruangsupapichat N, Parschau M, Maciá B, Katsonis N, Harutyunyan SR, Ernst KH, Feringa BL. Electrically driven directional motion of a four-wheeled molecule on a metal surface. Nature 2011, 479, 208–11.

[53] Feynman RP. There's plenty of room at the bottom. Eng. Sci. 1960, 23, 22–36.

9 Mediating molecular transformations

CONSPECTUS: *The potential scope of a receptor goes far beyond its ability to bind a molecule. A receptor can also alter the structure of the bound substrate, especially if functional groups along the binding site approach the substrate in the complex and actively participate in the transformation. Such receptors therefore accelerate reactions and may even control their outcome by imposing structural constraints on the substrate. If they bind the product with similar or greater affinity than the starting material, stoichiometric amounts are required to achieve full conversion, but receptors can also act catalytically, making their mode of action closely related to that of enzymes. In this chapter we will see examples of these cases. In addition, catalysts from the field of asymmetric organocatalysis will be discussed to show how important supramolecular concepts are in this context. At the end of the chapter, we will learn how to design molecules that mediate their own formation.*

9.1 Introduction

Two molecules with functional groups A and B must meet in solution before they can react. The rate of their reaction depends on a number of parameters, such as temperature, the concentrations of the reacting molecules, and the Gibbs free energy of activation $\Delta G^{\ddagger}$. The latter includes contributions from the activation enthalpy $\Delta H^{\ddagger}$, which correlates with the extent to which bonds must be broken and the reacting molecules distorted on the way to the transition state, and the activation entropy $\Delta S^{\ddagger}$, which reflects the organization of the transition state and the degrees of freedom that the reaction partners must give up for the reaction to occur.

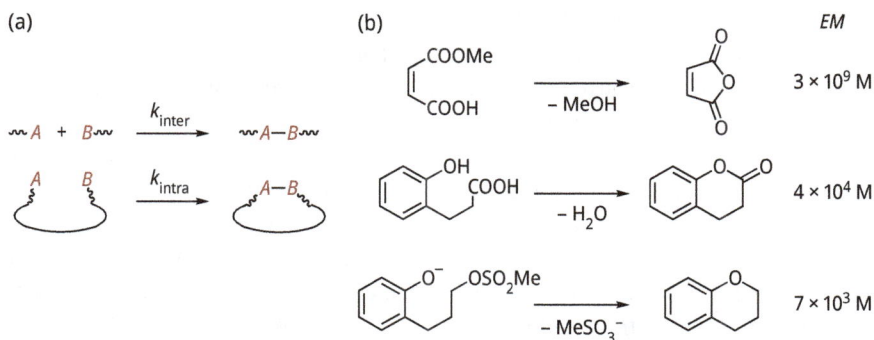

Figure 9.1: Schematic comparison of an intermolecular and an intramolecular reaction between two functional groups A and B (a) and examples of effective molarities of cyclization reactions (b).

When A and B are in the same molecule, they usually react much faster than when they are part of separate reactants (Figure 9.1a). The exact extent to which the intra-

https://doi.org/10.1515/9783111315171-009

molecular reaction is faster is quantified by relating the rate constants of the intramolecular and intermolecular reactions using the following equation:

$$EM = k_{\text{intra}}/k_{\text{inter}} \tag{9.1}$$

Since k_{intra} is associated with a first-order reaction and k_{inter} with a second-order reaction, the two rate constants have different dimensions and the ratio $k_{\text{intra}}/k_{\text{inter}}$ thus gives a concentration. In practice, equation (9.1) specifies the often unrealistically high concentration of a reactant required for the intermolecular reaction to have a pseudo-first-order rate constant identical to the rate constant of the intramolecular reaction ($k_{\text{intra}} = EM\ k_{\text{inter}}$). In other words, the ratio $k_{\text{intra}}/k_{\text{inter}}$ represents the kinetic equivalent of the effective molarity EM, which we encountered in its thermodynamic expression in Section 5.1 on self-assembly. EM was used in this context to assess the extent to which the intramolecular interaction of two binding partners is favored over the intermolecular interaction. In its kinetic form, the effective molarity describes how much faster an intramolecular reaction proceeds with respect to the corresponding intermolecular counterpart, with large EM values indicating particularly effective intramolecular pathways. To illustrate this aspect, effective molarities of several cyclization reactions are given in Figure 9.1b [1].

One reason for the high rates of many intramolecular reactions is the proximity of the reacting groups when they are part of the same molecule. This proximity increases the probability that the two groups will meet, making the activation entropy of the intramolecular reaction less unfavorable than that of the intermolecular reaction. An alternative explanation is that the loss of translational entropy associated with the intermolecular reaction is paid for in the intramolecular counterpart by positioning the reacting groups within the same molecule. Thus, the intramolecular reaction will usually have a lower Gibbs free energy of activation, unless it is associated with the buildup of strain, which has a detrimental effect on $\Delta H^{\ddagger}$.

Similar effects are operative when the proximity of the reacting groups is enforced by organizing them within a noncovalently stabilized complex (Figure 9.2a). In this way, the activation entropy $\Delta S^{\ddagger}$ also becomes less unfavorable compared to the reaction outside the receptor binding site because the reacting molecules give up their translational degrees of freedom in forming the complex. Since the kinetics of an intermolecular reaction changes from a second-order reaction in the absence of the receptor to a pseudo-first-order reaction in the complex, the extent to which complex formation influences the reaction rate can again be estimated from EM values [2]. An example is the CB[6]-induced azide–alkyne cycloaddition between **9.1** and **9.2** (Figure 9.2b). This reaction proceeds *via* a ternary complex in which both reaction partners are incorporated into the cucurbituril cavity with their ammonium groups binding to the carbonyl groups [3]. The resulting proximity of the azide and the alkyne group leads to a fast reaction. Specifically, the cycloaddition in HCOOH (88%)/H$_2$O, 1:1 (*v/v*) at 40 °C has a rate constant k_{inter} of 1.16×10^{-6} M^{-1} s^{-1} in the absence of CB[6], while the pseudo-first order rate constant k_{intra} is 0.019 s^{-1} when CB[6] is present, giving a remarkably high *EM* of 1.6×10^4 M [2]. This value is still two orders of magnitude lower than the upper limit of 4×10^6 M predicted for the *EM* of an intramolecular reaction in which the reacting groups are optimally arranged so that no strain develops during the reaction and the entropy changes are limited to the actual bond-breaking and bond-forming processes [4]. Nevertheless, it is much higher than many other receptor-mediated *EM* values, which span a range of seven orders of magnitude, from 10^{-3} to 10^4 M [2]. Lower values typically indicate that the reaction involves a significant reduction in the number of rotatable bonds in the reaction partners, or that the two reactants spend a large amount of time in an orientation within the cavity that is unsuitable for the reaction.

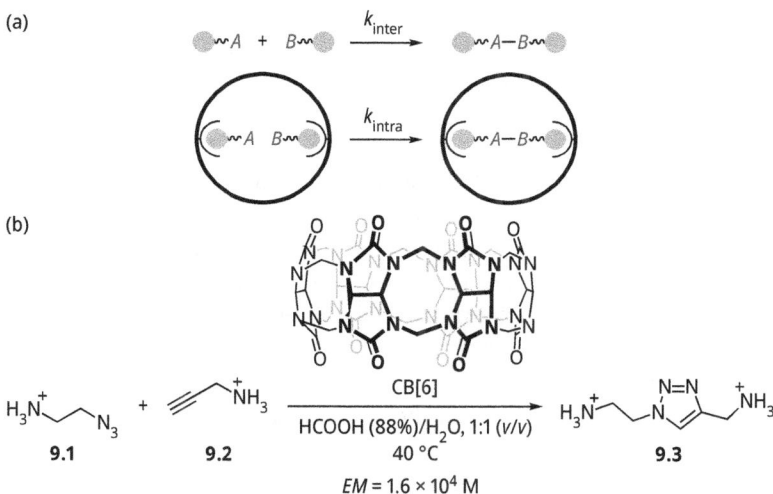

Figure 9.2: Comparison of the intermolecular reaction between two molecules with functional groups *A* and *B* and their pseudo-intramolecular reaction within a receptor cavity (a). The CB[6]-mediated 1,3-dipolar cycloaddition between azide **9.1** and alkyne **9.2** to give the 1,4-disubstituted 1,2,3-triazole **9.3** in (b) is an example of a reaction that proceeds significantly faster in the presence of a receptor.

Another important aspect of receptor-mediated reactions is that they can proceed with altered regioselectivity or stereoselectivity compared to the same transformation in the absence of the receptor. For example, in the reaction shown in Figure 9.2b, CB[6] induces the exclusive formation of the 1,4-disubstituted 1,2,3-triazole **9.3**, completely suppressing the formation of the corresponding 1,5-disubstituted isomer, which is normally also formed. The structure of the product thus reflects the preferred mutual arrangement of the two substrates within the CB[6] cavity.

A closer look at the first-order rate equation of a receptor-mediated reaction shows that the rate at which the complex is consumed depends not only on k_{intra} but also on the extent to which the starting materials are complexed, that is, on c_C (Figure 9.3).

Therefore, for effective substrate conversion, complex formation must not only affect the activation parameters of the reaction, but the complex must also be present in significant amounts. If it does not form at all, no rate enhancement can be expected, whereas high stability is advantageous.

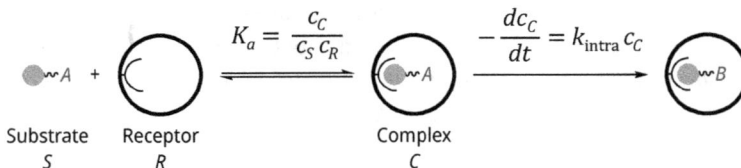

Figure 9.3: Both steps of a receptor-induced substrate conversion. The extent to which the intermediate receptor–substrate complex is formed in the first step depends on its thermodynamic stability, that is, on K_a, and the rate of the second step correlates with the concentration of the complex c_C and the rate constant k_{intra}.

Since binding precedes substrate conversion, any effect that affects the complex concentration will influence the rate of the reaction. An example is an increase in temperature, which will negatively affect the degree of complexation, although it may be beneficial for the actual transformation. Therefore, it is usually necessary to find a temperature at which complex formation is still guaranteed and the reaction is fast enough. Additional components in the reaction mixture that compete with the substrate for the receptor cavity inhibit the reaction. Conversely, the absence of an effect of a strongly binding substrate on the rate of the reaction often indicates that the reaction is not taking place in the receptor cavity. If the reaction proceeds more slowly in the presence of the receptor, the receptor itself may be an inhibitor, as in the reaction shown in Figure 5.68.

The bimolecular *fusion* of two molecules, shown in Figure 9.4a, is only one possible type of reaction a receptor-mediated transformation [5]. For this reaction to occur, there must be a driving force to form the ternary complex. The resulting proximity of the two substrate molecules then promotes their reaction without the need for further functional group participation in the bond-forming reaction. However, the arrangement of the reaction partners within the cavity may cause the reaction to proceed with a characteristic regioselectivity or stereoselectivity. A disadvantage of such reactions is that a ternary complex is converted into a binary complex, which is usually more stable for entropic reasons. Thus, the product tends to remain in the cavity, preventing the receptor from mediating further transformations. When product inhibition occurs, the receptor must be present in stoichiometric amounts to achieve full conversion. Only when the product is bound weaker than the transition state will turnover be observed.

A second type of reaction that can be induced by a receptor is the *fission* of the substrate into smaller fragments. This reaction could involve a functional group in the receptor that actively participates in the reaction by accepting one fragment of the sub-

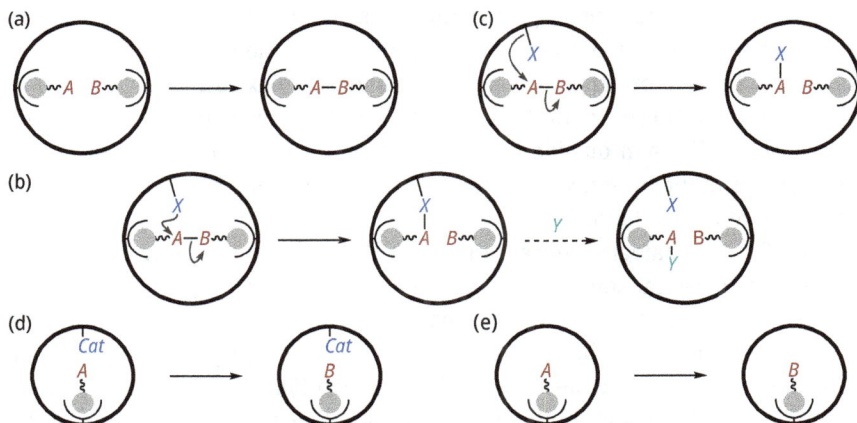

Figure 9.4: Schemes illustrating the types of reactions mediated by a receptor. In addition to the fusion of two substrate molecules (a), a receptor can also induce the cleavage of a substrate into smaller fragments, either by reacting directly with the incorporated substrate (b) or by transferring a simultaneously bound reactive species to it (c). The third mode of action involves the conversion of the substrate into another compound, with the receptor actively participating in the conversion (d) or not (e).

strate while releasing the other (Figure 9.4b). If the product resulting from the covalent receptor modification is stable under the reaction conditions, the reaction will go to completion only if stoichiometric amounts of the receptor are present. If the receptor is regenerated by cleavage of the bond formed in the first step, it acts as a catalyst. Alternatively, cleavage of the substrate is initiated by the transfer of a noncovalently bound reactive species to the substrate (Figure 9.4c). In this case, the receptor is not modified. Since the resulting ternary complex is usually less stable than that of the intact substrate because the products are smaller and therefore bound by fewer interactions, turnover is possible.

Finally, receptors can induce the *transformation* of a substrate molecule. Such a reaction may again involve a prosthetic group in the receptor that would mediate the same transformation in the absence of the receptor (Figure 9.4d). The role of the receptor is then to facilitate the conversion and make it more effective. Alternatively, complex formation could lead to stabilization of the transition state of the reaction, in which case confinement of the substrate in the receptor cavity alone is sufficient to achieve a rate increase without direct involvement of functional groups (Figure 9.4e). Regardless of the actual mode of action, the reaction within the receptor cavity may proceed with a different selectivity than that observed in the absence of the receptor. Catalytic transformations are possible because the substrate and product are often bound with similar affinity.

Do synthetic enzymes exist?

The way synthetic receptors mediate reactions is strikingly similar to the way enzymes work. In both cases, the reaction takes place within the confines of a binding pocket and may involve prosthetic groups that, after complex formation, end up in close proximity to the substrate. In this environment, the conversion of the substrate proceeds more rapidly than outside the cavity, often with characteristic selectivity. The reaction stops in the presence of a competitive inhibitor that binds to the binding site more effectively than the substrate molecule(s).

Because of these analogies, work on catalytically active receptors often aims at the development of enzyme mimics. However, few of the known supramolecular catalysts achieve the activity of enzymes [6]. Probably the most important reason for their often poorer performance is that enzymes are much more finely tuned to stabilize transition states than can be achieved by rational design [5]. Synthetic systems therefore often suffer from product inhibition and/or an unfavorable arrangement of the substrate in the cavity.

Cram had already encountered these problems in his pioneering work on enzyme mimics, which aimed to find a mimic for chymotrypsin, a protease responsible for cleaving peptide bonds [7]. The enzyme mediates this reaction by bringing a peptide group in the substrate close to a serine side chain in the binding site (Figure 9.5). The hydroxy group in this side chain is part of a charge-relay system that also includes the imidazole moiety of a histidine and the carboxylate group of an aspartate residue. In this catalytic triad, the carboxylate group causes partial deprotonation of the imidazole NH group, which increases the basicity of the other imidazole nitrogen atom to the point where it is capable to deprotonate the serine OH group. The resulting alkoxide initiates cleavage of the peptide bond by reacting with the peptide carbonyl group. An acylated enzyme is formed, which is ultimately hydrolyzed to regenerate the active form of the enzyme.

Cram used Corey–Pauling–Koltun models to design compound **9.4** (Figure 9.6a), which he expected to hydrolyze amino acid esters in a manner similar to how chymotrypsin cleaves peptide bonds [7]. In **9.4**, the spherand-type subunit should induce affinity for the ammonium group of the substrate by serving as a hydrogen bond acceptor (Section 4.1.3). The benzylic OH group takes over the role of the serine side chain, and the interaction of this group with the imidazole residue flanked by a carboxylate group completes the charge-relay system.

Due to the challenging synthesis of **9.4**, the work initially focused on simpler compounds whose structural complexity was gradually increased. First, the receptor scaffold **9.5** alone was shown to interact efficiently with ammonium groups in CDCl$_3$ (e.g., log K_a = 9.7 for (H$_3$C)$_3$CNH$_3^+$). Compound **9.5** was then converted to **9.6**, which contains the benzyl OH group as a nucleophile. The ability of **9.6** to induce transesterification of amino acid esters was tested in CDCl$_3$ containing *N,N*-diisopropylethylamine/*N,N*-diisopropylethylammonium perchlorate as buffer using the *p*-nitrophenol esters of L-alanine or other amino acids. Such esters are convenient substrates because their carbonyl group is activated for nucleophilic attack by the electron-withdrawing nature of the phenol group and because the released *p*-nitrophenolate anion allows the rate of reaction to be followed by UV–vis

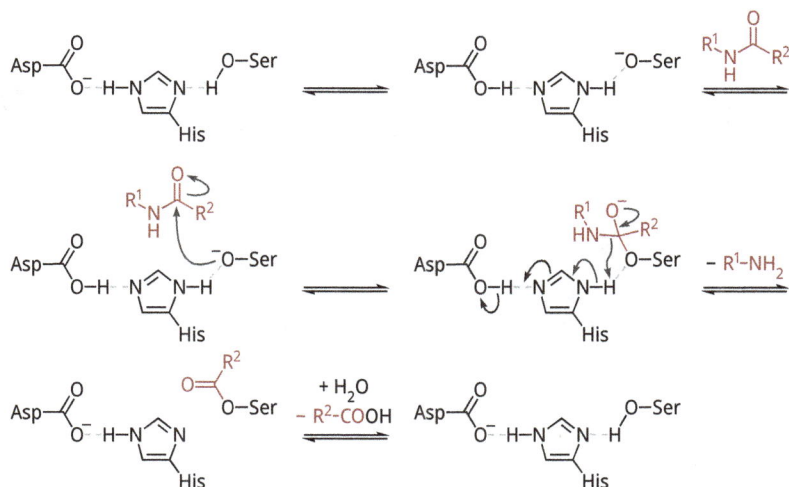

Figure 9.5: Reaction scheme illustrating the mechanism of peptide bond cleavage mediated by the charge-relay system of the enzyme chymotrypsin.

spectroscopy (Figure 9.6b) [8]. According to the results, the acylation of the benzyl alcohol is first order in the ratio of the buffer components, indicating that the deprotonated OH group is indeed the active nucleophile. Furthermore, the acetylation is approximately 10^{11} times faster than that of 3-phenylbenzyl alcohol, a model compound lacking the receptor moiety. Finally, the presence of an excess of $NaClO_4$ as a competitive inhibitor significantly reduces the reaction rate. These results demonstrate that the complexation of the substrate by **9.6** and the resulting proximity of the alkoxide to the ester group cause the expected rate enhancement.

Motivated by these results, the imidazole-containing receptor **9.7** was prepared in a 30-step synthesis [9]. According to kinetic studies, this compound is acylated approximately 10^5 times faster than a noncomplexing model compound, even in the absence of an externally added base, indicating that the imidazole residue in **9.7** does indeed contribute to the reaction. Unexpectedly, however, the reaction does not occur at the OH group, but at a nitrogen atom of the imidazole residue (Figure 9.6c) [10]. Accordingly, a derivative with a protected OH group also shows activity. The originally expected O-acylated product is only produced in a subsequent slow step by migration of the acyl group from the nitrogen to the oxygen atom. Since the mode of action of **9.7** differs from that of chymotrypsin, the synthesis of **9.4** was not attempted.

Although Cram's work demonstrated that the efficient transformation of suitable substrates is possible with properly designed receptors, it also showed that the deliberate design of enzyme mimics is far from trivial. The main reason is that efficient catalytic properties require a finely tuned interplay of binding properties, receptor structure, and other parameters such as solvation. Most importantly, high catalytic activity requires the

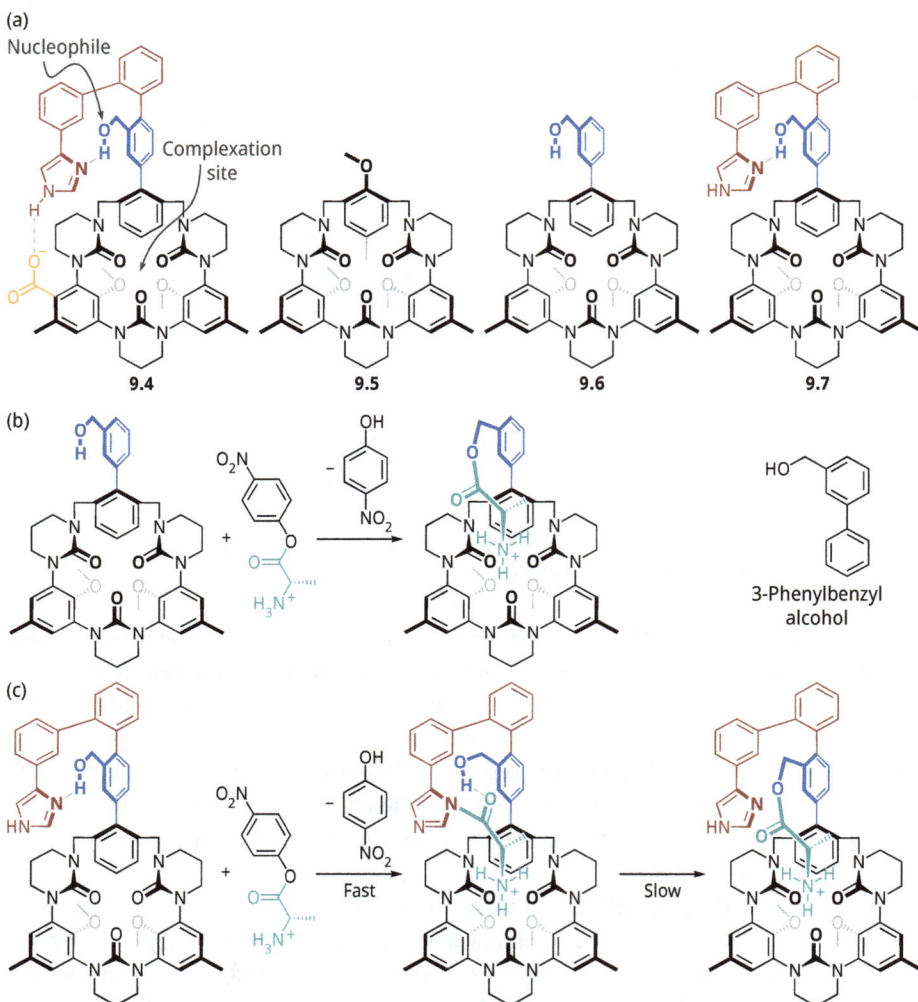

Figure 9.6: Molecular structures of chymotrypsin mimic **9.4** proposed by Cram, the structurally simplified analogs **9.5**, **9.6**, and **9.7**, and the model compound 3-phenylbenzyl alcohol (a). The reactions of **9.6** and **9.7** with the *p*-nitrophenol ester of L-alanine are shown in (b) and (c), respectively.

receptor to bind to the transition state of the reaction, which makes targeted structural design challenging. Nevertheless, many potent supramolecular catalysts have been developed since Cram's early achievements [11], a selection of which will be presented in the following sections. We begin the discussion with systems that mostly suffer from product inhibition and therefore operate in a stoichiometric manner. Of particular interest in this context are systems in which the confinement of substrate molecules within the cavity of a receptor, capsule, or cage induces reactivity not otherwise observed [12, 13]. We then look at catalytic systems, the strategies used to design them, and their mode of action. These catalysts cover a wide range of structures, from covalently and noncovalently assembled receptors to coordination cages, including switchable systems and asymmetric organocatalysts. The final part of the chapter introduces self-replicating molecules that catalyze their own formation. The use of noncovalent interactions to mediate the assembly of conventional catalysts or to optimize their properties is not considered [14].

9.2 Stoichiometric transformations

9.2.1 Transformation by functional group participation

Since enzymes operate in water, receptors active in the same environment should be a good starting point for the development of enzyme mimics. Cyclodextrins (CDs) (Section 4.1.4) are particularly attractive in this context, not only because of their ability to bind a variety of hydrophobic substrates in water, but also because they can be structurally varied over a wide range and thus tailored to the requirements of the substrate and its transformation. In addition, CDs are chiral and therefore potentially capable of inducing enantioselective reactions. However, many of the known CD-based enzyme mimics are not catalytically active, either because they are irreversibly modified during the transformation or because of product inhibition.

Even native CDs accelerate the rate of certain reactions, most notably the cleavage of phenol esters. The first reports of this reactivity were published in the early 1960s by the group of Friedrich Cramer, followed shortly thereafter by important contributions from Myron L. Bender, who showed that CDs have a marked effect on the cleavage of phenyl acetates [15]. For example, 3-nitrophenyl acetate is cleaved about 100 times faster in water at pH 10.6 and 25 °C when an excess of α-CD is present. This effect is even more pronounced for 3-*tert*-butylphenyl acetate, where the reaction is accelerated by a factor of 226. The reason for these rate enhancements is the partial deprotonation of 2-OH groups along the wider rim of the CD cavity at the high pH. The resulting nucleophilic alkoxide groups approach the bound substrate in the complex and induce ester cleavage by reacting with its carbonyl group (Figure 9.7). The reaction results in an acetylated CD, and the mode of action is therefore similar to that of chymotrypsin, with the secondary CD hydroxy group assuming the role of the serine side chain. However, the process is not catalytic since CD acetylation is irreversible.

Indications that the initial complexation of the substrate does indeed affect the reaction are the much smaller effects of β-CD and γ-CD on the reaction rate, as well as the influence of the substitution pattern of the substrate on the rate enhancement. For example, α-CD enhances the cleavage of 2-nitro and 4-nitrophenyl acetate only by a factor of 10 and 3, respectively, suggesting that the arrangement of these substrates does not allow an efficient reaction with a deprotonated 2-OH group. However, even

(a)

(b)

9.8   **9.9**

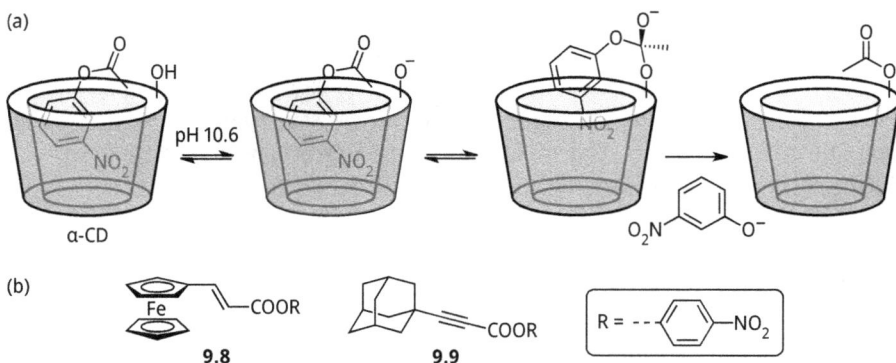

Figure 9.7: Schematic representation of the α-cyclodextrin-mediated cleavage of 3-nitrophenyl acetate in water at pH 10.6 (a), and structures of the ferrocene and adamantane-derived substrates **9.8** and **9.9** (b).

factors of 100–200 are still modest considering that the bond cleavage is mediated by a highly nucleophilic alkoxide, which Ronald Breslow attributed to the fact that 3-nitrophenyl acetate must partially leave the cavity to form the tetrahedral intermediate [16]. This is not the case for ferrocene derivative **9.8,** which was designed to orient the ester group in close proximity to a secondary OH group when bound by β-CD, and this ester is indeed cleaved 330,000 times faster in the presence of this CD. Rate enhancements of up to 5,900,000 have been observed for even better designed substrates [17].

The arrangement of the ester near the alkoxide group influences the first step of ester cleavage, the formation of the tetrahedral intermediate. Product formation in the next step depends not only on the propensity of the leaving group to depart, but also on the ease with which the resulting planar ester group is formed in the environment of the CD cavity. If the substrate is too rigid, the corresponding structural rearrangement is difficult and the second step of the reaction becomes rate-determining. More flexible substrates, such as the adamantane derivative **9.9**, do not have this disadvantage [18]. In this case, the tetrahedral intermediate forms less rapidly than that of **9.8** due to the rotation of the ester group around the ethinyl group. However, this inherent flexibility facilitates the second step, making this step much less dependent on the type of leaving group than in the case of the ferrocene derivative. Breslow concluded that it is necessary to consider the structural changes associated with the entire reaction when designing enzyme mimics to prevent an otherwise fast step from becoming rate-determining [16].

Although native CDs also mediate reactions other than transesterifications [16], CD-based enzyme mimics are more often based on substituted derivatives whose reactivity is controlled by the attached functional group. An example is the β-CD derivative **9.10** (Figure 9.8a) with a pyridoxalamine group attached to the 6-position of a glucose moiety [19]. Pyridoxal phosphate is the active form of vitamin B_6 and a cofactor in many biochemical processes, including transamination, decarboxylation, deamination, and race-mization reactions. Specifically, pyridoxalamine reacts with α-ketocarboxylic acids to form imines that, upon hydrogen shift and subsequent hydrolysis, yield the correspond-ing amino acids and pyridoxal (Figure 9.8b). Similarly, 3-(1*H*-indol-3-yl)-2-oxopropanoic acid is converted to tryptophan by **9.10** in a reaction that is 200 times faster than the same conversion performed in the presence of a pyridoxalamine derivative lacking the CD ring. This result and the fact that α-ketocarboxylic acids with substituents smaller than the indole group are less readily converted to the corresponding amino acids are strong indications that the incorporation of the substrate into the CD cavity plays a role in the conversion. Despite the chirality of **9.10**, the enantioselectivity of product forma-tion is low. Furthermore, there is no turnover because the pyridoxalamine is not regen-erated. Catalytically active substituted CDs are discussed in Section 9.3.1.

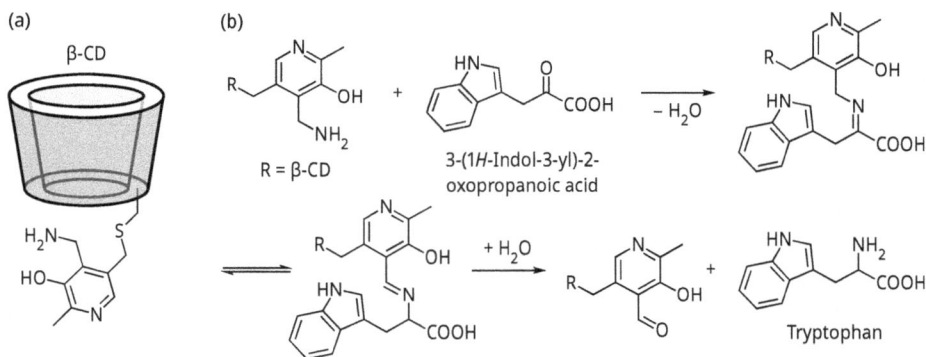

Figure 9.8: Molecular structure of pyridoxalamine-containing β-CD **9.10** (a) and mechanism of the transamination reaction that converts an α-ketocarboxylic acid to an amino acid (b).

Proximity effects also operate in the self-folding cavitands developed by Julius Rebek Jr. if they contain functional groups directed toward the cavity interior (Section 4.1.9) [20]. An example is **9.11** (Figure 9.9) with a Kemp's triacid incorporated into one of the walls. In the complexes with suitable amines, the methyl ester group ends up in close proxim-ity to the nitrogen atom of the guest. This allows the methyl group to be transferred in a nucleophilic substitution reaction, yielding the free carboxylate group and the corre-sponding ammonium ion [21]. The reaction is particularly rapid for quinuclidine, which is converted to the *N*-methylquiniclidinium ion with a half-life of less than three mi-nutes at room temperature. Amines that are less well bound by **9.11** react more slowly, and those that are preferentially bound with the nitrogen atom toward the cavity bot-tom, such as morpholine, do not react at all. A related Kemp's triacid lacking the cavi-tand residue does not react with the same amines even at 100 °C, demonstrating that the main reason for the efficiency of **9.11** is the proximity of the reacting groups in the complex and their good mutual orientation. An additional advantage is the absence of solvent molecules in the complex that shield the reacting functional groups. The high affinity of the cavitand to the positively charged product and the irreversible modification of the carboxyl group make the process noncatalytic.

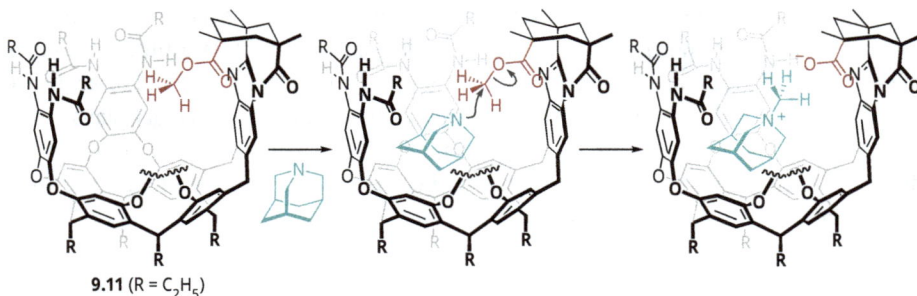

9.11 (R = C$_2$H$_5$)

Figure 9.9: Molecular structure of the functionalized self-folding cavitand **9.11** and its reaction with quinuclidine to give the corresponding free acid and the *N*-methylquinuclidinium ion. The front panel of the cavitand is not shown for reasons of clarity.

The structurally related deep cavitand **9.12**, which carries a 2-pyridone group, mediates the aminolysis of a choline-derived activated carbonate (Figure 9.10) [22]. The reaction involves the incorporation of the quaternary ammonium group into the cavity, placing the carbonate group close to the 2-pyridone moiety. This group stabilizes the tetrahedral intermediate by hydrogen bonding, thereby promoting product formation. Although the reaction is only about twice as fast when 10 mol% of **9.12** is present, this cavitand shows turnover.

9.2.2 Transformation by confinement

Bimolecular reactions between two substrates often proceed much faster when they take place in the confined space of a receptor cavity, even in the absence of functional groups that mediate bond formation. This is due to the proximity of the bound molecules, which makes the activation entropy of the reaction less unfavorable than that of the analogous reaction outside the cavity. The receptor could also alter the selectivity of the transformation as a consequence of the complexation-induced orientation of the guests [12]. Finally, the isolation of reactive intermediates in the cavity could prevent unwanted side reactions, resulting in fewer byproducts. In any case, the receptor serves primarily as a reaction vessel without actively participating in the transformation [23]. Capsules or coordination cages, in which the substrates are completely encapsulated and thus tightly held together, are particularly useful for these purposes

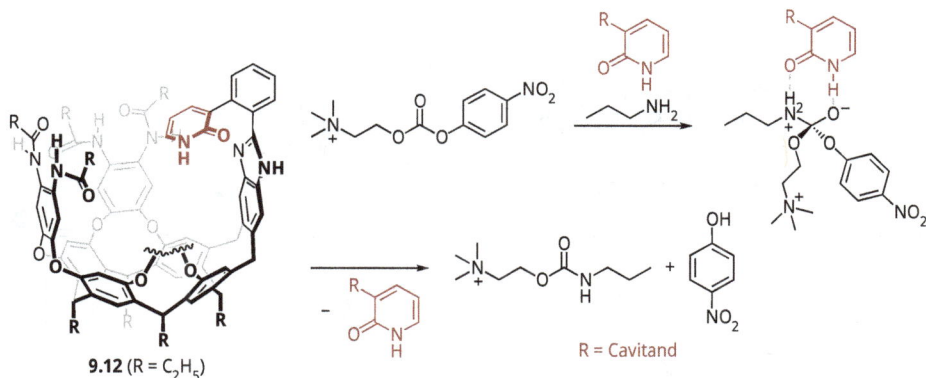

Figure 9.10: Molecular structure of the functionalized 2-pyridone-containing cavitand **9.12** and mechanism of the aminolysis of a choline-derived 4-nitrophenylcarbonate, illustrating the stabilization of the tetrahedral intermediate by the 2-pyridone moiety. The front panel of **9.12** is not shown for reasons of clarity.

because they are often assembled from simple building blocks and do not require elaborate synthesis.

Since the reaction between the complexed molecules must proceed without the need for additional reagents or catalysts, for which there is usually no space left in the cavity, thermal cycloadditions such as Diels–Alder reactions or 1,3-dipolar cyclo-additions are attractive reactions to study. In these reactions, the ternary complex between the receptor and the two substrate molecules is converted into the typically more stable binary complex of the product. Accordingly, product inhibition is usually observed, preventing catalysis.

An early example is the cycloaddition between benzoquinone and 1,3-cyclohexa-diene inside the cavity of Rebek's molecular softball (Section 5.3.3), which is assembled from two molecules of **9.13** (Figure 9.11a) [24]. Product formation in this case is 200 times faster than outside the softball and is associated with a high *EM* of 2.4 M.

CB[7] was shown by the Nau group to mediate the dimerization of cyclopenta-diene [25]. With a rate enhancement k_{cat}/k_{uncat} of 1.4×10^5 M, the reaction is only one order of magnitude slower than the estimated maximum rate achievable by confine-ment [26]. A factor contributing to this rate enhancement is the reduction of the pack-ing coefficient from 63% for the tightly packed ternary complex to the near-optimal packing coefficient of 58% of the product complex. Smaller and larger cucurbiturils are much less effective because those with less than seven subunits cannot bind two guest molecules simultaneously, while the larger ones bind two cyclopentadiene mol-ecules too loosely. Product inhibition can be prevented by the addition of small amounts of methanol, which weakens the stability of the CB[7] complex sufficiently to allow turnover. Neither softball **9.13** nor CB[7] affects the intrinsic selectivity of the Diels–Alder reaction, both preferentially inducing the formation of the *endo* product.

(a)

9.13 (Ar = 4-*n*-heptylphenyl)

(b)

CB[7]

(c)

9.14

(d)

9.15

(e)

9.16 (R = C₁₁H₂₃)

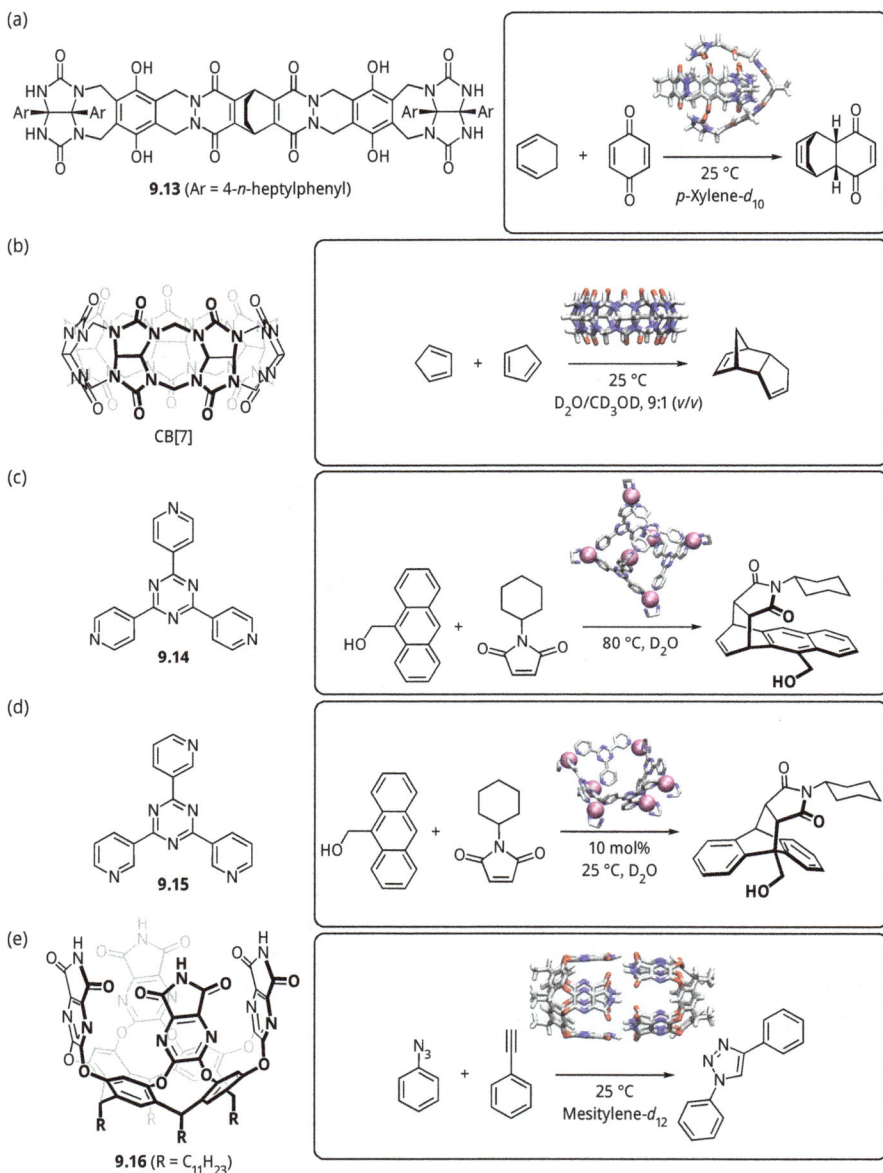

Figure 9.11: Diels–Alder reactions between benzoquinone and 1,4-cyclohexadiene inside the molecular softball **9.13** (a), between two molecules of cyclopentadiene in the presence of CB[7] (b), and between *N*-cyclohexylmaleimide and 9-hydroxymethylanthracene in the presence of the coordination cages formed from [Pd(en)(NO₃)₂] and the tripodal ligands **9.14** (c) and **9.15** (d). The reaction scheme in (e) shows the 1,3-dipolar cycloaddition between phenylacetylene and phenylazide inside the cavity of the self-assembled capsule formed from **9.16**.

This is different when the Diels–Alder reaction between *N*-cyclohexylmaleimide and 9-hydroxymethylanthracene is performed in the presence of Fujita's coordination cage assembled from [Pd(en)(NO$_3$)$_2$] and the tripodal ligand **9.14** (Section 5.5.5) [27]. While anthracene normally reacts with the center ring acting as the dienophile because the loss of aromatic stabilization is less than when one of the other rings reacts, this reaction is not observed when the reactants are incorporated into the coordination cage. Instead, a product is formed that contains the bridge in a terminal ring (Figure 9.11b). In contrast, the open and solvent-exposed cavity of the bowl-shaped receptor derived from ligand **9.15** allows the cycloaddition to proceed with normal regioselectivity, and since guest exchange is also possible, this receptor acts as a catalyst (Figure 9.11c) [27]. It mediates the complete conversion of the substrates within 5 h at 10 mol%, whereas 24 h are required when the receptor concentration is reduced to 1 mol%.

The hydrogen-bonded dimer of deep cavitand **9.16** (Section 5.3.3) simultaneously binds phenylacetylene and phenylazide, driving the 1,3-dipolar cycloaddition between these two molecules (Figure 9.11d) [28]. This reaction is 30,000-fold faster inside the cavity than outside and yields exclusively the 1,4-disubstituted 1,2,3-triazole, whereas in the absence of the capsule a mixture of the 1,4- and 1,5-disubstituted products is formed.

When molecules encapsulated in capsules or cages do not react spontaneously, their reaction can be induced by irradiation. These reactions may be concerted, such as the cycloadditions discussed earlier, or they may involve stepwise processes *via* intermediates whose reactivity is characteristically influenced by encapsulation. For example, irradiation of α-(*n*-alkyl)-dibenzyl ketones bound in the dimer of Gibb's octa acid **9.17** (Section 5.2) produces radical species by homolytic cleavage of the α-carbon bond, whose subsequent reactions are influenced by their orientation within the cavity (Figure 9.12) [29]. For example, the substrate with the methyl group (*n* = 1) fills the capsule with the two aromatic residues occupying the hemispheres and the alkyl group located in the equatorial region. In this case, decarbonylation followed by recombination is favored (Norrish type I reaction), leading to the exclusive formation of heterocoupled products. This is different from the reaction in solution, where homocoupled products are unavoidable. The arrangement of substrates with longer alkyl groups (*n* = 2–5) inside the cavity is less well defined because their alkyl and aryl groups compete for space in one of the hemispheres. As a result, additional reaction pathways become available, causing a progressive shift in the product ratio from Norrish type I products to rearranged and Norrish type II products. Once the length of the alkyl group reaches six or more carbon atoms, rearranged products are no longer formed, and the strong preference of the octyl group to occupy one hemisphere leads to the preferential formation of the Norrish type II product. In contrast, the same Norrish type II product is formed in solution in much smaller amounts. Thus, characteristic modes of uptake of these substrates cause them to react *via* distinct pathways that differ from those outside the cavity.

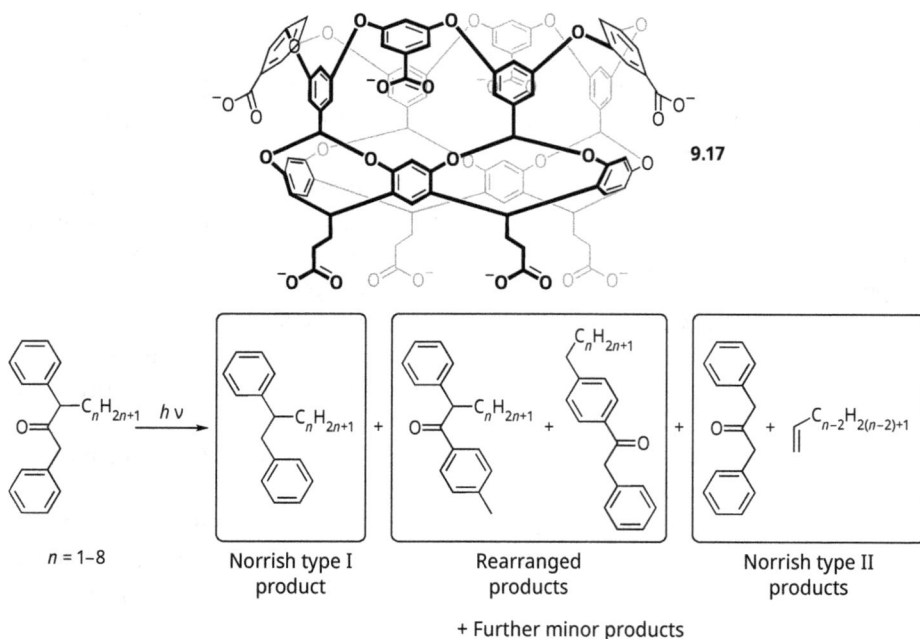

Figure 9.12: Photochemical reactions of α-(n-alkyl) dibenzyl ketones incorporated into the dimer of the octa acid **9.17**. The methyl derivative gives predominantly the decarbonylated heterocoupled Norrish type I product. The amount of this product decreases with increasing chain length in favor of rearranged and Norrish type II products. Rearranged products are not observed for the substrates with hexyl, heptyl, and octyl chains. The octyl derivative gives predominantly the Norrish type II product.

Other examples of photochemical reactions within the capsule formed from **9.17** are the photoisomerization of (*E*)-4,4'-dimethylstilbene and the photochemical dimerization of 4-methylstyrene [30]. (*E*)-4,4'-Dimethylstilbene can be photochemically isomerized to the *Z*-isomer, resulting in a photostationary state in the absence of the capsule with an *E*/*Z* ratio of approximately 1:3. When (*E*)-4,4'-dimethylstilbene is incorporated into the capsule and irradiated, much less (*Z*)-4,4'-dimethylstilbene is formed because this bulky isomer is a poor guest (Figure 9.13a). Consequently, the capsule also prevents the formation of the phenanthrene derivative, which is formed in small amounts from (*Z*)-4,4'-dimethylstilbene in solution. In the case of 4-methylstyrene, two molecules dimerize upon irradiation, giving preferentially 1,2-disubstituted cyclobutanes. The same reaction performed inside the capsule leads to an almost 1:1 mixture of two other products, one of which is the 1,3-disubstituted cyclobutane, again illustrating that the arrangement of the guests inside the capsule controls product formation (Figure 9.13b).

Effects on the outcome of photochemical [2 + 2]cycloadditions are also observed for substrates incorporated into the coordination cage derived from **9.14** [31]. For example, two acenaphthylene molecules dimerize inside this cage upon irradiation, yielding only

(a)

Without capsule	18%	76%	6%
With capsule	85%	15%	0%

(b)

Without capsule	83%	10%	0%	7%
With capsule	0%	0%	55%	45%

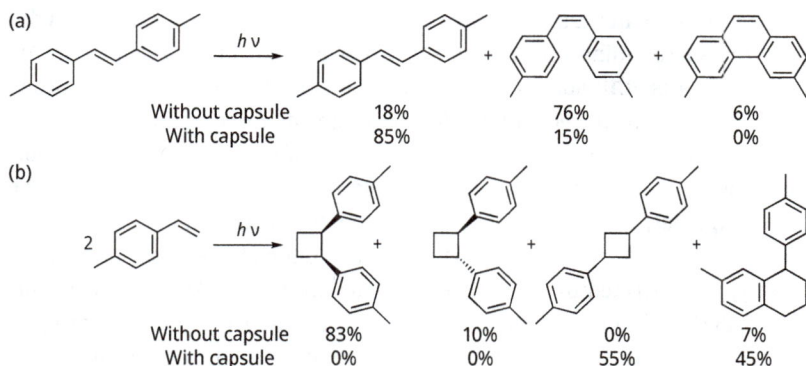

Figure 9.13: Effect of the dimer of octa acid **9.17** on the photoisomerization of (*E*)-4,4′-dimethylstilbene (a) and the photoinduced dimerization of 4-methylstyrene (b).

the *syn* product because the mutual arrangement of the reaction partners does not allow the formation of the more extended *anti*-isomer (Figure 9.14a). Of the four stereo-isomers that result from the photodimerization of 1-methylacenaphthylene, only the *syn*-isomer with the 1,3-*cis*-dimethylated cyclobutane ring is formed in the presence of the cage (Figure 9.14b).

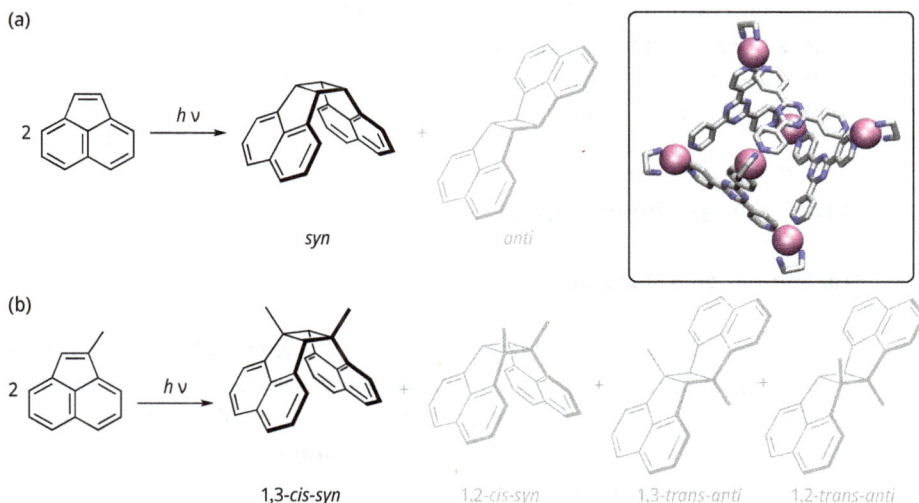

(a)

syn   anti

(b)

1,3-*cis-syn*   1,2-*cis-syn*   1,3-*trans-anti*   1,2-*trans-anti*

Figure 9.14: Effect of the coordination cage derived from **9.14** on the photochemical [2 + 2]cycloaddition of acenaphthylene (a) and 1-methylacenaphthylene (b). The products shown in gray are not formed in the presence of the coordination cage whose structure is shown in the inset.

The water-soluble deep cavitand **9.18** (Figure 9.15a) demonstrates that receptors with cavities open to the surrounding solvent can also serve as reaction vessels [32]. In the absence of suitable guests, **9.18** adopts the kite conformation in water, which is prone to self-assembly, forming an unreceptive velcrand dimer as discussed in Section 5.2. Suitable substrates, such as long-chain alkanes, shift the equilibrium to the vase conformation of the monomeric cavitand, whose deep cavity allows substrate binding. In the corresponding complexes, the alkanes adopt U-shaped conformations with the chain ends exposed to the solvent while the atoms near the center of the chain are buried deep in the cavity, as illustrated by the calculated structure of the 12-aminododecanoic acid complex of **9.18** (Figure 9.15b). Accordingly, α,ω-difunctionalized substrates are preorganized for cyclization. In the case of 12-aminododecanoic acid, ring closure is induced by treatment of the complex with a coupling reagent, giving the cyclic product in a yield approximately three times higher than in the absence of the cavitand (Figure 9.15c) [33]. Similarly, bis(lactams) are formed from α,ω-diamines and suitable activated diesters in up to 10 times higher yields when the reaction is performed under the influence of **9.18** (Figure 9.15d) [34]. Since the products remain in the cavity, there is no turnover.

In many of the examples discussed in this section, the receptors preorganize the bound substrates, thereby reducing the Gibbs free energy of activation of a particular reaction and making it faster than potential other transformations. The reactions are thus controlled by kinetic template effects, similar to those of external templates in macrocyclization reactions (Section 5.1). However, while templates in macrocyclization reactions often facilitate the synthesis of a receptor, in the examples above the receptors themselves take on the role of the template, mediating the transformations that take place within their cavities.

9.3 Catalytic transformations

9.3.1 Transformation by functional group participation

Functionalized receptors are capable of promoting the chemical conversion of a bound substrate, but for a receptor to act catalytically, several conditions must be met. First, the structural integrity of the receptor must be maintained throughout the catalytic cycle. If substrate conversion leads to structural changes, there must be a way to reverse these changes and thus regenerate the receptor in its catalytically active form. Second, the receptor should ideally bind more strongly to the transition state than to the substrate or the product to prevent product inhibition. Although this situation would most closely mimic the behavior of enzymes, it is usually very difficult to achieve by design, as we saw when discussing Cram's chymotrypsin mimic. Therefore, the receptor should at least bind more strongly to the substrate than to the

(a)

9.18 R =

(b)

(c)

(d)

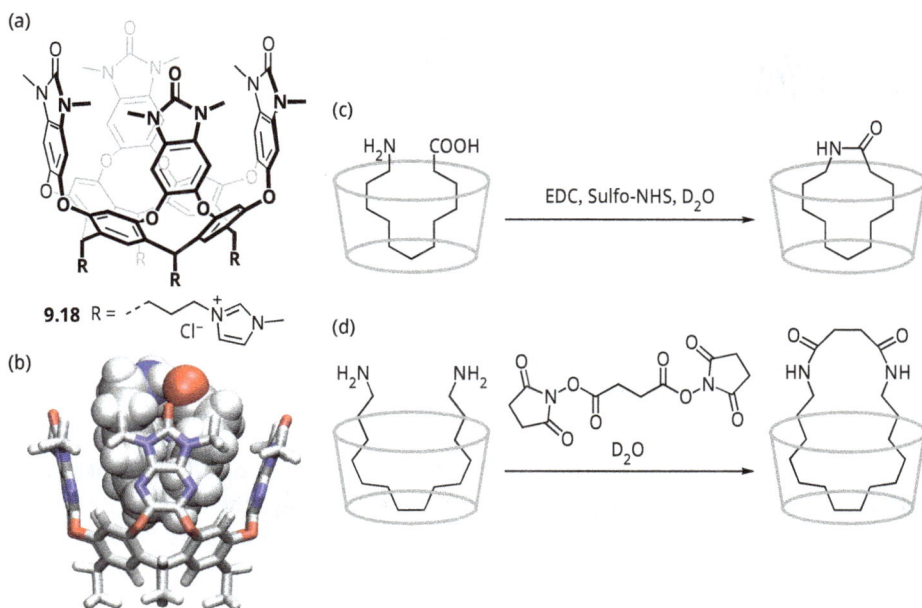

Figure 9.15: Molecular structure of the deep cavitand **9.18** (a), calculated structure of the 2-aminododecanoic acid complex of **9.18** (b), and schematic illustration of the cyclization of 12-aminododecanoic acid (c) and an α,ω-diamine (d) under the influence of **9.18**. The side chains of the cavitand in the calculated structure are omitted for reasons of clarity (EDC = 1-ethyl-3-(3-dimethylaminopropyl)carbodiimide hydrochloride, Sulfo-NHS = *N*-hydroxysulfosuccinimide).

product(s), allowing new substrate molecules to replace the product(s) from the complex after the reaction is complete.

The latter is the case in an early CD-based hydrolase mimic developed by the Breslow group [35]. In the corresponding bis(cyclodextrin) **9.19** (Figure 9.16a), two β-CD rings are linked by a 2,2′-bipyridyl moiety that coordinates to a copper(II) ion. This metal center serves to activate water molecules, thereby inducing the rapid hydroxide-mediated hydrolysis of esters such as **9.20**, whose terminal hydrophobic groups are incorporated into the CD cavities simultaneously placing the central carbonyl group near the metal ion (Figure 9.16b). The resulting ternary complex is less stable than that of the intact ester, allowing further substrate molecules to displace the products. Since the copper complex is also not consumed during the reaction, the overall process is catalytic, with a marked rate enhancement of 220,000 over the uncatalyzed reaction.

Other CD-derived catalysts are the tetrakis(cyclodextrin) derivative **9.21** (Figure 9.17a) with four CD moieties surrounding a porphyrin-manganese(III) complex and the bis(imidazole) **9.22** (Figure 9.17b). Both compounds exhibit turnover because the substrate and product of the respective reactions are bound with comparable affinities. In

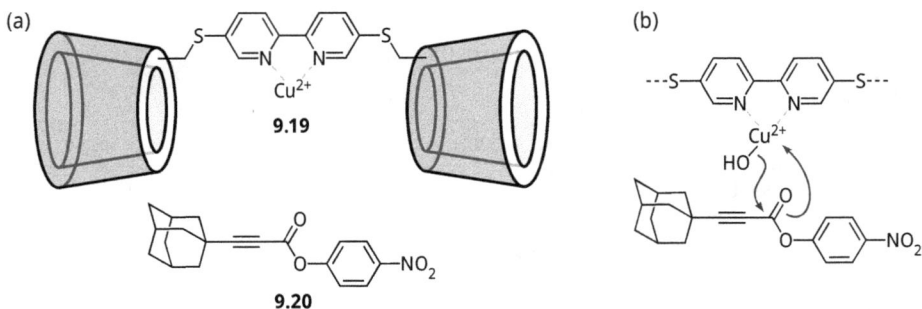

Figure 9.16: Molecular structures of bis(cyclodextrin) **9.19** and its substrate **9.20** (a), and proposed mechanism of ester cleavage mediated by the central copper(II) complex (b).

the case of **9.21**, the manganese(III) complex must first be activated by treatment with iodosobenzene [36]. The resulting oxo complex then mediates the epoxidation of stilbene derivatives or the oxidation of a nonactivated C–H bond in dihydrostilbene [37]. Note that the substrates used in these reactions contain bulky end groups to ensure binding to the CD moieties and positioning of the reacting groups close to the metal center. The same strategy allows the regioselective oxidation of C–H bonds in steroid-derived substrates.

Bis(imidazole) **9.22** facilitates the hydrolysis of a cyclic phosphate, with one imidazole unit acting as a general base to activate a water molecule and the protonated form of the other imidazole group transferring its proton to the substrate [38]. The reaction is most efficient when the two imidazole units are located in adjacent glucose units. In addition, the product with the phosphate group on the oxygen atom in *meta* position to the *tert*-butyl group is almost exclusively obtained, possibly because the other oxygen atom is in the apical position of the pentacoordinated intermediate, making it the better leaving group.

Another example of a receptor that catalyzes the cleavage of phosphate esters, specifically the dephosphorylation of adenosine triphosphate (ATP), is the polyamine-based anion receptor **9.23** (Figure 9.18a). The proposed catalytic cycle is shown in Figure 9.18b [39]. At the neutral pH at which the reaction takes place, **9.23** contains four to five protonated amino groups to which ATP binds by forming salt bridges. Complex formation brings the remaining unprotonated amino group close to the terminus of the triphosphate group, where it can induce cleavage by nucleophilic substitution. The final steps in the catalytic cycle are hydrolysis of the resulting phosphoramidate and dissociation of adenosine diphosphate (ADP), not necessarily in this order. Because the product has fewer negative charges than ATP, it binds less strongly to **9.23** and is therefore replaced by an unreacted substrate molecule. The intermediate phosphoramidate is also labile enough under the reaction conditions to be hydrolyzed, thereby restoring the active receptor.

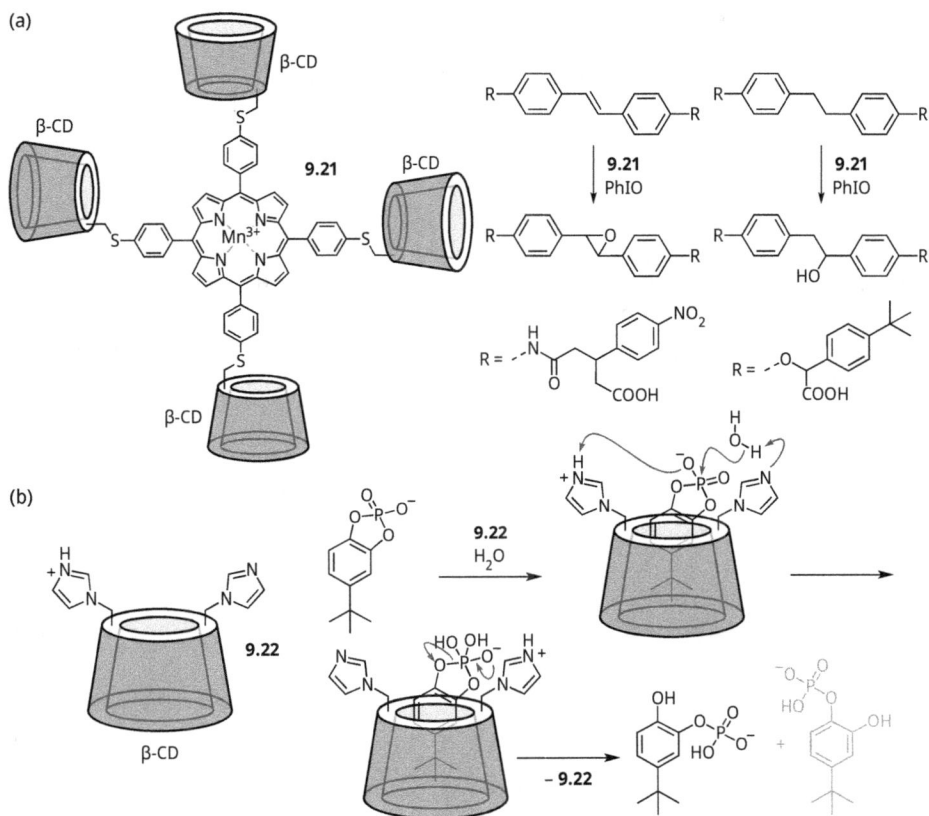

Figure 9.17: Molecular structure of tetrakis(cyclodextrin) **9.21** along with reaction schemes showing the transformations catalyzed by this compound (a). In (b), the schematic structure and mode of action of β-CD bis(imidazole) **9.22** is shown. The phosphate ester shown in gray is practically not formed (<1%).

The last receptor-based catalysts to be mentioned are compounds **9.24** and **9.25**. The calix[4]arene **9.24** with its two zinc(II) centers mimics the action of phosphodiesterases by inducing the transesterification of the RNA model 2-hydroxypropyl 4-nitrophenyl phosphate [40]. This transesterification, which is 23,000 times faster than the uncatalyzed one, requires the cooperative action of both metal centers and benefits from the presence of the calix[4]arene ring as a recognition unit (Figure 9.19).

Cyclophane **9.25** (Figure 9.20), developed by François Diederich, is a model for the enzyme pyruvate oxidase. It contains two prosthetic groups, a thiazolium moiety that activates the substrate and a flavin group that accepts a hydride equivalent from the reaction intermediate. The catalytic cycle involves the reaction of the thiazolium ion with an aldehyde, followed by a formal hydride transfer and methanolysis, yielding the oxidized substrate in the form of the methyl ester [41]. The cycle is completed by electrochemical oxidation of the reduced flavin moiety. In this way, naphthalene-2-

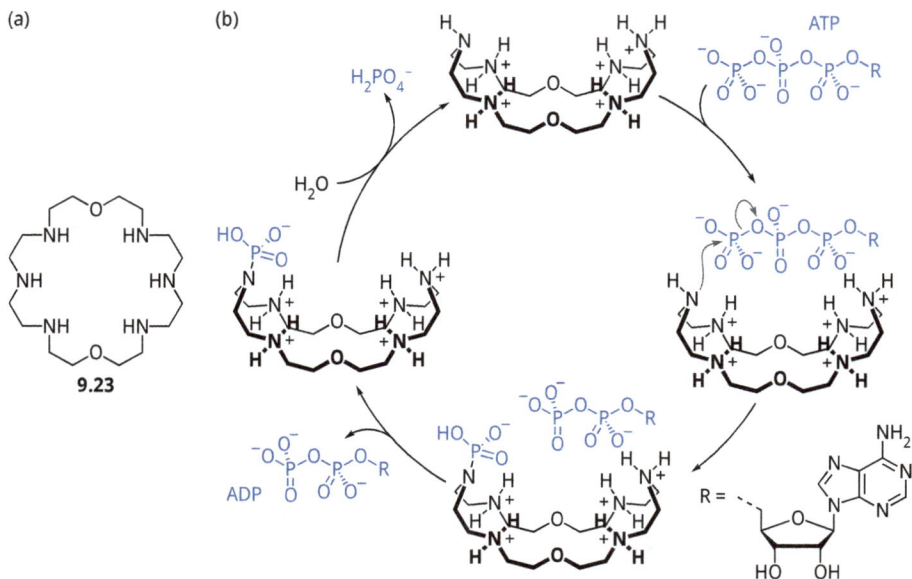

Figure 9.18: Molecular structure of polyazamacrocycle **9.23** (a) and catalytic cycle explaining the ability of **9.23** to mediate the dephosphorylation of ATP (b).

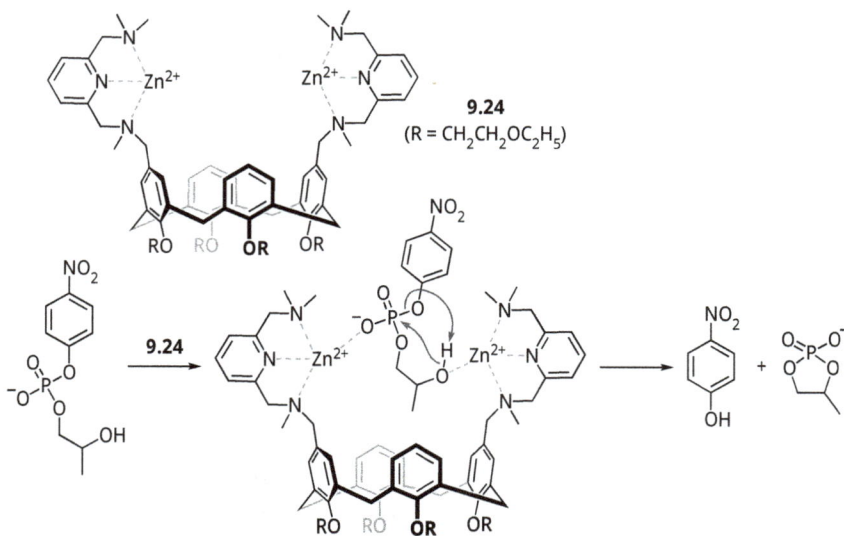

Figure 9.19: Molecular structure of the calix[4]arene dizinc(II) complex **9.24** and the proposed mechanism of intramolecular transesterification of the RNA model 2-hydroxypropyl 4-nitrophenyl phosphate.

carbaldehyde is converted to methyl 2-naphthoate in 16 h with a yield of 78% using 3 mol% of **9.25**.

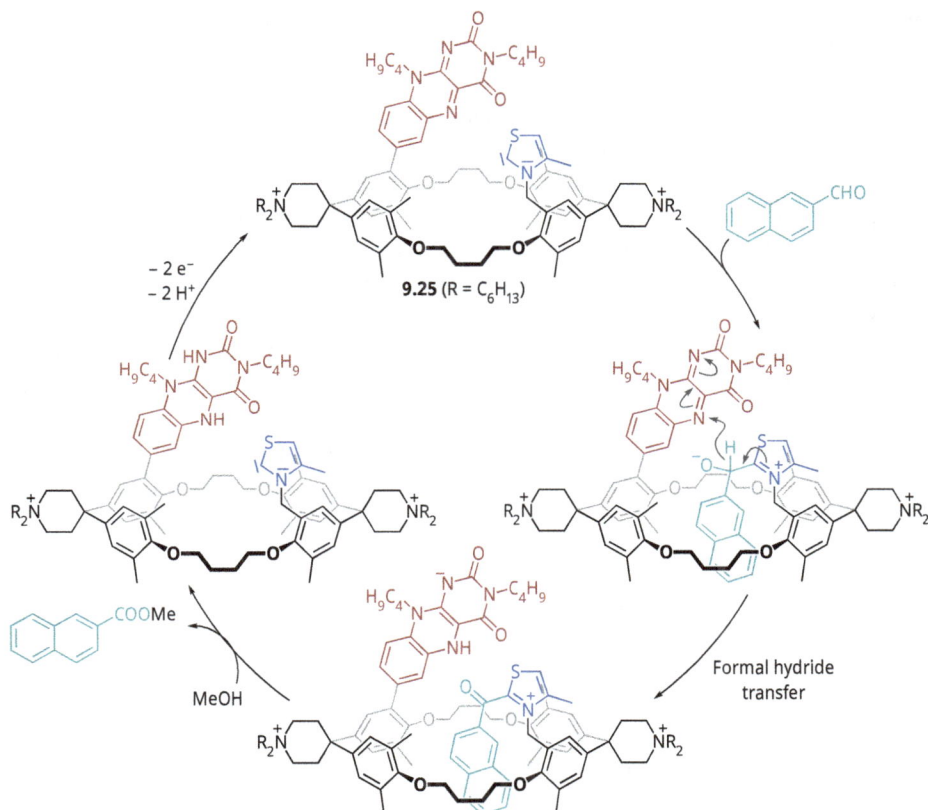

Figure 9.20: Molecular structure of cyclophane **9.25** and catalytic cycle showing the conversion of naphthalene-2-carbaldehyde to methyl 2-naphthoate.

As diverse as the reactions mediated by this selection of supramolecular catalysts may be, the principles underlying their mode of action are not too different from those of enzymes. In all receptors, substrate binding and conversion occurs at or near a binding site that is structurally defined by the receptor backbone [42]. Functional groups along this binding site are involved in substrate conversion, and the positioning of the substrate within the cavity and/or relative to the prosthetic groups affects the course of the reaction. Because of the receptor subunits, there is no doubt that the above catalysts deserve the label supramolecular, but are the ways in which selective transformations are mediated by catalysts that are not typically associated with supramolecular chemistry really fundamentally different?

As an example, consider the 1,1'-bi-2-naphthol (BINOL)-derived phosphate **9.26** (Figure 9.21a), which belongs to an important class of catalysts in the field of asymmetric organocatalysis [43]. One of the many reactions efficiently and enantioselectively mediated by such BINOL derivatives is the Mannich reaction between aldimines and ketene silyl acetals to form β-amino esters [44]. The reaction is mainly promoted by the phosphate group, which serves a dual purpose by acting as a Brønsted acid to activate the imine, while also providing a Lewis basic site for further interactions with the substrate. Since the hydroxy group in substrate is critical for achieving high enantioselectivity, it is likely that substrate activation involves the simultaneous interaction of the phosphate group with the imine nitrogen atom and the OH group (Figure 9.21b). These interactions take place within a sterically shielded cleft that controls the mutual arrangement of the reaction partners, thereby transferring the chirality of the catalyst to the product. The rate enhancement and chiral induction thus benefit from several types of noncovalent interactions, not only from salt bridges between the phosphate group and the iminium ion but also from aromatic and dispersion interactions between the substituents of the catalyst and the reaction partners sandwiched between them. Supramolecular principles therefore contribute to the performance of **9.26**.

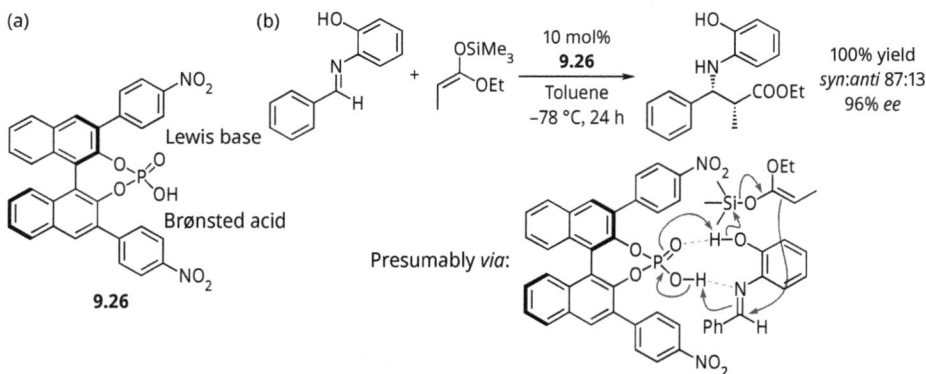

Figure 9.21: Molecular structure of BINOL-derived phosphate **9.26** (a) and Mannich reaction mediated by this organocatalyst (b).

The situation is similar for other organocatalysts, many of which have characteristic concave surfaces, clefts, or cavities. Examples are the phase-transfer catalysts **9.27** and **9.28** (Figure 9.22a, b), which form ion pairs with their substrates in nonpolar media, thereby inducing enantioselective transformations [45].

Even more obvious is the relationship of crown ether **9.29** to supramolecular chemistry. This catalyst binds to the cation of potassium enolates so that the addition to a Michael acceptor occurs in the confined environment of an ion pair (Figure 9.22c). An even larger class of organocatalysts have anion recognition sites [45]. These catalysts tightly interact with the anionic species released from the substrate or reagent, thereby controlling the course of the reaction. An example is the thiourea-derived catalyst **9.30** (Figure 9.22d), which promotes enantioselective acyl-Mannich reactions. Less conventional types of interactions have also found use in anion-binding catalysts, such as C–H hydrogen bonding in the helical tetrakis(triazole) **9.31** (Figure 9.22e) [46] and halogen bonding in the bis(5-iodobenzimidazolium) derivative **9.32** (Figure 9.22f) [47].

Catalyst **9.33** in Figure 9.22g is a chiral variant of the Kemp's triacid derivatives introduced in Section 4.1.12 [48]. When bound to this catalyst by hydrogen bonding, the substrate is oriented above the plane of the xanthone moiety in an arrangement defined by the position of the carbonyl group. The xanthone moiety serves as a photosensitizer that, upon excitation, undergoes intersystem crossing and subsequently transfers its triplet energy to the substrate. This energy transfer facilitates the intramolecular [2 + 2]cycloaddition, which yields two regioisomeric products due to the different orientations of the reacting double bonds in the transition state of the reaction. Both products are formed enantioselectively since the blocking of one side of the substrate by the xanthone moiety directs the intramolecular reaction to the opposite side. Finally, catenanes are also used as organocatalysts. For example, the bis(phosphate) **9.34** efficiently mediates the enantioselective transfer hydrogenation of 2-arylquinolines (Figure 9.23) [49].

The role of supramolecular chemistry in catalysis therefore goes beyond the development of enzyme mimics. Supramolecular aspects also contribute to the activity of many catalysts used in synthetic chemistry, and a deep understanding of the relevant noncovalent interactions could help understand their behavior or improve their performance [42]. However, since selectivity is a kinetic phenomenon that depends on the precise structure of the transition state, it is not trivial to understand or even manipulate the interactions that contribute to catalyst performance. For example, although binding studies provide insight into the affinity of a receptor for a particular substrate, the results are not necessarily transferable to the behavior of a catalyst whose efficiency depends on the stabilization of the transition state rather than the ground state. It is understandable, then, that catalyst development often involves the systematic fine-tuning of reaction conditions and catalyst structure. Nevertheless, it can be useful to take inspiration from supramolecular chemistry when designing a catalyst.

Supramolecular chemistry also plays a role in catalysts whose reactivity is controlled by external stimuli. An instructive example is the nanoswitch **9.35** (Figure 9.24a), developed in the group of Michael Schmittel, which allows the alternate activation and deactivation of two different catalytic processes [50]. One reaction is the 1,3-dipolar cycloaddition between a phenylalkyne and benzylazide, catalyzed by the copper(I) complex of the phenanthroline derivative **9.36** (Figure 9.24b), and the other is the piperidine-catalyzed Knoevenagel reaction between 4-nitrobenzaldehyde and diethyl malonate (Figure 9.24c).

When the four substrates of these reactions and **9.35** are mixed together with one equivalent of **9.36**, one equivalent of a copper(I) source, and one equivalent of piperidine, self-assembly selectively leads to State I (Figure 9.24d). In this state, piperidine

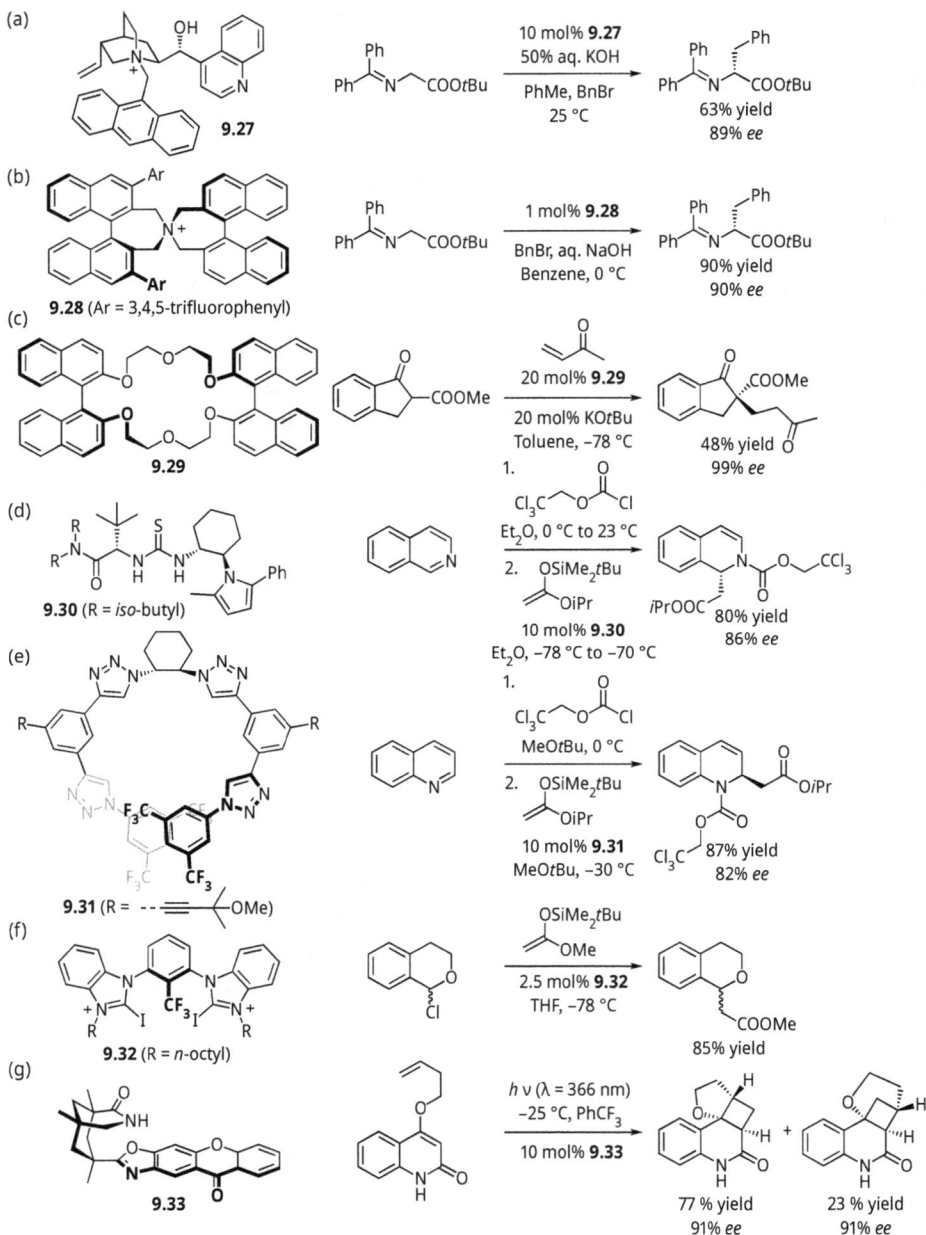

Figure 9.22: Examples of organocatalysts with concave surfaces, clefts or cavities, and/or binding sites for cations or anions, along with examples of reactions mediated by them.

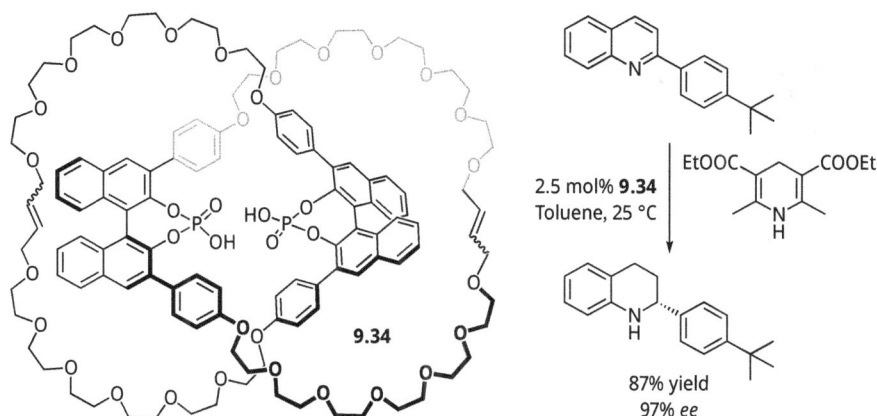

Figure 9.23: Molecular structure of the catenated BINOL-derived bis(phosphate) **9.34** and the transfer hydrogenation reaction mediated by this organocatalyst.

coordinates to the zinc ion in the porphyrin moiety and the additional equivalent of copper ions coordinates to **9.36**. The binding of piperidine shuts off the Knoevenagel reaction, while the phenanthroline-copper(I) complex is catalytically active and mediates the formation of the triazole. This reaction is stopped by adding another equivalent of **9.36**, which leads to the formation of the catalytically inactive $[Cu(9.36)_2]^+$ complex (State 0). A third equivalent of **9.36** displaces the arm from the copper complex in **9.35**, causing it to rotate and bind to the zinc ion (State II). The associated release of piperidine activates the Knoevenagel reaction. Returning from State II to State I requires the addition of one equivalent of a copper(I) salt, which releases **9.36** from the complex with the nanoswitch and restores the original structure of **9.35**. The entire process thus produces one equivalent of the $[Cu(9.36)_2]^+$ complex as waste.

This sophisticated system demonstrates that the design of supramolecular catalysts is not limited to achieving high reaction rates. Supramolecular chemistry also offers several strategies for designing catalysts whose activity can be controlled. Another one is the use of interlocked molecules, such as the rotaxane **9.37** developed by Leigh (Figure 9.25) [51]. The amino group in the axle promotes the Michael addition of an aliphatic thiol to (*E*)-cinnamaldehyde by iminium organocatalysis, but only if it is not blocked by the ring. This is the case when the amine is not protonated and the ring prefers to bind to the triazolium station. Once the amino group is protonated, it is a better binder for the crown ether, which consequently moves to the center of the axle, blocking the catalytically active group. Accordingly, changing the protonation state of the amine not only induces the movement of the ring but also turns the Michael addition off and on. Importantly, the protonated state of the axle alone still has a weak catalytic effect, clearly showing that the ring is required to completely shut down the reaction.

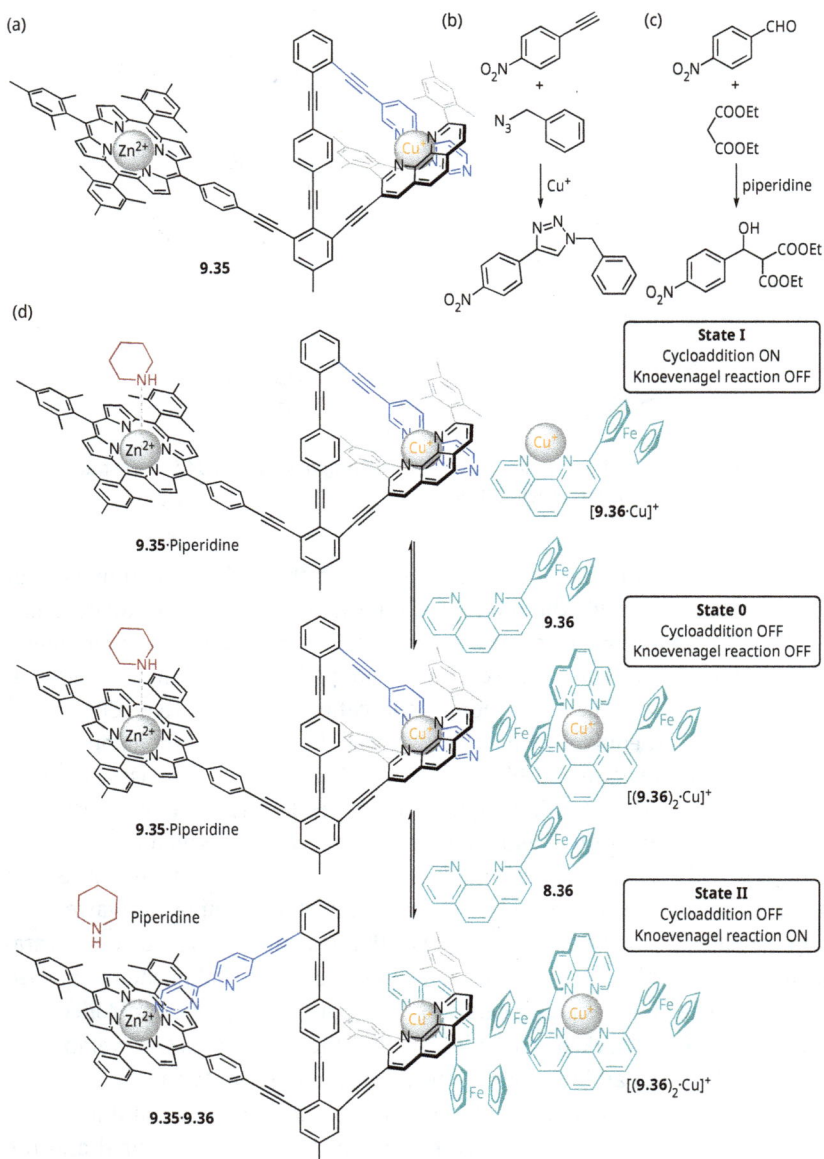

Figure 9.24: Molecular structure of nanoswitch **9.35** (a), 1,3-dipolar cycloaddition (b) and Knoevenagel (c) reactions activated and deactivated by this nanoswitch, and different states of the switch (d). State I mediates only the cycloaddition, State 0 is inactive, and State II mediates the Knoevenagel reaction. The addition of one equivalent of phenanthroline **9.36** allows switching State I into State 0, and another equivalent of **9.36** produces State II.

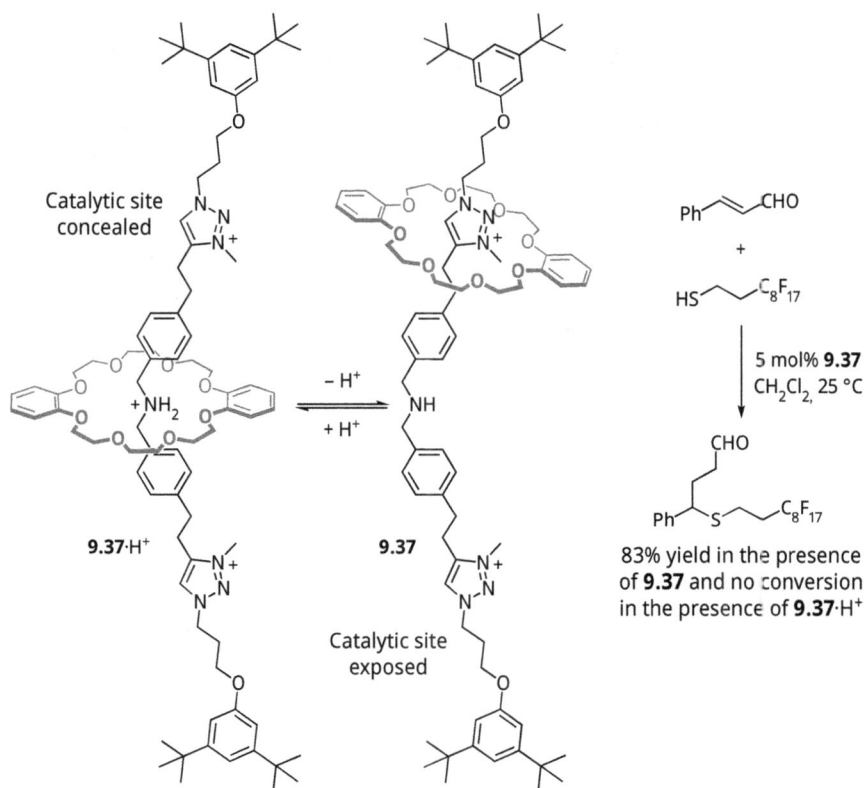

Catalytic site
concealed

– H⁺
+ H⁺

9.37·H⁺

9.37

Catalytic site
exposed

Ph ⟋⟍ CHO

+

HS ⟋⟍ C₈F₁₇

5 mol% **9.37**
CH₂Cl₂, 25 °C

CHO

Ph ⟋ S ⟋⟍ C₈F₁₇

83% yield in the presence
of **9.37** and no conversion
in the presence of **9.37·H⁺**

Figure 9.25: Protonation-/deprotonation-induced switching of the rotaxane **9.37**. The state with the ring bound to one of the triazolium stations allows the central amino group to promote the Michael addition of an aliphatic thiol to (*E*)-cinnamaldehyde. The rotaxane with the ring bound to the protonated amino group is catalytically inactive.

Ben L. Feringa showed that switchable molecules can be used to control the enantio-selectivity of a transformation. The corresponding catalyst **9.38** (Figure 9.26) is based on the sterically congested alkenes that are also the basis of Feringa's molecular mo-tors [52]. Compound **9.38** contains anion-binding motifs in the form of bis(triazole) units whose affinity for chloride ions allows them to activate 1-chlorochromane. The resulting ion pair reacts with a silyl ketene, with the outcome of the reaction depend-ing on the configuration of the double bond and the helicity of the flanking substitu-ents in **9.38** (Figure 9.26). In the *E*-isomer, the two side chains of **9.38** diverge and only bind individually to chloride anions. In this state, the catalyst is unable to transfer its chirality to the product, which is formed as a racemate. Upon irradiation and transi-tion to the *Z* state, the side chains of **9.38** converge, allowing them to bind the anion simultaneously. The resulting more compact complex, in which the helices have the *M* configuration, mediates the formation of the *R*-enantiomer of the product (with 48% enantiomeric excess, *ee*). The helix configuration is thermally reversed so that the catalyst now preferentially produces the *S*-enantiomer (with 80% *ee*). After irradiation and switching back to the *E*-form, the reaction becomes nonselective again. Thus, the unidirectionality of the switching process allows the transition from a nonselective catalyst to an enantioselective one, producing first the *R*-enantiomer and then the *S*-enantiomer of the product. This example illustrates the high degree of control over a reaction that can be achieved with a carefully designed catalyst.

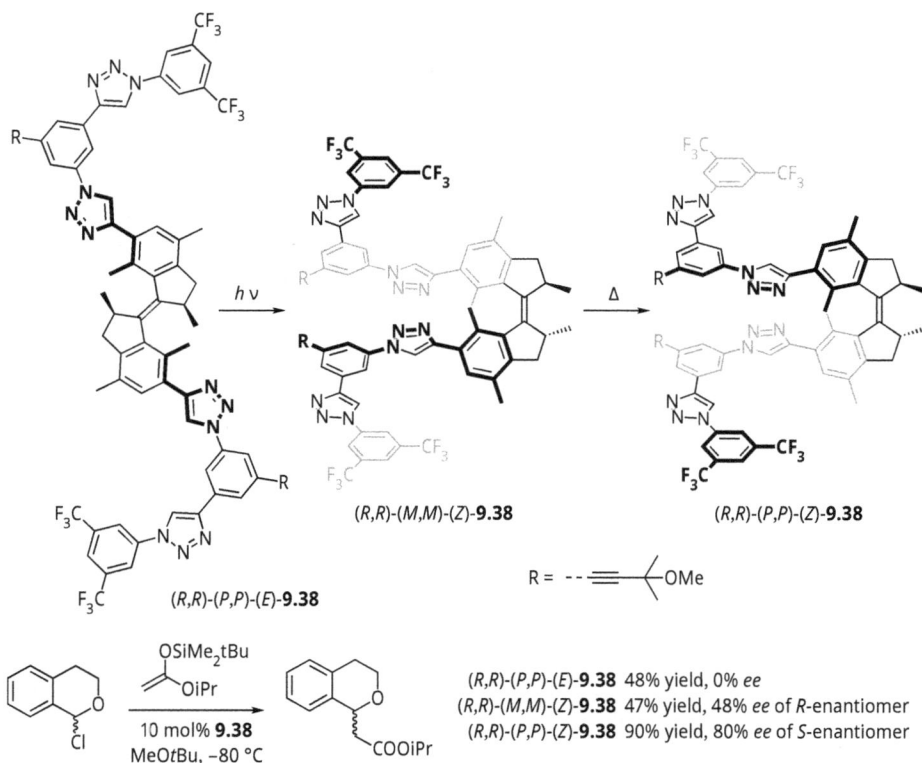

Figure 9.26: Structures of the three interconvertible states of organocatalyst **9.38** and effects of these structures on the enantioselectivity of the reaction between a silyl ketene acetal and 1-chlorochroman. (*R,R*)-(*P,P*)-(*E*)-**9.38** is nonselective, (*R,R*)-(*M,M*)-(*Z*)-**9.38** favors the formation of the *R*-enantiomer of the product, while (*R,R*)-(*P,P*)-(*Z*)-**9.38** leads to the preferential formation of the *S*-enantiomer.

9.3.2 Transformation by confinement

We now come to supramolecular catalysts based on coordination cages or capsules. These catalysts lack functional groups that participate in substrate conversion but owe their activity to the presence of a shielded cavity. They typically bind cationic substrates and promote reactions that proceed *via* positively charged transition states, often with characteristic selectivity. Some of the most efficient supramolecular catalysts known today belong to this category. One example is the 12-fold negatively charged coordination cage **9.39** (Figure 9.27a) developed by Kenneth N. Raymond.

We have seen in Section 5.5.5 that this cage has a high affinity for ammonium ions and related positively charged guests in water. Protonated amines, in particular, are so effectively stabilized that they persist in the cavity even when the external solution has a pH that would normally lead to deprotonation. Complex formation thus promotes the protonation of basic substrates, allowing **9.39** to catalyze reactions under conditions that would normally require acidic media [53]. An example is the acid-catalyzed hydrolysis of orthoformates (Figure 9.27b) [54].

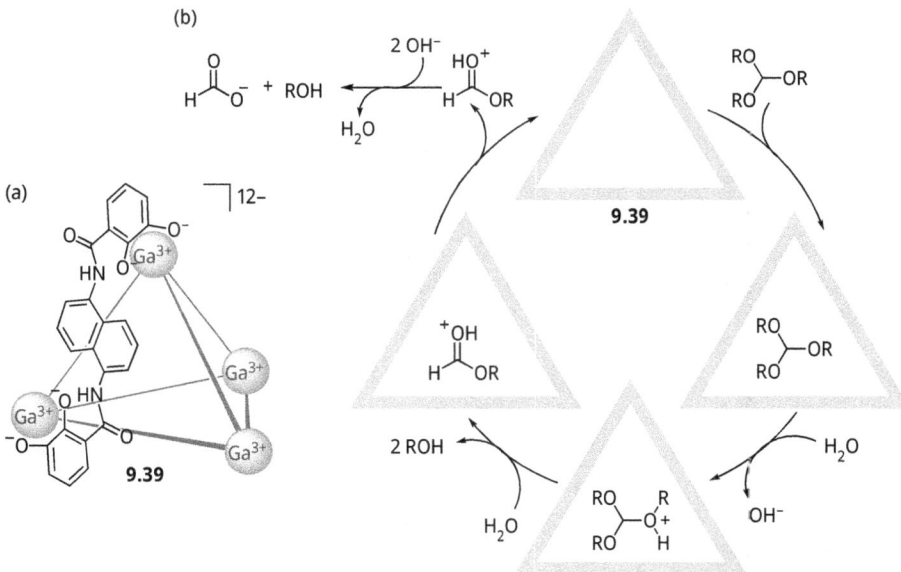

Figure 9.27: Structure of coordination cage **9.39** (a) and mechanism of orthoformate hydrolysis mediated by **9.39** (b).

In this reaction, the substrate is activated by the protonation of one of its oxygen atoms, which facilitates the loss of the first alcohol group. A water molecule then leads to the replacement of the second alcohol residue to yield the ester. Due to the efficient stabilization of the cationic intermediates by the coordination cage, the same reaction takes place in the cavity of **9.39** even when the external solution has a pH of 11, at which orthoformates are stable and uncatalyzed hydrolysis does not occur. Thus, there is no background reaction, and the product is rapidly hydrolyzed and removed from equilibrium as it leaves the cage, ensuring turnover. Catalysis is particularly efficient for substrates that fit tightly into the cavity. In addition, competitive binding of quaternary ammonium ions to **9.39** inhibits the reaction, demonstrating that it occurs in the cavity.

Other reactions mediated by **9.39** include the Nazarov reaction of penta-1,4-dien-3-ols (Figure 9.28a) [55] and the 3-aza-Cope rearrangement of allyl enammonium cations (Figure 9.28b) [56]. Both transformations benefit from the stabilization of the cationic intermediates. In addition, the substrates adopt well-organized conformations inside the cavity with converging end groups that resemble the cyclic transition states of the reactions. Complex formation thus leads to rate enhancements approaching those observed for enzymes. For example, the Nazarov reaction is over a million times faster than the uncatalyzed reaction. In the case of the 3-aza-Cope reaction, the product is rapidly hydrolyzed to an aldehyde as soon as it leaves the cavity. Therefore, product inhibition does not occur, but the additional presence of quaternary ammonium ions, which prevent the initial product from ion-pairing with the cage, proves beneficial. Product inhibition in the Nazarov reaction is prevented by the addition of maleimide, which captures the product as a Diels–Alder adduct and prevents it from reentering the cavity.

An example of an acid-catalyzed reaction that yields different stereoisomers in the absence and presence of **9.39** is the solvolysis of 1-phenylethyl 2,2,2-trichloroacetimidate in methanol to the corresponding methyl ether [57]. This nucleophilic substitution generally requires strong acids to activate the leaving group. The rear attack of methanol as the incoming nucleophile then preferentially affords the product with the inverted stereocenter (Figure 9.29). When the same reaction is performed in the presence of 2.5 mol% of **9.39**, it proceeds rapidly even at neutral pH. In addition, the absolute configuration of the substrate is maintained during the substitution, regardless of whether the reaction is performed in the presence of racemic or enantiomerically pure cages. Any potential chiral induction of the cage is thus overcompensated by the inherent effect of substrate confinement. This confinement presumably orients a naphthalene sidewall of the cage close to one side of the substrate. As a result, the positive charge generated upon cleavage of the leaving group is stabilized by cation–π interactions. In addition, the backside attack of the incoming nucleophile is prevented, so that the exit of the leaving group and the entry of the nucleophile must occur from the same side. The high degree of retention indicates that the intermediate does not have time to rotate between the two steps.

The resorcin[4]arene-derived hexameric capsules discussed in Section 5.3.3 also exhibit catalytic activity (Figure 9.30a) [58]. Like the coordination cage **9.39**, they contain electron-rich aromatic subunits along their inner surface that interact with and stabilize positively charged guest molecules. In addition, the cavity is surrounded by

(a)   (b)

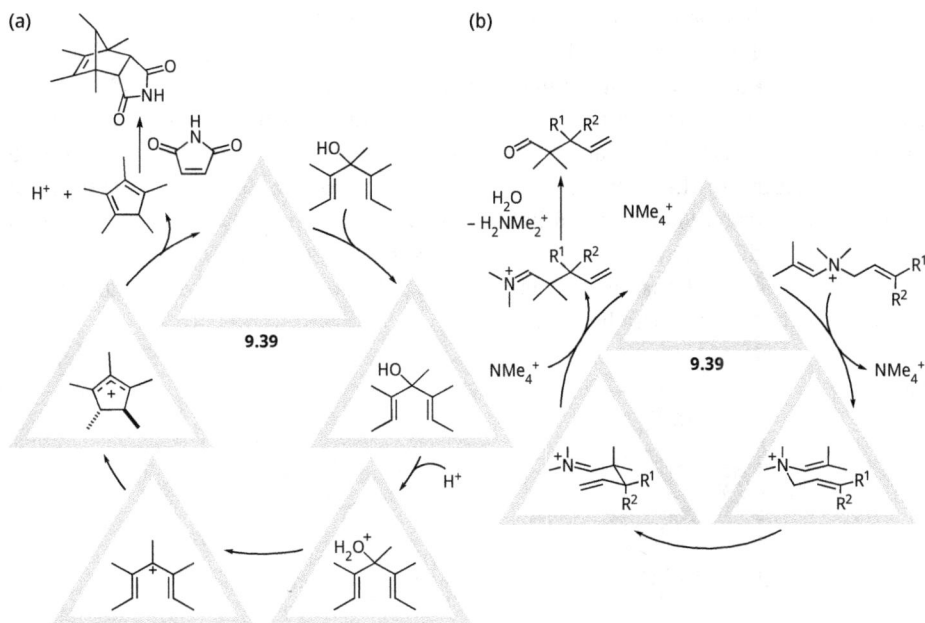

Figure 9.28: Catalytic cycles of the Nazarov reaction (a) and the 3-aza-Cope rearrangement (b) mediated by **9.39**.

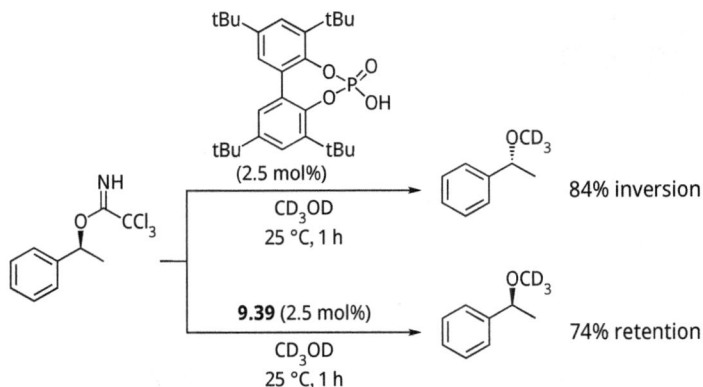

Figure 9.29: Stereoselectivity of the solvolysis of 1-phenylethyl 2,2,2-trichloroacetimidate in methanol mediated by an achiral Brønsted acid or by cage **9.39**.

weakly acidic phenolic OH groups, which potentially serve as Brønsted acids to protonate substrate molecules and thus initiate their reactions. Although these capsules have no permanent openings through which substrates can pass, their dynamic nature allows rapid guest exchange.

Konrad Tiefenbacher showed that these capsules act as catalysts in a variety of reactions, one of which is the hydrolysis of acetals (Figure 9.30b) [59]. For example, an 85% conversion to the corresponding aldehyde is observed within 1 h at 25 °C when 1,1-diethoxyethane is treated in water-saturated CDCl$_3$ with 10 mol% of **9.40**. Mechanistic studies indicate that trace amounts of HCl, which are almost unavoidable in CHCl$_3$, contribute to the conversion. The reaction is suppressed by the presence of salts such as tetrabutylammonium halides, whose cations occupy the cavity. When a mixture of 1,1-diethoxyethane and 1,1-diethoxydodecane is treated with **9.40**, the smaller substrate is almost exclusively hydrolyzed, showing that the reaction is limited to the substrate that can be bound.

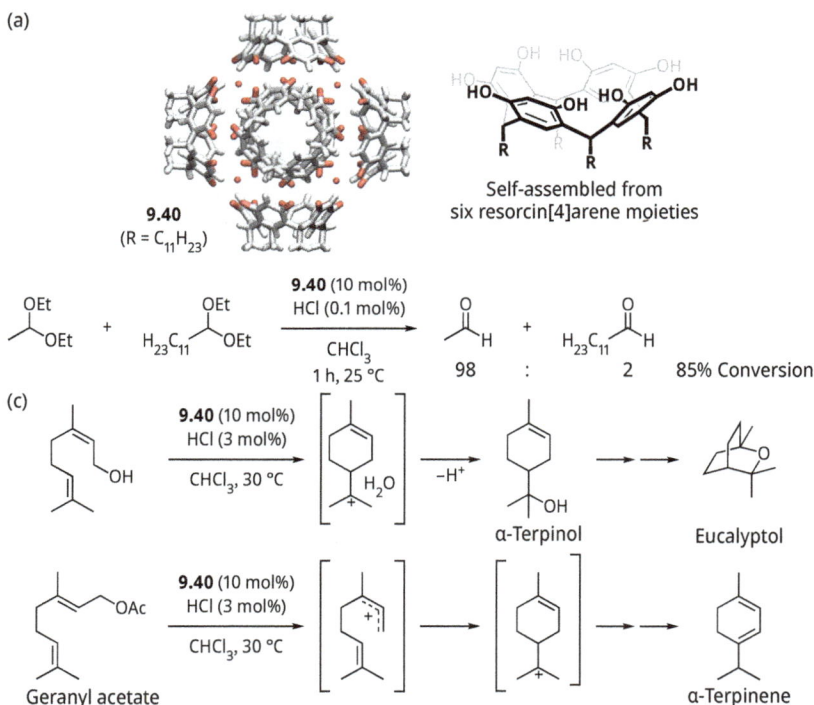

Figure 9.30: Structure of the hexameric capsule **9.40**, assembled from the corresponding resorcin[4] arene derivative (a), and acetal hydrolysis (b) and terpene cyclization (c) mediated by **9.40**. The reaction scheme in (b) shows that when using a mixture of acetals, only the acetal that fits into the cavity of **9.40** is consumed.

More impressive transformations catalyzed by **9.40** are the cyclizations of terpenes. The structural diversity of this class of natural products ultimately derives from the many ways in which a few acyclic precursors can be cyclized. In nature, these transformations take place within the active sites of enzymes that mediate selective product formation. Achieving the same selectivity with a conventional synthetic catalyst is challenging, especially when it requires an appropriate preorganization of the substrate. Supramolecular catalysts, on the other hand, are able to achieve this preorganization, as we have seen for the Nazarov reaction and the 3-aza-Cope rearrangement. In the context of terpene cyclizations, **9.40** catalyzes the selective formation of eucalyptol from nerol and α-terpinene from geranyl acetate (Figure 9.30c) [60]. Both reactions involve the initial acid-mediated cleavage of a leaving group to yield cations that selectively cyclize tail-to-head and then undergo further reactions. Because the products are less polar than the starting materials, they are also less efficiently bound, ensuring turnover. The same reactions performed in the absence of **9.40** yield only traces of complex mixtures of cyclic terpenes, demonstrating the pronounced confinement effect. This work demonstrates that supramolecular catalysts need not be elaborately designed and structurally complex to be useful. Enzyme mimics can have surprisingly simple structures or can be assembled noncovalently from suitable building blocks and still mediate transformations for which molecular chemistry has not yet provided effective solutions.

9.4 Self-replication

Living systems must be able to reproduce and accurately pass on their genetic information from one generation to the next. This information transfer is accomplished at the molecular level by DNA replication, where one strand of DNA serves as a template for the *de novo* synthesis of the complementary strand. The next cycle of replication then produces the exact copy of the original strand. Although DNA replication in living systems requires a complex network of different proteins, at its core it involves many concepts of supramolecular chemistry, such as molecular recognition, template effects, and supramolecular reaction control. This raises the question of whether small molecules can be designed to act in a similar way.

Is only DNA able to replicate?

Mimicking the principles of DNA replication is actually not that difficult. All that is needed is a molecule that dimerizes noncovalently and contains two or more covalently linked subunits [61]. Figure 9.31a shows schematically how such a minimal replicator works. This replicator R contains two complementary subunits A and B connected by a linker that is rigid enough to prevent the two subunits from interacting intramolecularly. R thus binds the individual but not yet connected components A and B, leading to the formation of the ternary complex $R \cdot A \cdot B$. In this complex, A and B are positioned close enough to react and form another molecule of R. The replicator R thus acts as a (kinetic) template in the reaction between A and B. Since the number of R doubles with each round of reaction, the concentration of the product is expected to increase exponentially with time until all the starting molecules are consumed. The overall process should therefore exhibit the typical profile of an autocatalytic reaction, characterized by a slow initial phase during which replicator molecules are formed by the uncatalyzed and therefore slow reaction between A and B. As the concentration of R increases, the catalyzed pathway gradually takes over, leading to an increase in the reaction rate, until the production of R levels off due to the consumption of the starting materials (Figure 9.31b).

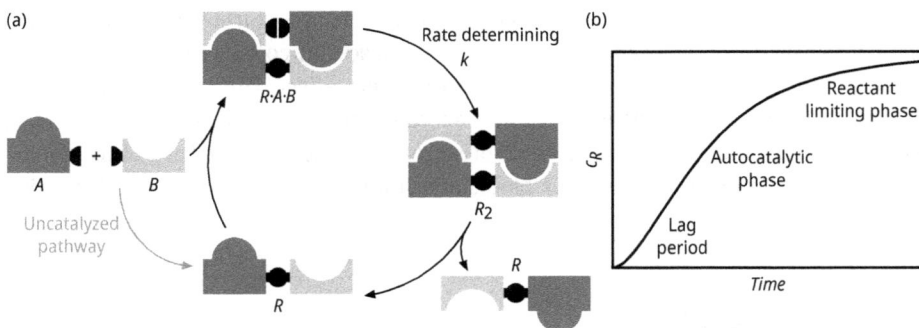

Figure 9.31: Catalytic cycle illustrating the formation of the replicator R from the two complementary subunits A and B (a) and the increase of the concentration of R with time, showing the typical sigmoidal profile of an autocatalytic process (b).

Although conceptually related to DNA replication, there is a fundamental difference between the replication of DNA and that of R. In the case of DNA, the exact copy of the original strand is obtained only after two rounds of reaction. In contrast, the self-complementarity of R leads directly to another copy of the same molecule, hence the term self-replication.

A critical factor in the catalytic cycle shown in Figure 9.31a that prevents many synthetic replicators from exhibiting the expected kinetic profile of an autocatalytic reaction is product inhibition. Once formed, the new replicator tends to remain in the dimer R_2, which for entropic reasons usually has a higher thermodynamic stability than the ternary complex $R \cdot A \cdot B$. As a consequence, R only mediates the formation of a single copy, so that the number of replicator molecules does not increase exponentially. There are still ways to experimentally assess whether product formation proceeds under the influence of the template. One possibility is to add small amounts of R to the mixture of the individual components. If the rate of product formation increases in response to this addition, the replicator is promoting the reaction. In quantitative terms, the reaction rate immediately after template addition should be proportional to the square root of the template concentration. This square root dependence, first observed experimentally by Günter von Kiedrowski, one of the pioneers in the development of self-replicating systems, can be derived from the catalytic cycle in Figure 9.31a [62]. It indicates that the rate-determining step is the reaction between A and B in the ternary complex $R \cdot A \cdot B$, yielding the dimer R_2. The corresponding rate equation is given by the following equation:

$$\frac{dc_{R_2}}{dt} = k \, c_{R \cdot A \cdot B} \tag{9.2}$$

The concentrations $c_{R \cdot A \cdot B}$ and c_{R_2} correlate with the stability constants $K_{a(R \cdot A \cdot B)}$ and $K_{a(R_2)}$, which are given by the laws of mass action (9.3) and (9.4). Of these equations, (9.4) describes the degree of product inhibition:

$$K_{a(R \cdot A \cdot B)} = \frac{c_{R \cdot A \cdot B}}{c_R \, c_A \, c_B} \tag{9.3}$$

$$K_{a(R_2)} = \frac{c_{R_2}}{c_R^2} \tag{9.4}$$

Combining equations (9.3) and (9.4) with equation (9.2) leads to the square root law (9.5). Assuming that the concentrations of reactants A and B immediately after the addition of R are approximately equal to the initial concentrations c_A^0 and c_B^0, and that c_{R_2} equals $c_R^0/2$, this equation shows that the rate at which R_2 is formed is proportional to the square root of the concentration of R:

$$\frac{dc_{R_2}}{dt} = k \, K_{a(R \cdot A \cdot B)} \, K_{a(R_2)}^{-0.5} \, c_A \, c_B \, c_{R_2}^{0.5} \tag{9.5}$$

Equation (9.5) thus reflects the kinetics of a self-replication process suffering from product inhibition, and it allows to verify experimentally whether a replicator acts as a template even if it is not catalytically active.

Taking inspiration from nature, von Kiedrowski designed the terminal-protected hexadeoxynucleotide **9.41** as the first synthetic replicator. This nucleotide has the palindromic base sequence 5′-CCGCGG-3′, meaning that each base finds a complementary partner when it dimerizes to form an antiparallel double strand (Figure 9.32) [62]. In addition, **9.41** contains only cytosine and guanine subunits to take advantage of the higher stability of the corresponding base pairs compared to those formed from adenine and thymine, which are stabilized by only two instead of three hydrogen bonds. Replicator **9.41** does indeed template the carbodiimide-mediated coupling of the two trinucleotide fragments **9.42** and **9.43**, although the reaction exhibits product inhibition and is therefore not autocatalytic. However, other nucleotide-derived replicators subsequently developed in the von Kiedrowski group showed the sigmoidal increase in replicator concentration expected for an autocatalytic system [61, 63].

Abiotic replicators were developed in the group of Julius Rebek Jr. [64]. These systems, an example of which is **9.44** (Figure 9.33), are based on the Kemp's triacid-based adenine receptors introduced in Section 4.1.12. In **9.44**, the imide group in the Kemp's triacid moiety serves to bind adenine *via* either a Watson–Crick or a Hoogsteen base

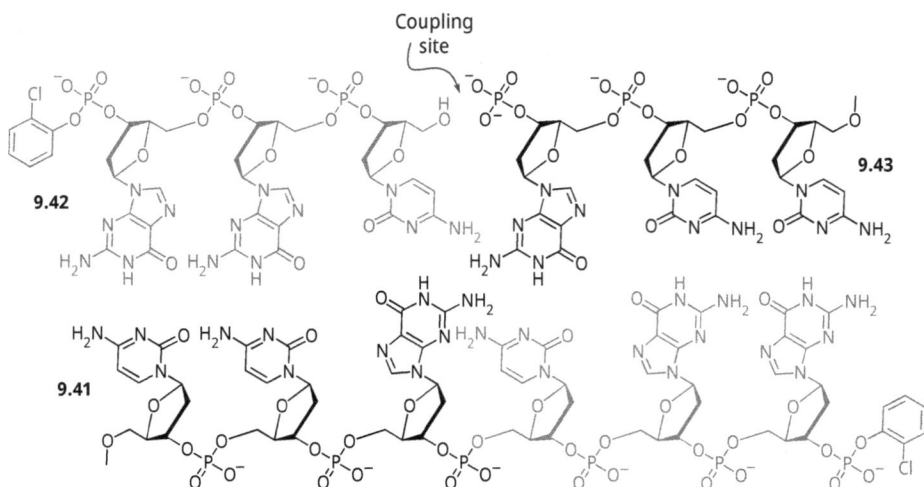

Figure 9.32: Hexanucleotide replicator **9.41**, which promotes the coupling of the two fragments **9.42** and **9.43**, resulting in the formation of another copy of **9.41**.

pairing mode. The other binding site is the adenine residue in the protected adenosine subunit. Thus, the two binding sites in **9.44** are self-complementary, but the molecule is unable to fold into a conformation in which the two subunits interact intramolecularly because this folding would require the central amide unit to adopt an unfavorable *cis* conformation. Replicator **9.44** promotes the coupling of the two components **9.45** and **9.46** by bringing the amino group in **9.45** and the activated ester in **9.46** together [65]. While this replicator also suffers from product inhibition, careful kinetic studies and examination of the behavior of structural analogs demonstrated that a template effect is operative. For example, a derivative of **9.44** with a methylated imide group is unable to promote coupling of the two fragments because base pairing is not possible.

Derivatives of **9.44** have been used in cleverly designed experiments to mimic principles that may have played a role in the origin of life. One such experiments examined the behavior of replicators in which the amino group of the adenine moiety was protected with either a benzyloxycarbonyl group or a 2-nitrophenylmethyloxycarbonyl group (Figure 9.34a) [66].

Both compounds **9.47a** and **9.47b** act as replicators, but they are inferior to the unprotected analog **9.47c** because the protecting groups induce a conformation of the adenine amino group that prevents the replicators from Watson–Crick base pairing (Section 4.1.12). In addition, the replicators are nonselective since they cannot promote only their own formation but also that of the other replicator. An important difference between the two receptors is that the 2-nitrophenylmethyloxycarbonyl group is photochemically cleavable, whereas the benzyloxycarbonyl group is not.

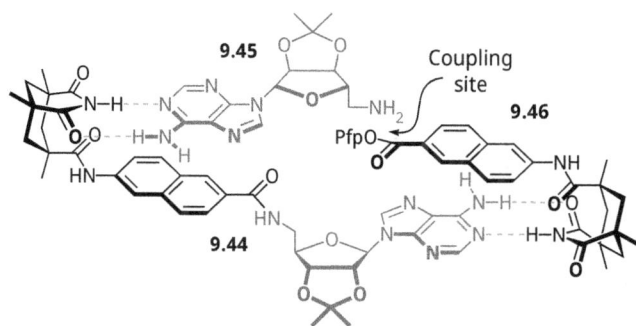

Figure 9.33: Rebek's replicator **9.44**, which mediates the coupling between **9.45** and **9.46** to form another replicator molecule (Pfp = pentafluorophenyl).

Thus, irradiation selectively converts replicator **9.47b** to the better replicator **9.47c**, which takes over control of product formation once it is formed. Consequently, a mutation in one of the replicators produces a derivative that is more fit than the others and thus suppresses self-replication of the inferior compounds.

In another experiment, the four building blocks **9.45**, **9.48**, **9.49**, and **9.50** were mixed. Of these, **9.45** and **9.48** are the building blocks of **9.47c**, and **9.49** and **9.50** are those of the independently designed replicator **9.51** (Figure 9.34b) [67]. All four building blocks combine to produce the two known replicators, as well as crossover species of which **9.52** is not a replicator because its convergent binding sites are not suitable for arranging two precursors in a way that would allow a reaction. Compound **9.53**, however, is an even better replicator than **9.47c** or **9.51**. In this early excursion into systems chemistry, the Rebek group thus demonstrated that new properties can emerge from complex systems. Although these replicators certainly did not play a role in the emergence of life, they demonstrate how useful model compounds can be in addressing fundamental questions.

In a similar vein, Douglas Philp explored the emergence of a replicator in a dynamic combinatorial library. He used *trans*-**9.54** (Figure 9.35) as the replicator, which is derived from nitrone **9.55** and maleimide **9.56**. These building blocks react by 1,3-dipolar cycloaddition to give either *trans*- or *cis*-**9.54**, but since the latter compound prefers a conformation in which the carboxyl and the 2-aminopyridine group interact intramolecularly, only the *trans* cycloadduct is capable of self-replication [68]. Indeed, mixing **9.55** and **9.56** under appropriate conditions results in a significantly higher amount of the *trans*-isomer in the product mixture than when using a maleimide derivative incapable of hydrogen bonding, suggesting that *trans*-**9.54** controls the outcome of the reaction. The concentration of *trans*-**9.54** increases in a sigmoidal fashion as expected for autocatalytic self-replication. The absence of product inhibition is surprising given the high stability of the dimer of *trans*-**9.54** ($\log K_a$ *ca.* 5–6), but is probably due to

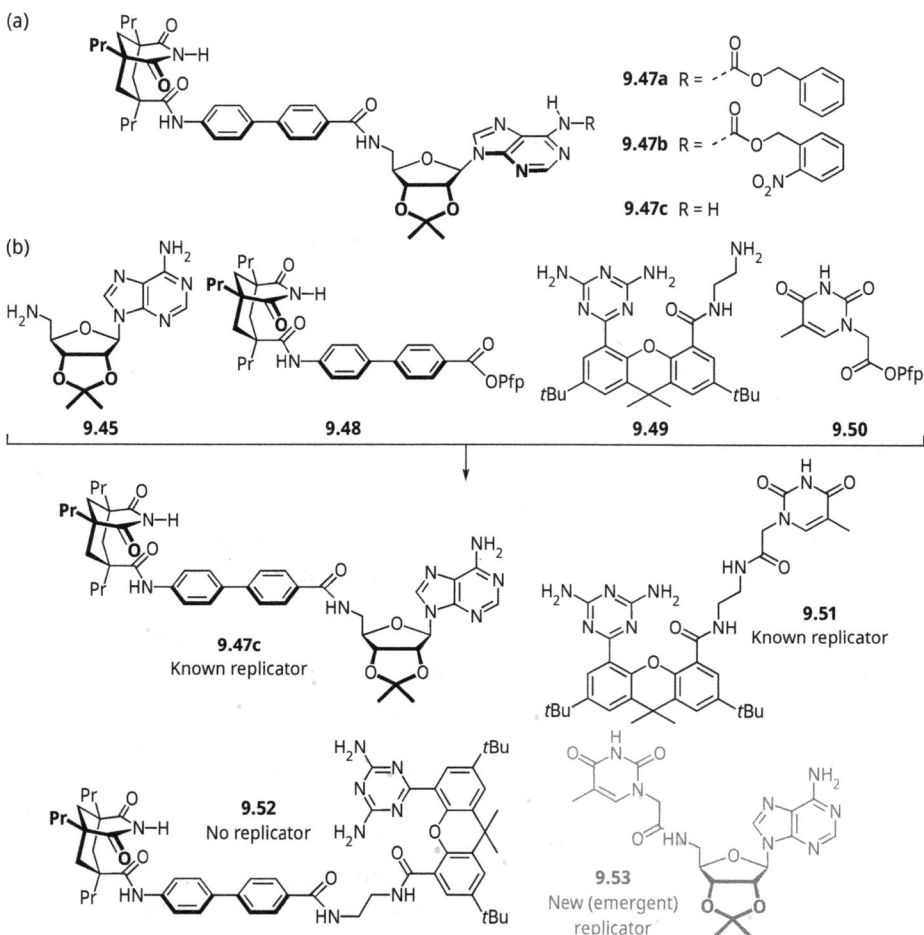

Figure 9.34: Molecular structures of the replicators **9.47a** and **9.47b**, of which the latter is selectively converted into the more efficient replicator **9.47c** upon irradiation (a), and structures of the four building blocks **9.45**, **9.48**, **9.49**, and **9.50**, whose coupling yields the replicators **9.47c** and **9.51** and the crossover products **9.52** and **9.53**, of which **9.52** is inactive while **9.53** is a new replicator of this series (b).

the stronger binding of *trans*-**9.54** to the transition state of the cycloaddition than to the product.

The nitrone **9.55** is formed from the hydroxylamine **9.57** and the aldehyde **9.58**. This reaction makes it possible to start the above self-replication process not from the nitrone itself, but from its two components. In this case, **9.55** is formed in a fast equilibrium and then reacts in a slower intermolecular reaction with **9.56** to give a mixture of *trans*-**9.54** and *cis*-**9.54**. Once *trans*-**9.54** is present in sufficient amounts, it

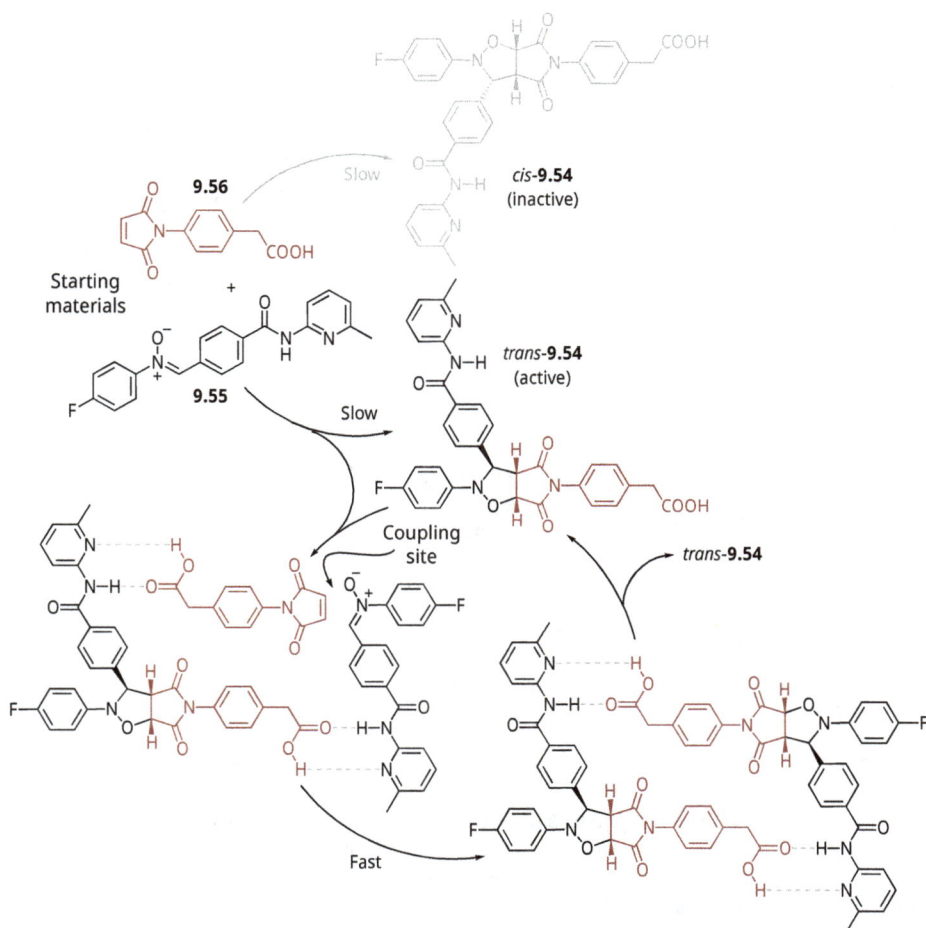

Figure 9.35: Catalytic cycle illustrating the self-replication of *trans*-**9.54**. This compound is formed along with the corresponding *cis*-diastereomer, which is unable to act as a replicator.

takes control of the further reaction by autocatalytically mediating its own formation (Figure 9.36a).

A further level of complexity is added by using a mixture of two aldehydes, an amine, and a hydroxylamine together with maleimide **9.56** (Figure 9.36b) [69]. Four of the starting materials now react to give the four rapidly interconverting products **9.55**, **9.59**, **9.60**, and **9.61**, one of which is nitrone **9.55**. In the absence of maleimide **9.56**, all four products are formed. In the presence of **9.56**, the two nitrones **9.55** and **9.59** react irreversibly to give four products, the *trans*- and *cis*-diastereomers of **9.54** and **9.62**, while the two imines cannot undergo cycloaddition. Of the possible cycloadducts, only *trans*-**9.54** is a replicator. Once enough of this replicator is present, its con-

(a)

Pool of starting materials   Pool of products

(b)

Pool of starting materials   Pool of products

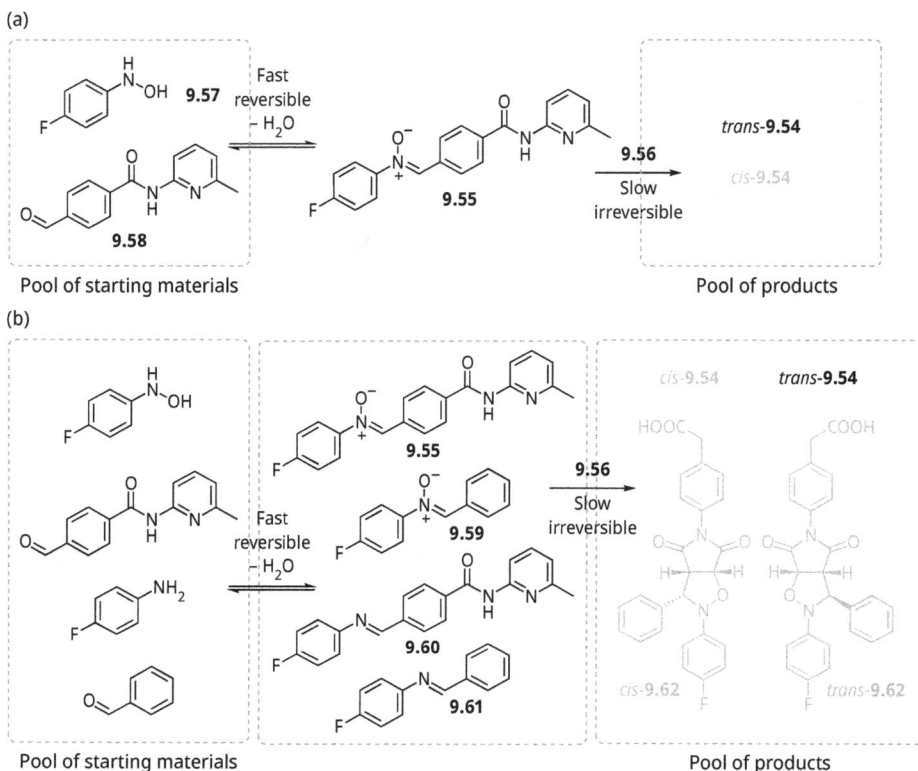

Figure 9.36: Condensation of **9.57** and **9.58** to nitrone **9.55** (a), and a dynamic library generated from two aldehydes, an amine, and a hydroxylamine (a). This library serves as a source for nitrone **9.55** and, in turn, for trans-**9.54**, which initiates the self-replicating process shown in Figure 9.35. Nitrone **9.59** gives a cycloadduct upon reaction with maleimide **9.56**, which cannot self-replicate. The imines cannot undergo cycloaddition. The inactive cycloadducts shown in gray do not strongly affect the self-replication of trans-**9.54** because the template-mediated cycloaddition is much faster than the competing reactions.

centration increases exponentially, simultaneously depleting the nitrone **9.55** in the reaction mixture. The equilibrium by which **9.55** is formed must therefore adapt by producing more of the required nitrone at the expense of the other condensation products.

   This work shows that a simple synthetic replicator promotes its own formation even when the required building blocks must first be generated from simpler components. When these components are part of a larger library of interconverting compounds, self-replication causes a shift in the library composition to ensure a supply of starting material. Thus, mixtures of molecules can exhibit predictable behavior when they do not behave independently [70]. For other examples, recall the emergence of self-replicating molecules from a pool of interconverting macrocycles discussed in Section 5.7.4, and the discussion of systems chemistry in Section 5.8.

Given that a crucial step in the origin of life must have been the emergence of replicating molecules, the above results provide insight into mechanisms that may have been relevant in this context. Although we will never know the exact details of how life evolved, model systems provide important clues. Another fundamental question is whether other biomolecules might have served as replicators before nucleotides were selected during chemical evolution. In fact, nucleotides are unlikely to have been information carriers and replicators in early life because of their structural complexity. It is more likely that the first replicators were based on simpler compounds, such as small catalytically active peptides, which also carry information in their amino acid sequence and are easily formed under prebiotic conditions. Yet, replicating peptides do not exist in nature. Does this mean that there is a fundamental reason why they are unknown, or could it be that they became extinct during chemical evolution? M. Reza Ghadiri addressed this important question by applying the principles of self-replication to synthetic peptides.

Ghadiri used the α-helical peptide **9.63** as a replicator (Figure 9.37a) [71]. In this peptide, polar amino acid chains are arranged on one side of the helix to mediate water solubility and nonpolar residues on the other side. Peptide **9.63** thus tends to self-assemble in aqueous solution to minimize solvent-exposed hydrophobic surfaces, resulting in aggregates in which two or three peptide strands coil around each other with a left-handed superhelical twist. These coiled coils are well known in nature and are found, for example, in the leucine zipper domain of a yeast transcription factor [72]. The presence of the cysteine units approximately in the middle of the 32-mer **9.63** allows the coupling of two fragments of this peptide by native peptide ligation. These fragments are the 17-mer **9.64** with a thioester group in the C-terminal alanine moiety and the corresponding 15-mer **9.65** with a cysteine subunit at the N-terminus, which couple autonomously in aqueous solution by the mechanism shown in Figure 9.37b. Coupling is significantly faster in the presence of **9.63** than in its absence, indicating that the 32-mer templates its own formation. Although product inhibition occurs, preventing autocatalysis, the rate of product formation depends on the square root of the concentration of **9.63**, consistent with a self-replication process. In addition, reagents such as guanidinium hydrochloride, which suppress peptide self-assembly and thus prevent **9.63** from acting as a template, cause the rate enhancement to disappear.

Ghadiri thus demonstrated that self-replicating peptides exist. Further work even showed that the sequence of amino acids in such peptides influences their reactivity [70]. Peptides should therefore not be excluded from the list of biomolecules that could have served as the first primitive replicators in the development of life. This work clearly shows that synthetic replicators are more than just a special class of supramolecular systems that happen to promote their own formation. They also help answer the fundamental question of how we all began.

(a)

Sequences 32-mer: ArCONH–RMKQLEEKVYELLSKVA-CLEYEVARLKKLVGE–CONH$_2$
17-mer: ArCONH–RMKQLEEKVYELLSKVA-COSCH$_2$Ph
15-mer: H$_2$N-CLEYEVARLKKLVGE–CONH$_2$

Color code ● Hydrophobic side chain
            ● Side chain mediating electrostatic interactions
            ● Side chain exposed to the solvent

(b)

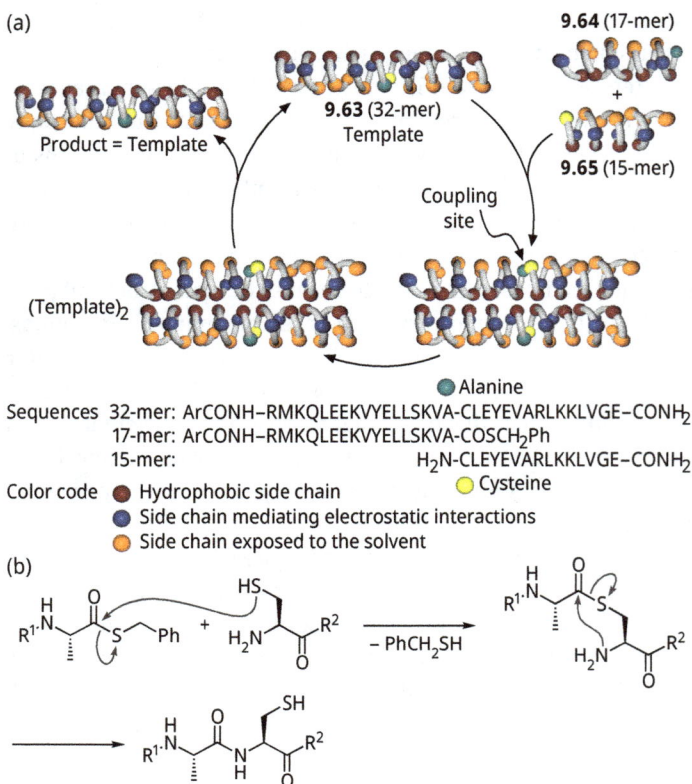

Figure 9.37: Schematic representation of the self-replication of peptide **9.63** (a), and mechanism of the native peptide ligation underlying the coupling of fragments **9.64** and **9.65** (b). The peptide sequences are indicated by the one letter code.

Bibliography

[1] Kirby AJ. Effective molarities for intramolecular reactions. Adv. Phys. Org. Chem. 1980, 17, 183–278.
[2] Cacciapaglia R, Di Stefano S, Mandolini L. Effective molarities in supramolecular catalysis of two-substrate reactions. Acc. Chem. Res. 2004, 37, 113–22.
[3] Mock WL, Irra TA, Wepsiec JP, Adhya M. Catalysis by cucurbituril. The significance of bound-substrate destabilization for induced triazole formation. J. Org. Chem. 1989, 54, 5302–8.
[4] Mandolini L. Intramolecular reactions of chain molecules. Adv. Phys. Org. Chem. 1986, 22, 1–111.
[5] Sanders JKM. Supramolecular catalysis in transition. Chem. Eur. J. 1998, 4, 1378–83.
[6] Raynal M, Ballester P, Vidal-Ferran A, van Leeuwen PWNM. Supramolecular catalysis. Part 2: artificial enzyme mimics. Chem. Soc. Rev. 2014, 43, 1734–87.
[7] Cram DJ. The design of molecular hosts, guests, and their complexes (Nobel Lecture). Angew. Chem. Int. Ed. Engl. 1988, 27, 1009–20.

[8]   Cram DJ, Katz HE, Dicker IB. Host–guest complexation. 31. A transacylase partial mimic. J. Am. Chem. Soc. 1984, 106, 4987–5000.

[9]   Cram DJ, Lam PYS. Host–guest complexation. 37. Synthesis and binding properties of a transacylase partial mimic with imidazole and benzyl alcohol in place. Tetrahedron 1986, 42, 1607–15.

[10]   Cram DJ, Lam PYS, Ho SP. A transacylase partial mimic. J. Am. Chem. Soc. 1986, 108, 839–41.

[11]   Olivo G, Capocasa G, Del Giudice D, Lanzalunga O, Di Stefano S. New horizons for catalysis disclosed by supramolecular chemistry. Chem. Soc. Rev. 2021, 50, 7681–724.

[12]   Wang R, Yu Y. Site-selective reactions mediated by molecular containers. Beilstein J. Org. Chem. 2022, 18, 309–24.

[13]   Otte, M. Reactions in endohedral functionalized cages. Eur. J. Org. Chem. 2023, 26, e202300012.

[14]   Raynal M, Ballester P, Vidal-Ferran A, van Leeuwen PWNM. Supramolecular catalysis. Part 1: non-covalent interactions as a tool for building and modifying homogeneous catalysts. Chem. Soc. Rev. 2014, 43, 1660–733.

[15]   VanEtten RL, Sebastian JF, Clowes GA, Bender ML. Acceleration of phenyl ester cleavage by cycloamyloses. A model for enzymic specificity. J. Am. Chem. Soc. 1967, 89, 3242–53.

[16]   Breslow R, Dong SD. Biomimetic reactions catalyzed by cyclodextrins and their derivatives. Chem. Rev. 1998, 98, 1997–2012.

[17]   Breslow R, Trainor G, Ueno A. Optimization of metallocene substrates for β-cyclodextrin reactions. J. Am. Chem. Soc. 1983, 105, 2739–44.

[18]   Breslow R, Chung S. Additional flexibility solves the leaving group problem in cyclodextrin acylation. Tetrahedron Lett. 1990, 31, 631–4.

[19]   Breslow R, Hammond M, Lauer M. Selective transamination and optical induction by a β-cyclodextrin-pyridoxamine artificial enzyme. J. Am. Chem. Soc. 1980, 102, 421–2.

[20]   Hooley RJ, Rebek Jr. J. Chemistry and catalysis in functional cavitands. Chem. Biol. 2009, 16, 255–64.

[21]   Purse BW, Ballester P, Rebek Jr. J. Reactivity and molecular recognition: amine methylation by an introverted ester. J. Am. Chem. Soc. 2003, 125, 14682–3.

[22]   Gissot A, Rebek Jr. J. A functionalized, deep cavitand catalyzes the aminolysis of a choline derivative. J. Am. Chem. Soc. 2004, 126, 7424–5.

[23]   Yoshizawa M, Klosterman JK, Fujita M. Functional molecular flasks: new properties and reactions within discrete, self-assembled hosts. Angew. Chem. Int. Ed. 2009, 48, 3418–38.

[24]   Kang J, Rebek Jr. J. Acceleration of a Diels–Alder reaction by a self-assembled molecular capsule. Nature 1997, 385, 50–2.

[25]   Tehrani FN, Assaf KI, Hein R, Jensen CME, Nugent TC, Nau WM. Supramolecular catalysis of a catalysis-resistant Diels–Alder reaction: almost theoretical acceleration of cyclopentadiene dimerization inside cucurbit[7]uril. ACS Catal. 2022, 12, 2261–9.

[26]   Page MI, Jencks WP. Entropic contributions to rate accelerations in enzymic and intramolecular reactions and the chelate effect. Proc. Natl. Acad. Sci. U. S. A. 1971, 68, 1678–83.

[27]   Yoshizawa M, Tamura M, Fujita M. Diels–Alder in aqueous molecular hosts: unusual regioselectivity and efficient catalysis. Science 2006, 312, 251–4.

[28]   Chen J, Rebek Jr. J. Selectivity in an encapsulated cycloaddition reaction. Org. Lett. 2002, 4, 327–9.

[29]   Gibb CLD, Sundaresan AK, Ramamurthy V, Gibb BC. Templation of the excited-state chemistry of α-(n-alkyl) dibenzyl ketones: how guest packing within a nanoscale supramolecular capsule influences photochemistry. J. Am. Chem. Soc. 2008, 130, 4069–80.

[30]   Parthasarathy A, Kaanumalle LS, Ramamurthy V. Controlling photochemical geometric isomerization of a stilbene and dimerization of a styrene using a confined reaction cavity in water. Org. Lett. 2007, 9, 5059–62.

[31]   Yoshizawa M, Takeyama Y, Kusukawa T, Fujita M. Cavity-directed, highly stereoselective [2+2] photodimerization of olefins within self-assembled coordination cages. Angew. Chem. Int. Ed. 2002, 41, 1347–9.

[32] Yu Y, Rebek Jr. J. Reactions of folded molecules in water. Acc. Chem. Res. 2018, 51, 3031–40.

[33] Mosca S, Yu Y, Gavette JV, Zhang KD, Rebek Jr. J. A deep cavitand templates lactam formation in water. J. Am. Chem. Soc. 2015, 137, 14582–5.

[34] Shi Q, Masseroni D, Rebek Jr. J. Macrocyclization of folded diamines in cavitands. J. Am. Chem. Soc. 2016, 138, 10846–8.

[35] Breslow R, Zhang B. Very fast ester hydrolysis by a cyclodextrin dimer with a catalytic linking group. J. Am. Chem. Soc. 1992, 114, 5882–3.

[36] Breslow R, Zhang X, Xu R, Maletic M, Merger R. Selective catalytic oxidation of substrates that bind to metalloporphyrin enzyme mimics carrying two or four cyclodextrin groups and related metallosalens. J. Am. Chem. Soc. 1996, 118, 11678–9.

[37] Breslow R, Zhang X, Huang Y. Selective catalytic hydroxylation of a steroid by an artificial cytochrome p-450 enzyme. J. Am. Chem. Soc. 1997, 119, 4535–6.

[38] Breslow R, Schmuck C. Goodness of fit in complexes between substrates and ribonuclease mimics: effects on binding, catalytic rate constants, and regiochemistry. J. Am. Chem. Soc. 1996, 118, 6601–5.

[39] Hosseini MW, Lehn JM, Maggiora L, Mertes KB, Mertes MP. Supramolecular catalysis in the hydrolysis of ATP facilitated by macrocyclic polyamines: mechanistic studies. J. Am. Chem. Soc. 1987, 109, 537–44.

[40] Molenveld P, Engbersen JFJ, Reinhoudt DN. Dinuclear metallo-phosphodiesterase models: application of calix[4]arenes as molecular scaffolds. Chem. Soc. Rev. 2000, 29, 75–86.

[41] Mattei P, Diederich F. Catalytic Cyclophanes. Part XI. A flavo-thiazolio-cyclophane as a biomimetic catalyst for the preparative-scale electro-oxidation of aromatic aldehydes to methyl esters. Helv. Chim. Acta. 1997, 80, 1555–88.

[42] Mitschke B, Turberg M, List B. Confinement as a unifying element in selective catalysis. Chem 2020, 6, 2515–32.

[43] Mitra R, Niemeyer J. Dual Brønsted-acid organocatalysis: cooperative asymmetric catalysis with combined phosphoric and carboxylic acids. ChemCatChem 2018, 10, 1221–34.

[44] Akiyama T, Itoh J, Yokota K, Fuchibe K. Enantioselective Mannich-type reaction catalyzed by a chiral Brønsted acid. Angew. Chem. Int. Ed. 2004, 43, 1566–8.

[45] Brak K, Jacobsen EN. Asymmetric ion-pairing catalysis. Angew. Chem. Int. Ed. 2013, 52, 534–61.

[46] Beckendorf S, Asmus S, Mück-Lichtenfeld C, García Mancheño O. "Click" bis-triazoles as neutral C–H⋯anion–acceptor organocatalysts. Chem. Eur. J. 2013, 19, 1581–5.

[47] Bulfield D, Huber SM. Halogen bonding in organic synthesis and organocatalysis. Chem. Eur. J. 2016, 22, 14434–50.

[48] Müller C, Bauer A, Bach T. Light-driven enantioselective organocatalysis. Angew. Chem. Int. Ed. 2009, 48, 6640–2.

[49] Mitra R, Zhu H, Grimme S, Niemeyer J. Functional mechanically interlocked molecules: asymmetric organocatalysis with a catenated bifunctional Brønsted acid. Angew. Chem. Int. Ed. 2017, 56, 11456–9.

[50] De S, Pramanik S, Schmittel M. A toggle nanoswitch alternately controlling two catalytic reactions. Angew. Chem. Int. Ed. 2014, 53, 14255–9.

[51] Blanco V, Carlone A, Hänni KD, Leigh DA, Lewandowski B. A rotaxane-based switchable organocatalyst. Angew. Chem. Int. Ed. 2012, 51, 5166–9.

[52] Dorel R, Feringa BL. Stereodivergent anion binding catalysis with molecular motors. Angew. Chem. Int. Ed. 2020, 59, 785–9.

[53] Pluth MD, Bergman RG, Raymond KN. Proton-mediated chemistry and catalysis in a self-assembled supramolecular host. Acc. Chem. Res. 2009, 42, 1650–9.

[54] Pluth MD, Bergman RG, Raymond KN. Acid catalysis in basic solution: a supramolecular host promotes orthoformate hydrolysis. Science 2007, 316, 85–8.

[55] Hastings CJ, Pluth MD, Bergman RG, Raymond KN. Enzymelike catalysis of the Nazarov cyclization by supramolecular encapsulation. J. Am. Chem. Soc. 2010, 132, 6938–40.

[56] Fiedler D, van Halbeek H, Bergman RG, Raymond KN. Supramolecular catalysis of unimolecular rearrangements: substrate scope and mechanistic insights. J. Am. Chem. Soc. 2006, 128, 10240–52.

[57] Zhao C, Toste FD, Raymond KN, Bergman RG. Nucleophilic substitution catalyzed by a supramolecular cavity proceeds with retention of absolute stereochemistry. J. Am. Chem. Soc. 2014, 136, 14409–12.

[58] Zhang Q, Catti L, Tiefenbacher K. Catalysis inside the hexameric resorcinarene capsule. Acc. Chem. Res. 2018, 51, 2107–14.

[59] Zhang Q, Tiefenbacher K. Hexameric resorcinarene capsule is a Brønsted acid: investigation and application to synthesis and catalysis. J. Am. Chem. Soc. 2013, 135, 16213–9.

[60] Zhang Q, Tiefenbacher K. Terpene cyclization catalysed inside a self-assembled cavity. Nat. Chem. 2015, 7, 197–202.

[61] Robertson A, Sinclair AJ, Philp D. Minimal self-replicating systems. Chem. Soc. Rev. 2000, 29, 141–52.

[62] von Kiedrowski G. A self-replicating hexadeoxynucleotide. Angew. Chem. Int. Ed. Engl. 1986, 25, 932–5.

[63] Bag BG, von Kiedrowski G. Templates, autocatalysis and molecular replication. Pure Appl. Chem. 1996, 68, 2145–52.

[64] Wintner EA, Conn MM, Rebek Jr. J. Studies in molecular replication. Acc. Chem. Res. 1994, 27, 198–203.

[65] Nowick JS, Feng Q, Tjivikua T, Ballester P, Rebek Jr. J. Kinetic studies and modeling of a self-replicating system. J. Am. Chem. Soc. 1991, 113, 8831–9.

[66] Hong JI, Feng Q, Rotello V, Rebek Jr. J. Competition, cooperation, and mutation: improving a synthetic replicator by light irradiation. Science 1992, 255, 848–50.

[67] Feng Q, Park TK, Rebek Jr. J. Crossover reactions between synthetic replicators yield active and inactive recombinants. Science 1992, 256, 1179–80.

[68] Kassianidis E, Philp D. Design and implementation of a highly selective minimal self-replicating system. Angew. Chem. Int. Ed. 2006, 45, 6344–8.

[69] Sadownik JW, Philp D. A simple synthetic replicator amplifies itself from a dynamic reagent pool. Angew. Chem. Int. Ed. 2008, 47, 9965–70.

[70] Kosikova T, Philp D. Exploring the emergence of complexity using synthetic replicators. Chem. Soc. Rev. 2017, 46, 7274–305.

[71] Lee DH, Granja JR, Martinez JA, Severin K, Ghadiri MR. A self-replicating peptide. Nature 1996, 382, 525–8.

[72] O'Shea EK, Klemm JD, Kim PS, Alber T. X-ray structure of the GCN4 leucine zipper, a two-stranded, parallel coiled coil. Science 1991, 254, 539–44.

10 Transporting molecules

CONSPECTUS: *Biological membranes effectively protect the cytoplasm of a cell and all its components from the surrounding medium. To ensure that nutrients or ions that cannot freely diffuse across the hydrophobic interior of a lipid bilayer can move from one side to the other, special transporter molecules exist. These molecules can act as carriers, capturing a target ion or molecule, transporting it across the membrane in the form of its complex, and releasing it again. Alternatively, transport may involve channels through which ions or polar molecules migrate. Both transport mechanisms can also rely on abiotic molecules, as shown in this chapter for cation, anion, and water transport.*

10.1 Introduction

Biological membranes are complex systems composed of several components. The actual bilayer is formed from phospholipids, which are natural amphiphiles consisting of two fatty acids covalently linked to adjacent hydroxy groups of glycerol. The third glycerol OH group carries a phosphate residue that can be esterified with ethanolamine, choline, or serine (Figure 10.1a).

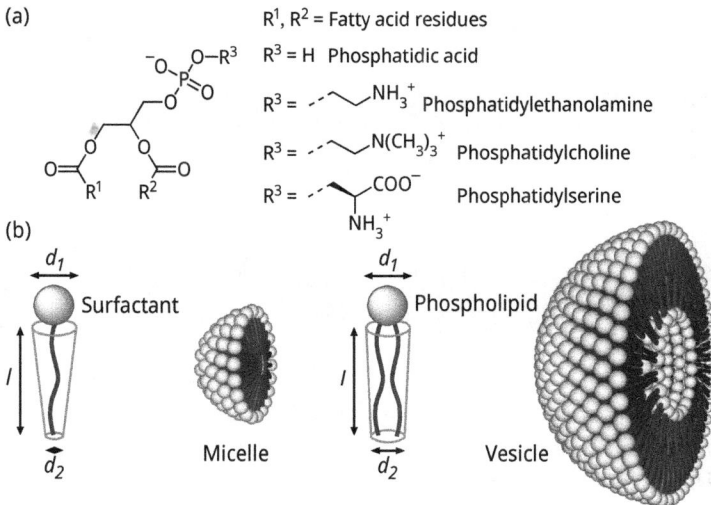

Figure 10.1: General structure of a phospholipid (a) and schematic illustration of the correlation between the structure of an amphiphile and the outcome of the self-assembly process (b). Surfactants with one hydrophobic chain and a polar head group assemble into micelles, whereas phospholipids with two hydrophobic chains form vesicles because the d_1/d_2 ratio is smaller than that of surfactants.

https://doi.org/10.1515/9783111315171-010

Like surfactants, phospholipids self-assemble in water to shield their hydrophobic residues from the surrounding medium, but unlike surfactants, they do not form spherical micelles. This is because the two hydrophobic chains limit the curvature of the self-assembled products, causing phospholipids to form leaflets that organize into bilayers (Figure 10.1b). These membranes have a hydrophobic interior containing the alkyl residues of the fatty acids, and hydrophilic outer and inner surfaces containing the phospholipid head groups. The resulting systems are called vesicles or, when formed from synthetic amphiphiles, also liposomes [1].

How do polar species cross cell membranes?

Phospholipid membranes are efficient barriers between the inside and outside of a cell, allowing only hydrophobic molecules or small neutral polar molecules such as water and urea to pass. Ions or polar organic compounds are incompatible with the hydrophobic interior of lipid bilayers, and these substrates must therefore rely on special transport systems. One way of transport is based on carriers that move relatively freely within the bilayer. These molecules bind to the substrate and thereby create an apolar layer around it, transport it from one side of the membrane to the other, and release it again (Figure 10.2a). The other transport mechanism involves channel-forming proteins that allow the substrate to move through a pore (Figure 10.2b).

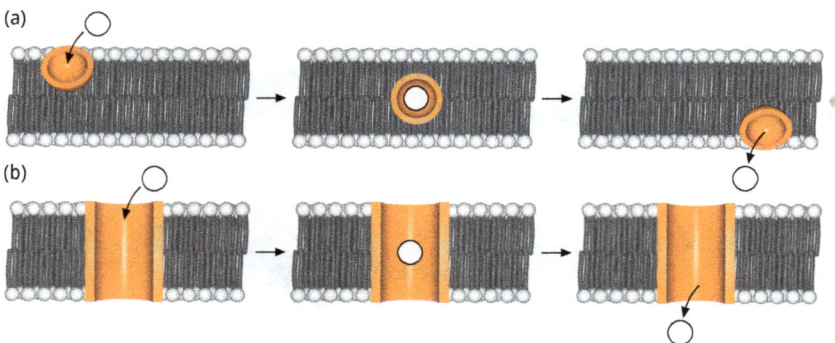

Figure 10.2: Schematic representation of membrane transport mediated by a carrier (a) or a channel-forming protein (b). A carrier forms a complex with the substrate that diffuses across the membrane, whereas a channel spans the membrane and allows the substrate to pass through.

Carriers are typically selective and transport the substrate along a concentration gradient from higher to lower concentrations. Channels, on the other hand, are less selective, but can be opened and closed by appropriate triggers, such as a change in membrane potential or the binding of a ligand to the extracellular segment of the transmembrane protein. Thus, channel-mediated transport can be controlled. It is also faster than carrier-mediated transport and occurs either with (passive transport) or against (active transport) a concentration gradient, with the energy required for active transport usually supplied by ATP hydrolysis.

There are several ways in which a charged substrate can be translocated across a membrane. Uniport involves the transport of either a cation or an anion, resulting in a change in membrane potential (Figure 10.3a). The transport of a cation together with an anion in the same direction is called symport (Figure 10.3b), whereas the transport of two ions with the same charge in opposite directions is called antiport (Figure 10.3c). In the latter two cases, the membrane potential is not affected.

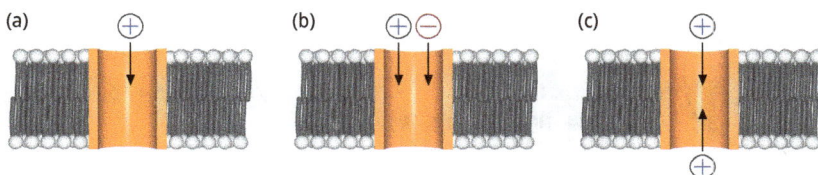

Figure 10.3: Schematic representation of ion translocation across a membrane by uniport (a), symport (b), and antiport (c). In the first case, only one ion moves. In the second, a cation and an anion move simultaneously in the same direction, and in the third, two ions with the same charge move in opposite directions.

How is membrane transport studied?

Studies of membrane transport t must be based on analytical techniques that provide information about the flow of the analytes of interest across a membrane. Among the various techniques available, the following are the most important.

U-tube experiments: These experiments are used to characterize the ability of receptor molecules to act as carriers. The experimental setup involves the use of a U-shaped glass tube or of a vial with a barrier in the middle that leaves an opening at the bottom (Figure 10.4a). In both cases, the vessel contains a solution of the carrier molecule in a hydrophobic organic solvent with a higher density than water (usually chloroform), which is intended to mimic the hydrophobic interior of a lipid bilayer. It is therefore called a bulk membrane or bulk liquid membrane. This solution is layered by aqueous solutions in the two separate compartments of the vessel, one of which, the source phase, contains the compound to be transported by the carrier. The other solution is the receiving phase into which the substrate is to be transported. Transport is driven by the concentration gradient from the source to the receiving phase. If ions are to be transported and the transport involves a symport mechanism, the receiving phase will be water. In the case of antiport, the receiving phase must contain an ion that is transported back to the source phase. The rate of transport is assessed by monitoring the appearance of the analyte in the receiving phase over time. A convenient way to do this for symport is to use a picrate salt and follow the increase in cation concentration by the simultaneous appearance of the yellow color of the picrate counterion. Other analytical techniques are available to detect the incoming ions. One application of a U-tube experiment is the separation of amino acid enantiomers, as discussed in Section 4.1.1.

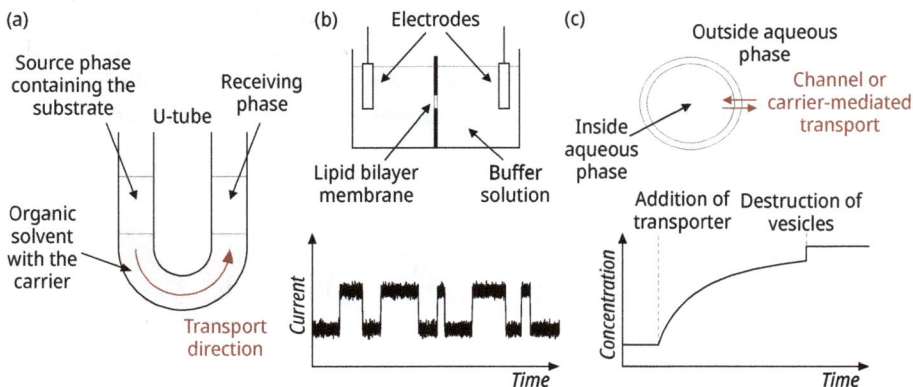

Figure 10.4: Experimental setups to study membrane transport. The U-tube experiment in (a) allows the characterization of the transport properties of molecules dissolved in the organic bulk liquid membrane separating the aqueous source and receiving phases. Bilayer conductance measurements are useful to study channels (b). They involve measuring the electrical current between two aqueous buffer solutions separated by a bilayer membrane. Vesicle-based experiments involve studying the transport of analytes from the aqueous solution inside a vesicle to the outside or in the opposite direction (c).

Bilayer conductance experiments: This method allows the characterization of ion channel properties and involves measuring the current between two chambers containing aqueous buffer solutions separated by a barrier with a micrometer-sized hole. This hole supports a lipid bilayer membrane deposited from a solution containing the corresponding phospholipid. The membrane acts as an insulator, preventing the flow of current after an electrical potential is applied. Once a channel is present in the membrane, characteristic steps are observed in a plot of current versus time. These steps correlate with the opening and closing of the channel. If the channel concentration in the membrane is sufficiently low, each step represents the activity of a single channel, allowing quantification of the single-channel current and the time the channel is open. An example of such a plot, along with a schematic representation of the underlying experiment, is shown in Figure 10.4b.

Vesicle-based experiments: Vesicles with a single bilayer membrane (unilamellar vesicles) are conveniently prepared by extruding aqueous dispersions of phospholipid films through polycarbonate filters with a defined pore size. During extrusion, the initially present lipid multilayers disintegrate, resulting in solutions containing unilamellar vesicles with a rather narrow size distribution. Depending on their size, the vesicles are classified as small (<100 nm), large (100–1,000 nm), or giant (>1,000 nm) unilamellar vesicles. These vesicles contain the solvent used during extrusion, making it possible to encapsulate salts, dye molecules, or other compounds. The vesicle solution is then subjected to dialysis or size exclusion chromatography to separate nonencapsulated compounds and establish a concentration gradient across the vesicle membranes (Figure 10.4c). After incorporating a transport system into the bilayer, usually by adding a concentrated solution of a transporter in a water-miscible solvent such as DMSO, membrane transport is initiated and followed by monitoring the change in analyte concentration inside or outside the vesicles. If the transporter causes the release of ions from the vesicle interior, the increase in the concentration of the respective ion in the surrounding solution is monitored electrochemically, for example using ion-selective electrodes. Conversely, it is possible to incorporate dyes into the vesicles, which report the pH change of the solution inside the vesicle or the presence of certain ions by a change in fluorescence. These experiments are typically performed by first recording the signal in the absence of the transporter. The addition of the transporter at a given time then causes a progressive increase in the concentration of the analyte, which is recorded. Finally, a detergent is added to destroy the vesicles and obtain the maximum observable signal under the chosen conditions. The resulting curves provide quantitative information about the transport rates and activities of the systems under study. Special strategies are available to obtain information about the transport mechanism. For example, whether a transporter acts as a carrier or channel is often assessed by adding compounds to the lipid bilayer that reduce fluidity. This addition should decrease the rate of ion transport if a carrier mechanism is operative, whereas the rate of transport through a channel should be unaffected. A counterion effect on the transport rate is indicative of a symport mechanism.

10.2 Cation transport

10.2.1 Channels

The transport of monovalent metal ions such as sodium or potassium across lipid membranes is crucial in biological systems for signal transduction or for maintaining the membrane potential. Membranes therefore contain specialized proteins that mediate cation transport and ensure that it is selective. Conversely, compounds that disrupt the membrane potential are harmful to cells and often have antibacterial activity. A protein responsible for maintaining the membrane potential is the potassium channel protein. Its structure is known at atomic resolution, providing insight into the structural basis of the transport properties, including the 10^5-fold selectivity for potassium over sodium ions [2]. The potassium channel protein is a tetramer of smaller subunits surrounding a central pore through which potassium ions pass. The pore is characterized by wider tunnels on either side, lined with negatively charged amino acid side chains that serve to increase the local cation concentration relative to the surrounding environment. In these regions, the cations are still fully solvated. The narrow region of the pore, the so-called selectivity filter, has a radius exactly matching the diameter of a potassium ion and is lined with carbonyl groups, as shown in the corresponding section of the crystal structure in Figure 10.5a. Potassium ions thus pass through the pore after having shed their solvation shell and are efficiently stabilized by ion–dipole interactions. Slightly larger rubidium ions can also pass, but sodium ions are either too small to interact efficiently with the carbonyl groups when desolvated or too large when solvated. These ions are therefore rejected, explaining the high K^+ selectivity of the channel.

(a)   (b)

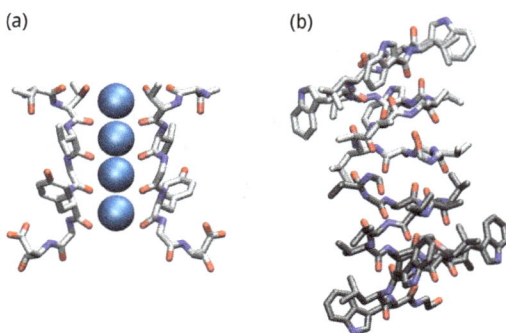

Figure 10.5: Section of the crystal structure of the potassium channel protein from *Streptomyces lividans* showing the environment in the filter region of the pore (a) and the NMR structure of the head-to-head dimer of gramicidin A in sodium dodecyl sulfate micelles (b). Only two of the four chains surrounding the potassium ions in the selectivity filter are shown in (a). The protons in (b) are omitted for reasons of clarity, and the carbon atoms of the protein backbones of the two gramicidin A subunits are shown in different shades of gray.

A channel protein that causes the collapse of the membrane potential is gramicidin A. This linear α-helical peptide is composed of 15 alternating L- and D-amino acids. When two gramicidin A molecules self-assemble within a lipid bilayer, a channel is formed through which up to 10^7 monocations pass per second. The structure of the head-to-head dimer of gramicidin A is shown in Figure 10.5b to illustrate its tubular structure [3].

Both structures provide insight into the concepts on which the design of artificial cation channels could be based. Such channels could involve either a single compound spanning the entire width of a lipid bilayer or an assembly of multiple subunits. Furthermore, pore-forming compounds alone are sufficient to achieve transport, with transport selectivity depending mainly on pore diameter. The additional presence of cation binding sites within the pore, such as anionic residues or oxygen atoms with which the cations can interact, is beneficial. All of these concepts have been realized in synthetic systems [4].

Synthetic channels are formed from the tube-forming *p*-octiphenyl-peptide conjugates developed by Stefan Matile and discussed in Section 5.3.4. These compounds assemble into tetrameric tubes whose height is comparable to the hydrophobic core
thickness of lipid bilayers. If the sequence of amino acids in the peptide side chains is
chosen so that hydrophobic residues end up on the outer surface of the tube, self-assembly within lipid bilayers is possible. The resulting structures resemble natural β-
barrels, pore-forming proteins containing cyclically arranged β-sheets of interacting peptide strands [5]. The pore diameter of these rigid-rod β-barrels can be controlled by varying the length of the peptide residues, while the transport properties depend on the
nature of the substituents arranged along the inner pore, with negatively charged
residues providing a suitable environment for cation transport. For example, pores
formed from *p*-octiphenyl derivative **10.1a** (Figure 10.6) have a relatively large diameter
(about 10 Å) and contain aspartate residues along the inner surface, which is why they
preferentially transport potassium over chloride ions [6]. According to bilayer conductance measurements, the lifetime of these channels is rather short ($\tau < 0.5$ ms), indicating that they are not very stable. The stability increases in the presence of magnesium
ions, which presumably cross-link the carboxylate groups along the pore interior, but
once these cations are present, the cation/anion selectivity is almost lost.

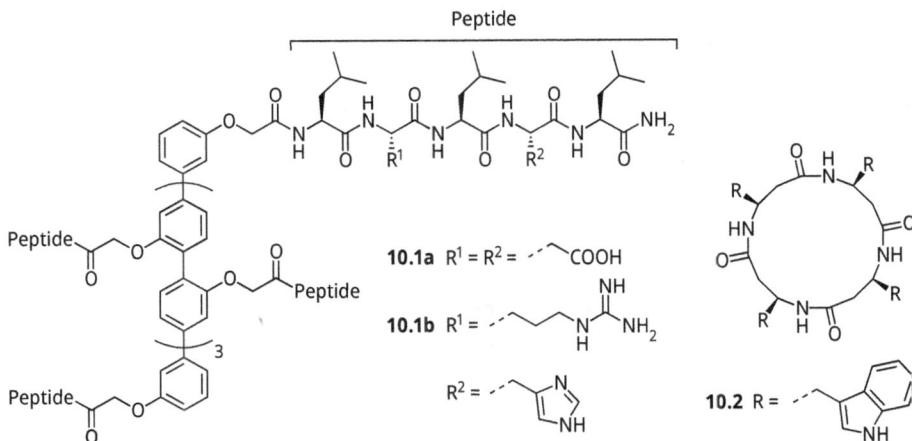

Figure 10.6: Molecular structures of the channel-forming *p*-octiphenyl-peptide conjugates **10.1a,b** and
cyclopeptide **10.2**. A schematic representation of the structure of the tetrameric assembly formed from
10.1a,b is shown in Figure 5.34.

Counterintuitively, the rigid-rod β-barrel derived from **10.1b** with internal arginine
and histidine residues is also cation selective, although it should contain positively
charged guanidinium groups along the inner surface. This is due to the simultaneous
presence of phosphate ions in the aqueous solution, which interact with the guanidinium residues and render the pore interior negatively charged. This selectivity is re-

versed by protonation of the phosphate and imidazole groups, resulting in an excess of positive charges in the pore. Other applications of this versatile family of β-barrel mimics include the transport of larger (charged) organic molecules, as well as sensing and catalysis [5].

Ghadiri's self-assembling cyclopeptides (Section 5.3.4) have also been used to mediate cation transport. The cyclic peptide **10.2** (Figure 10.6), for example, self-assembles within a lipid bilayer to form stacks of about six rings that act as discrete ion channels [7]. According to single-channel conductance measurements, the rate of potassium transport is 1.9×10^7 ions s^{-1}, which is greater than that induced by gramicidin A under similar conditions.

The arrangement of oxygen atoms along the inner surface of a synthetic ion channel is often based on the use of pore-forming crown ether derivatives, such as the α-helical peptide **10.3** or the *p*-octiphenyl derivative **10.4** (Figure 10.7) [4]. When these compounds are incorporated into lipid bilayers, their crown ether moieties stack to form channels through which cations can pass [8]. An alternative design of unimolecular ion channels is based on the so-called hydraphiles, introduced by George W. Gokel [9]. These large rings, an example of which is **10.5**, contain crown ether units, two of which face the membrane surfaces and form the channel gates when incorporated into a lipid bilayer. The alkyl chains and remaining crown ethers span the bilayer, creating a hydrophilic region that is critical for transport activity. Central residues that interact with water molecules are particularly beneficial, while those that bind to ions reduce transport activity. The propensity of oligoethylene chains to interact with cations also has a negative effect on transport activity, which is why hydraphiles with hydrocarbon chains tend to be more active. Somewhat related to the design of **10.5** is the hexasubstituted crown ether **10.6** described by Thomas M. Fyles, in which the crown ether ring acts as a central relay while the six substituents fix the channel within the bilayer [10]. This compound has a transport activity comparable to that of gramicidin A. Although it is tempting to ascribe its activity to the presence of the crown ether, the fact that the structurally simpler analog **10.7** mediates transport similarly well suggests that the crown ether ring has a limited functional role and serves mainly to properly arrange the substituents [4].

10.2.2 Carriers

Natural macrocyclic products such as valinomycin **10.8**, nonactin **10.9**, other members of the nactin family, or the acyclic monesin **10.10** (Figure 10.8a) transport cations across lipid membranes by a carrier mechanism. They induce the collapse of the transmembrane electrochemical gradient, which is harmful to cells and makes these ionophores potent antibiotics. Their mode of action involves complexation of the cation within a cleft or cavity where it binds to surrounding oxygen atoms by ion–dipole interactions. To illustrate this binding mode, the structure of the K$^+$ complex of valinomycin, which was already discussed in Section 4.1.1, is shown in Figure 10.8b. Complex formation pro-

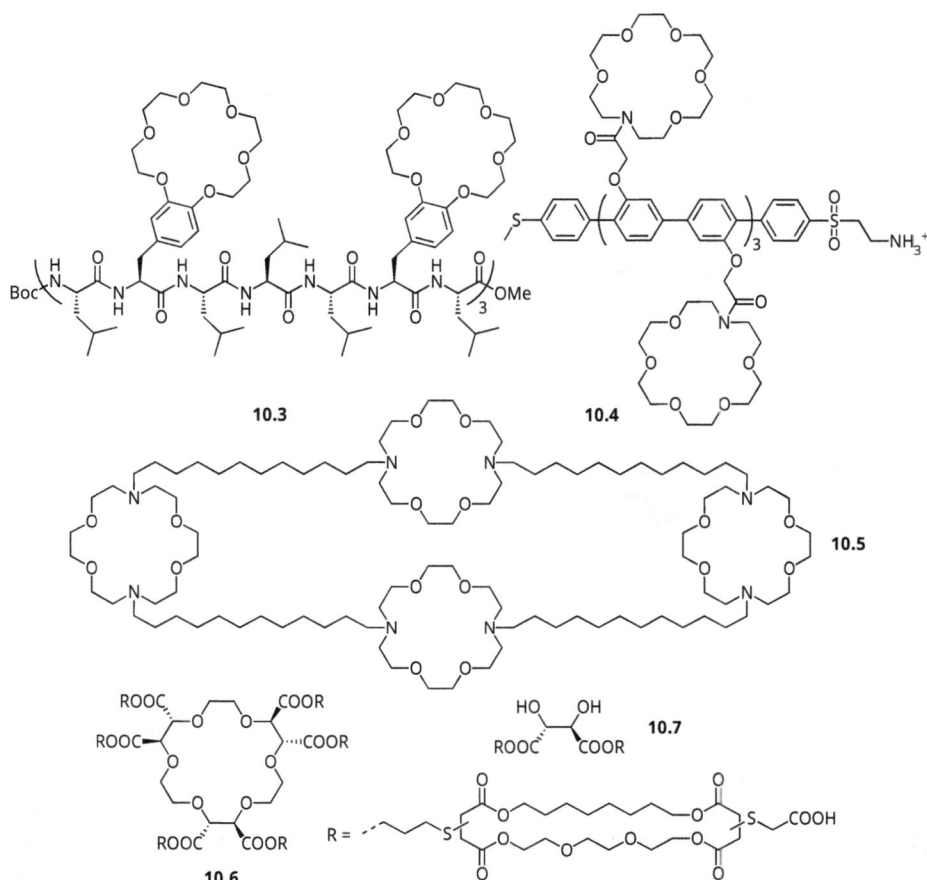

Figure 10.7: Molecular structures of the crown ether-derived unimolecular ion channels **10.3**, **10.4**, **10.5**, and **10.6**. The tartaric acid derivative **10.7** is a structurally simpler analog of **10.6** with similar ion transport activity. Compound **10.5** is a hydraphile.

duces a hydrophilic layer around the potassium ion, characterized by a cyclic arrangement of the isopropyl groups of the amino acid and α-hydroxycarboxylic acid subunits. This layer allows the complex to partition into the hydrophobic environment of a lipid bilayer without the need for an anion to compensate the positive charge. Valinomycin thus mediates the uniport of potassium ions along the concentration gradient selectively over slightly smaller and therefore less strongly bound sodium ions. Other ionophores form similar cation complexes but are often less selective.

The structures of these ionophores can serve as inspiration for the development of cation carriers. For example, there is a whole family of macrocyclic peptides and depsipeptides with structures and cation transport properties related to valinomycin

(a)   (b)

10.8

10.9   **10.10**

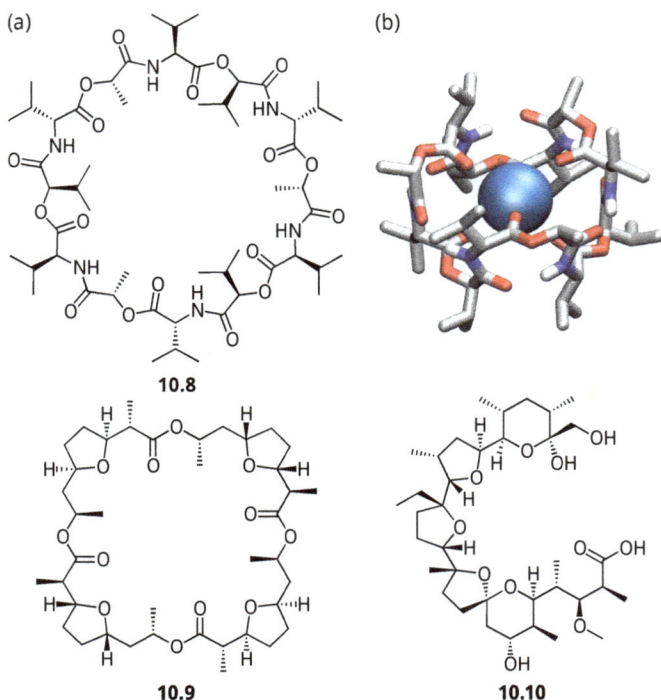

Figure 10.8: Molecular structures of valinomycin **10.8**, nonactin **10.9**, and monesin **10.10** (a), and crystal structure of the potassium complex of **10.8** (b).

[11]. The number of crown ether derivatives that have been studied for their cation transport properties is even larger. This large body of work, which often involved U-tube transport experiments, has provided detailed information on how the structure of crown ethers and the properties of their complexes influence the extraction of metal ions from the aqueous to the organic phase [12]. Less is known about the ability of crown ethers to mediate cation transport across membranes, although the toxicity of some crown ethers suggests that they are able to disrupt the transmembrane potential in a manner similar to valinomycin [13].

Crown ether-mediated cation transport can also be achieved using interlocked molecules [14]. An example is the rotaxane **10.11** developed by He Tian. This so-called molecular cable car contains an 18-crown-6 moiety moving along a symmetrical axle with two secondary amino groups near the end groups and a central triazolium moiety acting as a weak intermediate station (Figure 10.9) [15]. The length of the axle is comparable to the thickness of a phospholipid membrane, allowing **10.11** to be incorporated into vesicle membranes. In the presence of **10.11**, a pH gradient, resulting from the addition of KOH to the external solution, rapidly disappears, suggesting that the potassium ions are shuttled by the crown ether from the outside into the interior of the vesicle with the concomitant transport of hydroxide ions in the same direction or protons in the opposite direction. The decrease in transport rate when the triazolium subunit in the center of **10.11** is replaced by a less efficient station and the loss of transport activity when the amino groups are covalently modified suggest that the potassium transport is indeed due to the shuttling motion of the ring along the axle.

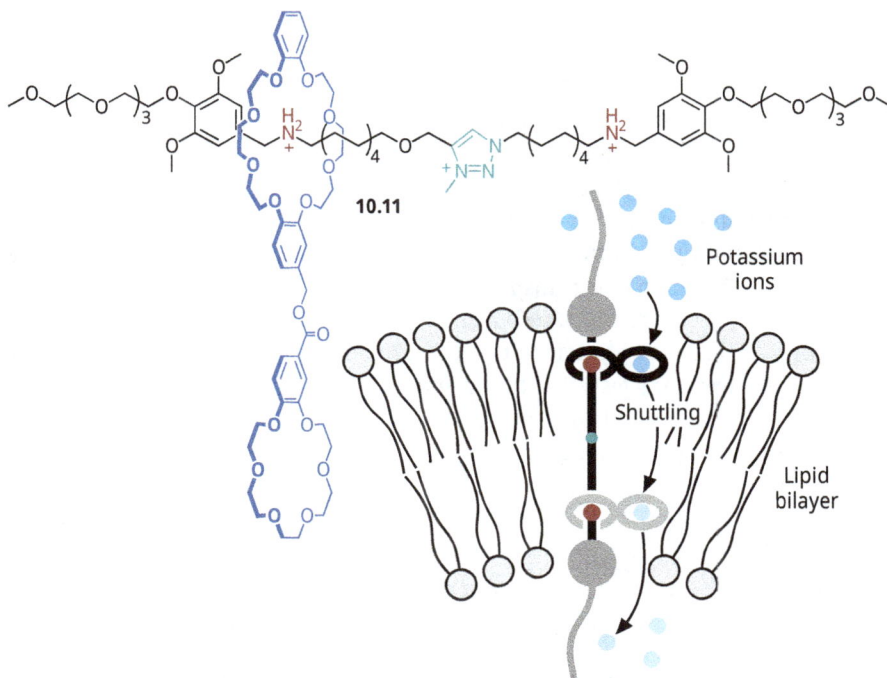

Figure 10.9: Structure of the molecular cable car **10.11** and schematic representation of potassium transport mediated by **10.11** from one side of a bilayer membrane to the other side.

10.3 Anion transport

10.3.1 Channels

A structurally diverse family of proteins is responsible for the selective transport of chloride anions across cell membranes. These so-called chloride ion channels (ClC) are involved in stabilizing the membrane resting potential and in mediating the flow of salt and water across epithelial barriers. According to crystal structures, they form homodimers in the membrane, with the two subunits arranged in opposite orientations [16]. Each subunit has its own selectivity filter in which a desolvated chloride anion is stabilized by a combination of electrostatic interactions with the dipoles of the α-helices converging on the binding site and direct hydrogen bonding interactions with two backbone NH groups and the OH groups of a serine and a tyrosine residue. While anions find an ideal environment within this filter region, cations would experience repulsive interactions, explaining the pronounced ion selectivity of these channels. The pore allows chloride and bromide ions to pass but is blocked by larger anions such as iodide or thiocyanate, which therefore inhibit chloride transport.

Synthetic systems that allow the incorporation of pores into lipid bilayers with an excess of positive charges along the inner surface can again be based on Matile's rigid-rod β-barrels [5]. In the context of anion transport, another family of transporters introduced by Matile is also noteworthy, consisting of rigid rod-like molecules with electron-deficient aromatic subunits of the right length to span a lipid bilayer. An example is oligo(p-phenylene)-N,N-naphthalenediimide **10.12** (Figure 10.10) [17, 18]. The aromatic subunits of this so-called π-slide create regions of substantial positive electrostatic potential within the membrane, allowing anion–π interactions to guide the movement of the anion. The transport activity depends on the peripheral substituents, with the singly charged **10.12a** showing the highest activity, while the activities of **10.12b** with charges at both termini and **10.12c** with two uncharged residues are much lower. This trend suggests that the orientation of the rods within the membrane is important for the activity [19]. These systems transport chloride more efficiently than the other halides and, among the oxoanions, have a selectivity for nitrate.

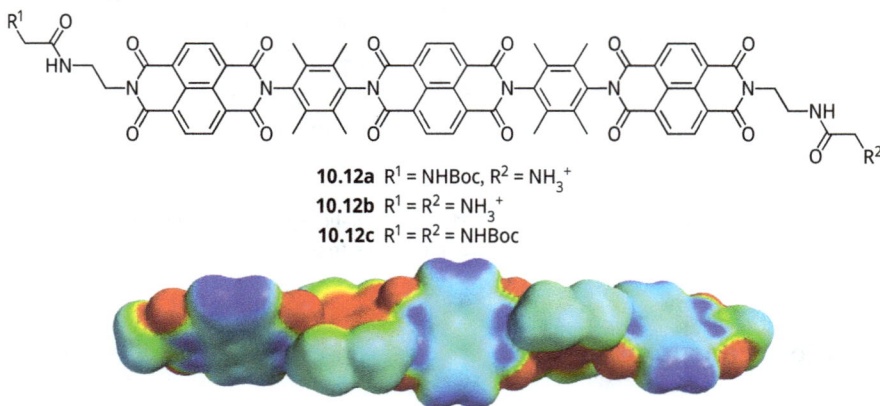

10.12a R^1 = NHBoc, R^2 = NH_3^+
10.12b R^1 = R^2 = NH_3^+
10.12c R^1 = R^2 = NHBoc

Figure 10.10: Molecular structures of π-slides **10.12a–c** and electrostatic potential surface of a truncated analog of **10.12a**, illustrating the positive electrostatic potentials generated by the naphthalenediimide subunits. The color coding covers a potential range from −75 to +75 kJ mol^{-1}, with red and blue indicating values greater than or equal to the absolute maximum in negative and positive potentials, respectively (Boc = *tert*-butyloxycarbonyl).

Structurally more closely related to natural anion channels are the amphiphilic peptides developed by the Gokel group of which **10.13** is an example (Figure 10.11) [20]. The two octadecyl chains at the *N*-terminus help anchor this compound in the membrane. The CH_2OCH_2 residue resembles the glyceryl residue of phospholipids, and the $(Gly)_3$ Pro(Gly)$_3$ subunit mimics peptide sequences found in natural chloride channels.

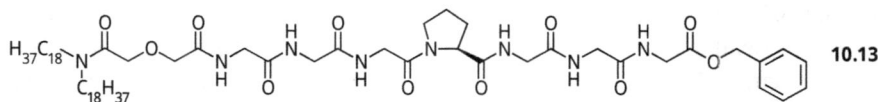

Figure 10.11: Molecular structure of the channel-forming amphiphilic peptide **10.13**, which promotes chloride transport.

Compound **10.13** dimerizes within the lipid bilayer to form pores with a diameter of 7–8 Å that promote chloride transport with approximately 10-fold selectivity over potassium. Each structural element has a characteristic influence on the transport activity. For example, a derivative of **10.13** with a leucine instead of the proline subunit is significantly less active, whereas a derivative with decyl instead of the octadecyl residues is more active but essentially unselective [20].

10.3.2 Carriers

Important natural anion carriers are the prodigiosins, pyrrole alkaloids produced by microorganisms such as *Serratia marcescens* [21]. These alkaloids are structurally quite diverse, but the pyrrolylopyrromethene moiety with a methoxy group in the 4-position of the central ring is found in all derivatives (Figure 10.12). Prodigiosins possess a number of biological activities, some of which are related to their ability to act as antiporters for chloride and bicarbonate or as symporters for protons and chloride ions across liposomal membranes. Proton-coupled chloride transport is somewhat related to phase-transfer catalysis in that protonation of the weakly basic nitrogen atom of the prodigiosin azafulvene subunit converts the entire molecule into a large lipophilic cation that associates with the chloride counterion and mediates its transport across the membrane. Chloride binding is mainly due to electrostatic interactions, but additional stabilizing effects of hydrogen bonding and/or charge-transfer interactions may also play a role.

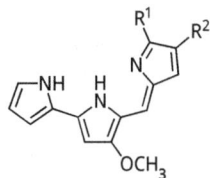

Figure 10.12: General structure of the prodigiosins. The different members of the prodigiosin family differ in the substituents R^1 and R^2.

The structural simplicity of prodigiosins and their cleft-like structure with a convergent arrangement of hydrogen bond donors suggest that typical anion receptors with appropriately placed binding sites should be able to act as anion carriers. The number of synthetic compounds that mediate anion transport is indeed large, as demonstrated by the work of Philip A. Gale [22]. The examples in Figure 10.13 illustrate that some transporters resemble natural carriers, while others have completely different structures with squaramide or urea groups for anion binding, or the binding sites attached to a tripodal scaffold.

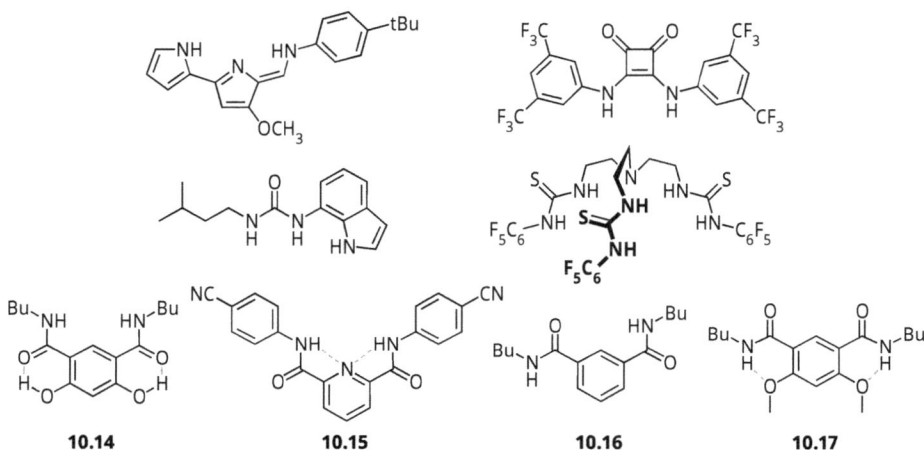

Figure 10.13: Examples of anion receptors mediating anion transport include the derivatives **10.14**–**10.17** of which **10.14** and **10.15** are preorganized for anion binding, rendering them carriers, whereas the lack of preorganization renders **10.16** and **10.17** inactive.

All of these compounds contain hydrogen bond donors for interaction with the substrate, with the isophthalamide derivatives **10.14**, **10.16**, and **10.17** being particularly instructive because they illustrate the importance of available hydrogen bond donors for anion transport [23]. The two hydroxy groups in **10.14** stabilize a conformation in which both NH groups can simultaneously interact with an anion. Thus, despite its structural simplicity, this compound efficiently mediates chloride/nitrate and chloride/bicarbonate exchange across lipid bilayers. For similar reasons, pyridine-2,6-dicarboxamides such as **10.15** are also efficient chloride carriers [24]. In contrast, the diamides **10.16** and **10.17**, which are either not well preorganized (**10.16**) or stabilized in a conformation in which the NH groups are not available for anion binding (**10.17**), are inactive.

Anthony P. Davis introduced cholic acid-based anion receptors, so-called cholapods, as chloride transporters [25]. A member of this family is **10.18** with three thiourea groups along the concave side of the steroidal scaffold (Figure 10.14). These binding sites are too far apart to interact intramolecularly, but have the optimal distance to interact with a chloride anion. As a result, cholapods have very high chloride affinities in chloroform. For example, the chloride complex of **10.18** has a $\log K_a$ of 11.3 according to binding studies using Cram's picrate extraction method. The lipophilicity of these receptors also renders their anion complexes soluble in the hydrophobic environment of lipid bilayers, allowing cholapods to act as efficient chloride/nitrate antiporters. Transport activity generally correlates with chloride affinity, but there appears to be an affinity limit beyond which no increase in transport rate is possible and any further increase in affinity becomes unproductive [26]. The steroidal scaffold is not required for activity, as the *trans*-decalin derivative **10.19** and the cyclohexane derivative **10.20** are also very effective [25].

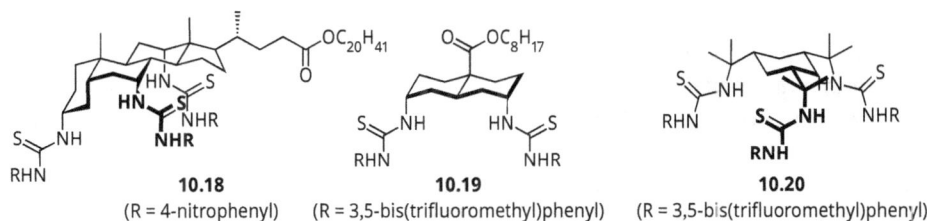

Figure 10.14: Molecular structures of cholapod **10.18** and the structurally simpler analogs **10.19** and **10.20** derived from *trans*-decaline and cyclohexane, respectively.

Another family of steroid-derived transporters are the molecular umbrellas developed by Steven L. Regen [27]. These compounds typically arrange multiple cholic acid subunits or other steroid-derived residues on a flexible backbone. The disubstituted spermidine derivative **10.21** (Figure 10.15) is an example. These compounds adopt different conformations depending on the polarity of the environment in which the hydrophilic faces of the cholic acids are either exposed to the solvent or form a polar cavity shielded by the hydrophobic outer faces of the steroidal hydrocarbon scaffolds. Accordingly, molecular umbrellas solubilize polar substrates such as nucleotides, oligonucleotides, and hydrophilic peptides within lipid bilayers and transport them from one side of the membrane to the other by passive diffusion.

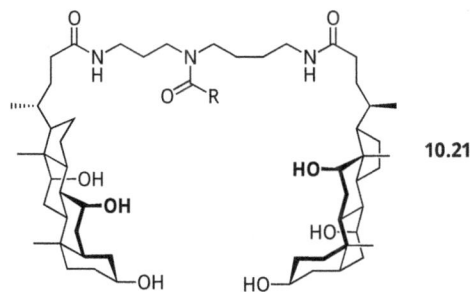

Figure 10.15: Molecular structure of the molecular umbrella **10.21**. The substituent R in this compound is variable and allows tuning the properties.

10.4 Water transport

Although water molecules can directly penetrate the hydrophobic interior of lipid bilayers, specialized proteins exist that mediate the rapid transport of water molecules from one side of the membrane to the other, thereby regulating osmotic pressure [28]. These so-called aquaporins consist of bundles of transmembrane α-helices surrounding a channel through which water molecules can pass but ions and other solutes cannot. The channel consists of larger conical cavities on either side connected by a narrow pore about 20 Å long, giving it an hourglass-like shape. In the larger cavities, the water molecules are hydrogen bonded intermolecularly, as in the bulk solution, while the pore is so narrow that the water molecules must pass through it in a single file. They enter the pore with the oxygen atom first but change orientation as they proceed to allow hydrogen bonding to asparagine side chains within the pore region. This reorientation prevents the water molecules inside the pore from forming a continuous chain of hydrogen bonds, which, in combination with the presence of positively charged or polarized residues positioned along the pore surface, prevents the transfer of protons between the water molecules. Thus, aquaporins selectively transport water molecules but block protons.

Examples of synthetic systems capable of transporting water across lipid bilayers are the ureido-imidazole derivative **10.22** and the pillar[5]arene **10.23** (Figure 10.16). Compound **10.22** crystallizes from water to form a layered structure, stabilized by intermolecular hydrogen bonds between the urea groups [29]. This structure contains channels surrounded by four imidazole units containing chains of water molecules. Each water molecule donates one hydrogen bond to its neighbor and one to a nitrogen atom of a surrounding imidazole residue, similar to the arrangement of water molecules in the pore of aquaporins (Figure 10.16a). Since liposomes dispersed in water and filled with aqueous sodium chloride begin to swell rapidly in the presence of **10.22**, it is likely that similar structures can form in the hydrophobic interior of lipid bilayers and mediate the passage of water molecules from the outside to the inside of a vesicle.

The pillar[5]arene **10.23** behaves similarly [30]. This compound contains substituents on either side of the pillar[5]arene ring with hydrazide subunits that stabilize, by intramolecular hydrogen bonding, a tube-like structure long enough to span a bilayer. Such tubes accommodate linear chains of water molecules in the solid state (Figure 10.16b), suggesting that they are capable of allowing the passage of water molecules when incorporated into a lipid bilayer, which is indeed observed.

Since the water molecules incorporated into the channels of **10.22** and **10.23** form an uninterrupted chain of hydrogen bonds, both compounds also mediate proton transfer. This transfer involves the rapid formation and cleavage of O–H bonds from one water molecule to the next (Grotthuss mechanism), which is not possible in aquaporins due to the reorientation of the water molecules in the pore. True mimics of natural water transporters must therefore prevent this proton hopping, which can be

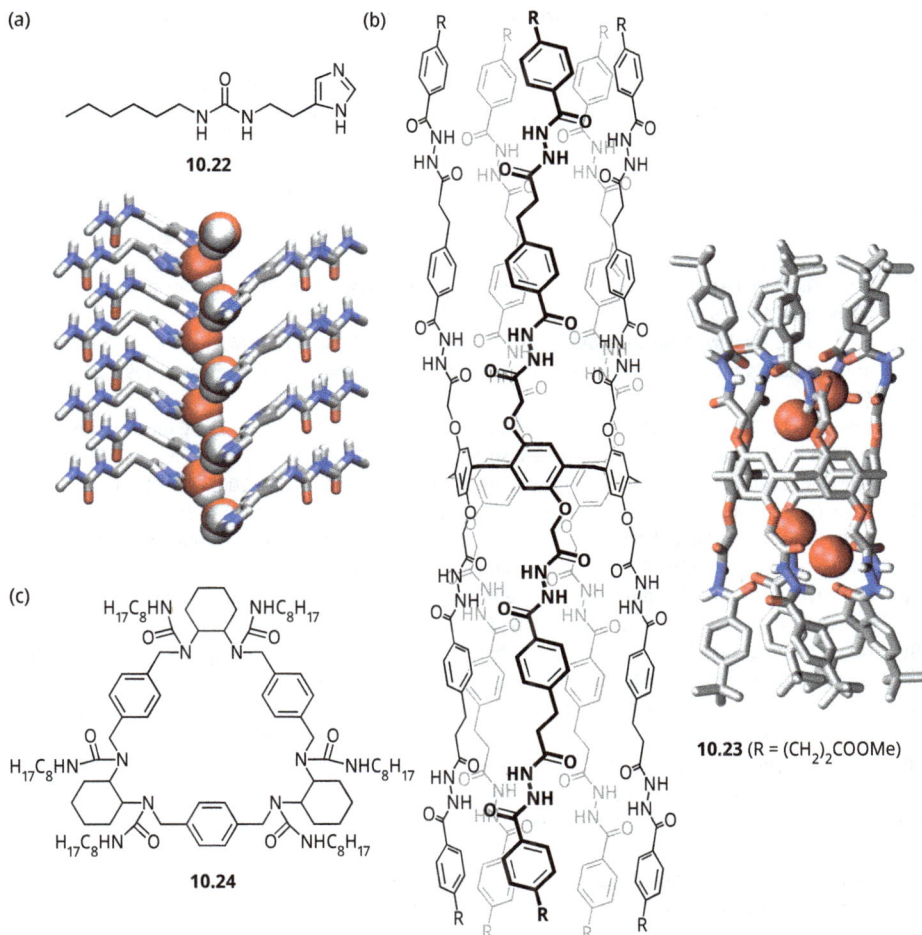

Figure 10.16: Molecular structure of the self-assembling ureido-imidazole **10.22** and section of its crystal structure showing the arrangement of water molecules in the channel formed from stacked subunits (a), structure of the pillar[5]arene **10.23** and crystal structure of an analog with a chain of four water molecules in the cavity, and structure of trianglamine **10.24**. Hydrogen atoms in the crystal structures, except those at nitrogen atoms, are omitted for reasons of clarity. The side chains of **10.22** are truncated in the crystal structure in (a). The water molecules are shown as space-filling models. The protons of the water molecules in the crystal structure in (b) could not be located.

achieved using alkyl-ureido-trianglamines such as **10.24** [31]. This macrocycle forms tubular superstructures in the solid state and presumably also in lipid bilayers, whose permeability approaches 10^8 water molecules per second per channel, which is only one order of magnitude lower than the permeability of natural water transporters. The pores are characterized by an alternating pattern of polar regions with a diameter similar to the pore size of aquaporins, where hydrogen bonding between the pore walls and water molecules is possible, and larger hydrophobic regions that interrupt the water channels. Since water transport is therefore discontinuous and does not involve a chain of hydrogen-bonded water molecules, proton transport is not possible.

Bibliography

[1]   Sessa G, Weissmann G. Phospholipid spherules (liposomes) as a model for biological membranes. J. Lipid Res. 1968, 9, 310–8.

[2]   Doyle DA, Cabral JM, Pfuetzner RA, Kuo A, Gulbis JM, Cohen SL, Chait BT, MacKinnon R. The structure of the potassium channel: molecular basis of K^+ conduction and selectivity. Science 1998, 280, 69–77.

[3]   Townsley LE, Tucker WA, Sham S, Hinton JF. Structures of gramicidins A, B, and C incorporated into sodium dodecyl sulfate micelles. Biochemistry 2001, 40, 11676–86.

[4]   Fyles TM. Synthetic ion channels in bilayer membranes. Chem. Soc. Rev. 2007, 36, 335–47.

[5]   Sakai N, Mareda J, Matile S. Artificial β-barrels. Acc. Chem. Res. 2008, 41, 1354–65.

[6]   Sakai N, Sordé N, Das G, Perrottet P, Gerard D, Matile S. Synthetic multifunctional pores: deletion and inversion of anion/cation selectivity using pM and pH. Org. Biomol. Chem. 2003, 1, 1226–31.

[7]   Clark TD, Buehler LK, Ghadiri MR. Self-assembling cyclic β^3-peptide nanotubes as artificial transmembrane ion channels. J. Am. Chem. Soc. 1998, 120, 651–6.

[8]   He L, Zhang T, Zhu C, Yan T, Liu J. Crown ether-based ion transporters in bilayer membranes. Chem. Eur. J. 2023, 29, e202300044.

[9]   Gokel GW, Daschbach MM. Coordination and transport of alkali metal cations through phospholipid bilayer membranes by hydraphile channels. Coord. Chem. Rev. 2008, 252, 886–902.

[10] Fyles TM, James TD, Kaye KC. Biomimetic ion transport: on the mechanism of ion transport by an artificial ion channel mimic. Can. J. Chem. 1990, 68, 976–8.

[11] Ovchinnikov YA, Ivanov VT. Conformational states and biological activity of cyclic peptides. Tetrahedron 1975, 31, 2177–209.

[12] Inoue Y, Gokel GW. Cation Binding by Macrocycles. Marcel Dekker: New York, 1990.

[13] Hendrixson RR, Mack MP, Palmer RA, Ottolenghi A, Ghirardelli RG. Oral toxicity of the cyclic polyethers – 12-crown-4, 15-crown-5, and 18-crown-6 – in mice. Toxicol. Appl. Pharm. 1978, 44, 263–8.

[14] Johnson TG, Langton MJ. Molecular machines for the control of transmembrane transport. J. Am. Chem. Soc. 2023, 145, 27167–84.

[15] Chen S, Wang Y, Nie T, Bao C, Wang C, Xu T, Lin Q, Qu DH, Gong X, Yang Y, Zhu L, Tian H. An artificial molecular shuttle operates in lipid bilayers for ion transport. J. Am. Chem. Soc. 2018, 140, 17992–8.

[16] Dutzler R, Campbell EB, Cadene M, Chait BT, MacKinnon R. X-ray structure of a ClC chloride channel at 3.0 Å reveals the molecular basis of anion selectivity. Nature 2002, 415, 287–94.

[17] Gorteau V, Bollot G, Mareda J, Perez-Velasco A, Matile S. Rigid oligonaphthalenediimide rods as transmembrane anion-π slides. J. Am. Chem. Soc. 2006, 128, 14788–9.

[18] Jentzsch AV, Hennig A, Mareda J, Matile S. Synthetic ion transporters that work with anion-π interactions, halogen bonds, and anion–macrodipole interactions. Acc. Chem. Res. 2013, 46, 2791–800.

[19] Gorteau V, Bollot G, Mareda J, Matile S. Rigid-rod anion-π slides for multiion hopping across lipid bilayers. Org. Biomol. Chem. 2007, 5, 3000–12.

[20] Yamnitz CR, Gokel GW. Synthetic, biologically active amphiphilic peptides. Chem. Biodiv. 2007, 4, 1395–412.

[21] Davis JT. Anion binding and transport by prodigiosin and its analogs. Top. Heterocycl. Chem. 2010, 24, 145–76.

[22] Busschaert N, Caltagirone C, Van Rossom W, Gale PA. Applications of supramolecular anion recognition. Chem. Rev. 2015, 115, 8038–155.

[23] Santacroce PV, Davis JT, Light ME, Gale PA, Iglesias-Sánchez JC, Prados P, Quesada R. Conformational control of transmembrane Cl⁻ transport. J. Am. Chem. Soc. 2007, 129, 1886–7.

[24] Yamnitz CR, Negin S, Carasel A, Winter RK, Gokel GW. Dianilides of dipicolinic acid function as synthetic chloride channels. Chem. Commun. 2010, 46, 2838–40.

[25] Valkenier H, Davis AP. Making a match for valinomycin: steroidal scaffolds in the design of electroneutral, electrogenic anion carriers. Acc. Chem. Res. 2013, 46, 2898–909.

[26] Edwards SJ, Valkenier H, Busschaert N, Gale PA, Davis AP. High-affinity anion binding by steroidal squaramide receptors. Angew. Chem. Int. Ed. 2015, 54, 4592–6.

[27] Janout V, Regen SL. Bioconjugate-based molecular umbrellas. Bioconjugate Chem. 2009, 20, 183–92.

[28] Agre P. Aquaporins water channels (Nobel lecture). Angew. Chem. Int. Ed. 2004, 43, 4278–90.

[29] Le Duc Y, Michau M, Gilles A, Gence V, Legrand YM, van der Lee A, Tingry S, Barboiu M. Imidazole-quartet water and proton dipolar channels. Angew. Chem. Int. Ed. 2011, 50, 11366–72.

[30] Hu HB, Chen Z, Tang G, Hou JL, Li ZT. Single-molecular artificial transmembrane water channels. J. Am. Chem. Soc. 2012, 134, 8384–7.

[31] Andrei IM, Chaix A, Benkhaled BT, Dupuis R, Gomri C, Petit E, Polentarutti M, van der Lee A, Semsarilar M, Barboiu M. Selective water pore recognition and transport through self-assembled alkyl-ureido-trianglamine artificial water channels. J. Am. Chem. Soc. 2023, 145, 21213–21.

11 Detecting molecules

CONSPECTUS: *One drop of an indicator solution is enough to determine whether water is acidic or basic. Supramolecular probes work in a similar way, changing color or some other easily measurable physical property when they sense the presence of an analyte. Sensing is based on interactions between the analyte and a receptor unit in the probe, which in turn affects the optical or redox properties of a reporter unit. Both units can be part of the same molecule or can be assembled noncovalently. Either way, the selectivity of detection is determined and limited by the selectivity of the receptor. This limitation is overcome by combining several different receptors, each of which interacts with the substrate in a slightly different way. The resulting response pattern allows differentiation of structurally related analytes or simultaneous detection of multiple analytes. This sensing strategy is similar to the way we perceive odors or tastes, demonstrating that supramolecular chemistry can mimic yet another fundamental principle of natural systems.*

11.1 Introduction

When two molecules come together to form a complex, the environment of the protons or chromophores in the binding partners changes, which can be detected using the techniques discussed in Section 2.2. Spectral changes induced by complex formation can thus be used to characterize receptor–substrate complexes in terms of structure and stability, but once the properties of a complex are known, a receptor can also be used as a probe to analyze whether and how much of the substrate is present in a sample of unknown composition. If the addition of the receptor to a solution does not produce the signal normally associated with complex formation, for example, one can conclude that the substrate is absent (if complex formation is not prevented by competing analytes). A measurable effect, on the other hand, not only qualitatively confirms the presence of the substrate, but also allows quantification of its concentration using a previously established relationship between the measured effect and the extent of complex formation. Such measurements form the basis of supramolecular analytical chemistry [1].

This branch of supramolecular chemistry is concerned with the development of strategies for the reliable and convenient detection of one or more analytes, ideally without the use of elaborate equipment. Receptors that change color upon complex formation, for example, signal the presence of an analyte in the same way that an indicator signals a change in pH. Quantitative measurements can be made using photometers, which are indeed widely used in supramolecular analytical chemistry. Of the other techniques discussed in Section 2.2, fluorescence spectroscopy and electrochemical methods are relevant when the emission or redox properties of the binding partners are affected by complex formation.

As attractive as supramolecular approaches are for achieving the sensitive and selective detection of a given analyte, their role in analytical chemistry should not be

https://doi.org/10.1515/9783111315171-011

overestimated. With the vast array of sophisticated analytical instrumentation available today, it is rare that a problem can be solved with a supramolecular approach but not with another existing technique. The simplicity and selectivity of supramolecular systems are still advantageous. In addition, by using the concepts of supramolecular chemistry, it is possible to design systems that work in a similar way to how smells or tastes are perceived. As a result, complex mixtures of compounds can be analyzed in a relatively simple way and with an accuracy that is sometimes difficult to achieve with other techniques. Supramolecular methods therefore have a firm place in analytical chemistry, as this chapter will show, but before we look at specific examples, it is important to introduce and define relevant terms.

What is a sensor?

A good term for the receptor mentioned above, which changes color when it binds to an analyte, would be *indicator*. Although very appropriate, this term is rarely used, perhaps to avoid confusion with pH indicators. Alternatively, the term *probe* is correct and well established. The most commonly used term for a receptor that can sense a substrate is *sensor*, or more precisely *chemosensor*, but these terms are somewhat problematic and their use has therefore been criticized [2]. The reason is that a true sensor is more than just a device that signals an event. A sensor also transmits information and triggers a response. For example, a motion detector that simply detects someone moving in your yard is not very useful, but it does serve a purpose when a light or alarm is subsequently turned on. In combination with feedback loops, sensors can even be used for more sophisticated applications such as process control. Chemosensors do not have these capabilities and therefore perform only half the tasks of a real sensor. Nevertheless, the term chemosensor is so firmly established in supramolecular analytical chemistry that it is not helpful to avoid it [3]. After all, it is the context that determines whether the term refers to a supramolecular or a real sensor, just as it is usually no problem for words like *bat* to have different meanings in different contexts.

The signal generated by a chemosensor is a direct consequence of complex formation or, in other words, reversible interaction with the analyte. Thus, like most other supramolecular systems, chemosensors operate under equilibrium conditions, that is, under thermodynamic control (Figure 11.1a). They are therefore able to respond in real time to changes in analyte concentration, allowing them to monitor reactions or other processes. In this respect, they differ fundamentally from another class of molecular probes known as *chemodosimeters* [4]. A chemodosimeter is a compound whose irreversible, kinetically controlled reaction with the analyte, which may involve cleavage and/or formation of covalent bonds, triggers a change in optical properties (Figure 11.1b). Such a compound is therefore also a useful optical probe, but because it is irreversibly modified when sensing the analyte, it only reports the total number of analyte molecules to which it has been exposed and cannot monitor

(a)

The signal intensity reflects the concentration of the complex at equilibrium

(b)

The signal intensity reflects the number of analyte molecules consumed in the reaction

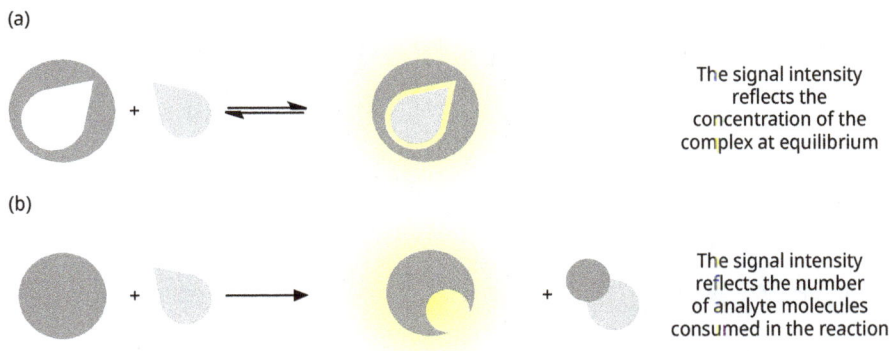

Figure 11.1: Schematic illustration of the working principles of a chemosensor (a) and a chemodosimeter (b).

changes in analyte concentration. In addition, the operating principle of a chemo-dosimeter does not involve a molecular recognition event, which is why these probes are outside the realm of supramolecular chemistry and will not be discussed here.

Analyte detection can be based on a single chemosensor molecule or an ensemble of molecules. Single molecules contain both a receptor and a reporter moiety and are typically designed to selectively target one analyte or a family of analytes such as amino acids or nucleotides. Sensing ensembles serve similar purposes, but consist of mixtures of noncovalently interacting receptor and reporter molecules. Each receptor–reporter pair in these mixtures typically generates a signal similar to a conventional chemosensor, but analysis of the composite information requires special methods. In addition, such assays can include any number of components, with increasing complexity usually having a positive effect on the fidelity of detection. Chemosensor arrays therefore allow analyses that are more difficult to perform with similar precision using other approaches. Thus, a variety of strategies exist to achieve selective and sensitive supramolecular detection of molecules, which will be discussed separately in the following parts of this chapter.

11.2 Single analyte sensing

11.2.1 Direct optical sensing

A probe that responds to the presence of an analyte by changing its optical properties must contain two components: a receptor moiety that interacts with the analyte and a chromophore that acts as a reporter by signaling the binding event. The integration of both groups into the same molecule usually requires the additional presence of a spacer, leading to the receptor–spacer–reporter paradigm proposed by A. Prasanna de Silva for the development of single molecule chemosensors (Figure 11.2) [5].

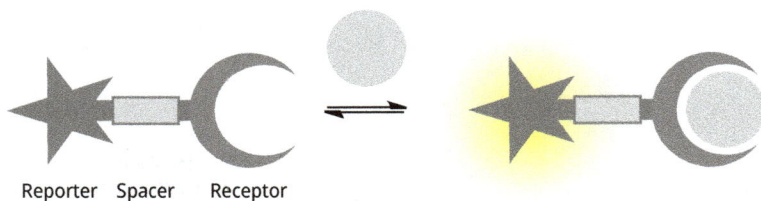

Reporter   Spacer   Receptor

Figure 11.2: General structure of a chemosensor combining a receptor and a reporter unit in the same molecule, covalently linked by a suitable spacer.

The design of such chemosensors is very modular and can include a wide range of different receptors and reporters to achieve the desired properties. One criterion to consider is analyte selectivity, which is controlled by the receptor. For example, when designing a probe for cations, this moiety can be based on any type of cation receptor that binds to the analyte of interest. Other considerations include the medium in which the assay will be performed and the stability of the complex, which determines the minimum analyte concentration that can be detected. Since selective detection is usually desired, the extent to which the analyte is preferred over competing substrates is also important. If selective binding cannot be ensured because the analyte is structurally too closely related to other components in the mixture, single molecule chemosensors reach their limits.

The reporter unit determines the technique required to follow the binding event and the sensitivity of the probe. When chromophores that absorb or emit in the visible region of the electromagnetic spectrum are used as reporters, the presence of the analyte can be conveniently detected with the naked eye. Spectral changes that are not immediately visible can be followed using UV–vis or more sensitive fluorescence techniques, which can also be used for quantitative measurements.

Finally, the spacer subunit ensures the proper mutual arrangement of the receptor and reporter units. If sensing requires direct electronic coupling of the two components, the spacer may be absent. Its presence is critical if too close proximity has a negative impact on the functionality of the receptor and/or reporter group. For example, direct coupling of the two subunits may interfere with analyte binding, which must be avoided.

Several photophysical processes can lead to an optical signal, with early chemosensors mostly relying on internal charge-transfer (ICT), sometimes called photoinduced charge-transfer (PCT). The reporter unit of such systems is usually a push-pull π-electron system, as in one of the first chemosensors, the crown ether **11.1** described by Fritz Vögtle (Figure 11.3) [6]. Photoexcitation facilitates the transfer of an electron between the conjugated subunits from one side to the other, resulting in an excited state charge distribution that is significantly different from the ground state. The positioning of an ion close to one side of the chromophore stabilizes or destabilizes this charge distribution, depending on the charge of the ion, and thus influences the frequency at which photoexcitation occurs. In the case of **11.1**, complexation of a potassium ion causes a 20 nm hypsochromic shift (shorter wavelength, higher frequency) in the absorption band, while sodium or rubidium ions have smaller effects, consistent with the generally lower affinity of 18-crown-6 derivatives for ions smaller and larger than potassium (Section 4.1.1). The effect of the cations is due to their interaction with the lone pair of the crown ether nitrogen atom, which prevents conjugation with the aromatic π-system and causes the excitation to require more energy.

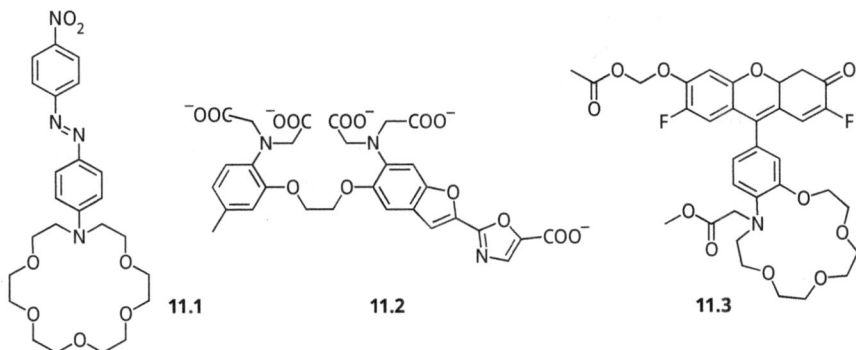

Figure 11.3: Molecular structures for the three ICT chemosensors **11.1** (for K^+), **11.2** (for Ca^{2+}), and **11.3** (for Na^+).

In general, a cation interacting with the donor moiety of a push–pull system (the amino group in **11.1**) will decrease the transition dipole moment and thus shift the ICT band to higher frequencies, while an anion interacting with the acceptor moiety will have the opposite effect. Other examples of ICT-based chemosensors include the chelating receptor **11.2**, which is used for Ca^{2+} sensing, and the xanthene-derived crown ether **11.3** (Figure 11.3), which is commercially available as CoroNa™ Green and exhibits an increase in green fluorescence intensity upon Na^+ binding. Both chemosensors are useful for cellular imaging (Section 12.2.3) [1, 5].

Fluorescence chemosensors alternatively rely on photoinduced electron-transfer (PET) for signal generation. These chemosensors contain a fluorescent chromophore and at least one functional group in close proximity that can donate an electron. Once the fluorophore is excited, this group transfers an electron to the now singly occupied ground state HOMO. As a result, the excited electron cannot return to this orbital by emitting a photon, and the molecule must choose nonradiative pathways to dissipate its excess energy (Figure 11.4a). Fluorescence is turned on by preventing the electron transfer that causes quenching. The actual behavior therefore depends on whether quenching occurs in the uncomplexed or complexed state. If the interactions with the analyte prevent quenching, the chemosensor goes from the nonfluorescent state to the fluorescent state upon analyte binding, making it a so-called turn-on chemosensor (Figure 11.4b). Conversely, if analyte binding enables PET, the chemosensor will respond to the presence of the analyte by turning off its fluorescence.

Both types of chemosensors exist, but a turn-on behavior is preferable. The reason is that it is easier to detect light against a dark background than it is to detect the disappearance of light in a bright environment. Consider an office building where all the lights behind the windows are on. If the lights in one room are turned off, it will take a while to find the dark window when looking at the building from the outside. On the other hand, if all the windows are dark, the room in which the lights are turned on can be spotted immediately. The important aspect in the context of sensing is that turn-off chemosensors become darker as the concentration of the analyte increases, and reliably measuring the faint light emitted by the sample as the analyte concentration changes against the background noise is challenging. Conversely, the signal from a turn-on chemosensor becomes more intense as the analyte concentration increases, with variations in background noise having little effect on the measurement.

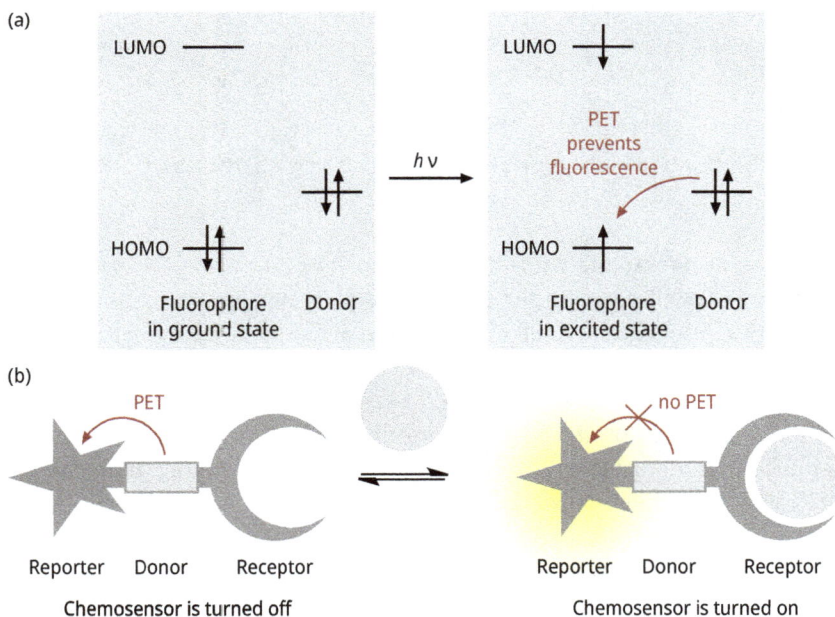

Figure 11.4: Schematic illustration of photoinduced electron-transfer (PET) (a), and the working principle of a turn-on chemosensor (b). During PET, the donor transfers an electron to the depleted orbital resulting from photoexcitation, preventing the system from relaxing to the ground state by emitting a photon.

Figure 11.5 shows examples of PET chemosensors. All of these systems contain one or more benzylic amino groups attached to the anthracene fluorophore. The chemosensors differ in the functional groups responsible for analyte recognition, with **11.4** containing a crown ether for metal ion binding, **11.5** containing ammonium groups to interact with a dihydrogen pyrophosphate anion, and **11.6** containing boronic acid groups to form esters with carbohydrates (Section 4.2.4). All free receptors are nonfluorescent due to the free electron pairs of the benzylic amino groups that act as donors (provided that the pH of the solution is sufficiently high to ensure that the amino groups are not protonated). Once the analytes are bound, electron transfer is no longer possible because the amino groups participate in stabilizing the complexes. In **11.4**, the lone pair of the amino group coordinates to the bound metal ion, while the corresponding electron pairs in **11.5** accept protons from the substrate. In **11.6**, the lone pairs of the nitrogen atoms coordinate to and stabilize the tetrahedral form of the boronic acid esters formed with the sugar. All of these compounds therefore begin to fluoresce upon analyte binding.

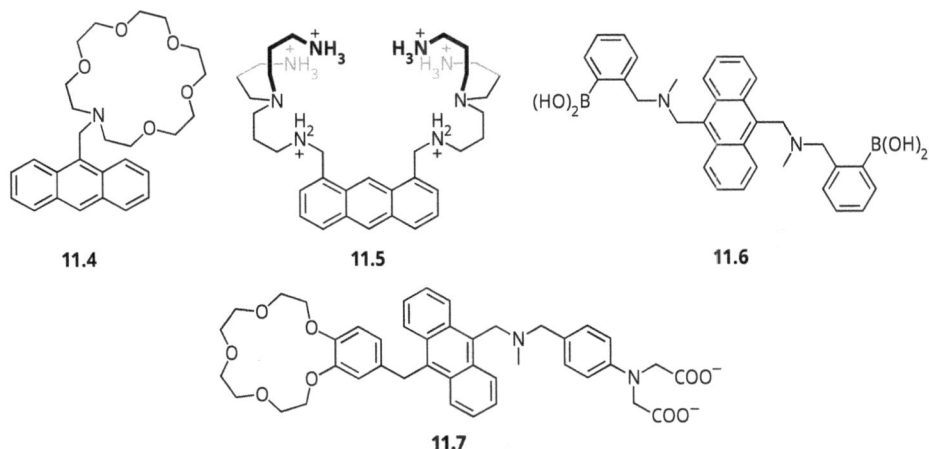

11.4   **11.5**   **11.6**

11.7

Figure 11.5: Molecular structures of the PET chemosensors **11.4** (for K$^+$), **11.5** (for H$_2$P$_2$O$_7$$^{2-}$), and **11.6** (for glucose), and of the *lab-on-a-molecule* **11.7**.

PET chemosensors such as **11.7** respond to more than one analyte. This compound contains a crown ether moiety to interact with Na$^+$, a phenyliminodiacetate group as a chelating group for Zn^{2+}, and a tertiary amino group that can accept a proton [7]. In the absence of any of these analytes, **11.7** will not fluoresce due to PET. Each analyte alone, or even any combination of two analytes, will not turn on fluorescence because there will always be a donor that quenches the anthracene chromophore. For example, when Na$^+$ binds to the crown ether and a proton binds to the amino group, the electron pair remaining in the phenyliminodiacetate moiety causes quenching. When

Na$^+$ and Zn^{2+} are bound simultaneously, the crown ether moiety is responsible for PET. The fluorescence of **11.7** is therefore only turned on when all three analytes are present. This ability to simultaneously analyze three different inputs is why **11.7** has been called a *lab-on-a-molecule*. Other such multifunctional chemosensors are known. Their behavior is often described using Boolean algebra, with **11.7** representing an AND logic gate [8].

Conformational changes upon analyte binding result in an optical signal when these changes affect the distance or orientation of the chromophores in an optical chemosensor, as in the case of the calix[4]arene derivative **11.8**, whose two pyrene units are in close proximity in the uncomplexed form (Figure 11.6) [9]. The typical pyrene excimer band is therefore observed in the UV–vis spectrum at about 480 nm, resulting from the interaction of a ground state and an excited state pyrene ring. Both rings are pushed apart when a sodium ion binds to the oxygen atoms (Section 4.1.7), causing the excimer band to disappear. Chemosensors that report analyte binding by the appearance of the excimer band also exist.

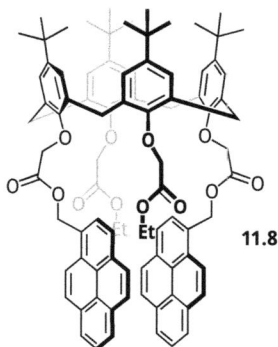

Figure 11.6: Molecular structure of the calix[4]arene derivative **11.8**, which responds to the presence of Na$^+$ by the disappearance of the pyrene excimer band.

When a conformationally flexible chemosensor contains two different fluorophores, Förster resonance energy transfer (FRET) is possible. In this case, the donor fluorophore is selectively excited and its energy is nonradiatively transferred to the acceptor chromophore, which emits a photon. This process requires an overlap of the donor emission band and the acceptor absorption band and depends on the relative orientation of the donor emission and acceptor absorption dipole moments. The efficiency of the energy transfer correlates with the distance r between the donor and acceptor with a dependence of r^{-6}. As a consequence, the acceptor emission progressively increases as the distance between the two groups becomes smaller, while the donor emission decreases. In addition to applications in chemosensing [10], FRET is used to probe the distances between different sites in biomolecules.

Other mechanisms that cause an optical response in a chemosensor are the rigidification of the molecule upon substrate binding, which usually results in an increase in fluorescence intensity due to the reduced number of nonradiative pathways for the excited state to dissipate its energy, and the effect of analyte binding on the absorption properties of metal-containing receptors [11, 12]. In addition, a number of cyclodextrin–dye conjugates are known whose interaction with the analyte causes the expulsion of the chromophore from the cyclodextrin cavity. The change in environment from apolar, when the chromophore is inside the cavity, to polar, when it is exposed to the external aqueous solution, causes a decrease in fluorescence intensity in the presence of the analyte [13].

11.2.2 Indirect optical sensing

The second approach to achieving optical sensing is to probe not the properties of the receptor, but those of an additional component that interacts with the receptor and is displaced from the receptor binding site by the analyte [14]. This strategy, illustrated in Figure 11.7, is called *indicator displacement assay* (IDA) and was popularized by Eric V. Anslyn, who demonstrated its practical use in many applications [15].

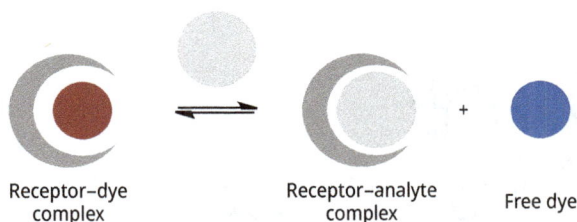

| Receptor–dye complex | Receptor–analyte complex | Free dye |

Figure 11.7: General principle of an indicator displacement assay (IDA). Sensing involves measuring the changes in the optical properties of the dye as it is displaced from its complex with the receptor by the analyte.

The actual probe consists of a mixture of a receptor and a reporter molecule. This reporter is usually a dye, often a conventional pH indicator, whose optical properties change upon binding to the receptor. The UV–vis spectrum in the absence of the analyte therefore reflects the optical properties of the receptor–dye complex and the extent to which it is formed. When the analyte is added, it competes with the dye for the binding site and since an optical signal is only expected if the analyte wins this competition, the receptor–analyte complex must be more stable than the receptor–dye complex. If this is the case, increasing the concentration of the analyte will cause a progressive displacement of the dye from the binding site. The stability constant of the receptor–analyte complex can be calculated from the associated spectral changes using the mathematical formalism explained in Appendix 13.2. This procedure differs from that used to determine the stability of a 1:1 complex because three binding partners must be considered instead of two.

Two examples from the Anslyn group will illustrate this concept. The first involves the use of receptor **11.9** (Figure 11.8), in which three 2-aminoimidazoline units are arranged around a 1,3,5-triethylbenzene ring [16]. The six substituents have an alternating up–down arrangement so that the three heterocyclic units end up on the same side of the ring (Section 4.1.12). Since 2-aminoimidazolines are also protonated at physiological pH, **11.9** has a preorganized binding site for anionic substrates. This receptor interacts strongly with tricarboxylates such as citrate, but also binds dicarboxylates, phosphates, and the trianionic dye 5-carboxyfluorescein, although with much lower affinity. For example, in 25 vol% water/methanol (buffered to pH 7.4) the $\log K_a$ of the citrate complex is 5.5, while the 5-carboxyfluorescein complex has a $\log K_a$ of only 3.7, which is optimal for an IDA. Binding of 5-carboxyfluorescein is associated with an increase in the absorbance and emission bands, so displacement of the dye by the more strongly bound citrate has the opposite effect. Plotting these optical changes against citrate concentration yields calibration curves that allow the quantification of the citrate content of unknown samples, including various sports drinks [16].

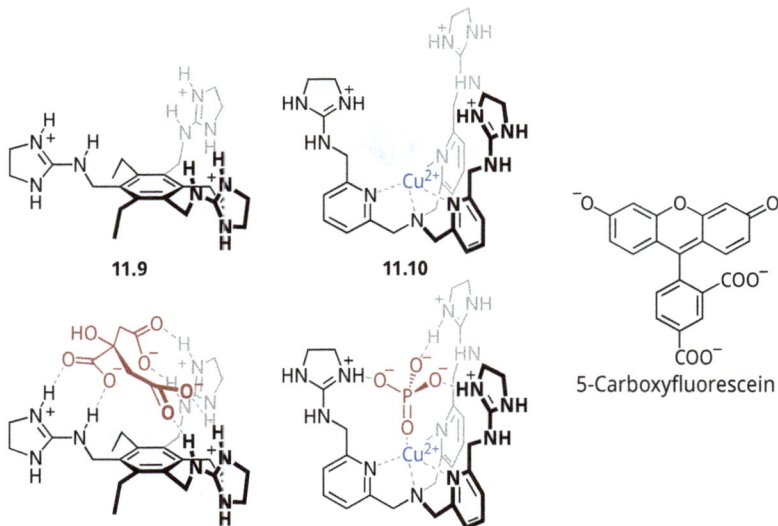

Figure 11.8: Molecular structures of receptors **11.9** and **11.10**, schematic structures of their complexes with citrate and phosphate, respectively, and molecular structure of 5-carboxyfluorescein.

The orthophosphate receptor **11.10** (Figure 11.8) has three 2-aminoimidazoline units arranged around a binding site containing a copper(II) ion [17]. It also interacts with 5-carboxyfluorescein in 50 vol% water/methanol (buffered to pH 7.4), resulting in a color change of the solution from yellow to light orange. Subsequent addition of phosphate restores the original color and the associated changes in the UV–vis spectrum can be used to determine phosphate in saliva.

IDAs are also useful for monitoring enzymatic reactions. Such so-called supramolecular tandem enzyme assays are based on receptor–dye complexes as sensing ensembles that respond differently to the product of an enzymatic reaction than to the substrate [18]. Product formation can thus be followed by optical measurements and even tracked in real time without having to label the compounds involved in the reaction. The example from the Nau group in Figure 11.9 illustrates the concept [19]. In this reaction, lysine is converted to the diammonium form of 1,5-diaminopentane by a decarboxylase. The receptor–dye ensemble consists of CB[7] and the fluorescent dye dapoxyl. This dye is weakly fluorescent in the medium in which the assay is performed, but becomes strongly fluorescent when incorporated into the CB[7] cavity. The corresponding complex is more stable than the complex between CB[7] and lysine, but dissociates in the presence of diammonium ions, which are better guests. The formation of the decarboxylation product therefore causes a decrease in fluorescence which, when followed, allows the determination of the rate of the reaction or the activity of the enzyme.

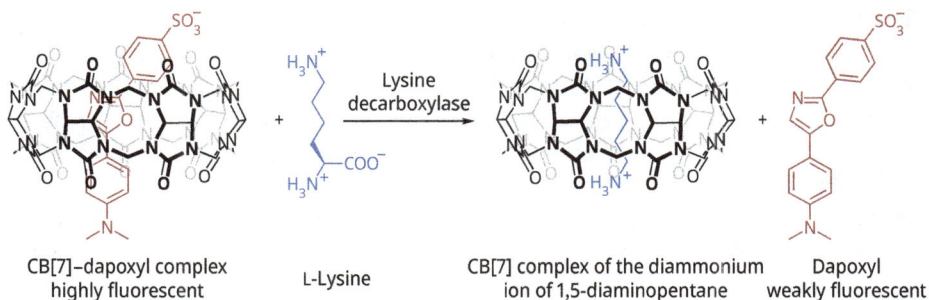

| CB[7]–dapoxyl complex highly fluorescent | L-Lysine | CB[7] complex of the diammonium ion of 1,5-diaminopentane | Dapoxyl weakly fluorescent |

Figure 11.9: Example of a supramolecular tandem enzyme assay for monitoring the decarboxylation of lysine using the CB[7]–dapoxyl complex as the sensing ensemble. This complex is stable in the presence of lysine, but releases the dye upon formation of the more strongly bound reaction product.

It should be noted that the concept of indicator displacement also allows for optical sensing when the receptor and indicator are covalently linked rather than noncovalently assembled. For example, the tripodal chemosensor **11.11** contains two thiourea groups as anion recognition moieties and a naphthyl carboxylate as an anionic chromophore (Figure 11.10) [20]. In the absence of analyte, **11.11** prefers a conformation in which the anionic substituent interacts with the urea groups. These intramolecular interactions are no longer possible in the glyphosate complex, so that the conformational rearrangement induced by the analyte leads to a pronounced reduction in fluorescence. Thus, indicator displacement plays a role in the sensing mechanism, but since the reporter and dye are covalently connected, the probe behaves strictly as a direct chemosensor, similar to the cyclodextrin derivatives mentioned at the end of the previous chapter.

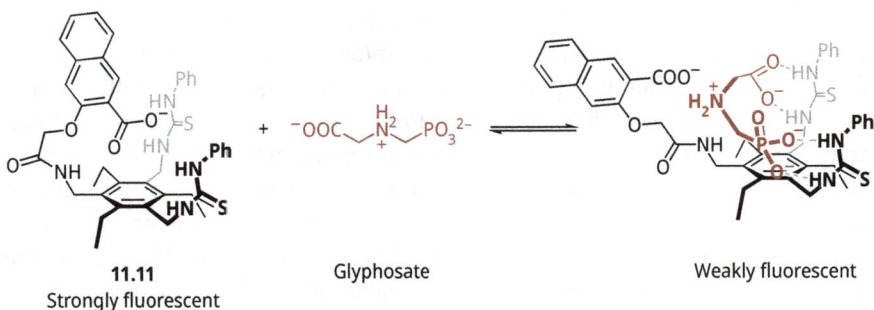

11.11
Strongly fluorescent

Glyphosate

Weakly fluorescent

Figure 11.10: Sensing of glyphosate with the receptor–dye conjugate **11.11**. Analyte binding in this case prevents the anionic chromophore from interacting with the thiourea moieties.

11.2.3 Direct electrochemical sensing

The use of electrochemical methods is well established in analytical chemistry. For example, nonredox-active inorganic cations or anions are reliably detected using ion-selective electrodes, many of which are commercially available [21]. These electrodes typically contain ionophores embedded in a polymeric membrane, which are responsible for the selective recognition of the charged guests. Alternatively, substrate sensing can be achieved using receptors immobilized on gold electrodes. The underlying principles will not be discussed here. Instead, we will focus on molecular chemosensors that allow electrochemical detection of the binding event. The design of such probes is based on the receptor–spacer–reporter paradigm mentioned earlier. In this case, the spacer connects the receptor to a redox-active group that serves as the reporter. Analyte binding affects the electrode potential at which redox processes occur, which can be measured by cyclic voltammetry.

Although the structures of optical and redox chemosensors are closely related, there is a fundamental difference in their behavior. In optical systems, analyte binding affects the properties of the reporter unit, and whether this unit is in the ground or excited state has little effect on the binding event because the photophysical processes underlying sensing are typically faster than the binding equilibrium. This is different in electrochemical chemosensors, where a change in the redox state of the reporter unit has a direct impact on the strength with which the receptor and analyte interact. The correlation is best understood using the square scheme in Figure 11.11a.

The vertical equilibria in this scheme describe the binding of the substrate S to the receptor R, which forms the complexes C_{red} and C_{ox}, depending on whether it is in the reduced or oxidized state R_{red} or R_{ox}. The horizontal equilibria describe the oxidation and reduction of R and C. The cyclic arrangement of these equilibria implies that the Gibbs free energy change of the entire cycle is zero. Thus, using the expressions $\Delta G^0 = -RT \ln K_a$ for the binding equilibria and $\Delta G^0 = -nF \Delta E^0$ for the electron-transfer

(a)

(b)

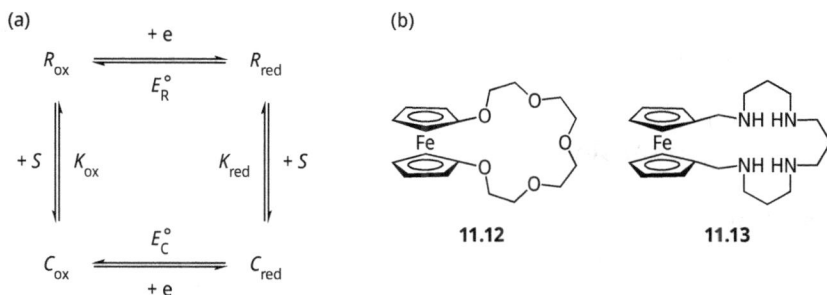

Figure 11.11: Cycle of equilibria associated with the binding of a substrate S to the oxidized and reduced versions of the receptor R to form the corresponding complexes C, and with the interconversion of the oxidized and reduced receptor and complex species (a). K is the association constant of the complexes and E^0 is the formal potential of the electron-transfer reactions. The molecular structures of the electrochemical chemosensors **11.12** (for Na^+) and **11.13** (for $HPO_4^{2-}/H_2PO_4^-$) are shown in (b).

processes, equation (11.1) can be derived, which shows that the shift in the electrode potential of a redox chemosensor $E_C^0 - E_R^0$ induced by analyte binding is a direct consequence of the different affinities with which the oxidized and reduced versions of the receptor bind to the analyte:

$$nF\left(E_C^0 - E_R^0\right) = RT \ln \frac{K_{red}}{K_{ox}} \tag{11.1}$$

Although the exact mechanism by which the redox state of the reporter unit affects binding varies from chemosensor to chemosensor [22], the correlation is straightforward: the generation of a positive charge in the reporter unit upon oxidation will negatively affect the binding of a nearby cation but will favor the binding of an anion. Electrochemical chemosensors for cations therefore typically signal the presence of the substrate by a positive (anodic) shift of the redox potential ($E_C^0 - E_R^0 > 0$ oxidation becomes more difficult), while the opposite is observed for anion binding.

The modular design of electrochemical chemosensors leads to a similar high structural variability as in optical systems. The receptor units are mostly based on established recognition elements, while a few reporter units have proven to be particularly useful. These reporters are based on either metal complexes or organic redox-active moieties, with ferrocene probably being the most commonly used metal-containing reporter group and tetrathiafulvalene the metal-free counterpart. Chemosensors **11.12** and **11.13** (Figure 11.11b) are examples of ferrocene-based systems. They allow the detection of sodium ions (**11.12**) [23] and hydrogen phosphate/dihydrogen phosphate ions (**11.13**) [24]. Their electrochemical responses are consistent with the above trends in that **11.12** shows a shift to a higher potential upon sodium binding, while the presence of the phosphate ions in the case of **11.13** causes a shift to a lower potential. Of the many known electrochemical chemosensors, those for anions were primarily developed in the group of Paul D. Beer [25].

11.3 Multi-analyte sensing

Taste and smell are senses that allow us to detect molecules in our environment. Although these senses differ in their exact biochemical mechanisms, the working principles are similar. When we smell coffee, for example, our olfactory system must detect and analyze the ratios of at least forty volatile compounds. Obviously, this task cannot be performed by a single nerve cell specialized for coffee. Instead, we have a large number, probably thousands, of different sensory neurons, each of which responds to the components of the coffee aroma by recognizing characteristic properties, such as structure, solubility, diffusion rate, and so on. Thus, a single component triggers responses in many neurons but typically to varying degrees. All of the signals combined result in a characteristic response pattern that is analyzed by the brain and associated with the aroma of coffee. To perceive taste, we have specialized receptors on our tongue that typically work in concert with olfactory receptors to produce such signal patterns. Thus, smell and taste receptors do not have to be highly selective to perceive aromas. Instead, the selectivity comes from the combination of many receptors, each of which recognizes specific features in a molecule without necessarily being selective for a single molecule.

The principles underlying our ability to smell and taste inspired the use of mixtures of receptors for sensing purposes [26]. This differential sensing strategy has several advantages. First, it allows the detection of multiple analytes with a single assay. Second, even structurally closely related analytes, for which the development of a selective receptor would be challenging, can be differentiated. Sensing relies on the combination of nonselective receptors, similar to how smell and taste receptors work, because the characteristic fingerprints produced by cross-reactive receptors that bind to the same analyte with different affinities provide more information about the identity of the analyte than the yes/no response of a single selective receptor. The corresponding chemosensors are either mixed in solution or spatially separated in an array, as shown schematically in Figure 11.12. Both approaches can involve direct or indirect sensing, but the use of sensing ensembles is often more convenient because it requires only a pool of receptors and indicators and no additional synthetic effort.

When multiple chemosensors are combined in the same solution, the UV–vis spectrum of the mixture reflects the number and relative concentrations of all the absorbing components (Figure 11.12a). These components include either the individual chemosensors or, if the assay is based on sensing ensembles, the receptors, indicators, and their complexes. The addition of an analyte causes characteristic perturbations in the composite UV–vis spectrum due to the different interactions of the analyte with the chemosensors or the extent to which the analyte shifts the equilibria in which the sensing ensembles are involved. Regardless of the sensing mechanism, the spectral changes are highly specific and allow discrimination of even structurally closely related compounds.

The spatial separation of different chemosensors or sensing ensembles is generally based on the use of microtiter plates containing a single probe in each well (Figure 11.12b). The presence of an analyte produces a characteristic response in each well, the combination of which is analyzed. Since the resulting patterns are again analyte-specific, this method also allows differentiation of multiple analytes.

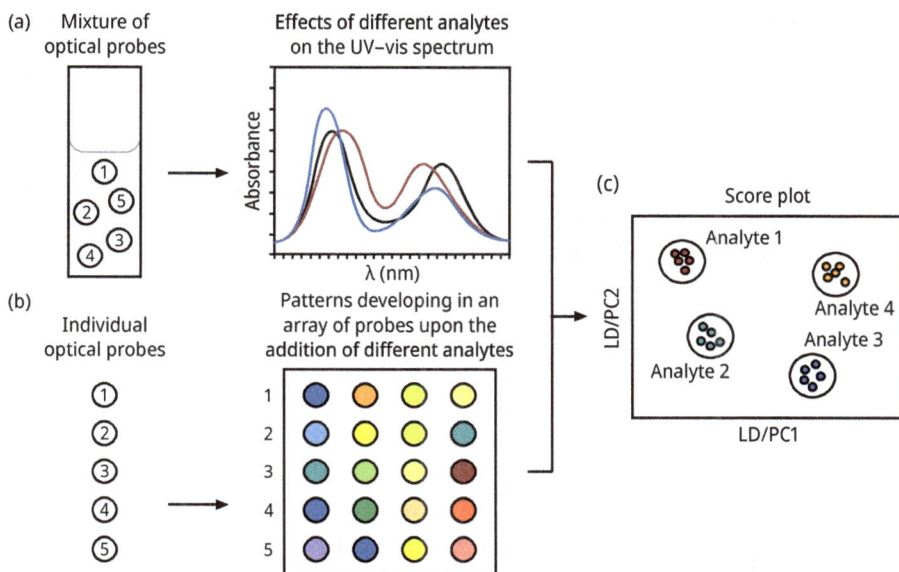

Figure 11.12: Schematic illustration of strategies to achieve multiple analyte detection. In (a), multiple optical probes are combined in the same solution and the characteristic effects of the analyte on the UV–vis spectrum of the mixture are analyzed. In (b), individual optical probes are spatially separated in an array. The pattern produced when the analyte is added to each probe is then used to obtain information about its nature. An example of a score plot resulting from such assays is shown in (c).

Since the amount of data resulting from such measurements is significantly larger than that generated by a single chemosensor, special chemometric methods such as principal component analysis (PCA) or linear discriminant analysis (LDA) must be used for data analysis. These methods find a correlation between the structure of the analyte and its effect on the recorded UV–vis spectrum over the entire wavelength range, or on the color pattern generated on the microtiter plates, by reducing the dimensionality of the data to the fraction that carries the relevant information. While the exact mathematical formalism of the two methods is different, the general strategy for establishing the correlation between the experimental result and the nature of the analyte is the same: the first step is always to train the method by performing a series of measurements with a selection of different analytes. Based on this training set, two- or three-dimensional score plots are generated that graphically illustrate analyte discrimination. An example is shown in Figure 11.12c. Each point in this plot represents a single measurement, with measurements involving identical analytes ideally resulting in closely spaced clusters of points. The further apart different clusters are, the more accurately the corresponding analytes can be distinguished. Once these plots are available, the actual samples can be analyzed.

The following examples illustrate successful applications of such assays. In an early attempt at multi-analyte sensing, the Anslyn group used the two receptors **11.14** and **11.15** together with bromopyrogallol red and pyrocatechol violet to detect malate and tartrate (Figure 11.13) [27]. Both receptors are structurally derived from citrate receptor **11.9**, but contain one (**11.14**) or two (**11.15**) boronic acid groups to confer affinity for diols. The binding properties of the two receptors are different, with **11.14** having similar affinity for tartrate and malate, whereas **11.15** binds tartrate more strongly. When increasing amounts of malate or tartrate are added to solutions containing constant concentrations of the receptors and dyes in methanol/water, 3:1 (v/v)

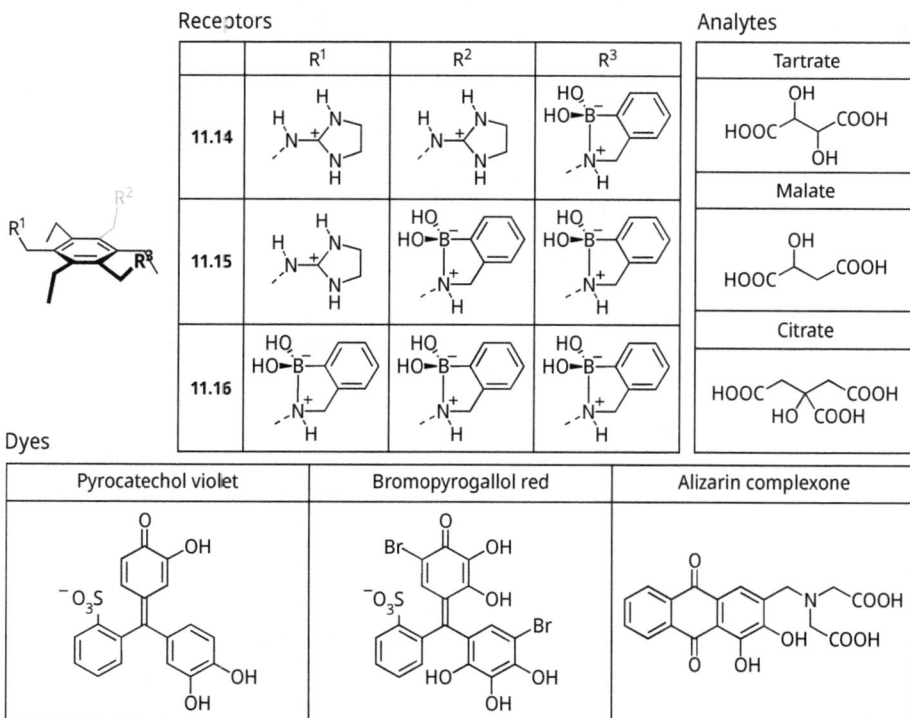

Figure 11.13: Molecular structures of the three receptors **11.14**, **11.15**, and **11.16**, the dyes bromopyrogallol red, pyrocatechol violet, and alizarin complexone, and the analytes malate, tartrate, and citrate.

characteristic changes in the UV–vis spectrum are observed, which can be used as a training set for pattern recognition. Based on the results, highly accurate determination of malate and tartrate concentrations is possible, even in mixtures.

When tris(boronic acid) **11.16** and alizarin complexone are also included in the assay (Figure 11.13), malate, tartrate, and citrate can be detected simultaneously [28]. Since these three acids are constituents of wine, UV–vis spectroscopic analysis of the assay response to different wine samples, followed by linear discriminant analysis, allows pattern-based discrimination of wine varieties.

The array-based assay shown in Figure 11.14, described by the group of Pavel Anzenbacher Jr. [29], detects different phosphate species in water and blood serum. This assay is based on direct sensing using **11.17a–f** as chemosensors, which respond to the presence of phosphate anions by characteristic color changes. A chemosensor array in a hydrophilic polyurethane matrix exhibits color patterns when exposed to the analytes, which are used as a training set for a PCA. The resulting score plot illustrates that it is possible to differentiate phosphate, pyrophosphate, adenosine monophosphate, and adenosine triphosphate.

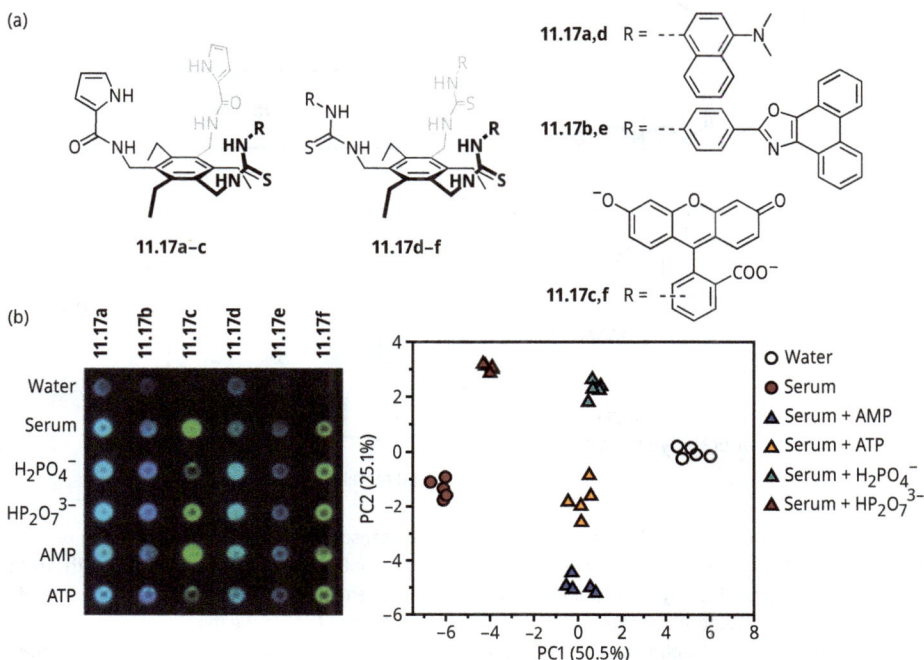

Figure 11.14: Molecular structures of receptors **11.17a–f** (a), differential responses of polymer-embedded chemosensors in an array in which the spots of the six receptors in each row are treated with the same analyte (b), and score plot illustrating the differentiation of analytes after principal component analysis (c). Image and plot adapted with permission from [29]. Copyright Wiley-VCH, 2007.

The Severin group demonstrated that setting up an effective assay can be surprisingly simple, requiring only the dissolution of three commercial dyes, arsenazo I, methyl calcein blue, and glycine cresol red (Figure 11.15a), along with $CuCl_2$ and $NiCl_2$, in aqueous buffer at pH 8.4 [30]. Under these conditions, the five components form a dynamic library of interconverting metal complexes. For the assay to work, it is not important to know how many complexes are present or what they look like. The only requirement is that the interconversion of the complexes is rapid and fully reversible. This approach thus demonstrates the usefulness of dynamic combinatorial chemistry (Section 5.7) for sensing.

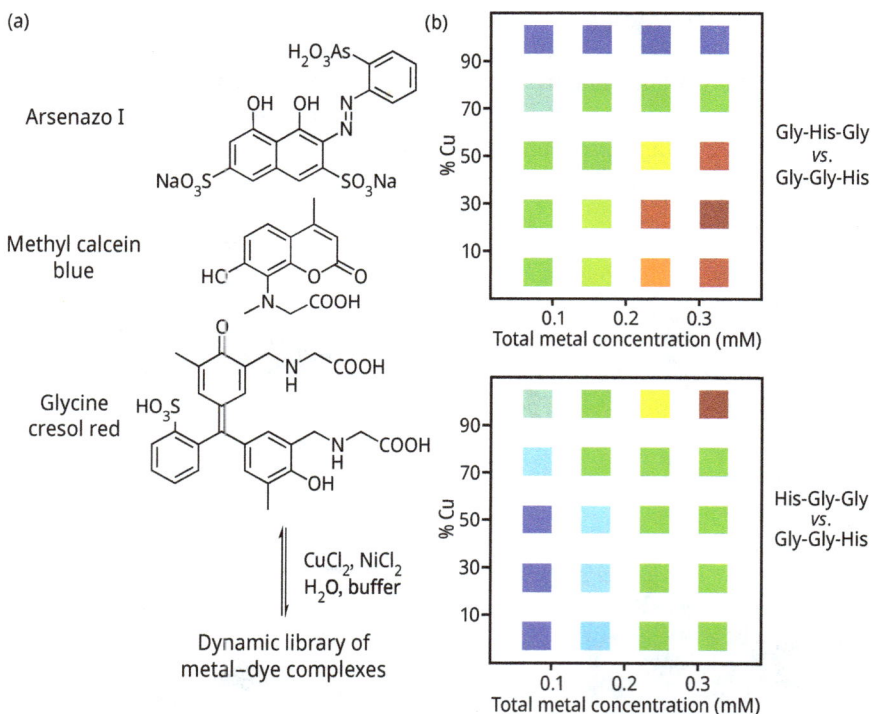

Figure 11.15: Components of a dynamic combinatorial chemosensor library for the differentiation of dipeptides and tripeptides (a) and the effect of the composition of the mixture on the accuracy with which the tripeptides Gly-Gly-His, Gly-His-Gly, and His-Gly-Gly are differentiated (b). In these graphs, the sum of the Cu^{2+} and Ni^{2+} concentrations is plotted against the percentage of Cu^{2+} in the mixture. A deep red color indicates optimal tripeptide differentiation. In both cases, the best differentiation is achieved at the highest total metal concentration, but at different metal ratios. Plots adapted with permission from [31]. Copyright American Chemical Society, 2006.

Sensing is based on the spectral changes caused by analytes that shift the equilibrium state of the dynamic library. These optical changes are structure sensitive, as demonstrated by the fact that each of the six dipeptides Val-Phe, Gly-Ala, His-Ala, Ala-His, Phe-Pro, and Pro-Gly produces a characteristic UV–vis spectrum when added to the library. The most pronounced changes are caused by His-Ala, probably because the imidazole moiety located at the N-terminus of this dipeptide competes particularly well with the dyes for binding to the metal ions. The spectral changes induced by structurally closely related dipeptides such as Gly-Ala, Val-Phe, Ala-Phe, Phe-Ala, and D-Phe-Ala are sufficiently characteristic to allow differentiation by LDA.

An advantage of this strategy is that it can be easily adapted to a specific analytical problem by varying the concentrations and ratios of the components [31]. For example, when using the tripeptides Gly-Gly-His, Gly-His-Gly, and His-Gly-Gly as analytes, a Cu^{2+}/Ni^{2+} ratio of 1:3 is optimal for distinguishing Gly-His-Gly from Gly-Gly-His, while a library containing only Cu^{2+} provides the best discrimination between His-Gly-Gly and Gly-Gly-His (Figure 11.15b). There are certainly other ways to analyze mixtures of these tripeptides. Liquid chromatography coupled to mass spectrometry might work if isomeric tripeptides give rise to characteristic fragmentation patterns. However, the above examples demonstrate that supramolecular analytical chemistry can also provide useful solutions to challenging analytical problems.

Bibliography

[1]   Anslyn EV. Supramolecular analytical chemistry. J. Org. Chem. 2007, 72, 687–99.
[2]   Wolfbeis OS. Probes, sensors, and labels: why is real progress slow? Angew. Chem. Int. Ed. 2013, 52, 9864–5.
[3]   Czarnik AW. Chemical communication in water using fluorescent chemosensors. Acc. Chem. Res. 1994, 27, 302–8.
[4]   Kaur K, Saini R, Kumar A, Luxami V, Kaur N, Singh P, Kumar S. Chemodosimeters: an approach for detection and estimation of biologically and medically relevant metal ions, anions and thiols. Coord. Chem. Rev. 2012, 256, 1992–2028.
[5]   de Silva AP, McCaughan B, McKinney BOF, Querol M. Newer optical-based molecular devices from older coordination chemistry. Dalton Trans. 2003, 1902–13.
[6]   Dix JP, Vögtle F. Ion-selective crown ether dyes. Angew. Chem. Int. Ed. Engl. 1978, 17, 857–9.
[7]   Magri DC, Brown GJ, McClean GD, de Silva AP. Communicating chemical congregation: a molecular AND logic gate with three chemical inputs as a "lab-on-a-molecule" prototype. J. Am. Chem. Soc. 2006, 128, 4950–1.
[8]   Erbas-Cakmak S, Kolemen S, Sedgwick AC, Gunnlaugsson T, James TD, Yoon J, Akkaya EU. Molecular logic gates: the past, present and future. Chem. Soc. Rev. 2018, 47, 2228–48.
[9]   Jin T, Ichikawa K, Koyama T. A fluorescent calix[4]arene as an intramolecular excimer-forming Na$^+$ sensor in nonaqueous solution. J. Chem. Soc., Chem. Commun. 1992, 499–501.
[10] Wu L, Huang C, Emery BP, Sedgwick AC, Bull SD, He XP, Tian H, Yoon J, Sessler JL, James TD. Förster resonance energy transfer (FRET)-based small-molecule sensors and imaging agents. Chem. Soc. Rev. 2020, 49, 5110–39.
[11] de Silva AP, Gunaratne HQN, Gunnlaugsson T, Huxley AJM, McCoy CP, Rademacher JT, Rice TE. Signaling recognition events with fluorescent sensors and switches. Chem. Rev. 1997, 97, 1515–66.

[12] You L, Zha D, Anslyn EV. Recent advances in supramolecular analytical chemistry using optical sensing. Chem. Rev. 2015, 115, 7840–92.

[13] Ogoshi T, Harada A. Chemical sensors based on cyclodextrin derivatives. Sensors 2008, 8, 4961–82.

[14] Wu J, Kwon B, Liu W, Anslyn EV, Wang P, Kim JS. Chromogenic/fluorogenic ensemble chemosensing systems. Chem. Rev. 2015, 115, 7893–943.

[15] Wiskur SL, Ait-Haddou H, Lavigne JL, Anslyn EV. Teaching old indicators new tricks. Acc. Chem. Res. 2001, 34, 963–72.

[16] Metzger A, Anslyn EV. A chemosensor for citrate in beverages. Angew. Chem. Int. Ed. 1998, 37, 649–52.

[17] Tobey SL, Anslyn EV. Determination of inorganic phosphate in serum and saliva using a synthetic receptor. Org. Lett. 2003, 5, 2029–31.

[18] Dsouza RN, Hennig A, Nau WM. Supramolecular tandem enzyme assays. Chem. Eur. J. 2012, 18, 3444–59.

[19] Hennig A, Bakirci H, Nau WM. Label-free continuous enzyme assays with macrocycle-fluorescent dye complexes. Nat. Meth. 2007, 4, 629–32.

[20] Minami T, Liu Y, Akdeniz A, Koutnik P, Esipenko NA, Nishiyabu R, Kubo Y, Anzenbacher Jr. P. Intramolecular indicator displacement assay for anions: supramolecular sensor for glyphosate. J. Am. Chem. Soc. 2014, 136, 11396–401.

[21] Privett BJ, Shin JH, Schoenfisch MH. Electrochemical sensors. Anal. Chem. 2008, 80, 4499–517.

[22] Beer PD, Gale PA, Chen GZ. Electrochemical molecular recognition: pathways between complexation and signalling. J. Chem. Soc., Dalton Trans. 1999, 1897–910.

[23] Tetsuo S. Electrochemically switched cation binding in pentaoxa[13]ferrocenophane. Chem. Lett. 1986, 15, 275–6.

[24] Beer PD, Cadman J, Lloris JM, Martínez-Máñez R, Padilla ME, Pardo T, Smith DK, Soto J. Selective electrochemical recognition of sulfate over phosphate and phosphate over sulfate using polyaza ferrocene macrocyclic receptors in aqueous solution. J. Chem. Soc., Dalton Trans. 1999, 127–34.

[25] Hein R, Beer PD, Davis JJ. Electrochemical anion sensing: supramolecular approaches. Chem. Rev. 2020, 120, 1888–935.

[26] Lavigne JJ, Anslyn EV. Sensing a paradigm shift in the field of molecular recognition: from selective to differential receptors. Angew. Chem. Int. Ed. 2001, 40, 3118–30.

[27] Wiskur SL, Floriano PN, Anslyn EV, McDevitt JT. A multicomponent sensing ensemble in solution: differentiation between structurally similar analytes. Angew. Chem. Int. Ed. 2003, 42, 2070–2.

[28] Gallagher LT, Heo JS, Lopez MA, Ray, BM, Xiao J, Umali AP, Zhang A, Dharmarajan S, Heymann H, Anslyn EV. Pattern-based discrimination of organic acids and red wine varietals by arrays of synthetic receptors. Supramol. Chem. 2012, 24, 143–8.

[29] Zyryanov GV, Palacios MA, Anzenbacher Jr. P. Rational design of a fluorescence-turn-on sensor array for phosphates in blood serum. Angew. Chem. Int. Ed. 2007, 46, 7849–52.

[30] Buryak A, Severin K. Dynamic combinatorial libraries of dye complexes as sensors. Angew. Chem. Int. Ed. 2005, 44, 7935–8.

[31] Buryak A, Severin K. Easy to optimize: dynamic combinatorial libraries of metal-dye complexes as flexible sensors for tripeptides. J. Comb. Chem. 2006, 8, 540–3.

12 Applying supramolecular systems

CONSPECTUS: The previous chapters were mainly concerned with explaining fundamental concepts, raising the question of whether supramolecular chemistry can be of practical use. In this chapter, examples from various fields will show that many of the concepts underlying supramolecular chemistry are indeed relevant to applications, and that receptors can also have practical purposes and commercial value. It is therefore quite possible that you have unknowingly encountered a supramolecular system already in your daily life.

12.1 Introduction

What is all of this good for?

After the serendipitous discovery of crown ethers, much work in supramolecular chemistry was devoted to understanding noncovalent interactions and developing new receptors. A major stimulus has also been the development of systems to help understand and mimic biological processes. Much of this work was probably driven by curiosity, with no specific application in mind. However, as supramolecular chemistry gradually developed into a mature field, the question naturally arose as to whether applications were conceivable. For a while, the lack of applications in supramolecular chemistry was indeed criticized, and supramolecular systems were considered interesting but mostly useless. But this criticism missed the point. In a field as diverse and multidisciplinary as supramolecular chemistry, there is no *single* application or field of use. Rather, supramolecular systems can play a role in many areas [1, 2], and it is this general relevance that this chapter aims to convey.

The first applications of supramolecular systems developed early. Crown ethers, for example, have long been used as alternatives to phase-transfer catalysts to promote chemical reactions (Section 12.5). This application is so well established that the fact that it is actually based on supramolecular concepts is often overlooked. The most widely used receptors are cyclodextrins, of which many tons are produced each year to meet demand. This success story has not been without its obstacles. When early toxicological studies suggested that consumption of cyclodextrins could be problematic, interest in these compounds initially waned. It was only when it became clear that cyclodextrins were toxicologically safe that their use was explored in a wider range of applications. Today, cyclodextrins and their derivatives are used as drugs, components of drug formulations, food additives, etc. Several other supramolecular concepts or systems also have or may soon have practical applications. In the following sections, we take a brief look at the various possibilities to illustrate that supramolecular chemistry can even have an impact on our daily lives.

https://doi.org/10.1515/9783111315171-012

12.2 Applications in medicine

12.2.1 Drugs

Any supramolecular receptor can have bioactivity if it binds to the species involved in a biological process under physiological conditions [3]. If the effect is harmful, the receptor is toxic. Conversely, if the receptor has a beneficial effect, it has the potential to be used as a drug. An example is the γ-CD derivative **12.1** (Figure 12.1), known as sugammadex and sold under the trade name Bridion™. This compound reverses the effects of rocuronium and vecuronium, drugs used in general anesthesia to induce neuromuscular blockade and muscle relaxation. Both anesthetics are active for a relatively long period of time, which can be problematic if the patient is difficult to intubate. In addition, both anesthetics cause side effects such as postoperative muscle weakness and respiratory problems. Sugammadex helps speed recovery of muscle function by acting as a sequestration agent, binding to the drug molecules and preventing them from interacting with their biological target. The efficiency of sequestration is due in part to the perfect fit of the large hydrophobic steroid systems of rocuronium and vecuronium into the cavity of the γ-CD ring. In addition, complex formation positions the quaternary ammonium groups of the drug molecules close to the carboxylate groups along the sugammadex cavity, which further stabilizes the complex. For example, rocuronium is complexed by sugammadex in water with a $\log K_a$ of 7.3 [4], which is one of the largest known stability constants for a γ-CD complex. This example demonstrates that the interactions between a supramolecular receptor and a suitable target can be sufficient to induce a biological effect, provided that the binding is strong and selective under physiological conditions.

The approval of sugammadex as a drug motivated research into whether other receptors could also serve as sequestration agents. Among the compounds tested in this context, the acyclic cucurbiturils developed by Lyle Isaacs show particularly promising results (Section 4.1.12) [5]. An example is **12.2** (Figure 12.1), which binds to rocuronium about two orders of magnitude more strongly than sugammadex ($\log K_a = 9.5$) and, like sugammadex, induces reversal of neuromuscular block in anesthetized rats [6]. Acyclic cucurbiturils also effectively bind to various drugs of abuse and show promising activity in the treatment of methamphetamine or other amphetamine intoxications. They also effectively bind to synthetic opioids such as fentanyl and carfentanil and may therefore be useful to treat addiction or overdose. In this context, the sulfated pillar[6]arene **12.3** also has promising activity, binding to fentanyl in 20 mM NaH_2PO_4 buffered water at pH 7.4 with a $\log K_a$ of 8.0 [7]. Because of its high affinity, especially for quaternary ammonium ions, **12.3** has been named Pillar[6]MaxQ. Since many of these receptors also show no signs of toxicity and activity *in vivo*, they have the potential to develop into the next generation of pharmaceutically useful sequestration agents [5].

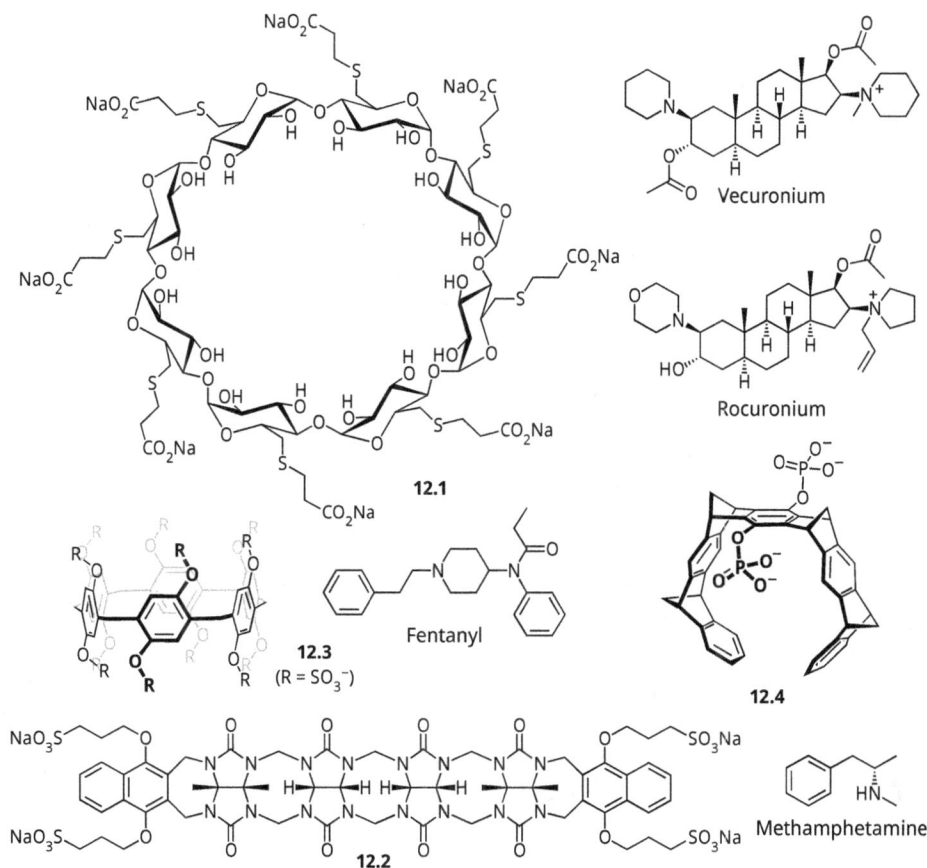

Figure 12.1: Molecular structures of sugammadex **12.1**, rocuronium and vecuronium, the acyclic cucurbituril **12.2** and methamphetamine, the pillar[6]arene derivative **12.3** and fentanyl, and the molecular clip **12.4**.

Other supramolecular receptors with biological activities are substituted calixarenes (Section 4.1.7), with positively charged calixarenes acting as gene transfection agents [8], while sulfonatocalix[4]arene and other negatively charged calixarenes have antiviral, antibacterial, and antifungal activity [9]. Sulfonated calix[4]arenes also allow the detection of posttranslational modifications in histone proteins and disrupt protein recognition events by binding to positively charged biomolecules or residues in biomolecules, such as the methylated amino groups in lysine side chains formed during posttranslational histone modification [10]. In addition, they reduce mortality in mice poisoned with the herbicides paraquat or diquat [11]. Similar effects have been observed for the larger sulfonatocalix[5]arene, carboxymethylated pillar[6]arenes, and CB[7] [5]. Supramolecular receptors, many based on cyclodextrins [12], have also been developed to detoxify chemical warfare agents.

The molecular clips developed by Klärner and Schrader (Section 4.1.12), of which **12.4** (Figure 12.1) is an example, inhibit the oligomerization and aggregation of proteins involved in diseases such as Alzheimer's and Parkinson's disease [13]. The development of Alzheimer's disease is accompanied by the appearance of Aβ, a 36–46 amino acid peptide derived from the amyloid precursor protein. Misfolded forms of the Aβ peptide aggregate to form oligomers that are toxic to neurons. These oligomers grow autocatalytically, with the initial aggregates inducing the misfolding of further peptides, ultimately leading to plaques. Clip **12.4** not only prevents this process by interacting with lysine and arginine residues in the Aβ peptide but also disassembles existing fibrils. Importantly, **12.4** is active *in vivo* where it shows no toxicity, making it a promising drug candidate for the treatment of a disease that is likely to affect more and more people in the future. Remarkably, **12.4** also disrupts the membrane that surrounds the HIV virus, which may allow its use in the treatment of viral infections.

Supramolecular drugs are also useful for mitigating the harmful effects of malfunctioning ion channels that cause incurable genetic diseases such as cystic fibrosis (CF). This disease affects approximately one in 3,000 newborns and is caused by a mutation in the gene for the CF transmembrane conductance regulator protein, an anion channel in epithelial cell membranes that controls the flux of chloride and bicarbonate anions. Dysregulation of this channel affects the transport of water across the membrane, leading to the formation of sticky mucus in organs such as the lungs. A potential strategy to treat this disorder is the administration of a carrier that helps restore anion transport [14]. Much work in this area has been inspired by the anion transport properties of prodigiosins (Section 10.3.2), and a number of synthetic anion carriers, such as those developed in the groups of Philip A. Gale and Anthony P. Davis, have indeed shown promise in mediating anion transport in epithelial cells [15]. Importantly, some of the most potent carriers have almost no toxicity, demonstrating that anion transport alone does not necessarily induce cell death.

Finally, nucleotides such as DNA are important drug targets. The anticancer drug cisplatin, for example, cross-links DNA strands and induces apoptosis of the affected cell, suggesting that molecules that interact specifically with DNA have the potential to influence metabolic processes such as the transcription of specific genes. Useful anchor points for such molecules are the edges of base pairs in the major and minor grooves of the DNA double helix, which contain sequences of hydrogen bond donors and acceptors that mirror the nucleobase sequence within the double helix. The DNA nucleobase sequence can thus be read without separating the strands, potentially allowing the control of DNA function and activity by blocking specific sites. In an approach that builds on the close links between supramolecular, bioorganic, and medicinal chemistry, the group of Peter B. Dervan deciphered the correlation between the hydrogen bonding pattern in the minor groove of DNA and the nucleobase sequence, and designed curved molecules containing amide and heterocyclic subunits with hydrogen bonding patterns complementary to specific minor groove sequences [16]. These compounds are expected to be useful as anti-infectives or for transcription therapy.

12.2.2 Drug formulations

The efficacy of a drug depends primarily on how well it interferes with the biological process that causes the disease, but there are other important parameters. One is bioavailability, which describes the fraction of the administered drug that reaches the systemic circulation. If this fraction is low, as is the case when a bioactive molecule with low water solubility is administered orally, the effect is usually weak, even if this molecule is intrinsically very active. Another problem arises when a drug is too unstable to ensure a sufficiently long shelf life. These problems can be overcome by mixing the drug with additional components that have no biological activity themselves but improve performance. Drug formulation, which results in the tablets, capsules, liquids, or creams sold in pharmacies, ensures that drugs are safely and easily administered at the correct dosage, are stable, and produce the desired effects.

A widely used strategy to improve the water solubility of the active drug ingredient is the use of cyclodextrins, which form water-soluble inclusion complexes with the bioactive molecule (Section 4.1.4) [17]. This approach benefits from the lack of toxicity of cyclodextrins when taken orally or applied to the skin. It is also quite versatile, not only because three different cyclodextrins are available to adapt the binding properties to the structural parameters of the drug, but also because the availability of different methylated, hydroxypropylated, or sulfobutylated cyclodextrin derivatives allows tuning of the properties of the complex.

In addition to increasing the solubility and bioavailability of a drug, complex formation has other beneficial effects, such as improving chemical stability. In addition, the slow release of active molecules from the complex prolongs the biological effect. In this context, cross-linked cyclodextrin-based materials can serve as useful drug depots. Finally, cyclodextrins enhance the transport of certain drugs across biological membranes, although the exact mechanism of this process is not yet fully understood. As a result, cyclodextrins are versatile components in drug formulations, which is why they are found in many commercial products ranging from tablets and capsules to ointments and drop solutions. For example, a sulfobutylated β-CD derivative is marketed under the name Captisol® and is currently used in nine FDA-approved drugs. It is important to note that there are alternatives to cyclodextrins in drug formulations, such as cyclic and acyclic cucurbiturils, which have also been shown to increase the solubility of bioactive molecules [18, 19].

12.2.3 Sensing

Supramolecular receptors can be used not only as therapeutic or sequestration agents or as additives to drug formulations, but also as chemosensors for the selective detection of biorelevant analytes in biofluids or in *in vitro* assays [20]. For example, the OPTI LION™ Electrolyte Analyzer, marketed by Optimedical Inc., is used in intensive care units and ambulances to quantify Na^+, K^+, and Ca^+ concentrations in blood serum. It contains the immobilized fluorescent probes **12.5a**, **12.5b**, and **12.5c** developed in the group of A. Prasanna de Silva (Figure 12.2a) [21].

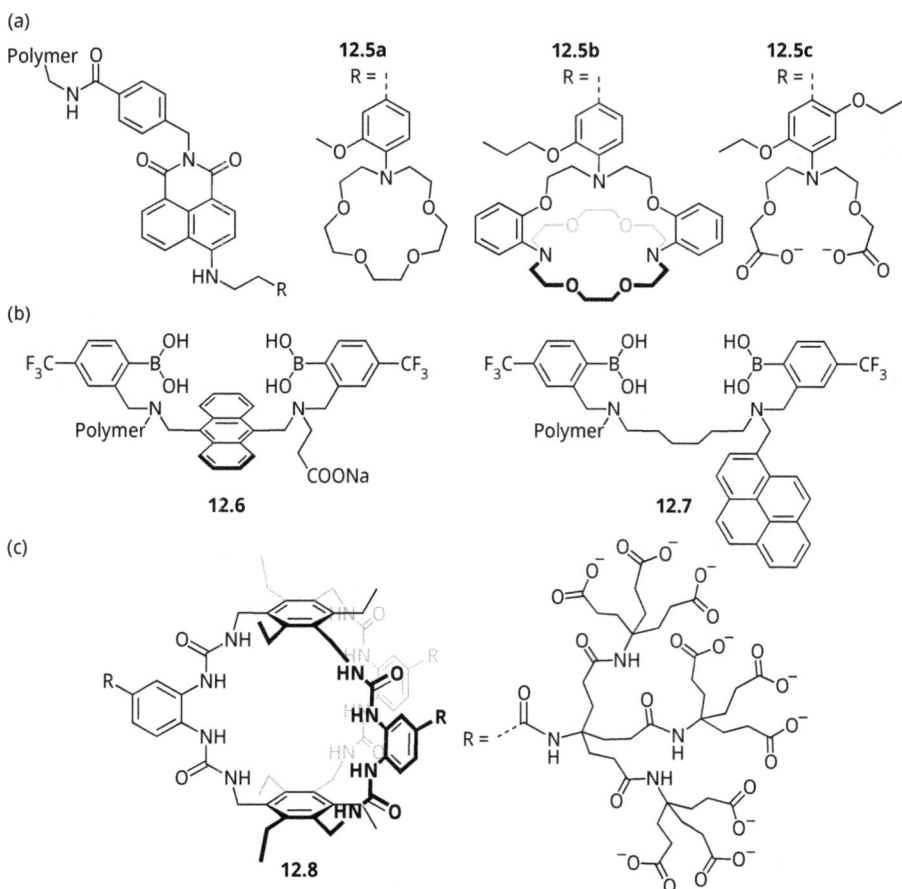

Figure 12.2: Structures of the fluorescent probes **12.5a**, **12.5b**, and **12.5c** for the detection of Na^+, K^+, and Ca^{2+} in blood serum, respectively (a), the bis(boronic acid)-based Eversense® and Glysure chemosensors **12.6** and **12.7** (b), and the glucose-selective temple receptor **12.8** (c).

Another important target in medical diagnostics is glucose, which needs to be monitored regularly in patients with diabetes to ensure that blood glucose levels are at healthy levels. In addition, careful control of blood glucose levels can reduce morbidity and mortality in hospital intensive care units, and there is evidence that glycated proteins may be important biomarkers for sugar-related diseases such as atherosclerosis, autoimmune diseases, and Alzheimer's disease. While it is important to detect and quantify glucose and other carbohydrates in biofluids such as blood, the use of synthetic receptors is challenging because carbohydrates are among the most difficult substrates to detect in the aqueous environment, as we have seen in Section 4.2.4. Nevertheless, sufficient progress has been made in supramolecular carbohydrate recognition and sensing to bring synthetic carbohydrate receptors to the market [22]. For example, the boronic acid-based chemosensor **12.6** (Figure 12.2b) developed by Tony D. James and Seiji Shinkai [23] has been implemented in the continuous glucose monitors marketed by Senseonics (Eversense®). Another glucose sensor, marketed by Glysure Ltd., contains bis(boronic acid) **12.7**, which was developed by the group of Tony D. James [24]. Both bis(boronic acids) are turn-on PET chemosensors that begin to fluoresce upon glucose binding (Section 11.2.1). In the commercial devices, these compounds are immobilized on various types of polymeric supports, integrated into an optical fiber system, and combined with a light-emitting diode, photodiodes, and a transmitter. The resulting systems are small enough to be implanted subcutaneously, and the data generated can be recorded and stored using an externally worn device or a mobile phone. In this way, patients can continuously monitor their blood glucose levels without having to externally analyze a drop of blood taken by pricking the skin.

Control of blood glucose concentrations may also be achieved by synthetic receptors capable of tightly and selectively binding glucose noncovalently in biological media. The currently most promising candidate for use as synthetic insulin for the treatment of diabetes is the *temple receptor* **12.8** (Figure 12.2b) developed by Anthony P. Davis (Section 4.2.4). This receptor is commercialized by Ziylo, a spin-out company from the University of Bristol, which was acquired by the global healthcare company Novo Nordisk in 2018.

These examples demonstrate that receptors and chemosensors can have commercial value. However, considering the large number of chemosensors described in the scientific literature, surprisingly few have actually reached the market [25]. The reason for this is probably the availability of various conventional analytical methods to reliably and sensitively detect any type of analyte. As a result, the incentive to introduce new methods is low. In addition, chemosensors often do not meet the requirements for practical use because they cannot be used continuously or for long periods of time under different environmental conditions. Nevertheless, supramolecular sensing is a powerful approach not only for the detection of biorelevant targets such as phosphate or phosphorylated species, drugs of abuse, hormones, toxins, and chemical warfare agents, and even biomolecules such as proteins or nucleotides, but also for environmentally relevant compounds such as pesticides, industrial products released into the environment, gases, and many other analytes. Sensing can be based on single receptors or arrays of receptors and can be optical or electrochemical. The techniques can be integrated into devices that can be miniaturized so that only small amounts of receptors are needed, making the use of structurally more complex systems not prohibitive. Therefore, it is probably only a matter of time before supramolecular sensors become more widely used.

12.2.4 Imaging

Imaging modalities are indispensable in modern medicine for the noninvasive diagnosis of disease. The use of X-rays is a classic method, but there are many other options such as ultrasound imaging (US), computed tomography (CT), magnetic resonance imaging (MRI), positron emission tomography (PET), single-photon emission CT (SPECT), and optical imaging (OI). Some of these methods require the simultaneous administration of a molecular agent that produces the measured signal or aids in the imaging process. In PET, for example, positron-emitting radioligands are introduced into the body to follow their binding to a protein or participation in a metabolic process. Optical imaging is based on probes in which a chromophore is attached to a binding motif that serves to recognize the biological target. In MRI, paramagnetic ions or molecules serve as contrast agents to reduce the relaxation times of the hydrogen nuclei of nearby water molecules, thereby increasing signal intensity and improving image resolution. The structural design of these agents and the tuning of their properties can be based on supramolecular concepts as the following examples illustrate.

Although MRI contrast agents are mostly based on gadolinium(III) complexes with chelating ligands such as diethylenetriaminepentaacetic acid (DTPA) or 1,4,7,10-tetraazacyclododecane-*N,N,N,N*-tetraacetic acid (DOTA) (Figure 12.3a), paramagnetic organic agents have advantages such as low cytotoxicity and good biodegradability. However, organic radicals often produce a weaker contrast (only one unpaired electron instead of the seven of Gd^{3+}) and suffer from low stability. The latter problem can be overcome by complexing them with a supramolecular receptor. For example, the 2,2,6,6-tetramethylpiperidinyloxyl (TEMPO) derivative **12.9** (Figure 12.3a) forms a complex with CB[8] with a log K_a of 6.2. CB[8] thus acts as a supramolecular protecting group that prevents the aminoxyl radical from decomposing. By attaching multiple copies of the CB[8] complex of **12.9** to the outer surface of a tobacco mosaic virus, a contrast agent was obtained that is much more stable than an analogous system with unprotected aminoxyl radicals and has a contrast strength close to that of the gadolinium(III)-DOTA complex [26].

Optical imaging in living organisms requires that the light used to excite the probe and the emitted light penetrate the tissue without significant loss of intensity due to interactions with biomolecules or heating of water. The optimal wavelength range, known as the phototherapeutic window, is between 650 and 1,000 nm, in the

(a)

Gd³⁺-DTPA

Gd³⁺-DOTA

12.9

(b)

Squaraine

12.10

$R = $

(c)

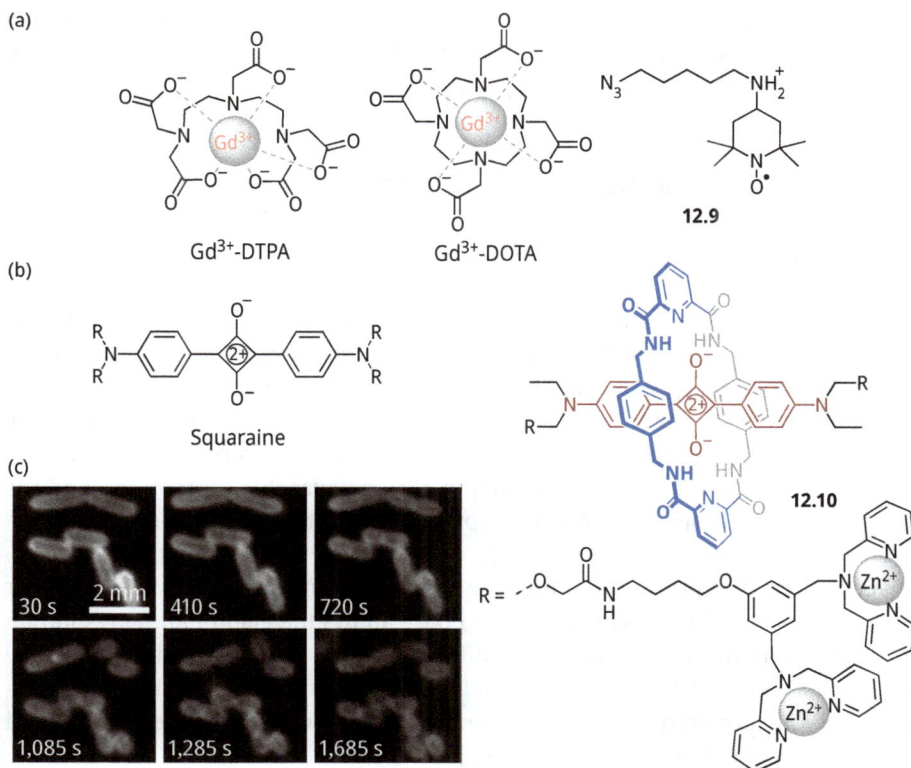

Figure 12.3: Structures of the Gd³⁺ complexes of DTPA and DOTA and of the TEMPO derivative **12.9** (a), general structure of a squaraine dye and molecular structure of the probe **12.10** (b), and images showing binary fission of *E. coli* cells stained with **12.10**. After the start of the experiment, the cells were imaged by fluorescence microscopy at the times indicated in the images. Image adapted with permission from [28]. Copyright Wiley-VCH, 2007.

deep red and near-infrared (NIR) regions. The intensely colored squaraines of the general structure shown in Figure 12.3b are in principle well suited for use as optical probes because of their absorption and emission properties, but they have the disadvantage of aggregating in aqueous media and reacting with nucleophiles. A strategy introduced by the group of Bradley D. Smith to overcome these drawbacks involves threading squaraines through a tetralactam ring [27]. The resulting rotaxanes have greatly improved solubility and stability, making them useful for imaging purposes after conjugation with appropriate target moieties. For example, rotaxane **12.10**, whose zinc–dipicolylamine complexes selectively bind to the surface of bacterial cells, allows staining and visualization of bacteria even within living organisms such as mice [28]. According to the images of *E. coli* in Figure 12.3c, which show cells stained with **12.10**, the probe is not harmful because binary fission still occurs within

30 min of measurement. Other applications of interlocked molecules in medicinal chemistry have been proposed, such as their use to enhance or inhibit the action of therapeutic agents, or as transporters or sensors [29].

12.3 Applications in separation processes

12.3.1 Chromatography

Chromatographic separations are mostly based on physisorption processes between compounds present in the mobile phase and selectors dissolved in or attached to the stationary phase. The underlying interactions are noncovalent in nature and subject to thermodynamic control, making them closely related to the interactions that stabilize receptor–substrate complexes. As a consequence, supramolecular receptors are useful selectors in chromatographic separations, and the properties of the respective columns typically correlate with the binding properties of the immobilized hosts. An early example reported by Donald J. Cram involves the use of a polymeric stationary phase containing the chiral crown ether **12.11** as a selector (Figure 12.4) [30]. This phase allows the enantiomeric resolution of the ammonium salts of various α-amino acids and α-amino acid esters. In all cases, the L-enantiomer elutes before the D-enantiomer, indicating that the immobilized (*R,R*)-configured crown ether interacts more strongly with the D-configured amino acids, consistent with the results of solution studies (Section 4.1.1).

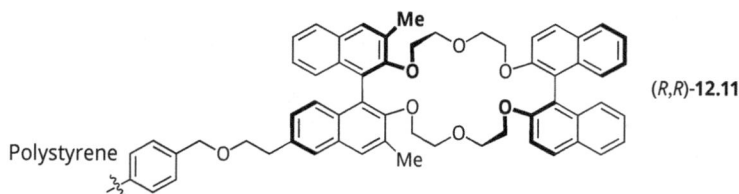

Figure 12.4: Molecular structure of the immobilized chiral crown ether (*R,R*)-**12.11**, which allows the separation of enantiomers of α-amino acids and α-amino acid esters.

Many other crown ether-based selectors have been developed and commercialized, mostly for enantiomeric separations, and columns with receptor-based chiral stationary phases are available from different suppliers. Calixarenes can also be useful, but the most important selectors are cyclodextrins for several reasons. First, cyclodextrins are available in different ring sizes and therefore allow the separation of a wide range of compounds, especially hydrophobic organic compounds with residues that can be incorporated into the cyclodextrin cavity. Second, cyclodextrins can be easily immobilized and structurally modified to tune their separation properties and adapt them to a specific application. Native cyclodextrins or those with polar functional groups are typically used in liquid chromatography and capillary electrophoresis, while nonpolar derivatives such as alkylated cyclodextrins are more suitable for gas chromatography. Finally, cyclodextrins are chiral and therefore capable of separating enantiomers, which is probably the most important application of cyclodextrin-based columns. Most suppliers of chromatography equipment have such columns in their portfolios.

12.3.2 Adsorption

Materials such as zeolites, metal organic frameworks (MOFs), covalent organic frameworks (COFs), and other porous solids are useful for separating gases, hydrocarbons, and other industrial products due to their ordered porous structure combined with a high surface area. The use of these materials is not only more economical than energy-intensive distillation, but can also be more efficient, especially when the components to be separated have similar boiling points.

Alternatively, separation processes can be based on crystals of macrocyclic receptors [31]. Such crystals are often nonporous despite the presence of macrocyclic cavities, but can structurally adapt to the presence of guest molecules. The first example of such nonporous adaptive crystals (NACs) was described by the group of Tomoki Ogoshi [32]. He showed that crystallization of the perethylated pillar[5]arene **12.12** (Figure 12.5) from acetone yields crystals in which the pillararene cavities form channels occupied by acetone molecules. When these crystals are heated to 50 °C under reduced pressure, the acetone molecules are desorbed and the crystals are converted to a nonporous structure. Exposure of these activated crystals to alkanes triggers another structural transformation, but only if the alkane is longer than butane. Importantly, cyclic or branched alkanes are also rejected, so that adsorption is selective for linear alkanes with chain lengths from n-pentane onward. This selectivity mirrors the binding selectivity of pillar[5]arenes in solution. The crystal structure of the n-hexane complex shows the formation of a herringbone structure in which individual n-hexane molecules are surrounded by two rings of **12.12**, demonstrating that the crystals are adaptive.

Since then, many other examples of this concept have been described [31]. For example, the separation of ethylbenzene and styrene, which is difficult by distillation because the boiling points of these compounds differ by only 9 °C, can be achieved with activated crystals of perethylated pillar[6]arene **12.13** (Figure 12.5), which selectively adsorb styrene from a 50:50 v/v mixture with >99% purity in a single adsorption cycle [33]. Structural studies showed that the crystals reorganize upon styrene binding, forming a structure with channels between individual pillar[6]arene molecules in which the styrene molecules reside. The macrocycles themselves are deformed and have cavities that are too small to accommodate a guest.

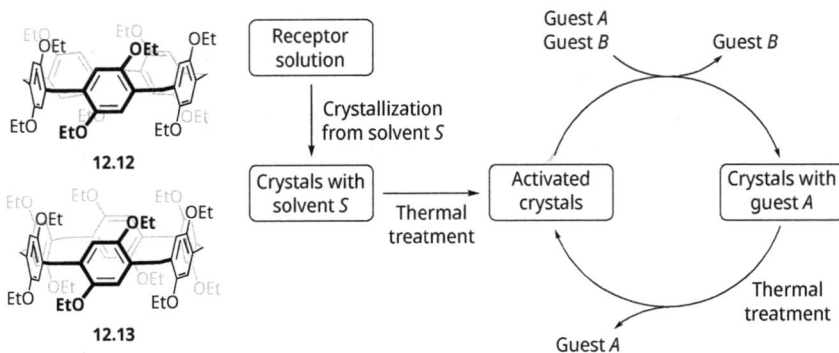

Figure 12.5: Molecular structure of perethylated pillar[5]arene **12.12** and pillar[6]arene **12.13**, and schematic representation of the adaptive behavior of the crystals of these macrocycles, which allows the selective adsorption of linear alkanes longer than butane in the case of **12.12**, and the separation of styrene and ethylbenzene in the case of **12.13**.

The process is fully reversible, allowing desorption of the bound styrene and reuse of the crystals for another round of separation.

Although this concept is promising, several challenges need to be overcome before it can be put to practical use. Aspects such as selectivity or long-term stability can most likely be overcome by proper selection and structural fine-tuning of the receptors. Slow adsorption and desorption kinetics also need to be addressed, but the biggest challenge will probably be to perform these separations on an industrially useful scale and to ensure the availability of sufficient amounts of the required receptors. The necessary development effort will be worthwhile for industrially relevant separations that cannot be performed with conventional methods.

12.3.3 Extraction

Hydrophobic organic compounds released into the environment such as herbicides or pesticides, or other chemicals such as polycyclic aromatic hydrocarbons (PAHs), polychlorinated biphenyls, dioxins or benzopyrenes, are dangerous and often persistent pollutants. They are also good guests for cyclodextrins, making the use of cyclodextrins to extract these pollutants from water or soil an attractive remediation strategy. For example, soil containing polycyclic aromatic hydrocarbons such as phenanthrene and pyrene can be decontaminated by extraction with an aqueous solution of permethylated β-CD [34]. After reextraction of the hydrophobic contaminants from the aqueous phase, the cyclodextrin solution can be reused. Cyclodextrins and cyclodextrin derivatives are also useful for extracting dibenzodioxins and polychlorinated dibenzofurans from water [35], or PAHs from oil at spill sites [36]. The ability of cyclodextrins to operate in these complex environments and their lack of toxicity make their use in remediation applications particularly attractive.

Metals such as copper, nickel, cobalt, and zinc can be separated from complex matrices by extractive metallurgy using appropriate cation receptors [37]. While the recognition of these transition metal ions is mainly based on the principles of coordination chemistry, the extraction of main group metal ions can be achieved with supramolecular receptors such as those marketed by IBC Advanced Technologies Inc. [38]. For example, the crown ether–calixarene receptor **12.14** (Figure 12.6a), also known as BOBcalix and marketed by IBC under the trade name MacroLig® 209, has been used for the selective removal of the radioactive isotope ^{137}Cs from nuclear waste at the Savannah River Site and radioactive storage sites in Germany [39, 40]. Extraction of the positively charged radioactive uranyl cation UO_2^{2+} can be achieved with receptors containing pyrrole moieties [41] or with a tripodal receptor derived from Kemp's triacid with three carboxylate groups [42].

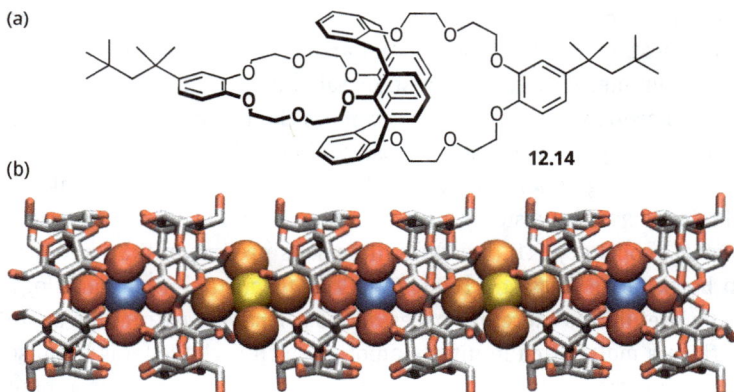

(a)

12.14

(b)

Figure 12.6: Molecular structure of BOBcalix **12.14**, a crown ether–calix[4]arene-derived receptor for Cs^+ (a), and crystal structure of the K[AuBr$_4$]–α-CD coprecipitate (b). Hydrogen atoms are not shown in the crystal structure for reasons of clarity. The cyclodextrin rings are shown as stick models, whereas the $AuBr_4^-$ and the hydrated K^+ ions are shown as space-filling models.

An important source of precious metals is the large amount of electrical and electronic waste generated today [43]. Among its many valuable constituents is gold, the extraction of which is economically attractive but should avoid the use of highly toxic and environmentally damaging reagents such as cyanide. In this context, the discovery by the Stoddart group that α-CD can be used to selectively precipitate KAuBr$_4$ from an aqueous solution, even in the presence of other square-planar noble metal complexes such as [PtBr$_4$]$^{2-}$ and [PdBr$_4$]$^{2-}$ is particularly noteworthy [44]. X-ray crystallography showed that the α-CD units are arranged in a columnar head-to-head manner in the crystals, stabilized by hydrogen bonds between the secondary hydroxy groups of adjacent rings (Figure 12.6b). The [AuBr$_4$]$^-$ anions and hydrated potassium ions are alternately arranged along the columns, with the anions surrounded by the primary hydroxy groups of two α-CD rings and the cations surrounded by the secondary hydroxy groups. The precipitate can be decomposed by adding reducing agents such as Na$_2$S$_2$O$_5$ to yield metallic gold, and the α-CD can be recrystallized and reused. Based on this finding, a cyanide-free gold extraction method has been proposed in which the gold is dissolved with concentrated HBr/HNO$_3$, the solution is neutralized with aqueous KOH, and K[AuBr$_4$] is precipitated with α-CD. Precipitates can also be obtained from K[AuBr$_4$] and β-CD. In this case, however, additives such as diethylene glycol dibutyl ether (dibutyl carbitol) are required to induce binding of the tetrabromoaurate anions to the primary faces of the cyclodextrin ring, resulting in the formation of columnar structures similar to those observed with α-CD [45]. This technology is being commercialized by Cycladex Inc., a company specializing in the development of environmentally friendly methods for the isolation of important metals such as gold, silver, and lithium.

Inorganic anions of environmental concern are nitrate and phosphate, which are responsible for eutrophication, or toxic cyanide and arsenate. These species occur in different matrices – sometimes dissolved in fresh or salt water, sometimes adsorbed on solids – making the development of synthetic receptors challenging. Sulfate anions, although not environmentally relevant *per se*, pose problems for the long-term disposal of radioactive waste, large quantities of which are currently stored in tanks. It has been proposed to convert this waste by vitrification into glass logs for subsequent disposal in a geological repository [41], but the sulfate salts present in the waste are not well soluble in borosilicate glass and reduce the long-term stability of such logs. For vitrification to be useful, it is therefore necessary to selectively extract the small amounts of sulfate from a matrix rich in nitrate anions, which is challenging because nitrate anions are much easier to transfer from a polar to a nonpolar environment than doubly charged and efficiently hydrated sulfate anions. A solution to this problem was proposed by the Sessler group [46]. They showed that exposing a solution of the nitrate salt of the doubly protonated form of the lipophilic cyclo[8]pyrrole **12.15** (Figure 12.7a) in toluene (containing additional trioctylammonium nitrate) to 0.02 M aqueous Na$_2$SO$_4$ and varying concentrations of NaNO$_3$ causes the macrocycle to selectively take up sulfate from the aqueous phase and release the nitrate counterions (Figure 12.7b).

The group of Katrina A. Jolliffe showed that the separation of sulfate anions from an aqueous salt mixture can also be achieved by selective transport across a liquid membrane [47]. The actual experiment involved a U-tube containing receptor **12.16** (Figure 12.7a) and tetrabutylammonium nitrate in the organic phase. This receptor picks up sulfate anions from the source phase, with the tetrabutylammonium ions ensuring charge neutralization, and transports them to the receiving phase, even when the source solution contains a mixture of salts typically found in nuclear water solutions. The presence of barium chloride in the receiving phase improves the efficiency of

(a)

12.15 (R = C$_{11}$H$_{23}$)

12.16

(b)

2 [(H$_{17}$C$_8$)$_3$NH$^+$]NO$_3^-$ [(H$_{17}$C$_8$)$_3$NH$^+$]$_2$SO$_4^{2-}$

Toluene

Water

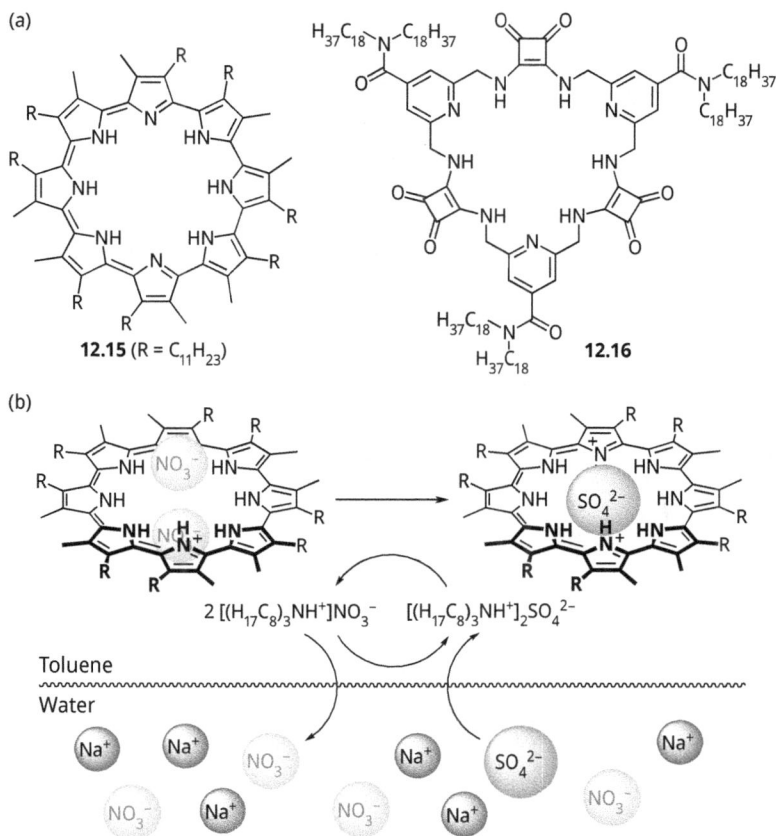

Figure 12.7: Molecular structures of cyclo[8]pyrrole **12.15** and the macrocyclic squaramide **12.16** (a) and schematic representation of sulfate extraction mediated by **12.15** (b).

the transport by precipitating sulfate as BaSO$_4$. While the performance of these processes is unlikely to be of practical use, they demonstrate that high sulfate-over-nitrate selectivity can be achieved under solvent extraction conditions. A different approach to the same problem is presented in the following section.

The tetrahedral pertechnetate anion TcO$_4^-$ is radioactive, with the added hazard that it readily migrates in the surface layers of the Earth's crust and thereby enters the food chain. One isotope of pertechnetate is used in diagnostic imaging, while those produced in the nuclear fuel cycle have very long half-lives. Strategies to remove pertechnetate from water include solvent extraction and ion exchange, but these methods suffer from low selectivity. Strategies based on supramolecular receptors potentially allow selective removal, but although pertechnetate looks similar to sulfate, it is actually larger and much less strongly coordinating, making the development of suitable receptors difficult [48]. Nevertheless, several classes of receptors have shown promise. Examples include the cyclotriveratrylene derivative **12.17** (Figure 12.8), which can almost completely remove TcO_4^- by a single extraction from a 3 mM aqueous solution into nitromethane even in the presence of competing anions [49], the sapphyrin derivative **12.18**, whose monoprotonated form binds to pertechnetate in 2.5 vol% methanol/water with a $\log K_a$ of 3.6 [50], and the polyazacryptand **12.19** [51]. The hexaprotonated form of the latter receptor binds a pertechnetate anion in water with a $\log K_a$ of 5.5, two orders of magnitude stronger than nitrate. The only anion that is bound with similar affinity is perrhenate ReO_4^-, which is structurally very similar to the pertechnetate anion.

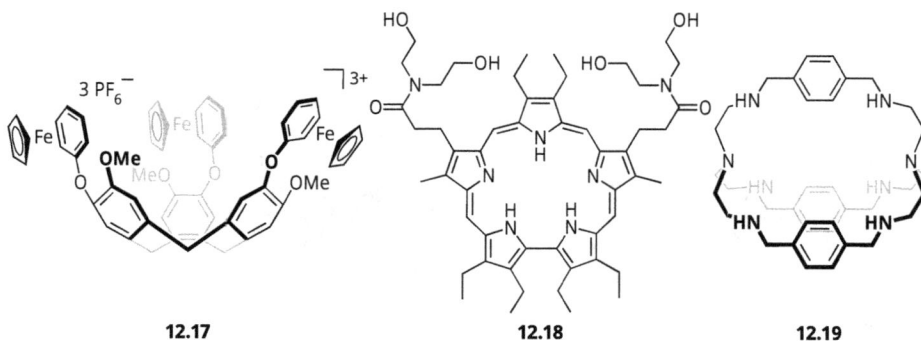

Figure 12.8: Molecular structures of pertechnetate binders **12.17**, **12.18**, and **12.19**.

12.3.4 Precipitation

Another strategy to remove sulfate from radioactive waste prior to vitrification is to selectively precipitate it from solution. Ideally, this process should be reversible to allow recycling of the precipitant. The feasibility of this strategy was demonstrated by Radu Custelcean using the tripodal tris(urea) **12.20** (Figure 12.9a), which binds to sulfate by forming a complex in which the anion is sandwiched between two interdigitating subunits of **12.20** and held in place by 12 hydrogen bonds [52]. To illustrate this arrangement, the crystal structure of the sulfate complex of **12.20** is shown in Figure 12.9b. This complex crystallizes from an aqueous solution containing 5 M $NaNO_3$, 1.25 M NaOH, and 0.044 M Na_2SO_4, which closely resembles the radioactive waste mixture in terms of ionic strength, alkalinity, and sulfate concentration [53]. In the precipitate formed, the sodium cations bridge the individual capsules by binding to the nitrogen atoms in the pyridine units. The yield is 90%, meaning that only 10% of the sulfate anions remain in solution after one round of precipitation. The precipitate decomposes upon dissolution in fresh water, from which the uncomplexed **12.20** subsequently crystallizes, allowing it to be recycled.

A second strategy is similar but involves the use of the bis(guanidinium) salt **12.21**, which is obtained in one step from the condensation of aminoguanidinium chloride

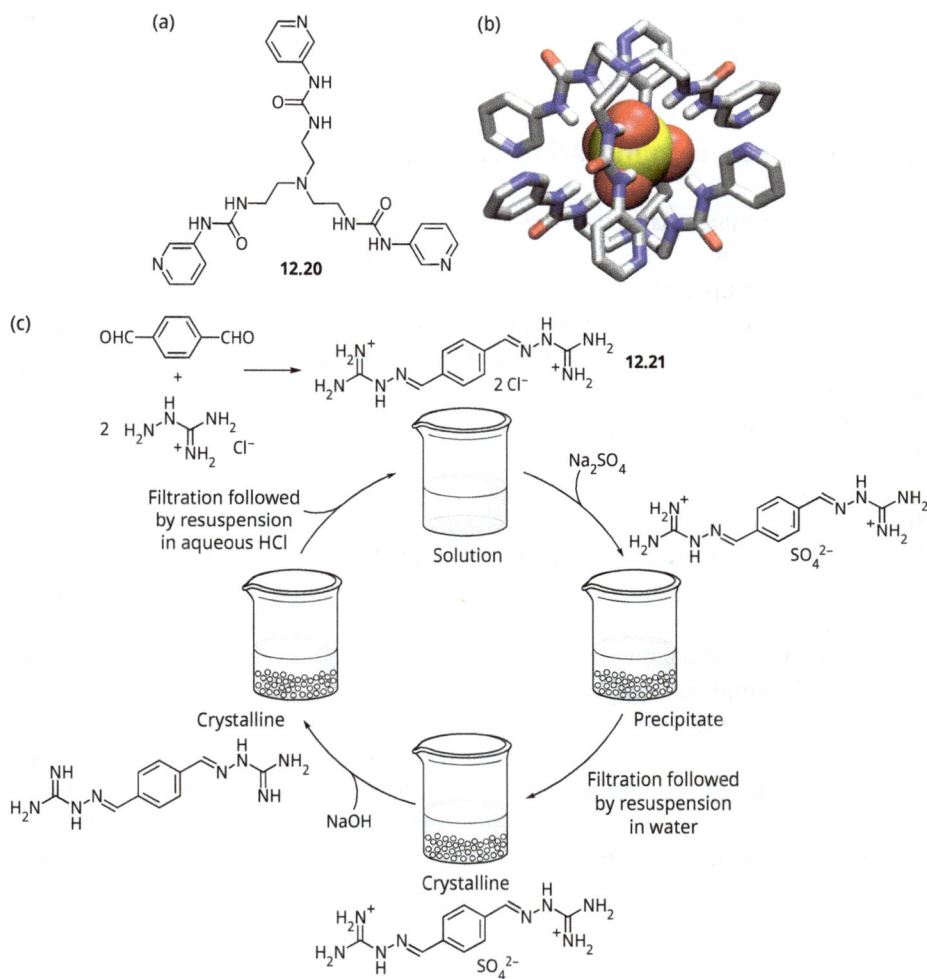

Figure 12.9: Molecular structure of sulfate binder **12.20** (a), crystal structure of the complex between two molecules of **12.20** and a sulfate anion (b), and scheme showing the synthesis of **12.21** and the cycle of sulfate precipitation and recovery of the precipitation agent. Hydrogen atoms other than those on NH groups are not shown in the crystal structure for reasons of clarity.

and terephthalaldehyde. While the chloride salt of **12.21** is well soluble in water, the corresponding sulfate salt has a solubility product pK_{SP} of 9.6, only slightly lower than that of $BaSO_4$ ($pK_{SP} = 10.0$) [54]. The addition of **12.21** to an aqueous solution of Na_2SO_4 thus results in the formation of a precipitate that can be filtered off. The precipitation agent is recycled by decomposing the sulfate salt with NaOH, filtering off the neutral salt-free form, and dissolving it in aqueous HCl, as shown in Figure 12.9c. The process is selective because only the sulfate complex of **12.21** precipitates, even if the solution contains a mixture of chloride (0.1 M), nitrate (0.07 M), and sulfate (0.034 M) salts.

12.4 Applications in materials chemistry

Supramolecular materials formed or stabilized by reversible interactions have a number of properties that make them interesting for applications. One is the ability to self-heal, because the weak interactions that are preferentially broken under mechanical stress can be easily reformed. Therefore, fractures or damage can heal autonomously or upon a mild (thermal) stimulus, resulting in a material with nearly the same mechanical properties as before. A prerequisite for self-healing is that the polymer network is flexible enough to allow the binding partners involved in the reversible interactions to find each other.

The first example of a self-healing polymeric material whose elastomeric properties are mediated by noncovalent interactions was described by Ludwik Leibler [55]. This material is obtained by sequential treatment of a mixture of di- and tricarboxylic acids containing long-chain alkyl groups with diethylenetriamine and urea, yielding oligomers containing amide, urea, and imidazolidone subunits along the chain that interact by hydrogen bonding. At temperatures above 160 °C, this material behaves as a viscoelastic liquid that can be extruded, while it has rubber-like properties above the glass transition temperature of 28 °C. Lowering the glass transition temperature by adding a plasticizer results in a polymer that is an elastomer at room temperature and can self-heal at ambient conditions. This product, marketed under the brand name Reverlink™, has been used to make a self-healing wire from a tube of the polymer containing a liquid gallium–indium alloy [56]. When this wire is cut and the two ends are pressed back together, the original state is restored to its mechanically stable and fully functional form.

The 2-ureido-4[1H]-pyrimidinone (UPy) unit introduced by Bert Meijer for the preparation of supramolecular polymers (Section 6.2.1) has been used to prepare self-healing hydrogels. An example is the copolymer **12.22**, which contains 2-(dimethylamino)ethyl and UPy units in the side chains (Figure 12.10a) [57]. This polymer is water-soluble at acidic pH where the tertiary amino groups are protonated. When the pH is increased to about 8 by the addition of NaOH, gelation occurs, indicating the formation of a neutral polymer with cross-links formed by UPy dimers. An incision made at 50 °C does not heal at that temperature, but lowering the temperature to 20 °C results in rapid self-healing. Many other reversible interactions are used to produce such self-healing materials, including aromatic interactions, coordinative interactions, or host–guest interactions, as in the hydrogels discussed in Section 6.3 [58].

UPy-containing polymers can also be used for tissue engineering [59]. In this context, the Meijer group showed that oligocaprolactone, a biocompatible and biodegradable but brittle polymer, yields a strong, elastic material when modified with two

(a)

(b)

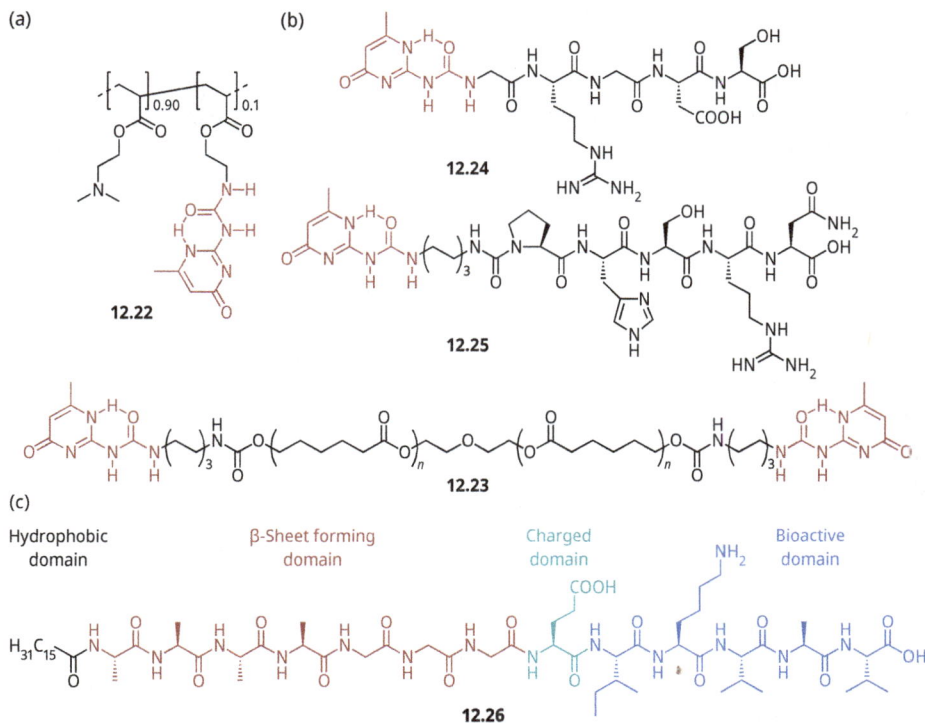

12.22

12.24

12.25

12.23

(c)

Hydrophobic domain	β-Sheet forming domain	Charged domain	Bioactive domain

12.26

Figure 12.10: Molecular structure of copolymer **12.22** (a), the UPy-containing building blocks **12.23**, **12.24**, and **12.25** for a biocompatible, bioactive supramolecular polymer (b), and the peptide amphiphile **12.26** (c).

terminal UPy moieties that mediate self-assembly [60]. This polymer can be processed into films, fibers, grids, or meshes when heated above 80 °C, where depolymerization occurs, yielding a low-viscosity solution. The resulting material is biocompatible and stable not only in buffered aqueous solution but also when implanted subcutaneously, despite the fact that the self-assembly is due to hydrogen bonding, which is typically weak in polar media with high ionic strength. Advantageously, the properties and biological activity of this material can be tuned by simply mixing **12.23** with bioactive peptides such as **12.24** and **12.25** (Figure 12.10b). The resulting peptide-containing polymers exhibit strong and specific cell adhesion *in vitro* and *in vivo*, which is not observed for the polymer containing only **12.23**.

The commercial value of supramolecular polymers is being exploited by a company called SupraPolix [61]. In addition to biomedical applications, the thermally reversible depolymerization–polymerization equilibria allow these materials to be used as ink-jet inks, which must have low viscosity when ejected through the nozzle of the printer head, but should solidify progressively as they reach the paper to produce a sharp image. Xerox and other companies have filed patents for the use of UPy-containing polymers in this application.

The peptide amphiphiles introduced by Samuel I. Stupp also produce materials with promising biomedical applications [62]. These molecules, an example of which is **12.26** (Figure 12.10c), consist of a hydrophobic alkyl chain attached to a short peptide sequence capable of inducing self-assembly by hydrogen bonding. A charged amino acid in the chain is responsible for mediating solubility in water, and the final structural peptide domain, exposed on the outside of the resulting aggregates, is structurally variable to induce bioactivity and interaction with cells or proteins. At the first level of organization, these compounds self-assemble into cylindrical micelles in water, which further aggregate into a dense network of nanofibers, even under physiological conditions and *in vivo*. Potential applications of the resulting materials include encapsulation of hydrophobic drugs or use as a matrix to template inorganic mineralization processes. Cells encapsulated in this network are viable and continue to proliferate, suggesting that these materials can also be used as biomimetic matrices in tissue engineering and regenerative medicine.

Hydrogels, that is, cross-linked polymer networks with high water adsorption capacity, are usually obtained from water-compatible polymers by covalent or noncovalent cross-linking. We have seen several examples in Section 6.3, and the self-healing copolymer **12.22** is also a hydrogel. These polymers have many applications, ranging from medicine and cosmetics to food and agriculture. They can also be obtained from low molecular weight compounds that self-assemble in solution into fibrillar structures and further into networks of fibers that trap the surrounding solvent and prevent bulk flow of the material. A typical scanning electron microscope (SEM) image of such a gel network is shown in Figure 12.11.

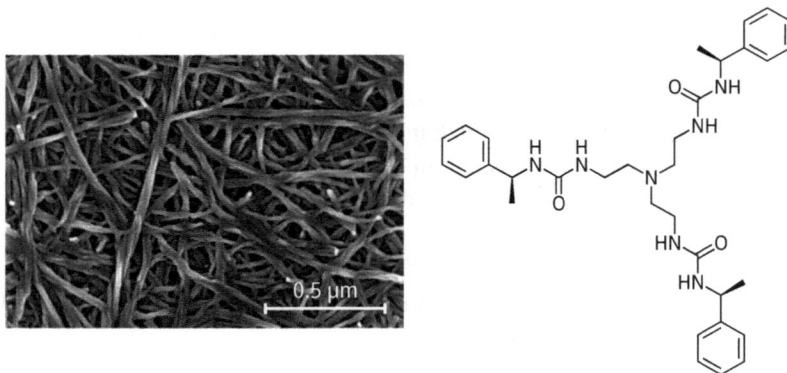

Figure 12.11: SEM image of a gel formed from a low molecular weight gelator, in this case the tris(urea) shown on the right. Image adapted with permission from [63]. Copyright The Royal Chemical Society, 2006.

The advantage of such self-assembled gels over gels made from cross-linked polymers is their responsiveness to various stimuli and the ease with which their properties can be tuned by changing the structure and ratio of the building blocks. Many gelators are also biocompatible, allowing their use in applications similar to those of conventional hydrogels [64].

Unique mechanical properties are observed for a class of gels, the so-called slide-ring gels introduced by Kohzo Ito (Section 6.3), in which the cross-links are due to interconnected rings threaded onto the polymer chains. When a mechanical force is applied to such polymers, the rings connecting the chains and the chains themselves respond by moving apart, and when the force is removed, the original state is easily restored. These materials are therefore elastic and soft, so that scratches on the surface quickly disappear [65]. As a result, coatings containing such slide-ring gels have scratch-resistant properties that are attractive for use as automotive paints. For example, Nissan, in collaboration with the University of Tokyo and Advanced Softmaterials Inc., has developed the Scratch Shield coating for several cars. Nissan is now also marketing a scratch-resistant mobile phone case based on the same technology, and another application for these gels is their use as vibration-proof and sound-proof insulation materials for sound speakers.

12.5 Applications in catalysis

Although impressive progress has been made in the development of supramolecular catalysts, as we saw in Chapter 9, such sophisticated systems are not yet in use. However, in addition to accelerating reactions by bringing together reaction partners or stabilizing transition states, receptors can also modulate the solubility or reactivity of a reagent and mediate mass transport. Even commercially available receptors are useful in this context, such as crown ethers, which are efficient phase-transfer catalysts.

In phase-transfer catalysis, the catalyst promotes the reaction by facilitating the transfer of an anionic reagent or reactant dissolved in an aqueous phase to the organic phase, where it reacts with the substrate. Conventional phase-transfer catalysts are salts of quaternary ammonium ions that exchange their counterion at the liquid–liquid interface, thereby mediating the dissolution of the anionic reagent in the organic phase. A similar effect can be achieved with crown ethers, which interact with the positively charged counterion of the anionic reagent. As the complex migrates into the organic phase, the anion follows to maintain charge neutrality and subsequently reacts. The catalytic cycle of a crown ether-mediated phase transfer is illustrated in Figure 12.12a for a nucleophilic substitution reaction.

Crown ethers are efficient catalysts for a variety of reactions, including substitutions, eliminations, or oxidations [66], and chiral crown ethers also mediate enantioselective reactions [67]. These compounds have the additional advantage that, unlike quaternary ammonium salts, they also promote the dissolution of solid salts in or-

(a)

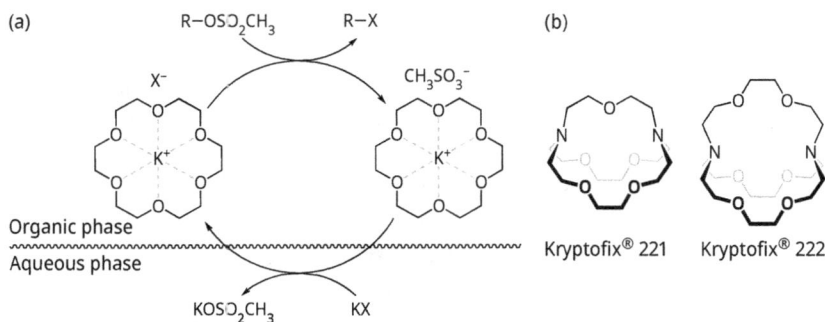

(b)

Figure 12.12: General illustration of the course of a nucleophilic substitution under phase-transfer conditions mediated by a crown ether (a) and example of cryptands commercialized under the brand name Kryptofix® (b).

ganic solvents, allowing direct phase transfer from the solid to the organic phase. For example, 18-crown-6 dissolves potassium permanganate in benzene, yielding a purple solution that acts as an efficient oxidation reagent. Similarly, KCN can be dissolved in dichloromethane or KOAc in acetonitrile.

Anions are more reactive in organic solvents because they lack the shell of water molecules that surrounds them in water, but the extent to which they are truly *naked* when dissolved in an organic solvent depends on how strongly they interact with the counterion. In the case of crown ethers, the exposed arrangement of the cation allows it to pair with the anion in the complex, reducing its reactivity. Cryptands, on the other hand, completely surround the cation and are therefore often more effective phase-transfer catalysts [68]. Many cryptands are commercially available and sold under the trade name Kryptofix®. Examples are shown in Figure 12.12b.

The ability of supramolecular receptors to promote phase transfer is also attractive for industrial processes. An example is the hydroformylation of higher alkenes under biphasic conditions. In this process, the substrates and products are in the organic phase, while the catalyst is dissolved in water so that it can be separated and recycled (Figure 12.13a) [69]. An additional phase-transfer agent mediates the transport of the substrate from the organic phase to the catalyst phase and of the product back to the organic phase, which is why this strategy is referred to as *inverse phase-transfer catalysis*. Several supramolecular receptors are useful for this purpose, with calixarenes and methylated cyclodextrins being particularly effective. Their interaction with the substrate ensures mass transfer. In addition, the geometry with which the substrate is incorporated into the receptor cavity influences the selectivity of the conversion, that is, the ratio of linear to branched products in the hydroformylation. Receptors with additional Lewis-basic sites can interact directly with the catalytically active metal center, thereby also influencing the solubility of the catalyst and its selectivity.

(a)

The branched aldehyde

CHO

is typically also formed

(b)

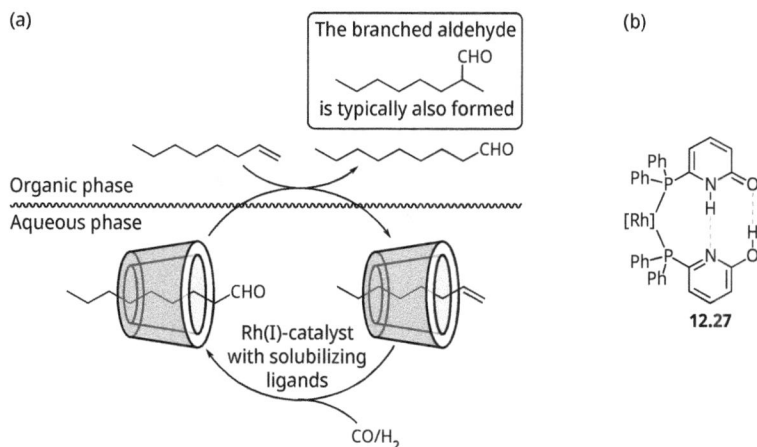

Figure 12.13: Schematic representation of biphasic hydroformylations (a), and structure of the self-assembled rhodium(I) complex **12.27** (b).

Another modular approach to ligand design involves the use of noncovalent interactions to structurally stabilize chelating ligands. Complex **12.27** (Figure 12.13b) is an example of such a catalyst in which hydrogen bonds between the two heterocyclic rings help bring the two phosphine residues together, facilitating their simultaneous interaction with the metal ion [70].

12.6 Applications in molecular electronics

Molecular electronics is an interdisciplinary field of research that focuses on the use of molecules to design and assemble electronic devices [71]. One way to obtain these devices is the controlled positioning of molecules with extended π-systems on a support or in a film by self-assembly or supramolecular polymerization to achieve the desired function [72]. This functionality is the result of the electronic and/or optical properties of the collection of molecules brought into close proximity and oriented in a specific arrangement in the device.

Alternatively, the function of an electronic device can be based on the properties of individual molecules, as in the rotaxane-based memory device developed by J. Fraser Stoddart. In this system, the 0 and 1 states of a bit are stored in the position of the ring in a bistable [2]rotaxane **12.28**, which has a structure similar to the molecular valve discussed in Section 8.2 (Figure 12.14) [73]. By sandwiching rotaxanes between 16 nm wide rows of orthogonally oriented Si and Ti nanowires, a crossbar architecture is obtained in which approximately 100 molecules are present at each junction, with their hydrophilic end groups preferentially oriented toward the Si wires. Switching is achieved by applying high positive or negative voltage pulses that cause oxidation and reduction of the tetrathiafulvalene groups. In the reduced state, the blue box surrounds the tetrathiafulvalene moiety, as shown in Figure 12.14, while oxidation causes it to move to the naphthalene station. Both states persist after the trigger is turned off. In addition, a lower, nonperturbing voltage allows probing where the ring is located, with the crossbar architecture ensuring that each junction can be addressed separately.

Although the first prototype of this device suffered from defects that reduced the reliability of the readout, a remarkably high storage density of 10^{11} bits per cm^2 was achieved. Such bistable rotaxanes have also been used in other experimental setups [74].

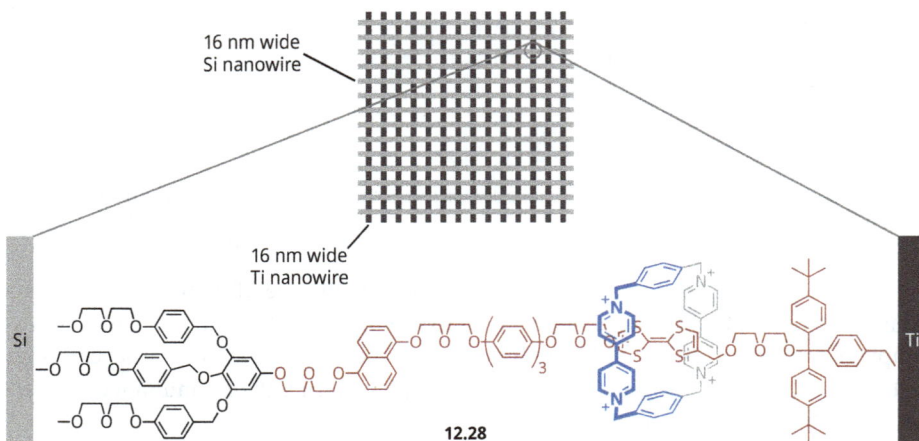

Figure 12.14: Molecular structure of the bistable [2]rotaxane **12.28** and schematic representation of the memory device assembled therefrom. In the corresponding crossbar architecture, rows of orthogonally arranged Si and Ti nanowires sandwich approximately 100 molecules of **12.28** at each junction.

12.7 Applications in consumer products

12.7.1 Textiles

Now we come to the applications where supramolecular chemistry reaches our daily lives. Since it is essential that the compounds used in this context are nontoxic, available in large quantities, and inexpensive, only a few types of receptors are actually used, the most important of which are cyclodextrins.

In the textile industry, the intensity and distribution of color on the fabric can be improved by introducing the dye molecules as their cyclodextrin complexes. Supramolecular receptors can also be used to reduce the residual color in the effluent from dyeing dying processes. Cucurbiturils are particularly useful in this context because they form insoluble complexes with many dyes. Solid cucurbiturils allow the decoloration of wastewater when used as a stationary phase in a column. The adsorbed dye molecules are selectively destroyed with ozone, while cucurbiturils are stable under these conditions.

Cyclodextrins are also used to impart special properties to finished textiles. For example, by attaching cyclodextrins to fibers, materials can be produced that adsorb unpleasant odors and reduce the development of body odor. Textiles with cyclodextrins also serve as filters to remove organic compounds from the air, leading to protective clothing. Finally, complexes of cyclodextrins with drugs or fragrances immobilized on the fiber act as depots. When the treated garment is not worn, the complexes are stable, and the bound substances do not evaporate. When worn, the water on the skin and the slightly elevated temperature cause the complexes to dissociate and the active ingredients to be released [75].

12.7.2 Food

The only supramolecular receptors currently approved for use in foods are cyclodextrins. They stabilize emulsions, which is useful in the manufacture and long-term stability of sauces, creams, and desserts. Cyclodextrins are also used to protect flavors and other active food ingredients from oxygen, heat, or light. Complexation of volatile flavors with cyclodextrins improves the shelf life of instant meals and reduces unpleasant tastes and odors.

Many food products containing cyclodextrins are on the market, especially in Japan. Examples include powdered green tea or powdered flavors, spices, and herbs. In these products, cyclodextrins not only facilitate processing and act as a protective agent, but also influence the actual taste, since complexed flavor molecules are generally released slowly. This slow release also prolongs the flavor of chewing gum.

12.7.3 Household

Various other products containing cyclodextrins are available in supermarkets. Examples include cosmetics that contain cyclodextrins to solubilize, stabilize, and suppress the volatility of fragrances. In addition, containers and packaging materials are sometimes coated with cyclodextrins to bind preservatives or compounds with antibacterial properties. A common cyclodextrin-based supermarket product is manufactured by Procter & Gamble under the name Febreze™. This odor eliminator and air freshener is applied by spraying an aqueous solution containing hydroxypropyl β-CD as the active ingredient onto fabrics or furniture. It works by masking volatile molecules through complexation, which reduces evaporation and the associated odors. So the easiest way to conduct an experiment in supramolecular chemistry is to spray Febreze™ on your sofa.

Is this the end?

Yes, this concludes our journey through the field of supramolecular chemistry. We started with the basics and then discussed many different topics that characterize the field. Hopefully, one thing that has become clear is that supramolecular chemistry is a modern, diverse, and multidisciplinary field of research, driven by the curiosity and creativity of the many contributing scientists. Some, but by no means all, have been mentioned. If this book has sparked your interest, you are welcome to join the community.

Bibliography

[1]   Kolesnichenko IV, Anslyn EV. Practical applications of supramolecular chemistry Chem. Soc. Rev. 2017, 46, 2385–90.

[2]   Williams GT, Haynes CJE, Fares M, Caltagirone C, Hiscock JR, Gale, PA. Advances in applied supramolecular technologies Chem. Soc. Rev. 2021, 50, 2737–63.

[3]   Pan YC, Tian JH, Guo DS. Molecular recognition with macrocyclic receptors for application in precision medicine. Acc. Chem. Res. 2023, 56, 3626–39.

[4]   Cameron KS, Clark JK, Cooper A, Fielding L, Palin R, Rutherford SJ, Zhang MQ. Modified γ-cyclodextrins and their rocuronium complexes. Org. Lett. 2002, 4, 3403–6.

[5]   Deng CL, Murkli SL, Isaacs LD. Supramolecular hosts as in vivo sequestration agents for pharmaceuticals and toxins. Chem. Soc. Rev. 2020, 49, 7516–32.

[6]   Haerter F, Simons JCP, Foerster U, Moreno-Duarte I, Diaz-Gil D, Ganapati S, Eikermann-Haerter K, Ayata C, Zhang B, Blobner M, Isaacs L, Eikermann M. Comparative effectiveness of calabadion and sugammadex to reverse non-depolarizing neuromuscular-blocking agents. Anesthesiology 2015, 123, 1337–49.

[7]   Brockett AT, Xue W, King D, Deng CL, Zhai C, Shuster M, Rastogi S, Briken V, Roesch MR, Isaacs L. Pillar[6]MaxQ: a potent supramolecular host for in vivo sequestration of methamphetamine and fentanyl. Chem 2023, 9, 881–900.

[8]   Bagnacani V, Franceschi V, Fantuzzi L, Casnati A, Donofrio G, Sansone F, Ungaro R. Lower rim guanidinocalix[4]arenes: macrocyclic nonviral vectors for cell transfection. Bioconjugate Chem. 2012, 23, 993–1002.

[9]   Perret F, Lazar AN, Coleman AW. Biochemistry of the *para*-sulfonato-calix[n]arenes. Chem. Commun. 2006, 2425–38.

[10] Hof F. Host–guest chemistry that directly targets lysine methylation: synthetic host molecules as alternatives to bio-reagents. Chem. Commun. 2016, 52, 10093–108.

[11] Guo DS, Liu Y. Supramolecular chemistry of p-sulfonatocalix[n]arenes and its biological applications. Acc. Chem. Res. 2014, 47, 1925–34.

[12] Letort S, Balieu S, Erb W, Gouhier G, Estour F. Interactions of cyclodextrins and their derivatives with toxic organophosphorus compounds. Beilstein J. Org. Chem. 2016, 12, 204–28.

[13] Schrader T, Bitan G, Klärner FG. Molecular tweezers for lysine and arginine – powerful inhibitors of pathologic protein aggregation. Chem. Commun. 2016, 52, 11318–34.

[14] Busschaert N, Gale PA, Small-molecule lipid-bilayer anion transporters for biological applications. Angew. Chem. Int. Ed. 2013, 52, 1374–82.

[15]   Li H, Valkenier H, Thorne AG, Dias CM, Cooper Ja, Kieffer M, Busschaert N, Gale PA, Sheppard DN, Davis AP. Anion carriers as potential treatments for cystic fibrosis: transport in cystic fibrosis cells, and additivity to channel-targeting drugs. Chem. Sci. 2019, 10, 9663–72.

[16]   Dervan PB. Molecular recognition of DNA by small molecules. Bioorg. Med. Chem. 2001, 9, 2215–35.

[17]   Muankaew C, Loftsson T. Cyclodextrin-based formulations: a non-invasive platform for targeted drug delivery. Basic Clin. Pharmacol. Toxicol. 2018, 122, 46–55.

[18]   Zhao Y, Pourgholami MH, Morris DL, Collins G, Day AI. Enhanced cytotoxicity of benzimidazole carbamate derivatives and solubilisation by encapsulation in cucurbit[n]uril. Org. Biomol. Chem. 2010, 8, 3328–37.

[19]   Hettiarachchi G, Samanta SK, Falcinelli S, Zhang B, Moncelet D, Isaacs L, Briken V. Acyclic cucurbit[n] uril-type molecular container enables systemic delivery of effective doses of albendazole for treatment of SK-OV-3 xenograft tumors. Mol. Pharmaceutics 2016, 13, 809–18.

[20]   Krämer J, Kang R, Grimm LM, De Cola L, Picchetti P, Biedermann F. Molecular probes, chemosensors, and nanosensors for optical detection of biorelevant molecules and ions in aqueous media and biofluids. Chem. Rev. 2022, 122, 3459–636.

[21]   de Silva AP, Vance TP, West, MES, Wright GD. Bright molecules with sense, logic, numeracy and utility. Org. Biomol. Chem. 2008, 6, 2468–80.

[22]   Sun X, James TD. Glucose sensing in supramolecular chemistry. Chem. Rev. 2015, 115, 8001–37.

[23]   James TD, Sandanayake KRAS, Iguchi R, Shinkai S. Novel saccharide-photoinduced electron transfer sensors based on the interaction of boronic acid and amine. J. Am. Chem. Soc. 1995, 117, 8982–7.

[24]   Arimori S, Bell ML, Oh CS, Frimat KA, James TD. Modular fluorescence sensors for saccharides. Chem. Commun. 2001, 1836–7.

[25]   Wolfbeis OS. Probes, sensors, and labels: why is real progress slow? Angew. Chem. Int. Ed. 2013, 52, 9864–5.

[26]   Lee H, Shahrivarkevishahi A, Lumata JL, Luzuriaga MA, Hagge LM, Benjamin CE, Brohlin OR, Parish CR, Firouzi HR, Nielsen SO, Lumata LL, Gassensmith JJ. Supramolecular and biomacromolecular enhancement of metal-free magnetic resonance imaging contrast agents. Chem. Sci. 2020, 11, 2045–50.

[27]   Gassensmith, JJ, Baumes JM, Smith BD. Discovery and early development of squaraine rotaxanes. Chem. Commun. 2009, 6329–38.

[28]   Johnson JR, Fu N, Arunkumar E, Leevy WM, Gammon ST, Piwnica-Worms D, Smith BD. Squaraine rotaxanes: superior substitutes for Cy-5 in molecular probes for near-infrared fluorescence cell imaging. Angew. Chem. Int. Ed. 2007, 46, 5528–31.

[29]   Riebe J, Niemeyer J. Mechanically interlocked molecules for biomedical applications. Eur. J. Org. Chem. 2021, 5106–16.

[30]   Sogah GDY, Cram DJ. Host–guest complexation. 14. Host covalently bound to polystyrene resin for chromatographic resolution of enantiomers of amino acid and ester salts. J. Am. Chem. Soc. 1979, 101, 3035–42.

[31]   Zhang G, Lin W, Huang F, Sessler J, Khashab NM. Industrial separation challenges: how does supramolecular chemistry help? J. Am. Chem. Soc. 2023, 145, 19143–63.

[32]   Ogoshi T, Sueto R, Yoshikoshi K, Sakata Y, Akine S, Yamagishi T. Host–guest complexation of perethylated pillar[5]arene with alkanes in the crystal state. Angew. Chem. Int. Ed. 2015, 54. 9849–52.

[33]   Jie K, Liu M, Zhou Y, Little MA, Bonakala S, Chong SY, Stephenson A, Chen L, Huang F, Cooper AI. Styrene purification by guest-induced restructuring of pillar[6]arene. J. Am. Chem. Soc. 2017, 139, 2908–11.

[34]   Petitgirard A, Djehiche M, Persello J, Fievet P, Fatin-Rouge N. PAH contaminated soil remediation by reusing an aqueous solution of cyclodextrins. Chemosphere 2009, 75, 714–8.

[35] Cathum SJ, Dumouchel A, Punt M, Brown CE. Sorption/desorption of polychlorinated dibenzo-*p*-dioxins and polychlorinated dibenzo furans (PCDDs/PCDFs) in the presence of cyclodextrins. J. Soil Sediment Contamin. 2007, 16, 15–27.

[36] Serio N, Levine M. Solvent effects in the extraction and detection of polycyclic aromatic hydrocarbons from complex oils in complex environments. J. Inclusion Phenom. Macrocyclic Chem. 2016, 84, 61–70.

[37] Wilson AM, Bailey PJ, Tasker PA, Turkington JR, Grant RA, Love JB. Solvent extraction: the coordination chemistry behind extractive metallurgy. Chem. Soc. Rev. 2014, 43, 123–34.

[38] Izatt RM, Izatt SR, Bruening RL, Izatt NE, Moyer BA. Challenges to achievement of metal sustainability in our high-tech society. Chem. Soc. Rev. 2014, 43, 2451–75.

[39] Walker DD, Norato MA, Campbell SG, Crowder ML, Fink SD, Fondeur FF, Geeting MW, Kessinger GF, Pierce RA. Cesium removal from Savannah River Site radioactive waste using the caustic-side solvent extraction (CSSX) process. Sep. Sci. Technol. 2005, 40, 297–309.

[40] Simonnet M, Sittel T, Weßling P, Geist A. Cs extraction from chloride media by calixarene crown-ethers. Energies, 2022, 15, 7724.

[41] Rambo BM, Sessler JL. Oligopyrrole macrocycles: receptors and chemosensors for potentially hazardous materials. Chem. Eur. J. 2011, 17, 4946–59.

[42] Sather AC, Berryman OB, Rebek Jr. J. Selective recognition and extraction of the uranyl ion. J. Am. Chem. Soc. 2010, 132, 13572–4.

[43] Rao MD, Singh KK, Morrison CA, Love JB. Challenges and opportunities in the recovery of gold from electronic waste. RSC Adv. 2020, 10, 4300–9.

[44] Liu Z, Frasconi M, Lei J, Brown ZJ, Zhu Z, Cao D, Iehl J, Liu G, Fahrenbach AC, Botros YY, Farha OK, Hupp JT, Mirkin CA, Stoddart JF. Selective isolation of gold facilitated by second-sphere coordination with α-cyclodextrin. Nat. Commun. 2013, 4, 1855.

[45] Wu H, Wang Y, Tang C, Jones LO, Song B, Chen XY, Zhang L, Wu Y, Stern CL, Schatz GC, Liu W, Stoddart JF. High-efficiency gold recovery by additive-induced supramolecular polymerization of β-cyclodextrin. Nat. Commun. 2023, 14, 1284.

[46] Eller LR, Stępień M, Fowler CJ, Lee JT, Sessler JL, Moyer BA. Octamethyl-octaundecylcyclo[8]pyrrole: a promising sulfate anion extractant. J. Am. Chem. Soc. 2007, 129, 11020–1.

[47] Qin L, Vervuurt SJN, Elmes RBP, Berry SN, Proschogo N, Jolliffe KA. Extraction and transport of sulfate using macrocyclic squaramide receptors. Chem. Sci. 2020, 11, 201–7.

[48] Katayev EA, Kolesnikov GV, Sessler JL. Molecular recognition of pertechnetate and perrhenate. Chem. Soc. Rev. 2009, 38, 1572–86.

[49] Gawenis JA, Holman KT, Atwood JL, Jurisson SS. Extraction of pertechnetate and perrhenate from water with deep-cavity [CpFe(arene)]$^+$-derivatized cyclotriveratrylenes. Inorg. Chem. 2002, 41, 6028–31.

[50] Gorden AEV, Davis J, Sessler JL, Král V, Keogh DW, Schroeder NL. Monoprotonated sapphyrin-pertechnetate anion interactions in aqueous media. Supramol. Chem. 2004, 16, 91–100.

[51] Alberto R, Bergamaschi G, Braband H, Fox T, Amendola V. ^{99}TcO$_4^-$: selective recognition and trapping in aqueous solution. Angew. Chem. Int. Ed. 2012, 51, 9772–6.

[52] Custelcean, R. Urea-functionalized crystalline capsules for recognition and separation of tetrahedral oxoanions. Chem. Commun. 2013, 49, 2173–82.

[53] Rajbanshi A, Moyer BA, Custelcean R. Sulfate separation from aqueous alkaline solutions by selective crystallization of alkali metal coordination capsules. Cryst. Growth Des. 2011, 11, 2702–6.

[54] Custelcean R, Williams NJ, Seipp CA, Ivanov AS, Bryantsev VS. Aqueous sulfate separation by sequestration of [(SO$_4$)$_2$(H$_2$O)$_4$]$^{4-}$ clusters within highly insoluble imine-linked bis-guanidinium crystals. Chem. Eur. J. 2016, 22, 1997–2003.

[55] Cordier P, Tournilhac F, Soulié-Ziakovic C, Leibler L. Self-healing and thermoreversible rubber from supramolecular assembly. Nature 2008, 451, 977–80.

[56]   Palleau E, Reece S, Desai SC, Smith ME, Dickey MD. Self-healing stretchable wires for reconfigurable circuit wiring and 3D microfluidics. Adv. Mater. 2013, 25, 1589–92.

[57]   Cui J, del Campo A. Multivalent H-bonds for self-healing hydrogels. Chem. Commun. 2012, 48, 9302–4.

[58]   Yang Y, Urban MW. Self-healing of polymers via supramolecular chemistry. Adv. Mater. Interfaces 2018, 5, 1800384.

[59]   Goor OJGM, Hendrikse SIS, Dankers PYW, Meijer EW. From supramolecular polymers to multi-component biomaterials. Chem. Soc. Rev. 2017, 46, 6621–37.

[60]   Dankers PYW, Harmsen MC, Brouwer LA, Van Luyn MJA, Meijer EW. A modular and supramolecular approach to bioactive scaffolds for tissue engineering. Nat. Mater. 2005, 4, 568–74.

[61]   Bosman AW, Sijbesma RP, Meijer EW. Supramolecular polymers at work. Mater. Today 2004, 7, 34–9.

[62]   Cui H, Webber MJ, Stupp SI. Self-assembly of peptide amphiphiles: From molecules to nanostructures to biomaterials. Pept. Sci. 2010, 94, 1–18.

[63]   Stanley CE, Clarke N, Anderson KM, Elder JA, Lenthall JT, Steed JW. Anion binding inhibition of the formation of a helical organogel. Chem. Commun. 2006, 3199–201.

[64]   Hirst AR, Escuder B, Miravet JF, Smith DK. High-tech applications of self-assembling supramolecular nanostructured gel-phase materials: from regenerative medicine to electron c devices. Angew. Chem. Int. Ed. 2008, 47, 8002–18.

[65]   Noda Y, Hayashi Y, Ito K. From topological gels to slide-ring materials. J. Appl. Polym. Sci. 2014, 131, 40509.

[66]   Gokel GW, Durst HD. Principles and synthetic applications in crown ether chemistry. Synthesis 1976, 168–84.

[67]   Oliveira MT, Lee JW. Asymmetric cation-binding catalysis. ChemCatChem 2017, 9, 377–84.

[68]   Landini D, Maia A, Montanari F, Pirisi FM. Crown ethers as phase-transfer catalysts. A comparison of anionic activation in aqueous–organic two-phase systems and in low polarity anhydrous solutions by perhydrodibenzo-18-crown-6, lipophilic quaternary salts, and cryptands. J. Chem. Soc., Perkin Trans. 2 1980, 46–51.

[69]   Obrecht L, Kamer PCJ, Laan W. Alternative approaches for the aqueous-organic biphasic hydroformylation of higher alkenes. Catal. Sci. Technol. 2013, 3, 541–51.

[70]   Breit B, Seiche W. Hydrogen Bonding as a construction element for bidentate donor ligands in homogeneous catalysis: regioselective hydroformylation of terminal alkenes. J. Am. Chem. Soc. 2003, 125, 6608–9.

[71]   Mathew PT, Fang F. Advances in molecular electronics: a brief review. Engineering 2018, 4, 760–71.

[72]   Jain A, George SJ. New directions in supramolecular electronics. Mater. Today 2015, 18, 206–14.

[73]   Green JE, Choi JW, Boukai A, Bunimovich Y, Johnston-Halperin E, DeIonno E, Luo Y, Sheriff BA, Xu K, Shin YS, Tseng HR, Stoddart JF, Heath JR. A 160-kilobit molecular electronic memory patterned at 10^{11} bits per square centimetre. Nature 2007, 445, 414–7.

[74]   Yu H, Luo Y, Beverly K, Stoddart JF, Tseng HR, Heath JR. The molecule-electrode interface in single-molecule transistors. Angew. Chem. Int. Ed. 2003, 42, 5706–11.

[75]   Ferreira L, Mascarenhas-Melo F, Rabaça S, Mathur A, Sharma A, Giram PS, Pawar KD, Rahdar A, Raza F, Veiga F, Mazzola PG, Paiva-Santos AC. Cyclodextrin-based dermatological formulations: dermopharmaceutical and cosmetic applications. Colloids Surf. B, 2023, 221, 113012.

13 Appendices

13.1 Concentrations in a 1:2 binding equilibrium

How do I do the math?

The fundamental equations describing the formation of a 1:2 receptor–substrate complex were introduced in Section 2.1, but a strategy for solving these equations was not explained. One approach, which was used to generate the graphs in Figure 2.3, is presented here [1], although alternative methods exist [2]. We start with the laws of mass action in equations (13.1a, b), which describe the stepwise formation of the 1:2 complex:

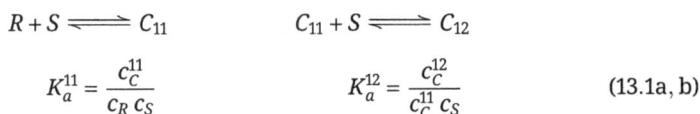

$$R + S \rightleftharpoons C_{11} \qquad\qquad C_{11} + S \rightleftharpoons C_{12}$$

$$K_a^{11} = \frac{c_C^{11}}{c_R\, c_S} \qquad\qquad K_a^{12} = \frac{c_C^{12}}{c_C^{11}\, c_S} \qquad\qquad (13.1a, b)$$

The mass balances for these equilibria are specified in the following equations:

$$c_R = c_R^0 - c_C^{11} - c_C^{12} \qquad\qquad (13.2a)$$

$$c_S = c_S^0 - c_C^{11} - 2\, c_C^{12} \qquad\qquad (13.2b)$$

To reduce the number of unknowns, we first seek an expression for c_C^{11}, starting from equations (13.1a, b) and (13.2a):

$$c_R^0 - c_C^{11} - \frac{c_C^{11}}{K_a^{11} c_S} - K_a^{12} c_C^{11} c_S = 0 \qquad\qquad (13.3)$$

An expression for c_S results by combining equations (13.1b) and (13.2b):

$$c_S = c_S^0 - c_C^{11} - 2\, K_a^{12} c_C^{11} c_S \qquad\qquad (13.4)$$

The rearrangement of equation (13.4) affords equation (13.5):

$$c_S = \frac{c_S^0 - c_C^{11}}{1 + 2\, K_a^{12} c_C^{11}} \qquad\qquad (13.5)$$

When combining this equation with equation (13.3), the following equation is obtained:

$$c_R^0 - c_C^{11} - \frac{c_C^{11}}{K_a^{11}} \frac{1 + 2\, K_a^{12} c_C^{11}}{c_S^0 - c_C^{11}} - K_a^{12} c_C^{11} \frac{c_S^0 - c_C^{11}}{1 + 2\, K_a^{12} c_C^{11}} = 0 \qquad\qquad (13.6)$$

https://doi.org/10.1515/9783111315171-013

Rearranging equation (13.6) affords the cubic equation (13.7):

$$K_a^{12}\left(1 - \frac{4K_a^{12}}{K_a^{11}}\right)\left(c_C^{11}\right)^3 + \left[1 - 2K_a^{12}\left(c_R^0 + \frac{2}{K_a^{11}}\right)\right]\left(c_C^{11}\right)^2$$
$$- \left[c_R^0 + c_S^0 + \frac{1}{K_a^{11}} + K_a^{12}c_S^0\left(c_S^0 - 2c_R^0\right)\right]c_C^{11} + c_R^0 c_S^0 = 0 \tag{13.7}$$

A solution for c_C^{11} can be found by using Newton's iterative method. For this purpose, equation (13.7) is defined as $f\left(c_C^{11}\right)$. The corresponding derivative $f'\left(c_C^{11}\right)$ has the following form:

$$f'(x) = 3K_a^{12}\left(1 - \frac{4K_a^{12}}{K_a^{11}}\right)\left(c_C^{11}\right)^2 + 2\left[1 - 2K_a^{12}\left(c_R^0 + \frac{2}{K_a^{11}}\right)\right]\left(c_C^{11}\right)$$
$$- \left[c_R^0 + c_S^0 + \frac{1}{K_a^{11}} + K_a^{12}c_S^0\left(c_S^0 - 2c_R^0\right)\right] = 0 \tag{13.8}$$

We now calculate c_C^{11} by using the following equation:

$$c_C^{11}(\text{new}) = c_C^{11}(\text{old}) - \frac{f\left(c_C^{11}(\text{old})\right)}{f'\left(c_C^{11}(\text{old})\right)} \tag{13.9}$$

The first iteration requires a guess for $c_C^{11}(\text{old})$ and using $c_R^0/2$ is often a good start. The calculation is repeated by using the resulting value $c_C^{11}(\text{new})$ as the new starting value $c_C^{11}(\text{old})$ for the next iteration until convergence is reached, typically requiring 5 to 10 steps. With the c_C^{11} thus obtained, c_C^{12} is calculated by using equation (13.11), which is obtained by combining equations (13.1b) and (13.2b) to give equation (13.10), followed by rearrangement:

$$c_C^{12} = K_a^{12}c_C^{11}\left(c_S^0 - c_C^{11} - 2\ c_C^{12}\right) \tag{13.10}$$

$$c_C^{12} = \frac{K_a^{12}c_C^{11}\left(c_S^0 - c_C^{11}\right)}{1 + 2K_a^{12}c_C^{11}} \tag{13.11}$$

Analogous expressions can be derived if the formation of the complex involves the binding of two receptor molecules to one substrate molecule. The easiest way to approach this situation is to replace c_S^0 by c_R^0 and *vice versa* in the above equations.

13.2 Concentrations in an indicator displacement assay

In an indicator displacement assay (IDA), the substrate S (analyte) is added to a solution of a receptor R and an indicator I. Both the substrate and the indicator bind to the same binding site of the receptor, so that the two complexes C_S and C_I, of which C_S should be more stable than C_I, coexist in the equilibrium. If only 1:1 complexes are formed, the overall equilibrium can be broken down into two equilibria, describing the formation of the substrate complex C_S and the indicator complex C_I with the associated laws of mass action (13.12a, b):

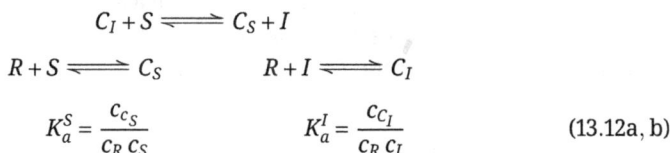

$$C_I + S \rightleftharpoons C_S + I$$

$$R + S \rightleftharpoons C_S \qquad\qquad R + I \rightleftharpoons C_I$$

$$K_a^S = \frac{c_{C_S}}{c_R\, c_S} \qquad\qquad K_a^I = \frac{c_{C_I}}{c_R\, c_I} \qquad\qquad (13.12a, b)$$

The concentrations c_R, c_S, c_I, c_{C_S}, and c_{C_I} correlate with the total concentrations c_R^0, c_S^0, and c_I^0 according to the mass balances (13.13a–c):

$$c_R = c_R^0 - c_{C_S} - c_{C_I} \qquad\qquad (13.13a)$$

$$c_S = c_S^0 - c_{C_S} \qquad\qquad (13.13b)$$

$$c_I = c_I^0 - c_{C_I} \qquad\qquad (13.13c)$$

We use these equations to derive an expression that leaves only one unknown concentration. If we chose c_R as the unknown, we Must define all other unknown concentrations in terms of c_R. To do so, we combine the above equations and arrive at the following expressions:

$$c_{C_S} = \frac{K_a^S c_R}{1 + K_a^S c_R}\, c_S^0 \qquad\qquad (13.14a)$$

$$c_{C_I} = \frac{K_a^I c_R}{1 + K_a^I c_R}\, c_I^0 \qquad\qquad (13.14b)$$

$$c_I = \frac{c_I^0}{1 + K_a^I c_R} \qquad\qquad (13.14c)$$

Substituting these equations into (13.13a) leads to the following equation:

$$c_R = c_R^0 - \frac{K_a^S c_R}{1 + K_a^S c_R}\, c_S^0 - \frac{K_a^I c_R}{1 + K_a^I c_R}\, c_I^0 \qquad\qquad (13.15)$$

Rearranging this equation gives the cubic equation (13.16):

$$K_a^S K_a^I (c_R)^3 + \left[K_a^S + K_a^I + K_a^S K_a^I \left(c_S^0 + c_I^0 - c_R^0\right)\right](c_R)^2$$
$$+ \left[1 + K_a^S c_S^0 + K_a^I c_I^0 - \left(K_a^S + K_a^I\right) c_R^0\right] c_R - c_R^0 = 0 \qquad\qquad (13.16)$$

A solution to this equation is again found using Newton's method, as explained in the previous section. Once c_R is known, the other concentrations are calculated using equations (13.14a–c).

In an IDA, the total concentrations of the host and indicator are normally held constant while the concentration of the substrate (analyte) is increased. The change in absorbance of the solution A_{obs} during the titration is recorded and then plotted against the substrate concentration, c_S^0. This measurement is carried out at a wavelength at which neither the receptor nor the substrate absorbs, so that A_{obs} reflects only the contributions of the free indicator and its complex according to expression (13.17) (length of the cuvette = 1 cm):

$$A_{obs} = A_I + A_{c_{C_I}} = \varepsilon_I c_I + \varepsilon_{C_I} c_{C_I} \tag{13.17}$$

Substituting the above-derived equations (13.14c) and (13.14b) into (13.17) gives equation (13.18), which allows K_a^S to be obtained by nonlinear regression. The parameters ε_I, ε_{C_I}, and K_a^I should be determined prior to the titration to reduce the number of variables and thereby increase the reliability of the fit:

$$A_{obs} = \frac{c_I^0}{1 + K_a^I c_R} \left(\varepsilon_I + \varepsilon_{C_I} K_a^I c_R \right) \tag{13.18}$$

13.3 Python scripts

This section contains selected Python scripts to illustrate the strategies used to calculate the binding isotherms and various other graphs presented throughout the book. All scripts have the same structure for ease of understanding. They were not written to be particularly elegant or fast, but they can serve as a basis for fitting routines.

13.3.1 Binding isotherm of 1:1 complex formation

For an example, see Figure 2.2.

```python
import numpy as np
from matplotlib import pyplot as plt


# Equation
def equation(c, e, K):
    t1 = (c + c*e + 1/K)/2
    cc = t1 - (t1**2 - c*c*e)**0.5
    ct = cc/c
    return(ct)
```

```
# Start Program
try:

  # Input
  conc = float(input('Concentration of the receptor in mol/L: '))
  equi = float(input(Equivalents of the substrate added: '))
  Ka_1 = float(input('Stability constant of the 1:1 complex in L/mol: '))

  # Calculate function
  x = np.linspace(0, equi, int(equi*50))
  y = equation(conc, x, Ka_1)

  # Plot function
  plt.clf()
  plt.xlabel('c(S)/c(R)')
  plt.ylabel('Degree of complexation')
  title = 'Binding isotherm for a 1:1 complex\n(receptor concentration:
'+str(conc)+' mol/L,\nstability constant: '+str(Ka_1)+' L/mol)'
  l1 = plt.plot(x, y, label = title)
  plt.figlegend(loc='lower right', frameon = 0, borderaxespad = 5)
  plt.show()

except:

  print('Unable to parse input.\n')
```

13.3.2 Job plot of a 1:1 complex

For an example, see Figure 2.7.

```
import numpy as np
from matplotlib import pyplot as plt

# Equation
def equation(c, e, K):

  t1 = (c + 1/K)/2
  cc = t1 - (t1**2 - e*c*(1-e)*c)**0.5
  ct = 2*cc/c
  return(ct)
```

```
# Start Program
try:

  # Input
  conc = float(input('Sum of receptor and substrate concentration in
mol/L: '))
  Ka_1 = float(input('Stability constant of the 1:1 complex in L/mol: '))

  # Calculate function
  x = np.linspace(0, 1, 50)
  y = equation(conc, x, Ka_1)

  # Plot function
  plt.clf()
  plt.xlabel('X')
  plt.ylabel('Degree of complexation')
  title = 'Job plot for a 1:1 complex\n(sum of receptor and substrate
\nconcentration: '+str(conc)+' mol/L,\nstability constant: '+str(Ka_1)+'
L/mol)'
  l1 = plt.plot(x, y, label = title)
  plt.figlegend(loc='lower center', frameon = 0, borderaxespad = 5)
  plt.show()

except:

  print('Unable to parse input.\n')
```

13.3.3 Binding isotherm of 1:2 complex formation

For an example, see Figure 2.3.

```
import numpy as no
from matplotlib import pyplot as plt

# Equation
def equation(c, e, K1, K2):

  t1 = K2*(1-4*K2/K1)
  t2 = 1-2*K2*(c+2/K1)
  t3 = c+c*e+1/K1+K2*c*e*(c*e-2*c)
  t4 = c*c*e
```

```
    c1 = c/10
    ca = c

    while abs(np.sum(c1)-np.sum(ca)) > 1e-6:
        ca  = c1
        cx  =   t1*c1**3 +   t2*c1**2 - t3*c1 + t4
        cxa = 3*t1*c1**2 + 2*t2*c1    - t3
        c1  = c1 - cx/cxa

    c2 = K2*c1*(c*e-c1)/(1+2*K2*c1)

    c1 = c1
    c2 = c2
    c3 = c1+c2
    ct = np.array([[c1],[c2],[c3]])
    return(ct)

# Start Program
try:

    # Input
    conc = float(input('Concentration of the receptor in mol/L: '))
    equi = float(input('Equivalents of the substrate added: '))
    Ka_1 = float(input('Stability constant of the 1:1 complex in L/mol: '))
    Ka_2 = float(input('Stability constant of the 1:2 complex in L/mol: '))

    # Calculate function
    x = np.linspace(0, equi, int(equi*50))
    y = equation(conc, x, Ka_1, Ka_2)

    # Plot function
    plt.clf()
    plt.xlabel('c(S)/c(R)')
    plt.ylabel('Degree of complexation')
    title1 = 'Binding isotherm of the 1:1 complex'
    l1 = plt.plot(x, y[0,0], label = title1)
    title2 = 'Binding isotherm of the 1:2 complex'
    l2 = plt.plot(x, y[1,0], label = title2)
    title3 = 'Sum of both isotherms\n(receptor concentration: '+str(conc)+'
mol/L,\nstability constants - Ka(1:1):
'+str(Ka_1)+' L/mol),\nKa(1:2): '+str(Ka_2)+' L/mol)'
    l3 = plt.plot(x, y[2,0], label = title3)
```

```
    plt.figlegend(loc='lower right', frameon = 0, borderaxespad = 5)
    plt.show()

except:

    print('Unable to parse input.\n')
```

13.3.4 Job plot of a 1:2 complex

For an example, see Figure 2.8.

```
import numpy as np
from matplotlib import pyplot as plt


# Equation
def equation(c, e, K1, K2):

    t1 = K2*(1-4*K2/K1)
    t2 = 1-2*K2*(e*c+2/K1)
    t3 = c+1/K1+K2*c*(1-e)*(c*(1-e)-2*e*c)
    t4 = c*e*c*(1-e)
    c1 = c/10
    ca = c

    while abs(np.sum(c1)-np.sum(ca)) > 1e-6:
        ca  = c1
        cx  =   t1*c1**3 +   t2*c1**2 - t3*c1 + t4
        cxa = 3*t1*c1**2 + 2*t2*c1    - t3
        c1  = c1 - cx/cxa

    c2 = K2*c1*(c*(1-e)-c1)/(1+2*K2*c1)

    c1 = 2*c1/c
    c2 = 2*c2/c
    c3 = c1+c2
    ct = np.array([[c1],[c2],[c3]])

    return(ct)


# Start Program
try:
```

```python
  # Input
  conc = float(input('Sum of receptor and substrate concentration in
mol/L: '))
  Ka_1 = float(input('Stability constant of the 1:1 complex in L/mol: '))
  Ka_2 = float(input('Stability constant of the 1:2 complex in L/mol: '))

  # Calculate function
  x = np.linspace(0, 1, 50)
  y = equation(conc, x, Ka_1, Ka_2)

  # Plot function
  plt.clf()
  plt.xlabel('X')
  plt.ylabel('Degree of complexation')
  title1 = 'Job plot of the 1:1 complex'
  l1 = plt.plot(x, y[0,0], label = title1)
  title2 = 'Job plot of the 1:2 complex'
  l2 = plt.plot(x, y[1,0], label = title2)
  title3 = 'Sum of Job plots\n(receptor concentration: '+str(conc)+'
mol/L,\nstability constants - Ka(1:1): '+str(Ka_1)+' L/mol),\nKa(1:2):
'+str(Ka_2)+' L/mol)'
  l3 = plt.plot(x, y[2,0], label = title3)
  plt.figlegend(loc='lower center', frameon = 0, borderaxespad = 5)
  plt.show()

except:

  print('Unable to parse input.\n')
```

13.3.5 ITC binding isotherm of 1:1 complex formation

For an example, see Figure 2.15.

```python
import numpy as np
from matplotlib import pyplot as plt

# Equation
def equation(cc, cs, e, H, K):

  v0 = 1e-3
  vi = v0*e*cc/cs
```

```
   i  = (v0*cc)/(cs*10)
   Li = cs*(1-(1-i/v0)**(vi/i))
   Mi = cc*(1-i/v0)**(vi/i)
   t1 = (Li+Mi+1/K)/2
   t2 = t1-(t1*t1-Li*Mi)**0.5
   vj = (v0*e*cc/cs)-i
   Lj = cs*(1-(1-i/v0)**(vj/i))
   Mj = cc*(1-i/v0)**(vj/i)
   u1 = (Lj+Mj+1/K)/2
   u2 = u1-(u1*u1-Lj*Mj)**0.5
   ht = (v0*H)/(i*cs)*(t2-u2*(1-i/v0))
   return(ht)

# Start Program
try:

   # Input
   conc_c = float(input('Concentration in the cell in mol/L: '))
   conc_s = float(input('Concentration in the syringe in mol/L: '))
   equi   = float(input('Equivalents of the binding partner added: '))
   heat   = float(input('Complexation enthalpy in kJ/mol: '))
   Ka_1   = float(input('Stability constant of the 1:1 complex in
L/mol: '))

   # Calculate function
   x = np.linspace(0, equi, int(equi*50))
   y = equation(conc_c, conc_s, x, heat, Ka_1)

   # Plot function
   plt.clf()
   plt.xlabel('c(Syringe)/c(Cell)')
   plt.ylabel('Complexation enthalpy / kJ/mol')
   title = 'Binding isotherm for a 1:1 complex\n(concentration in cell:
'+str(conc_c)+'\nconcentration in syringe: '+str(conc_s)+'\ncomplexation
enthalpy: '+str(heat)+' kJ/mol,\nstability constant: '+str(Ka_1)+' L/mol)'
   l1 = plt.plot(x, y, label = title)
   plt.figlegend(loc='lower right', frameon = 0, borderaxespad = 5)
   plt.show()
```

except:

```
  print('Unable to parse input.\n')
```

13.3.6 Binding isotherm of dimerization

For an example, see Figure 6.3.

```python
import numpy as np
from matplotlib import pyplot as plt

# Equation
def equation(c, K):

  alpha = 1-((-1+(1+8*K*c)**0.5)/(4*K*c))
  return(alpha)

# Start Program
try:

  # Input
  conc_l = float(input('Lower concentration limit in mol/L: '))
  conc_h = float(input('Upper concentration limit in mol/L: '))
  Ka     = float(input('Stability constant of the dimer in L/mol: '))

  # Calculate function
  x = np.logspace(np.log10(conc_l), np.log10(conc_h), 100)
  y = equation(x, Ka)

  # Plot function
  plt.clf()
  plt.xlabel('c')
  plt.ylabel('Degree of dimerization')
  title = 'Binding isotherm for a dimerization\n(stability constant: '+str(Ka)+' L/mol)'
  l1 = plt.semilogx(x, y, label = title)
  plt.figlegend(loc='lower right', frameon = 0, borderaxespad = 5)
  plt.show()
```

```
except:

  print('Unable to parse input.\n')
```

13.3.7 Binding isotherm of isodesmic supramolecular polymerization

For an example, see Figure 6.4a.

```python
import numpy as npfrom matplotlib import pyplot as plt

# Equation
def equation(Kc):

  alpha = 1-((1+2*Kc-(4*Kc+1)**0.5)/(2*(Kc)**2))
  return(alpha)

# Start Program
try:

  # Input
  conc_l = float(input('Lower limit of Ka*cM0: '))
  conc_h = float(input('Upper limit of Ka*cM0: '))

  # Calculate function
  x = np.logspace(np.log10(conc_l), np.log10(conc_h), 100)
  y = equation(x)

  # Plot function
  plt.clf()
  plt.xlabel('Ka*cM0')
  plt.ylabel('Degree of polymerization')
  title = 'Binding isotherm for an\nisodesmic supramolecular
\npolymerization'
  l1 = plt.semilogx(x, y, label = title)
  plt.figlegend(loc='lower right', frameon = 0, borderaxespad = 5)
  plt.show()

except:

  print('Unable to parse input.\n')
```

13.3.8 Number and weight average degrees of polymerization of isodesmic supramolecular polymerization

For an example, see Figure 6.4b.

```python
import numpy as np
from matplotlib import pyplot as plt


# Equation
def equation(Kc):

    KcM1 = (1+2*Kc-(4*Kc+1)**0.5)/(2*Kc)
    dpn  = 1/(1-KcM1)
    dpw  = (1+KcM1)/(1-KcM1)
    dpt  = np.array([[dpn],[dpw]])
    return(dpt)


# Start Program
try:

    # Input
    conc_l = float(input('Lower limit of Ka*cM0: '))
    conc_h = float(input('Upper limit of Ka*cM0: '))

    # Calculate function
    x = np.logspace(np.log10(conc_l), np.log10(conc_h), 100)
    y = equation(x)

    # Plot function
    plt.clf()
    plt.xlabel('Ka*cM0')
    plt.ylabel('Number and weight averages of\ndegree of polymerization')
    title1 = 'Number average of degree\nof polymerization'
    l1 = plt.semilogx(x, y[0,0], label = title1)
    title2 = 'Weight average of degree\nof polymerization'
    l2 = plt.semilogx(x, y[1,0], label = title2)
    plt.figlegend(loc='lower left', frameon = 0, borderaxespad = 6)
    plt.show()
```

```
except:

  print('Unable to parse input.\n')
```

13.3.9 Binding isotherm of ring–chain supramolecular polymerization

For an example, see Figure 6.6a.

```
import sys
import numpy as np
from matplotlib import pyplot as plt

# Equation
def equation(Kc, e, K):

  c  = 1
  ca = 2

  while abs(np.sum(c)-np.sum(ca)) > 1e-6:

    t1  = 0
    t1a = 0
    t2  = 0
    t2a = 0
    m   = 1
    ta  = 1
    tn  = 2

    while tn/ta - 1 > 1e-6:
      ta  = tn
      t1  = t1  + m          * c**m
      t1a = t1a + m**2       * c**(m-1)
      t2  = t2  + m**(-3/2) * c**m
      t2a = t2a + m**(-1/2) * c**(m-1)
      m   = m + 1
      tn  = np.sum(t1) + np.sum(t1a) + np.sum(t2) + np.sum(t2a)

    print('.', end="")
    sys.stdout.flush()
    ca = c
    c1 = t1 + e * K * t2 - Kc
```

```
      c1a = t1a + e * K * t2a
      c   = ca  - (c1/c1a)

   t1  = 0
   t2  = 0
   m   = 1
   ta  = 1
   tn  = 2

   while tn/ta - 1 > 1e-6:
      ta = tn
      t1 = t1 + m          * c**m
      t2 = t2 + m**(-3/2) * c**m
       m = m  + 1
      tn = np.sum(t1) + np.sum(t2)

   alpha = t1/(t1+e*K*t2)
   return(alpha)

# Start Program
try:

   # Input
   conc_l = float(input('Lower limit of Ka*cM0: '))
   conc_h = float(input('Upper limit of Ka*cM0: '))
   EM     = float(input('Effective molarity: '))
   Ka     = float(input('Stability constant: '))

   # Calculate function
   x = np.logspace(np.log10(conc_l), np.log10(conc_h), 100)
   y = equation(x, EM, Ka)

   # Plot function
   plt.clf()
   plt.xlabel('Ka*cM0')
   plt.ylabel('Fraction of monomers in chains')
   title = 'Binding isotherm for a\nring-chain supramolecular
polymerization'
   l1 = plt.semilogx(x, y, label = title)
   plt.figlegend(loc='lower right', frameon = 0, borderaxespad = 5)
   plt.show()
```

```
except:

  print('Unable to parse input.\n')
```

13.3.10 Number average degree of polymerization of ring–chain supramolecular polymerization

For an example, see Figure 6.6b.

```python
import sys
import numpy as np
from matplotlib import pyplot as plt

# Equation
def equation(Kc, e, K):

  c  = 1
  ca = 2

  while abs(np.sum(c)-np.sum(ca)) > 1e-6:

    t1  = 0
    t1a = 0
    t2  = 0
    t2a = 0
    m   = 1
    ta  = 1
    tn  = 2

    while tn/ta - 1 > 1e-6:
      ta  = tn
      t1  = t1  + m         * c**m
      t1a = t1a + m**2      * c**(m-1)
      t2  = t2  + m**(-3/2) * c**m
      t2a = t2a + m**(-1/2) * c**(m-1)
      m   = m + 1
      tn  = np.sum(t1) + np.sum(t1a) + np.sum(t2) + np.sum(t2a)

    print('.', end="")
    sys.stdout.flush()
    ca = c
```

```
    c1  = t1  + e * K * t2 - Kc
    c1a = t1a + e * K * t2a
    c   = ca  - (c1/c1a)

  t1 = 0
  t2 = 0
  m  = 1
  ta = 1
  tn = 2

  while tn/ta - 1 > 1e-6:
    ta  = tn
    t1  = t1 +                c**m
    t2  = t2 + m**(-5/2) * c**m
    m   = m + 1
    tn  = np.sum(t1) + np.sum(t2)

  dpn = Kc/(t1+e*K*t2)
  return(dpn)

# Start Program
try:

  # Input
  conc_l = float(input('Lower limit of Ka*cM0: '))
  conc_h = float(input('Upper limit of Ka*cM0: '))
  EM     = float(input('Effective molarity: '))
  Ka     = float(input('Stability constant: '))

  # Calculate function
  x = np.logspace(np.log10(conc_l), np.log10(conc_h), 100)
  y = equation(x, EM, Ka)

  # Plot function
  plt.clf()
  plt.xlabel('Ka*cM0')
  plt.ylabel('Number average of\ndegree of polymerization')
  title = 'Number average\ndegree of polymerization for a\nring-chain
supramolecular\npolymerization (EM: '+str(EM)+')'
  l1 = plt.loglog(x, y, label = title)
  plt.figlegend(loc='lower left', frameon = 0, borderaxespad = 6)
  plt.show()
```

```
except:

  print('Unable to parse input.\n')
```

13.3.11 Binding isotherm of cooperative supramolecular polymerization

For an example, see Figure 6.7a.

```python
import numpy as np
from matplotlib import pyplot as plt

# Equation
def equation(Kc, s):

  t1 = s-1
  t2 = Kc+2-2*s
  t3 = 1+2*Kc
  t4 = Kc
  c  = 1
  ca = 2

  while abs(np.sum(c)-np.sum(ca)) > 1e-6:
    ca  = c
    cx  =   t1*c**3 +   t2*c**2 - t3*c + t4
    cxa = 3*t1*c**2 + 2*t2*c    - t3
    c   = ca - cx/cxa

  alpha = 1-c/Kc
  return(alpha)

# Start Program
try:

  # Input
  conc_l = float(input('Lower limit of Ka*cM0: '))
  conc_h = float(input('Upper limit of Ka*cM0: '))
  Sigma  = float(input('Sigma: '))

  # Calculate function
  x = np.logspace(np.log10(conc_l), np.log10(conc_h), 100)
  y = equation(x, Sigma)
```

```
# Plot function
plt.clf()
plt.xlabel('Ka*cM0')
plt.ylabel('Degree of Polymerization')
title = 'Binding isotherm for a\ncooperative supramolecular
\npolymerization (sigma: '+str(Sigma)+')'
l1 = plt.semilogx(x, y, label = title)
plt.figlegend(loc='lower right', frameon = 0, borderaxespad = 5)
plt.show()

except:

print('Unable to parse input.\n')
```

13.3.12 Number average degree of polymerization of cooperative supramolecular polymerization

For an example, see Figure 6.7b.

```
import numpy as np
from matplotlib import pyplot as plt

# Equation
def equation(Kc, s):

  t1 = s-1
  t2 = Kc+2-2*s
  t3 = 1+2*Kc
  t4 = Kc
  c  = 1
  ca = 2

  while abs(np.sum(c)-np.sum(ca)) > 1e-6:
    ca  = c
    cx  =   t1*c**3 +   t2*c**2 - t3*c + t4
    cxa = 3*t1*c**2 + 2*t2*c    - t3
    c   = ca - cx/cxa

  dpn = ((1-s)*(1-c)**2+s)/((1-(1-s)*c)*(1-c))
  return(dpn)
```

```
# Start Program
try:

    # Input
    conc_l = float(input('Lower limit of Ka*cM0: '))
    conc_h = float(input('Upper limit of Ka*cM0: '))
    Sigma  = float(input('Sigma: '))

    # Calculate function
    x = np.logspace(np.log10(conc_l), np.log10(conc_h), 100)
    y = equation(x, Sigma)

    # Plot function
    plt.clf()
    plt.xlabel('Ka*cM0')
    plt.ylabel('Number average of\ndegree of polymerization')
    title = 'Number average\ndegree of polymerization for a\ncooperative
supramolecular\npolymerization (sigma: '+str(Sigma)+')'
    l1 = plt.loglog(x, y, label = title)
    plt.figlegend(loc='lower right', frameon = 0, borderaxespad = 5)
    plt.show()

except:

    print('Unable to parse input.\n')
```

Bibliography

[1]   Hargrove, AE, Zhong Z, Sessler JL, Anslyn EV. Algorithms for the determination of binding constants
      and enantiomeric excess in complex host:guest equilibria using optical measurements. New
      J. Chem. 2010, 34, 348–54.
[2]   Thordarson, P. Determining association constants from titration experiments in supramolecular
      chemistry. Chem. Soc. Rev. 2011, 40, 1305–23.

Index

https://doi.org/10.1515/9783111315171-014